Biology and Management of the Formosan Subterranean Termite and Related Species

Biology and Management of the Formosan Subterranean Termite and Related Species

Edited by:

Nan-Yao Su, University of Florida

and

Chow-Yang Lee, University of California, Riverside

CABI

CABI is a trading name of CAB International

CABI
Nosworthy Way
Wallingford
Oxfordshire OX10 8DE
UK

Tel: +44 (0)1491 832111
E-mail: info@cabi.org
Website: www.cabi.org

CABI
200 Portland Street
Boston
MA 02114
USA

Tel: +1 (617)682-9015
E-mail: cabi-nao@cabi.org

A catalogue record for this book is available from the British Library, London, UK.

ISBN-13: 9781800621572 (hardback)
9781800621589 (ePDF)
9781800621596 (ePub)

DOI: 10.1079/9781800621596.0000

Commissioning Editor: Ward Cooper
Editorial Assistant: Lauren Davies
Production Editor: James Bishop

Typeset by Straive, Pondicherry, India
Printed and bound in the UK by CPI Group (UK) Ltd, Croydon, CR0 4YY

Contents

Dedication

We dedicate this book to the following six researchers whose work over the past 110 years has enabled us to advance the taxonomy, biology, and management of the Formosan subterranean termite, *Coptotermes formosanus* Shiraki.

Jeffery P. La Fage (1945–1989) received his PhD in 1976 from the University of Arizona under the tutelage of the renowned termite ecologist, W.L. Nutting. La Fage's doctoral research focused on the nutrition and bioenergetics of a desert drywood termite. His dissertation included an extensive review of the nutrient ecology of termites, which was later published as a highly influential chapter titled "Nutrient dynamics of termites" in the book *Production Ecology of Ants and Termites*, edited by Brian (1978). This chapter continues to be widely cited and serves as essential literature for those studying termite ecology. In 1977, La Fage joined the faculty of Louisiana State University, where he established a program to study the biology and control of *C. formosanus*, a termite species wreaking havoc in New Orleans and especially the historic French Quarter. One of his notable innovations was the utilization of rolled corrugated cardboard as an efficient method to extract termites from infested logs or lumbers, providing a swift and effective means of termite collection. La Fage's life was tragically cut short in 1989 when he intervened in an attempted robbery on the streets of the French Quarter. Nevertheless, despite the brevity of his professional career, La Fage made remarkable contributions to the field, publishing a total of 55 peer-reviewed papers in the short 17 years.

Masamitsu Oshima (1884–1965) visited Taiwan in 1907 during his tenure as a Zoology student at the Tokyo Imperial University (today's University of Tokyo) to investigate the termite problem. Following his graduation in 1908, he was hired by the Agricultural Experiment Station, Government of Formosa, to conduct an in-depth study on termites and their detrimental effects in Taiwan. Oshima's significant contributions to termite research in Taiwan were reflected in a series of "Investigation report on termites" published between 1909 and 1917, which laid the groundwork for future studies of termites. During his time on the island, Oshima diligently collected termites from various locations, providing valuable specimens that Shiraki used to describe *C. formosanus*. Oshima's collection of *Coptotermes* specimens likely included both *C. formosanus* and *C. gestroi*. However, entomologists at that time mistakenly assumed the presence of a single *Coptotermes* species on the island, leading to the initial confusion regarding the identity of *C. formosanus* in Taiwan. Subsequently,

Oshima's research interest shifted toward reptiles and fish, culminating in the completion of his PhD in 1920. His doctoral work encompassed detailed descriptions of over 77 new species of freshwater fish in Taiwan.

After graduating from Sapporo Agricultural College (later Hokkaido Imperial University and today's Hokkaido University) in 1906, **Tokuichi Shiraki** (1882–1970) taught mineralogy, agriculture production, and entomology at his alma mater. In 1907, he moved to Taiwan to assume the position of the head of the entomology department at the Agricultural Experiment Station, Government of Formosa. Despite being the author of *C. formosanus*, Shiraki's contribution extended far beyond his studies on termites during his tenure in Taiwan. He was entrusted with the responsibility of combating the yellow stem borer (*Scirpophaga incertulas*), a significant pest affecting rice crops in Taiwan. Shiraki spearheaded a groundbreaking biological control project by introducing the ladybird beetle, *Novius* (= *Rodolia*) *cardinalis*, effectively controlling the cottony cushion scale, *Icerya purchase*, that almost destroyed the citrus industry on the island. In 1913, Shiraki spent three years at the British Museum of Natural History to sort out the taxonomy of insects found in Taiwan. His meticulous research on *S. incertulas* earned him a PhD from Hokkaido Imperial University in 1917. He authored over 150 scientific articles and numerous entomological volumes, including *Medical Entomology*, *Entomological Dictionary*, *Insect Taxonomy*, and others. Shiraki's pivotal role in advancing the field of entomology during Taiwan's colonial era cemented his position as a central figure in its development.

Upon returning from World War II as a highly decorated soldier in 1945, **Minoru Tamashiro** (1923–2020) attended the University of Hawai'i, where he earned a BSc (1952) and an MSc (1954) degree in Entomology. Later, he went to the University of California, Berkeley, and received a PhD in 1959 under the guidance of E.A. Steinhaus, the father of insect pathology. Joining the faculty of the University of Hawai'i, he attempted a biological control of *C. formosanus* with entomopathogenic fungi but soon realized the challenges in locating, collecting, and studying this cryptic insect species both in the laboratory and field. To overcome these challenges, Tamashiro developed ingenious techniques such as stake surveys and on-the-ground monitoring stations. His method of using wooden ramps to separate field-collected termites from debris remains one of the most efficient procedures for manipulating a large number of subterranean termites in the laboratory. Together with his student, P.-Y. Lai, Tamashiro also devised a dye marker to establish the interconnection among monitoring stations and to delineate the foraging territory of a *C. formosanus* colony. Researchers studying subterranean termites have widely adopted these methods, and his survey stake and monitoring station techniques formed the foundation of today's baiting technology.

Kunio Tsunoda (1948–2011) received his PhD from Kyoto University in 1978, investigating the ecology of shipworm. Before joining Kyoto University, he pursued research under the mentorship of Ruth Turner (Harvard University), one of the foremost marine scientists of the 20th century. His experience at Harvard inculcated him for scientific excellence in his career. Tsunoda joined the Wood Research Institute, Kyoto University, in 1979 and spent the next 32 years researching wood-destroying organisms such as termites and fungi. He is best known for research on termite management in Japan, preservatives for wood and wood-based composites, molds, and fungal decay, test methodology and standardization, fumigants, natural durability, and coatings. Tsunoda founded the Pacific-Rim Termite Research Group, an academic society dedicated to researching the biology and management of termites in the Pacific region. He received the Honorary Life Membership of the International Research Group on Wood Protection (IRG) in 2007.

Tsuyoshi Yoshimura (1960–2021) received his PhD from Kyoto University in 1995. He was a faculty member of the Research Institute for Sustainable Humanosphere (RISH) (formerly Wood Research Institute) at Kyoto University from 1990 to 2021. He significantly advanced research on the nutritional physiology of termites by revealing the contribution of protozoan fauna in the hindgut to wood decomposition in termites. His research on termite feeding ecology, feeding behavior, and biological control using insect–pathogen led to a comprehensive termite management system in Japan. He authored 187 peer-reviewed papers, 20 books, and 24 review articles in academic journals. He received numerous accolades, including awards from the Japan Wood Research Society (1996), the Japan Termite Control Association (1996, 2000, 2007, 2009), and the Japanese Society of Environmental Entomology and Zoology (2009).

List of Contributors

Thomas Chouvenc
Entomology and Nematology Department, Ft. Lauderdale Research and Education Center, 3205 College Avenue, University of Florida, FL 33314, USA. Email: tomchouv@ufl.edu

J. Kenneth Grace
Department of Plant and Environmental Protection Sciences, University of Hawai'i, 3050 Maile Way, Room 310, Honolulu, HI 95822-2271, USA. Email: kennethg@hawaii.edu

Claudia Husseneder
Department of Entomology, Louisiana State University, A543 Life Sciences Annex, Baton Rouge, LA 70803, USA. Email: chusseneder@agcenter.lsu.edu

Shuji Itakura
Department of Applied Biological Chemistry, Faculty of Agriculture, Kindai University, 3327-204 Naka-machi, Nara 631-8505, Japan. Email: itakura@nara.kindai.ac.jp

Chow-Yang Lee
Department of Entomology, University of California, 900 University Avenue, Riverside, CA 92521, USA. Email: chowyang.lee@ucr.edu

Hou-Feng Li
Department of Entomology, National Chung Hsing University, 145 Xingda Road, South District, Taichung City 402, Taiwan. Email: houfeng@nchu.edu.tw

Nathan Lo
School of Life and Environmental Sciences, F22 Life Earth & Environmental Science Building, University of Sydney, Australia. Email: nathan.lo@sydney.edu.au

Kok-Boon Neoh
Department of Entomology, National Chung Hsing University, 145 Xingda Road, South District, Taichung City 402, Taiwan. Email: neohkokboon@nchu.edu.tw

Wakako Ohmura
Laboratory of Innovative Humano-habitability, Research Institute for Sustainable Humanosphere, Kyoto University, Gokasho, Uji 611-0011, Japan. Email: murasan@rish.kyoto-u.ac.jp

Faith M. Oi
Entomology and Nematology Department, University of Florida, PO Box 110620, Building 970, Natural Area Drive, Gainesville, FL 32611, USA. Email: foi@ufl.edu

Rudolf H. Scheffrahn
Entomology and Nematology Department, Ft. Lauderdale Research and Education Center, 3205 College Avenue, University of Florida, FL 33314, USA. Email: rhsc@ufl.edu

Nan-Yao Su
Entomology and Nematology Department, Ft. Lauderdale Research and Education Center, 3205 College Avenue, University of Florida, FL 33314, USA. Email: nysu@ufl.edu

Qian Sun
Department of Entomology, Louisiana State University, A511 Life Sciences Annex, Baton Rouge, LA 70803, USA. Email: qsun@agcenter.lsu.edu

Gaku Tokuda
Tropical Biosphere Research Center, COMB, University of the Ryukyus, 1 Senbaru, Nishihara, Okinawa, Japan. Email: tokuda@comb.u-ryukyu.ac.jp

Edward L. Vargo
Department of Entomology, Texas A&M University, 370 Olsen Blvd, College Station, TX 77843, USA. Email: edward.vargo@ag.tamu.edu

Chia-Chien Wu
Department of Entomology, National Chung Hsing University, 145 Xingda Road, South District, Taichung City 402, Taiwan. Email: daisy100406@gmail.com

Koichi Yamamoto
Japan Wood Protection Association, Dai. 3 Matsuzaka Bldg. 2-5, 4-chome, Toranomon, Minato-ku, Tokyo 105-0001, Japan. Email: jwpa-4@mokuzaihozon.org

Preface

The Formosan subterranean termite, *Coptotermes formosanus*, holds the notorious title of being the most destructive termite species, while its close relative, the Asian subterranean termite, *C. gestroi*, has gained infamy as the most invasive termite species worldwide. *Coptotermes formosanus* stands out as the only termite species listed among the world's 100 worst invasive alien species of the Global Invasive Species Database (2023). Annually, its infestation costs more than US$4 billion in control and damage repairs of structures in the U.S. alone. Surprisingly, despite its economic significance, no dedicated book has been published on *C. formosanus*.

Previous literature has provided authoritative treatises on termite biology and control. One notable work, *Termites and Termite Control*, edited by C.A. Kofoid (1934), comprised 54 comprehensive chapters covering a broad range of topics such as biology and distribution of many known termite species in the continental U.S., Mexico, West Indies, Panama, Hawai'i, and the Philippines. This groundbreaking work also addressed management options, prevention, and damage repair. S.F. Light and T. Snyder, regarded as pioneers of termitology in North America, collectively contributed 23 chapters, and some of the information documented in 1934 remains relevant today. The seminal volumes *Biology of Termites*, edited by Krishna and Weesner (1969, 1970), have been indispensable textbooks for students of termites, delving into intricate aspects of fundamental biology, termite associations with microbes and symbionts, phylogeny, and descriptions of termites in various regions. A concise yet informative book titled *Termites in Buildings* by Edwards and Mill (1986) offered a succinct description of basic termite biology and summarized available control measures, proving to be a valuable resource for researchers and termite control professionals. *Termites: Evolution, Sociality, Symbioses, Ecology*, edited by Abe *et al.* (2000), provided a much-needed update to the *Biology of Termites*, bridging a 30-year gap and primarily focusing on basic science in termite biology. Subsequently, *Biology of Termites: a Modern Synthesis*, edited by Bignell *et al.* (2010), quickly incorporated results brought about by advancements in molecular tools. Despite recent changes in termite taxa, the most comprehensive work on termite taxonomy remains the *Treatise on the Isoptera of the World* by Krishna *et al.* (2013).

While these earlier treatises provided a vital foundation in termite biology and control, this volume takes a specific focus on *C. formosanus*. It serves as the first comprehensive reference book compiling literature on this important termite pest species since Tokuichi Shiraki described it in 1909. The book covers the global distribution of *C. formosanus*, its biogeography, and the mechanism by which it has spread from its native range in southern

China and Taiwan to other parts of the world, drawing upon recorded infestations and population genetics studies. A chapter reviews its present taxonomic status and the recent progress made in *Coptotermes* taxonomy. Subsequent chapters delve into the biology, ecology, foraging behavior, physiology, and chemical ecology of *C. formosanus*. It explores the termite's association with endosymbionts, particularly the gut microbiota. The book also addresses inspection and monitoring and various control options against this species, including baits, liquid termiticides, wood preservatives, alternative and experimental management methods, as well as the integrated pest management (IPM) and area-wide population management approach. Additionally, two chapters are dedicated to discussing another highly destructive species, *C. gestroi*, and the recently discovered *C. formosanus/C. gestroi* hybrids. *Coptotermes gestroi* represents the primary destructive subterranean termite species in tropical Asia and South America, with introductions to other parts of the world, such as the U.S. and Africa. The final two chapters offer valuable laboratory techniques to study rhinotermitids, including *C. formosanus*, and highlight unsolved questions that warrant further investigation.

Control of *C. formosanus* has been challenging due in part to their large colony size. With the advent of new technologies and strategies, there is a need to review the pros and cons of all available options. The book will be an essential and valuable reference for researchers, graduate students, pest management professionals, chemical manufacturer personnel, building and property managers, and others. It will provide readers with a comprehensive understanding of the biology and management of the Formosan subterranean termite and the Asian subterranean termite.

The editors express their sincere gratitude to all 17 authors for their contributions in making this book an authoritative reference of the Formosan subterranean termite. They also want to record their appreciation to the following colleagues who reviewed one or more chapters of the book: Hunter Bergbower (University of Florida), Tom Bourguignon (Okinawa Institute of Science and Technology Graduate University), Tiago Carrijo (Universidade Federal do ABC, Brazil), Thomas Chouvenc (University of Florida), Brian Forschler (University of Georgia), J. Kenneth Grace (University of Hawai'i), Sang-Bing Lee (University of Florida), Vernard Lewis (University of California, Berkeley), Hongjie Li (Ningbo University, China), Houfeng Li (National Chung Hsing University, Taiwan), Nobuaki Mizumoto (Auburn University), Aaron Mullins (University of Florida), Christine Nalepa (North Carolina State University), Alain Robert (University of Paris XIII, France), Yves Roisin (Université Libre de Bruxelles, Belgium), Michael Scharf (University of Florida), Michael K. Rust (University of California, Riverside), Nicholas Saleh (University of Florida), Rudolf Scheffrahn (University of Florida), Jan Šobotník (Czech Academy of Science, Czech Republic), Edward Vargo (Texas A&M University), and Xuguo Zhou (University of Kentucky).

Nan-Yao Su
Chow-Yang Lee
June 2023

References

Abe, T., Bignell, D.E. and Higashi M. (eds) (2000) *Termites: Evolution, Sociality, Symbioses, Ecology*. Springer, Dordrecht, Netherlands.

Bignell, D.E, Roisin, Y. and Lo, N. (eds) (2010) *Biology of Termites: a Modern Synthesis*. Springer, Dordrecht, Netherlands.

Edwards, R. and Mill, A.E. (1986) *Termites in Buildings: Their Biology and Control*. Rentokil Ltd, East Grinstead, UK

Global Invasive Species Database (2023) Species profile: *Coptotermes formosanus*. Downloaded from http://www.iucngisd.org/gisd/speciesname/Coptotermes+formosanus on 09-10-2023.

Kofoid, C. (ed.) 1934 *Termites and Termite Control*. University of California Press, Berkeley, California.

Krishna, K. and Weesner, F.M. (eds.) (1969) *Biology of Termites*. Vol. 1. Academic Press, New York.
Krishna, K. and Weesner, F.M. (eds) (1970) *Biology of Termites*, Vol. 2. Academic Press, New York.
Krishna, K., Grimaldi, D.A., Krishna, V. and Engel, M.S. (2013) Treatise on the Isoptera of the World: Vol. 1. *Bulletin of the American Museum of Natural History* 377, 200 pp.
Shiraki, T. (1909) On the termites of Japan. *Transactions of the Entomological Society of Japan* 2, 229–242 [in Japanese]

1 Introduction

Nan-Yao Su[1] and Chow-Yang Lee[2]
[1]*University of Florida, nysu@ufl.edu;* [2]*University of California, Riverside, chowyang.lee@ucr.edu*

1.1 Formosan subterranean termite and related species

As implied by its Chinese and Japanese names, 家白蟻 or "house termite," the Formosan subterranean termite, *Coptotermes formosanus* Shiraki, has been known to associate with manmade structures throughout history. It is considered the most economically important termite pest globally, yet no comprehensive book dedicated to this serious pest species has been published. Nonetheless, the pest status of *C. formosanus* has spurred a substantial amount of literature on this species, covering both basic biology and management measures. It would not be an overstatement to assert that a significant portion of our knowledge about termites originates from research on *C. formosanus*, given its prominence as the number one termite pest.

Historically, *Coptotermes* has taxonomically been placed in the "lower" termites (Rhinotermitidae). However, it is now established that *Coptotermes* belongs to the sister clade to Termitidae (Bucek *et al.*, 2019), explaining why the biological and behavioral traits of *Coptotermes* species share some similarities with "higher" termites (Termitidae), including the presence of separate nest structures, where termites venture out from the central nest to search for food, bifurcated developmental pathways resulting in the production of terminal "true" workers, and the formation of large colonies. The large colony size of *C. formosanus*, as revealed through early excavation studies (Erhorn, 1934; King and Spink, 1969), left a lasting impression on those studying this species.

A significant milestone in the field study of *C. formosanus* colonies occurred with the development of stake survey and monitoring stations by Tamashiro *et al.* (1973). These innovative devices provided unprecedented access to the underground colonies of this cryptic species. Colony boundaries could be delineated when combined with the use of dye markers (Lai *et al.*, 1983a), leading to the practical definition of a colony as "a group of termites sharing an interconnected gallery system" (Su and Scheffrahn, 1998). Moreover, the monitoring devices served as means to introduce control agents such as microbes and bait toxicants. The dye marker facilitated the establishment of interconnection among monitoring stations, and some stations were left untreated to serve as control stations for unbiased assessment of treatment efficacy. The field protocol of Tamashiro *et al.* (1973), which involved using survey stakes to detect termites before installing monitoring stations, later became the operation model for the early-generation bait products that employed the monitoring–baiting procedure.

 Biology and Management of the Formosan Subterranean Termite and Related Species (eds Nan-Yao Su and Chow-Yang Lee)
DOI: 10.1079/9781800621596.0001

In recent years, our knowledge of *C. formosanus* biology has made significant strides, primarily driven by the rapid progress in molecular techniques. A major development in the management of subterranean termites occurred with the introduction of commercial baits in 1995. Moreover, all living organisms are now facing the challenge of climate change. The northward expansion range of *Coptotermes gestroi* (Wasmann) has increased its distribution overlap with *C. formosanus* and is now creating opportunities for hybridization between these two invasive termite pest species. The past few decades of research have fundamentally changed our understanding of subterranean termite pest species, and such changes have necessitated the publication of this comprehensive volume, which consolidates the most up-to-date information on the biology and management of *C. formosanus* and related species.

1.2 Biology

Genetic evidence indicated that *C. formosanus* originated in southern China and Taiwan when these regions were land-connected during the late Pleistocene (Li *et al.*, 2009) (Chapter 2 of this volume). Over time, it spread to other parts of the world through human activities, although the exact pathway of its dispersal remains unknown (Chapter 9 of this volume). Specimens of *C. formosanus* were collected at the type locality in Taiwan and designated as the neotype to replace the lost holotype, and this species has been re-described (Chen *et al.*, 2020) (Chapter 3 of this volume). Recent advancements in genetic research, such as the availability of partial mitochondria gene sequences and whole-genome sequences of *C. formosanus* (Itakura *et al.*, 2020), have provided valuable insights. Combining these genetic findings with morphological studies, 15 junior synonyms of *C. formosanus* were proposed (Chapter 3 of this volume). Utilization of molecular markers has drastically enhanced our understanding of colony breeding structure (Chapter 9 of this volume), which will provide unparalleled insights into the dynamics of colony population structure and demographics (Chapter 4 of this volume).

The caste differentiation and developmental biology of termites present a complex subject. Although there is a general trend, each termite family (and sometimes genus) follows its distinct developmental pathway, which can be influenced by various factors such as sex, social context, colony age, nutrition, and surrounding environment. In the past, treatises of termite caste differentiation often listed multiple pathways specific to different termite families, which can lead to confusion unless readers possess a deep understanding of the subject matter. To address this issue, Chapter 4 of this volume focuses primarily on the individual developmental pathway and the colony development of *C. formosanus* and is essential reading for students of termite biology. However, it is crucial to recognize that this represents only one among numerous pathways observed in different termite species, and termites of other families may take different pathways to caste formation (Roisin, 2000). Throughout a colony's lifespan, it progresses through a series of stages, including incipient colony, immature colony, juvenile colony, mature colony, and ultimately senescent colony. Laboratory and field data revealed that the duration of each stage depends on multiple factors, and each colony undertakes a unique progression through its lifespan (Chouvenc *et al.*, 2022). These findings provide valuable insights into the intricate dynamics of termite colonies and their developmental progress.

A mature *C. formosanus* colony typically comprises millions of termites and may forage up to 100 m in search of food. *Coptotermes formosanus* propagates tunnels away from the tunnel origin following a Global Away Vector (GAV) generated through path integration (Bardunias and Su, 2009). This process allows termites to geolocate their current location relative to tunnel origin. A laboratory study showed that the location of newly discovered food is georeferenced by path integration, and *C. formosanus* constructs new tunnels heading directly toward the food to optimize transportation efficiency (Michael *et al.*, 2023). The recent progress in our understanding of *Coptotermes* behavioral ecology is reviewed here (Chapter 5 of this volume), along with the recent advances in *C. formosanus* chemical ecology (Chapter 7 of this volume).

The ability of termites to digest lignocellulose in wood represents a unique attribute of these fascinating insects. Many have long aspired to replicate this digestive process outside termite guts, transforming wood debris into alcohol as a biofuel. Achieving this goal remains elusive. *Coptotermes formosanus* is known to secrete endogenous cellulases in the salivary glands and the midgut, but studies indicated that cellulases produced by hindgut protists play a primary role in the total cellulase activity against crystalline cellulose (Chapter 6 of this volume). It is worth noting that the contribution of protists in lignocellulose digestion may only represent a fraction of the functions performed by the complex microbe community residing in termite guts. The symbiotic microbes found in *C. formosanus* guts comprise a quadripartite system consisting of protists, archaea, bacteria, and viruses such as bacteriophages (Chapter 8 of this volume). Traditionally three protist species are known from *C. formosanus*, namely *Pseudotrichonympha grassii*, *Holomastigotoides hartmanni*, and *Cononympha (Spirotrichonympha) leidyi* (Koidzumi, 1921; Lai *et al.*, 1983b; Kitade *et al.*, 2013). Recent genetic and microscopy studies have identified two additional species, *H. minor*, and *Con. koidzumii* (Jasso-Selles *et al.*, 2020; Nishimura *et al.*, 2020). While the role of protists in lignocellulose digestion is well established, the contributions of other microbes to termite hosts remain relatively obscure. Bacteria likely play supporting roles in nutrient acquisition, and earlier studies suggested their involvement in nitrogen fixation to compensate for limited nitrogen content in wood diets (Breznak *et al.*, 1973). However, recent studies have shown that *C. formosanus* rely more on dietary nitrogen from the soil than intrinsic nitrogen fixation by diazotrophic bacteria in their hindguts (Mullins and Su, 2018; Mullins *et al.*, 2021, 2022). This ability to acquire nitrogen from soil organic matter potentially enables *C. formosanus* and other subterranean termites to develop and sustain large colonies.

1.3 Management

Management of *C. formosanus* requires skill and knowledge on how to monitor, inspect, and treat for structural infestation from the pest control industry (Chapter 10 of this volume). Over the past century, the pest management industry has widely used soil termiticides to control subterranean termites. While their use has generally declined, 33% of termite control services still relied exclusively on liquid termiticides, as of 2022, whereas 62% utilized a combination of liquid termiticides and baits (Chapter 12 of this volume). The primary objective of soil termiticide is to create an unbroken insecticide barrier that effectively prevents soilborne subterranean termites from accessing structures. As of 2023, non-repellent termiticides such as fipronil are the most commonly used by the industry. When a minimum rate of 0.06% fipronil is employed, ~0.745 kg of the termiticide active ingredient (AI) is required to form a complete barrier beneath and surrounding a 230 m^2 home, amounting to 32.4 kg/hectare (Chapter 15 of this volume).

In contrast to soil termiticides, the primary objective of bait application is to address termite infestation at its source by achieving colony elimination. Following the successful demonstration of a chitin synthesis inhibitor (CSI), hexaflumuron, to eliminate colonies of *C. formosanus* and *Reticulitermes flavipes* (Kollar) (Su, 1994), the first subterranean termite bait, Recruit® of Sentricon® system, was commercially launched in 1995 (Chapter 11 of this volume). Since then, significant progress has been made in improving the performance of bait, and all commercial baits now utilize CSIs as active ingredients. Field studies have demonstrated that less than 1 g of CSI is sufficient to eliminate colony(s) of subterranean termites in the vicinity of a house. This translates to a remarkably low pesticide application rate of ~0.043 kg/hectare, which is ~753 times lower than the application rate of soil termiticide treatments. The significant reduction in pesticide usage achieved by baits underscores their effectiveness and environmentally conscious approach to termite control.

Su and Scheffrahn (1998) emphasized that the mere combination of available control options does not constitute integrated pest management (IPM) for subterranean termites. Instead, a subterranean termite IPM must consider the cost and benefits of a termite management program. Control measures should minimize risks to human health and the environment while effectively reducing termite damage potential. Comparing the two major control options, baits offer distinct advantages over soil termiticide treatments. Baits utilize 33-fold less toxic insecticide and are applied at lower rates (753-fold) than soil insecticide treatments. This translates to a staggering 24,849-fold reduction in environmental impact when baits are employed instead of soil termiticides. Moreover, eliminating termite colonies through baiting effectively reduces termite damage potential. Based on this premise, baits have been implemented in large-scale area-wide (AW) projects. These initiatives have demonstrated that key factors contributing to the success of AW projects include well-defined objectives, a simple managerial structure, and the use of control measures supported by reliable data that substantiate claims of colony elimination. On the other hand, political interference and overly complex managerial structures tend to lead to the failures of AW projects, highlighting the importance of maintaining a streamlined and focused approach to termite management.

While applications of liquid termiticides or termite baits have dominated the pest management standard since the late 1990s, other research avenues continue to be investigated. Alternative approaches for *C. formosanus* management have continued to receive attention over the decades, despite a relative lack of commercial successes (Chapter 14 of this volume). In addition, wood damage prevention with preservatives remains an active research area (Chapter 13 of this volume).

1.4 Climate change: *C. gestroi* and hybrids with *C. formosanus*

Coptotermes gestroi is another congeneric species to *C. formosanus* with strong invasive capabilities and is a major pest around the tropics (Chapter 16 of this volume). The reality of climate change and its impact on global warming is unequivocal, and this is further substantiated by the observable northward movement of tropical species like *C. gestroi* in recent years (Su *et al.*, 2017). First discovered in the continental U.S. from Miami, Florida (25° 45′ 42.0516″ N) in 1996 (Su *et al.*, 1997), *C. gestroi* has gradually expanded its range northward, with reports of its presence in Tampa, Florida (27° 57′ 50.9652″ N) as of 2022. The current distribution of *C. gestroi* in Florida can be found at https://flrec.ifas.ufl.edu/termites-in-florida/termite-distribution/, which is routinely updated from the records of the University of Florida Termite Collection.

When *C. formosanus* was first documented in Taiwan in the early 20th century (Oshima, 1909; Shiraki, 1909), *C. gestroi* was already established in southern Taiwan, most likely below the Tropic of Cancer (23° 27′ N) (Chapter 3 of this volume). A comprehensive island-wide survey conducted between 2015 and 2020 revealed that *C. gestroi* distributed from southern regions of Taiwan to Taichung (24° 9′ 19.6776″ N), with a few isolated samples collected in Taipei (25° 6′ 19.7892″ N) (Huang *et al.*, 2022). In areas where *C. formosanus* and *C. gestroi* coexist, instances of heterospecific mating and resultant hybrid colonies have been observed in the laboratory (Chouvenc *et al.*, 2015). Moreover, hybrid alates have been collected since 2019 in both Taiwan and Florida (Chen, 2022). Although the hybrid F1 and some F2 alates appeared viable, their overall fitness and potential gene flow between *C. formosanus* and *C. gestroi* remain uncertain and require further investigation (Chapter 17 of this volume). The interactions and potential hybridization between *C. formosanus* and *C. gestroi*, possibly influenced by changing climatic conditions, raise important questions regarding their long-term impact on the ecosystem and the genetic integrity of each species. Ongoing research and monitoring efforts will be crucial in understanding the consequences of these interactions and developing effective strategies for their management in the face of changing environmental conditions.

1.5 Experimental protocols

In his seminal chapter, Becker (1969) stressed the significance of preserving termite social behavior to ensure the biological relevance of experimental findings. However, subsequent laboratory studies on subterranean termites have predominantly utilized isolated foraging groups collected from field colonies. While this approach may be adequate for examining individual termite responses to stimuli or control agents, it falls short of replicating the complex dynamics of a whole colony with a complete caste composition. Unfortunately, many of these simplified laboratory studies, conducted with only a few termites in a Petri dish, are often extrapolated to a real-world condition, despite the potential limitations and differences.

Using incipient colony(s) initiated from founding pairs, particularly when coupled with planar arenas, has yielded unprecedented insights into termite responses to experimental inputs. Notably, the discoveries of molting site affinity and its implication for CSI bait applications (Kakkar *et al.*, 2017, 2018) would not have been possible without employing these laboratory protocols. While these protocols may initially appear intricate and potentially costly, their benefits are highly valuable. On this premise, a comprehensive, step-by-step guide to these experimental protocols is presented in Chapter 18 of this volume. Embracing these protocols will further enhance our knowledge of termite biology, facilitate the development of novel management strategies, and promote greater compatibility of results across different laboratories.

1.6 Conclusion

This book offers comprehensive and up-to-date knowledge of *C. formosanus*, providing readers with a profound understanding of this termite species. It delves into aspects such as basic biology, ecology, symbiosis, and evolution, creating an engaging introduction to termite biology for academic and non-academic readers. What sets this book apart is its practicality. It incorporates chapters focused on pest management, making it a versatile resource for researchers, educators, those involved in extension activities, and industry. The book bridges the gap between theoretical knowledge and practical solutions by addressing real-world applications.

The diverse perspectives offered by the different contributing authors also provide a range of expertise and opinions that would not have been possible otherwise. In Chapter 19 of this volume, contributing authors muse on the potential future avenues of research on *Coptotermes* by pointing to outstanding questions to be answered. Despite significant progress in termite science in recent decades, this final chapter underscores the persistent gaps in our understanding and unveils numerous opportunities for future research endeavors.

In summary, this book is an indispensable resource, consolidating the latest knowledge on *C. formosanus*. Its comprehensive coverage, practical applications, diverse author perspectives, and the prospect of future research make it an invaluable asset for anyone interested in termite biology.

References

Bardunias, P. and Su, N.-Y. (2009) Dead reckoning in the tunnel propagation of the Formosan subterranean termite (Isoptera: Rhinotermitidae). *Annals of the Entomological Society of America* 102, 158–165.

Becker, G. (1969) Rearing of termites and testing methods used in the laboratory. In: Krishna, K. and Weesner, F.M. (eds) *Biology of Termites*. Vol. 1. Academic Press, New York. pp. 351–385.

Breznak, J.A., Brill, W.J., Mertins, J.W. and Coppel, H.C. (1973) Nitrogen fixation in termites. *Nature* 244, 577–580.

Bucek, A., Šobotník, J., He, S., Shi, M., McMahon, D.P. *et al.* (2019) Evolution of termite symbiosis informed by transcriptome-based phylogenies. *Current Biology* 29, 3728–3734.

Chen, G.Y. (2022) Hybridization between Formosan subterranean termite and Asian subterranean termite in Taiwan. MSc dissertation, National Chung Hsing University, Taichung, Taiwan.

Chen, G.-Y., Ke, Y.-L., Liang, W.-R. and Li, H.-F. (2020) Redescription of Formosan subterranean termite, *Coptotermes formosanus* (Blattodea: Rhinotermitidae), with three new synonyms from China. *Acta Entomologica Musei Nationalis Pragae* 60, 599–608.

Chouvenc, T., Helmick, E.E. and Su, N.-Y. (2015) Hybridization of two major termite invaders as a consequence of human activity. *PLoS ONE* 10, e0120745.

Chouvenc, T., Ban, P.M. and Su, N.-Y. (2022) Life and death of termite colonies, a decades-long age demography perspective. *Frontiers in Ecology and Evolution* 10, 911042.

Erhorn, E.M. (1934) The termites of Hawaii, their economic significance and control, and the distribution of termites by commerce. In: Kofoid, C.A. (ed.) *Termites and Termite Control*. University of California Press, Berkeley, California, pp. 321–333.

Huang, S.-Y., Chiu, C.-I, Tsai, Y.-Y., Li, W.-J., Wu, C.-C. *et al.* (2022) Nationwide termite pest survey conducted in Taiwan as a citizen science project. *Journal of Economic Entomology* 115, 1650–1658.

Itakura, S., Yoshikawa, Y., Togami, Y. and Umezawa, K. (2020) Draft genome sequence of the termite, *Coptotermes formosanus*: genetic insights into the pyruvate dehydrogenase complex of the termite. *Journal of Asia-Pacific Entomology* 23, 666–674.

Jasso-Selles, D.E., De Martini, F., Velenovsky IV, J.F., Mee, E.D., Montoya, S.J. *et al.* (2020) The complete protist symbiont communities of *Coptotermes formosanus* and *Coptotermes gestroi*: morphological and molecular characterization of five new species. *Journal of Eukaryotic Microbiology* 67, 626–641.

Kakkar, G., Osbrink, W. Mullins, A. and Su, N.Y. (2017) Molting site fidelity in workers of Formosan subterranean termites (Isoptera: Rhinotermitidae). *Journal of Economic Entomology* 110, 2512–2517. doi: 10.1093/jee/tox246

Kakkar, G., Osbrink, W. and Su, N.Y. (2018) Molting site fidelity accounts for colony elimination of the Formosan subterranean termites (Isoptera: Rhinotermitidae) by chitin synthesis inhibitor baits. *Science Reports* 8, 1259. doi:10.1038/s41598-018-19603-8

King, E.G. and Spink, W.T. (1969) Foraging galleries of the Formosan termite, *Coptotermes formosanus*, in Louisiana. *Annals of the Entomological Society of America* 62, 536–542.

Kitade, O., Hayashi, Y. and Noda, S. (2013) Symbiotic protist communities in the termite *Coptotermes formosanus* in Japan and a comparison of community structures between workers and soldiers. *Japanese Journal of Protozoology* 46, 21–29.

Koidzumi, M. (1921) Studies on the intestinal protozoa found in the termites of Japan. *Parasitology* 13, 235–309.

Lai, P.Y., Tamashiro, M., Fujii, J.K., Yates J. and Su, N.-Y. (1983a) Sudan Red 7B, a dye marker for *Coptotermes formosanus*. *Proceedings of Hawaiian Entomological Society* 24, 277–282.

Lai, P.Y., Tamashiro, M. and Fujii, J.K. (1983b) Abundance and distribution of the three species of symbiotic protozoa in the hindgut of *Coptotermes formosanus* (Isoptera: Rhinotermitidae). *Proceedings of the Hawaiian Entomological Society* 24, 271–276.

Li, H.F., Ye, W., Su, N.Y. and Kanzaki, N. (2009) Phylogeography of *Coptotermes gestroi* and *Coptotermes formosanus* (Isoptera: Rhinotermitidae) in Taiwan. *Annals of the Entomological Society of America* 102, 684–693.

Michael, Z., Chouvenc, T., Su, N.-Y. and Lee, S.-B. (2023) Finding shortcuts through collective tunnel excavations in a subterranean termite. *Behavioral Ecology* 34, 354–362. doi: 10.1093/beheco/arad007

Mullins, A. and Su, N.-Y. (2018) Parental nitrogen transfer and apparent absence of N_2 fixation during colony foundation in *Coptotermes formosanus* Shiraki. *Insects* 9, 37.

Mullins, A., Chouvenc T. and Su, N.-Y. (2021) Soil organic matter is essential for colony growth in subterranean termites. *Scientific Report* 11, 21252. doi: 10.1038/s41598-021-00674-z

Mullins, A., Sheffhrahn, R.H. and Su, N.Y. (2022) Nitrogen inventories and nitrogenase expression rates of a drywood and a subterranean termite. *Annals of the Entomological Society of America* 115, 387–390. doi: 10.1093/aesa/saac014

Nishimura, Y., Otagiri, M., Yuki, M., Shimizu, M., Inoue, J.-I. *et al.* (2020). Division of functional roles for termite gut protists revealed by single-cell transcriptomes. *The ISME Journal* 14, 2449–2460.

Oshima, M. (1909) Taiwanese termites. In: Oshima, M. (ed.) *The First Official Report on Termites*. Taiwan Sōtokufu, Taihoku, Japan, pp. 29–40.

Roisin, Y. (2000) Diversity and evolution of caste patterns. In: Abe, T., Bignell, D.E. and Higashi, M. (eds) *Termites: Evolution, Sociality, Symbioses, Ecology*. Kluwer Academic Dordrecht, Netherlands, pp. 95–119.

Shiraki, T. (1909) On the termites of Japan. *Transactions of the Entomological Society of Japan* 2, 229–242 [in Japanese]

Su, N.-Y. (1994) Field evaluation of a hexaflumuron bait for population suppression of subterranean termites (Isoptera: Rhinotermitidae). *Journal of Economic Entomology* 87, 389–397.

Su, N.-Y. and Scheffrahn, R.H. (1998) A review of subterranean termite control practices and prospects for integrated pest management programs. *Integrated Pest Management Reviews* 3, 1–13.

Su, N.-Y., Scheffrahn R.H. and Weissling, T. (1997) A new introduction of a subterranean termite, *Coptotermes havilandi* Holmgren (Isoptera: Rhinotermitidae) in Miami, Florida, *Florida Entomologist* 80, 408–411.

Su, N.-Y., Chouvenc, T. and Li, H.-F. (2017) Potential hybridization between two invasive termite species, *Coptotermes formosanus* and *C. gestroi* (Isoptera: Rhinotermitidae), and its biological and economic implications. *Insects* 8, 14.

Tamashiro, M., Fujii J.K. and Lai P.Y. (1973) A simple method to observe, trap, and prepare large numbers of subterranean termites for laboratory and field experiments. *Environmental Entomology* 2, 721–722

2 Biogeography of *Coptotermes formosanus*

Rudolf H. Scheffrahn
Entomology and Nematology Department, Ft. Lauderdale Research and Education Center, University of Florida, rhsc@ufl.edu

Abstract
Modern records of the Formosan subterranean termite (FST) in Asia include southeastern China, Hainan Island, northern Vietnam, Taiwan, the Ryukyus Arc, and southern mainland Japan (Kyushu, Shikoku, and Honshu). Geologic and paleoclimatic reconstructions during the Pleistocene support various subaerial connections and FST-suitable climates among Asian land masses. Localities with both FST and its staphylinid termitophiles may also point to an endemic range but it is unknown how flightless termitophiles advance with their host range. The cold conditions in southern Japan during the Last Glacial Maximum (LGM) bring into doubt the viability of FST on the main islands of Japan and suggest possible anthropogenic origin from southern Asian seaports as early as 1691. The occurrence of FST on the deep-water islands of Lanyu, Chichijima, and Hachijojima point to shipboard infestations. The first establishment of FST outside of Asia was discovered in Hawai'i before 1907 followed by the first Nearctic discovery in 1957 from coastal North Carolina. So far, genetic evidence has given general trends in global dispersal of FST from Asia to Hawai'i and then to the U.S. mainland, but more precise pathways and timing remain obscure. The global latitudinal distribution of FST is 18° N to 36 °N and is primarily limited to temperate climates with hot summers and with or without dry seasons. Some native and introduced climates are either tropical or arid. Before 1983, all North American establishments of FST were located near marine dockage, confirming the importance of alate dispersal from infested boats. Later inland infestations were probably the result of roadway transport of infested goods. Future global range expansion of FST is discussed.

2.1 Introduction

The biogeography of organisms is driven by natural and, in some cases, human forces. Before the Pleistocene, the dispersal of non-avian terrestrial organisms was directed by geologic and climatic forces and over-water rafting events (Antonelli, 2017; Darlington, 1957; De Queiroz, 2014; Lindo, 2020). About 70,000 years ago (70 kya), humans began their migration out of Africa as the first dominant invasive species (Marean, 2015). Mice and rats are well-known examples of very early human-aided transoceanic dispersals from about 2 kya (Searle *et al.*, 2009; Matisoo-Smith and Robins, 2004; Zeng *et al.*, 2018). Although natural forces continued to influence the biogeography of species following the Last Glacial Maximum (LGM) (Hewitt, 2003), the anthropogenic introduction of invasive or alien species probably began when regional maritime

DOI: 10.1079/9781800621596.0002

trade (ca. 2 kya) (Dennell, 2017; Yongjing *et al.*, 2013) and global maritime trade began to flourish ca. 0.5 kya (Hulme, 2009; Boivin *et al.*, 2016).

The natural dispersal of termites is short-ranged and gradual, being measured at rates of hundreds of meters per quinquennial, due to slow colony maturation time, short alate flight distance, and post-flight colony inception (Tonini *et al.*, 2013, 2014; Borges *et al.*, 2014; Mullins *et al.*, 2015; Scheffrahn *et al.*, 2017). These factors preclude natural flight dispersal over small (> 250 m) bodies of water, after which a viable mate must be encountered, followed by sequestration in a favorable microclimate. Although natural transoceanic rafting shaped the historical biogeography of extant termite taxa, successful continental establishments by these extremely rare events are measured in million-year timescales (Bourguignon *et al.*, 2016, 2017; Buček *et al.*, 2022). Incidents of natural transoceanic rafting by termites have never been directly observed.

The first recorded anthropogenic transport of a subterranean termite occurred sometime before 1837 when Kollar (1837) described the imago of the Nearctic *Reticulitermes flavipes* (Kollar) from hothouses in Schönbrunn, Austria. The mechanism for this landlocked introduction is unknown but the termites were found infesting wooden plant tubs imported from America (Gay, 1967). *Reticulitermes flavipes* did not establish in the Austrian landscape but was followed by separate successful establishments in Southern Europe (Austin *et al.*, 2005; Dedeine *et al.*, 2016). As with rodents, anthropogenic overwater dispersal of termites may have occurred much earlier than first detected (Scheffrahn *et al.*, 2009).

The earliest encounters of FST originate in Asia. Shiraki (1909) described FST from specimens collected in Taipei, on the island of Taiwan (formerly Formosa). Oshima (1913) was the first to report FST from Japan (north to south: Honshu, Shikoku, Kyushu, Hachijojima, and the Ryukyu Archipelago). Repeating Oshima's 1913 localities, Yano (1915) was the first to report *C. formosanus* (FST) from mainland China ("southern part of China"). The pre-human distribution of FST, however, cannot be inferred from its first encounters, but must be hypothesized from paleographic and paleoclimatic events in this region.

2.2 Range of FST during the Pleistocene, the last Glacial Maximum, and the Mid-Holocene

The most recent major east Asian geological event began 2 million years ago (mya) when the Ryukyus Archipelago formed from the eastern margin of the Chinese mainland (Xu *et al.*, 2016). At this time, the Okinawa Trough began its rift between the mainland and the proto-Ryukyu islands. Deep sediments in the rift from the Yangtze River (Park *et al.*, 1998) were eroded by the warm Kuroshio Current, thereby separating the limestone proto-Ryukyus from the mainland. By 1.5 mya, several deep gaps (from the south: Yonaguni, Kerama, and the Tokara Gap) divided the Ryukyu archipelago (Kimura, 2000; Osozawa *et al.*, 2012) to form the southern, central, and northern Ryukyu Island groups (Fig. 2.1). Osozawa *et al.* (2013) gave numerous examples of speciation by volant insect biota between China and the Ryukyus. This suggests that FST may also have reached the Ryukyus before the Holocene but later anthropogenic introductions to the Ryukyus may have happened (see below). At this time, the Tsushima current, an extension of the Kuroshio current, flowed into the Japan Sea through the Tsushima Strait. The Tsushima Strait may have been a biological barrier between Japan, Korea, and northern China during warmer interglacial periods (Osozawa *et al.*, 2012); however, Jiang *et al.* (2021) suggested that Kyushu, the southernmost main island of Japan, was subaerial with Korea and eastern China in the East China Sea from 600 kya to the LGM. Today the Korean peninsula remains free of FST (Lee *et al.*, 2021). Taiwan and Hainan joined the mainland of China numerous times during the late Pleistocene (Voris, 2000) during several interglacial periods from 0.2 million kya to the LGM (Zhang *et al.*, 2022). A genetic comparison of

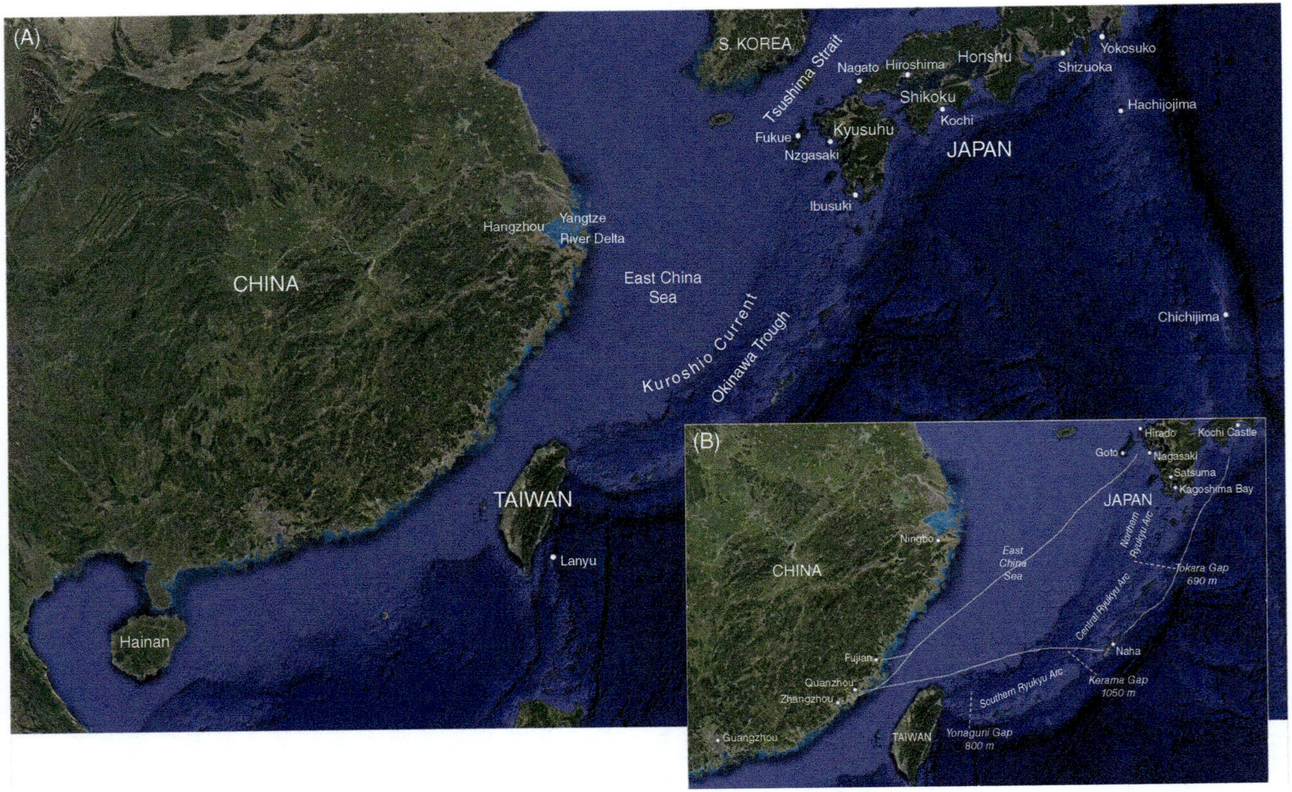

Fig. 2.1. (A) Locations referred to in text and bathymetry. (B) Approximate trading routes between China, Ryukyu, and Japan adapted from the Selden map (Batchelor, 2013).

FST (Li *et al.*, 2009) confirmed that Taiwan and China constituted part of the natural range of the termite. Similarly, Hainan and the southern coast of China were connected during glacial cycles (Yao *et al.*, 2009) and genetic differences between Hainan and mainland China FST populations could not be detected (Li *et al.*, 2009).

The native range of FST during the LGM (late Pleistocene, ca. 20 kya) could be inferred from climate reconstructions. At 18 kya, pollen records show that southernmost China, Taiwan, and Hainan shared a warm mixed forest habitat while further north (present-day Hangzhou, 30° N) the biomes consisted of steppe and desert (Yu *et al.*, 2000), suggesting exclusion of FST northward. Low sea levels bound the three as a single landmass for the last time during the LGM (Lambeck *et al.*, 2014). By the early Holocene (ca. 10 kya), the sea level, measured from the central coast of China, was 40 m below present (Liu *et al.*, 2021), at which time both Taiwan (Voris, 2000) and Hainan (Yao *et al.*, 2009) once again emerged as continental islands. The Ryukyus remained an archipelago due to the deep-water Okinawa Trough and interisland gaps (up to 1050 m, Fig. 2.1) surrounding these islands (Kaifu *et al.*, 2020). During the LGM, the temperatures of central China and southern Japan (Kyushu) were estimated to be much colder: 4–10°C (Zhuo *et al.*, 1998), 5–6°C (Tsukada, 1983), 7°C (Gotanda and Yasuda, 2008) below Holocene temperatures. During the LGM, Hiroshima was 8.3°C colder than during the mid-Holocene (Kato *et al.*, 2021). Today, the coldest non-endemic locality for FST (Vargo *et al.*, 2003) is in Rutherford County, North Carolina (January mean high, mean, and mean low were 11°C, 5°C, and 1°C, respectively). When decreasing the current January temperatures by 6°C to approximate LGM temperatures, the Japanese FST-inhabited localities would have January temperatures as follows: Nagasaki, Kyushu (4°, 1°, –2°C), Ibusuki, Kyushu (7°, 2°, 0°C), Kochi, Shikou (5°, 0°, –5°C), Nagato, Honshu (3°, 0, ° –3°C), and Tokyo (Nanei), Honshu (3°, 0°, –3°C). This may have been too cold to sustain the FST in Japan before temperatures rebounded in the Early Holocene. Blumenfeld *et al.* (2021) found that the populations of FST from eastern China clustered together with those from Okinawa but not mainland China. Temperature was shown to be the main driver for ant biodiversity patterns during the LGM in Japan (Wepfer *et al.*, 2016).

By the mid-Holocene (6 kya), thermal maxima were attained (Otto-Bliesner *et al.*, 2006) and the northern hemisphere's climate approximated current conditions (Wanner *et al.*, 2008). The forest biomes of eastern China and the southern Japanese Archipelago were also similar to present-day forests based on pollen analyses (Lin *et al.*, 2019). The Köppen climate classifications for eastern China and southern Japan were also congruent during the mid-Holocene (Chen and Chen, 2013; Oh and Shin, 2016). Therefore, climate and land cover in southern Japan would likely have been conducive for successful anthropogenic introduction of the FST as early as the mid-Holocene were it not for the absence of seaworthy vessels and frequent trade at that time.

2.3 Do termitophiles mark the endemic boundaries of their host?

It has been hypothesized that the co-occurrence of obligate termitophiles and their termite hosts marks the endemic range of both. Emerson (1955) speculated that *Thyreoxenus*, a staphylinid termitophile, accompanied its *Nasutitermes* host to the Solomon Islands via the pre-tectonic landbridge hypotheses. Even after Tectonic Theory, Emerson (1971) maintained that the geographic localities of termitophiles gave strong evidence for the endemic origins of its host. He further stated:

> No termitophilous beetle has ever been found with dealates or small incipient colonies of living termites. Host-specific termitophilous beetles must find their hosts sometime after the maturity of the termite colony, so that geographical occurrence together indicates continuous land masses or close proximity of land masses under appropriate climatic conditions.

When Kistner (1985) described *Sinophilus xiai*, an obligate staphylinid termitophile of FST from mainland China, he also adopted

Emerson's argument and concluded that only mainland China was the endemic origin of FST. The discovery of two termitophilous staphylinids of FST, *Sinophilus yukoae* Maruyama & Iwata and *Japanophilus hojoi* Maruyama & Iwata, on mainland Japan and the Ryukyu islands (Maruyama and Iwata, 2002, Maruyama *et al.*, 2012; Kanao *et al.*, 2019) expanded the "native" range of FST based on Emerson's and Kistner's hypotheses. Finally, Liang *et al.* (2020) collected both *S. yukoae* and *J. hojoi* from FST colonies in Taiwan, completing the hypothesized native range of FST to include all major islands bordering the East China Sea between 22° N and 34° N.

The reasoning for coupling termitophile distribution with the native range of the termite host does not consider that anthropogenically transported colonies on seagoing vessels (see next section) might also host termitophiles as they do on land. When discussing the transport of *Sinophilus xiai*, Kistner (1985) could not envision the existence of an FST nest on board a boat. He wrote that:

> ... while termites, such as *Coptotermes* can be transported relatively easily in shipments of wood and wood products, it is very difficult to transport the termitophiles because this usually demands that the entire nest be transported. This probably would not happen because the nests are located underground or in rotten stumps of no commercial value.

No termitophiles have yet been found on *Coptotermes*-infested vessels. The tiny (1.2 mm long) first instar of *J. hojoi*, a flightless adult (T. Kanao, Japan, 2023, personal communication), is adorned with four anal hooks (Liang *et al.*, 2020). Could these hooks be used to cling to FST alates? Although alate phoresy by termitophiles has not been observed, it has been implied from amber fossils (Ninon *et al.*, 2019) and suggests that the endemic termitophile/host hypothesis could be argued against. To date, however, no termitophiles have been found on *Coptotermes*-infested vessels.

2.4 Boats, not cargo, as primary overwater dispersal platforms for FST

Many reports suggest that long-distance transport of termites is facilitated by onboard shipping cargo such as infested crates, packaging, pallets, lumber, or shipping containers (La Fage, 1987; Mori, 1987; Su and Tamashiro, 1987). Although evidence is convincing that incipient colonies of FST were spread throughout the southeastern U.S. in infested railroad cross-ties (Jenkins *et al.*, 2002; Austin *et al.*, 2008), there is no compelling evidence that cargo of any type unloaded from ships resulted in colony establishment on land. In his treatise on introduced termites, Gay (1967) stated that:

> ... ships have played a significant role in the dispersal of termites, not merely by carrying termite-infested cargoes from place to place but also through infestations established in the woodwork of the ships themselves.

But Gay (1967), citing cargo interceptions of *Coptotermes*, did not provide evidence that, had these been overlooked, they would have resulted in terrestrial establishments. Brown (1934) reported *Coptotermes* "nesting in large numbers in the woodwork of an ocean liner lying at dock on the San Francisco waterfront" in 1927. Kalshoven (1962) gave a personal account of a 1937 incident of a "thriving colony" of *Coptotermes gestroi* (Wasmann) causing destruction to the woodwork of a small tanker sailing between Sumatra and other Asian seaports. Li *et al.* (2008) reported an FST-infested sport-fishing yacht on Lanyu Island, Taiwan. Dhang (2014) noted numerous infestations of *C. gestroi* and *Microcerotermes losbanosensis* Oshima aboard yachts in Manila harbor. Although distant oceanic travel by yachts and sailboats is uncommon, these vessels can be carried worldwide by large ships designed for that purpose (DYT Yacht Transport, 2023).

All early FST discoveries in the southeastern U.S. were in the proximity of marine-accessible dockage (Beal, 1987), pointing to boats as vehicles for introduction and establishment. Boats, ranging from ancient junks (see next section) to modern yachts, contain the microenvironment needed for FST colony foundation and maturation: wood, voids, and moisture directly from rainfall or seawater accumulations in the bilge. The importance of boats as carriers of mature FST colonies became clear from observations in Florida and Australia, with the first structural boat infestation of a boat docked on Islamorada, Florida Keys in 1986 (Scheffrahn and Crowe, 2011). Scheffrahn and Crowe (2011) reported 28 yacht and sailboat infestations of FST in Florida between 1986 and 2008, including two that originated from Hong Kong. Between 1994 and 2009, they reported 13 infested boat interceptions of FST in Australian waters. These boats were almost exclusively private vessels (Fig. 2.2). In 1986 a wooden yacht from Taiwan was found to be infested with FST at a marina in Hypoluxo, Florida (R.H. Scheffrahn, Florida, 1986, personal observation). In 2008, Lee *et al.* (2009) reported that a 14 ft fiberglass fishing boat, trailered 5 years earlier from New Orleans to the inland community of Poplarville, Mississippi, contained a mature FST colony of some 150,000 individuals. Hochmair and Scheffrahn (2010) observed a strong correlation between locations of land-based infestations of *Coptotermes* and dockage of private vessels (yachts, sailboats, etc.). When infested boats are docked during the late spring dispersal season, the crepuscular FST alates are readily attracted to lights on land or other vessels. The University of Florida Termite Collection database (Scheffrahn, 2019) currently lists 104 termite infestations onboard private marine vessels, including 74 *Coptotermes* spp. and, of those, 44 FST infestations, none involving cargo.

Modern construction of container ships, shipping methods, quarantine inspections, and shipping regulations for cargo virtually preclude the commercial transport of viable termite colonies. Since 1952, the International Plant Protection Convention (IPPC) under the Food and Agriculture Organization of the United Nations have placed regulations on the trade of plants and plant products, including raw wood and lumber, to

Fig. 2.2. Boat infestations of Formosan subterranean termites. (A) Nest in houseboat in Florida; (B) drydock boat in Florida; (C) quarantined drydock yacht in Australia (courtesy W. Crowe); and (D) 27 m yacht under fumigation in Florida.

prevent the spread of pests (FAO, 2011). In 2002, the International Standards for Phytosanitary Measures No. 15 (ISPM15) was adopted to provide treatment standards for wood packaging materials used in international trade (International Plant Protection Convention, 2002). Examples of wood packaging materials regulated by ISPM15, primarily focused on wood-boring beetles (Haack *et al.*, 2014), include dunnage, crates, reels, collars, and pallets; and other wooden items such as bracing and all solid wood and wood packaging materials (WPM) over 6 mm in thickness. All wood products must be treated using heat (core 56°C for at least 30 min) or methyl bromide/sulfuryl fluoride fumigation (Allen *et al.*, 2017). Both Canada and the U.S. are members of the IPPC and have adopted ISPM-15 regulations to maintain international obligations along with international trading partners (IPPC, 2018).

Developed in the 1950s, standardized metal containers are now universally used to transport practically all non-bulk cargo on ships (Vis and De Koster, 2003; Heins, 2016). The containers themselves lack cellulose and termite refugia, and commercial cellulosic cargo within containers are kept free of water that could degrade cargo quality. Therefore, using commercial log, lumber, and timber import records as proxies for future overseas transport of FST to a warming Korea (Lee *et al.*, 2021) would be unconvincing compared with data on the travels of private vessel.

Blumenfeld and Vargo (2020) evaluated 906 international non-native termite interceptions recorded by the USDA APHIS PPQ from 1924 to 2017 at commercial U.S. ports of entry. Presumably, international seaports constituted the vast bulk of interceptions, with land border inspections from Mexico accounting for 48 interceptions. The USDA data used by Blumenfeld and Vargo (2020) listed 45 FST interceptions between 1932 and 2017, but 16 originated from regions that do not support FST (e.g., Europe, tropical Pacific, and southeast Asia). The authors did not specify the condition of the termite samples (e.g., caste or number) at the time of interception, so it could not be determined if any of these cases could have resulted in viable establishments had they not been detected. Many of the intercepted taxa, especially the Termitidae (e.g., *Amitermes*, *Anoplotermes*, *Cahualitermes*, *Odontotermes*, *Parvitermes*, *Subulitermes*, *Termes*, etc.), have never become established outside their endemic habitats, because of their unique microclimate and food requirements.

The USDA APHIS PPQ cargo inspections are limited to commercial carriers at U.S. seaports, airports, and U.S.–Mexico land border crossings (McCullough *et al.*, 2006). Operators of pleasure boats originating from foreign ports must report to U.S. Customs and Border Protection (CBP) within 48 h of their arrival. Passengers must present identification documents and a CBP officer may inspect the vessel and its cargo for contraband; however, unlike commercial cargo, there is no mechanism in place for pest inspections (CBP, 2023). Unlike container vessels that have an unloading/loading turnaround time of less than one day (Ducruet and Merk, 2013), sailboats and yachts may remain docked for months or years, allowing ample opportunity for alates to disperse to land. Hochmair and Scheffrahn (2010) and Scheffrahn and Crowe (2011) argued that the lack of focus on shipboard inspection of termites had allowed termites to spread domestically and internationally. Unless focus is placed on inspecting a vessel's internal structure for termites, FST and other species will continue to expand their range.

Mature shipboard FST colonies add an additional dimension to long-distance dispersal not realized by other invasive taxa. Firstly, boat infestations are comparable to "floating bridgeheads." Normally, the "bridgehead effect" refers to an exotic pest population that becomes the source of additional, usually distant, new invasions (Bertelsmeier and Keller, 2018). Blumenfeld and Vargo (2020) limited bridgehead populations of invasive termites to terrestrial localities. Boat infestations supplement or replace terrestrial bridgeheads as intermediary platforms for invasive range expansion. Large dispersal flights of FST (Su and Scheffrahn, 1987) from ships (Scheffrahn and Su, 2005) produce overwhelming propagule pressure on nearby climatically suitable terrestrial habitats on which trees or buildings occur.

In the context of a mature termite colony aboard a boat, propagule pressure, i.e., the number of reproducing individuals released into a non-native region, is uniquely distinct for termites compared with invasion by solitary organisms per Lockwood *et al.* (2005) or as interception of a few specimens at or from foreign locations (Blumenfeld and Vargo, 2020). As exemplified by FST, a single colony can release thousands of alates during periodic bursts in the flight season (Su and Scheffrahn, 1987). When docked, FST alates will be attracted to nearby lighted residential areas or other illuminated boats where many habitats exist for successful incipient colony development. Furthermore, there is no evidence that the diversity of invasive populations of FST differs from those of endemic origin, or that FST populations must undergo adaptations for propagule invasiveness (Blumenfeld *et al.*, 2021). Thus, they are not influenced to reduced fitness from the Allee Effect (Taylor and Hasting, 2005).

The FST does not fit the normal paradigms for most invasive species (Sakai *et al.*, 2001; Beck *et al.*, 2008). It has the same biological traits in its Asian range (China, Taiwan, Okinawa, Japan) as it does in introduced non-Asian geographies. In both hemispheres, FST is an economic pest (Sugio and Miyaguni, 2019; Mori, 1987; Lin, 1987; Li *et al.*, 2009) and established colonies have no natural enemies (Chouvenc *et al.*, 2011).

2.5 Early anthropogenic dispersal within Asia

If Kyushu was indeed too cold to support FST during the LGM, anthropogenic establishment by shipboard infestations after Holocene warming is plausible. Records of Chinese sailing vessels (junks) that were used for navigating coastal routes from China to Korea and Japan (Schottenhammer, 2012) go back 2 millennia to the Han Dynasty (Guan *et al.*, 2013). The oldest Chinese shipwreck, a 13th century Song Dynasty vessel known as the "Quanzhou Ship", was discovered near the city of Quanzhou in Fujian Province (Green, 1983; Green and Burningham, 1998). It was during this time that trade from China to Korea, Ryukyus, and Japan was expanding (Schottenhammer, 2012). The "Quanzhou Ship" was a three-masted vessel, 34.4 m in overall length and 9.84 m at the beam, with an estimated loaded draught of 3 m and an estimated displacement of about 374 tons (Green, 1983; Li, 1989). It was estimated that some Song period junks had a displacement of up to 1900 tons, which is "in all probability the largest trading vessels anywhere in the world at the time" (Wake, 1997). The oldest tangible evidence (ca. 1619) of early Chinese trading routes with the southern Japanese archipelago was discovered in a British library in 2008. This "Selden Map of China" (Batchelor, 2013) depicts merchant shipping routes between southern Chinese seaports, including Quanzhou, to the Ryukyus (Okinawa) and the main southern Japanese islands of Kyushyu (Nagasaki) and Shikoku (Fig. 2.1). Mori (1987) reports the earliest reliable record of FST in Japan was that of Dr. Engelbert Kaempfer (1712). During his visit to Kyushu in 1690–1691, Kaempfer was restricted to the artificial island of Deshima, at the seaport of Nagasaki (Bowers, 1966) where he presumably collected the termites. Early shipborne introductions may have introduced FST to the deep-water islands of Lanyu (Taiwan), Chichijima, and Hachijojima (Fig. 2.1).

Mori (1987) reported that, before World War II, the range of FST in Japan extended only to the Pacific Coast west of Shizuoka Prefecture, probably from Sodeshi (now Shizuoka City), which was where Okada (1912) found FST. Mori (1987) argued that the FST was introduced further to the north and west (Yokosuka Naval Base) by U.S. military activities associated with the war. However, the current January mean high, mean, and mean low temperatures for Yokosuka (10°, 7°, 4°C) (Weather Spark, 2023) are actually slightly warmer than those of Shizuoka (10°, 5°, 2°C), meaning that the climate was suitable for FST in Yokosuka before the war and that FST may have been present there before the war.

Genetic studies to date support that FST was introduced to Japan but do not provide a precise time of occurrence. Vargo *et al.* (2003) demonstrated that Japanese FST populations from Kyushu and Fukue islands had no significant isolation and no evidence of a recent bottleneck, suggesting they were introduced by humans hundreds of years ago. Husseneder *et al.* (2012) reanalyzed FST populations from Kyushu and Fukue (Vargo *et al.*, 2003) and compared their genetic results with populations from China, Hawai'i, and the continental U.S., again pointing to an early introduction of FST to Japan. Blumenfeld *et al.* (2021) found that the populations of FST from eastern China clustered together with those from Okinawa but not mainland Japan, suggesting anthropomorphic introduction of FST between China and Okinawa, and a separate introduction to Japan.

2.6 Anthropogenic dispersal beyond Asia

The earliest published account of FST beyond Asia was recorded on 5 December 1913 from an infestation in the Kamehameha Chapel, Honolulu, Hawai'i (Swezey, 1914). Tamashiro *et al.* (1987) noted that the first FST alates in Honolulu were caught in 1907 and were deposited in the Bishop Museum. Su and Tamashiro (1987) suggested that the Hawaiian establishments were the result of 19th century sandalwood trade (see above regarding cargo) and the arrival of Asian immigrants. The sandalwood logs, however, were exported from Hawai'i to China (St John, 1947), suggesting that the trade boats, not the wood, were the vehicles of FST introduction to Hawai'i. The FST is now widespread in the Hawai'ian Archipelago (Husseneder *et al.*, 2012).

The first report of the presence of FST in the New World was that of "a well-established colony with thousands of workers ... recently discovered in a vessel in San Francisco harbor" (Light, 1929). To date, no establishment of FST has occurred in northern California (Fig. 2.3). Specimens of the first established population in the New World were collected in Charleston, South Carolina, in 1957 (Chambers *et al.*, 1988). Since then, FST has proliferated throughout the southeastern U.S. (Fig. 2.3). La Fage (1987) speculated that the FST, first reported in 1966 in Louisiana, originated with the influx of post-WWII shipments from Asia to Lake Charles and New Orleans. Beal (1987) reported FST infestations at Texas shipyards in Houston (1965) and Galveston (1966), while Thompson (1985) reported an infestation in Hallandale, Florida, in 1980, and Howell *et al.* (1987) reported a population in Beaumont, Texas, in 1981.

It was suspected that in the 1970s an infested yacht from Taiwan, docked on the intercoastal waterway in Hallandale, Florida, initiated the original land-borne FST infestation in the state. Alates, preparing for dispersal flights, have been collected from boats in marinas in Fort Lauderdale, Hollywood, Marathon, Madeira Beach, Ponce Inlet (18 m Hatteras yacht from Hong Kong), Port Orange, and Tampa, Florida (data in Scheffrahn, 2019). Insular infestations of FST were discovered on residential islands east of Miami (Biscayne Bay), Grand Bahama Island, Bahamas (Scheffrahn *et al.*, 2015), and on Marathon, Florida Keys (data in Scheffrahn, 2019).

The first inland FST infestation was reported by Thompson (1985) in Orlando, Florida, in 1983. Sponsler *et al.* (1988) recorded the second inland infestation in Auburn, Alabama, in 1987. Even recently, most FST infestations are within 2 km of marine dockage (Hochmair and Scheffrahn, 2010) in Florida as well as globally (Fig 2.2). There is considerable circumstantial evidence that inland FST populations were spread by movement of infested trees, reused old railroad crossties (sleepers), etc. throughout the southeastern U.S. (Sponsler *et al.*, 1988; Henderson, 2001; Jenkins *et al.*, 2002; Messenger *et al.*, 2002; Austin *et al.*, 2006; Evans *et al.*, 2019). It is entirely likely that a royal pair or young incipient FST colony could be sustained in the crevices or cracks on the

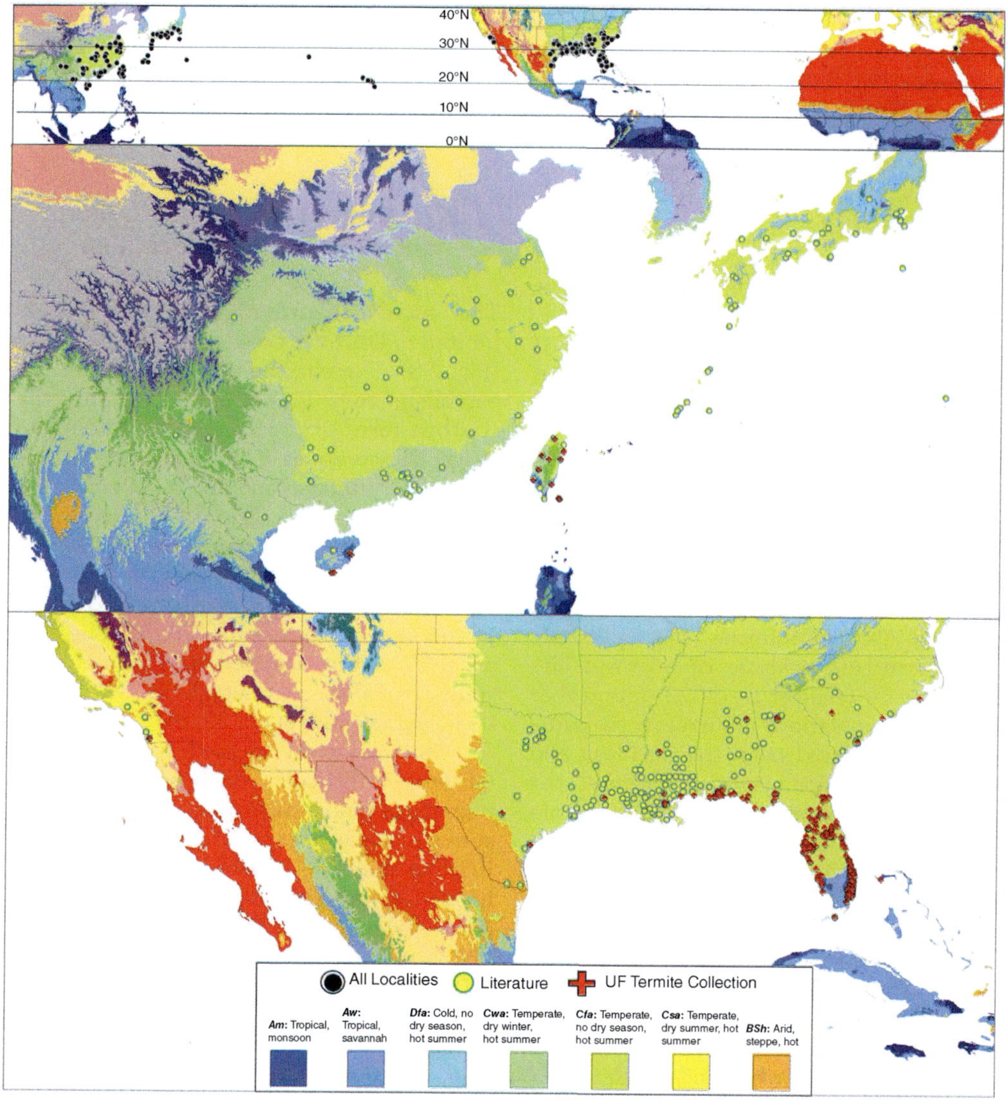

Fig. 2.3. Current global distribution of the Formosan subterranean termite from published localities (Krishna *et al.*, 2013) and University of Florida Termite Collection data (Scheffrahn, 2019). Base map of Köppen-Geiger climate classifications from Beck *et al.* (2018).

surface of weathered and wet railroad ties before installation in outdoor landscapes. J.W. Austin (North Carolina, 2022, personal communication) noted that two cases in Texas were neighborhood FST populations probably originating from FST-infested ties transported from a storage depot to Lakeway and Rockport in 2005. The hypothesis that FST can infest commercial shipments of wood products, e.g., palettes, crates, and other wooden material (Su and Tamashiro, 1987; Husseneder *et al.*, 2012; Lee *et al.*, 2021) has not been directly observed. It is unlikely that an FST colony could survive transport in dry commodities for more than a few days.

The 1992 discovery of a mature population of FST in La Mesa, California, was probably the result of a direct shipment of nursery goods from Hawai'i in June 1976 (Atkinson *et al.*, 1993). Establishment of an FST population in this Mediterranean climate was unexpected, yet as early as 1993 the population released hundreds of alates up to 270 m from the infested neighborhood. (Haagsma *et al.*, 1995). By 1997, FST alates were trapped up to 1,100 m from the original source, and additional adjacent infestations were discovered (Rust *et al.*, 1998). In 2018, FST again surfaced in La Mesa, a likely descendent of the original 1992 population (Tseng *et al.*, 2021). In 2019, FST was found in a residential neighborhood in Canyon Lake, California, but the haplotype of this population differed from that of the La Mesa populations (Tseng *et al.*, 2021). In 2020 FST was found to be established in Israel (Scheffrahn *et al.*, 2020). In 2021, two additional discoveries appeared in Rancho Santa Fe and Highland Park, California (Tseng *et al.*, 2022). Structure analyses of microsatellite genotypes by Tseng *et al.* (2022) found that both 2021 populations occurred in the same cluster but differed from the La Mesa and Canyon Lake populations. Therefore, the California FST populations originated from three independent introduction events outside the state. The discovery of localized populations of FST in California and Israel predicts future establishments in rather hostile Mediterranean urban/suburban environments, albeit with limited or no dispersal capabilities but supported by irrigation as a substitute for rainfall.

Recently, Blumenfeld and Vargo (2020) argued that termite introductions, including FST, are aided by mechanisms of propagule pressure and bridgehead effects. Using USDA APHIS records for termite interceptions at U.S. ports of entry, they found a significant association between the frequency of foreign species interceptions (propagule pressure proxy) and the likelihood of establishment. They also found that "bridgehead" interceptions, e.g., FST from Hawai'i (a "secondary" location where FST had previously invaded) were more common than those from Asia (a "primary" location). Using phylogenetic analyses, Blumenfeld *et al.* (2021) reconstructed the invasive global routes of FST. The clearest invasion route by FST was Hawai'i to Texas and Louisiana. The Florida populations arose from the Asian and the Texas/Louisiana populations.

2.7 Current global distribution and projected new localities

The current distribution of the FST is limited to latitudes between 18° N and 36° N (Fig. 2.3). Therein, the vast majority of FST establishments in eastern Asia and the southeastern U.S. occur in the Köppen–Geiger climate zones of *Cfa* (temperate, no dry season, hot summer) and *Cwa* (temperate, dry, hot summer, mostly China). Adjacent localities with tropical climates include Hainan Island (Li *et al.*, 2009), Hawai'ian Archipelago, southern Florida, and Grand Bahama Island (Scheffrahn *et al.*, 2015). The FST occupies a few isolated arid climates in southern California (Tseng *et al.*, 2022) and Israel (Scheffrahn *et al.*, 2020).

Morphological identification of *Coptotermes* is difficult, which has resulted in numerous synonymies of species within the genus (Chouvenc *et al.*, 2016). For example, Krishna *et al.* (2013) listed nine synonymies for FST alone. Reports of FST in the Philippines (Light, 1921; Okonya and Kroschel, 2013) and the Marshall Islands (Sugerman, 1972) are almost certainly attributable to the misidentification of *C. gestroi.* The identification of FST in east Africa (Wanyonyi *et al.*, 1984) is likely an endemic African *Coptotermes* species. Guam was reported as an FST locality (Hromada, 1970) but was confirmed to be *C. gestroi* when the island was surveyed (Su and Scheffrahn, 1998). Populations in South Africa (Coaton, 1950) were apparently extirpated (V. Uys, South Africa, 2005, personal communication) and the record of FST from Sri Lanka (Ahmad, 1953a) is also a likely misidentification of *C. emersoni* (Ahmad, 1953b).

The status of FST in Vietnam has become more certain. Vu *et al.* (2007) reported FST in residential areas of Hanoi. Nguyen *et al.* (2014) used genetic markers to identify *C. gestroi* in urban areas around

Hanoi but did not cite the work of Vu *et al.* (2007). Nam *et al.* (2018) collected both FST and *C. gestroi* in the southern highlands of Vietnam but cited neither Vu *et al.* (2007) nor Nguyen *et al.* (2014). Considering the climate similarities of Hainan, southeastern China, and northern Vietnam (Fig. 2.3), it is indeed likely that both *Coptotermes* species are sympatric residents of Vietnam. Reported Vietnam localities of FST (Nguyen *et al.*, 2014) are included in Fig. 2.3. Alates and soldiers of *C. gestroi* and FST can now be readily identified morphologically (Scheffrahn *et al.*, 1990; Scheffrahn *et al.*, 2015).

Future establishments of FST will be driven by local climate, coastal accessibility, and the effectiveness of official quarantine regulations. Based on the expansive *Cfa* biome in the southeastern U.S. (Fig. 2.3) (Beck *et al.*, 2018), it is possible that FST will be found in southern Arkansas and Oklahoma. Other northern hemisphere localities in the less favorable biome could include northeastern India, the Azores, and Jeju Island, S. Korea. Given that tropical *Am* and *Aw* climates support FST in Hainan, Hawai'i, southern Florida, and Grand Bahama Island, future proximal tropical localities might include other Bahamian Islands, Cuba, and along the southeastern coast of Mexico, including the Yucatan Peninsula. Given that Mediterranean climates (*Csa*) can support FST populations, albeit under marginal growth and expansion conditions, the northern coast of the African continent might be a plausible establishment zone. Future southern hemisphere establishments, most likely between 18°S and 36°S, might occur along the southeast coast of South Africa, far southeastern Brazil, Easter Island, Reunion Island, and eastern coastal Australia.

References

Ahmad, M. (1953a) Two new cases of introduction of termites. *Spolia Zeylanica* 27, 35–36.

Ahmad, M. (1953b) New termites and hitherto unknown caste from Ceylon. *Spolia Zeylanica* 27, 37–41.

Allen, E., Noseworthy, M., and Ormsby, M. (2017). Phytosanitary measures to reduce the movement of forest pests with the international trade of wood products. *Biological Invasions* 19, 3365–3376.

Antonelli, A. (2017) Biogeography: drivers of bioregionalization. *Nature Ecology and Evolution* 1, 1–2.

Atkinson, T.H., Rust, M.K. and Smith J.L. (1993) The Formosan subterranean termite, *Coptotermes formosanus* Shiraki (Isoptera: Rhinotermitidae), established in California. *Pan-Pacific Entomologist* 69, 111–113.

Austin, J.W., Szalanski, A.L., Scheffrahn, R.H., Messenger, M.T., Dronnet, S. *et al.* (2005) Genetic evidence for the synonymy of two *Reticulitermes* species: *Reticulitermes flavipes* and *Reticulitermes santonensis*. *Annals of the Entomological Society of America* 98, 395–401.

Austin, J.W., Szalanski, A.L., Scheffrahn, R.H., Messenger, M.T., McKern, J.A. *et al.* (2006) Genetic evidence for two introductions of the Formosan subterranean termite, *Coptotermes formosanus* (Isoptera: Rhinotermitidae), to the United States. *Florida Entomologist* 89, 183–193.

Austin, J.W., Glenn, G.J. and Gold, R.E. (2008) Protecting urban infrastructure from Formosan termite (Isoptera: Rhinotermitidae) attack: a case study for United States railroads. *Sociobiology* 51, 231–247.

Batchelor, R. (2013) The Selden map rediscovered: a Chinese map of East Asian shipping routes, c. 1619. *Imago Mundi* 65, 37–63.

Beal, R.H. (1987) Practical considerations of the Formosan subterranean termite in Louisiana. In: Tamashiro, M. and Su, N.Y. (eds) *Proceedings of the International Symposium on the Formosan subterranean termite.* Hawaii Institute of Tropical Agriculture and Human Resources. University of Hawaii, Honolulu, Hawaii, pp. 48–53.

Beck, H.E., Zimmermann, N.E., McVicar, T.R., Vergopolan, N., Berg, A. *et al.* (2018) Present and future Köppen-Geiger climate classification maps at 1-km resolution. *Scientific Data* 5, 1–12.

Beck, K.G., Zimmerman, K., Schardt, J.D., Stone, J., Lukens, R. *et al.* (2008) Invasive species defined in a policy context: Recommendations from the Federal Invasive Species Advisory Committee. *Invasive Plant Science and Management* 1, 414–421.

Bertelsmeier, C., and Keller, L. (2018) Bridgehead effects and role of adaptive evolution in invasive populations. *Trends in Ecology and Evolution* 33, 527–534.

Blumenfeld, A.J. and Vargo, E.L. (2020) Geography, opportunity and bridgeheads facilitate termite invasions to the United States. *Biological Invasions* 22, 3269–3282.

Blumenfeld, A.J., Eyer, P.A., Husseneder, C., Mo, J., Johnson, L., Wang, C. *et al.* (2021) Bridgehead effect and multiple introductions shape the global invasion history of a termite. *Communications Biology* 4, 1–12.
Boivin, N.L., Zeder, M A., Fuller, D.Q., Crowther, A., Larson, G. *et al.* (2016) Ecological consequences of human niche construction: Examining long-term anthropogenic shaping of global species distributions. *Proceedings of the National Academy of Sciences* 113, 6388–6396.
Borges, P.A., Guerreiro, O., Ferreira, M.T., Borges, A., Ferreira, F. *et al.* (2014) *Cryptotermes brevis* (Isoptera: Kalotermitidae) in the Azores: Lessons after 2 yr of monitoring in the Archipelago. *Journal of Insect Science* 14, 1–7.
Bourguignon, T., Lo, N., Šobotník, J., Sillam–Dussès, D., Roisin, Y. and Evans, T.A. (2016) Oceanic dispersal, vicariance and human introduction shaped the modern distribution of the termites *Reticulitermes*, *Heterotermes* and *Coptotermes*. *Proceedings of the Royal Society B: Biological Sciences* 283, 20160179.
Bourguignon, T., Lo, N., Šobotník, J., Ho, S. Y., Iqbal, N., Coissac, E. *et al.* (2017) Mitochondrial phylogenomics resolves the global spread of higher termites, ecosystem engineers of the tropics. *Molecular Biology and Evolution* 34, 589–597.
Bowers, J.Z. (1966) Engelbert Kaempfer: Physician, explorer, scholar and author. *Journal of the History of Medicine and Allied Sciences* 21, 237–259.
Brown, A.A. (1934) Introduction. In: Kofoid, C. (ed.) *Termites and Termite Control*. University of California Press, Berkeley, California, pp. xxi–xxiv.
Buček, A., Wang, M., Šobotník, J., Hellemans, S., Sillam-Dussès, D. *et al.* (2022) Molecular phylogeny reveals the past transoceanic voyages of drywood termites (Isoptera, Kalotermitidae). *Molecular Biology and Evolution* 39, 1–20.
CBP (2023) *Pleasure boats – Requirements for entering the United States*. U.S. Customs and Border Protection. Available at: https://help.cbp.gov/s/article/Article-67?language=en_US) (accessed 7 September 2023)
Chambers, D.M., Zungoli, P.A. and Hill Jr, H.S. (1988) Distribution and habitats of the Formosan subterranean termite (Isoptera: Rhinotermitidae) in South Carolina. *Journal of Economic Entomology* 81, 1611–1619.
Chen, D. and Chen, H.W. (2013) Using the Köppen classification to quantify climate variation and change: An example for 1901–2010. *Environmental Development* 6, 69–79.
Chouvenc, T., Su, N.Y., and Grace, J.K. (2011) Fifty years of attempted biological control of termites – Analysis of a failure. *Biological Control* 59, 69–82.
Chouvenc, T., Li, H.F., Austin, J., Bordereau, C., Bourguignon, T., *et al.* (2016) Revisiting *Coptotermes* (Isoptera: Rhinotermitidae): a global taxonomic road map for species validity and distribution of an economically important subterranean termite genus. *Systematic Entomology* 41, 299–306.
Coaton, W.G.H. (1950) Infestation of buildings in South Africa by subterranean wood-destroying termites. *Bulletin of the Department of Agriculture and Forestry, South Africa* 299, 1–89.
Darlington, P. (1957) *Zoogeography. The Geographical Distributions of Animals*. Harvard University Press, Cambridge, Massachusetts.
De Queiroz, A. (2014) *The Monkey's Voyage: How Improbable Journeys Shaped the History of Life*. Basic Books, New York.
Dedeine, F., Dupont, S., Guyot, S., Matsuura, K., Wang, C. *et al.* (2016) Historical biogeography of *Reticulitermes* termites (Isoptera: Rhinotermitidae) inferred from analyses of mitochondrial and nuclear loci. *Molecular Phylogenetics and Evolution* 94, 778–790.
Dennell, R. (2017) Human colonization of Asia in the Late Pleistocene: the history of an invasive species. *Current Anthropology* 58, 383–396.
Dhang, P. (2014) Managing subterranean termites in ships and yachts across Philippine ports. *International Pest Control* 56, 262–263.
Ducruet, C., and Merk, O. (2013) Examining container vessel turnaround times across the world. *Port Technology International* 59, 1–3.
DYT Yacht Transport (2023) *Routes and schedules*. Available at: https://www.yacht-transport.com/schedules/ (accessed 7 September 2023)
Emerson, A.E. (1955) Geographical origins and dispersions of termite genera. *Fieldiana: Zoology* 37, 465–521.
Emerson, A.E. (1971) Tertiary fossil species of the Rhinotermitidae (Isoptera), phylogeny of genera, and reciprocal phylogeny of associated Flagellata (Protozoa) and the Staphylinidae (Coleoptera). *Bulletin of the American Museum of Natural History* 146, 243–303.

Evans, T.A., Forschler, B.T. and Trettin, C.C. (2019) Not just urban: The Formosan subterranean termite, *Coptotermes formosanus*, is invading forests in the southeastern USA. *Biological Invasions* 21, 1283–1294.

FAO (2011) *Guide to implementation of phytosanitary standards in forestry*. FAO Forestry Paper No. 164. Food and Agriculture Organization of the United Nations, Rome.

Gay, F.J. (1967) A world review of introduced species of termites. *Commonwealth Scientific and Industrial Research Organization Bulletin* 286, 1–88.

Gotanda, K. and Yasuda, Y. (2008) Spatial biome changes in southwestern Japan since the Last Glacial Maximum. *Quaternary International* 184, 84–93.

Green, J. (1983) The Song Dynasty shipwreck at Quanzhou, Fuji Province, People's Republic of China. *International Journal of Nautical Archaeology* 12, 253–261.

Green, J. and Burningham, N. (1998) The ship from Quanzhou, Fujian Province, People's Republic of China. *International Journal of Nautical Archaeology* 27, 277–301.

Guan, Y., Xiong, Z., Ruan, X., Wang, H. and Terrasi, F. (2013) Evidences for the view of the importance of Hepu seaport in ancient China. *Nuclear Instruments and Methods in Physics Research Section B: Beam Interactions with Materials and Atoms* 294, 688–691.

Haack, R.A., Britton, K.O., Brockerhoff, E.G., Cavey, J.F., Garrett, L.J. *et al.* (2014) Effectiveness of the International Phytosanitary Standard ISPM No. 15 on reducing wood borer infestation rates in wood packaging material entering the United States. *PLoS ONE* 9, e96611.

Haagsma, K., Rust, M.K, Reierson, D.A., Atkinson, T.H. and Kellum, D. (1995) Formosan subterranean termite established in California. *California Agriculture* 49, 30–33.

Heins, M. (2016) *The Globalization of American Infrastructure: The Shipping Container and Freight Transportation.* Routledge. New York.

Henderson, G. (2001) Practical considerations of the Formosan subterranean termite in Louisiana: a 50-year-old problem. *Sociobiology* 37, 281–292.

Hewitt, G. (2003) Ice ages: Species distributions and evolution. In: Rothschild, L.J and Lister, A.M. (eds) *Evolution on Planet Earth*, Elsevier, Amsterdam, Netherlands.

Hochmair, H.H. and Scheffrahn, R.H. (2010) Spatial association of marine dockage with land-borne infestations of invasive termites in urban South Florida. *Journal of Economic Entomology* 103, 1338–1346.

Howell, H.N., Hamman, P.J. and Granovsky, T.A. (1987) The geographical distribution of the termite genera *Reticulitermes, Coptotermes and Incisitermes* in Texas. *The Southwestern Entomologist* 12, 119–125.

Hromada, C.J. (1970) Formosan termites on Guam. A massive control job for Terminix. *Pest Control* 38, 10–25.

Hulme, P.E. (2009) Trade, transport, and trouble: managing invasive species pathways in an era of globalization. *Journal of Applied Ecology* 46, 10–18.

Husseneder, C., Simms, D.M., Delatte, J.R., Wang, C., Grace, J.K. and Vargo, E.L. (2012) Genetic diversity and colony breeding structure in native and introduced ranges of the Formosan subterranean termite, *Coptotermes formosanus. Biological Invasions* 14, 419–437.

IPPC (International Plant Protection Convention) (2002) *International standards for phytosanitary measures: guidelines for regulating wood packaging material in international trade*. Publication No. 15. Food and Agriculture Organization of the United Nations, Rome.

IPPC (International Plant Protection Convention) (2018) International standards for phytosanitary measures: revision of ISPM No. 15, regulation of wood packaging material in international trade. Food and Agriculture Organization of the United Nations, Rome.

Jenkins, T.M., Dean, R.E. and Forschler, B.T. (2002) DNA technology, interstate commerce and the likely origin of Formosan subterranean termite (Isoptera: Rhinotermitidae) infestation in Atlanta, Georgia. *Journal of Economic Entomology* 95, 381–389.

Jiang, K., Tong, X., Ding, Y., Qi., Wang, Z.W., Miao, L.-Y. *et al.* (2021) Shifting roles of the East China Sea in the phylogeography of red nanmu in East Asia. *Journal of Biogeography* 48, 2486–2501.

Kaempfer, E. (1712) *Amoenitatum exoticarum politico-physico-medicarum fasciculi V: quibus continentur variae relationes, observationes & descriptiones rerum Persicarum & ulterioris Asiae : multâ attentione in peregrinationibus per universum orientem collectae*. Printed by H.W. Meyer (1657–1722). doi: 10.5962/bhl.title.147012

Kaifu, Y., Kuo, T.H., Kubota, Y. and Jan, S. (2020) Palaeolithic voyage for invisible islands beyond the horizon. *Scientific Reports* 10, 1–12.

Kalshoven, L.G.E. (1962) Observations on *Coptotermes havilandi* Holmgr. (*javanicus* Kemn.) (Isoptera). *Beaufortia: Zoological Museum Amsterdam Miscellaneous Publications* 9, 121–137.

Kanao, T., Iwata, R., Sasaki, T., Hirose, H. and Yoshimura, M. (2019) New distributional records of two termitophilous rove beetles (Coleoptera, Staphylinidae) in Japan. *Elytra, New Series* 9, 39–40.

Kato, H., Amekawa, S., Hori, M., Shen, C.C., Kuwahara, Y. *et al.* (2021) Influences of temperature and the meteoric water δ18O value on a stalagmite record in the last deglacial to middle Holocene period from southwestern Japan. *Quaternary Science Reviews* 253, 106746.

Kimura, M. (2000) Paleogeography of the Ryukyu Islands. *Tropics* 10, 5–24.

Kistner, D.H. (1985) A new genus and species of termitophilous Aleocharinae from mainland China associated with *Coptotermes formosanus* and its zoogeographic significance (Coleoptera: Staphylinidae). *Sociobiology* 10, 93–104.

Kollar, V. (1837) Naturgeschichte der schadlichen Insekten. *Verhandlungen Landwirtschaftsrats Gesellschaft in Wien* 5, 411–413.

Krishna, K., Grimaldi, D.A., Krishna, V. and Engel, M.S. (2013) Treatise on the Isoptera of the World: Vol 3. Neoisoptera excluding Termitidae. *Bulletin of the American Museum of Natural History* 377, 623–973.

La Fage, J.P. (1987) Practical considerations of the Formosan subterranean termite in Louisiana: a 30-year-old problem. In: Tamashiro, M. and Su, N.-Y. (eds) *Biology and control of the Formosan subterranean termite. Proceedings of the International Symposium on the Formosan Subterranean Termite.* Research Extension Series 083, Hawaii Institute of Tropical Agriculture and Human Resources, pp. 37–42.

Lambeck, K., Rouby, H., Purcell, A., Sun, Y. and Sambridge, M. (2014) Sea level and global ice volumes from the Last Glacial Maximum to the Holocene. *Proceedings of the National Academy of Sciences* 111, 15296–15303.

Lee, K.C., Sun, J.Z., Zhu, Y. and Mallette, E.J. (2009) A case study of the Formosan subterranean termite, *Coptotermes formosanus* (Isoptera: Rhinotermitidae) transported with a non-cellulosic commercial carrier in south Mississippi. *Sociobiology* 53, 619–630.

Lee, S.B., Tong, R.L., Kim, S.H., Im, I.G. and Su, N.Y. (2021) Potential pest status of the Formosan subterranean termite, *Coptotermes formosanus* Shiraki (Blattodea: Isoptera: Rhinotermitidae), in response to climate change in the Korean Peninsula. *Florida Entomologist* 103, 431–437.

Li, G.Q (1989) Archaeological evidence for the use of 'chu-nam' on the 13th century Quanzhou ship, Fujian Province, China. *International Journal of Nautical Archaeology* 18, 277–283.

Li, H.F., Scheffrahn, R.H., Su, N.Y., Kanzaki, N., and Yang, R.L. (2008) Survey of the termites (Isoptera: Kalotermitidae, Rhinotermitidae, Termitidae) of Lanyu Island, Taiwan. *Florida Entomologist* 91, 472–473.

Li, H.F., Ye, W., Su, N.Y. and Kanzaki, N. (2009) Phylogeography of *Coptotermes gestroi* and *Coptotermes formosanus* (Isoptera: Rhinotermitidae) in Taiwan. *Annals of the Entomological Society of America* 102, 684–693.

Liang, W.R., Maruyama, M., Kanao, T., Iwata, R. and Li, H.F. (2020) Discovery of termitophilous rove beetles associated with Formosan subterranean termite *Coptotermes formosanus* in Taiwan, with the first larval description for the tribe Termitohospitini (Coleoptera: Staphylinidae). *Acta Entomologica Musei Nationalis Pragae* 60, 77–87.

Light, S.F. (1921) Notes on Philippine termites, I. *Philippine Journal of Science* 18, 243–257.

Light, S.F. (1929) Present status of our knowledge of the termites of China. *Lingnan Science Journal* 7, 581–600.

Lin, S.Q. (1987) Present status of *Coptotermes formosanus* and its control in China. In: Tamashiro, M. and Su, N.Y. (eds.) *Proceedings of the International Symposium on the Formosan subterranean termite.* Hawaii Institute of Tropical Agriculture and Human Resources, University of Hawaii, Honolulu, Hawaii, pp. 31–37.

Lin, Y., Ramstein, G., Wu, H., Rani, R., Braconnot, P., *et al.* (2019) Mid-Holocene climate change over China: model–data discrepancy. *Climate of the Past* 15, 1223–1249.

Lindo, Z. (2020) Transoceanic dispersal of terrestrial species by debris rafting. *Ecography* 43, 1364–1372.

Liu, Y., Deng, L., He, J., Zhao, X., Wang, H., Feng, D., Chen, J., Li, M. and Sun, Q. (2021) Holocene geomorphological evolution and the Neolithic occupation in South Hangzhou Bay, China. *Geomorphology* 389, 107827.

Lockwood J.L., Cassey P. and Blackburn T. (2005) The role of propagule pressure in explaining species invasions. *Trends in Ecology and Evolution* 20, 223–228.

Marean, C.W. (2015) The most invasive species of all. *Scientific American* 313, 32–39.

Maruyama, M. and Iwata, R. (2002) Two new termitophiles of the tribe Termitohospitini (Coleoptera: Staphylinidae: Aleocharinae) associated with *Coptotermes formosanus* (Isoptera: Rhinotermitidae). *The Canadian Entomologist* 134, 419–432.

Maruyama, M., Kanao, T. and Iwata, R. (2012) Discovery of two Aleocharine Staphylinid Species (Coleoptera) associated with *Coptotermes formosanus* (Isoptera: Rhinotermitidae) from central

Japan, with a review of the possible natural distribution of *C. formosanus* in Japan and surrounding countries. *Sociobiology* 59, 605–616.

Matisoo-Smith, E. and Robins, J.H. (2004) Origins and dispersals of Pacific peoples: evidence from mtDNA phylogenies of the Pacific rat. *Proceedings of the National Academy of Sciences, USA* 101, 9167–9172.

McCullough, D.G., Work, T.T., Cavey, J.F., Liebhold, A.M. and Marshall, D. (2006) Interceptions of nonindigenous plant pests at US ports of entry and border crossings over a 17-year period. *Biological Invasions* 8, 611–630.

Messenger M.T., Su, N.Y. and Scheffrahn, R.H. (2002) Current distribution of the Formosan subterranean termite and other termite species (Isoptera: Rhinotermitidae, Kalotermitidae) in Louisiana. *Florida Entomologist* 85, 580–587.

Mori, H. (1987) The Formosan subterranean termite in Japan: its distribution, damage and current and potential control measures. In: Tamashiro, M. and Su, N.Y. (eds.) *Proceedings of the International Symposium on the Formosan subterranean termite.* Hawaii Institute of Tropical Agriculture and Human Resources, University of Hawaii, Honolulu, pp. 23–26.

Mullins, A.J., Messenger, M.T., Hochmair, H.H., Tonini, F., Su, N.-Y. *et al.* (2015) Dispersal flights of the Formosan subterranean termite (Isoptera: Rhinotermitidae). *Journal of Economic Entomology* 108, 707–719.

Nam, N.X., Huy, N.Q. and Le Quang Thinh, N.T.H. (2018) Species composition and distribution of termites according to altitude in the Central Highlands, Vietnam. In: Forschler, B.T., Li, H.F., Foo, F.K., Yusuf, S., Himmi, S.K. and Ismayati, M. (eds) *Proceedings of the 12th Pacific Rim Termite Research Group Conference*, Yogyakarta, Indonesia, pp. 16–23.

Nguyen, T.T., Do, T.H., Duong, T.H., Le, Q.G., Dao, T.-K. *et al.* (2014) Identification of Vietnamese *Coptotermes* pest species based on the sequencing of two regions of 16S rRNA gene. *Bulletin of Insectology* 67, 131–136.

Ninon, R., D'Haese, C. and Barden, P. (2019) Fossil amber reveals springtails' longstanding dispersal by social insects. *BMC Evolutionary Biology* 19, 1–12.

Oh, H. and Shin, H.J. (2016) Climatic classification over Asia during the middle Holocene Climatic Optimum based on PMIP models. *Journal of Earth Science* 27, 123–129.

Okada, T. (1912) On the distribution of *Coptotermes formosanus* in the Shizuoka Prefecture. *Konchiu-Sekai* 16, 11–12.

Okonya, J.S. and Kroschel, J. (2013) Indigenous knowledge of seasonal weather forecasting: a case study in six regions of Uganda. *Agricultural Sciences* 4, 641–648.

Oshima, M. (1913) Notes on the termites of Japan, with description of one new species. *Philippine Journal of Science, Section D, General Biology, Ethnology and Anthropology* 8, 271–281.

Osozawa, S., Shinjo, R., Armid, A., Watanabe, Y., Horiguchi, T., *et al.* (2012) Palaeogeographic reconstruction of the 1.55 Ma synchronous isolation of the Ryukyu Islands, Japan, and Taiwan and inflow of the Kuroshio warm current. *International Geology Review* 54, 1369–1388.

Osozawa, S., Su, Z.H., Oba, Y., Yagi, T., Watanabe, Y., *et al.* (2013) Vicariant speciation due to 1.55 Ma isolation of the Ryukyu islands, Japan, based on geological and GenBank data. *Entomological Science* 16, 267–277.

Otto-Bliesner, B.L., Brady, E.C., Clauzet, G., Tomas, R., Levis, S., *et al.* (2006) Last Glacial Maximum and Holocene climate in CCSM3. *Journal of Climate* 19, 2526–2544.

Park, J.O., Tokuyama, H., Shinohara, M., and Suyehiro, K. (1998) Seismic record of tectonic evolution and backarc rifting in the southern Ryukyu Island arc system. *Tectonophysics* 94, 21–42.

Rust, M., Reierson, D., Paine, E., Kellum, D. and Haagsma, K. (1998) Ravenous Formosan subterranean termites persist in California. *California Agriculture* 52, 34–37.

Sakai, A.K., Allendorf, F.W., Holt, J.S., Lodge, D.M., Molofsky, J., *et al.* (2001) The population biology of invasive species. *Annual Review of Ecology and Systematics* 32, 305–332.

Scheffrahn, R.H. (2019) University of Florida Termite Collection (UFTC). Available at: https://www.termitediversity.org/ (accessed 1 February 2023).

Scheffrahn, R.H. and Crowe, W. (2011) Ship-borne termite (Isoptera) border interceptions in Australia and onboard infestations in Florida, 1986–2009. *Florida Entomologist* 94, 57–63.

Scheffrahn, R.H. and Su, N.Y. (2005) Distribution of the termite genus *Coptotermes* (Isoptera: Rhinotermitidae) in Florida. *Florida Entomologist* 88, 201–203.

Scheffrahn, R.H., Su, N.Y. and Diehl, B. (1990) Native, introduced, and structure-infesting termites of the Turks and Caicos Islands, BWI (Isoptera: Kalotermitidae, Rhinotermitidae, Termitidae). *Florida Entomologist* 73, 622–627.

Scheffrahn, R.H., Křeček, J., Ripa, R. and Luppichini, P. (2009) Endemic origin and vast anthropogenic dispersal of the West Indian drywood termite. *Biological Invasions* 11, 787–799.

Scheffrahn, R.H., Carrijo, T F., Křeček, J., Su, N.Y., Szalanski, A.L., *et al.* (2015) A single endemic and three exotic species of the termite genus *Coptotermes* (Isoptera, Rhinotermitidae) in the New World. *Arthropod Systematics and Phylogeny* 73, 333–348.

Scheffrahn, R.H., Carrijo, T.F., Postle, A.C. and Tonini, F. (2017) *Disjunctitermes insularis*, a new soldierless termite genus and species (Isoptera, Termitidae, Apicotermitinae) from Guadeloupe and Peru. *ZooKeys* 665, 71–84.

Scheffrahn, R.H., Bahder, B.W. and Lu, T. (2020) *Coptotermes formosanus* (Blattodea: Rhinotermitidae) established in Israel and world distribution of a major termite pest. *Check List* 16, 1537–1543.

Schottenhammer, A. (2012) The "China Seas" in world history: A general outline of the role of Chinese and East Asian maritime space from its origins to c. 1800. *Journal of Marine and Island Cultures* 276, 201–207.

Searle, J.B., Jamieson, P.M., Gündüz, I., Stevens, M.I., Jones, E.P., Gemmill, C.C.E. and King, C.M. (2009) The diverse origins of New Zealand house mice. *Proceedings of the Royal Society B: Biological Science* 276, 209–217.

Shiraki, T. (1909) On the termites of Japan. *Transactions of the Entomological Society of Japan* 2, 229–242 [in Japanese]

Sponsler R. C., Jordan, K. S. and Appel, A. G. (1988) New distribution record of the Formosan Subterranean Termite, *Coptotermes formosanus* (Isoptera: Rhinotermitidae), in Auburn, Alabama. *Entomological News* 99, 87–89.

St John, H. (1947) The history, present distribution and abundance of sandalwood on Oahu, Hawaiian Islands: Hawaiian Plant Studies 14. *Pacific Science* 1, 5–20.

Su, N.Y. and Scheffrahn, R.H. (1987) Alate production of a field colony of the Formosan subterranean termite (Isoptera: Rhinotermitidae). *Sociobiology* 13, 167–172.

Su, N.Y. and Scheffrahn, R.H. (1998) *Coptotermes vastator* Light (Isoptera: Rhinotermitidae) in Guam. *Proceedings of the Hawaiian Entomological Society* 33, 13–18.

Su, N.Y. and Tamashiro, M. (1987) An overview of the Formosan subterranean termite (Isoptera: Rhinotermitidae) in the world. In: Tamashiro, M. and Su, N.Y. (eds) *Proceedings of the International Symposium on the Formosan subterranean termite*. Hawaii Institute of Tropical Agriculture and Human Resources. University of Hawaii, Honolulu, Hawaii, pp. 3–15.

Sugerman, B.B. (1972) Insects and other arthropods from Kwajalein Atoll (Marshall Islands). *Proceedings Hawaiian Entomological Society* 21, 271–286.

Sugio, K. and Miyaguni, Y. (2019) Characteristics of dispersal flight of the Formosan subterranean termite, *Coptotermes formosanus* Shiraki (Isoptera: Rhinotermitidae) in Okinawa. *Japanese Journal of Environmental Entomology and Zoology* 30, 63–69.

Swezey, O.H. (1914) Notes and exhibitions. *Proceedings of the Hawaiian Entomological Society* 3, 27.

Tamashiro, M., Yates, J.R and Ebesu, R.H. (1987) The Formosan subterranean termite in Hawaii: problems and control. In: Tamashiro, M. and Su, N.Y. (eds) *Proceedings of the International. Symposium on the Formosan subterranean termite.* Hawaii Institute of Tropical Agriculture and Human Resources, University of Hawaii, Honolulu, Hawaii, pp. 15–20.

Taylor, C.M. and Hastings, A. (2005) Allee effects in biological invasions. *Ecology Letters* 8, 895–908.

Thompson, C.R. (1985) Detection and distribution of Formosan termites (Isoptera: Rhinotermitidae) in southeastern Florida. *Journal of Economic Entomology* 78, 528–530.

Tonini, F., Hochmair, H.H., Scheffrahn, R.H. and Deangelis, D.L. (2013) Simulating the spread of an invasive termite in an urban environment using a stochastic individual-based model. *Environmental Entomology* 42, 412–423.

Tonini, F., Hochmair, H.H., Scheffrahn, R.H. and DeAngelis, D.L. (2014) Stochastic spread models: a comparison between an individual–based and a lattice–based model for assessing the expansion of invasive termites over a landscape. *Ecological Informatics* 24, 222–230.

Tseng, S.P., Boone, J., Boone, N., King, S., Taravati, S. *et al.* (2021) Genetic analysis of Formosan subterranean termite (Blattodea: Rhinotermitidae) populations in California. *Journal of Economic Entomology* 114, 1264–1269.

Tseng, S.P., Taravati, S., Choe, D.H., Rust, M.K. and Lee, C.Y. (2022) Genetic evidence for multiple invasions of *Coptotermes formosanus* (Blattodea: Rhinotermitidae) in California. *Journal of Economic Entomology* 115, 1251–1256.

Tsukada, M. (1983) Vegetation and climate during the last glacial maximum in Japan. *Quaternary Research* 19, 212–235.

Vargo, E.L., Husseneder, C., and Grace, J.K. (2003) Colony and population genetic structure of the Formosan subterranean termite, *Coptotermes formosanus*, in Japan. *Molecular Ecology* 12, 2599–2608.

Vis, I.F. and De Koster, R. (2003) Transshipment of containers at a container terminal: an overview. *European Journal of Operational Research* 147, 1–16.

Voris, H.K. (2000) Maps of Pleistocene Sea levels in Southeast Asia: shorelines, river systems and time durations. *Journal of Biogeography* 27, 1153–1167.

Vu, Q.M., Nguyen, H.H. and Smith, R.L. (2007) The termites (Isoptera) of Xuan Son National Park, northern Vietnam. *The Pan-Pacific Entomologist* 83, 85–94.

Wake, C. (1997) The great ocean-going ships of southern China in the age of Chinese maritime voyaging to India, twelfth to fifteenth centuries. *International Journal of Maritime History* 9, 51–81.

Wanner, H., Beer, J., Bütikofer, J., Crowley, T.J., Cubasch, U. *et al.* (2008) Mid–to Late Holocene climate change: an overview. *Quaternary Science Reviews* 27, 1791–1828.

Wanyonyi, K., Darlington, J.P.E.C. and Bagine, R.K.N. (1984) Checklist of the species of termites (Isoptera) recorded from East Africa. *Journal of the East Africa Natural History Society and National Museum* 181, 1–10.

Weather Spark (2023) *The Weather Year Round Anywhere on Earth*. Cedar Lake Ventures, Minneapolis, Minnesota. Available at: https://weatherspark.com (accessed 1 February 2023)

Wepfer, P.H., Guénard, B. and Economo, E.P. (2016) Influences of climate and historical land connectivity on ant beta diversity in East Asia. *Journal of Biogeography* 43, 2311–2321.

Xu, X., Liu, F., Chen, J., Ono, H., Agnarsson, I. *et al.* (2016) Pre-Pleistocene geological events shaping diversification and distribution of primitively segmented spiders on East Asian margins. *Journal of Biogeography* 43, 1004–1019.

Yano, M. (1915) White ants in Japan. *Bulletin of the Forest Experiment Station (Meguro, Tokyo)* 9, 134–142.

Yao, Y., Harff, J., Meyer, M. and Zhan, W. (2009) Reconstruction of paleocoastlines for the northwestern South China Sea since the Last Glacial Maximum. *Science in China Series D: Earth Sciences* 52, 1127–1136.

Yongjing, G., Zhaoming, X., Xiangdong, R., Huijuan, W. and Terrasi, F. (2013) Evidences for the view of the importance of Hepu seaport in ancient China. *Nuclear Instruments and Methods in Physics Research Section B: Beam Interactions with Materials and Atoms* 294, 688–691.

Yu, G., Chen, X., Ni, J., Cheddadi, R., Guiot *et al.* (2000) Palaeovegetation of China: a pollen data-based synthesis for the mid-Holocene and last glacial maximum. *Journal of Biogeography* 27, 635–664.

Zeng, L., Ming, C., Li, Y., Su, L.Y., Su, Y.H. *et al.* (2018) Out of Southern East Asia of the brown rat revealed by large-scale genome sequencing. *Molecular Biology and Evolution* 35, 149–158.

Zhang, C., Yang, S., Huang, X., Dou, Y., Li, F. *et al.* (2022) Sea level change and Kuroshio intrusion dominated Taiwan sediment source-to-sink processes in the northeastern South China Sea over the past 244 kyrs. *Quaternary Science Reviews* 287, 107558.

Zhuo, Z., Baoyin, Y. and Petit-Maire, N. (1998) Paleoenvironments in China during the Last Glacial Maximum and the Holocene optimum. *Episodes Journal of International Geoscience* 21, 152–158.

3 Taxonomic status of *Coptotermes formosanus* and related species

Hou-Feng Li[1] and Chia-Chien Wu[2]
Department of Entomology, National Chung Hsing University, [1]houfeng@nchu.edu.tw; [2]daisy100406@gmail.com

Abstract
The taxonomy of *Coptotermes formosanus* Shiraki, 1909 and congeneric species has been challenging since the early 20th century. To date, 15 junior synonyms of *C. formosanus* have been documented. Recent taxonomic treatments to validate *C. formosanus* are references for further clarification of uncertain congeneric species. In this chapter, the general morphology of the alate, soldier, and worker castes of *C. formosanus* is provided, and the progress of *Coptotermes* taxonomy, including new species and junior synonyms, is reviewed. A checklist of 58 *Coptotermes* species shows the species validation status and describes castes, gene sequence availability, and their putative native areas. Of the 35 uncertain species, 31 are in Asia. To accelerate the overall revision of *Coptotermes*, we recommend an intensive collection at type localities, obtaining gene sequences to group termite samples into operational taxonomic units, analyzing the intraspecific morphometric variation, and validating the species by redescribing the species.

3.1 Introduction

The genus *Coptotermes* Wasmann, 1896 is a well-recognized termite taxon due to its economic importance and invasive tendency (Rust and Su, 2012; Evans *et al.*, 2013; see Chapter 2 of this volume). One hundred and twelve names of *Coptotermes* have been proposed (Krishna *et al.*, 2013; Chouvenc *et al.*, 2016; Lee *et al.*, 2017), and 49 of them have been designated as junior synonyms or *nomina nuda*. To date, 59 living and four fossil species names remain. The revision of *Coptotermes* has been in progress in the past two decades, but more effort is required to complete this task. This chapter focuses on the taxonomic works of the Formosan subterranean termite, *Coptotermes formosanus* Shiraki, 1909. The difficulty and source of controversy of *C. formosanus* taxonomy in the early 20th century is introduced. The effort to validate the taxonomic status of *C. formosanus* is described, which is an example of further taxonomic treatments of other congeneric species. The general morphology of the alate, soldier, and worker caste of *C. formosanus* are described. In addition to *C. formosanus*, we review the progress of *Coptotermes* taxonomy in the past decade. The strategies to complete the taxonomic revision of *Coptotermes* are also discussed.

3.2 Background of *C. formosanus* discovery in Taiwan

After 1895, Taiwan became a territory of Japan until 1945. Because wooden constructions in

 Biology and Management of the Formosan Subterranean Termite and Related Species (eds Nan-Yao Su and Chow-Yang Lee)
DOI: 10.1079/9781800621596.0003

Taiwan suffered severe termite damage, the Japanese government tried to reduce the termite problem. At the time, the termite pest species was known colloquially as "house termite", イエシロアリ (in Japanese), or 家白蟻 (in Chinese). In 1909, a Japanese researcher, Tokuichi Shiraki, described the house termite as *C. formosanus* based on samples collected in Taiwan by Masamitsu Oshima (Shiraki, 1909). The morphology of male and female alates, nymph, soldier, worker, and queen of *C. formosanus* were described by using traditional Japanese literary language in the original description. No figures were included, and no holotype was designated. In the same year, Oshima (1909) provided more detailed information on the morphological and behavioral characters of *C. formosanus*, including drawings of a dealate, soldier, worker, and wing. However, in the following year, Oshima (1910) suggested that *C. formosanus* is a junior synonym of *Coptotermes gestroi* (Wasmann 1896) because soldiers of *C. formosanus* and those of *C. gestroi* bear superficial resemblance. At the time, the alate morphologies of the two species were not compared, because the *C. gestroi* alate had not been described (Wasmann, 1896). Between 1909 and 1912, several other Japanese and European researchers were involved in the debate of *Coptotermes* identification in Taiwan, including Munemoto Yano (1883–1970), Shozaburo Watasé (1862–1929), Sanji Hozawa (1885–1947), Eric Wasmann (1859–1931), and Nils Holmgren (1877–1954). Oshima (1912) collected *Coptotermes* samples from Taipei and southern Japan and did a detailed morphological measurement of the soldier caste. With these results, Oshima (1912) determined that the samples represented a new termite species, *C. formosanus*, which was agreed upon by other researchers.

After 1912, *C. gestroi* in Taiwan was not mentioned until 2003 (Tsai and Chen, 2003). Detailed soldier morphological information supported that *C. gestroi* did occur in Taiwan. Li *et al.* (2009) confirmed that both *C. formosanus* and *C. gestroi* occurred in Taiwan based on morphological data of both soldier and alate castes and genetic analysis. *Coptotermes formosanus* is found throughout Taiwan, and *C. gestroi* is limited to the southern area. Li *et al.* (2010) reviewed the taxonomic controversy which occurred in the early 20th century and pointed out two main causes.

The first was a false assumption. Researchers involved in the debate believed only one *Coptotermes* species existed in Taiwan. Each researcher obtained samples of either or both species, did morphological observations and lumped measurements together, and reached different conclusions. According to the current distribution of *Coptotermes* spp. in Taiwan, samples collected by Oshima in 1909 and used by Shiraki for the original description of *C. formosanus* likely included both species. The coloration of a dealate painted by Oshima (1909) resembles that of *C. gestroi*. However, Oshima (1912) collected samples in Taipei and Japan, where only *C. formosanus* occurs, which led to the conclusion that only one novel *Coptotermes* species occurs in Taiwan.

The second reason was an overreliance on soldier morphology for species identification. *Coptotermes* soldiers display active defensive behavior by guarding the opening of a shelter tube, and collecting the soldier caste of *Coptotermes* is easy. The shape and length of soldier heads, such as the head width-to-length ratio and elongated or circular head shapes, were the characters used to differentiate *C. formosanus* and *C. gestroi* (Yano, 1911; Oshima, 1912). However, the length, width, and shape of the soldier head of *C. formosanus* vary (Chouvenc and Su, 2014; Chen *et al.*, 2020). The intra- and interspecific variation of soldier morphometric data were underestimated. On the other hand, the seasonal occurrence of alates impedes the simultaneous collection of both soldier and alate castes. Even though the alate caste provided much more diagnostic characters than the soldier caste, the alates of the two species were not compared at the time.

3.3 Validation of *C. formosanus*

To date, 15 junior synonyms of *C. formosanus* have been proposed, including: *C. formosae* Holmgren, 1911; *C. hongkonensis* Oshima,

1914; *C. intrudens* Oshima, 1920; *C. remotus* Silvestri, 1927; *C. eucalyptus* Ping, 1984; *C. xiaoliangensis* Ping, 1984; *C. guangzhouensis* Ping, 1985; *C. heteromorphus* Ping, 1985; *C. rectangularis* Ping and Gong, 1986; *C. ochraceus* Ping and Xu, 1986; *C. commnuis* Xia and He, 1986; *C. changtaiensis* Xia and He, 1986; *C. hekouensis* Xia and He, 1986; *C. suzhouensis* Xia and He, 1986; and *C. guizhouensis* He and Qiu, 1992 (Li *et al.*, 2012; Krishna *et al.*, 2013; Ke *et al.*, 2016; Chen *et al.*, 2020). The number of junior synonyms of *C. formosanus* was highest among other congeneric species. To validate the taxonomic status of *C. formosanus* and to prevent further taxonomic confusion, several works have been carried out in the past two decades.

Mass collection of *Coptotermes* samples was conducted at the type locality, Taiwan. To confirm the sympatric distribution of *C. formosanus* and *C. gestroi* in Taiwan, 220 *Coptotermes* samples were collected (Li *et al.*, 2010), and 1,429 *Coptotermes* samples were used to map the current distribution of both species (Huang *et al.*, 2022).

A male alate collected by H.-F. Li at Taoyuan City, Taoyuan District, Taiwan (24.99033° N, 121.30314° E; 20-VI-2006 (TW49)) was designated as neotype (Li *et al.*, 2010) and deposited in the National Museum of Natural Science, Taichung, Taiwan (NMNS). Several female alates, soldiers, and workers collected within the same colony of the neotype were preserved in the five institutes, including NMNS; Taiwan Forestry Research Institute, Taipei, Taiwan; National Taiwan University, Taipei, Taiwan; National Chung Hsing University, Taichung, Taiwan; and the University of Florida Termite Collection, Fort Lauderdale Research and Education Center.

The gene sequence information for *C. formosanus* is now available for public access on GenBank. The partial mitochondrial gene sequences, including COII, 12S rRNA, and 16S rRNA, were obtained from workers collected within the same colony of the neotype. The sequence data were deposited in GenBank with accession numbers EU805758, EU805712, and EU805735 (Li *et al.*, 2009). The whole genome sequence of *C. formosanus* is also available (Itakura *et al.*, 2020).

Coptotermes formosanus was redescribed using the type specimen (male alate), 35 alates of 14 colonies, and 34 soldiers of 12 colonies from Taiwan (Chen *et al.*, 2020). In addition to detailed morphological information, the intraspecific variations, including head shape, fontanelle shape, and morphometric information, were presented. Individuals collected from multiple colonies provided information on the intraspecific morphological variation of *C. formosanus*.

3.4 General morphology and diagnostic characters of *C. formosanus*

This section describes the external morphology of the alate, queen, king, soldier, and worker. Their color was compared with the Munsell color system (Munsell, 1967). Living or fresh individuals were used to determine the hue, value, and chroma of each character.

Alate: Head capsule brownish yellow (Munsell color system: 10YR 5/10, Fig. 3.1A). Head subcircular. Antennal spots inconspicuous. Labrum liguiform, broadest at middle (Fig. 3.1A). Antenna with 19–21 articles; 1st article cylindrical, longest; 3rd article shortest; 3rd to 20th articles moniliform; last article oblong. Eye circular to subcircular (Fig. 3.1B). Ocellus oval. Pronotum dark goldenrod (10YR), trapezoidal, broadest at anterior, anterior margin concave, sides converging posteriorly, posterior margin with middle notched (Fig. 3.1C). Forewing scale overlapping the basal part of hindwing scale (Fig. 3.1C). Scale of the forewing broader than the stump of the hindwing. Abdomen oblong, hairy. Legs pale yellow (5Y 9/6). Femur shorter than the tibia. Tibia spurs formula 3:2:2. Tarsi 4-jointed. Wing with membrane whitish, semitransparent, and hairy (Fig. 3.1D; Fig. 3.2A,B); costal margin and radial sector sclerotized, media and cubitus faint; cubitus with 7–11 branches; average setae length 78±5 μm (71–85 μm), average setae density 502.8±70.8/mm^2 (384–636/mm^2) (Fig. 3.2C).

The wing scales on the thoraxes of male and female dealates indicate that they are

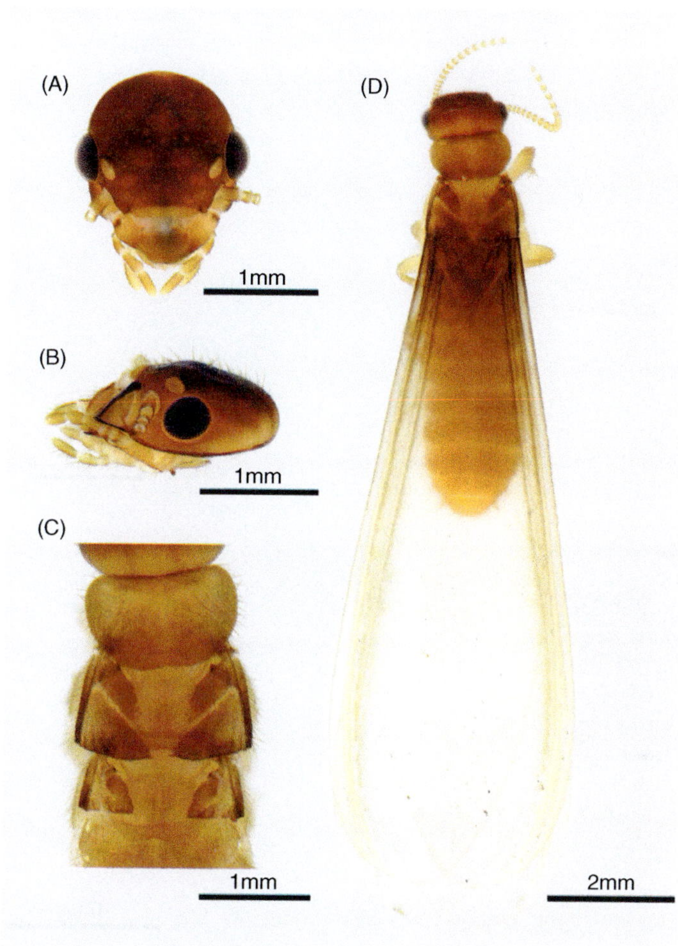

Fig 3.1. Alate of *Coptotermes formosanus*. (A) Head in frontal view; (B) in lateral view; (C) thorax in dorsal view; (D) habitus of alate in dorsal view.

primary reproductives (Fig. 3.3A). The morphology of the male dealate (king) is not much different from male alate, but the abdomen of the female dealate (queen) swollen significantly along colony development (Fig. 3.3B).

Soldier: Head capsule yellowish (2.5Y 8.5/10, Fig. 3.4A). Head typically pear-shaped, some individual heads rounded, elliptical, oval, rounded rectangle, or rounded trapezoid (Chen *et al.*, 2020). Labrum acuminates, broadest at the base (Fig. 3.4A). Antenna with 14–17 articles; 1st article cylindrical, longest; 2nd to 15th articles moniliform; last article oblong (Fig. 3.4 B). Fontanelle shaped triangular to arched (Fig. 3.4C,D), opening frontally, with two pairs of setae around the fontanelle (Fig. 3.4D). Postmentum long, broadest at anterior one-third, narrowest at the posterior one-third (Fig. 3.4E). Pronotum yellowish (2.5Y 8.5/10, Fig. 3.4F), trapezoid, narrower than the head, broadest at anterior, anterior and posterior margin slightly bilobed, sides converging posteriorly. Abdomen oblong, hairy. Leg whitish. Tibial spurs formula 3:2:2. Tarsi 4-jointed.

Worker: Head capsule whitish (2.5Y 9/4, Fig. 3.5A). Head oval. Labrum liguiform. Antenna

Fig 3.2. Wing of *Coptotermes formosanus*. (A) Forewing; (B) hindwing; (C) proximal sections of the forewing. Costal margin (Co), radial sector, (Rs), median (M), cubitus (Cu).

Fig 3.3. (A) King (upper left) and queen (bottom right) of a two-year-old colony of *C. formosanus* raised in the laboratory. (B) Physogastric queen collected from an infested manmade construction. Body length of the queen is ~25mm.

Fig 3.4. Soldier of *Coptotermes formosanus*. (A) Head in dorsal view, (B) in lateral view, and (C) in frontal view; (D) fontanelle surrounded with four setae pointed by arrows; (E) head in ventral view; (F) habitus of soldier in dorsal view.

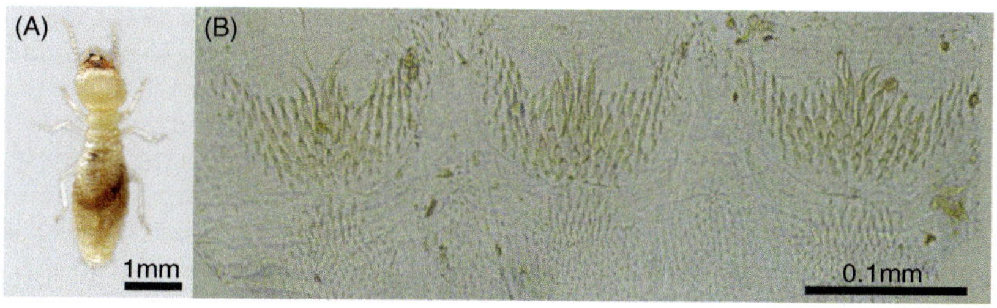

Fig 3.5. Worker of *Coptotermes formosanus*. (A) Habitus of worker in dorsal view; (B) enteric valve.

with 11–16 articles; 1st article cylindrical, longest; 2nd to 14th articles moniliform; last article oblong. Pronotum trapezoid, narrower than the head, broadest at anterior part, anterior and posterior margins slightly concave, lateral margin converging posteriorly (Fig. 3.5A). Enteric valve with three cushions; each cushion with two parts (Fig. 3.5B); larger part crescent-shaped, with 35–45 short spines, middle with 8–11 long spines; smaller part half-moon-like, with 35–40 short spines. Abdomen oblong, hairy. Tibia spurs formula 3:2:2. Tarsi 4-jointed.

The whitish and semitransparent wings of *C. formosanus* could be distinguished from the species with yellowish, brownish, or darkish wings, like *C. acinaciformis acinaciformis* (yellowish), *C. amanii* (pale brown), *C. brunneus* (pale fuscous), *C. curvignathus* (light brownish yellow), *C. frenchi* (light fuscous),

C. grandiceps (yellowish), *C. lacteus* (dark brown), *C. michaelseni* (light smoky), *C. dreghorni* (brown), *C. sjostedti* (brown), and *C. testaceus* (shiny-black, or dark brown) (Haviland, 1898; Holmgren 1913; Hill, 1942; Harris, 1966; Gay, 1955; Thapa, 1982; Bourguignon and Roisin, 2011; Scheffrahn *et al.*, 2015). The alates of *C. formosanus* look similar to five species, including *C. acinaciformis raffrayi*, *C. elisae*, *C. gestroi*, *C. intermedius*, and *C. travians*. So far, no reliable characters have been used to differentiate *C. formosanus* from *C. acinaciformis raffrayi* and *C. elisae*. The antennal spots and the color between the head and abdomen could be used to distinguish *C. formosanus* and *C. gestroi*. *Coptotermes formosanus* with antenna spots inconspicuous, the color of the head darker than the abdomen in dorsal view, and *C. gestroi* with antenna conspicuous, the color of the head as dark as the abdomen in dorsal view (Li *et al.*, 2010). In addition to antennal spots and body pigmentation, *C. formosanus* and *C. gestroi* alates could be distinguished by the basal coloration of wing veins (more pigmented in *C. formosanus*) and wing membrane setae are longer and denser in *C. formosanus* than that of *C. gestroi* (Scheffrahn *et al.*, 2015). The length of the hind tibia of *C. formosanus* is 1.33–1.60 mm, and *C. intermedius* is 1.04–1.14 mm (Harris, 1966). The 1st antennal article of *C. formosanus* is longest, compared with the *C. travians* with the 2nd antennal article longest (Haviland, 1898).

Very few characters of the soldier caste can be used to identify the *Coptotermes* species. The shape, size, slope of the fontanelle, and the number of the setae around the fontanelle are good characters to distinguish the species; however, previous studies seldom described the fontanelle clearly. The shape and size of the fontanelle of *C. formosanus* (triangular to arched, 0.15–0.22 mm) are different from *C. curvignathus* (0.21–0.25 mm), *C. gestroi* (oval) and *C. travians* (round) (Holmgren, 1913; Thapa, 1982; Scheffrahn *et al.*, 2015). The slope of the fontanelle could be used to distinguish the *C. formosanus* and other species; for example, the fontanelle of *C. formosanus* could only be observed from the frontal view, and the fontanelle of *C. elisae* could be well visible from the dorsal view (Bourguignon and Roisin, 2011). In profile, the anterior vertex of *C. formosanus* ends in an abrupt angle where it meets the dorsum of the fontanelle. In contrast, in *C. gestroi*, the vertex is rounded near the fontanellar junction (Scheffrahn *et al.*, 1990). The arrangement of two pairs of setae around the fontanelle of *C. formosanus* is different from those of *C. gestroi* (one pair of setae), *C. sjostedti* (one pair), or *C. testaceus* (three pairs) (Scheffrahn *et al.*, 2015).

3.5 Recent progress of *Coptotermes* taxonomy

In the worldwide termite catalog (Krishna *et al.*, 2013), 66 living species and four fossil species of *Coptotermes* were listed. In the past ten years, two new described species and nine junior synonyms were proposed, resulting in 59 living *Coptotermes* species (Table 3.1). The two new species, *C. nanus* Lee *et al.*, 2017 and *C. cooloola* Lee *et al.*, 2017 from Western Australia, were supported with morphological information and molecular data, including 12S, 16S rRNA, COI, COII sequences, and nuclear internal transcribed spacer 1 gene (ITS1) (Lee *et al.*, 2017). Nine taxonomic treatments of junior synonyms were conducted through multiple methods and supported by various evidence. Five junior synonyms of Chinese species were proposed based on a re-examination of the original description and type specimens. Li *et al.* (2012) proposed that *C. quizhouensis* He and Qui, 1992 is a junior synonym of *C. formosanus* by comparing morphological descriptions of the two original descriptions. *Coptotermes changtaiensis* Xia and He, 1986, *C. hekouensis* Xia and He, 1986, *C. suzhouensis* Xia and He, 1986 are proposed as the junior synonyms of *C. formosanus* after re-examining the type specimens of the three species in both winged imago and soldier castes (Chen *et al.*, 2020). *Coptotermes cochlearus* Xia and He, 1986 was confirmed as the junior synonym of *C. gestroi* based on the soldier caste's morphological data of original

Table 3.1. Checklist of *Coptotermes* worldwide (updated from Krishna *et al.*, 2013; Chouvenc *et al.*, 2016).

	Species name[a]	Alate	Soldier	Gene sequences	Putative native area
01	***C. acinaciformis acinaciformis* (Froggatt, 1898)**	Y	Y	Y	Australia
02	***C. acinaciformis raffrayi* Wasmann, 1900**	Y	Y	Y	Western Australia
03	***C. amanii* (Sjöstedt, 1911)**	Y	Y		Ethiopian region
04	*C. amboinensis* Kemner, 1931	Y	Y		Indonesia: Maluku
05	*C. bannaensis* Xia and He, 1986		Y		China: Yunnan
06	*C. beckeri* Mathur and Chhotani, 1969		Y		South India
07	*C. bentongensis* Krishna, 1956		Y		Malaysia
08	*C. boetonensis* Kemner, 1934		Y		Indonesia: Java
09	***C. brunneus* Gay, 1955**	Y	Y	Y	Western Australia
10	*C. ceylonicus* Holmgren, 1911	Y	Y		South India, Sri Lanka
11	*C. chaoxianensis* Huang and Li, 1985	Y	Y		China: Anhui
12	***C. cooloola* Lee et al., 2017**		Y	Y	Australia
13	***C. curvignathus* Holmgren, 1913**	Y	Y	Y	Indonesia, Vietnam
14	*C. cyclocoryphus* Zhu *et al.*, 1984	Y	Y		China: Guangdong
15	*C. dimorphus* Xia and He, 1986	Y	Y		China: Yunnan
16	*C. dobonicus* Oshima, 1914		Y		Papua New Guinea
17	***C. dreghorni* Hill, 1942**	Y	Y	Y	Queensland
18	***C. elisae* (Desneux, 1905)**	Y	Y		Indonesia, Malaysia
19	*C. emersoni* Ahmad, 1953		Y		Sri Lanka
20	***C. formosanus* Shiraki, 1909**	Y	Y	Y	Mainland China, Taiwan
21	***C. frenchi* Hill, 1932**	Y	Y	Y	Australia
22	*C. fumipennis* (Walker, 1853)	Y			Unknown
23	***C. gambrinus* Bourguignon and Roisin, 2011**		Y		Papua New Guinea
24	*C. gaurii* Roonwal and Krishna, 1955		Y		Sri Lanka, Nicobar
25	***C. gestroi* (Wasmann, 1896)**	Y	Y	Y	Indonesia, Malaysia, Philippines
26	***C. grandiceps* Snyder, 1925**	Y	Y		Papua New Guinea
27	*C. grandis* Li and Huang, 1985		Y		China: Fujian
28	*C. guangdongensis* Ping, 1985		Y		China: Guangdong
29	*C. gulangyuensis* Li and Huang, 1986		Y		China: Fujian
30	*C. hainanensis* Li and Tsai, 1985		Y		China: Hainan
31	*C. heimi* (Wasmann, 1902)	Y	Y	Y	India, Pakistan, Nepal
32	***C. intermedius* Silvestri, 1912**	Y	Y	Y	West Africa
33	***C. kalshoveni* Kemner, 1934**		Y	Y	Indonesia, Malaysia

Continued

Table 3.1. Continued.

	Species name[a]	Alate	Soldier	Gene sequences	Putative native area
34	*C. kishori* Roonwal and Chhotani, 1962		Y		India
35	***C. lacteus* (Froggatt, 1898)**	Y	Y	Y	Eastern Australia
36	*C. longignathus* Xia and He, 1986		Y		China: Yunnan
37	*C. longistriatus* Li and Huang, 1985		Y		China: Guangdong
38	*C. mauricianus* (Rambur, 1842)	Y			Mauritius
39	*C. melanoistriatus* Gao *et al.*, 1995		Y		China: Hong Kong
40	*C. menadoae* Oshima, 1914	Y			Indonesia: Sulawesi
41	***C. michaelseni* Silvestri, 1909**	Y	Y	Y	Western Australia
42	*C. minutissimus* Kemner, 1934		Y		Indonesia: Sulawesi
43	*C. monosetosus* Tsai and Li, 1985	Y	Y		China: Yunnan
44	***C. nanus* Lee *et al.*, 2017**		Y	Y	Australia
45	*C. oshimai* Light and Davis, 1929		Y		Indonesia: Sulawesi
46	***C. pamuae* Snyder, 1925**		Y		Papua New Guinea
47	*C. paradoxus* (Sjöstedt, 1911)	Y			Ethiopian region
48	*C. peregrinator* Kemner, 1934		Y		Indonesia: Sulawesi
49	*C. premrasmii* Ahmad, 1965	Y	Y		Thailand
50	*C. remotus* Hill, 1927	Y	Y		Papua New Guinea
51	*C. sepangensis* Krishna, 1956		Y	Y	Indonesia, Malaysia
52	*C. shanghaiensis* Xia and He, 1986	Y	Y		China: Shanghai
53	*C. silvaticus* Harris, 1968		Y		Ethiopian region
54	*C. sinabangensis* Oshima, 1923		Y		Sumatra, Malaysia
55	***C. sjostedti* Holmgren, 1911**	Y	Y	Y	Ethiopian region to Senegal
56	***C. testaceus* (Linnaeus, 1758)**	Y	Y	Y	Neotropics
57	***C. travians* (Haviland, 1898)**	Y	Y	Y	Sumatra, Malaysia
58	***C. truncatus* (Wasmann, 1897)**		Y		Madagascar
59	*C. varicapitatus* Tsai and Li, 1985		Y		China: Guangdong

[a]**Bold** species denotes that there is robust evidence supporting its validity; non-bold species indicates that additional taxonomic information required to confirm its validity.

descriptions and type specimens (Ke *et al.*, 2022). Molecular tools were employed to extract the DNA of these Chinese type specimens for gene sequencing, but no results were obtained, likely due to specimen decomposition.

Confirming junior synonyms with molecular and morphometric data has been a favorable methodology in recent years. *Coptotermes vastator* Light, 1929 was confirmed as the junior synonym of *C. gestroi* based on the molecular phylogenetic analysis (12S, 16S rRNA, COII sequences) and morphometric data of soldier castes (Yeap *et al.*, 2007). *Coptotermes crassus* Snyder, 1922 and *C. niger* Snyder, 1922 were confirmed as the junior synonyms of *C. testaceus* (Linnaeus, 1758) based on a comparison of the 16S rRNA sequences and the morphological characters of *Coptotermes* species collected in Central and South America (Scheffrahn *et al.*, 2015). *Coptotermes ochraceus* Ping and Xu, 1986 was proposed as a junior synonym of *C. formosanus*, according to the 16S rRNA, COII sequences, and nine morphological characters of soldier castes in the samples collected from the type locality, Guizhou, China (Ke *et al.*, 2016).

In addition to the nine synonyms mentioned above, Yeap *et al.* (2009) considered that *C. dimorphus* Xia and He, 1986 is a junior synonym of *C. formosanus*, based on genetic evidence (12S and 16S rRNA). However, according to the original descriptions, the soldier caste of *C. dimorphus* has only one pair of setae around the fontanelle, but that of *C. formosanus* has two pairs. Because Yeap *et al.* (2009) did not provide detailed information on the collection site of *C. dimorphus* in China and no morphological evidence was provided, the taxonomic status of *C. dimorphus* requires further study. Jamil *et al.* (2018) proposed that *C. sepangensis* Krishna, 1956 was the junior synonym of *C. kalshoveni* Kemer, 1934, according to the COII sequences and the 17 morphological characters of the soldier caste. However, samples used to represent *C. sepangensis* and *C. kalshoveni* in Jamil *et al.* (2018) were collected from Sarawak, Malaysia, not from their type localities, Sepang, Malaysia, and Semarang, Indonesia, respectively. We suggest that the collection of *C. sepangensis* and *C. kalshoveni* from their type localities and a comparison of their sequences are necessary to further confirm the taxonomic status of *C. sepangensis*.

3.6 Toward *Coptotermes* revision

Following the *Coptotermes* revision roadmap proposed by Chouvenc *et al.* (2016), the 59 names of *Coptotermes* spp. listed in Table 3.1 are divided into two categories, i.e., 23 valid species (bold species names) and 36 uncertain species (non-bold species names). The valid species included a detailed morphological description, interspecific comparison, intraspecific morphological variability, and genetic information. The uncertain species were seldom studied, less so than other species, or without available molecular data supporting their validities. In the past ten years, in addition to the two newly described species and nine junior synonyms, validity of three species had been changed. First, the validity of *C. acinaciformis raffrayi* Wasmann, 1980 and *C. curvignathus* Holmgren, 1913 were confirmed based on new morphological and molecular information (Yeap *et al.*, 2009; Lee *et al.*, 2015). However, the data to differentiate *C. gestroi* and *C. heimi* (Wasmann, 1902) were insufficient to conclude. Yeap *et al.* (2010) proposed *C. heimi* as a junior synonym of *C. gestroi*, based on the mitochondrial gene (16S rRNA and COII sequences) and 17 morphometric data of the soldier caste. The type locality of *C. heimi* is Ahmednagar district, India, but the collection sites of *C. heimi* of Yeap *et al.* (2010) were not from the type localities (Pakistan and unknown location in India). Additional DNA sequences from type locality specimens and alate morphological data are required to confirm the validity of *C. heimi*, which is now listed as an uncertain species (Table 3.1).

Historically, the description of *Coptotermes* species mainly relied on the soldier caste. Soldier morphological description is available for 55 of 59 species (Table 3.1). The four species described based solely on alate caste are *C. fumipennis* (Walker, 1853), *C. mauricianus*

(Rambur, 1842), *C. menadoae* Oshima, 1914, and *C. paradoxus* (Sjöstedt, 1911). All four species had very few collection records, and some had only one individual. *C.fumipennis* might be the junior synonym of *C. testaceus* (Linnaeus, 1758), according to the morphology of two syntypes, which were considered as non-Australian and species (Watson and Gay, 1980). *C. mauricianus* was only recorded in Mauritius, which is regarded as the senior synonym of *C. truncates* (Wasmann, 1897) or *C. gestroi* (Krishna *et al.*, 2013, Chouvenc *et al.*, 2016). *C. paradoxus* may be the synonym of *C. intermedius* Silvestri, 1912 according to the unpublished file card from Emerson (Krishna *et al.*, 2013). Although alate caste provides more identification characters (Li *et al.*, 2010; Scheffrahn *et al.*, 2015), the alate caste of 27 species (27/59, 45.8%) remains unknown. In addition to morphological information, gene sequence data are available only from 19 of the 59 species (32.2%).

From a geographic perspective of *Coptotermes* taxonomy, Australian and American species have been well revised (Scheffrahn *et al.*, 2015; Lee *et al.*, 2017). Few African species have been described and studied (Harris, 1966). Asian species comprise the majority and require the most attention. The *Coptotermes* revision in Asia has been in progress for the past decade (Li *et al.*, 2012; Ke *et al.*, 2016; Jamil *et al.*, 2018; Chen *et al.*, 2020; Ke *et al.*, 2022). Among the 36 uncertain *Coptotermes* species, the putative native areas of 32 species are in Asia (Table 3.1). Further studies in China (14 uncertain species), Indonesia (seven uncertain species), India, and Sri Lanka (six uncertain species) will lead to significant improvement in *Coptotermes* revision. To accelerate the revision of *Coptotermes*, we suggest the following strategies.

3.6.1 Intensive collection around type localities

A sample size of ten colonies for each *Coptotermes* sp. is a conservative number to present intraspecific variation. A collection of 100 individuals, including > 10 soldiers, would be sufficient for further morphometric measurement and genetic studies of each termite colony. Termite individuals found in the same tunnel or mud tube could be considered intra-colonial individuals. Termite samples collected at infested sites > 100 m apart from each other could be regarded as different colonies (see Chapter 5 of this volume).

Dispersal flights of multiple species and multiple colonies may occur at the same time in each area. Alates collected during their dispersal flights could be preserved together in the same vial with ethanol, but each alate should be treated as an independent sample. Soldiers and mature alates collected from the same colony are ideal samples to represent the species and to be selected as type specimens.

Researchers could focus on collecting *Coptotermes* soldier and worker castes at feeding sites in natural environments, such as logs, rotten wood, and trunks of living trees. Mud tubes built on living trees are another clue for searching *Coptotermes* spp. (Chiu *et al.*, 2022). Setting up Malaise traps and light traps eases alate collection and estimates dispersal flight seasons (Chiu *et al.*, 2016). Because access to private buildings in urban environments is often limited, launching citizen science projects to invite the public and pest management professionals to collect termites would be a practical method to collect many samples over a long period (Huang *et al.*, 2022).

3.6.2 Obtaining gene sequences

Mitochondrial genes 16S, COI, and COII are the most frequently sequenced genes for species identification. So far, gene sequences of 19 *Coptotermes* species are available (Table 3.1). Gene sequences obtained from multiple colonies of an uncertain species in its type locality could be aligned with sequences of the other valid species for further synonymization or validation. Gene sequence comparison is also helpful for matching conspecific soldier and alate castes when they are collected separately.

So far, all *Coptotermes* species have been described based on the soldier or alate morphology instead of gene sequences. To confirm the linkage between gene sequences and named species requires morphological and collection information. Hence, assuming species identification in any molecular data platform is unreliable without confirming its morphological description and collection information.

3.6.3 Confirmation of the linkage between gene sequences and morphometric data

Intraspecific morphological variation may result from colony age (Chouvenc *et al.*, 2014), nutrition availability (Mullins *et al.*, 2021), and climatic and geographic factors. For example, measurements of 30 individuals of *C. formosanus* soldier collected from 10 locations in Taiwan indicated a considerable intraspecific variation, with the head length ranging between 1.048 and 1.694 mm, which overlapped with all described *Coptotermes* species from China (Chouvenc *et al.*, 2016).

Using multiple gene sequences, termite samples could be grouped into operational taxonomic units (OTU). Within each OTU, morphometric measurements could be conducted to obtain the intraspecific variation. Observation of multiple colony samples can reveal body color variation due to colony age, post-molting time, and preservation condition. Submitting both genetic and morphological data for species identification will make it more reliable to describe new species, redescribe uncertain species, and synonymize some species.

3.6.4 Cautions with species hybridization

Hybridization of *C. formosanus* and *C. gestroi* has been documented in the lab (Chouvenc *et al.*, 2015) and field (Chen, 2022). The hybrid soldier and alate present intermediate morphology (Chen, 2022; Patel *et al.*, 2023). Since mitochondria are inherited from female linkage, detecting the hybridization of a population would require both mitochondrial and nuclear markers or microsatellite markers (Chouvenc *et al.*, 2017). Even though the hybridization of *C. formosanus* and *C. gestroi* is the only proven case within the genus *Coptotermes*, it may occur in sympatric areas which are inhabited by multiple *Coptotermes* species. Hybridization should be considered a potential factor for taxonomic confusion.

3.7 Conclusion

Previous taxonomic studies of the Formosan subterranean termite present examples of the challenge of *Coptotermes* identification. This chapter has reviewed the current taxonomic status of all *Coptotermes* spp. and provided practical methods for future research. Revision of the *Coptotermes* genus was not and will not be an easy task, but this complex work could be divided into many simple ones and conducted by researchers of respective geographical areas. International cooperation, especially in Asia, will accelerate the overall revision of this economically important termite genus.

References

Bourguignon, T. and Roisin, Y. (2011) Revision of the termite family Rhinotermitidae (Isoptera) in New Guinea. *ZooKeys* 148, 55–103.

Chen, G.Y. (2022) Hybridization between Formosan subterranean termite and Asian subterranean termite in Taiwan. MSc dissertation, National Chung Hsing University, Taichung, Taiwan.

Chen, G.-Y., Ke, Y.-L., Liang, W.-R. and Li, H.-F. (2020) Redescription of Formosan subterranean termite, *Coptotermes formosanus* (Blattodea: Rhinotermitidae), with three new synonyms from China. *Acta Entomologica Musei Nationalis Pragae* 60, 599–608.

Chiu, C.-I, Yeh, H.-T., Tsai, M.-J. and Li, H.-F. (2016) Naturalization and control of *Coptotermes gestroi* in a Taiwanese forest. *Journal of Economic Entomology* 109, 1317–1325.

Chiu, C.-I, Chen, B.-Y., Chang, F.-C., Kuan, K.-C. and Li, H.-F. (2022) Functional plasticity of foraging shelter tubes built by termites. *Environmental Entomology* 51, 649–659.

Chouvenc, T. and Su, N.-Y. (2014) Colony age-dependent pathway in caste development of *Coptotermes formosanus* Shiraki. *Insectes Sociaux* 61, 171–182.

Chouvenc, T., Basille, M., Li, H.-F. and Su, N.-Y. (2014) Developmental instability in incipient colonies of social insects. *PLoS ONE* 9(11), e113949.

Chouvenc, T., Helmick, E. E. and Su, N.-Y. (2015) Hybridization of two major termite invaders as a consequence of human activity. *PLoS ONE* 10(3), e0120745.

Chouvenc, T., Li, H.-F., Austin, J., Bordereau, C., Bourguignon, T. *et al.* (2016). Revisiting *Coptotermes* (Isoptera: Rhinotermitidae): a global taxonomic road map for species validity and distribution of an economically important subterranean termite genus. *Systematic Entomology* 41, 299–306.

Chouvenc, T., Osorio, S., Chakrabarti, S., Helmick, E.E., Li, H.-F. *et al.* (2017) Assessment of genetic markers for the determination of *Coptotermes formosanus* × *Coptotermes gestroi* (Isoptera: Rhinotermitidae) F1 hybrids. *Florida Entomologist* 100, 657–659.

Evans, T. A., Forschler, B. T. and Grace, J. K. (2013) Biology of invasive termites: a worldwide review. *Annual Review of Entomology* 58, 455–474.

Gay, F.J. (1955) A new *Coptotermes* and *Ahamitermes* (Isoptera) from Australia. *Proceedings of the Linnean Society of New South Wales* 79 (5-6) [1954], 177–181.

Harris, W.V. (1966) On the genus *Coptotermes* in Africa (Isoptera: Rhinotermitidae). *Proceedings of the Royal Entomological Society of London, Series B, Taxonomy* 35, 161–171.

Haviland, G. D. (1898) Observations on termites; with descriptions of new species. *Journal of the Linnean Society of London, Zoology* 26, 358–442 + 4 pls.

Hill, G. F. (1942) *Termites (Isoptera) from the Australian Region*. Council for Scientific and Industrial Research, Melbourne, 479 pp. + 24 pls.

Holmgren, N. (1913) Termitenstudien. 4. Versuch einer systematischen Monographie der Termiten der orientalischen Region. *Kungliga Svenska Vetenskaps-Akademiens Handlingar* 50, 1–276 + 8 pls.

Huang, S.-Y., Chiu, C.-I, Tsai, Y.-Y., Li, W.-J., Wu, C.-C. *et al.* (2022) Nationwide termite pest survey conducted in Taiwan as a citizen science project. *Journal of Economic Entomology* 115, 1650–1658.

Itakura, S, Yoshikawa, Y., Togami, Y. and Umezawa, K. (2020) Draft genome sequence of the termite, *Coptotermes formosanus*: Genetic insights into the pyruvate dehydrogenase complex of the termite. *Journal of Asia-Pacific Entomology* 23, 666–674.

Jamil, N., Ismail, W.N.W. and Singham, G.V. (2018) Morphometric variation and genetic relationship of *Coptotermes* spp. (Blattodea: Rhinotermitidae) in Sarawak, Malaysia. *Malaysian Applied Biology* 47, 113–125.

Ke, Y., Zhang, S. and Li, Z. (2022) Phylogenetic relationships of Chinese *Coptotermes* (Blattodea: Isoptera: Rhinotermitidae) termites and a new synonym inferred from morphological data. *Journal of Entomological Science* 57, 64–81.

Ke, Y., Wu, W., Liu, G. and Li, Z. (2016) Genetic relationship between *Coptotermes ochraceus* and *Coptotermes formosanus* (Isoptera: Rhinotermitidae). *Journal of Environmental Entomology* 38, 341–347.

Krishna, K., Grimaldi, D.A., Krishna, V. and Engel, M.S. (2013) Treatise on the Isoptera of the World: Neoisoptera Excluding Termitidae. *Bulletin of the American Museum of Natural History* 377, 623–973.

Lee, T.R., Cameron, S.L., Evans, T.A., Ho, S.Y. and Lo, N. (2015) The origins and radiation of Australian *Coptotermes* termites: from rainforest to desert dwellers. *Molecular Phylogenetics and Evolution* 82, 234–244.

Lee, T.R., Evans, T.A., Cameron, S.L., Ho, S.Y., Namyatova, A.A. *et al.* (2017) A review of the status of *Coptotermes* (Isoptera: Rhinotermitidae) species in Australia with the description of two new small termite species from northern and eastern Australia. *Invertebrate Systematics* 31, 180–190.

Li, H.-F., Su, N.-Y. and Wu., W.-J. (2010) Solving the hundred-year controversy of *Coptotermes* taxonomy in Taiwan. *American Entomologist* 56, 222–227.

Li, H.-F., Ye, W., Su, N.-Y. and Kanzaki, N. (2009) Phylogeography of *Coptotermes gestroi* and *Coptotermes formosanus* (Isoptera, Rhinotermitidae) in Taiwan. *Annals of the Entomological Society of America* 102, 684–693.

Li, Z.-Q., Zhong, J.-H. and Xiao, W.-L. (2012) A new synonymy of *Coptotermes formosanus* (Isoptera: Rhinotermitidae). *Sociobiology* 59, 1223–1227.

Mullins, A., Chouvenc, T. and Su, N.-Y. (2021) Soil organic matter is essential for colony growth in subterranean termites. *Scientific Reports* 11(1), 1–11.

Munsell, A.H. (1967) *Munsell book of color: Glossy finish collection., vol.* Munsell Color Company, Baltimore, Maryland.

Oshima, M. (1909) *Taiwanese termites*. In Oshima, M. (ed.) *The First Official Report on Termites.* Taiwan Sōtokufu, Taihoku, Japan, pp. 29–40

Oshima, M. (1910) Taiwanese termites. *Dobutsugaku Zasshi*, 22, 343–346.

Oshima, M. (1912) *Third Official Report on Termites* [Dai Sankai Shiroari Chosa Hokoku]. Institute of Science, Government of Formosa Taihoku [Taipei], 186 pp.

Patel, J.S., Chouvenc, T., Wu, C.-C., Li, H.-F. *et al.* (2023) Phenotypic differences among mating combinations of two *Coptotermes* species and their hybrids (Blattodea: Rhinotermitidae). *Journal of Insect Science*, In press.

Rust, M.K. and Su, N.-Y. (2012) Managing social insects of urban importance. *Annual Review of Entomology* 57, 355–375.

Scheffrahn, R.H., Su, N.-Y. and Diehl, B. (1990) Native, introduced, and structure-infesting termites of the Turks and Caicos Islands, BWI (Isoptera: Kalotermitidae, Rhinotermitidae, Termitidae). *Florida Entomologist* 73, 622–627.

Scheffrahn, R.H., Carrijo, T.F., Křeček, J., Su, N.-Y., Szalanski, A.L. *et al.* (2015) A single endemic and three exotic species of the termite genus *Coptotermes* (Isoptera, Rhinotermitidae) in the New World. *Arthropod Systematics and Phylogeny* 73, 333–348.

Shiraki, T. (1909) On the termites of Japan. *Transactions of the Entomological Society of Japan* 2, 229–242. [in Japanese]

Thapa, R. S. (1982) Termites of Sabah. *Sabah Forest Record* 12, 1–374.

Tsai, C.-C. and Chen, C.-S. (2003) First record of *Coptotermes gestroi* (Isoptera: Rhinotermitidae) from Taiwan. *Formosan Entomologist* 23, 157–161.

Wasmann, E. (1896) Viaggio di Leonardo Fea in Birmania e regioni vicine LXXII. Neue Termitophilen und Termiten aus Indien. I–III. *Annali del Museo Civico di Storia Naturale di Genova* 16, 613–630.

Watson, J. and Gay, F. (1980) The identities of *Termes austraus* Walker and *Termes fumipennis* Walker (Isoptera). *Australian Journal of Entomology* 19, 19–25.

Yano, M. (1911) On the scientific names of Japanese termites. *Dobutsugaku Zasshi* 23, 364–368.

Yeap, B.K., Othman, A.S., Lee, V.S. and Lee, C.Y. (2007) Genetic relationship between *Coptotermes gestroi* (Wasmann) and *Coptotermes vastator* Light (Isoptera: Rhinotermitidae). *Journal of Economic Entomology* 100, 467–474.

Yeap, B.-K., Othman, A.S. and Lee, C.-Y. (2009) Molecular systematics of *Coptotermes* (Isoptera: Rhinotermitidae) from East Asia and Australia. *Annals of the Entomological Society of America* 102, 1077–1090.

Yeap, B.-K., Dugal, F.M., Othman, A.S. and Lee, C.-Y. (2010) Genetic relationship between *Coptotermes heimi* and *Coptotermes gestroi* (Isoptera: Rhinotermitidae). *Sociobiology* 56, 291–312.

4 A primer to termite biology: *Coptotermes* colony life cycle, development, and demographics

Thomas Chouvenc
Entomology and Nematology Department, Ft. Lauderdale Research and Education Center, University of Florida, tomchouv@ufl.edu

Abstract
In the past five decades, major advancements in the understanding of subterranean termite biology have been made, partially thanks to the research that focused on *Coptotermes* pest species. In this chapter, a comprehensive overview of the *Coptotermes* colony life cycle is placed in the broader context of termite biology, to assemble the interconnected pieces of a complex sociobiological entity. As eusocial cockroaches, termite biology is deeply rooted in its ancestral hemimetabolous developmental constraints, and the life cycle of a termite colony reflects this aspect throughout its lifetime. Each individual termite develops in a socio-environmental context that influences its caste determination, its function within the colony and its potential longevity. A *Coptotermes* colony goes from foundation to death by passing through major steps of colony growth and maturation, while optimizing reproductive output and high colony fitness over time. This chapter provides a systemic approach to this *Coptotermes* biology, with first a focus on individual termite development, physiology, symbiosis, and behavior, to the colony-wide implications of individual differentiation over time, and finally, with important considerations for colony demographic alteration throughout the lifetime of a *Coptotermes* colony.

4.1 Introduction

Coptotermes formosanus Shiraki is one of the most damaging termite pest species in the world and has received extensive research efforts, especially after it established beyond its native range (Lax and Osbrink, 2003; Blumenfeld *et al.*, 2021). This historic biological invasion has led the bulk of subterranean termite research to focus on *C. formosanus* in the past five decades, as testified by the very existence of this book (Chapter 1 of this volume). *Coptotermes* as a genus is considered a pest globally in the tropics, subtropics, and warm temperate parts of the world (Chapter 2 of this volume). Most of the 23 estimated valid species have a tangible economic impact in their native and invasive ranges (Chouvenc *et al.*, 2016). Consequently, sizable research activity also focused on the biology of *Coptotermes* species beyond *C. formosanus*. Several aspects of the biology of *Coptotermes* are relatively conserved within the genus, which implies that observations made in *C. formosanus* often apply to most *Coptotermes* species, and vice versa. Of course, each *Coptotermes* species displays unique traits that set them apart, such as their invasive potential, nest architecture, and environmental extremes tolerance, but the discussion of such traits is outside of the scope of the current chapter.

 Biology and Management of the Formosan Subterranean Termite and Related Species (eds Nan-Yao Su and Chow-Yang Lee)
DOI: 10.1079/9781800621596.0004

Therefore, while this chapter primarily focuses on *C. formosanus* general biology, several aspects of the *Coptotermes* biology covered herein were studied in other *Coptotermes* species and are assumed to be similar or identical in *C. formosanus*, to provide a comprehensive overview of the complexity of *Coptotermes* genus biology.

Coptotermes is a subterranean termite genus which has historically been placed in the Rhinotermitidae family (Krishna *et al.*, 2013). However, it belongs to a clade (*Reticulitermes* + *Heterotermes* + *Coptotermes*) that is a sister group to the Termitidae family, making the Rhinotermitidae[1] a paraphyletic group (Bourguignon *et al.*, 2016a; Chouvenc *et al.*, 2021; Hellemans *et al.*, 2022) (Fig. 4.1). As a result, this clade displays some biological traits that depart from the rest of the Rhinotermitidae but also shares some traits found in Termitidae (e.g., large colonies, bifurcated developmental pathways). Furthermore, while genera within the *Heterotermes* + *Reticulitermes* + *Coptotermes* clade share common ancestral traits, *Coptotermes* morphology, biology, and ecology sometimes depart from *Reticulitermes* and *Heterotermes* (Lo *et al.*, 2004; Vargo and Husseneder, 2009; Bourguignon *et al.*, 2016a; Wang *et al.*, 2019). Therefore, as a word of caution, some of the observations on *Coptotermes* summarized herein may not always apply to other subterranean termite genera.

This chapter first provides an overview of the *Coptotermes* life cycle and highlights how the inherent termite molting process drives major aspects of the biology of *Coptotermes* colonies. Next, it provides step-by-step details of the complex colony foundation process and the successive transitions toward colony maturity. Finally, it discusses the implications of colony demographic changes over time within the context of potential strategies toward colony reproductive perennity. Ultimately, this chapter aims at synthesizing the current understanding of the complexity of the biology of *Coptotermes* colonies and how future research avenues can further inform aspects of the biology of an often-misunderstood termite model system, when compared with other social insects.

4.2 General termite biology

4.2.1 Termite developmental biology and its evolutionary context

Termites form a clade of eusocial cockroaches (= Blattodea: Isoptera[2]), which is the sister group to *Cryptocercus* woodroaches (Lo *et al.*, 2000, 2003, 2007; Inward *et al.*, 2007; Nalepa, 2010a; Wang *et al.*, 2022) (Fig. 4.1). The transition from a woodroach ancestor to termites marked a major shift toward complex degrees of social organizations, which resulted in the emergence of many traits that facilitated termites' successful dominance of various ecological niches over the past 150 million years of diversification (Bignell, 2016; Chouvenc *et al.*, 2021). Approximately 3,000 extant termites have been described, with a wide diversity in soldier morphology, diet specialization, breeding systems, nest architecture, and caste differentiation mechanisms (Watson *et al.*, 1985; Eggleton, 2000; Krishna *et al.*, 2013). Eusociality in organisms is classically represented by a biological entity (the colony) which displays cooperative brood care, overlapping generations, and reproductive division of labor (Wilson, 1971). A termite ancestor acquired eusociality independently from eusocial hymenopterans as a diplo-diploid organism through different selective pressures (Howard and Thorne, 2010; Nalepa, 2010a, 2011; Kronauer and Libbrecht, 2018). Termites are hemimetabolous insects that underwent a strong "altricial development–biparental care" codependence through evolutionary processes (Nalepa, 2011). Along with a relationship with

[1] The Rhinotermitidae family is currently under phylogenomic scrutiny and will most likely be revised to resolve the monophyly of each termite family in a near future. As a result, the *Heterotermes–Reticulitermes–Coptotermes* clade will most likely be elevated to its own family: Heterotermitidae (Chouvenc *et al.*, 2021).

[2] Although Isoptera is no longer accepted as an insect order (Inward *et al.*, 2007), it can still be used as an unranked clade name that includes all termites, as their biology departs from all other cockroaches (Lo *et al.*, 2007), although the Infraorder rank was suggested (Krishna *et al.*, 2013).

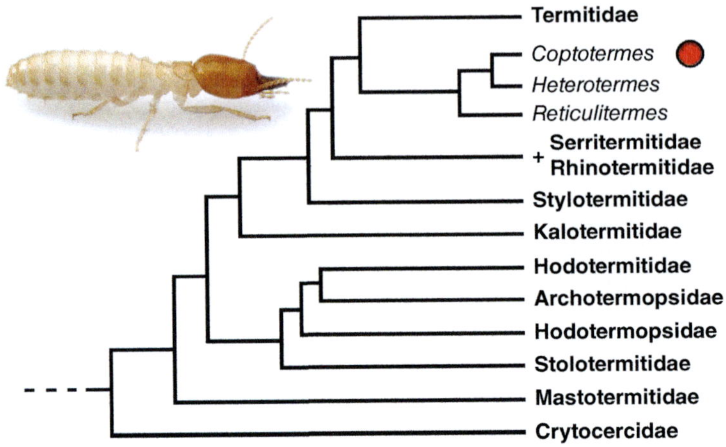

Fig. 4.1. Phylogenetic position of *Coptotermes* within the extant termite diversity. Modified from Chouvenc *et al.* (2021) with permission. (T. Chouvenc.)

obligate gut mutualists and the emergence of epigenetically regulated developmental processes, it ultimately resulted in nymphs expressing alloparental care behaviors, moving these ancestors toward eusociality (Nalepa, 2015; Matsuura, 2020; Chouvenc *et al.*, 2021; Chouvenc, 2022).

Contrary to eusocial Hymenoptera in which the brood care is inherently provided by adult individuals that went through holometabolous metamorphosis, termite brood care is provided by individuals that are indefinitely maintained in a juvenile state (Nalepa, 2010a). This difference is critical, as it implies that termites must engage in periodic molting that may never lead to the imaginal molt (adult alate). In "lower"[3] termite families (i.e., non-termitid families) as well as in all cockroaches, many genera display a linear developmental pathway, as each juvenile helper (often referred to as "pseudergate") has the opportunity to proceed toward the imaginal molt and participate in a dispersal flight (Watson *et al.*, 1985; Roisin and Korb, 2010; Korb, 2015; Oguchi *et al.*, 2021). Such pseudergate individuals retain some degrees of developmental totipotency, where the indefinite developmental stasis in juvenile morphology is not permanent. However, such ancestral hemimetabolous linear development was replaced in several termite clades by a bifurcated developmental ground plan, as observed in Mastotermitidae, Hodotermitidae, some Rhinotermitidae, and all Termitidae (Watson *et al.*, 1985; Legendre *et al.*, 2013; Revely *et al.*, 2021). Bifurcated developments imply that the imaginal developmental pathway (alates) is irrevocably detached from the sterile worker pathway, as workers have lost their developmental totipotency (= "true" workers). It is most likely that the general bifurcated development scheme observed in *Coptotermes* was already present in the ancestor common to the *Heterotermes* + *Reticulitermes* + *Coptotermes* + Termitidae clade (Fig. 4.1) (Bourguignon *et al.*, 2016b; Chouvenc *et al.*, 2021; Revely *et al.*, 2021).

4.2.2 Termite terminology

The termite literature has historically used terms that are specific to termite biology

[3] The term "lower" termite has historically been used to represent all termite families, minus Termitidae (= "higher" termites). The commonality among all "lower" termites is their mode of cellulose digestion that relies on cellulolytic protists in their hindgut, a mutualism ancestral to all termites. This symbiosis was lost in Termitidae, as their diet and aspects of their biology shifted away from some ancestral conditions (Chouvenc *et al.*, 2021). However, one should abstain from ranking lower/higher termites, as the use of these terms, while convenient, ignores that all extant lineages have independently evolved novel traits on their own since the emergence of eusociality. Therefore, "lower termites" does not imply conditions that are less complex or more primitive than in "higher termites." (Carrijo *et al.*, 2023).

and are often context-dependent in their meaning. As a result, such terms have been, and still are, a source of confusion for readers new to termite biology. For clarity purposes in this chapter, terms that relate to *Coptotermes* biology primarily follow definitions revisited from Grassé (1982), Thorne (1996), and Revely *et al.* (2021).

Alates: mature winged individuals that proceeded through imaginal molt (= imago, adult phenotype), after molting through several nymphal instars. Alates leave their natal nest during dispersal flights.

Dealates: imagoes that have dropped their wings after a dispersal flight, in search of a mating partner, or in the process of establishing an incipient colony.

Primary reproductives: the queen and king that initially established the colony after a dispersal flight event (alates → dealates → primary reproductives). The primary queen and king in *Coptotermes* start as the sole reproductively competent individuals in the colony, mate and lay eggs, and can maintain reproductive functions for more than a decade. Wing scales are present in primary reproductives.

Secondary reproductives (= nymphoids): individuals that proceeded through part of the nymphal pathway but did not engage in imaginal molt (brachypterous neoteny). As a result, compound eyes are underdeveloped, cuticle sclerotization is weak, and wing buds are often short. They remain in the colony to supplement or replace egg production functions.

Physogastry: mature reproductive state in queens (primary and secondary) that expanded their abdomen to accommodate their increased individual egg-laying capacity.

Nymphs: Individuals engaged in the reproductive development pathway, which display wing buds, and successively molt toward the imago ($L_2 \rightarrow N_1 \rightarrow \ldots \rightarrow N_6 \rightarrow$ alate) or as secondary reproductives ($N_n \rightarrow$ SR). The older the nymphal instar, the more the wing buds and compound eyes are developed.

Eggs: oocytes produced by sexual reproduction (= zygote), laid by a queen, and undergoing embryonic development (to date, parthenogenesis has not been described in *Coptotermes*). Eggs (E) rely on competent caregivers to prevent desiccation and to complete embryonic development.

Larvae: the first two instars after hatching ($E \rightarrow L_1 \rightarrow L_2$), which remain fully dependent on competent colony members to continue their post-embryonic development. While "larva" is a term specific to holometabolous insects, it was historically co-opted in termites (hemimetabolous insects), as an analogy to eusocial Hymenoptera, where poorly sclerotized larval instars are nutritionally dependent on other members of the colony.

Workers: undifferentiated individuals beyond the second instar larva that are not engaged in the nymphal pathway. The worker caste in *Coptotermes* represents the majority of individuals in the colony, with 85–90% of individuals in foraging populations (Haverty, 1977). Workers are nutritionally competent individuals that are in charge of finding and processing food, brood care, nest maintenance, and most other miscellaneous chores within the colony. In *Coptotermes*, individuals reaching the first instar worker (W_1) can no longer engage in the nymphal pathway but continue to molt indefinitely toward the next worker instar (W_{n+1}) until individual senescence, or molt toward a soldier morph. *Coptotermes* is not known to have the ability to produce fertile workers (ergatoid reproductives = "tertiary" or third-form apterous reproductives) (Myles, 1999; Chouvenc and Su, 2014). In *Coptotermes*, workers do not have a terminal molt but rarely survive beyond the W_7 instar.

Soldiers: Individuals displaying defensive traits, with a highly sclerotized head capsule, slashing mandibles, and a fontanelle connected to a large frontal gland reservoir that expands into the abdomen. Soldiers are nutritionally dependent on workers and are primarily in charge of defending the colony from predators and competitors. The *Coptotermes* soldier caste represents 10–15% of individuals in a colony. The soldier developmental pathway is accessible from relatively young worker instars through an intermediate molt (= pre-soldier or "white soldier"). The soldier is a terminal stage, unable to molt.

4.2.3 *Coptotermes* life cycle overview

Like the vast majority of termite species (Nutting, 1969), *Coptotermes* initiate new colonies after a dispersal flight event from established mature colonies (Fig. 4.2). Male and female alates (winged imagoes) leave their natal colony to find a mate and establish an incipient termite colony as primary queen and king. The first few eggs laid (E) hatch into first instar larvae (L_1) which then molt into second instar larvae (L_2). At this point, the incipient colony is under biparental care. As larvae molt into first instar workers (W_1), they acquire nutritional symbionts from their parents and progressively take over brood care functions while continuing to molt into further worker instars (W_{n+1}). Soldiers (S) are produced to provide defense capabilities for the young colony. It then takes several years for the colony to grow to millions of individuals, as the queen becomes physogastric, and nymphs (N) are finally produced and can advance to the alate stage (A), which later engages in dispersal flights that result in the next generation of established colonies. Secondary reproductives (SRs) can be produced to replace a lost primary reproductive and maintain egg-laying activity. While the *Coptotermes* life cycle appears relatively simple at first glance (Fig. 4.2), the developmental pathways for various castes are not always accessible to all individuals and are often dependent on the social context (Chouvenc and Su, 2014).

4.3 Developmental pathways and caste determination mechanisms

4.3.1 Developmental pathways are colony-age dependent

While descriptions of the developmental pathways in *Coptotermes* have historically relied on observations of termites sampled from mature field colonies (Shimizu, 1962; Roy-Noël, 1968; King and Spink, 1975; Costa-Leonardo *et al.*, 1999; Roisin and Lenz, 1999; Raina *et al.*, 2004; Barsotti and Costa-Leonardo, 2005), this has long obscured the complexity of how individual development is

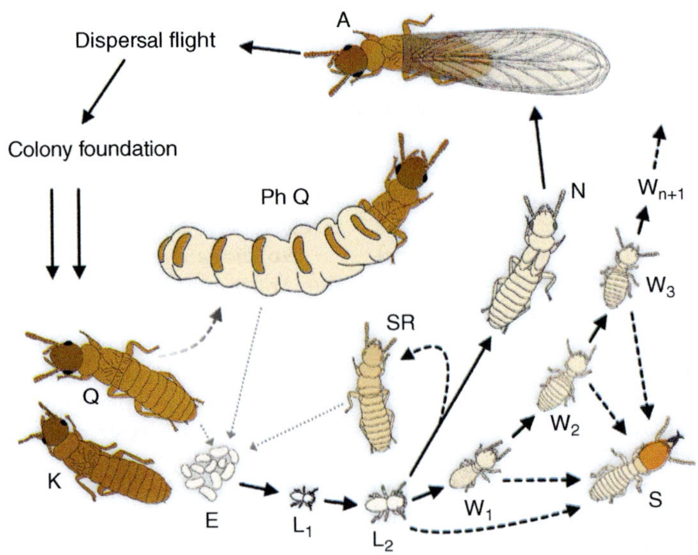

Fig. 4.2. *Coptotermes* life cycle. E = Eggs, L_n = Larval instar, W_n = Worker instar, S = Soldier, N_n = Nymphal instar, A = Alate, K = King, Q = Queen, PhQ = Physogastric queen. SR = Secondary Reproductive. Full-line arrow: primary pathways. Dotted arrow: conditional or temporarily dependent pathway. (T. Chouvenc.)

regulated depending on the social context (Chouvenc and Su, 2014). This social context primarily correlates with the colony maturity status, for which colony age is often a reliable proxy in the early years of colony growth (< 5 years, prior to colony maturity). Instar identification can be estimated by counting the number of antennomeres, as the 3rd article of the antenna duplicates during each molt. Chouvenc and Su (2014) identified five distinct colony maturity phases, where some individual developmental pathways are only accessible when a critical colony capacity is reached, and where some pathways are abandoned (temporarily through seasonality or permanently through colony growth). Progressively, a colony transitions through stages: incipient colony → immature colony → juvenile colony → mature colony → senescent colony. For each stage of colony maturity, the possible developmental pathway for individuals differs (Fig. 4.3A). For example, incipient colonies can only produce soldiers through the $L_2 \rightarrow PS_1 \rightarrow S_1$ pathway (nanitic soldier). As the colony develops toward maturity, this nanitic soldier pathway is no longer available, as soldiers are progressively produced through the $W_1 \rightarrow PS_2 \rightarrow S_2$, then the $W_2 \rightarrow PS_3 \rightarrow S_3$ pathway, and finally the $W_3 \rightarrow PS_4 \rightarrow S_4$ pathway. S_5 soldiers have been suspected to be possible but have not been confirmed in *Coptotermes*, raising the possibility that once a worker reaches the fourth instar (W_4), it can only continue molting toward the subsequent worker instar (W_{n+1}) (Barsotti and Costa-Leonardo, 2005; Chouvenc and Su, 2014). In addition, individual worker longevity is dependent on colony maturity, as living conditions and resource accessibility progressively improve with increasing colony size and age. Workers are unlikely to reach W_3 in incipient colonies, as they usually die early due to poor living conditions. As conditions improve, workers have an increasingly higher probability of survival until the next molting opportunity, with some W_7 workers reported in mature colonies (Chouvenc and Su, 2014), with a maximum age speculated at 3–4 years old. The production of nymphs (N_n) through the opening of the imaginal developmental pathway defines when a colony has reached reproductive maturity (= mature colony), which ultimately results in the production of alates (A). Finally, with the eventual complete loss of reproduction, a colony is doomed toward senescence, which is reflected by an accumulation of old instar workers in the absence of population renewal over several years of colony decline (Chouvenc *et al.*, 2022).

4.3.2 Molting process in *Coptotermes*

The periodic molting of individual termites drives major aspects of termite biology. It allows individuals to engage in caste differentiation, to improve individual physiological and behavioral efficiency, to renew mandible sharpness, and to adjust individual functionality to inherently variable colony needs (Xing *et al.*, 2013; Roisin, 2016; Lee *et al.*, 2022). However, regardless of the caste differentiation pathway an individual is engaged in, the molting process in *Coptotermes* follows a distinct series of events that is common to most castes. As a general example, this section reviews the molting process for a worker instar (W_n) molting into the next worker instar (W_{n+1}), revised (Fig. 4.3B) from a handful of studies that specifically made observations on the *C. formosanus* molting process (Raina *et al.*, 2008; Xing *et al.*, 2013; Kakkar *et al.*, 2016a,b; Tong *et al.*, 2021).

Termite workers that are in their intermolt period remain active and involved with various tasks within the colony (Du *et al.*, 2017a). This period may be as short as less than a month for young instar workers but can last for several months for older instar workers, which was speculated to be up to ~6 months (Shimizu, 1962; Nakajima *et al.*, 1963; Kakkar *et al.*, 2016b). Eventually, workers periodically engage in the molting process, which lasts approximately 14 days. Around 10 days prior to ecdysis (= shedding of the old cuticle), individual workers start reducing their activity (Raina *et al.*, 2008), and if located at a remote foraging site (Chapter 5 of this volume), return back to the central part of the nest, where the brood and reproductives are located (Kakkar *et al.*, 2017, 2018). At 6 days prior to ecdysis, workers enter a fasting period during which

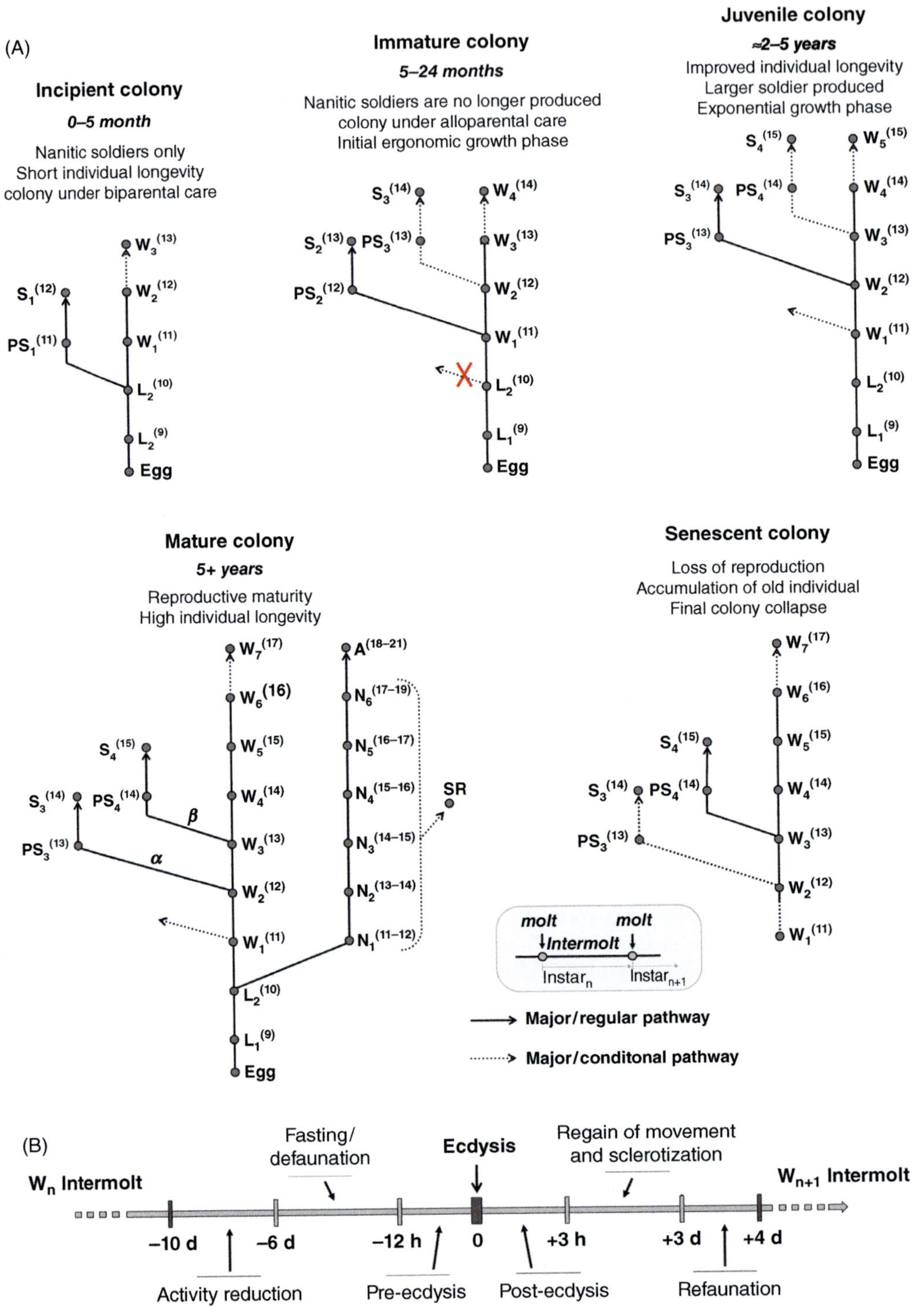

Fig. 4.3. (A) Developmental pathways in *Coptotermes* according to colony developmental stage. Numbers in parentheses indicate the usual number of antennomeres. (B) Individual molting sequence in a *Coptotermes* worker. (T. Chouvenc.)

the paunch is defaunated from symbiotic protists (Cleveland, 1925; Buchli, 1958; Soltani-Mazouni and Bordereau, 1987; Raina *et al.*, 2008). Although it was initially thought that protists die out of starvation or gut environmental changes as a consequence of feeding cessation (Raina *et al.*, 2008), it is more likely that protists rapidly die directly as a result of changes in hormonal titers in the termite host (Nalepa, 2017).

After initial defaunation, workers progressively form a new cuticle under the existing one (Raina *et al.*, 2008; Xing *et al.*, 2013). Around 12 h prior to ecdysis, the old cuticle and the new cuticle are slowly separated via dorsoventral contractions as the old cuticle thins out (apolysis) and the new cuticle expands underneath, creating wrinkles visible through the transparency of the old cuticle, resulting in a "mottled" appearance (Raina *et al.*, 2008; Xing *et al.*, 2013). The actual ecdysis process takes approximately 25–30 min to complete and is facilitated through the help of nearby workers (Xing *et al.*, 2013; Du *et al.*, 2016). Ecdysis is initiated with abdominal peristalsis that leads to the dorsal breach of the old cuticle, which expands antero-posteriorly. As the exuvial sac protrudes from the abdomen, the molting worker lies on its side in a jack-knife position (Su and Scheffrahn, 1993), with the legs, head, and antennae pulled away from their respective exuviae. Finally tracheal exuviae are separated from spiracle openings (Xing *et al.*, 2013). Within the first hour post-ecdysis, the newly emerged termite regains some degree of mobility, while the cuticle continues to expand and to unwrinkle for 3 h post-ecdysis. It then takes an additional 36 h for the sclerotization of the cuticle to complete, which includes the hardening of the new mandibles (Kakkar *et al.*, 2016a). Remarkably, the paunch remains devoid of protozoa for the first 3 days post-ecdysis but starts to be refaunated through proctodeal trophallaxis from nestmates by day 4 (Raina *et al.*, 2008). The complete refaunation process may take several additional days (Nalepa, 2017), but beyond 4 days after ecdysis, individuals are involved with some of the tasks within the colony as new functional workers (Fig. 4.3B).

From this general succession of events of the current understanding of the molting process in *Coptotermes*, exceptions and unknowns remain. First, the molting process in larval instars is unclear regarding its duration and the series of steps involved. The L_1 larval instar duration is ~18 days, while the L_2 instar is ~25 days, and the length of the fasting phase (if any) is unknown. This suggests that the molting process in larvae may be far less than the ~14-day period observed in workers, but this remains to be investigated. Second, the duration of the intermolt period for nymphs is mostly unknown. Nymphs can be found year-round at foraging sites of mature colonies (Raina *et al.*, 2004; Albino and Costa-Leonardo, 2011), and it was suggested that the $N_1 \rightarrow \ldots \rightarrow N_6 \rightarrow$ alate pathway is initiated seasonally and may take more than one year to complete (Chouvenc *et al.*, 2022). Third, while nymphs also lose their gut symbionts during their successive molts, alates retain a small protist population during the imaginal molt (Grassé and Noirot, 1945; Honigberg, 1970; Nalepa, 2017). Fourth, *Coptotermes* individuals are known to proceed only through progressive molts. Regressive or stationary molts may be restricted to termite clades that follow the linear development, and therefore those with totipotent pseudergates (Lebrun, 1991; Korb, 2015), which are absent in *Coptotermes* (= "true" worker caste). Fifth, while the production of secondary reproductives (SR) originates from nymphal instars, it remains unclear which N_n is the primary source for such differentiation, as secondary reproductives with highly variable wing bud sizes, number of antennomeres and eye formations were observed in *Coptotermes* colonies (Lenz *et al.*, 1988; Costa-Leonardo *et al.*, 2004, 2022; Chouvenc and Su, 2014). Finally, the differentiation toward the soldier caste involves an intermediate molt (= pre-soldier, PS, or "white" soldier), as the head capsule must undergo critical morphological changes (Park and Raina, 2004; Sugime *et al.*, 2015). In *Coptotermes*, the developmental processes in soldier differentiation remain to be further investigated, as soldiers display an extreme modification of the frontal gland among non-Termitidae taxa (Quennedey, 1984)

Each *Coptotermes* worker develops independently of the timing of the molting of their nestmates. Such asynchrony results in a colony-wide daily rate of molting for workers that varies from 0.6% to 2.2% (Raina *et al.*, 2008; Kakkar *et al.*, 2016b; Tong *et al.*, 2021), which may depend on temperature (Kakkar *et al.*, 2016b), resource availability (Su and La Fage, 1986; Chouvenc, 2020) and ongoing colony demography, i.e., different cohort size and age of caregivers (Du *et al.*, 2016; Chouvenc *et al.*, 2022). The periodicity of molting for all workers also implies that for a consequential part of their individual life (~14 days per molting cycle), workers are not actively participating in colony tasks, which would inherently impact colony functioning. The asynchronous molting of workers therefore inherently prevents a potential accumulation of negative effects from the temporary loss of a major part of the workforce, minimizing the underlying cost of molting on the colony at any given time (Kakkar *et al.*, 2016a). Finally, the physiological and physical conditions of workers early in their intermolt may differ from their conditions late in intermolt, including mandible sharpness (Roisin, 2016). Therefore, in addition to an existing age polyethism in *Coptotermes* (Du *et al.*, 2016, 2017b; Lee *et al.*, 2022), it is possible that task allocation may also be driven by intermolt polyethism, which remains to be investigated.

4.3.3 Caste determination mechanisms

Termites undergoing caste differentiation during their molting process are subjected to peculiar post-embryonic developmental alterations, which often deviate from the ancestral egg-to-imago developmental pathway of a hemimetabolous insect. The phenotypic plasticity observed in social insects has long been a source of interest in sociobiological circles, particularly how a single genotype would result in distinct phenotypes and functions. The factors involved in caste determination are often mixed from endogenous and exogenous sources, are variable across taxa, and are often difficult to disentangle. Such factors include sex, nutrition, social context, and environmental context, which interact through complex epigenetic regulations, which, in turn, modulate caste determination (see extensive review material for details: Miller, 1969; Watson *et al.*,1985; Roisin, 2000; Hayashi *et al.*, 2007; Miura and Scharf, 2010; Watanabe *et al.*, 2014; Korb, 2015; Korb and Belles, 2017; Jongepier *et al.*, 2018; Lo *et al.*, 2018; Suzuki *et al.*, 2019; Matsuura, 2020; Miura and Maekawa, 2020; Maekawa *et al.*, 2022).

The bulk of the research about caste determination in termites has primarily been performed on taxa with linear developments (with totipotent pseudergates), including *Hodotermopsis* (Miura *et al.*, 1999; Koshikawa *et al.*, 2001; Miura, 2001; Cornette *et al.*, 2008, 2013; Oguchi *et al.*, 2016, 2022), *Zootermopsis* (Miura *et al.*, 2003; Masuoka and Maekawa, 2016a, 2016b; Glastad *et al.*, 2016; Yaguchi *et al.*, 2019), *Kalotermes* (Lüscher and Springhetti, 1960; Lüscher, 1972; Soltani-Masouni and Bordereau, 1987; Lebrun, 1985, 1991) and *Cryptotermes* (Korb *et al.*, 2003, 2009, 2021; Korb, 2015). Studies on termite species with bifurcated developments have primarily focused on *Reticulitermes* (Hrdý and Křeček, 1972; Okot-Kotber *et al.*, 1993; Scharf *et al.*, 2003, 2005; Zhou *et al.*, 2006; Elliott and Stay, 2008; Tarver *et al.*, 2010, 2011). In comparison, relatively little has been done on *Coptotermes* despite its major pest status; these studies have primarily focused on mechanisms toward soldier differentiation through experimentation with juvenile hormones or juvenile hormone analogues (Park and Raina, 2004, 2005; Liu *et al.*, 2005a; Mao *et al.*, 2005; Tarver *et al.*, 2012; Du *et al.*, 2020). However, some observations made in *Reticulitermes* may also apply (but not always) to *Coptotermes*. Here is a short overview of what is known in *Coptotermes*.

As found in all termites (Okot-Kotber *et al.*, 1993; Cornette *et al.*, 2008; Nalepa, 2017; Su and Monteagudo, 2017), ecdysone or ecdysteroid hormones are directly involved in triggering the molting process in *Coptotermes* (Raina *et al.*, 2004). Soldier differentiation from workers is accompanied by a rise in juvenile hormone titers prior to the pre-soldier molt (Park and Raina, 2003, 2004, 2005; Mao *et al.*, 2005, 2010; Tarver *et al.*, 2012; Maekawa *et al.*, 2014; Du *et al.*, 2019, 2020), a mechanism conserved in most termite clades (Hrdý, 1976; Korb, 2015). In addition, inherent colony-level juvenile hormone titers can vary within the year through seasonal changes, as caste regulation may also be altered during nymph production and alate dispersal events (Liu *et al.*, 2005a,b; Mao and Henderson, 2007). Finally, while ecdysone may not be involved in physiologically functioning mature primary reproductives, as they are a terminal caste, it was suggested that the ecdysone pathway may still be used toward egg development in *Coptotermes* queens through the corpora allata (Raina *et al.*, 2003a).

Observations about soldier caste ratio in *Coptotermes* have varied widely from ~5% to 25% (Haverty 1977), but such observations may greatly depend on sampling protocols from the field and/or laboratory conditions. Overall, across the exhaustive *C. formosanus* literature, a consensus about soldier caste ratios converged toward 10–15%, but inherent colony variability may persist. The role of the genetic sex of individuals on worker/soldier caste determination can vary across and within termite taxa (Watson *et al.*, 1985; Miura and Maekawa, 2020), and this appears to hold true in *Coptotermes*. In *C. formosanus*, workers, soldiers and nymphs can be either male or female (Henderson and Rao, 1993; Raina *et al.*, 2004), while all soldiers produced in *C. lacteus* are female (Roisin and Lenz, 1999). The potential of sex determination factors for soldier castes among all other *Coptotermes* spp. remains to be investigated.

Nutrition and physical interactions were shown to affect the production of soldiers in *Zootermopsis* especially during early colony development (Maekawa *et al.*, 2012; Watanabe *et al.*, 2014; Yaguchi *et al.*, 2016). Low nutrition can also result in reduced soldier ratios in *C. formosanus* (Su and La Fage, 1986; Waller and La Fage, 1988; Song *et al.*, 2006), potentially due to the inherent mortality of failed molting pre-soldiers, as the soldier production pathway remains active regardless of resource limitations (Chouvenc *et al.*, 2015a; Chouvenc, 2020). In *Coptotermes*, soldier caste ratio regulation appears to be mitigated by a negative feedback mechanism (Haverty, 1979; Mao *et al.*, 2005; Dong *et al.*, 2009; Mao and Henderson, 2010), as observed in other termite taxa (Springhetti, 1973; Bordereau, 1985; Watanabe *et al.*, 2011). As *Coptotermes* soldier proportion may surpass a specific threshold, it may inhibit the rise of juvenile hormones in workers about to molt, preventing the accumulation of superfluous soldiers (Haverty, 1979; Su and La Fage, 1986; Park and Raina, 2004, 2005). Conversely, the removal of soldiers results in higher juvenile hormone titers in molting workers, resulting in the colony compensating the loss of soldiers by replacing them up to the specific threshold (Park and Raina, 2005; Chouvenc *et al.*, 2015a).

In *Coptotermes*, L_2 remain developmentally totipotent and can therefore molt toward a functional caste that would be required by the colony, either by engaging in the nymphal pathway, or by engaging in the sterile pathway (Chouvenc and Su, 2014). However, it is unclear how the $L_2 \rightarrow N_1$ differentiation is triggered. Regulatory pathways through juvenile hormone expression may be involved (Liu *et al.*, 2005a), but socio-environmental factors are most likely incorporated, as the nymphal pathway becomes available to larvae when colonies have reached a critical maturity (Raina *et al.*, 2004; Albino and Costa-Leonardo, 2011; Chouvenc and Su, 2014; Chouvenc *et al.*, 2022), or when *Coptotermes* colonies are orphaned, so that replacement neotenic reproductives are produced (or attempted) (Lenz and Runko, 1993; Costa-Leonardo *et al.*, 2004; Chouvenc *et al.*, 2015b), although it remains to be further investigated in *C. formosanus* specifically. The latter observation implies that there may be royal factors that could inhibit the production of nymphs in the early years of the development of the colony, which, if lifted through orphaning, may result in the production of secondary reproductives. However, with limited access to resources, it may directly impact the potential reproductive functionality of such individuals (Costa-Leonardo *et al.*, 2004, 2022; Chouvenc *et al.*, 2015b). In *Coptotermes*, such putative factors remain to be investigated, but it could be challenging. Indeed, studies on kalotermitids using totipotent pseudergates may allow for the manipulation of relatively small groups of termites for the forced production of secondary reproductives (Weil *et al.*, 2007, 2009; Korb *et al.*, 2009). Similarly, the high potential for the production of ergatoid reproductives in groups of *Reticulitermes* also allows for direct experimentation (Scharf *et al.*, 2005; Maekawa *et al.*, 2010; Matsuura *et al.*, 2010; Shimada and Maekawa, 2010; Matsuura and Yamamoto, 2011; Funaro *et al.*, 2019; Sun *et al.*, 2017, 2020; Takata *et al.*, 2023). In the absence of ergatoid and pseudergate morphologies in *Coptotermes*, the processes involved in $L_2 \rightarrow N_1$ and $N_n \rightarrow$ SR caste differentiations therefore remain challenging to investigate.

Finally, asexual queen succession has not been observed in *Coptotermes*. This mode of reproduction involves a queen laying parthenogenetic eggs with transgenerational epigenetic markers that lead to the development of a secondary reproductive. Asexual queen succession was first identified in some *Reticulitermes* species (Matsuura *et al.*, 2009; Vargo *et al.*, 2012) and has since been observed in several genera within Termitidae (Fougeyrollas *et al.*, 2015, 2017; Fournier *et al.*, 2016; Matsuura, 2017; Hellemans *et al.*, 2019). A few attempts to document parthenogenesis in *Coptotermes* concluded in its absence or unlikeliness (King and Spink, 1974; Higa, 1981; Roisin and Lenz, 2002; Raina *et al.*, 2003b; Ferraz and Cancello, 2004; Lee, 2015), although it was not investigated across *Coptotermes* species diversity. Therefore, the following description of the *Coptotermes* overall life cycle and growth may hold true for most, if not all, species within *Coptotermes*.

4.4 Dispersal flight and partner-finding behaviors

Colony foundation by alates is a critical step within the life cycle of termites. During this sensitive time, many factors at play can hinder the chance for a successful colony establishment, where often just a few lucky alates may attain this goal. Nutting (1969) provided a remarkable overview of the processes involved surrounding termite dispersal flights, including pre-flight colony activity, alate development, environmental conditions during flight, dispersal flight, tandem behavior, and initial colony foundation. To date, the comprehensive resource provided by Nutting's book chapter remains a universal reference to always return to when studying dispersal and early colony life of termites, as testified by being the most cited chapter from the irreplaceable *Biology of Termites* (Krishna and Weesner, 1969). It is therefore unnecessary to reiterate here Nutting's extensive effort, as this section instead covers a summary of the key aspects surrounding termite dispersal from Nutting (1969), and also expands from developments on *Coptotermes* since then.

4.4.1 Dispersal flights

The dispersal flight of termites is a relatively well-described aspect of their biology, as a sudden emergence of large numbers of flying alates is relatively easy to notice when compared with the cryptic activity of established colonies. These events are often referred to as "swarms" or "mating flights" in the literature and in the general press, as an analogy to honeybee swarms or ant mating flights, but for termites, the term "dispersal flights" is most appropriate when compared with other eusocial insects, as it depicts a functional dispersal of individual alates away from natal nests. With very few potential exceptions (Nutting, 1969), mating in termites only occurs later during colony foundation, and differs from most social Hymenoptera, which mate during dispersal flights.

When environmental conditions are favorable, *Coptotermes* engage in large synchronous dispersal flights, where many mature colonies within the local environment each contribute many thousands of alates (Su and Scheffrahn, 1987; Husseneder *et al.*, 2006; Simms and Husseneder, 2009; Mullins *et al.*, 2015). Both *C. formosanus* and

C. gestroi have vespertine flight activity, where alates can be observed flying in large densities during the evening twilight. However, *C. gestroi* can be observed as soon as sunset (early twilight), while *C. formosanus* will initiate flight closer to dusk (late twilight), resulting in ~15 min delay in their emergence if both species initiate their flights on the same day in co-occurring locations (Chouvenc *et al.*, 2017a). In comparison, several Australian *Coptotermes* species initiate flights mid-afternoon (Calaby and Gay, 1956; Ewart, 1988), and *C. testaceus* in South-Central America also participate in diurnal flights (Emerson, 1938). The distinct difference in the time of the day for dispersal flights among *Coptotermes* species can partially explain why *C. formosanus* and *C. gestroi* are both major invaders, while most of the other *Coptotermes* species have no-to-little invasive abilities (Chouvenc *et al.*, 2016). *Coptotermes* alates have functional compound eyes (Chang *et al.*, 2005) and, as both *C. formosanus* and *C. gestroi* engage in crepuscular dispersal flights, they have the opportunity to fly toward urban lights (see practical implications in Chapter 18 of this volume), facilitating their association with human activity (Scheffrahn *et al.*, 2015). In addition, alates can fly to lighted boats temporarily docked in ports or marinas, creating vessel opportunities for manmade dispersal to other parts of the world (Hochmair and Scheffrahn, 2010; Scheffrahn and Crowe, 2011; Su, 2013). In *Reticulitermes flavipes*, a day-flying species, its invasiveness may be instead partially due to its colony budding abilities with ergatoid reproductives (Vargo and Hussenener, 2009; Eyer *et al.*, 2021).

Coptotermes formosanus flight phenology has been studied in most geographic locations where the species is established, and the seasonality of dispersal flights varies depending on latitude and other climatic conditions at each location. In most locations, dispersal flights occur from mid-April to mid-June, with a peak of activity around mid-May in Taiwan (Huang *et al.*, 2022), in Florida (Chouvenc *et al.*, 2017a), in Texas (Puckett *et al.*, 2014), in Hawai'i (Bess, 1970; Higa and Tamashiro, 1983), and in Louisiana (King and Spink, 1974; Henderson, 1996; Messenger and Su, 2005a; Mullins *et al.*, 2015), but with several weeks delay in South Carolina and North Carolina (Chambers *et al.*, 1988) and in Japan (Oshima, 1919; Sugio and Miyaguni, 2019), and finally with flights in July–August in Israel (Scheffrahn *et al.*, 2020). In mainland China, *C. formosanus* peak of flight activity ranges from late April to mid-June, in a general south–north latitudinal gradient (Wang *et al.*, 2002a). While no longer observed in South Africa from the only previously known established population in the southern hemisphere (Chapter 2 of this volume), dispersal flights of *C. formosanus* were initially monitored in December–January (Coaton and Sheasby, 1976). In comparison, the peak of *C. gestroi* dispersal flights is in March in Florida (Chouvenc *et al.*, 2017a) and in Taiwan (Chiu *et al.*, 2016), in mid-April in Malaysia (Neoh and Lee, 2009), and in September in Brazil (Ferraz and Cancello, 2001; Albino and Costa-Leonardo, 2011) and in Réunion Island (Bordereau *et al.*, 1999).

All observations from the studies cited above confirmed that alate flight activity in *Coptotermes* follows a clear seasonal pattern, with several relatively small density flights early in the dispersal season, to a rise of activity with a handful of major dispersal flight events (peak activity), to then finish with additional small density flights after the peak (see Chapter 17 of this volume). However, multi-year surveys demonstrated that the seasonal flight activity can shift up to a couple of weeks earlier or later, depending on climatic conditions (Higa and Tamashiro, 1983; Puckett *et al.*, 2014; Chiu *et al.*, 2016; Chouvenc *et al.*, 2017a). These conditions at the time of major dispersal flights have been categorized for *Coptotermes* as relatively warm days within the season, low wind speed, and relatively high humidity (potentially following or preceding a rain event). However, there is a consensus that it is not possible to accurately predict the exact timing for the emergence of major dispersal flights (Leong *et al.*, 1983; Henderson and Delaplane, 1994; Chouvenc *et al.*, 2017a), for several potential reasons. First, the production of alates may depend on the asynchronous nymphal maturation process within colonies (Raina *et al.*, 2004; Albino and Costa-Leonardo, 2011; Chouvenc *et al.*, 2022). Second, climate condition in the months prior to the dispersal flight season may influence nymphal development toward alate maturation (Henderson and Delaplane, 1994; Chouvenc *et al.*, 2017a). Third, prolonged unfavorable environmental conditions during the flight season may delay major flight events, to the point that colonies accumulating ready-to-fly alates may engage in dispersal flights during less-than-optimal flight conditions (Chouvenc *et al.*, 2017a). Finally, microenvironmental changes influencing intrinsic social interactions within a colony may trigger or prevent alate exodus (Nutting, 1969; Leong *et al.*, 1983; Chouvenc *et al.*, 2017a).

The duration of a single active flight can last for as little as 15 min for relatively small density flights but can last for up to an hour for dense flight events (Chouvenc *et al.*, 2017a). Such relatively long flight duration can be attributed to two distinct factors that result in the temporary accumulation of flying alates in the air. First, alates leave the colony from a finite number of emergence holes, which results in thousands of alates waiting in line to disperse, so that it may take up to 20–30 min for all alates to be progressively released from the natal colony (Leong *et al.*, 1983). Second, each alate may fly for 20–30 min to disperse in the environment (Ikehara, 1966; Raina *et al.*, 2003c) for distances that can be relatively short (< 50 m), to a record distance of ~1,300 m, and with an average flight distance of 621 m, prior to falling onto the ground or landing on trees (Mullins *et al.*, 2015). With both factors combined, it results in the continuous visual observation of flying alates for up to one hour during major events, with an initial progressive increase in alate density to a final decrease in alate density (Chouvenc *et al.*, 2017a).

Mature *Coptotermes* colonies release ~80% of all alates produced annually in only four to six major synchronized flight events, where often > 35% of all alates can be released in a single large event (Chouvenc *et al.*, 2017a). While the sex ratio of dispersal flights may have important implications for population genetic structure (Kobayashi *et al.*, 2013; Hellemans *et al.*, 2019; Sinotte *et al.*, 2021), *Coptotermes* averages a 1:1 sex ratio during major synchronous dispersal flights, but the sex ratio can be biased toward females early in the season and biased toward males late in the season (Ferraz and Cancello, 2001; Chouvenc *et al.*, 2017a). When compared with most other "lower termite" clades that produce relatively small-to-moderate dispersal flight events, *Coptotermes* colonies conversely produce a handful of large inundative dispersal flights per annual cycle (Su and Scheffrahn, 1987; Chouvenc *et al.*, 2017a), more in line with many Termitidae (Nutting, 1969; Weesner, 1970). Such major events rarely occur on consecutive days, as colonies may have temporarily depleted their matured ready-to-fly alate reserves (Higa, 1981; Chouvenc *et al.*, 2017a). These major flights have deep implications for the probability of an individual alate surviving the event, finding a suitable mate, and establishing a successful incipient colony. It has been argued that it is most likely that the majority of successful colony establishments within the annual cycle occur during these few large dispersal events (Chouvenc *et al.*, 2017a; Chouvenc, 2019). The sheer number of alates synchronously produced by mature colonies creates a situation where, even if only a small fraction of alates survive the process, it would still result in a relatively high number of newly established colonies, under the constraints of a strong Allee effect (Kusaka and Matsuura, 2018), where alate individual fitness and natal colony fitness can both be positively influenced by the high density of dispersal flights.

Large synchronous flights can benefit individual alates as they participate in a numbers game, because inundative dispersal flights facilitate predation avoidance (Nutting, 1979; Lepage, 1991; Chouvenc *et al.*, 2017a). Major flight events provide predators from across the animal kingdom with a unique feasting opportunity, as several metric tons of highly nutritious defenseless insects are suddenly released into the environment (Silva *et al.*, 2021). A long list of potential predators was provided by

Nutting (1969), as any opportunistic predators present at the location during the flight can directly contribute to the demise of many alates (Chouvenc *et al.*, 2017a). However, despite such intense predatory pressure, inundative dispersal flights overwhelm the local predatory capacity during these ephemeral events. Therefore, alates can increase their individual chance of survival from predation by taking part in these few major flight events (Nutting, 1979). In addition, alate density directly impacts the chance for individuals to encounter a potential mate in an environment favorable for colony foundation (Husseneder *et al.*, 2006; Chouvenc, 2019).

Finally, alates can have additional ecological roles beyond the *Coptotermes* dispersal flight consequences on termite colonies' reproductive success and the temporary nutritional benefit of dispersal events for many opportunistic predators. During dispersal, *Coptotermes* alates not only carry their nutritional symbionts within their gut (Velenovsky *et al.*, 2021); they may also carry opportunistic organisms on their cuticle, such as nematodes (Foley *et al.*, 2018), bacteria (Chouvenc *et al.*, 2018), and mites (Wang *et al.*, 2002b), among potential others. These phoretic, parasitic or pathogenic associations may provide dispersal opportunities for a wide range of soil organisms and reflect that termite alates may have other potential effects on soil communities.

4.4.2 Dealation and tandem behavior

After their dispersal flight, *C. formosanus* alates can land on the ground, often after being attracted to urban lights, and it has since been assumed that *Coptotermes* primarily initiate colonies in the ground (Fei and Henderson, 1999; Raina *et al.*, 2003c; Park *et al.*, 2004; de Lima *et al.*, 2006; Mullins *et al.*, 2015). However, *Coptotermes* alates also aggregate on trees after their dispersal flight, and it was suggested that this behavior might be most relevant to *Coptotermes*' partner-finding strategy and to the colony foundation process (Waller and La Fage, 1988; Chouvenc *et al.*, 2017a, 2020; Mizumoto *et al.*, 2020), as *Coptotermes* colonies have the peculiar ability to infest live trees (Greaves, 1962; Lai *et al.*, 1983; Osbrink *et al.*, 1999; Chouvenc and Foley, 2018). *Coptotermes*' default flight directionality may be toward live trees, as a default rendezvous point for both sexes (Chouvenc *et al.*, 2020; Mizumoto *et al.*, 2020).

Alates do not always drop their wings as soon as they land. Sometimes, after a short walk, individuals may take flight again to pursue further dispersal. However, individuals eventually proceed to dealation by contracting thoracic muscles, which snaps the wings off at their basal suture (Nutting, 1969; Myles, 1988; Matsuura and Nishida, 2002). What triggers *Coptotermes* alates to dealate may depend on various factors, including time/distance of flight, alate density, and predation pressure. Laboratory observation of *Reticulitermes* demonstrated that when held in high density, it may take several hours for individuals to dealate, while when alone or in a pair, dealation occurs rapidly (Matsuura and Nishida, 2002). Such observation was confirmed in *Coptotermes* during light trapping, where small-density flights would result in rapid dealation of individuals in the container of the light trap, while highly dense flights would result in a virtual absence of dealation (Chapter 18 of this volume). In this situation, large numbers of alates sometimes arrange in a rosette (Myles, 1988) or *rosace* (Grassé, 1984) formation, i.e., an aggregation of alates in concentric rows where individuals hide under the folded wings of other alates. However, such observations in *Coptotermes* may be rather artificial, owing to the sudden containment of alates after a relatively brief flight period. The lack of dealation after light trapping, when contained in large numbers, may therefore be due to social cues that resemble the ones for alates, while waiting in the colony prior to a flight, when compartmentalized in special chambers within the nest (Calaby and Gay, 1956; Nutting, 1969; Grassé, 1984). Instead, as *Coptotermes* alates land on trees in high density, they rapidly dealate and actively seek a potential partner, implying that within the appropriate socio-environmental context, alates readily proceed to dealation shortly after they land from their individual dispersal effort (Raina *et al.*, 2003b; Chouvenc *et al.*, 2020; Mizumoto *et al.*, 2020).

After dealation, females run randomly on the substrate, while displaying their tergal gland to release a sex pheromone which facilitates males' ability to engage in tandem behavior (Raina *et al.*, 2003c; Bordereau and Pasteels, 2010), although *Coptotermes* females do not display a characteristic "call" posture (King and Spink, 1974; Higa, 1981; Raina *et al.*, 2003b). Instead, the exposure of the tergal glands is subtle and ephemeral while the female is actively running (Chouvenc *et al.*, 2020). The semiochemical profile of the *Coptotermes* female tergal glands involves dodecatrienol and trilinolein, which when combined, effectively results in a contact sex pheromone (Raina *et al.*, 2003c; Chouvenc *et al.*, 2020) (see Chapter 7 of this volume for additional details). It is necessary for a male to be within ~1 cm distance from a female to potentially detect its presence (Raina *et al.*, 2003c; Mizumoto *et al.*, 2020). It was suspected that *Coptotermes* females do not rely on long-distance volatile pheromones to attract males, because inundative dispersal flights readily result in high localized density of potential mates, often on live trees, where dealates repeatedly run into other individuals of both sexes in a chaotic manner (Grassé, 1984; Raina *et al.*, 2003c; Mizumoto and Dobata, 2019; Chouvenc *et al.*, 2020; Mizumoto *et al.*, 2020).

When a male encounters a female, the tandem forms as the male antennae and maxillary palps initiate contact with the abdominal tip of a female (Raina *et al.*, 2003c). The tandem can be maintained as long as the movements of the two partners are sufficiently coordinated (Mizumoto *et al.*, 2021) and the male closely follows the female, who takes the lead in search of a favorable nesting site. However, tandems often break up, either because the male loses contact, or if other individuals accidentally run into the tandem. When the separation occurs, the female temporarily pauses its movement, while the male walks frantically in a search pattern using sinusoidal movements, in attempt to relocate the female (Mizumoto and Dobata, 2019). In low dealate density, this sexual dimorphic movement results in increasing the chance for the female to be reconnected with the same male, while in high dealate density it increases the chance for the male to find another female (Mizumoto *et al.*, 2020), which results in repeated trial-and-error attempted tandems.

Under high predatory pressure, males and females must rapidly pair up and isolate themselves, leaving little room for the luxury of being nitpicky in mate selection (Nutting, 1979; Chouvenc *et al.*, 2017a), making active sexual selection unlikely (Hartke and Baer, 2011). However, dealates tend to pair up with partners of a similar size (Husseneder and Simms, 2008), which may result from individuals in a tandem being able to have relatively coordinated movements (Mizumoto and Dobata, 2019; Mizumoto *et al.*, 2020). In addition, individuals with low heterozygosity may not pair up as readily as individuals with high heterozygosity (Husseneder and Simms, 2008). as they may have an increased difficulty in maintaining tight movement coordination during tandem runs. Male–female dealate tandem formation and maintenance may therefore not have arisen from an active case of sexual selection, but instead from a passive case of natural selection. In fact, there is no kin discrimination during tandem formation (Husseneder and Simms, 2008), and it can be argued that there is also little-to-no species discrimination within *Coptotermes* during tandem formation (Chouvenc *et al.*, 2020).

Therefore, random partner encounter remains the predominant factor for initial tandem formation (Raina *et al.*, 2003c; Chouvenc *et al.*, 2017a, 2020; Mizumoto *et al.*, 2020, 2021), highlighting the potential role of luck for alate survival and pairing success prior to colony foundation. For example, alates may initiate colony foundation with an incomplete gut mutualist fauna (Michaud *et al.*, 2020; Velenovsky *et al.*, 2023). In addition, observations from Australian *Coptotermes* populations inferred that most mature field colonies display relatively low levels of homozygosity (Thompson *et al.*, 2007), potentially reflecting a low fitness from colonies established by inbred siblings. With such a reproductive strategy, it is inconsequential for a mature colony to produce alates for which a majority would die during dispersal flights or fail to establish a viable colony, as colony fitness is eventually assured through repeated inundative dispersal flight events, which are reiterated perennially over many years (Chouvenc *et al.*, 2022) as a high-input numbers game. For invasive populations of *Coptotermes* with limited initial genetic input, a relatively high degree of heterozygosity (> 2 alleles per loci) could be maintained (Blumenfeld *et al.*, 2021), as the same numbers game can weed out deleterious homozygote combinations through failed colony establishment of such pairings.

4.4.3 Nest site selection

If the tandem run persists between the female and the male, the female leads in the search for a suitable nesting site, as the male maintains tandem connection. The pair rapidly switches from positive phototaxis to negative phototaxis, and also displays high thigmotaxis, so as to quickly find a place to hide (Grassé, 1984; Mizumoto and Bourguignon, 2023). Studies that investigated nest site selection in *Coptotermes* showed that, when given time in artificial conditions for choice assays, dealates would preferentially establish in soils with organic matter instead of sand (de Lima *et al.*, 2006); they would also preferentially establish in relatively moist environments (Fei and Henderson, 1999), and differentially establish in the presence of mulch from various plant origins (Sun *et al.*, 2006). However, during the chaos of a dispersal flight and tandem formations, dealates have no time to be highly selective on where to establish the colony: while the tandem pair may skip potential nesting sites for any reason, each passed opportunity translates into an ever-increasing chance to fall prey to a predator. Therefore, while some factors may reduce the chance of tandem pair establishing in an unfavorable location, they often settle for the first available opportunity, as randomness again remains a prime factor.

During extensive observations (personal observation, flight seasons of 2013–2023 in South Florida), within 30 min after landing, dealates are no longer visible anywhere in the environment. At this point, tandems either have initiated the excavation of their claustral chamber or have been preyed upon. Exhaustive surveys in the hours following large dispersal flights revealed dealate pairs hiding under rocks, wood pieces on the ground, car tires, a loose shingle plate on a roof, and in the leaf litter, among many other examples, reflecting the opportunistic nature of site selection. Many of these sites are subject to desiccation (full sun) or flooding (water accumulation from rain) and often cannot be seen as "favorable sites" for initial colony establishment. However, after major dispersal events, trees are often surrounded by a thick layer of wings in a ring pattern on the ground around the trunk, resulting from the dealation of alates after they landed on the tree. Manual excavation of the dirt at the trunk–soil interface confirmed that many dealate pairs were in the process of excavating a claustral chamber, often within a few centimeters from each other. These sites often contained high moisture, organic soil matter and wood particles, implying that the combination of the "preferred" characteristics defined by previous studies (Fei and Henderson, 1999; de Lima *et al.*, 2006; Sun *et al.*, 2006) is inherently defined at the tree trunk–soil interface, among other potential favorable nesting sites. Although access to soil is not required for initial colony foundation (Raina *et al.*, 2003b; Ferraz and Cancello, 2004), access to soil later becomes critical during colony growth in *Coptotermes* (Mullins *et al.*, 2021).

4.5 Colony foundation and incipient colony development

4.5.1 Initial colony establishment

Colony foundation is initiated as the female and male isolate themselves from the surrounding environment. Within the first few hours after dispersal, the male and female dealates build a sealed chamber. This chamber has received many names, depending on the authors, across termite diversity, including: coplarium, nuptial chamber, nursery cavity, royal chamber, claustral chamber, incipient cell, etc. However, many of these terms may often reflect a particular period within the colony maturation process. Therefore, "claustral chamber" is here used as a generic term, as it applies to the incipient family unit (from 0 to ~150 days post-foundation), which remains isolated from the rest of the environment during this period. During the first day of establishment, tandem runs are no longer exhibited by the founding pair (Park *et al.*, 2004) and both individuals engage in active claustral chamber excavation and construction, which involves the manipulation of soil and/or wood particles and potentially the use of saliva for consolidation (King and Spink, 1975; Higa, 1981). The isolation reduces the risk of predation and desiccation. During the first 24 h, the two individuals also amputate 2–5 distal antennomeres from each other, extensively groom each other, and mate at least once (Higa, 1981; Raina *et al.*, 2003b). Antennal cropping was suggested to be involved with triggering a behavioral and physiological shift toward reproduction (Nalepa *et al.*, 2011; Miyaguni *et al.*, 2013) but remains to be investigated further.

As both the male and female lack external genitalia (Laranjo *et al.*, 2019), the mating behavior in *Coptotermes* involves the male and female positioning themselves in opposite directions, connecting the tips of their abdomens for ~30 seconds while displaying brief lateral waggling movements (Raina *et al.*, 2003b). The sperm, which is spherical and lacks flagella, is transferred from the male to the female genital chamber, and then moved to the relatively small spermatheca via muscular contractions (Costa-Leonardo and Patricio, 2005; Raina *et al.*, 2007; Laranjo *et al.*, 2019; da Silva and Costa-Leonardo, 2023). While *Coptotermes* sperm was suggested to live at least 60 days in the female's spermatheca, multiple early mating events improve the egg-laying capacity of the female, with one to two mating events per day (Higa, 1981; Raina *et al.*, 2003b; da Silva and Costa-Leonardo, 2023).

The pre-oviposition period can last from 5 days to 2 weeks, as this period can vary extensively depending on the rearing conditions, with relatively low temperature delaying initial oviposition (Bess, 1970; King and Spink, 1974; Higa, 1981; Raina *et al.*, 2003b; Chouvenc and Lee, 2021). In addition, if provided with a pre-formed claustral chamber, the female may initiate oviposition as early as day 3 post-foundation (Raina *et al.*, 2003b), implying that the excavation efforts during the initial claustral chamber construction come at a cost to the founding pair. The first handful of eggs laid by the female may be relatively big compared with subsequently laid eggs (Matsuura and Kobayashi, 2010), as increased size may accelerate the emergence of the first few larvae. Within 10 days after laying the initial egg, a dozen additional eggs are laid, and within a month, a total of 15–25 eggs can accumulate in the claustral chamber in a single cluster (= egg pile) (King and Spink, 1975; Higa, 1981; Raina *et al.*, 2003b). Egg incubation must be at temperatures above 21°C, as embryonic development fails below this temperature (King and Spink, 1974; Fei and Henderson, 2003). Within this first month, eggs develop under the extensive care of both the male and the female, as eggs are groomed extensively and sometimes moved around the claustral chamber. If no care is provided in the form of salivary deposition during grooming activity, eggs rapidly desiccate or display microbial growth (Raina *et al.*, 2003b; Ferraz and Cancello, 2004). The egg incubation period is temperature-dependent, as they hatch within ~30 days when kept at 26°C, or as early as ~17 days when kept at 30°C (Higa, 1981).

During the initial egg incubation period, the male and female continue to modify the claustral chamber, by chewing on wood and then spreading the wood pulp mixed with saliva on the chamber's walls (Fig. 4.4). They also periodically engage in reciprocal allogrooming, stomodeal and proctodeal trophallaxis, and additional mating events. In addition, physiological changes occur in both individuals within the first month.

Fig. 4.4. Initial claustral chamber within 3 weeks after foundation, where the dealates isolated themselves using masticated wood pulp to seal themselves from the environment. (T. Chouvenc.)

First, they both increase their individual biomass by ~33% by absorbing environmental moisture and by ingesting a small amount of wood (Chouvenc, 2022). Second, the increase of wet biomass is accompanied by a rise in protists abundance in the male's and female's hindguts (Shimada *et al.*, 2013; Velenovsky *et al.*, 2021). Third, the wing muscles are progressively atrophied, as both individuals most likely reallocate these resources toward reproduction (Zhang *et al.*, 2021). Finally, as both partners undergo behavioral and reproductive shifts from their previous alate condition, an alteration of gene expression over the course of the colony foundation most likely takes place (Husseneder *et al.*, 2012) but remains to be investigated further.

Within 30–50 days post-foundation, a few first instar larvae (L_1) emerge, as the hatching process is facilitated by the parents via careful manipulation of the eggs with their mandibles and maxillary palps. The parents provide grooming and limited trophallaxis to the neonate offspring, as larvae remain fully dependent on their parents to continue their post-embryonic development (Nalepa, 2010b, 2011). Again, developmental rates strongly depend on temperature, as first (L_1) and second (L_2) instar larvae take from 13 to 20 days at 26°C or below, and ~7 days at 30°C to molt into the next instar (Higa, 1981). As a result, first instar workers (W_1) and first pre-soldiers (PS_1) start emerging 50–75 days post-foundation; and within ~10 days after the first PS_1 emerged, the first nanitic soldiers (S_1) emerge (from 0 to 3 S_1 produced by 75 days) (King and Spink, 1975; Higa, 1981; Barsotti and Costa-Leonardo, 2005). At this point, the incipient colony still remains a woodroach-like family unit under biparental care, where the nurturing capacity of the colony is limited to just the parents (Chouvenc, 2022). When the first few W_1 emerge from molting L_2, both parents engage in extensive proctodeal trophallaxis toward W_1, which acquire the microbial gut community from their parents, making them nutritionally self-competent (Shellman-Reeve, 1990; Brossette *et al.*, 2019). The repeated transfers of gut fluids containing nutritional reserves, protozoa and other gut microbes result in a rapid loss of weight in both parents leading to a ~40% loss from its peak, finally reaching a minimal weight around 100 days post-foundation (Velenovsky *et al.*, 2021; Chouvenc, 2022).

With the progressive reserve exhaustion of the parents, an oviposition hiatus, or a reduced oviposition, is initiated around 60–75 days post-foundation in *C. formosanus* (King and Spink, 1974; Higa, 1981; Morales-Ramos and Rojas, 2003; Raina *et al.*, 2003b, 2007). The intensity of this reduction may vary among *Coptotermes* species, as it was observed to be minimal in *C. gestroi* (Ferraz and Cancello, 2004; Chouvenc, 2022). However, all experiments on *C. formosanus* did not provide the incipient colonies with organic soil material, which directly limits colony growth (Mullins and Su, 2018). It is therefore possible that the duration, intensity, and potential repeatability of oviposition hiatuses within the first year (King, 1971; Higa, 1981; Raina *et al.*, 2003b) may be directly influenced by the availability of dietary nitrogen, either trophically acquired (Mullins *et al.*, 2021), or recycled through exuviae consumption and cannibalism (Chouvenc, 2020; Tong *et al.*, 2023), and by the temperature (Higa, 1981), as subtropical *C. formosanus* may be seasonally exposed to colder temperatures than tropical *C. gestroi* (Patel *et al.*, 2019a, 2019b).

All these observations confirm that incipient *Coptotermes* colonies are under a tight resource budget (Chouvenc, 2019, 2022), and that the colony foundation process has evolved to optimize the initial resources investment into castes, as exemplified by the production of soldiers (Chouvenc *et al.*, 2015a). Soldiers produced during the incipient colony phase are all nanitic (S_1) through accelerated development (Fig. 4.3A). However, S_1 soldiers come as a trade-off as they are relatively "cheap" to produce, but are also poorly functional (Chouvenc *et al.*, 2014). First, as S_1 are being produced with at least two molts less than soldiers in mature colonies (S_3 or S_4), they are produced relatively quickly and require minimal resources. Being so remarkably small, their defense functions are relatively poor, which is mostly irrelevant in the context of the initial claustral chamber (Chouvenc and Su, 2014). Second, the rapid development under biparental care conditions, where the nurturing capacity is limited to the two parents, results in soldiers displaying strong fluctuating asymmetry, which reflects the stressful conditions in which the larvae and pre-soldiers were developing (Chouvenc *et al.*, 2017b). Third, their function is primarily to signal to the colony that the ~10% soldier ratio is reached, which inhibits the production of additional soldiers (Chouvenc *et al.*, 2015a). Therefore, nanitic soldiers are a compromise for incipient colonies, to rapidly reach the inherent species-specific soldier ratio, while limiting the underlying cost of producing fully developed soldiers. As a result, while incipient colonies limit their investment in soldier production, they can reallocate such resources toward the production of a few additional workers (Chouvenc *et al.*, 2015a).

By ~100 days post-foundation, several workers can reach the W_2 instar, and additional W_1 are also produced from larvae, with the total number of offspring ranging from just a handful of individuals for the least successful incipient colonies, to up to 50 individuals for the most successful ones (King and Spink, 1975; Higa, 1981; Fei and Henderson, 2003; Chouvenc, 2022), depending on many confounding factors. These factors may include temperature (King and Spink, 1975; Higa, 1981), access to wood, water and dietary nitrogen (Raina *et al.*, 2003b; Mullins *et al.*, 2021), initial internal resources of parents (Mullins and Su, 2018; Chouvenc, 2022), genetic profiles and partner compatibility (Fei and Henderson, 2003; Eyer and Vargo, 2022; Velenovsky *et al.*, 2023), microbial environment (Rosengaus and Traniello, 1993; Fei and Henderson, 2003; Calleri *et al.*, 2006; Cole and Rosengaus, 2019; Eyer and Vargo, 2022) and "luck" (Nutting, 1969; Chouvenc, 2019). Both W_1 and W_2 workers engage in brood care behaviors, as the nurturing capacity of the colony is the combined brood care behavior from both parents and the workers. Individuals developed under this shared brood care capacity undergo reduced developmental stress when compared with the first few emerged individuals (Chouvenc *et al.*, 2017b). In the following ~50 days, the incipient colony undergoes the critical shift from biparental care to alloparental care (Nalepa, 2015; Chouvenc, 2022).

4.5.2 Transition from biparental care to alloparental care

As workers progressively accumulate in the incipient colony, the nurturing capacity of the colony improves. The newer cohort of eggs and larvae developing under this extended brood care capacity becomes more and more reliant on the workers' contribution. Between ~100 and ~150 days, the responsibilities of grooming, egg tending, proctodeal and stomodeal trophallaxis, wood chewing, nest maintenance, assisting hatching eggs and molting individuals, progressively shift away from the parents toward the workers (Chouvenc and Su, 2017a; Brossette *et al.*, 2019; Chouvenc, 2022). While both parents still host protozoa in their hindguts (Velenovsky *et al.*, 2021), the workers progressively become the primary donors for the colony-wide transfer of nutritional mutualists (Noirot and Noirot-Timothée, 1969; Du *et al.*, 2017a). By ~150 days, the transfer of brood care duties to alloparents is completed and is irreversible, as the removal of all workers at this point would result in the colony's death (Chouvenc and Su, 2017a). Parents are now completely dependent on their offspring, have potentially lost the ability to provide brood care, and instead now focus on reproductive activities. The transition to alloparental care is also often accompanied by a renewed oviposition activity (Chouvenc, 2022), although the drop of temperature from seasonality in subtropical/warm temperate areas (King and Spink, 1975) and limited access to dietary nitrogen (Mullins *et al.*, 2021) may delay oviposition resumption. Ultimately, as the two parents receive nutritional support and care from their own offspring, oviposition resumption contributes to the initial ergonomic growth of the colony (Oster and Wilson, 1978).

Remarkably, individuals produced under biparental care have shorter life expectancy than the ones produced under alloparental care (Shimizu, 1962; King and Spink, 1974; Higa, 1981) as incipient colonies rarely display W_3 (Chouvenc and Su, 2014). Again, it reflects that the first cohort of offspring developed under stressful factors (Chouvenc et *al.*, 2014), despite the complete dedication of both parents to rearing this first cohort (Chouvenc, 2022). The environmental conditions inside the nest of incipient colonies are not yet stable, as they are removed from the homeostatic nest conditions observed in mature colonies (Chouvenc *et al.*, 2017b). While short-lived, these first few workers (= first alloparents), which developed under biparental care, are indispensable for the initial positive feedback on resource acquisition and improvement of colony nurturing capacity. If using the incipient colony as an analogy to an economic engine, this short-lived cohort of workers represents the first and unique biparental investment, to prime the system toward initial economic growth, with long-term positive feedback of returns on investment (Oster and Wilson, 1978; Chouvenc, 2022).

As the transition from biparental care to alloparental care is completed, division of labor, cooperative brood care, and overlapping cohorts of offspring are established. The incipient colony at this point is no longer a woodroach-like family unit, but instead has transitioned to a eusocial termite colony where the two parents have now become functional queen and king (Wilson, 1971; Nalepa, 2015; Chouvenc, 2022). Retrospectively, any attempts to determine the actual survival rate of an individual alate from dispersal flight to alloparental care in *Coptotermes* remains mostly speculative. However, it may be reasonable to expect that less than 0.01% of all alates engaging in dispersal flights may successfully become queens or kings of established incipient colonies. Therefore, if two alates are fortunate enough to have established a colony and reached the transition to alloparental care, they most likely have survived the most critical existential threats they would ever face over their lifetime.

4.6 Colony growth and maturation

4.6.1 Immature colony

Under the alloparental care regime, the colony has an improved nurturing capacity that further contributes to colony growth, which implies that the colony is running out of space in the initial claustral chamber. Thus, the colony starts to venture outside the claustral chamber, as workers initiate excavation activity with the formation of underground galleries (distances < 30 cm) (King and Spink, 1975; Du *et al.*, 2017b). In addition, gallery excavation may allow the immature *C. formosanus* colony to dig below the soil surface (20–30 cm deep) to retreat from potential winter air temperatures, as the colony must maintain the "nursery" above 21°C to maintain oviposition and egg hatching (King and Spink, 1975). Gallery excavation activity has important consequences for the colony at this life stage, as it now allows for foraging activities to locate additional wood and nitrogen resources (see Chapter 5 of this volume), but it also exposes the colony to novel predation and competition pressures. This timing coincides with the colony shifting the production of soldiers away from the $L_2 \rightarrow PS_1 \rightarrow S_1$ pathway, to the $W_1 \rightarrow PS_2 \rightarrow S_2$ pathway. These newly produced S_2 soldiers are still not as efficient in defense abilities as soldiers produced in mature colonies, but they are bigger and more robust than S_1, and therefore can actively participate in the defense of the nest and foraging locations

(Chouvenc and Su, 2014). Within one year of colony growth, the most successful colonies can accumulate up to ~200 individuals, and the queen starts showing early signs of physogastry as it expands ovarian capacity with the addition of functional ovarioles (Higa, 1981; Raina *et al.*, 2003b). In addition, the prolonged absence of direct wood chewing by the queen and king has resulted in the progressive atrophy of their jaw muscles (Cleveland, 1925).

In the second year of colony growth, competition with nearby immature termite colonies may increase, alongside predation exposures. Frequent encounter with neighboring colonies may result in direct agonism, where the role of the S_2 soldiers may become critical. Often, a few soldiers may engage in active defense behaviors against intruders within the gallery, providing time for the workers to seal access to the colony using soil, saliva, and feces. By sacrificing a few individuals, the colony may maintain territory boundaries, while limiting casualties during agonistic confrontations with ants (Chouvenc *et al.*, 2015c) or other subterranean termite colonies (Li *et al.*, 2010). Alternatively, if there is a lack of initial agonism due to the absence of colony recognition cues (Chouvenc and Su, 2017b; Gordon *et al.*, 2020), interaction between two young conspecific colonies may result in colony fusion, where ultimately one of the royal pair dies, while the other royal pair inherits the combined nurturing capacity of the two initial colonies (Guaraldo and Costa-Leonardo, 2009; Gordon *et al.*, 2020), or a large colony can integrate a smaller colony (Su and Scheffrahn, 1988a; Lee *et al.*, 2019).

By 2 years of colony growth, most successful colonies can reach up to 5,000 individuals, with an accumulation of W_2 and W_3 which can display improved longevity (Chouvenc and Su, 2014). A few individuals can reach the W_4 instar, again highlighting the ongoing improvement of the homeostatic conditions in the nest (Chouvenc *et al.*, 2017b). During this period, the accumulation of workers of different ages allows for a progressive establishment of age polyethism (Du *et al.*, 2016, 2017a). In addition, with the expansion of the foraging territory of underground galleries, a heterogenous distribution of individuals can be observed, where younger worker instars tend to remain in the proximity of the royal cell and egg pile, while older instar workers tend to venture to and forage at the periphery of the nest (Du *et al.*, 2017b). The age-based task division among workers allows for an optimization of colony functioning, which also improves colony resiliency from inherent environmental perturbations as observed in many social insect taxa (Lee *et al.*, 2022).

4.6.2 Juvenile colony

Between 2 to 5 years, the colony continues to grow exponentially, reaching up to ~20,000 individuals within 3 years, and up to ~80,000 individuals within 4 years. At this point, the queen is reaching high degrees of physogastry and, as in *C. gestroi*, queens go from an initial mass of ~7 mg to ~70 mg (Chouvenc, 2018). In *C. formosanus*, queens can go from an initial mass of ~11 mg to ~90 mg, as ovaries reach high reproductive output, with up to ~375 ovarioles, and between 20 to 40 oocytes in various degrees of maturation in each ovariole (Higa, 1981). Workers and soldiers continue to have an improved life expectancy, as W_4, W_5 and W_6 start accumulating in the colony among large cohorts of W_2 and W_3 (Chouvenc and Su, 2014; Su *et al.*, 2017). For soldier production, the use of the $W_1 \rightarrow PS_2 \rightarrow S_2$ pathway is reduced, with a primary use of the $W_2 \rightarrow PS_3 \rightarrow S_3$ pathway (Higa, 1981; Chouvenc and Su, 2014).

Foraging distances continue to expand as the colony requires more and more resources to maintain colony growth, from 20 m foraging distances in colonies with ~5,000 termites and up to 100 m in colonies with > 20,000 termites (King and Spink, 1969; Osbrink *et al.*, 2016; Su *et al.*, 2017). Colony-wide wood consumption activity is partially proportional to colony size, as *Coptotermes* colonies with ~40,000 workers can feed on ~5 g of wood per day (Patel *et al.*, 2020). The heterogeneous distribution of different worker instars initially observed in immature colonies (Du *et al.*, 2017b) continues to persist, with the oldest available instar workers at the far end of the foraging galleries, while youngest instars remain at the center of the nest (Su *et al.*, 2017; Lee *et al.*, 2022) (see Chapter 5 of this volume). Finally, as colonies grow in size and in territory, molting site fidelity is observed, as foraging termites come back to the central part of the nest to molt (Kakkar *et al.*, 2017).

4.6.3 Mature colony

As the colony approaches 1 million termites or beyond, it engages in the production of nymphs, which advance toward alate maturation. Mature colonies are able to contribute seasonally to the production of thousands of alates, which then engage in dispersal flight events. While the number of alates a mature colony can annually produce may vary extensively depending on resource availability and colony health, one study reported that ~70,000 alates dispersed from one mature *C. formosanus* colony during a single dispersal season (Su and Scheffrahn, 1987). It is possible that colonies early in their maturity produce far fewer alates, but over the years of dispersal perennity and ongoing colony growth that generate several millions of termites (Su and Scheffrahn, 1988b; Chouvenc *et al.*, 2022), alate production most likely increases accordingly (see Chapter 5 of this volume). Therefore, within the life of a *Coptotermes* colony, it is likely that over the years or decades it produces more than a million alates, implying that, even if the speculated 0.01% alate success rate is a conservative estimation, a single parental colony can be directly responsible for the establishment of 100+ daughter colonies throughout its lifetime, ensuring high reproductive success.

The physogastric primary queen of a mature *Coptotermes* colony (Fig. 4.5) may have over 1,000 active ovarioles and may weigh up to 300 mg (Higa, 1981). Egg-laying activity may vary over time (Bess, 1970; Higa, 1981), but the maximum output of a single primary queen from a mature colony has been speculated to range from a few hundred eggs to 1,000+ eggs per day (Higa, 1981; Lax and Osbrink, 2003; Lo *et al.*, 2012), although such estimates may still be conservative, as mature colonies can maintain several millions of termites over years (Su and Scheffrahn, 1988b). The high oviposition rate allows the colony to sustain population growth or to maintain maximal population size, as it compensates for inherent individual senescence, as workers may now live up to the W_6 or W_7 instar (Shimizu, 1962; Nakajima *et al.*, 1963; Chouvenc and Su, 2014). Soldiers in mature colonies are predominantly represented by S_3 (~64%), but a few S_2 soldiers are still produced (~12%), and the $W_3 \rightarrow PS_4 \rightarrow S_4$ pathway is also used (~24%) (Chouvenc and Su, 2014).

Mature *Coptotermes* colonies have extensive foraging territories (King and Spink, 1969; Su and Scheffrahn, 1988b). One of the consequences of such foraging ability is that mature colonies periodically engage in agonistic encounters with predators, predominantly ants (Cornelius and Grace, 1995; Chouvenc *et al.*, 2015b; Tuma *et al.*, 2020), or other termite colonies that compete for the same resources (Messenger and Su, 2005a,b; Messenger *et al.*, 2005; Uchima and Grace, 2009; Li *et al.*, 2010; Mullins *et al.*, 2011; Bernard *et al.*, 2017). Another consequence of such foraging ability is their capacity to locate and infest food resources within the environment (Lee *et al.*, 2006), which could be a new live tree to infest or a manmade structure. Contrary to *Reticulitermes*, which has the ability to establish

Fig. 4.5. Physogastric *Coptotermes* queen and king from a 5-year-old laboratory colony. (T. Chouvenc.)

many secondary reproductives across multiple satellite nests (Thorne *et al.*, 1999; Dronnet *et al.*, 2005), *Coptotermes* appear to primarily retain reproduction in a centralized location, which can be located at a food source (inside a tree) or remote from a foraging area (Kakkar *et al.*, 2017). For additional information about the foraging ecology in *Coptotermes*, see Chapter 5 of this volume.

Finally, as *Coptotermes* colonies have a characteristic ability to infest live trees (from initial foundation location, or from later infestation from foraging activity), trees are often piped from the inside (Greaves, 1962; Osbrink *et al.*, 1999; Chouvenc and Foley, 2018; Costa *et al.*, 2020). The void is then filled with "carton material" (Fig. 4.6), which is a sponge-like structure resulting from the building behavior of workers which reuse fecal material as a primary building block (King and Spink, 1969; Du *et al.*, 2016). The carton helps maintain homeostatic conditions within the nest (Wood, 1988) and possesses its own microbial community that helps mitigate potential entomopathogens from spreading within the colony (Chouvenc *et al.*, 2013b, 2018). When compared with other Rhinotermitidae, *Coptotermes* cartons are rather complex and extensive, and the nest structure elaboration often resembles the complexity that can be found in some Termitidae (Chouvenc *et al.*, 2021). The carton three-dimensional structure also potentially optimizes the negative spaces resulting from wood feeding, by improving the surface-to-volume ratio to allow for relatively high termite density in their living space. *Coptotermes* "carton nests" can sometimes be found in structures if a colony has been infesting it for long enough. Such "above-ground nests" have long been interpreted as "aerial nests," which could imply that the colony is not connected to the ground. However, in the vast majority of cases, such carton nests are connected through galleries to the ground, as colonies are unable to survive and thrive without a reliable source of environmental moisture, as they do in the case of boat infestation (Scheffrahn and Crowe, 2011).

Fig. 4.6. Carton material from *Coptotermes* sp. in a piped live oak tree (T. Chouvenc.)

Additional discussion of colony demographic including production of nymphs and secondary reproductives production can be found below in section 4.7 ("Mature colony demographic considerations").

4.6.4 Senescent colony

Ultimately, from a few years to several decades of colony maturity and alate production (Nakajima *et al.*, 1963; Grace *et al.*, 1995; Chouvenc *et al.*, 2022), the colony eventually loses reproductive capabilities. Such loss of reproduction may be the outcome of the death of the primary queen and/or king, in which case the absence of the establishment of functional secondary reproductives would lead to the death of the colony. Such a scenario may occur if the queen loses reproductive health, leading to the progressive absence of eggs and larvae, but remains alive (Bess, 1970), potentially inhibiting the production of replacement secondary reproductives (Chouvenc and Su, 2014). Alternatively, as several replacement reproductive queens/kings become established over several cycles of reproductive succession, long-term consequences of inbreeding cycles may eventually lead to colony health decline (Chouvenc *et al.*, 2022). Regardless of the cause of the loss of reproduction, the long-term disappearance of eggs and larvae leads to the absence of population replacement.

The consequence of the absence of the production of young individuals is the progressive accumulation of old individuals within the colony (Chouvenc and Su, 2014). Within a year of loss of reproduction, the colony may engage in a last dispersal flight season, as the terminal nymph cohort takes ~1 year to mature into alates (Chouvenc *et al.*, 2022). Within 3 years after the loss of reproduction, the last few remaining workers and soldiers reach old instars (W_5, W_6, W_7 for workers, and S_4 soldiers only are left) and eventually all die of individual senescence during the final colony collapse (Chouvenc and Su, 2014; Chouvenc *et al.*, 2022). Although this general pattern of colony death may hold true for many colonies, the description of the colony collapse process from several *C. formosanus* field colonies revealed that each colony displays a unique demographic trajectory toward colony death (Chouvenc *et al.*, 2022). Eventually, with the final colony collapse, the nest becomes a novel niche for a wide range of organisms, in a chaotic ecological succession (Chouvenc *et al.*, 2013b).

While mature colonies have high degrees of resiliency and can survive many potential selective pressures, colonies are periodically exposed to environmental threats that may lead to their demise, independently of their reproductive and health status (Lepage and Darlington, 2000; Chouvenc *et al.*, 2022). Such threats include predation, competition, environmental extremes (temperature, water, resource depletion) and pest management practices if the colony infests manmade structures (Chapter 15 of this volume). With the emergence of chitin synthesis inhibitor (CSI) baits as a major tool against *Coptotermes* subterranean termite infestations (Su, 2019), it was established that CSI bait consumption by a subterranean termite colony will result in the accelerated senescence of the colony (Chouvenc *et al.*, 2022) (Chapter 11 of this volume). Functionally, CSI triggers the death of relatively young individuals (Chouvenc, 2018) and prevents the queen from laying viable eggs (Rojas and Morales-Ramos, 2004; Chouvenc and Lee, 2021), leading to the accumulation of old individuals, the last to engage in the molting cycle (Nakajima *et al.*, 1963). Effectively, colony collapse resulting from feeding on a CSI bait recapitulates the succession of events during the natural senescence process that typically takes ~3 years, but instead occurs within a 3-month period (Chouvenc and Lee, 2021; Chouvenc *et al.*, 2022).

4.7 Mature colony demographic considerations

This last section focuses on the remaining aspects of mature colony functioning. First, it covers the nymph production process depending on the reproductive status of the colony. Second, it covers the emergence of secondary reproductives and their impact on the colony demography. Finally, it provides a heuristic perspective on colony functioning through the lens of demographic changes over the lifetime of a *Coptotermes* colony.

4.7.1 Production of nymphs through the life of the mature colony

Coptotermes colonies require larval instars ($L_2 \rightarrow N_1$) to initiate the nymphal developmental pathway (Lenz *et al.*, 1988; Roisin and Lenz, 1999; Costa-Leonardo and Barsotti, 2001; Barsotti and Costa-Leonardo, 2005; Chouvenc and Su, 2014). The suggested *de novo* production of nymphs from workers (Raina *et al.*, 2004) remains unlikely in *Coptotermes* (Chouvenc and Su, 2014). While it is unclear if the nymphal developmental pathway is triggered during L_2, or if factors are already involved in earlier stages (L_1 or E), the $L_2 \rightarrow N_1$ pathway is not accessible to young colonies despite an abundance of L_2. This confirms that nymph production depends on socio-environmental influences, such as its access to resources and the critical mass of the colony, which have yet to be investigated in *Coptotermes*. As a mature colony temporarily engages in the production of a large number of nymphs, it directly competes with the production of workers and soldiers, resulting in a temporary aging of worker and soldier cohorts. Such seasonal nymph investment is therefore reflected within the seasonal changes in worker and soldier demographic variation (Chouvenc *et al.*,

2022). However, environmental temperature fluctuation through seasonality may also influence egg production (Waller and La Fage, 1987) and the metabolic rates of individuals (Kakkar *et al.*, 2016b), adding potential confounding factors to the periodical age demographic oscillations of sterile castes (Chouvenc *et al.*, 2022).

It takes more than a year for a mature *Coptotermes* colony to produce a new cohort of alates. This implies that, as the colony proceeds to its seasonal alate dispersal flights, it is already engaged in the production of the next cohort of nymphs for the following year (Albino and Costa-Leonardo, 2011; Chouvenc *et al.*, 2022). Consequently, nymphs produced in mature colonies are found year-round at foraging sites but are limited to N_{3+}, as young nymphal instars apparently remain close to the central nest with larvae and young worker instars (Crosland and Su, 2006; Albino and Costa-Leonardo, 2011; Chouvenc *et al.*, 2022). Nymph production requires a substantial investment from the colony (Costa-Leonardo *et al.*, 2022), as the annual alate departure results in the significant loss of colony total biomass (here speculated to range from ~5% to 20%). From a nutritional perspective, nymph development and maturation may impose an important cost on the caregiving workforce. As N_1 and N_2 apparently remain close to the central nest and are presumably nutritionally dependent on workers, N_{3+} are primarily located at foraging sites and can actually feed on their own (Crosland and Su, 2006). This may be a strategy to limit the inherent cost of nymph production for the colony, by relocating them to where the food is, avoiding nutritional dependency on workers, and alleviating their potential metabolic burden on the nurturing capacity of the colony (Chouvenc *et al.*, 2022). During their consecutive molts toward the imago, it is unclear if nymphs move back to the central nest each time, as observed in workers (Kakkar *et al.*, 2017). In addition, just like workers, nymphs lose their gut symbionts during each molt and need to be refaunated. However, they retain a subset of their symbionts during their imaginal molt (May, 1941; Grassé and Noirot, 1945; Nalepa, 2017) which they will carry with them during the dispersal flight to seed the incipient colony with their obligate symbionts (Shimada *et al.*, 2013; Velenovsky *et al.*, 2021, 2023).

While nymphs are only produced during "mature colony" status, their inherent quality and quantity may vary as the colony undergoes reproductive shifts (Chouvenc *et al.*, 2022). First, during a temporary loss of reproduction (loss of primary queen/king), the resulting absence of eggs and larvae can prevent the colony from producing nymphs for the following year (Chouvenc *et al.*, 2022). Second, with the establishment of several replacement queens and the increase of oviposition, the number of nymphs produced can increase substantially in the following years compared with when it was under the primary reproductives. Third, as the colony may now be going through inbreeding cycles, these nymphs are much smaller than those initially produced by the colony, implying a shift in its reproductive trade-off, with a potential reduction of quality but an increase of quantity of alates (Chouvenc *et al.*, 2022). Finally, aberrant nymph-to-worker ratios may sometimes occur during initial colony senescence, as most, if not all, available larvae engage in the imaginal developmental pathway (Chouvenc *et al.*, 2022). This directly interferes with the replacement of workers, leading to a major portion of the colony turning into alates for a final dispersal flight, in an attempt to maximize terminal colony reproductive success (Su, 1994; Chouvenc *et al.*, 2022). Although the sex ratio of nymphs produced from healthy mature *Coptotermes* colonies is not expected to deviate much from 1:1 (Chouvenc *et al.*, 2017a), some unusual colony conditions may sometimes result in sex-biased production of alates (Crosland *et al.*, 1994; Roisin and Lenz, 2002).

4.7.2 Status of secondary reproductives

There is no evidence that *Coptotermes* can produce secondary reproductives in the presence of the primary queen and king. Instead, functional secondary reproductives (male or female) appear to be produced only as replacement reproductives in the case of the loss of a primary reproductive (Lenz *et al.*, 1988; Myles, 1999; Vargo *et al.*, 2003; Chouvenc and Su, 2014). Alternatively, non-functional neotenic individuals may be produced in the presence of live primary reproductives, but their function remains unclear (Costa-Leonardo and Arab, 2004; Costa-Leonardo *et al.*, 2004); they may serve as a standby strategy. In addition, after the loss of primary reproductives, some colonies fail to produce functional replacement reproductives (Bess, 1970; Lenz and Runko, 1993; Costa-Leonardo *et al.*, 2004), ultimately leading to the absence of population renewal and to colony senescence. Interestingly, immature colonies that have lost a primary reproductive have the ability to produce one or two non-functional replacement reproductives (Chouvenc *et al.*, 2015c). This observation was suggested to be a vestigial trait, but it directly demonstrated that non-mature colonies may still have a limited potential for a few individuals to engage in the $L_2 \rightarrow N_1 \rightarrow \ldots \rightarrow$ SR pathway, but only under peculiar social circumstances (loss of primary reproductives).

Consequently, successful production of secondary reproductives depends on the socio-environmental context, including current critical mass of the colony, its nutritional environment, its overall health, and its potential genetic makeup and population demographic structure (Nutting, 1969; La Fage and Nutting, 1977; Chouvenc and Su, 2014; Vargo, 2019; Chouvenc *et al.*, 2022; Costa-Leonardo *et al.*, 2022). *Coptotermes* secondary queens can accumulate within the colony, reaching remarkable degrees of physogastry, but not as developed as in primary queens (Fig. 4.7). Similar to primary reproductives, secondary reproductives also lose their gut protozoa over time (Velenovsky *et al.*, 2021; Inagaki *et al.*, 2022). The production of multiple functional secondary reproductives (queens and/or kings) can lead to a rapid rejuvenation of the colony. Over time, an accumulation and eventually a succession of secondary reproductives may potentially undergo cumulative inbreeding effects. From a demographic perspective, the addition of several functional secondary queens leads to a massive input of new individuals within the colony (high natality), but such individuals also potentially display relatively high mortality (short longevity) from inbreeding (Chouvenc *et al.*, 2022). As a result, a relatively old mature colony undergoing such inbreeding may display a demographic profile resembling a juvenile colony (relatively young average worker instar) but still maintain alate production and reproductive perennity (Chouvenc *et al.*, 2022).

4.7.3 Colony demographic changes through reproductive perennity

A termite colony represents a continuous succession of overlapping cohorts of individuals. An individual is born within the context of the existing colony, develops, serves its function, and eventually dies or disperses. From a colony perspective, its demographic makeup at any point in time is represented by the assemblage of all current overlapping cohorts of individuals, which is the inherent byproduct of two driving forces: the natality rate of individuals, and the mortality rate/dispersal of individuals, which both change over time. As a result, the demographic structure of a termite colony is highly dynamic and each colony demographic trajectory is unique (Chouvenc *et al.*, 2022). Here, the use of population pyramids (age structure diagrams) in the social study of human

Fig. 4.7. Secondary reproductives obtained from a single *Coptotermes gestroi* laboratory 7-year-old colony, with three physogastric secondary queens and one secondary king. (T. Chouvenc.)

populations (Goldstein, 2009) can be directly applied to a termite colony, where its instar composition (as a proxy to age composition) informs on its reproductive and health statuses, and on the overall life expectancy of individuals. The longevity of workers and soldiers also has a direct impact on the functionality of the colony, as they can maintain their respective functions over time resulting in a high return on investment per individual (Lemanski and Fefferman, 2018), which varies extensively over the lifetime of a *Coptotermes* colony (Chouvenc *et al.*, 2022).

To visualize the complexity of the dynamic nature of *Coptotermes* colony demography and to close this chapter on a "big picture" perspective, this final section offers a graphic representation of the potential demographic shifts a colony may go through over its lifetime. To do so, a generic *C. formosanus* colony is used (Fig. 4.8), by combining various elements observed over decades from multiple *C. formosanus* field colonies (Chouvenc *et al.*, 2022). For further details on the "demographic individuality" of a termite colony, it is suggested to refer directly to Chouvenc *et al.* (2022). In this generic (fictional) *C. formosanus* colony, the estimation of its relative population size over time is first provided (Fig. 4.8A). The demographic variation of the average instar cohort of workers is then provided as a proxy to the age structure of the colony over time (Fig. 4.8B). As soldier replacement is initiated from the pool of available workers, the changes of the S_2, S_3 and S_4 proportions within the colony closely follow the demographic variation of workers and are therefore not further discussed herein (but see Chouvenc *et al.*, 2022 for additional details). Finally, population pyramids from each time point reflect the demographic change over time for this generic colony (Fig. 4.8C). The following paragraphs therefore describe the series of events (numbers in the figure) this colony eventually goes through over its lifetime.

Phase 1: Foundation and initial colony growth. The colony is founded by a pair of alates, which produce their first cohort of offspring during the incipient colony phase. As the colony shifts from a biparental to an alloparental care regime, the colony enters its immature phase, initiates foraging outside of the claustral chamber and starts accumulating workers and soldiers. However, at this point, individual longevity remains limited as workers rarely survive past W_3.

Phase 2: Exponential colony growth. Exponential growth of the colony during the juvenile colony phase, where individuals' longevity improves, W_2, W_3, W_4 and W_5 can progressively be found at foraging sites. This results in a progressive accumulation of relatively old individuals, increasing the average worker instar within the colony. Age polyethism is well developed at this point and the colony expands its foraging territory.

Phase 3: Mature colony. The colony reaches maturity and optimal homeostatic conditions, allowing individuals to reach maximal longevity. The colony produces nymphs on a seasonal basis, which disperse as alates (green arrows), effectively reducing the number of termites within the colony. The colony enters seasonal cycles of reproductive perennity that may repeat for many years, sometimes decades. Here, worker instar demographic structure oscillates as a result of the subtle change in the ratios of instars from W_2 to W_{6+} over time, as the seasonal production of nymphs ($L_2 \rightarrow N_1$) may compete with the production of workers. It inherently leads to a temporary absence of worker replacement, which in turn temporarily increases the average age of workers.

Phase 4: Loss of primary reproductives. Eventually, the colony may lose reproduction with the death of the primary queen and/or king. The lack of oviposition leads to the absence of population replacement and to an accumulation of relatively older instar workers. In some cases, the loss of reproduction is permanent, which would directly lead the colony to phases 7 (senescence) and 8 (death). However, if the colony can resume oviposition with the production of secondary reproductives, the colony senescence may be temporary, and would lead to phase 5.

Phase 5: Inbreeding and colony rematuration. With the successful emergence of replacement reproductives, the sudden influx of eggs leads to the rejuvenation of the worker pool. It would result in a temporary bimodal

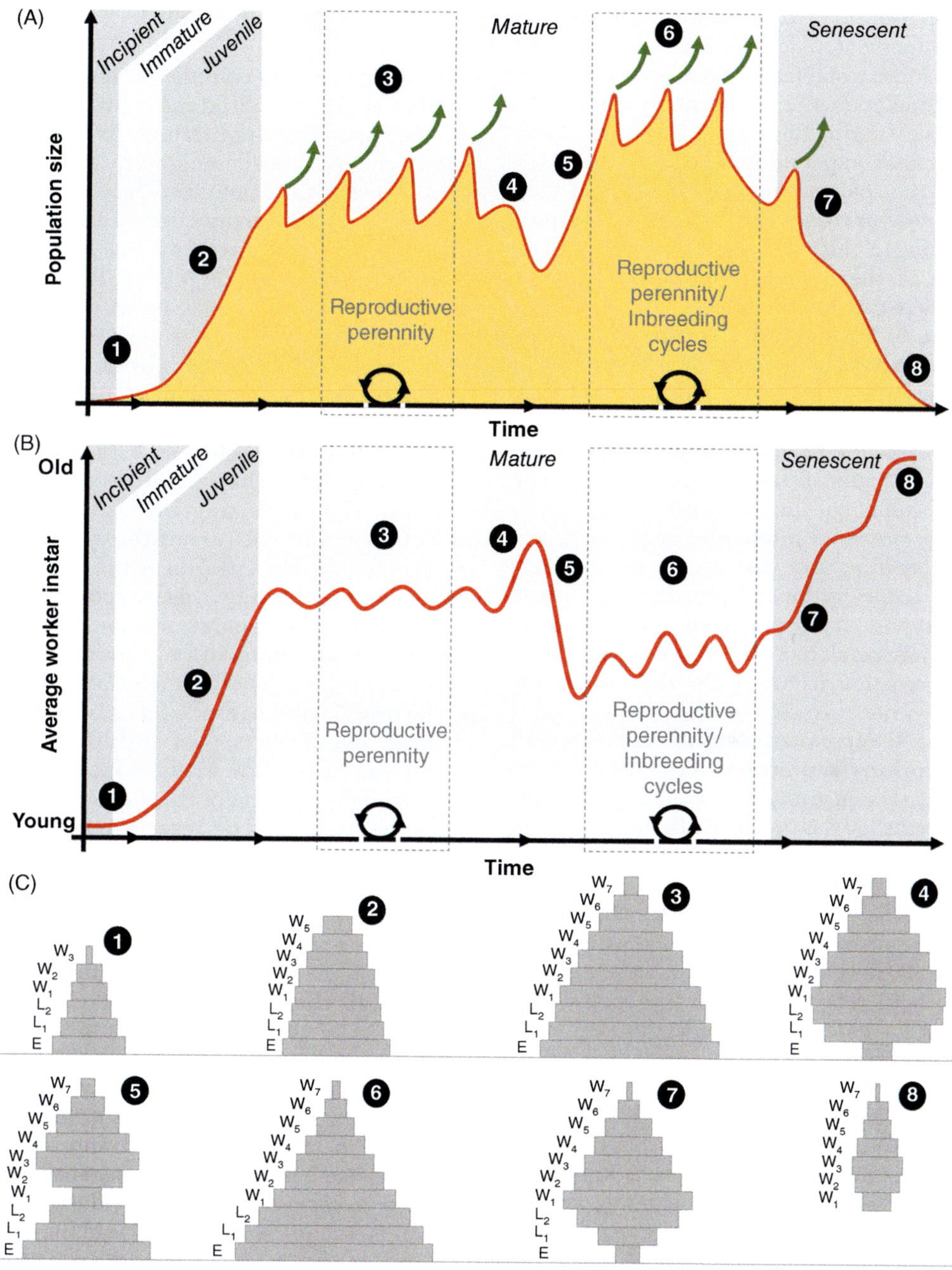

Fig. 4.8. Demographic changes in a generic *Coptotermes* colony over its lifetime. (A) Relative population size, (B) worker demography, and (C) age structure of the colony. Numbers in the figure represent a phase within the life of the colony, with a detailed explanation for each number in the main text. Phase 1: Foundation and initial colony growth. Phase 2: Exponential colony growth. Phase 3: Mature colony. Phase 4: Loss of primary reproductives. Phase 5: Inbreeding and colony rematuration. Phase 6: Mature colony with secondary reproductives. Phase 7: Permanent loss of reproduction and colony senescence. Phase 8: Colony death. (T. Chouvenc.)

distribution of two distinct worker age cohorts (demographic heterogeneity). In addition, it can result in the augmentation of colony population size, as natality would exceed mortality during this period. If the initial loss of oviposition period (phase 4) was relatively long, the colony may miss one or two consecutive years of seasonal nymphs/alates production. In this case, the temporary senescence of the colony would then lead to a colony re-maturation process.

Phase 6: Mature colony with secondary reproductives. Mature colony oviposition is now potentially led by multiple secondary queens/kings, which then may result in years/decades of inbreeding cycles. While the oviposition from multiple physogastric replacement queens may result in high natality, the potential cumulative negative effects of inbreeding can also result in relatively short longevity for individuals, and a lower quality of alates produced. These factors therefore result in a worker age demographic that remains relatively "young" for many years (when compared with phase 3).

Phase 7: Permanent loss of reproduction and colony senescence. Eventually, just as previously seen in phase 4, a loss of reproduction may occur, but this time as a definitive absence of oviposition. This permanent loss of reproduction leads to an accumulation of old workers during the early senescence of the colony. Most remaining larvae may engage in nymphal development for a last alate dispersal flight, while the absence of population replacement leads to an aging population.

Phase 8: Colony death. Finally, the colony collapses, with only a few remaining old instars individuals until final colony death. During this process, it is possible that a healthy neighboring *Coptotermes* colony may take over the territory of the collapsing colony.

4.8 Conclusion

A termite colony is a complex biological entity that engages in differential production of individuals over its lifetime. As a social insect colony, its demography changes upon intrinsic metabolic and reproductive priorities, and extrinsic environmental fluctuations. Once a colony has reached maturity, it can maintain reproductive perennity by seasonally engaging in dispersal flights, to maximize fitness output, as the colony may also resort to various reproductive strategies through its lifetime. This chapter provides a reflection of the proximate and ultimate mechanisms that drive such demographic changes in *Coptotermes* colonies, while providing extensive elements of the natural history of the species to set the colony life cycle in a robust biological context. While this chapter summarizes the advancement of knowledge of termite colony foundation, growth and maturation since the consequential chapter by Nutting (1969), here the focus is on *Coptotermes* owing to its pest status and its position as a biological model. This chapter also highlights aspects of the biology of termites that remain to be investigated. Paradoxically, such aspects often remain elusive as a direct result of the complexity of the biology of *Coptotermes* colonies, and where hypotheses and speculations may linger for a foreseeable future because of the limits imposed on their putative testability (see Chapter 18 of this volume). It is therefore optimistic to hope that the advancement achieved in *Coptotermes* biology over the past five decades will enable major breakthroughs in further understanding this complex, challenging and deeply fascinating termite genus.

References

Albino, E. and Costa-Leonardo, A.M. (2011) Nymphs in foraging populations of *Coptotermes gestroi* (Isoptera: Rhinotermitidae). *Animal Biology* 61, 427–439.

Barsotti, R.C. and Costa-Leonardo, A.M. (2005) The caste system of *Coptotermes gestroi* (Isoptera: Rhinotermitidae). *Sociobiology* 46, 87–103.

Bernard, S., Osbrink, W. and Su, N.-Y. (2017) Response of the Formosan subterranean termite to neighboring con-specific populations after baiting with noviflumuron. *Journal of Economic Entomology* 110, 575–583.

Bess, H.A. (1970) Termites of Hawaii and the oceanic islands. In: Krishna, K. and Weesner, F.M. (eds) *Biology of Termites*. Vol. 1. Academic Press, New York, pp. 449–476.

Bignell, D.E. (2016) The role of symbionts in the evolution of termites and their rise to ecological dominance in the tropics. In: Hurst, C.J. (ed.) *The Mechanistic Benefits of Microbial Symbionts*. Springer, Dordrecht, Netherlands, pp. 121–172.

Blumenfeld, A.J., Eyer, P.A., Husseneder, C., Mo, J., Johnson, L.N., *et al.* (2021) Bridgehead effect and multiple introductions shape the global invasion history of a termite. *Communications Biology* 4, e.196.

Bordereau, C. (1985) The role of pheromones in termite caste differentiation. In: Watson, J.A.L., Okot-Kotber, B.M. and Noirot, C. (eds) *Caste Differentiation in Social Insects*. Pergamon Press, Oxford, pp. 221–226.

Bordereau, C. and Pasteels, J.M. (2010) Pheromones and chemical ecology of dispersal and foraging in termites. In: Bignell, D.E, Roisin Y. and Lo, N. (eds) *Biology of Termites: A Modern Synthesis*. Springer, Dordrecht, Netherlands, pp. 279–320.

Bordereau, C., Peppuy, A., Connétable, S. and Robert, A. (1999) Les termites de l'île de la Réunion et leur importance économique. *Actes des Colloques Insectes Sociaux* 12, 159–164.

Bourguignon, T., Lo, N., Šobotník, J., Sillam-Dussès, D., Roisin, Y. *et al.* (2016a) Oceanic dispersal, vicariance and human introduction shaped the modern distribution of the termites *Reticulitermes*, *Heterotermes* and *Coptotermes*. *Proceedings of the Royal Society B: Biological Sciences* 283, 20160179.

Bourguignon, T., Chisholm, R.A. and Evans, T.A. (2016b) The termite worker phenotype evolved as a dispersal strategy for fertile wingless individuals before eusociality. *The American Naturalist* 187, 372–387.

Brossette, L., Meunier, J., Dupont, S., Bagnères, A.G. and Lucas, C. (2019) Unbalanced biparental care during colony foundation in two subterranean termites. *Ecology and Evolution* 9, 192–200.

Buchli, H. (1958) L'origine des castes et les potentialités ontogéniques des termites européens du genre *Reticulitermes* Holmgren. *Annales des Sciences Naturelles, Zoologie, 11e série* 20, 263–429.

Calaby, J.H. and Gay, F.J. (1956) The distribution and biology of the genus *Coptotermes* (Isoptera) in Western Australia. *Australian Journal of Zoology* 4, 19–39.

Calleri, D.V., McGrail Reid, E., Rosengaus, R.B., Vargo, E.L. and Traniello, J.F. (2006) Inbreeding and disease resistance in a social insect: effects of heterozygosity on immunocompetence in the termite *Zootermopsis angusticollis*. *Proceedings of the Royal Society B: Biological Sciences* 273, 2633–2640.

Carrijo, T.F., Engel, M.S., Chouvenc, T., Gile, G.H., Mikaelyan, A., Dedeine, F., Ware, J.L., Haifig, I., Arab, A., Constantini, J.P. and Souza, J.P. (2023) A call to termitologists: it is time to abandon the use of "lower" and "higher" termites. *Insectes Sociaux* 70, 295–299.

Chambers, D.M., Zungoli, P.A. and Hill Jr, H.S. (1988) Distribution and habitats of the Formosan subterranean termite (Isoptera: Rhinotermitidae) in South Carolina. *Journal of Economic Entomology* 81, 1611–1619.

Chang, L., Hsu, E.L. and Wu, W.J. (2005) The ultrastructure of compound eye of Formosan subterranean termite, *Coptotermes formosanus* (Isoptera: Rhinotermitidae). *Collection and Research* 18, 1–6.

Chiu, C.-I., Yeh, H.-T., Tsai, M.-J. and Li, H.-F. (2016) Naturalization and control of *Coptotermes gestroi* (Blattodea: Rhinotermitidae) in a Taiwanese forest. *Journal of Economic Entomology* 109, 1317–1325.

Chouvenc, T. (2018) Comparative impact of chitin synthesis inhibitor baits and non-repellent liquid termiticides on subterranean termite colonies over foraging distances: colony elimination versus localized termite exclusion. *Journal of Economic Entomology* 111, 2317–2328.

Chouvenc, T. (2019) The relative importance of queen and king initial weights in termite colony foundation success. *Insectes Sociaux* 66, 177–184.

Chouvenc, T. (2020) Limited survival strategy in starving subterranean termite colonies. *Insectes Sociaux* 67, 71–82.

Chouvenc, T. (2022) Eusociality and the transition from biparental to alloparental care in termites. *Functional Ecology* 36, 3049–3059.

Chouvenc, T. and Foley IV, J.R. (2018) *Coptotermes gestroi* (Wasmann) (Blattodea [Isoptera]: Rhinotermitidae), a threat to the southeastern Florida urban tree canopy. *Florida Entomologist* 101, 79–90.

Chouvenc, T. and Lee, S.-B. (2021) Queen egg laying and egg hatching abilities are hindered in subterranean termite colonies when exposed to a chitin synthesis inhibitor bait formulation. *Journal of Economic Entomology* 114, 2466–2472.

Chouvenc, T. and Su, N.-Y. (2014) Colony age-dependent pathway in caste development of *Coptotermes formosanus* Shiraki. *Insectes Sociaux* 61, 171–182.

Chouvenc, T. and Su, N.-Y. (2017a) Irreversible transfer of brood care duties and insights into the burden of caregiving in incipient subterranean termite colonies. *Ecological Entomology* 42, 777–784.

Chouvenc, T. and Su, N.-Y. (2017b) Testing the role of cuticular hydrocarbons on intercolonial agonism in two subterranean termite species (*Coptotermes*) and their hybrids. *Insectes Sociaux* 64, 347–355.

Chouvenc, T., Bardunias, P., Efstathion, C.A., Chakrabarti, S., Elliott *et al.* (2013a) Resource opportunities from the nest of dying subterranean termite (Isoptera: Rhinotermitidae) colonies: a laboratory case of ecological succession. *Annals of the Entomological Society of America* 106, 771–778.

Chouvenc, T., Efstathion, C.A., Elliott, M.L. and Su, N.-Y. (2013b) Extended disease resistance emerging from the faecal nest of a subterranean termite. *Proceedings of the Royal Society B: Biological Sciences* 280, 20131885.

Chouvenc, T., Basille, M., Li, H.-F. and Su, N.-Y. (2014) Developmental instability in incipient colonies of social insects. *PLoS ONE* 9, e113949.

Chouvenc, T., Basille, M. and Su, N.-Y. (2015a) The production of soldiers and the maintenance of caste proportions delay the growth of termite incipient colonies. *Insectes Sociaux* 62, 23–29.

Chouvenc, T., Mullins, A.J. and Su, N.-Y. (2015b) Territorial status-quo between the big-headed ant (Hymenoptera: Formicidae) and the Formosan subterranean termite (Isoptera: Rhinotermitidae). *Florida Entomologist* 98, 157–161.

Chouvenc, T., Mullins, A.J. and Su, N.-Y. (2015c) Rare production of nymphs in an Asian subterranean termite (Isoptera: Rhinotermitidae) incipient colony. *Florida Entomologist* 98, 972–973.

Chouvenc, T., Li H.-F., Austin, J., Bordereau, C., Bourguignon, T., *et al.* (2016) Revisiting *Coptotermes* (Isoptera: Rhinotermitidae): a global taxonomic road map for species validity and distribution of an economically important subterranean termite genus. *Systematic Entomology* 4, 299–306.

Chouvenc, T., Scheffrahn, R.H., Mullins, A.J. and Su, N.-Y. (2017a) Flight phenology of two *Coptotermes species* (Isoptera: Rhinotermitidae) in southeastern Florida. *Journal of Economic Entomology* 110, 1693–1704.

Chouvenc, T., Basille, M. and Su, N.Y. (2017b) Role of accelerated developmental pathway and limited nurturing capacity on soldier developmental instability in subterranean termite incipient colonies. *Insectes Sociaux* 64, 477–483.

Chouvenc, T., Elliott, M.L., Šobotník, J., Efstathion, C.A. and Su, N.-Y. (2018) The termite fecal nest: a framework for the opportunistic acquisition of beneficial soil *Streptomyces* (Actinomycetales: Streptomycetaceae). *Environmental Entomology* 47, 1431–1439.

Chouvenc, T., Sillam-Dussès, D. and Robert, A. (2020) Courtship behavior confusion in two subterranean termite species that evolved in allopatry (Blattodea, Rhinotermitidae, *Coptotermes*). *Journal of Chemical Ecology* 46, 461–474.

Chouvenc, T., Šobotník, J., Engel, M.S. and Bourguignon, T. (2021) Termite evolution: mutualistic associations, key innovations, and the rise of Termitidae. *Cellular and Molecular Life Sciences* 78, 2749–2769.

Chouvenc, T., Ban, P.M. and Su, N.-Y. (2022) Life and death of termite colonies, a decades-long age demography perspective. *Frontiers in Ecology and Evolution* 10, 911042.

Cleveland, L.R. (1925) The feeding habit of termite castes and its relation to their intestinal flagellates. *Biological Bulletin* 48, 295–308.

Coaton, W.G.H. and Sheasby, J.L. (1976) National survey of the Isoptera of southern Africa. 11. The genus *Coptotermes* Wasmann (Rhinotermitidae: Coptotermitinae). *Cimbebasia A* 3, 140–172.

Cole, E.L. and Rosengaus, R.B. (2019) Pathogenic dynamics during colony ontogeny reinforce potential drivers of termite eusociality: mate assistance and biparental care. *Frontiers in Ecology and Evolution* 7, 473.

Cornelius, M.L. and Grace, J.K. (1995) Laboratory evaluations of interactions of three ant species with the Formosan subterranean termite (Isoptera: Rhinotermitidae). *Sociobiology* 26, 291–298.

Cornette, R., Gotoh, H., Koshikawa, S. and Miura, T. (2008) Juvenile hormone titers and caste differentiation in the damp-wood termite *Hodotermopsis sjostedti* (Isoptera, Termopsidae). *Journal of Insect Physiology* 54, 922–930.

Cornette, R., Hayashi, Y., Koshikawa, S. and Miura, T. (2013) Differential gene expression in response to juvenile hormone analog treatment in the damp-wood termite *Hodotermopsis sjostedti* (Isoptera, Archotermopsidae). *Journal of Insect Physiology* 59, 509–518.

Costa, J.G., Santos, I.C.D.L., Peres Filho, O., Zanetti, R. and Santos, A.D. (2020) Characterization of injury caused by *Coptotermes testaceus* in *Tectona grandis* plantations in Brazil. *Cerne* 26, 310–314.

Costa-Leonardo, A.M. and Arab, A. (2004) Reproductive strategy of *Coptotermes gestroi* (Isoptera: Rhinotermitidae) in Brazil. *Sociobiology* 44, 123–126.

Costa-Leonardo, A.M. and Barsotti, R.C. (2001) Growth patterns of incipient colonies of *Coptotermes havilandi* (Isoptera, Rhinotermitidae) initiated in the laboratory from swarming alates. *Sociobiology* 37, 551–561.

Costa-Leonardo, A.M. and Patricio, G.B. (2005) Structure of the spermatheca in five families of Isoptera. *Sociobiology* 45, 659–670.

Costa-Leonardo, A.M., Barsotti, R.C. and Camargo-Dietrich, D. (1999) Review and update on the biology of *Coptotermes havilandi* (Isoptera, Rhinotermitidae). *Sociobiology* 33, 339–356.

Costa-Leonardo, A.M., Arab, A. and Casarin, F.E. (2004) Neotenic formation in laboratory colonies of the termite *Coptotermes gestroi* after orphaning. *Journal of Insect Science* 4, e10.

Costa-Leonardo, A.M., Janei, V. and da Silva, I.B. (2022) Comparative reproductive biology of pre-, imaginal, and neotenic castes of the Asian termite *Coptotermes gestroi* (Blattaria, Isoptera, Rhinotermitidae). *Bulletin of Entomological Research* 6, 827–836.

Crosland, M.W. and Su, N.-Y. (2006) Work allocation among castes in a rhinotermitid termite (Isoptera) - Are nymphs a working caste? *Sociobiology* 48, 585–598.

Crosland, M.W.J., Li, G.X., Huang, L.W. and Dai, Z.R. (1994) Switch to single sex alate production in a colony of the termite, *Coptotermes formosanus*. *Journal of Entomological Science* 29, 523–525.

da Silva, I.B. and Costa-Leonardo, A.M. (2023) Mating mediates morphophysiological changes in the spermathecae of *Coptotermes gestroi* queens. *Entomologia Experimentalis et Applicata* 171, 361–373.

de Lima, M.M., Gonçalves, T.T., DeSouza, O. and Reis, R. Jr (2006) Nesting site selection by *Coptotermes gestroi* (Insecta: Isoptera). *Sociobiology* 48, 246–251.

Dong, S.L., Mao, L. and Henderson, G. (2009) Physical contact between soldier and worker is essential in soldier self-regulation of *Coptotermes formosanus* (Isoptera, Rhinotermitidae). *Insectes Sociaux* 56, 28–34.

Dronnet, S., Chapuisat, M., Vargo, E.L., Lohou, C. and Bagnères, A.G. (2005) Genetic analysis of the breeding system of an invasive subterranean termite, *Reticulitermes santonensis*, in urban and natural habitats. *Molecular Ecology* 14, 1311–1320.

Du, H., Chouvenc, T., Osbrink, W.L.A. and Su, N.-Y. (2016) Social interactions in the central nest of *Coptotermes formosanus* juvenile colonies. *Insectes Sociaux* 63, 279–290.

Du, H., Chouvenc, T. and Su, N.-Y. (2017a) Development of age polyethism with colony maturity in *Coptotermes formosanus* (Isoptera: Rhinotermitidae). *Environmental Entomology* 46, 311–318.

Du, H., Chouvenc, T., Osbrink, W.L.A. and Su, N.-Y. (2017b) Heterogeneous distribution of castes/instars and behaviors in the nest of *Coptotermes formosanus* Shiraki. *Insectes Sociaux* 64, 103–112.

Du, H., Wu, W., Huang, X. and Li, Z. (2019) Screening of reference genes for expression analysis in the study of soldier caste differentiation of Formosan subterranean termite *Coptotermes formosanus* Shiraki. *PeerJ* 7, e7981.

Du, H., Tong, R.L., Huang, X., Liu, B., Huang, R. *et al.* (2020) Methoprene-induced genes in workers of Formosan subterranean termites (*Coptotermes formosanus* Shiraki). *Insects* 11, e71.

Eggleton, P. (2000) Global patterns of termite diversity. In: Abe, T., Bignell, D.E. and Higashi M. (eds) Termites: Evolution, Sociality, Symbioses, Ecology. Springer, Dordrecht, Netherlands, pp. 25–51.

Elliott, K.L. and Stay, B. (2008) Changes in juvenile hormone synthesis in the termite *Reticulitermes flavipes* during development of soldiers and neotenic reproductives from groups of isolated workers. *Journal of Insect Physiology* 54, 492–500.

Emerson, A.E. (1938) Termite nests-a study of the phylogeny of behavior. *Ecological Monographs* 8, 247–284.

Ewart, D. (1988) Aspects of the ecology of the termite *Coptotermes lacteus* (Froggatt). Doctoral dissertation, Ph.D thesis, Dept. Zool., La Trobe University, Bundoora, Victoria, Australia, 343 pp.

Eyer, P.A. and Vargo, E.L. (2022) Short and long-term costs of inbreeding in the lifelong-partnership in a termite. *Communications Biology* 5, e389.

Eyer, P.A., Blumenfeld, A.J., Johnson, L.N., Perdereau, E., Shults, P. *et al.* (2021) Extensive human-mediated jump dispersal within and across the native and introduced ranges of the invasive termite *Reticulitermes flavipes*. *Molecular Ecology* 30, 3948–3964.

Fei, H.X. and Henderson, G. (1999) Effect of moisture and two nitrogen sources on nest site choice by alates and dealates of *Coptotermes formosanus* in the laboratory (Isoptera: Rhinotermitidae). *Sociobiology* 34, 591–589.

Fei, H.X. and Henderson, G. (2003) Comparative study of incipient colony development in the Formosan subterranean termite, *Coptotermes formosanus* Shiraki (Isoptera, Rhinotermitidae). *Insectes Sociaux* 50, 226–233.

Ferraz, M.V. and Cancello, E.M. (2001) Swarming behaviour of the economically most important termite, *Coptotermes havilandi* (Isoptera: Rhinotermitidae), in Southeastern Brazil. *Sociobiology* 38, 683–694.

Ferraz, M.V. and Cancello, E.M. (2004) Strategies on the developmental biology of incipient colonies of *Coptotermes gestroi* (Isoptera: Rhinotermitidae) in different substrates. *Sociobiology* 44, 109–122.

Foley, J.R. IV, Chouvenc, T., Giblin-Davis, R.M., Su, N.-Y. and Kanzaki, N. (2018) Phoresy and within-colony transmission of nematodes associated with alates of subterranean termites (Isoptera: Rhinotermitidae). *Environmental Entomology* 47, 1107–1116.

Fougeyrollas, R., Dolejšová, K., Sillam-Dussès, D., Roy, V., Poteaux, C., *et al.* (2015) Asexual queen succession in the higher termite *Embiratermes neotenicus*. *Proceedings of the Royal Society B: Biological Sciences* 282, 20150260.

Fougeyrollas, R., Křivánek, J., Roy, V., Dolejšová, K., Frechault, S. *et al.* (2017) Asexual queen succession mediates an accelerated colony life cycle in the termite *Silvestritermes minutus*. *Molecular Ecology* 26, 3295–3308.

Fournier, D., Hellemans, S., Hanus, R. and Roisin, Y. (2016) Facultative asexual reproduction and genetic diversity of populations in the humivorous termite *Cavitermes tuberosus*. *Proceedings of the Royal Society B: Biological Sciences* 283, 20160196.

Funaro, C.F., Schal, C. and Vargo, E.L. (2019) Queen and king recognition in the subterranean termite, *Reticulitermes flavipes*: Evidence for royal recognition pheromones. *PLoS ONE* 14, e0209810.

Glastad, K.M., Gokhale, K., Liebig, J. and Goodisman, M.A. (2016) The caste-and sex-specific DNA methylome of the termite *Zootermopsis nevadensis*. *Scientific Reports* 6, e37110

Goldstein, J.R. (2009) How populations age. In: Uhienberg, P. (ed.) *International Handbook of Population Aging*. Springer, Dordrecht, Netherlands, pp. 7–18

Gordon, J.M., Šobotník, J. and Chouvenc, T., 2020. Colony-age-dependent variation in cuticular hydrocarbon profiles in subterranean termite colonies. *Ecology and Evolution*, 10(18), pp.10095-10104.

Gordon, J.M., Velenovsky, J.F. and Chouvenc, T. (2022) Subterranean termite colony elimination can be achieved even when only a small proportion of foragers feed upon a CSI bait. *Journal of Pest Science* 95, 1207–1216.

Grace, J.K., Yamamoto, R.T. and Tamashiro, M. (1995) Relationship of individual worker mass and population decline in a Formosan subterranean termite colony (Isoptera: Rhinotermitidae). *Environmental Entomology* 24, 1258–1262.

Grassé, P.-P. (1982) *Termitologia Vol. I: Anatomie-Physiologie-Reproduction des termites*. Masson, Paris.

Grassé, P.-P. (1984) *Termitologia Vol. II: Fondation des sociétés, construction*. Masson, Paris.

Grassé, P.-P. and Noirot, C. (1945) La transmission des flagellés symbiotiques et les aliments des termites. *Bullletin Biologique de la France et de la Belgique* 79, 273–297.

Greaves, T. (1962) Studies of foraging galleries and the invasion of living trees by *Coptotermes acinaciformis* and *C. brunneus* (Isoptera). *Australian Journal of Zoology* 10, 630–651.

Guaraldo, A.C. and Costa-Leonardo, A.M. (2009) Preliminary fusion testing between whole young colonies of *Coptotermes gestroi* (Isoptera: Rhinotermitidae). *Sociobiology* 53, 767–774.

Hartke, T.R. and Baer, B. (2011) The mating biology of termites: a comparative review. *Animal Behaviour* 82, 927–936.

Haverty, M.I. (1977) The proportion of soldiers in termite colonies: a list and a bibliography. *Sociobiology* 2, 199–216.

Haverty, M.I. (1979) Soldier production and maintenance of soldier proportions in laboratory experimental groups of *Coptotermes formosanus* Shiraki. *Insectes Sociaux* 26, 69–84.

Hayashi, Y., Lo, N., Miyata, H. and Kitade, O. (2007) Sex-linked genetic influence on caste determination in a termite. *Science* 318, 985–987.

Hellemans, S., Dolejšová, K., Křivánek, J., Fournier, D., Hanus, R. *et al.* (2019) Widespread occurrence of asexual reproduction in higher termites of the *Termes* group (Termitidae: Termitinae). *BMC Evolutionary Biology* 19, e131.

Hellemans, S., Wang, M., Hasegawa, N., Šobotník, J., Scheffrahn, R.H. *et al.* (2022) Using ultraconserved elements to reconstruct the termite tree of life. *Molecular Phylogenetics and Evolution* 173, 107520.

Henderson, G. (1996) Alate production, flight phenology, and sex-ratio in *Coptotermes formosanus* Shiraki, an introduced subterranean termite in New Orleans, Louisiana. *Sociobiology* 28, 319–326.

Henderson, G. and Delaplane, K.S. (1994) Formosan subterranean termite swarming behavior and alate sex-ratio (Isoptera: Rhinotermitidae). *Insectes Sociaux* 41, 19–28.

Henderson G. and Rao K.S. (1993) Sexual dimorphism in soldiers of Formosan subterranean termites (Isoptera: Rhinotermitidae). *Sociobiology* 21, 341–345.

Higa S.Y. (1981) Flight, colony foundation and development of the gonads of the primary reproductives of the Formosan subterranean termites, *Coptotermes formosanus* Shiraki. PhD dissertation, University of Hawaii, Honolulu, 173 pp.

Higa, S.Y. and Tamashiro, M. (1983) Swarming of the Formosan subterranean termite, *Coptotermes formosanus* Shiraki in Hawaii (Isoptera: Rhinotermitidae). *Proceedings of the Hawaiian Entomological Society* 24, 233–238.

Hochmair, H.H. and Scheffrahn, R.H. (2010) Spatial association of marine dockage with land-borne infestations of invasive termites (Isoptera: Rhinotermitidae: *Coptotermes*) in urban South Florida. *Journal of Economic Entomology* 103, 1338–1346.

Honigberg, B.M. (1970) Protozoa associated with termites and their role in digestion. In: Krishna, K. and Weesner, F.M. (eds) *Biology of Termites, Vol. 2*. Academic Press, New York, pp. 1–36.

Howard, K.J. and Thorne, B.L. (2010) Eusocial evolution in termites and Hymenoptera. In: Bignell, D.E, Roisin Y., and Lo, N. (eds) *Biology of Termites: A Modern Synthesis*. Springer, Dordrecht, Netherlands, pp. 97–132.

Hrdý, I. (1976) The influence of juvenile hormone analogues on caste development in termites. In: Lüscher, M. (ed.) *Phase and Caste Determination in Insects*, Pergamon, Oxford, p. 71.

Hrdý, I. and Křeček, J. (1972) Development of superfluous soldiers induced by juvenile hormone analogues in the termite, *Reticulitermes lucifugus santonensis*. *Insectes Sociaux* 19, 105–109.

Huang, S.-Y., Chiu, C.-I., Tsai, Y.-Y., Li, W.-J., Wu, C.-C. and Li, H.-F. (2022) Nationwide termite pest Survey conducted in Taiwan as a citizen science project. *Journal of Economic Entomology* 115, 1650–1658.

Husseneder, C. and Simms, D.M. (2008) Size and heterozygosity influence partner selection in the Formosan subterranean termite. *Behavioral Ecology* 19, 764–773.

Husseneder, C., Simms, D.M. and Ring, D.R. (2006) Genetic diversity and genotypic differentiation between the sexes in swarm aggregations decrease inbreeding in the Formosan subterranean termite. *Insectes Sociaux* 53, 212–219.

Husseneder, C., McGregor, C., Lang, R.P., Collier, R. and Delatte, J. (2012) Transcriptome profiling of female alates and egg-laying queens of the Formosan subterranean termite. *Comparative Biochemistry and Physiology Part D: Genomics and Proteomics* 7, 14–27.

Ikehara, S. (1966) Distribution of termites in the Ryukyu Archipelago. Bulletin of Arts & Science Division, University of the Ryukyus. *Mathematics and Sciences* 9, 49–178.

Inagaki, T., Nozaki, T. and Matsuura, K. (2022) Rapid elimination of symbiotic intestinal protists during the neotenic differentiation in a subterranean termite, *Reticulitermes speratus*. *Insectes Sociaux*, 69, 335–343.

Inward, D., Beccaloni, G. and Eggleton, P. (2007) Death of an order: a comprehensive molecular phylogenetic study confirms that termites are eusocial cockroaches. *Biology Letters* 3, 331–335.

Jongepier, E., Kemena, C., Lopez-Ezquerra, A., Belles, X., Bornberg-Bauer, E. *et al.* (2018) Remodeling of the juvenile hormone pathway through caste-biased gene expression and positive selection along a gradient of termite eusociality. *Journal of Experimental Zoology Part B: Molecular and Developmental Evolution* 330, 296–304.

Kakkar, G., Chouvenc, T. and Su, N.-Y. (2016a) Postecdysis sclerotization of mouthparts of the Formosan subterranean termites (Isoptera: Rhinotermitidae). *Journal of Economic Entomology* 109, 792–799.

Kakkar, G., Chouvenc, T., Osbrink, W. and Su, N.-Y. (2016b) Temporal assessment of molting in workers of Formosan subterranean termites (Isoptera: Rhinotermitidae). *Journal of Economic Entomology* 109, 2175–2181.

Kakkar, G., Osbrink, W., Mullins, A. and Su, N.-Y. (2017) Molting site fidelity in workers of Formosan subterranean termites (Isoptera: Rhinotermitidae). *Journal of Economic Entomology* 110, 2512–2517.

Kakkar, G., Osbrink, W. and Su, N.-Y. (2018) Molting site fidelity accounts for colony elimination of the Formosan subterranean termites (Isoptera: Rhinotermitidae) by chitin synthesis inhibitor baits. *Scientific Reports* 8, e1259.

King, E.G. Jr (1971) Biology of the Formosan subterranean termite, *Coptotermes formosanus*, Shiraki, with primary emphasis on young colony development. PhD Dissertation, Louisiana State University and Agricultural & Mechanical College, 225 pp.

King, E.G. Jr and Spink, W.T. (1969) Foraging galleries of the Formosan subterranean termite, *Coptotermes formosanus*, in Louisiana. *Annals of the Entomological Society of America* 62, 536–542.

King, E.G. Jr and Spink, W.T. (1974) Laboratory studies on the biology of the Formosan subterranean termite with primary emphasis on young colony development. *Annals of the Entomological Society of America* 67, 953–958.

King, E.G. Jr and Spink, W.T. (1975) Development of incipient Formosan subterranean termite colonies in the field. *Annals of the Entomological Society of America* 68, 355–358.

Kobayashi, K., Hasegawa, E., Yamamoto, Y., Kawatsu, K., Vargo, E.L. *et al.* (2013) Sex ratio biases in termites provide evidence for kin selection. *Nature Communications* 4, e2048.

Korb, J. (2015) Juvenile hormone: a central regulator of termite caste polyphenism. *Advances in Insect Physiology* 48, 131–161.

Korb, J. and Belles, X. (2017) Juvenile hormone and hemimetabolan eusociality: a comparison of cockroaches with termites. *Current Opinion in Insect Science* 22, 109–116.

Korb, J., Roux, E.A. and Lenz, M. (2003) Proximate factors influencing soldier development in the basal termite *Cryptotermes secundus* (Hill). *Insectes Sociaux* 50, 299–303.

Korb, J., Hoffmann, K. and Hartfelder, K. (2009) Endocrine signatures underlying plasticity in postembryonic development of a lower termite, *Cryptotermes secundus* (Kalotermitidae). *Evolution & Development* 11, 269–277.

Korb, J., Greiner, C., Foget, M. and Geiler, A. (2021) How can termites achieve their unparalleled postembryonic developmental plasticity? A test for the role of intermolt-specific high juvenile hormone titers. *Frontiers in Ecology and Evolution* 9, 619594.

Koshikawa, S., Matsumoto, T. and Miura, T. (2001) Regressive molt in the Japanese damp-wood termite *Hodotermopsis japonica* (Isoptera: Termopsidae). *Sociobiology* 38, 495–500.

Krishna, K. and Weesner, F.M. (1969) *Biology of Termites, Vol. 1*. Academic Press, New York, 598 pp.

Krishna, K., Grimaldi, D.A., Krishna, V. and Engel, M.S. (2013) Treatise on the Isoptera of the World: Vol. 1. *Bulletin of the American Museum of Natural History* 377, 1–200.

Kronauer, D.J. and Libbrecht, R. (2018) Back to the roots: the importance of using simple insect societies to understand the molecular basis of complex social life. *Current Opinion in Insect Science* 28, 33–39.

Kusaka, A. and Matsuura, K. (2018) Allee effect in termite colony formation: influence of alate density and flight timing on pairing success and survivorship. *Insectes Sociaux* 65, 17–24.

La Fage, J.P. and Nutting W.L. (1977) Nutrient dynamics of termites. In: Production ecology of ants and termites (Brian, M.V., Ed), Cambridge University Press, Cambridge. pp. 165–232.

Lai, P. Y., M. Tamashiro, J. R. Yates, J. K. Fujii, N.-Y. Su *et al.* (1983) Living plants in Hawaii attacked by *Coptotermes formosanus*. *Proceeding of the Hawaii Entomological Society* 24, 283–286.

Laranjo, L.T., Haifig, I. and Costa-Leonardo, A.M. (2019) A glycoproteinaceous secretion in the seminal vesicles of the termite *Coptotermes gestroi* (Isoptera: Rhinotermitidae). *Insects* 10, e428.

Lax, A.R. and Osbrink, W.L. (2003) United States Department of Agriculture—agriculture research service research on targeted management of the Formosan subterranean termite *Coptotermes formosanus* Shiraki (Isoptera: Rhinotermitidae). *Pest Management Science* 59, 788–800.

Lebrun, D. (1985) The role of hormones in social polymorphism and reproduction in *Kalotermes flavicollis* Fabr. In: Watson, J.A.L., Okot-Kotber, B.M. and Noirot, C. (eds) *Caste Differentiation in Social Insects*. Pergamon, Oxford, pp. 227–238.

Lebrun, D. (1991) Hormonal control of caste differentiation in termites. *Ethology Ecology and Evolution* 3 (supp1.), 9–13.

Lee, S.B., Mullins, A., Aguilera-Olivares, D., Chouvenc, T. and Su, N.-Y. (2019) Fused colonies of the Formosan subterranean termite (Blattodea: Rhinotermitidae) for laboratory experiments. *Journal of Economic Entomology* 112, 2311–2315.

Lee, S.-B., Chouvenc, T., Mizumoto, N., Mullins, A. and Su, N.-Y. (2022) Age-based spatial distribution of workers is resilient to worker loss in a subterranean termite. *Scientific Reports* 12, e7837.

Lee, S.H., Bardunias, P. and Su, N.Y. (2006) Food encounter rates of simulated termite tunnels with variable food size/distribution pattern and tunnel branch length. *Journal of Theoretical Biology* 243, 493–500.

Lee, T.R.C. (2015) Molecular phylogenetic, ecological, and evolutionary studies of Australian Rhinotermitid termites. PhD Dissertation, University of Sydney, Australia.

Legendre, F., Whiting, M.F. and Grandcolas, P. (2013) Phylogenetic analyses of termite post-embryonic sequences illuminate caste and developmental pathway evolution. *Evolution and Development* 15, 146–157.

Lemanski, N.J. and Fefferman, N.H. (2018) How life history shapes optimal patterns of senescence: implications from individuals to societies. *The American Naturalist* 191(6), 756–766.

Lenz, M. and Runko, S. (1993) Long-term impact of orphaning on field colonies of *Coptotermes lacteus* (Froggatt) (Isoptera: Rhinotermitidae). *Insectes Sociaux* 40, 439–456.

Lenz, M., Barrett, R.A. and Miller, L.R. (1988) Mechanisms of colony re-establishment after orphaning in *Coptotermes lacteus* (Froggatt) (Isoptera; Rhinotermitidae). *Sociobiology* 14, 245–268.

Leong, K.L.H., Tamashiro, M., Yates, J. and Su, N.-Y. (1983) Microenvironmental factors regulating the flight of *Coptotermes formosanus* Shiraki in Hawaii (Isoptera: Rhinotermitidae). *Proceedings of the Hawaiian Entomological Society* 24, 287–291.

Lepage, M. (1991) Predation on the termite *Macrotermes michaelseni* reproductives and post settlement survival in the field (Isoptera: Macrotermitinae). *Sociobiology* 18, 153–166.
Lepage, M. and Darlington, J.P. (2000) Population dynamics of termites. In: Abe, T., Bignell, D.E. and Higashi M. (eds) *Termites: Evolution, Sociality, Symbioses, Ecology*. Springer, Dordrecht, Netherlands, pp. 333–361.
Li, H.-F., Yang, R.-L. and Su, N.-Y. (2010) Interspecific competition and territory defense mechanisms of *Coptotermes formosanus* and *Coptotermes gestroi* (Isoptera: Rhinotermitidae). *Environmental Entomology* 39, 1601–1607.
Liu, Y., Henderson, G., Mao, L. and Laine, R.A. (2005a) Seasonal variation of juvenile hormone titers of the Formosan subterranean termite, *Coptotermes formosanus* (Rhinotermitidae). *Environmental Entomology* 34, 557–562.
Liu, Y., Henderson, G., Mao, L. and Laine, R.A. (2005b) Effects of temperature and nutrition on juvenile hormone titers of *Coptotermes formosanus* (Isoptera: Rhinotermitidae). *Annals of the Entomological Society of America* 98, 732–737.
Lo, N., Tokuda, G., Watanabe, H., Rose, H., Slaytor, M., *et al.* (2000) Evidence from multiple gene sequences indicates that termites evolved from wood-feeding cockroaches. *Current Biology* 10, 801–804.
Lo, N., Bandi, C., Watanabe, H., Nalepa, C. and Beninati, T. (2003) Evidence for cocladogenesis between diverse dictyopteran lineages and their intracellular endosymbionts. *Molecular Biology and Evolution* 20, 907–913.
Lo, N., Kitade, O., Miura, T., Constantino, R. and Matsumoto, T. (2004) Molecular phylogeny of the Rhinotermitidae. *Insectes Sociaux* 51, 365–371.
Lo, N., Engel, M.S., Cameron, S., Nalepa, C.A., Tokuda, G. *et al.* (2007) Save Isoptera: a comment on Inward *et al. Biology Letters* 3, 562.
Lo, N., Li, B. and Ujvari, B. (2012) DNA methylation in the termite *Coptotermes lacteus*. *Insectes Sociaux* 59, 257–261.
Lo, N., Simpson, S.J. and Sword, G.A. (2018) Epigenetics and developmental plasticity in orthopteroid insects. *Current Opinion in Insect Science* 25, 25–34.
Lüscher, M. (1972) Environmental control of juvenile hormone (JH) secretion and caste differentiation in termites. *General and Comparative Endocrinology* 3(Supplement), 509–514.
Lüscher, M. and Springhetti, A. (1960) Untersuchungen über die Bedeutung der Corpora allata für die Differenzierung der Kasten bei der Termite *Kalotermes flavicollis* F. *Journal of Insect Physiology* 5, 190–212.
Maekawa, K., Ishitani, K., Gotoh, H., Cornette, R. and Miura, T. (2010) Juvenile Hormone titre and vitellogenin gene expression related to ovarian development in primary reproductives compared with nymphs and nymphoid reproductives of the termite *Reticulitermes speratus*. *Physiological Entomology* 35, 52–58.
Maekawa, K., Nakamura, S. and Watanabe, D. (2012) Termite soldier differentiation in incipient colonies is related to parental proctodeal trophallactic behavior. *Zoological Science* 29, 213–217.
Maekawa, K., Hayashi, Y., Lee, T. and Lo, N. (2014) Presoldier differentiation of Australian termite species induced by juvenile hormone analogues. *Austral Entomology* 53, 138–143.
Maekawa, K., Hayashi, Y. and Lo, N. (2022) Termite sociogenomics: evolution and regulation of caste-specific expressed genes. *Current Opinion in Insect Science* 50, 100880.
Mao, L. and Henderson, G. (2007) A case for a free-running circannual rhythm in soldier developmental time of Formosan subterranean termites. *Insectes Sociaux* 54, 388–392.
Mao, L. and Henderson, G. (2010) Group size effect on worker juvenile hormone titers and soldier differentiation in Formosan subterranean termite. *Journal of Insect Physiology* 56, 725–730.
Mao, L., Henderson, G., Liu, Y. and Laine, R.A. (2005) Formosan subterranean termite (Isoptera: Rhinotermitidae) soldiers regulate juvenile hormone levels and caste differentiation in workers. *Annals of the Entomological Society of America* 98, 340–345.
Mao, L., Henderson, G. and Dong, S.L. (2010) Studies of precocene I, precocene II and biogenic amines on Formosan subterranean termite (Isoptera: Rhinotermitidae) presoldier/soldier formation. *Journal of Entomological Science* 45, 27–34.
Masuoka, Y. and Maekawa, K. (2016a) Gene expression changes in the tyrosine metabolic pathway regulate caste-specific cuticular pigmentation of termites. *Insect Biochemistry and Molecular Biology* 74, 21–31.
Masuoka, Y. and Maekawa, K. (2016b) Ecdysone signaling regulates soldier-specific cuticular pigmentation in the termite *Zootermopsis nevadensis*. *FEBS letters* 590, 1694–1703.
Matsuura, K. (2017) Evolution of the asexual queen succession system and its underlying mechanisms in termites. *Journal of Experimental Biology* 220, 63–72.

Matsuura, K. (2020) Genomic imprinting and evolution of insect societies. *Population Ecology* 62, 38–52.

Matsuura, K. and Kobayashi, N. (2010) Termite queens adjust egg size according to colony development. *Behavioral Ecology* 21, 1018–1023.

Matsuura, K. and Nishida, T. (2002) Mechanism, induction factors, and adaptive significance of dealation in the subterranean termite *Reticulitermes speratus* (Isoptera, Rhinotermitidae). *Insectes Sociaux* 49, 241–244.

Matsuura, K. and Yamamoto, Y. (2011) Workers do not mediate the inhibitory power of queens in a termite, *Reticulitermes speratus* (Isoptera, Rhinotermitidae). *Insectes Sociaux* 58, 513–518.

Matsuura, K., Vargo, E.L., Kawatsu, K., Labadie, P.E., Nakano, H. *et al.* (2009) Queen succession through asexual reproduction in termites. *Science* 323, 1687

Matsuura, K., Himuro, C., Yokoi, T., Yamamoto, Y., Vargo, E.L. *et al.* (2010) Identification of a pheromone regulating caste differentiation in termites. *Proceedings of the National Academy of Sciences of the USA* 107, 12963–12968

May, E. (1941) The behavior of the intestinal protozoa of termites at the time of the last ecdysis. *Transactions of the American Microscopical Society* 60, 281–292.

Messenger, M.T. and Su, N.-Y. (2005a) Colony characteristics and seasonal activity of the Formosan subterranean termite (Isoptera: Rhinotermitidae) in Louis Armstrong park, New Orleans, Louisiana. *Journal of Entomological Science* 40, 268–279.

Messenger, M.T. and Su, N.Y. (2005b) Agonistic behavior between colonies of the Formosan subterranean termite (Isoptera: Rhinotermitidae) from Louis Armstrong Park, New Orleans, Louisiana. *Sociobiology* 45, 331–346.

Messenger, M.T., Su, N.Y., Husseneder, C. and Grace, J.K. (2005) Elimination and reinvasion studies with *Coptotermes formosanus* (Isoptera: Rhinotermitidae) in Louisiana. *Journal of Economic Entomology* 98, 916–929.

Michaud, C., Hervé, V., Dupont, S., Dubreuil, G., Bézier, A.M., Meunier, J., Brune, A. and Dedeine, F. (2020) Efficient but occasionally imperfect vertical transmission of gut mutualistic protists in a wood-feeding termite. *Molecular Ecology* 29(2), 308–324.

Miller, E.M. (1969) Caste differentiation in the lower termites. In: Krishna K. and Weesner F.M. (eds) *Biology of Termites. Vol. 1*. Academic Press, New York. pp. 283–310.

Miura, T. (2001) Morphogenesis and gene expression in the soldier-caste differentiation of termites. *Insectes Sociaux* 48, 216–223.

Miura, T. and Maekawa, K. (2020) The making of the defensive caste: Physiology, development, and evolution of the soldier differentiation in termites. *Evolution and Development* 22, 425–437.

Miura, T. and Scharf, M.E. (2010) Molecular basis underlying caste differentiation in termites. In: Bignell, D.E, Roisin Y. and Lo, N. (eds) *Biology of Termites: A Modern Synthesis*. Springer, Dordrecht, Netherlands, pp. 211–253.

Miura, T., Kamikouchi, A., Sawata, M., Takeuchi, H., Natori, S. *et al.* (1999) Soldier caste-specific gene expression in the mandibular glands of *Hodotermopsis japonica* (Isoptera: Termopsidae). *Proceedings of the National Academy of Sciences of the USA* 96, 13874–13879.

Miura, T., Koshikawa, S. and Matsumoto, T. (2003) Winged presoldiers induced by a juvenile hormone analog in *Zootermopsis nevadensis*: implications for plasticity and evolution of caste differentiation in termites. *Journal of Morphology* 257, 22–32.

Miyaguni, Y., Sugio, K. and Tsuji, K. (2013) Antennal cropping in the Asian dry-wood termite, *Neotermes koshunensis*. *Insectes Sociaux* 60, 223–229.

Mizumoto, N. and Bourguignon, T. (2023) Light alters activity but does not disturb tandem coordination of termite mating pairs. *Ecological Entomology* 48, 145–153.

Mizumoto, N. and Dobata, S. (2019) Adaptive switch to sexually dimorphic movements by partner-seeking termites. *Science Advances* 5, 6108.

Mizumoto, N., Rizo, A., Pratt, S.C. and Chouvenc, T. (2020) Termite males enhance mating encounters by changing speed according to density. *Journal of Animal Ecology* 89, 2542–2552.

Mizumoto, N., Lee, S.B., Valentini, G., Chouvenc, T. and Pratt, S.C. (2021) Coordination of movement via complementary interactions of leaders and followers in termite mating pairs. *Proceedings of the Royal Society B* 288, 20210998.

Morales-Ramos, J.A. and Rojas, M.G. (2003) Nutritional ecology of the Formosan subterranean termite (Isoptera: Rhinotermitidae): growth and survival of incipient colonies feeding on preferred wood species. *Journal of Economic Entomology* 96, 106–116.

Mullins, A. and Su, N.-Y. (2018) Parental nitrogen transfer and apparent absence of N_2 fixation during colony foundation in *Coptotermes formosanus* Shiraki. *Insects* 9, 37.

Mullins, A.J., Su, N.Y. and Owens, C. (2011) Reinvasion and colony expansion of *Coptotermes formosanus* (Isoptera: Rhinotermitidae) after areawide elimination. *Journal of Economic Entomology* 104, 1687–1697.

Mullins, A.J., Messenger, M.T., Hochmair, H.H., Tonini, F., Su, N.Y. *et al.* (2015) Dispersal flights of the Formosan subterranean termite (Isoptera: Rhinotermitidae). *Journal of Economic Entomology* 108, 707–719.

Mullins, A., Chouvenc, T. and Su, N.-Y. (2021) Soil organic matter is essential for colony growth in subterranean termites. *Scientific Reports* 11, e21252.

Myles, T.G. (1988) Dealation in termites (Isoptera). *Sociobiology* 14, 61–88.

Myles, T.G. (1999) Review of secondary reproduction in termites (Insecta: Isoptera) with comments on its role in termite ecology and social evolution. *Sociobiology* 33, 1–43.

Nakajima, S., Shimizu, K. and Nakajima, Y. (1963) Analytical studies on the vitality of colonies of the Formosan termite, *Coptotermes formosanus* Shiraki. *Bulletin of the Faculty of Agriculture, University of Miyazaki* 8, 106–110.

Nalepa, C.A. (2010a) Altricial development in wood-feeding cockroaches: the key antecedent of termite eusociality. In: Bignell, D.E, Roisin Y. and Lo, N. (eds) *Biology of Termites: a Modern Synthesis*. Springer, Dordrecht, Netherlands, pp. 69–95.

Nalepa, C.A. (2010b) Altricial development in subsocial cockroach ancestors: foundation for the evolution of phenotypic plasticity in termites. *Evolution and Development* 12, 95–105.

Nalepa, C.A. (2011) Body size and termite evolution. *Evolutionary Biology* 38, 243–257.

Nalepa, C.A. (2015) Origin of termite eusociality: trophallaxis integrates the social, nutritional, and microbial environments. *Ecological Entomology* 40, 323–335.

Nalepa, C.A. (2017) What kills the hindgut flagellates of lower termites during the host molting cycle? *Microorganisms* 5, e82.

Nalepa, C.A., Evans, T.A. and Lenz, M. (2011) Antennal cropping during colony foundation in termites. *ZooKeys* 148, 185–196

Neoh, K.B. and Lee, C.Y. (2009) Flight activity and flight phenology of the Asian subterranean termite, *Coptotermes gestroi* (Blattodea: Rhinotermitidae). *Sociobiology* 54, 521–530.

Noirot, C. and Noirot-Timothée, C. (1969) The Digestive System. In: Krishna, K. and Weesner, F.M. (eds) *The Biology of Termites, Vol. 1*. Academic Press, New York, pp. 49–80.

Nutting, W.L. (1969) Flight and colony foundation. In: Krishna, K. and Weesner, F.M. (eds) *The Biology of Termites, Vol. 1*. Academic Press, New York, pp.233–282.

Nutting, W.L. (1979) Termite flight periods: strategies for predator avoidance? *Sociobiology* 4, 141–151.

Oguchi, K., Shimoji, H., Hayashi, Y. and Miura, T. (2016) Reproductive organ development along the caste differentiation pathways in the dampwood termite *Hodotermopsis sjostedti*. *Insectes Sociaux* 63, 519–529.

Oguchi, K., Maekawa, K. and Miura, T. (2021) Regulatory mechanisms underlying the differentiation of neotenic reproductives in termites: Partial release from arrested development. *Frontiers in Ecology and Evolution* 9, 635552.

Oguchi, K., Koshikawa, S. and Miura, T. (2022) Hormone-related genes heterochronically and modularly regulate neotenic differentiation in termites. *Developmental Biology* 485, 70–79.

Okot-Kotber, B.M., Prestwich, G.D., Strambi, A. and Strambi, C. (1993) Changes in morphogenetic hormone titers in isolated workers of the termite *Reticulitermes flavipes* (Kollar). *General and Comparative Endocrinology* 90, 290–295.

Oshima, M. (1919) Formosan termites and methods of preventing their damage. *Philippine Journal of Science* 15, 319–383.

Osbrink, W.L., Woodson, W.D. and Lax, A.R. (1999) Population of Formosan subterranean termite, *Coptotermes formosanus* (Isoptera: Rhinotermitidae), established in living urban trees in New Orleans, Louisiana, USA. In: *Proceedings, 3rd International Conference on Urban Pests*. Graficke zavody hronov, Prague, Czech Republic, pp. 19–22.

Osbrink, W.L.A., Cornelius, M.L. and Showler, A.T. (2016) Bionomics and formation of "Bonsai" colonies with long-term rearing of *Coptotermes formosanus* (Isoptera: Rhinotermitidae). *Journal of Economic Entomology* 109, 770–778.

Oster G.F. and Wilson, E.O. (1978) *Caste and Ecology in the Social Insects*. Princeton University Press, New Jersey.

Park, Y.I. and Raina, A. (2003) Factors regulating caste differentiation in the Formosan subterranean termite with emphasis on soldier formation. *Sociobiology* 41, 49–60.

Park, Y.I. and Raina, A.K. (2004) Juvenile hormone III titers and regulation of soldier caste in *Coptotermes formosanus* (Isoptera: Rhinotermitidae). *Journal of Insect Physiology* 50, 561–566.

Park, Y.I. and Raina, A.K. (2005) Regulation of juvenile hormone titers by soldiers in the Formosan subterranean termite, *Coptotermes formosanus*. *Journal of Insect Physiology* 51, 385–391.

Park, Y.-I., Bland, J.M. and Raina, A.K. (2004) Factors affecting post-flight behavior in primary reproductives of the Formosan subterranean termite, *Coptotermes formosanus* (Isoptera: Rhinotermitidae). *Journal of Insect Physiology* 50, 539–546.

Patel, J.S., Tong, R.L., Chouvenc, T. and Su, N.Y. (2019a) Comparison of temperature-dependent survivorship and wood-consumption rate among two invasive subterranean termite species (Blattodea: Rhinotermitidae: *Coptotermes*) and their hybrids. *Journal of Economic Entomology* 112, 300–304.

Patel, J.S., Chouvenc, T. and Su, N.Y. (2019b) Temperature preference of two invasive subterranean termite species and their hybrids (Blattodea: Rhinotermitidae: *Coptotermes*). *Journal of Economic Entomology* 112, 2888–2893.

Patel, J.S., Lee, S.B., Chouvenc, T. and Su, N.Y. (2020) Inferring termite colony size using wood consumption in subterranean termites (Blattodea: Rhinotermitidae) in laboratory-rearing conditions. *Journal of Economic Entomology* 113(2), 905–910.

Puckett, R.T., Espinoza, E.M. and Gold, R.E. (2014) Alate trap-based assessment of Formosan subterranean termite (Isoptera: Rhinotermitidae) dispersal flight phenology associated with an urbanized barrier island ecosystem. *Environmental Entomology* 43, 868–876.

Quennedey, A. (1984) Morphology and ultrastructure of termite defense glands. In: Hermann, H.R. (ed.) *Defensive Mechanisms in Social Insects*. Praeger, New York, pp. 151–200.

Raina, A.K., Park, Y.I. and Hruska, Z. (2003a) Ecdysone agonist halofenozide affects corpora allata and reproductive physiology of the Formosan subterranean termite, *Coptotermes formosanus*. *Journal of Insect Physiology* 49, 677–683.

Raina, A., Park, Y.-I. and Florane, C. (2003b) Behavior and reproductive biology of the primary reproductives of the Formosan subterranean termite (Isoptera: Rhinotermitidae). *Sociobiology* 41, 37–48.

Raina, A.K., Bland, J.M., Dickens, J.C., Park, Y.-I. and Hollister, B. (2003c) Premating behavior of dealates of the Formosan subterranean termite and evidence for the presence of a contact sex pheromone. *Journal of Insect Behavior* 16, 233–245.

Raina, A., Osbrink, W. and Park, Y.I. (2004) Nymphs of the Formosan subterranean termite (Isoptera: Rhinotermitidae): aspects of formation and transformation. *Annals of the Entomological Society of America* 97, 757–764.

Raina, A., Murphy, C., Florane, C., Williams, K., Park, Y.I. *et al.* (2007) Structure of spermatheca, sperm dynamics, and associated bacteria in Formosan subterranean termite (Isoptera: Rhinotermitidae). *Annals of the Entomological Society of America* 100, 418–424.

Raina, A., Park, Y.I. and Gelman, D. (2008) Molting in workers of the Formosan subterranean termite *Coptotermes formosanus*. *Journal of Insect Physiology* 54, 155–161.

Revely, L., Sumner, S. and Eggleton, P. (2021) The plasticity and developmental potential of termites. *Frontiers in Ecology and Evolution* 9, 552624.

Roisin, Y. (2000) Diversity and evolution of caste patterns. In: Abe, T., Bignell, D.E. and Higashi M. (eds) *Termites: Evolution, Sociality, Symbioses, Ecology*. Springer, Dordrecht, Netherlands, pp. 95–119.

Roisin, Y. (2016) What makes the cost of brood care important for the evolution of termite sociality? Its insignificance. *Ecological Entomology* 41, 31–33.

Roisin, Y. and Korb, J. (2010) Social organisation and the status of workers in termites. In: Bignell, D.E, Roisin Y. and Lo, N. (eds) *Biology of Termites: A Modern Synthesis*. Springer, Dordrecht, Netherlands, pp. 133–164.

Roisin, Y. and Lenz, M. (1999) Caste developmental pathways in colonies of *Coptotermes lacteus* (Froggatt) headed by primary reproductives (Isoptera, Rhinotermitidae). *Insectes Sociaux* 46, 273–280.

Roisin, Y. and Lenz, M. (2002) Origin of male-biased sex allocation in orphaned colonies of the termite, *Coptotermes lacteus*. *Behavioral Ecology and Sociobiology* 51, 472–479.

Rojas, M.G. and Morales-Ramos, J.A. (2004) Disruption of reproductive activity of *Coptotermes formosanus* (Isoptera: Rhinotermitidae) primary reproductives by three chitin synthesis inhibitors. *Journal of Economic Entomology* 97, 2015–2020.

Rosengaus, R.B. and Traniello, J.F. (1993) Disease risk as a cost of outbreeding in the termite *Zootermopsis angusticollis*. *Proceedings of the National Academy of Sciences of the USA* 90, 6641–6645.

Roy-Noël, J. (1968) Études biométrique et morphologique du couvain de *Coptotermes intermedius* Silvestri (Isoptère Rhinotermitidae). *Insectes Sociaux* 15, 389–394.

Scharf, M.E., Wu-Scharf, D., Pittendrigh, B.R. and Bennett, G.W. (2003) Caste-and development-associated gene expression in a lower termite. *Genome Biology* 4, R62.

Scharf, M.E., Wu-Scharf, D., Zhou, X., Pittendrigh, B.R. and Bennett, G.W. (2005) Gene expression profiles among immature and adult reproductive castes of the termite *Reticulitermes flavipes*. *Insect Molecular Biology* 14, 31–44.

Scheffrahn, R.H. and Crowe, W. (2011) Ship-borne termite (Isoptera) border interceptions in Australia and onboard infestations in Florida, 1986–2009. *Florida Entomologist* 94, 57–63.

Scheffrahn, R.H., Carrijo, T.F., Křeček, J., Su, N.Y., Szalanski, A.L. *et al.* (2015) A single endemic and three exotic species of the termite genus *Coptotermes* (Isoptera, Rhinotermitidae) in the New World. *Arthropod Systematics and Phylogeny* 73, 333–348.

Scheffrahn, R.H., Bahder, B.W. and Lu, T. (2020) *Coptotermes formosanus* (Blattodea: Rhinotermitidae) established in Israel and world distribution of a major termite pest. *Check List* 16, 1537–1543.

Shellman-Reeve, J.S. (1990) Dynamics of biparental care in the dampwood termite, *Zootermopsis nevadensis* (Hagen): response to nitrogen availability. *Behavioral Ecology and Sociobiology* 26, 389–397.

Shimada, K. and Maekawa, K. (2010) Changes in endogenous cellulase gene expression levels and reproductive characteristics of primary and secondary reproductives with colony development of the termite *Reticulitermes speratus* (Isoptera: Rhinotermitidae). *Journal of Insect Physiology* 56, 1118–1124.

Shimada, K., Lo, N., Kitade, O., Wakui, A. and Maekawa, K. (2013) Cellulolytic protist numbers rise and fall dramatically in termite queens and kings during colony foundation. *Eukaryotic Cell* 12, 545–550.

Shimizu, K. (1962) Analytical studies on the vitality of colonies of the Formosan subterranean termite, *Coptotermes formosanus* Shiraki. I. Analysis of the strength of vitality. *Bulletin of the Faculty of Agriculture, University of Miyazaki* 8, 106–110.

Silva, I.S., Lucena, E.F., Moura, F.M. and Vasconcellos, A. (2021) Termite flights seasonally promote nutrient pulses in the Caatinga dry forest in northeastern Brazil. *Applied Soil Ecology* 166, 104066.

Simms, D. and Husseneder, C. (2009) Assigning individual alates of the Formosan subterranean termite (Isoptera: Rhinotermitidae) to their colonies of origin within the context of an area-wide management program. *Sociobiology* 53, 631–650.

Sinotte, V.M., Conlon, B.H., Seibel, E., Schwitalla, J.W., de Beer, Z.W. *et al.* (2021) Female-biased sex allocation and lack of inbreeding avoidance in *Cubitermes* termites. *Ecology and Evolution* 11, 5598–5605.

Soltani-Mazouni, N. and Bordereau, C. (1987) Changes in the cuticle, ovaries and colleterial glands during the pseudergate and neotenic molt in *Kalotermes flavicollis* (Fabr.) (Isoptera: Kalotermitidae). *International Journal of Insect Morphology and Embryology* 16, 221–235.

Song, D., Hu, X.P. and Su, N.Y. (2006) Survivorship, cannibalism, body weight loss, necrophagy, and entombment in laboratory groups of the Formosan subterranean termite, *Coptotermes formosanus* under starvation (Isoptera: Rhinotermitidae). *Sociobiology* 47, 27–39.

Springhetti, A. (1973) Group effects in the differentiation of the soldiers of *Kalotermes flavicollis* Fabr. (Isoptera). *Insectes Sociaux* 20, 333–342.

Su, N.Y. (1994) Field evaluation of a hexaflumuron bait for population suppression of subterranean termites (Isoptera: Rhinotermitidae). *Journal of Economic Entomology* 87, 389–397.

Su, N.-Y. (2013) How to Become a Successful Invader. *Florida Entomologist* 96, 765–769.

Su, N.Y. (2019) Development of baits for population management of subterranean termites. *Annual Review of Entomology* 64, 115–130.

Su, N.Y. and La Fage, J.P. (1986) Effects of starvation on survival and maintenance of soldier proportion in laboratory groups of the Formosan subterranean termite, *Coptotermes formosanus* (Isoptera: Rhinotermitidae). *Annals of the Entomological Society of America* 79, 312–316.

Su, N.Y. and Monteagudo, E.J. (2017) Hyperecdysonism in the Formosan subterranean termite and eastern subterranean termite (Isoptera: Rhinotermitidae). *Journal of Economic Entomology* 110, 1736–1739.

Su, N.Y. and Scheffrahn, R.H. (1987) Alate production of a field colony of the Formosan subterranean termite (Isoptera: Rhinotermitidae). *Sociobiology* 13, 167–172.

Su, N.-Y., and Scheffrahn R.H. (1988a) Intra- and interspecific competition of the Formosan and eastern subterranean termite: Evidence from field observations (Isoptera: Rhinotermitidae). *Sociobiology* 14, 157–164.

Su, N.-Y. and Scheffrahn, R.H. (1988b) Foraging population and territory of the Formosan subterranean termite (Isoptera: Rhinotermitidae) in an urban environment. *Sociobiology* 14, 353–360.

Su, N.Y. and Scheffrahn, R.H. (1993) Laboratory evaluation of two chitin synthesis inhibitors, hexaflumuron and diflubenzuron, as bait toxicants against Formosan and eastern subterranean termites (Isoptera: Rhinotermitidae). *Journal of Economic Entomology* 86, 1453–1457.

Su, N.-Y., Osbrink, W., Kakkar, G., Mullins, A. and Chouvenc, T. (2017) Foraging distance and population size of juvenile colonies of the Formosan subterranean termite (Isoptera: Rhinotermitidae) in laboratory extended arenas. *Journal of Economic Entomology* 110, 1728–1735.

Sugime, Y., Ogawa, K., Watanabe, D., Shimoji, H., Koshikawa, S. *et al.* (2015) Expansion of presoldier cuticle contributes to head elongation during soldier differentiation in termites. *The Science of Nature* 102, e71.

Sugio, K. and Miyaguni Y. (2019) Characteristics of dispersal flight of the Formosan subterranean termite, *Coptotermes formosanus* Shiraki (Isoptera: Rhinotermitidae) in Okinawa. *Japanese Journal of Environmental Entomology and Zoology* 30, 63–69.

Sun, J.Z., Lockwood, M.E. and Etheridge, J.L. (2006) Evaluation of the preferences of dealates of *Coptotermes formosanus* (Isoptera: Rhinotermitidae) among landscape mulches. *Sociobiology* 48, 85–99.

Sun, Q., Haynes, K.F., Hampton, J.D. and Zhou, X. (2017) Sex-specific inhibition and stimulation of worker-reproductive transition in a termite. *The Science of Nature* 104, e79.

Sun, Q., Hampton, J.D., Merchant, A., Haynes, K.F. and Zhou, X. (2020) Cooperative policing behaviour regulates reproductive division of labour in a termite. *Proceedings of the Royal Society B* 287, 20200780.

Suzuki, R., Yaguchi, H. and Maekawa, K. (2019) Histone modifying genes are involved in the molting period during soldier differentiation in *Zootermopsis nevadensis*. *Journal of Insect Physiology* 117, 103892.

Takata, M., Nagai, S., Inagaki, T., Ohkubo, Y., Tasaki, E. *et al.* (2023) Heritable effects on caste determination and colony-level sex allocation in termites under field conditions. *Iscience* 26, e106207

Tarver, M.R., Zhou, X. and Scharf, M.E. (2010) Socio-environmental and endocrine influences on developmental and caste-regulatory gene expression in the eusocial termite *Reticulitermes flavipes*. *BMC Molecular Biology* 11, e28.

Tarver, M.R., Schmelz, E.A. and Scharf, M.E. (2011) Soldier caste influences on candidate primer pheromone levels and juvenile hormone-dependent caste differentiation in workers of the termite *Reticulitermes flavipes*. *Journal of Insect Physiology* 57, 771–777.

Tarver, M.R., Florane, C.B., Zhang, D., Grimm, C., Lax, A.R. *et al.* (2012) Methoprene and temperature effects on caste differentiation and protein composition in the Formosan subterranean termite, *Coptotermes formosanus*. *Journal of Insect Science* 12, e18.

Thompson, G.J., Lenz, M., Crozier, R.H. and Crespi, B.J. (2007) Molecular-genetic analyses of dispersal and breeding behaviour in the Australian termite *Coptotermes lacteus*: evidence for non-random mating in a swarm-dispersal mating system. *Australian Journal of Zoology* 55, 219–227.

Thorne, B.L. (1996) Termite terminology. *Sociobiology* 28, 253–263.

Thorne, B.L., Traniello, J.F.A., Adams, E.S. and Bulmer, M. (1999) Reproductive dynamics and colony structure of subterranean termites of the genus *Reticulitermes* (Isoptera Rhinotermitidae): a review of the evidence from behavioral, ecological, and genetic studies. *Ethology Ecology and Evolution* 11, 149–169.

Tong, R.L., Aguilera-Olivares, D., Chouvenc, T. and Su, N.-Y. (2021) Nitrogen content of the exuviae of *Coptotermes gestroi* (Wasmann) (Blattodea: Rhinotermitidae). *Heliyon* 7, e06697.

Tong, R.L., Patel, J.S., Gordon, J.M., Lee, S.B., Chouvenc, T. *et al.* (2023) Exuviae recycling can enhance queen oviposition and colony growth in subterranean termites (Blattodea: Rhinotermitidae: *Coptotermes*). *Environmental Entomology* 52, 254–258.

Tuma, J., Eggleton, P. and Fayle, T.M. (2020) Ant-termite interactions: an important but under-explored ecological linkage. *Biological Reviews* 95, 555–572.

Uchima, S.Y. and Grace, J.K. (2009) Interspecific agonism and foraging competition between *Coptotermes formosanus* and *Coptotermes gestroi* (Blattodea: Rhinotermitidae). *Sociobiology* 54, 765–776.

Vargo, E.L. (2019) Diversity of termite breeding systems. *Insects* 10, 52.

Vargo, E.L. and Husseneder, C. (2009) Biology of subterranean termites: insights from molecular studies of *Reticulitermes* and *Coptotermes*. *Annual Review of Entomology* 54, 379–403.

Vargo, E.L., Husseneder, C. and Grace, J.K. (2003) Colony and population genetic structure of the Formosan subterranean termite, *Coptotermes formosanus*, in Japan. *Molecular Ecology* 12, 2599–2608.

Vargo, E.L., Labadie, P.E. and Matsuura, K. (2012) Asexual queen succession in the subterranean termite *Reticulitermes virginicus*. *Proceedings of the Royal Society B: Biological Sciences* 279, 813–819.

Velenovsky, J.F., Gile, G.H., Su, N.Y. and Chouvenc, T. (2021) Dynamic protozoan abundance of *Coptotermes* kings and queens during the transition from biparental to alloparental care. *Insectes Sociaux* 68, 33–40.

Velenovsky, J.F., De Martini, F., Hileman, J., Gordon J.M., Su, N.-Y. *et al.* (2023) Vertical transmission of cellulolytic protists in termites is imperfect, but sufficient, due to biparental transmission. *Symbiosis* (in press)

Waller, D.A. and La Fage, J.P. (1987) Food quality and foraging response by the subterranean termite *Coptotermes formosanus* Shiraki (Isoptera: Rhinotermitidae). *Bulletin of Entomological Research* 77, 417–424.

Waller, D.A. and La Fage, J.P. (1988) Size variation in *Coptotermes formosanus* Shiraki (Rhinotermitidae): Consequences of host use. *American Midland Naturalist* 119, 436–440.

Wang, C., Powell, J.E. and Liu, Y. (2002a) A literature review of the biology and ecology of *Coptotermes formosanus* (Isoptera: Rhinotermitidae) in China. *Sociobiology* 40, 343–364.

Wang, C., Powell, J.E. and O'Connor, B.M. (2002b) Mites and nematodes associated with three subterranean termite species (Isoptera: Rhinotermitidae). *Florida Entomologist* 85, 499–506.

Wang, M., Buček, A., Šobotník, J., Sillam-Dussès, D., Evans, T.A. *et al.* (2019) Historical biogeography of the termite clade Rhinotermitinae (Blattodea: Isoptera). *Molecular Phylogenetics and Evolution* 132, 100–104.

Wang, M., Hellemans, S., Šobotník, J., Arora, J., Buček, A. *et al.* (2022) Phylogeny, biogeography and classification of Teletisoptera (Blattaria: Isoptera). *Systematic Entomology* 47, 581–590.

Watanabe, D., Gotoh, H., Miura, T. and Maekawa, K. (2011) Soldier presence suppresses presoldier differentiation through a rapid decrease of JH in the termite *Reticulitermes speratus*. *Journal of Insect Physiology* 57, 791–795.

Watanabe, D., Gotoh, H., Miura, T. and Maekawa, K. (2014) Social interactions affecting caste development through physiological actions in termites. *Frontiers in Physiology* 5, e127.

Watson, J.A.L., Okot-Kotber, B.M., Noirot, C. (1985) *Caste Differentiation in Social Insects*. Pergamon, Oxford.

Weesner, F.M. (1970) Termites of the Nearctic region. In: Krishna K. and Weesner F.M. (eds) *The Biology of Termites, Vol. 2*. Academic Press, New York, pp.477–522.

Weil, T., Rehli, M. and Korb, J. (2007) Molecular basis for the reproductive division of labour in a lower termite. *BMC Genomics* 8, e198.

Weil, T., Korb, J. and Rehli, M. (2009) Comparison of queen-specific gene expression in related lower termite species. *Molecular Biology and Evolution* 26, 1841–1850.

Wilson, E.O. (1971) *The Insect Societies*. Harvard University Press, Cambridge, Massachusetts.

Wood, T.G., 1988. Termites and the soil environment. *Biology and Fertility of Soils* 6, 228–236.

Xing, L., Chouvenc, T. and Su, N.-Y. (2013) Molting process in the Formosan subterranean termite (Isoptera: Rhinotermitidae). *Annals of the Entomological Society of America* 106, 619–625.

Yaguchi, H., Inoue, T., Sasaki, K. and Maekawa, K. (2016) Dopamine regulates termite soldier differentiation through trophallactic behaviours. *Royal Society Open Science* 3, 150574.

Yaguchi, H., Suzuki, R., Matsunami, M., Shigenobu, S. and Maekawa, K. (2019) Transcriptomic changes during caste development through social interactions in the termite *Zootermopsis nevadensis*. *Ecology and Evolution* 9, 3446–3456.

Zhang, Z.Y., Ren, J., Chu, F., Guan, J.X., Yang, G.Y. *et al.* (2021) Biochemical, molecular, and morphological variations of flight muscles before and after dispersal flight in a eusocial termite, *Reticulitermes chinensis*. *Insect Science* 28, 77–92.

Zhou, X., Tarver, M.R., Bennett, G.W., Oi, F.M. and Scharf, M.E. (2006) Two hexamerin genes from the termite *Reticulitermes flavipes*: Sequence, expression, and proposed functions in caste regulation. *Gene* 376, 47–58.

5 Ecology and foraging behavior

Nan-Yao Su
Entomology and Nematology Department, Ft. Lauderdale Research and Education Center, University of Florida, nysu@ufl.edu

Abstract
A single colony of the Formosan subterranean termite, *Coptotermes formosanus* Shiraki, may contain millions of termites that forage up to 100 m in search of food. In highly populated areas, inter-colony competition is fierce, and colony boundaries are delimited by intercepted underground tunnels. The wood consumption rate of *C. formosanus* is among the highest of most termite species investigated. Because wood contains little nitrogen, an earlier study suggested that *C. formosanus* acquired nitrogen through nitrogen fixation by the hindgut symbionts. However, recent studies showed that *C. formosanus* depend more on dietary nitrogen than previously assumed. *Coptotermes formosanus* construct underground tunnel systems with a Global Away Vector via path integration, and its geometric pattern is adopted to intercept food in a clumped distribution. Once food is discovered and its location is georeferenced by path integration, *C. formosanus* construct new tunnels heading directly toward the food to optimize transportation efficiency. Recent studies with planar arenas showed that age polyethism of lower termites such as *C. formosanus* might be more extensive and complex than previously thought.

5.1 Introduction

Studies on subterranean termites' relationship to other organisms and to their physical environment provide vital information on their roles in the ecosystem as well as to better management. Subterranean termites inhabit nesting structures below ground, and ecological studies on field colonies are challenging. However, due to their economic importance, underground foraging galleries of the Formosan subterranean termite, *Coptotermes formosanus* Shiraki, were meticulously excavated as early as 1928 (Erhorn, 1934). Following the excavation studies, on-the-ground monitoring stations were developed to study *C. formosanus* colony activity (Tamashiro *et al.*, 1973), and a dye marker was used to establish the interconnection among monitoring stations (Lai *et al.*, 1983a). These methods provided tools to study field biology of *C. formosanus* colonies, including colony foraging territory, forager movement, and population estimates. Laboratory studies using orphan foraging groups are common in studies of various ecological parameters of *C. formosanus*; however, due to limits imposed by artificial conditions, i.e., exposure to open air in Petri dishes, etc., interpretations of such data need caution. To remedy the shortcomings of earlier laboratory protocols with Petri

 Biology and Management of the Formosan Subterranean Termite and Related Species (eds Nan-Yao Su and Chow-Yang Lee)
DOI: 10.1079/9781800621596.0005

dish studies, planar arenas have been used in recent years, and observations through these arenas, especially when used in conjunction with incipient colonies containing the entire colony castes, provided unprecedented insights into inter-colony and interspecies relationships of subterranean termites (Bernard *et al.*, 2017).

Of the many aspects of termite ecology, wood-feeding that led to the evolution of termite eusociality is a unique and important subject. Lower termites such as *C. formosanus* depend on cellulosic enzymes in symbionts to digest wood that produce carbohydrates. Because of the poor nitrogen contents of wood, studies have suggested that termites also rely on microbes in guts for nitrogen fixation (Benemann, 1973; Breznak *et al.*, 1973; Noda *et al.*, 2002). Recent studies suggested that *C. formosanus*, at least during the first year of colony founding, did not acquire nitrogen from symbionts and instead relied on dietary sources resulting from microbial decomposition activity (Mullins and Su, 2018; Mullins *et al.*, 2021).

Foraging behavior includes two primary phases, food searching and transportation. Food searching in one-piece nest species (kalotermitids) is carried out by founding alate pairs, and once a colony is established in a piece of wood, foraging is unnecessary. Most subterranean termites (rhinotermitids) tunnel through soil in search of food. They may occasionally appear on the soil surface to access food, but most of the foraging gallery remains underground, and studies on the mechanism by which these tunnels intercept food in or on soil are the key to understanding the search phase of subterranean termite foraging behavior (Robson *et al.*, 1995). Recent studies with planar arenas revealed that subterranean termites are capable of geolocating their positions in soil by using path integration (Bardunias and Su, 2009a). Foods are intercepted by tunnels following a Global Away Vector (GAV) and transported via the shortest path by constructing a direct tunnel from the central nest to food (Campora and Grace, 2001; Michael *et al.*, 2023)

Division of labor during foraging was most extensively studied in higher termites, such as *Nasutitermes*, *Hospitalitermes*, and *Macrotermes* spp., that forage in the open (Traniello and Leuthold, 2000). In *Nasutitermes corniger* (Motschulsky), for example, soldiers functioned as scouts to discover food before workers were recruited to the site (Traniello, 1981), which presented a clear division of labor in foraging. Polyethism during foraging of lower termites, such as the subterranean termite, *Reticulitermes fukiensis* Light, is less distinct, as older workers performed most of the foraging activities with little involvement from younger workers (Crossland *et al.*, 1997). Recent studies with *C. formosanus*, however, provided more information on the foraging polyethism of this species. This chapter reviews studies on colony biology, nutrient ecology, and foraging behavior of *C. formosanus*.

5.2 Colony biology

5.2.1 Colony population size and foraging territory

The line drawing by an anonymous author of the Oahu Sugar Company in 1928 (Erhorn, 1934) remains one of the oldest records of the underground foraging gallery of the Formosan subterranean termite (Fig. 5.1A). The gallery of *C. formosanus* in Hawai'i extended more than 50 m and was probably longer, because the tunnel was not traced further when the depth reached > 3 m. King and Spink (1969) laboriously traced termite tunnels from cypress standing logs to reveal an elaborate underground tunneling system of *C. formosanus* in Louisiana (Fig. 5.1B). Their results in Louisiana agreed with that of Hawai'i in that the gallery system may extend at least 50 m and reach a depth of 3 m. These studies provided the first visual representations of the underground foraging galleries of *C. formosanus*.

Adopting the stake method described by Erhorn (1934), Tamashiro *et al.* (1973)

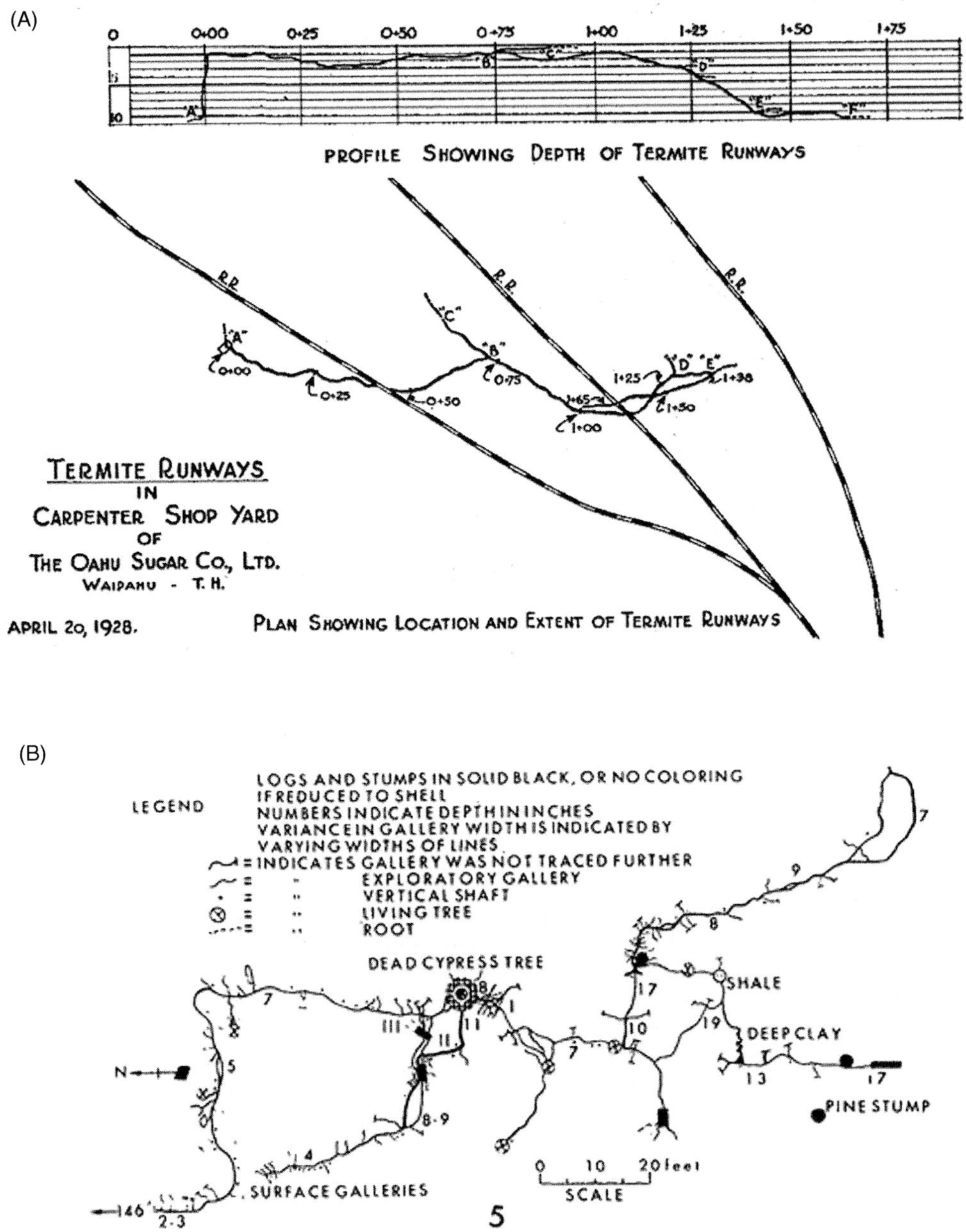

Fig. 5.1. Excavated underground foraging galleries of *C. formosanus* in (A) Hawai'i (reprinted with permission from Erhorn, 1934) and (B) Louisiana (reprinted with permission from King and Spink, 1969).

developed a monitoring system by first driving wooden stakes into the ground to detect *C. formosanus* presence in soil. Infested stakes were then replaced by on-the-ground monitoring stations. Because termites in the stations represent a subset of an underground termite population, the station serves as a window to study the belowground cryptic colonies. Following the development of the monitoring stations, Lai *et al.* (1983a) used a dye marker, Sudan Red 7B, to determine the interconnectivity among stations, which led to the definition of a colony as "a group of termites sharing interconnected foraging sites" (Su and Scheffrahn, 1998). Many have adopted and modified the monitoring station to advance our understanding of the field ecology of subterranean termites (Su and Scheffrahn, 1986; Grace *et al.*, 1989). In addition to Sudan Red 7B, other dyes such as Neutral Red and Nile Blue A were discovered (Su *et al.*, 1991) and have been widely used for studies of subterranean termites. In his study using Sudan Red 7B, Lai (1977) uncovered dyed *C. formosanus* works in a station 110 m away from the point of release, indicating a foraging distance that is more than double of those reported from excavation studies of ~50 m (Erhorn, 1934; King and Spink, 1969). Because termite tunnels were not traced further when they reached 3 m deep, foraging distances were invariably underestimated in these excavation studies. The dye marker studies provide more realistic data on colony foraging distance than excavation studies.

Colony foraging distances of *Coptotermes gestroi* (Wasmann) as determined by mark-recapture protocol were generally shorter than those of *C. formosanus*, i.e., 4–6 m in Thailand (Sornnuwat *et al.*, 1996), 2–57 m in Little Cayman Island (Su *et al.*, 2000), 24–133 m in Brazil (Arab *et al.*, 2005), 5–60 m in Malaysia (Majid and Ahmad, 2015; Umar and Majid, 2020), and 2–94 m in a natural forest in Taiwan (Chiu *et al.*, 2016).

In addition to the foraging distance and territory of a subterranean termite colony, the colony population size has been the interest of many who studied these cryptic insects. Howard *et al.* (1982), for example, extracted termites of *Reticulitermes flavipes* (Kollar) from logs within a 20 m radius with the assumption that they belonged to the same colony. The numbers of termites in logs were estimated by weight. Termites in soil within the colony territory were exhaustively trapped by pieces of cardboard, counted, and added to the number of termites extracted from logs to yield the total population size of a colony. Their results showed that a matured colony of *R. flavipes* may contain, on average ~250,000 (~50,000–350,000) termites, but these figures could have underestimated the actual population size because many live termites could have remained in soil or underground logs and escaped the cardboard sampling. Darlington (1984) sampled the population sizes of mound-building termites, *Macrotermes michaelseni* (Sjöstedt) and *Macrotermes subhyalinus* (Rambur) by excavations of live mounds and compared the results with mounds that had been fumigated. Her results showed that counts of live mounds yielded mean numbers of adult workers per mound (68,300 and 197,706 for *M. michaelseni* and M. *subhyalinus*, respectively) that were substantially lower than those of fumigated mounds (974,066 and 1,362,034 for *M. michaelseni* and *M. subhyalinus*, respectively). Because termites may escape the disturbance of excavations, direct samplings of live colonies of these mound-building termites underestimated the colony size by a factor of ~7–14, according to the results of Darlington (1984). Aside from being too labor intensive, a direct excavation of field colonies of subterranean termites renders the colonies useless for further studies.

Although initially intended for establishing interconnectivity among monitoring stations, Lai (1977) also recorded the number of dyed workers to estimate the worker population of a *C. formosanus* colony with a simple Lincoln index, and his results showed that a single colony of *C. formosanus* in Hawai'i contained ~1.3–1.6 million foragers. In the absence of another practical method to count the colony population size, the mark–release–recapture protocol has been widely used to estimate the population size of cryptic colonies of subterranean termites (Jones, 1990; Grace *et al.*, 1995; Myles, 1996). A triple mark–recapture protocol was later used in conjunction with the weighted model (Begon, 1979) to improve the estimated accuracy/precision of colony size (Su and Scheffrahn, 1988a; Haagsma and Rust, 1993; Sornnuwat *et al.*, 1996). Population estimates using the triple mark–recapture protocol showed that a single colony of *C. formosanus* may contain 1–6.9 million foragers in Florida (Su and Scheffrahn, 1988a; Su, 1994), which was compatible with those reported from Hawai'i (Lai, 1977). Notwithstanding the shorter foraging distance, population estimates of *C. gestroi* colonies, according to triple mark–recapture studies, are compatible with those of *C. formosanus*, i.e., 1.1–2.6 million in Little Cayman (Su *et al.*, 2000), 0.7–2.0 million in Brazil (Arab *et al.*, 2005), 0.1–4.2 million in Malaysia (Majid and Ahmad, 2015), and 0.1–0.2 million in Taiwan (Chiu *et al.*, 2016).

Mark–recapture estimates of subterranean termite colony populations have been criticized (Forschler and Townsend, 1996; Thorne *et al.*, 1996; Evans *et al.*, 1998). These studies pointed out possible violations of model assumptions, including the persistence of dye markers, mortality of marked individuals, and the equal capture probability of both marked and unmarked individuals (Begon, 1979). Marker persistence and dye-induced mortality are technical issues that can be resolved, i.e., the use of safer dyes such as Neutral Red, adopting different marking protocol (e.g., a lower concentration at longer marking time, or vice versa), or the discovery of new markers, but the most serious violation is probably the assumption of equal capture probability, which was often referred to as "equal mixing" because termites are "captured" from monitoring stations. Marked animals following a release in the wild usually do not meet the equal capture probability assumption, and most variants of capture–recapture models address the immigration/emigration (or birth/death) issues instead. A diffusion equilibrium model that accounts for the time-wise changes in the spatial distribution of marked termites was developed, and its accuracy was validated using laboratory arenas with known population sizes (Su and Lee, 2008), but this model has yet to be used in the field. Notwithstanding the debates over its accuracy or precision, the capture–recapture model remains the only practical method to estimate colony population sizes of subterranean termites, especially when the colonies have to be kept alive for other purposes. Among the majority of those studying *C. formosanus*, the current consensus of the population size of a mature colony is several million individuals.

5.2.2 Impact of climatic parameters on colonies

Temperature is the primary factor dictating the seasonality and distribution of an organism. In a laboratory test where termites were confined in a jar with moistened sand and a wooden block for 28 days, *C. formosanus* survived significantly better at 15–20°C than in other temperature ranges (Cao and Su, 2016). The distribution of *C. formosanus* can be best explained by its ability to survive at a suitable temperature. It should be noted, however, that temperatures used in the laboratory test are more representative of soil temperatures than ambient temperatures in the natural habitats of subterranean termites. Soil temperature is usually more stable than ambient temperature and is affected by soil cover, soil type, depth, moisture, and regions. The difference between soil and ambient temperature could be > 10°C (Zheng *et al.*, 1993) and probably even more in deeper soil. The 15–20°C range may appear to be narrow, but considering it represents soil temperature, this range may account for its current distribution of *C. formosanus* from Fort Lauderdale, Florida (26.1 °N, summer high ambient temperature: 32°C) to Wilmington, North Carolina (34.2° N, winter low ambient temperature: 5°C) in the U.S.

Another climatic parameter that affects *C. formosanus* distribution is humidity. In a laboratory study to test water loss tolerance and RH (relative humidity) preference among four termite species that inhabit distinct microhabitats, *C. formosanus* turned out to be the least tolerant to low RH (Zukowski and Su, 2017). None of the tested *C. formosanus* survived, even at the RH of 72.9% in 12 days. They survived only when RH was > 92.0% and avoided chambers with RH of < 70%. Even the dampwood termite, *Neotermes jouteli* (Banks), was more desiccation tolerant than *C. formosanus*. The inability of *C. formosanus* to survive in low RH environments may explain their absence in arid southwestern regions of the U.S., including western Texas, New Mexico, Arizona, and most of southern California except some isolated populations (Atkinson *et al.*, 1993; Tseng *et al.*, 2021, 2022) that were probably supported by landscape irrigation.

While the temperature- and RH-dependent survivability of *C. formosanus* account for its distribution, temperature preference and temperature-dependent activity levels such as wood consumption rate, tunneling, and food searching speed (Cao and Su, 2014, 2016) may account for its seasonality. Delaplane (1991) reported that the feeding rates of *C. formosanus* colonies confined in cypress trees in Louisiana were highest in summer (June–September) and correlated with daily maximum ambient temperature and day length. In New Orleans, Louisiana, Fei and Henderson (2004) reported that soil temperature at 10.2 cm from a building foundation ranged from 9°C to 24°C, significantly affected the seasonal foraging activity of *C. formosanus*. Cornelius and Osbrink (2011) showed that feeding rates of *C. formosanus* colonies (also in New Orleans) were significantly higher in June when soil temperature ranged from 23°C to 35°C, which agreed with the laboratory results of 20–35°C as reported by Cao and Su (2016). In subtropical south Florida, *C. formosanus* colonies were generally more active in June–October than in other months (Su, 1994; Su *et al.*, 1995), whereas in tropical Hawai'i (19.9° N) *C. formosanus* activity peaked in December–March (Hapukotuwa and Grace, 2012).

Alates of *C. formosanus* were known to fly on humid and still evenings of May and June in Hawai'i (Bess, 1970; Higa, 1981). In addition to high RH, Leong *et al.* (1983) reported that dispersal flights were initiated when light intensity dropped below 10.8 lumen/m^2 and wind velocity was below 3.7 km/h (30.8 m/second). Early on, Ikehara (1980) reported that *C. formosanus* alates flew 460 m at the wind velocity of 2.2 m/second. However, Higa and Tamashiro (1983) estimated the flight distance of ~100 m. Using orange fluorescent powder as a marker, Messenger and Mullins (2005) traced alates that flew across the Mississippi River in New Orleans with a flight distance up to 892 m when the wind velocity was below 1.0 m/second. Mullins *et al.* (2015) recorded the longest dispersal flight by capturing marked *C. formosanus* alates 1,300 m from the flight origin. This was the furthest extent of recapture light traps, and individuals may well have flown further but avoided capture. Data on the number of alates produced by a *C. formosanus* colony is scarce. Su and Scheffrahn (1987) reported a total count of over 68,000 *C. formosanus* alates from a large quantity of carton-encrusted wood of a colony during a swarming season, and even this was most likely an underestimate since there were probably other alates in the colony.

Coptotermes formosanus colonies not only prefer highly moist conditions, but they also appear to be able to survive in soil under prolonged inundation (Owens *et al.*, 2012). Following the floods of Hurricane Katrina in 2005, 80% of New Orleans was inundated for 1–3 weeks. After 2 weeks of inundation, when the water receded, *C. formosanus* returned to many in-ground monitoring stations (Cornelius *et al.*, 2007). By comparing the genotypic profiles, Owens *et al.* (2012) confirmed that, of the 17 *C. formosanus* colonies tested, 14 (82%) survived the inundation for up to 2 weeks. In a laboratory study, Cottone *et al.* (2015) reported that 60 workers of *C. formosanus* survived more than 3 weeks in an air pocket of 5.94 cm^3 at 21°C. Cottone and Su (2017) used artificial trees to demonstrate that *C. formosanus* workers built watertight galleries to trap air pockets to survive prolonged inundation. Presumably, *C. formosanus* may also be able to construct airtight foraging tubes extending above water levels for air access. In their native habitats of southern China, including deltas of Pearl River (Guangdong province) and Yangzhu River (Zhejiang province), flooding events are frequent, and *C. formosanus* is probably adapted to survive prolonged inundation events.

5.2.3 Inter-colony competition

Agonistic behavior between conspecific colonies is typical and often used to delimit colony boundaries (Nel, 1968; Thorne, 1982; Levings and Adams, 1984; Clément, 1986; Adams and Levings, 1987). *Coptotermes formosanus* colonies, however, displayed a varying degree of aggression. In newly introduced

Florida, for example, most *C. formosanus* colonies were not agonistic against each other (at least as of 1990), but in Hawai'i, where this species has been known since the late 19th century (and possibly with multiple introductions), some colonies displayed inter-colony aggression whereas other did not (Su and Haverty, 1991). Presumably, the agonistic interaction kept field colonies separated.

After establishing 54 in-ground monitoring stations in the 32-acre Armstrong Park, New Orleans, Lousiana, Messenger and Su (2005a) used mark–recapture protocol with three dye markers to map out colony boundaries of 13 *C. formosanus* colonies (Fig. 5.2). Their colony affinities were confirmed with genotype profiling (Husseneder *et al.*, 2003). *Coptotremes formosanus* were then collected from 11 of the 13 colonies, and individuals from each colony were divided into two groups for agonistic assay, one marked with Nile Blue A or Neutral Red and the other unmarked (Messenger and Su, 2005b). Two tests were carried out, a no-choice test in which termites from two colonies were placed together in a Petri dish, and a choice test in which termites of two colonies were placed in a planar arena through two chambers at the diagonally opposite end. Marked termites from each colony were paired with unmarked termites, totaling 121 possible pairs, including 11 intra-colony pairs. With the no-choice bioassay in which termites were forced to blend, agonistic encounters were observed in 46 of 110 inter-colony pairs. In the choice test where termites tunneled through moistened sand to encounter the other colony, they had a choice to invade the other colony's tunnels or to retreat. Under the choice test, agonistic encounters were observed in 88 out of 110 pairs. In 42 pairings, agonistic responses were absent in the no-choice Petri dish study but were present in the arena (choice) study.

The results suggested that field colonies of *C. formosanus* more aggressively maintain their colony boundaries than the no-choice Petri dish test might indicate. When tunnels from two hostile colonies intersected, a small skirmish would break out, followed by

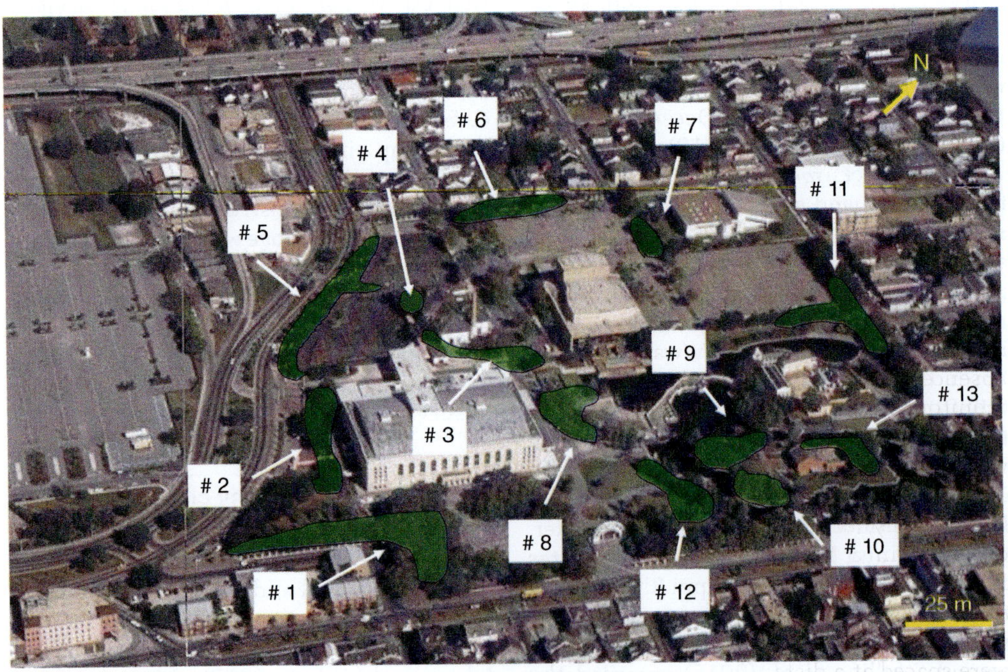

Fig. 5.2. Colony boundaries of 13 colonies of *C. formosanus* in the 32-acre Armstrong Park, New Orleans, Louisiana.

a larger number of workers and soldiers moving quickly to the point of encounter. They packed tightly in tunnels until they died (termed "suicide cramming"), after which the tunnels were sealed from behind (Messenger and Su, 2005b). In an area inhabited by many *C. formosanus* colonies, exploratory tunnels are probably constantly constructed from the main nest. When they encounter tunnels of hostile neighboring colonies, the intercepted tunnel tips would be closed after the "suicide cramming," resulting in the formation of stable boundaries among colonies. Another interesting observation from the study of Messenger and Su (2005b) was that seven pairs (Fig. 5.2, #1 vs #8, #1 vs #11, #2 vs #10, #3 vs #6, #4 vs #7, #4 vs #8, and #5 vs 8) did not display any aggression toward each other in both tests. These non-aggression pairs were typically far away from each other in the park or had other colony(s) between them and probably have never encountered each other in the field. This suggested that the agonistic behavior could be, in part, the result of learning in addition to inherited chemical cues associated with the colonies' genetic relatedness. Because many distant colonies are agonistic against each other when paired (for example, some colonies from Hawai'i vs Louisiana), the most likely explanation could be that closely related colonies, such as those neighboring each other in Armstrong Park (Messenger and Su, 2005b), may require frequent contact to develop an agonistic response toward each other.

Colony boundaries between the 13 colonies, as in Fig. 5.2, remained practically the same from 1998 to 2001, with temporary absence (and return) of termites in some monitoring stations. In fact, territories of *C. formosanus* colonies are remarkably stable, with some that remained the same for 10–20 years (see Chapter 4, this volume). Occasionally, *R. flavipes* were also briefly observed in a few stations in Armstrong Park, but they were always replaced by *C. formosanus*. Because the colony boundaries were defined by monitoring stations that were spaced at a distance, there appeared to be gaps between colonies (Fig. 5.2), but it was hypothesized that adjacent colonies were constantly interacting with each other through intercepting underground tunnels between them. To test this hypothesis, three colonies (Fig. 5.2, colonies #2, #3, #9) in Armstrong Park were selected for elimination by bait applications to observe the reactions of their adjacent colonies (Messenger *et al.*, 2005). As colony #2 began to decline with the application of hexaflumuron baits, one monitoring station was occupied by *R. flavipes,* and other stations were slowly occupied by *C. formosanus*. DNA genotyping indicated the newly arrived *C. formosanus* was different from colony #2, #5, or #1, and was probably originated from a colony outside of the park across the street (Fig. 5.2). A similar outcome was observed when colony #3 was eliminated and replaced by a previously unknown *C. formosanus* colony. Curiously enough, colonies #8 and #4 that were close to the eliminated colony #3 did not make a move. Results with colony #2 and #3 suggested that there were many more smaller colonies than the 13 identified colonies (Messenger and Su, 2005a) and that they were undetected because larger colonies may have outcompeted them. When the larger colonies were eliminated, these smaller and probably younger latercomers might take over the vacant territories. Elimination of colony #9 resulted in a more visual outcome. Colony #9 was located on an artificial island with colony #13 on the north and colony #10 on the east (Fig. 5.2). As the activity of colony #9 declined by baits, colony #13 moved in and took over close-by stations of colony #9 three months after bait application, forcing colony #9 into a few stations away from colony #13. Colony #13 continued to move in aggressively, and ten months after bait application, the entire territory of colony #9 was taken over by colony #13. In the meantime, territory of colony #10 remained the same. The results suggested that there were close underground contacts between colonies #9 and #13 before bait application, and colony #13 moved into the territory of colony #9 immediately after baits negatively impacted it. The field result was later confirmed by Bernard *et al.*

(2017), who used planar arenas to demonstrate that *C. formosanus* colonies readily invaded baited neighboring colonies, accessed the abandoned tunneling system, and were subsequently eliminated after feeding on the leftover baits.

After eliminating three selected colonies (Messenger *et al.*, 2005), all termite colonies in the park were eliminated by baits in 2002–2003 (Mullins *et al.*, 2011). Following the total elimination of all termite colonies, numbers of stations with termites were recorded monthly–quarterly between 2004 and 2009 (except for six months each in 2005 and 2007 due to the adverse impacts of Hurricane Katrina) to observe the population recovery process by *C. formosanus* and *R. flavipes* in the 32-acre park (Su *et al.*, 2016). Results showed a sigmoid population growth curve from zero to 109 active stations between 2004 and 2008 (Fig. 5.3A). One year later, in 2009, the activity plateaued to ~80 stations that were occupied by 14 *C. formosanus* colonies and one *R. flavipes* colony (Fig. 5.3B). The sigmoid mid-point was 1,202 days (40 months), and it was concluded that if surrounded by heavy termite pressure, a termite-free zone may be repopulated by termites from outside in 53 months (Su *et al.*, 2016). The population growth process and colony territory formation were further examined by a simulation study with a lattice model (Lee and Su, 2009). Alate seeding pairs were randomly assigned to a two-dimensional lattice space and allowed to grow cell-by-cell. When cells encounter those of other colony(s), they are given a choice to stop, retreat, or advance by a coin-toss rule. Results showed that those that arrived earlier and with less competition tended to occupy larger territories than latecomers. The simulation results suggested that in a large area such as Armstrong Park, there were probably many small colonies, i.e., latecomers, whose growth was constrained by neighboring larger colonies (early arrivals). They show themselves only after their larger neighbors are eliminated, as shown in the three-colony elimination study (Messenger *et al.*, 2005). Alternatively, some of the young and small colonies may be absorbed by nearby larger colonies, as evidenced by a field observation (Su and Scheffrahn, 1988b).

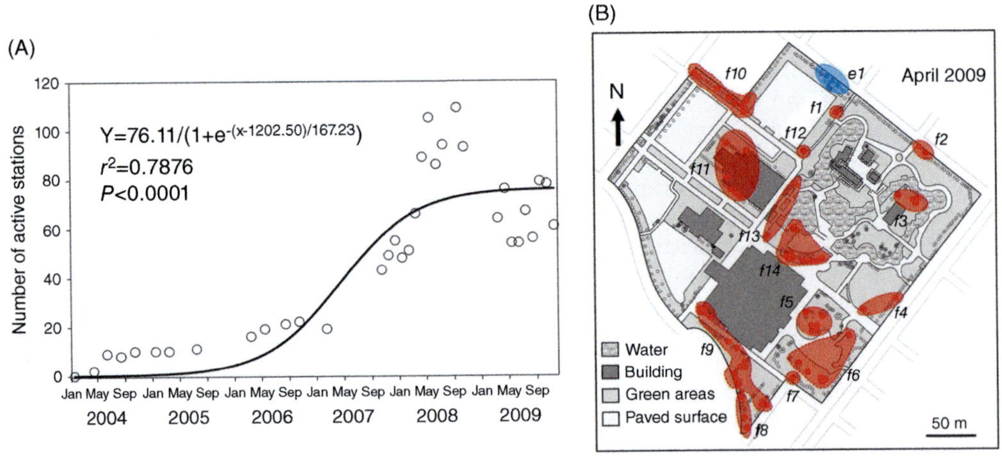

Fig. 5.3. Following the total elimination of all termite colonies in 2002–2003, numbers of active termite stations were periodically recorded to investigate the termite population recovery process. (A) Results showed that the termite population grew from zero (2004) to 109 active stations in five years. (B) By 2009, the once termite-free zone was occupied by 14 *C. formosanus* colonies and one *R. flavipes* colony. (Reprinted with permission from Su *et al.*, 2016.)

5.3 Nutrient ecology

5.3.1 Wood consumption rate

The Formosan subterranean termite is considered one of the most damaging structural pest species. The mean wood consumption rate of *C. formosanus* was estimated at ~50 mg wood/g termite/day (Su and La Fage, 1984), which is higher than most termite species (~10–30 mg/g/day) listed by Wood (1978). Wood consumption rates differ substantially among *C. formosanus* colonies (23–78 mg/g/day), and colonies with smaller workers (younger colonies) tend to consume more wood (Su and La Fage, 1984).

Coptotermes formosanus is known to damage a wide range of living plants (Lai *et al.*, 1983b), but when confined to force-feed on non-preferred wood species such as Western red cedar (*Thuja plicata* Donn) or redwood (*Sequoia sempervirens* (D. Don) Endl.), it suffered significant mortality (Su and Tamashiro, 1986). Large colony size also contributes to damaging potential of *C. formosanus*. Assuming an average worker weighs 3 mg, a *C. formosanus* colony with 1 million workers can potentially consume 150 g wood daily. It is tempting to use an estimate like this to project period of infestation in a structure, but since a field colony does not exclusively feed on a structure, it should be noted that such exercise is not valid.

5.3.2 Nitrogen acquisition

The unique ability of termites to feed on wood fibers is considered one factor for their evolution in eusociality. Wood diet is rich in carbohydrates but poor in nitrogen. Wood contains as little as 0.03–0.15% nitrogen. Yet, the nitrogen content of whole termites is similar to other animals, i.e., ~10% (La Fage and Nutting, 1978), and termite nitrogen acquisition has been one topic pursued by many. Early studies by Benemann (1973) on *Incisitermes minor* Hagen, and Breznak *et al.* (1973) on *C. formosanus*, concluded that these lower termites acquired nitrogen through nitrogen fixation by their hindgut symbionts. Their studies contributed to significant advances in our understanding of termite nutrient ecology, but the results of follow-up studies were inconsistent. Tayasu *et al.* (1998), for example, estimated that 30–60% of the nitrogen in termite *Neotermes koshunensis* Shiraki was of a recent atmospheric origin, whereas none of the nitrogen in termite tissue was of recent atmospheric origin in species such as *Macrotermes ukuzii* Holmgren (Rohrmann and Rossman, 1980).

In a study to measure the change in total nitrogen during the early growth of *C. formosanus* colonies, Mullins and Su (2018) did not detect an increase in total nitrogen over 12 months, even in the presence of functional workers. The results indicated the absence of nitrogen fixation by bacteria in worker hindguts. A recent study by Mullins *et al.* (2021) reported that 14-month-old incipient colonies of *C. formosanus* developed faster when reared on wood and organic soil than those reared on wood and inorganic sand, and no significant difference was found in nitrogenase expression in their hindguts between these two groups. The results indicated that *C. formosanus* colonies rely more on dietary nitrogen for colony growth than intrinsic nitrogen fixation by diazotrophic bacteria in their hindguts. In another study, Mullins *et al.* (2022) demonstrated that nitrogenase expression rates in hindguts of kalotermitids, e.g., *Cryptotermes* species, were significantly higher than those of rhinotermitids, e.g., *Coptotermes* species. Collectively these results suggested that subterranean species such as *C. formosanus* depend more on dietary nitrogen than nitrogen fixation by diazotrophic bacteria in hindguts, which is the primary source of nitrogen for kalotermitids that live in a piece of wood with no access to soil. The variabilities in nitrogen fixation rates among 32 termite species, as listed by Breznak (2000), may be explained, at least in part, by their access to soil organic matter.

5.3.3 Nitrogen conservation

The scarcity of nitrogen in a wood diet also demands termites to recycle and conserve nitrogen within their body. Termite gut bacteria are known to degrade uric acid, a common nitrogenous waste product, to be recycled and reabsorbed by termites (Breznak, 2000). Cannibalism is often listed as another mechanism of nitrogen recycling within a colony (La Fage and Nutting, 1978). Presumably, cannibalizing individuals which died of old age could provide a reliable source of nitrogen, but this has not been quantified. Exuviae is another source of nitrogen for recycling at the colony level. Laboratory observations showed that the exuviae of a molting termite were groomed off and immediately consumed by nearby assisting nestmates (Raina *et al.*, 2008; Xing *et al.*, 2013). Exuviae of *C. gestroi* contained 11.24% nitrogen, which is higher than that of termite whole bodies (~10%), wood (~0.1%), or organic soil (~2.5%) (Tong *et al.*, 2021).

Moreover, a study showed that *C. formosanus* workers always returned to a central nest to molt, which led to a hypothesis that exuviae at the central nest were fed to the queen, which requires extra nitrogen for egg production (Kakkar *et al.*, 2017). When exuviae marked with immunoglobulin G (IgG) were fed to 2-year-old *C. gestroi* colonies, the markers were always detected from queens, and marker concentrations were higher in queens and second or higher instar workers (W_{2+}) than the eggs and first instar larvae (Tong *et al.*, 2023a). In another follow-up study, the addition of exuviae in incipient colonies reared on nitrogen-poor medium, i.e., sand vs organic soil, significantly increased the colony biomass after 3.5 months. In contrast, queens from colonies in which exuviae were removed (by removing pre-molt individuals) laid fewer eggs than queens from colonies in which exuviae were not removed (Tong *et al.*, 2023b). These results confirmed the hypothesis that nitrogen sources from the exuviae shed at the central nest were passed on to queens and contributed to egg production.

5.4 Foraging behavior

The foraging behavior of termites is a collective activity of a group of individuals and in some epigeal species, such as *Nasutitermes* spp. or *Macrotermes* spp., is considered wholly dependent on semiochemicals (Traniello and Leuthold, 2000). With the exception of above-ground foraging (Reinhard *et al.*, 1997), however, semiochemicals may have limited roles in the food-searching phase of foraging behavior for subterranean termites, because chemicals do not travel far in soil. Once the food is found, semiochemicals may play an important role, especially in recruitment. In the fungus-growing termite, *Macrotermes carbonarius* (Hagen), underground tunnels are excavated from the mound, leading to foraging holes from which termites venture out to the soil surface in search of food (Sugio, 1995). Presumably, *M. carbonarius* foragers may rely on semiochemicals to locate food on the soil surface. In the arid landscape of New Mexico, the desert subterranean termites *Gnathamitermes tubiformans* (Buckley) and *Amitermes wheeleri* (Desneux) were known to locate surface foods such as cow dung by sensing the thermal shows of these objects (Etthershank *et al.*, 1980).

Underground foraging galleries of *C. formosanus* (Erhorn, 1934; King and Spink, 1969) showed that most of the tunnels were constructed at > 10 cm below ground (Fig. 5.1). They may occasionally emerge in the open to construct shelter tubes to access food, but the extensive subterranean gallery system suggests that subterranean termites such as *C. formosanus* mostly intercept food via underground tunnels in the soil. However, the mechanism by which they construct underground tunnels to locate food was mostly unknown until recently, when results of laboratory studies using planar arenas became available. Another aspect of foraging behavior in subterranean termites that is poorly understood is how foragers optimize food transportation efficiency. Unlike terrestrial insects such as ants and bees that use path integration to find the shortest return route along a homing vector, subterranean termites must either use the existing tunnels or excavate new tunnels with the shortest distance between the nest and newly discovered food.

5.4.1 Food searching phase

In planar arena studies, termites are placed in a thin layer (but thicker than termite height) of sand or soil sandwiched between two clear Plexiglas® sheets to allow observation of tunneling behaviors and tunnel geometry. With the planar arenas, Robson *et al.* (1995) reported that the food search pattern in *R. flavipes* was non-random, and termites systematically constructed tunnels to cover areas with minimum redundancy. Their pioneering work was the first to describe foraging behavior based on mechanical processes instead of chemical means. The results agreed with the earlier report of Ratcliffe and Greaves (1940), who described underground subterranean galleries of *Coptotermes lacteus* (Frogatt).

When *C. formosanus* groups were placed at the center of planar arenas, tunnels were initially excavated uniformly from the point of origin, but once the food was found, more tunnels were excavated toward the food source (Campora and Grace, 2001; Michael *et al.*, 2023). Campora and Grace (2001) also reported that the presence of wood did not affect tunnel distribution and concluded that *C. formosanus* responded to structural anomalies in the arena substrate rather than the presence of wood. Their conclusion, confirmed by Puche and Su (2001), was that sound wood embedded in the sand of planar arenas did not affect the tunnel directions of *C. formosanus* or *R. flavipes*. In fact, the results of Puche and Su (2001) showed that *C. formosanus* excavated a tunnel as close as 2.5 mm to a piece of sound wood without intercepting it (Fig. 5.4A). When wood discs decayed with the brown rot fungus, *Gloeophyllum trabeum* (Pers. ex Fr), were embedded in arena sand, however, tunnels were excavated toward the decayed wood (Su, 2005). *Gloeophyllum trabeum* is known to produce 3*Z*,6*Z*,8*E*-dodecatrienol that elicits trail-following responses by termites (Matsumura *et al.*, 1968). Both *C. formosanus* and *R. flavipes* showed the same responses toward decayed wood as well as polymer discs treated with acetone extracts of decayed wood. The attraction properties were probably water-soluble and permeated through wet sand to alter termite tunneling orientation toward the source at 12–18 cm (Fig. 5.4B). Another factor that influences termite tunneling behavior is moisture. When termites (both *C. formosanus* and *R. flavipes*) were allowed to tunnel through the sand with moisture gradients, more tunnels were excavated in the sand with higher moisture (Su and Puche, 2003).

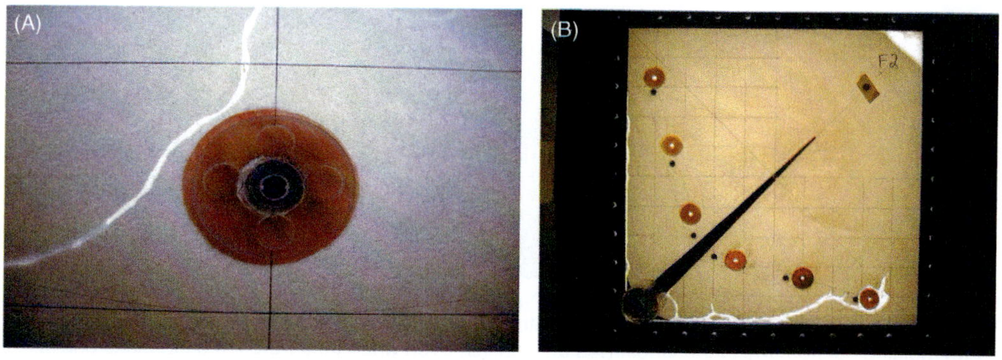

Fig. 5.4. (A) *Coptotermes formosanus* excavated a tunnel as close as 2.5 mm to a piece of sound wood without intercepting it. (Reprinted with permission from Puche and Su, 2001.) (B) Termites excavated tunnels toward a wood-disc decayed with brown rot fungus, which indicated that attractants permeated from the wood disc to altered tunnel orientation at the distance of 12–18 cm. (Reprinted with permission from Su, 2005.)

Structural anomaly, trail-following pheromone, and moisture gradient are extrinsic factors that influence termite tunnel behavior. The characteristic geometric pattern, however, is the product of the group behaviors of subterranean termites that follow a set of intrinsic rules. When *C. formosanus* workers were introduced in a sand-filled arena, excavated tunnels always radiated out from the tunnel origin (Fig. 5.5A) (Su *et al.*, 2004). Even when forced to make one or two 90° turns at various lengths, they invariably made necessary corrections to return to the vector away from the tunnel origin (Fig. 5.5B) (Bardunias and Su, 2009a). The results showed that the tunnel direction of *C. formosanus* was governed by the Global Away Vector (GAV), which is opposite to a homing vector of birds or bees, and GAV was generated by a path integration (or dead reckoning) through which termites were able to geolocate their current location in reference to the point of tunnel origin. When moving through an open chamber, more tunnels were propagated from the second chamber, suggesting that termites updated the GAV after encountering an empty space or food source (Bardunias and Su, 2010a).

Although groups of termites excavate the extensive gallery, not all termites in a colony participate in tunneling activity. Laboratory studies showed that in any termite group, only 16–20% of *C. formosanus* workers participated in tunnel excavation (Yang *et al.*, 2009; Bardunias *et al.*, 2010). Of the 16% reported by Bardunias *et al.* (2010), 20.6% (or 3.3% of the test group) performed more than half of tunnel excavation activities. This small number of termites may be responsible for the overall tunnel direction and branching formation, but the roles may be rotated to other individuals as they age. The active excavators usually congregate at the tunnel tip to form a queue behind those at the digging face (Bardunias and Su, 2010b). With a large population size, more termites are delayed in a queue and cannot reach the tunnel tip. Instead, those in a queue excavate laterally by taking soil from the tunnel walls at their current position, resulting in wider tunnels. Mizumoto *et al.* (2020) found that subterranean termite species that switched to lateral excavation more quickly when stuck in a queue characteristically showed a higher branching rate than species that were more likely to wait longer in the queue. This suggests that lateral excavation

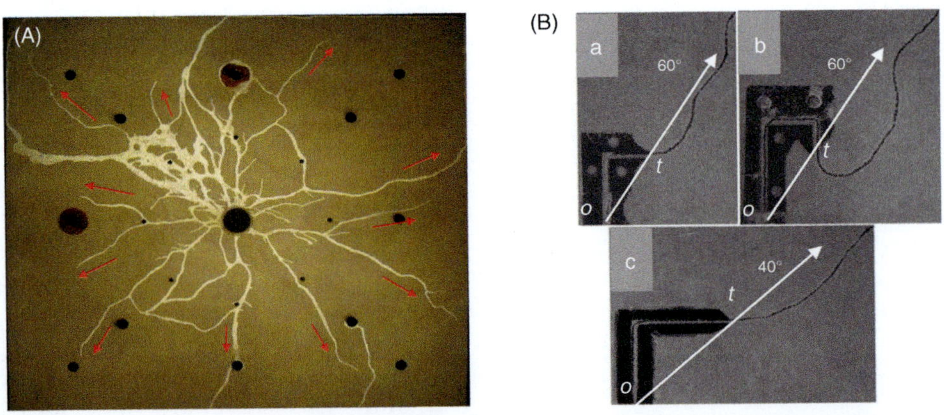

Fig. 5.5. (A) When *C. formosanus* workers were introduced into a sand-filled arena, excavated tunnels always radiated out from the tunnel origin (red arrows). (B) Even when they were forced to make one or two 90° turns at various lengths (*a, b,* and *c*), they invariably made necessary corrections at the tunnel tip (*t*) to return to the vector away from tunnel origin (*o*). The results showed that the tunnel direction of *C. formosanus* was governed by the Global Away Vector (GAV), which was generated by a path integration through which termites were able to geolocate their own current location (*t*) in reference to the point of origin (*o*). (Reprinted with permission from Bardunias and Su, 2009a.)

may begin at a depression, which may be further excavated to form a branch tunnel (Bardunias and Su, 2009b).

During tunnel construction, sand or soil removed from excavated tunnels is deposited in voids within the gallery system. In planar arenas, sand was deposited in the central container to introduce termites; and in the field colonies, soil removed from excavated tunnels was deposited in wooden voids created by feeding (Li and Su, 2008). When termites (both *C. formosanus* and *R. flavipes*) of various group sizes were placed in planar arenas, tunnel volumes increased and plateaued at the levels that were proportional to the group size, i.e., groups with more termites plateaued at higher volumes (Su and Lee, 2009). However, termites did not stop tunneling even when the tunnel volume reached the plateau. Instead, they continued constructing new tunnels in different areas of the arena, and sand removed from newly excavated tunnels was deposited in old and unused tunnels, resulting in an equilibrated tunnel volume (Su and Lee, 2009). The equilibrated tunnel volume was species-specific and inversely correlated with group size, suggesting that gallery systems of larger colonies are more crowded than smaller colonies.

Colony size also impacts foraging distance. When 6–8-year-old incipient colonies of *C. formosanus* (~1,000–25,000 workers per colony) were placed in extended arenas and allowed to continue foraging until reaching maximum distance, the total foraging distances ranged from ~15 m to 329 m and were positively correlated with the population size (Su *et al.*, 2017a).

In a field colony of *C. formosanus*, multiple foraging sites and nests are interconnected by extensive underground tunnels (Fig. 5.1B). The cost of constructing the elaborate tunnel system to encounter a food source (benefit) can be viewed from the conceptual framework of "optimal foraging theory" (Emlen, 1966; MacArthur and Pianka, 1966). Investigating tunnel geometry in relation to the cost–benefits in termite foraging may provide clues to understanding how termites use their tunneling system to optimize the foraging efficiency. However, studies of termite gallery systems at the colony level are labor-intensive and impractical. Simulation models of termite tunnels offer an alternative. To provide data for such models, critical components of termite tunnel geometry were identified using movement pathways and tested for their validity with a simulation program (Fig. 5.6A) (Su *et al.*, 2004). Eight critical components were initially identified, including: (1) the number of primary tunnels; (2) the linear length of primary tunnel segments; (3) the turning angle of each linear segment; (4) the branching angle of secondary tunnels; (5) the probability of branching; (6) termination probability of primary tunnels; (7) termination probability of secondary tunnels; and (8) the probability of continuing through an intersection when two tunnels meet. When they were incorporated into a simulation program to generate virtual tunnels, some tunnels looped back to the tunnel origin, which was not observed in empirical tunnels. To correct the discrepancy, two more components were added to the program, i.e.: (9) maximum deviation from the start angle; and (10) minimum angles between two primary tunnel origins. The "no loop back" feature of the termite tunnel system was later confirmed by the "Global Away Vector," as described by Bardunias and Su (2009a). The simulated tunnels based on these ten geometric components (Fig. 5.6B) were compared for their complexities with empirical tunnels using a fractal analysis (Su *et al.*, 2004).

The ten geometric components identified from empirical tunnels of *C. formosanus* were incorporated in a lattice gas model to test searching efficiency in relation to food distribution pattern, and the results showed that *C. formosanus* tunnels were most efficient in intercepting food particles spaced in clumped rather than random or uniformed distribution (Lee *et al.*, 2006). Before becoming closely associated with human activity, the native landscape inhabited by *C. formosanus* was probably littered with large fallen trees, and their large size, compared with termites, represents "clumped" and aggregated food source. The conclusion agreed with that of Robson *et al.* (1995) that the *R. flavipes* search gallery was adopted to intercept

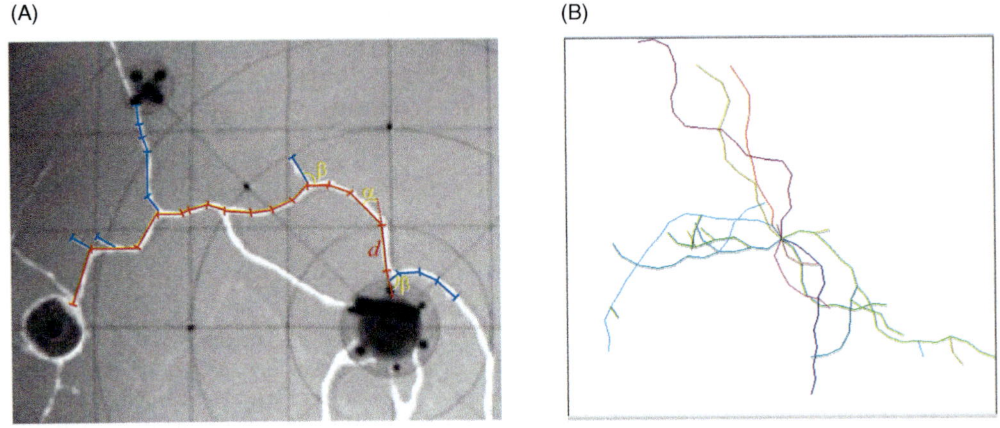

Fig. 5.6. (A) Some critical components of termite tunnel geometry were identified by using movement pathways, including the linear length of primary tunnel segments (d), the turning angle of each linear segment (α), and the branching angle of secondary tunnels (β). (B) Simulated tunnels with ten critical components of termite tunnel geometry. (Reprinted with permission from Su *et al.*, 2004.)

"clumped" decaying wood on the soil surface. Studies with other animal species also showed that collective foraging is more efficient than solitary foraging when food distribution is clumped (Zahavi, 1971; Krebs *et al.*, 1972).

Simulation studies with a lattice gas model were later used to investigate other geometric characteristics of termite tunnels in relation to foraging efficiency. The empirical tunnels, as reported by Su *et al.* (2004), showed that both primary and secondary tunnels were terminated when they reached a certain length. Lee *et al.* (2008) examined the food encounter rate of termite tunnels with various length ratios between primary and secondary tunnels, $\lambda = L_2/L_1$, where λ is the length ratio, and L_1 and L_2 are primary and secondary tunnel lengths, respectively. Their results showed that food encounter rate was a bimodal function of length ratio λ, with the encounter rate optimized by two values of $\lambda = 0.1$ and 0.6. When $\lambda = 0.1$, the gallery system is characterized by long primary tunnels with short branches, which are optimized to cover a long distance, i.e., a distance-searching strategy. When $\lambda = 0.6$, lengths of primary and branch tunnels were organized to cover a wider area, i.e., an area-searching strategy. Depending on the environmental constraints, the tunnel system of some termites may be configured to search for food at a distance, whereas others may be configured to explore an area extensively.

The model prediction of the "area" or "distance" searching strategy was realized by the finding of *C. gestroi* populations that appeared unable to intercept survey stakes or in-ground bait stations (Su *et al.*, 2023). Since the discovery of *C. gestroi* in Miami, Florida, in 1996 (Su *et al.*, 1997), and especially after it was found in Broward County in 2001 (Su *et al.*, 2017b), repeated attempts have been made to use stake survey and in-ground monitoring stations to study their field colonies, but none of these in-ground devices has been intercepted by *C. gestroi* populations in southeastern Florida. Survey stakes and in-ground monitoring stations are in widespread use to study *C. gestroi* in various regions, including Thailand (Sornnuwat *et al.*, 1996), Little Cayman (Su *et al.*, 2000), Brazil (Santos *et al.*, 2010), Hawai'i (Uchima and Grace, 2003), and Malaysia (Lee, 2002). However, a closer look at the literature showed that the interception rates of these devices by *C. gestroi* in some areas are low, i.e., 1.7% in Taiwan (Chiu *et al.*, 2016), and 4.5% in Brazil (Arab *et al.*, 2005), which are considerably lower than that of *C. formosanus*, ~17% (Su, 1994). A follow-up study with simulated structural infestations showed that *C. gestroi* constructed longer primary tunnels and shorter secondary tunnels with less branching frequency than *C. formosanus*, resulting in a significantly smaller length ratio than that of *C. formosanus* (Su and Lee, 2023). Because of the smaller length ratio, *C. gestroi* appears to adopt the "distance searching" strategy rather than the "area searching" strategy of *C. formosanus* (Fig. 5.7). With more branch tunnels and complex tunnel geometry, *C. formosanus* search for food in an area before moving on to other areas, which allowed them to eventually intercept even the sparsely placed (~3 m apart) survey stakes or in-ground stations. Due to the fewer primary tunnels that tend to travel straight at a distance, *C. gestroi* (at least those in southeastern Florida) may have bypassed the sparsely distributed survey stakes.

However, the tunnel geometry of *C. gestroi* in southeastern Florida is not ubiquitous to this species. Laboratory observations showed that *C. gestroi* in Hawai'i constructed tunnels with more branches than *C. formosanus* (Grace *et al.*, 2004; Hapukotuwa and Grace, 2012), and *C. gestroi* in Malaysia also propagated highly branched tunnels (Yeoh and Lee, 2007). As reported in the literature, the intra-specific dissimilarity in tunnel geometry and the discrepancies in their ability to intercept survey stakes or in-ground stations suggest that the differences may be due to environment or colony age instead of being species-specific. For example, the invasion of *C. gestroi* in Broward County is a recent event (since 2011), and most colonies in this area could be young and presumably small and may not venture out into the soil, or if they do, may only construct simple tunnels to search for food. Another reason could be the difference in laboratory protocols among studies. Field-collected orphan populations of *C. gestroi* were used in the studies in Hawai'i (Grace *et al.*, 2004; Hapukotuwa and Grace, 2012) and Malaysia (Yeoh and Lee, 2007), but in southeastern Florida, incipient colonies with a complete caste composition were used (Su and Lee, 2023). Further studies are needed to elucidate the differences.

5.4.2 Food transportation phase

Despite being one important element of foraging optimality, food transportation efficiency is not as well studied as food searching efficiency. Because subterranean termites locate and transport food back to

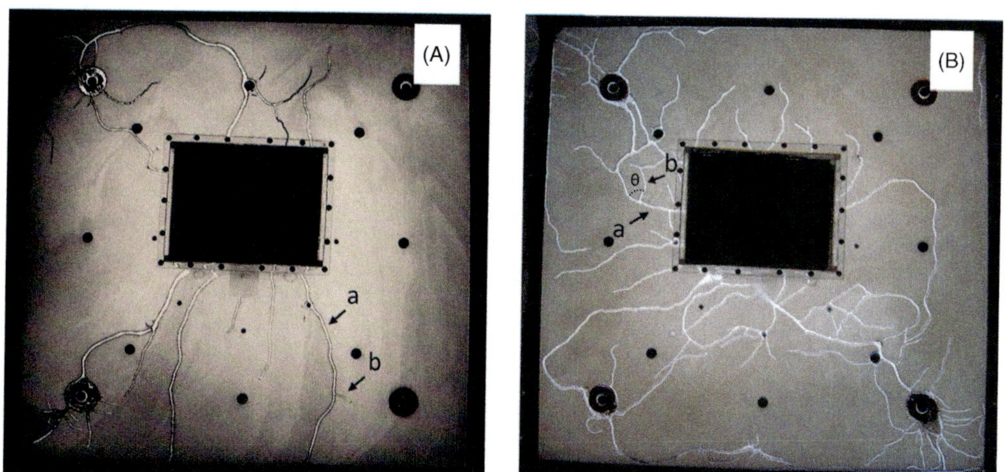

Fig. 5.7 (A) *Coptotermes gestroi* that constructs longer and more linear primary tunnels with shorter secondary tunnels adopts a "distance searching strategy," whereas (B) *C. formosanus* that propagates shorter primary tunnels and longer secondary tunnels adopts an "area searching strategy" in search of food. *a*, primary tunnel; *b*, secondary tunnel; θ, branching angle. (Reprinted with permission from Su and Lee, 2023.)

the central nest using the tunnel system, Lee *et al.* (2007) hypothesized that certain features of the tunnel pattern may be optimized for the ratio (ε) of food gain to loss due to transportation. They focused on the branch (secondary tunnel) length of termite tunnels and revealed that branch length distribution followed a negative exponential curve, $P(L) = e^{-\alpha L}$ (Fig. 5.8A), and α = 0.15 for *C. formosanus* according to empirical data (Su *et al.*, 2004). Virtual termite tunnels were generated using the lattice model with α values ranging from 0.05 to 0.25. As shown in Fig. 5.8B, smaller α generated tunnels with longer branches, and vice versa. The simulation results with these virtual tunnels showed that ε (ratio of food gain to transportation loss) was optimized when α ~ 0.15 (Fig. 5.8B), which confirmed the hypothesis that branch length distribution of *C. formosanus* tunnels was configured to maximize the transportation efficiency (Lee *et al.*, 2007).

Aside from the simulation study, a laboratory arena study showed that *C. formosanus* excavated shortcut tunnels to travel directly to a food source following its discovery (Michael *et al.*, 2023). A 2-year-old *C. formosanus* incipient colony (~3,000 workers) was connected to three planar arenas with preformed tunnels of different complexities, named "simple," "intermediate," and "complex" arenas (Fig. 5.9). The "simple arena" was provided with a straight tunnel from the tunnel origin (Fig. 5.9*a*) to the food source (moistened cellulose media pad, 2 × 2 cm) (Fig. 5.9F). In contrast, preformed tunnels for an "intermediate" and "complex" arena had three and eleven 90° turns, respectively (Fig. 5.9, 0 h). Immediately after entering the arenas, termites followed the preformed tunnels to reach the food source in all three arenas. Within 12 h, termites in the "simple" arena mostly used the straight preformed tunnel to reach food, whereas those in the "intermediate" arena excavated branch tunnels from the lower horizontal tunnel heading to the food source (Fig. 5.9*b*, 12 h). In the "complex" arena, one tunnel was propagated from the 1st turn and headed straight to the food source (Fig. 5.9*c*, 12 h). In addition, numerous branched tunnels were excavated from the lower horizontal tunnel (Fig. 5.9*d*, 12 h), and one connected to the 4th turns of the preformed tunnel, forming a shortcut. Branch tunnels were also excavated from the 9th turn (Fig. 5.9*e*, 12 h) that connected to the upper horizontal tunnel, forming another shortcut. Another branch tunnel originated from the 5th turn (Fig. 5.9*f*, 12 h) headed toward the food source. By 24 h, termites in the "simple" arena formed several short branches that appeared to follow the GAV, but they did not develop further or

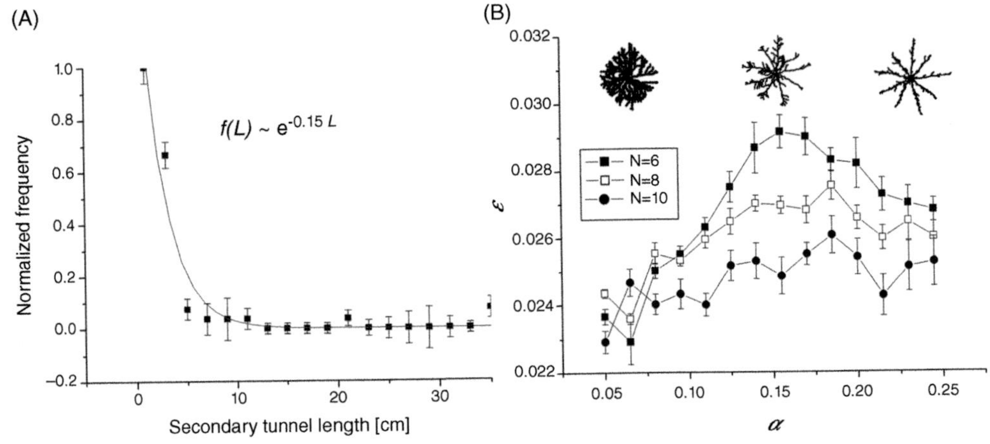

Fig. 5.8. (A) Length distribution of secondary tunnels of *C. formosanus* followed a negative exponential curve, $P(L) = e^{-\alpha L}$ with α = 0.15. (B) A simulation study showed that ε (ratio of food gain to transportation loss) was optimized when α ~ 0.15. (Reprinted with permission from Lee *et al.*, 2007.)

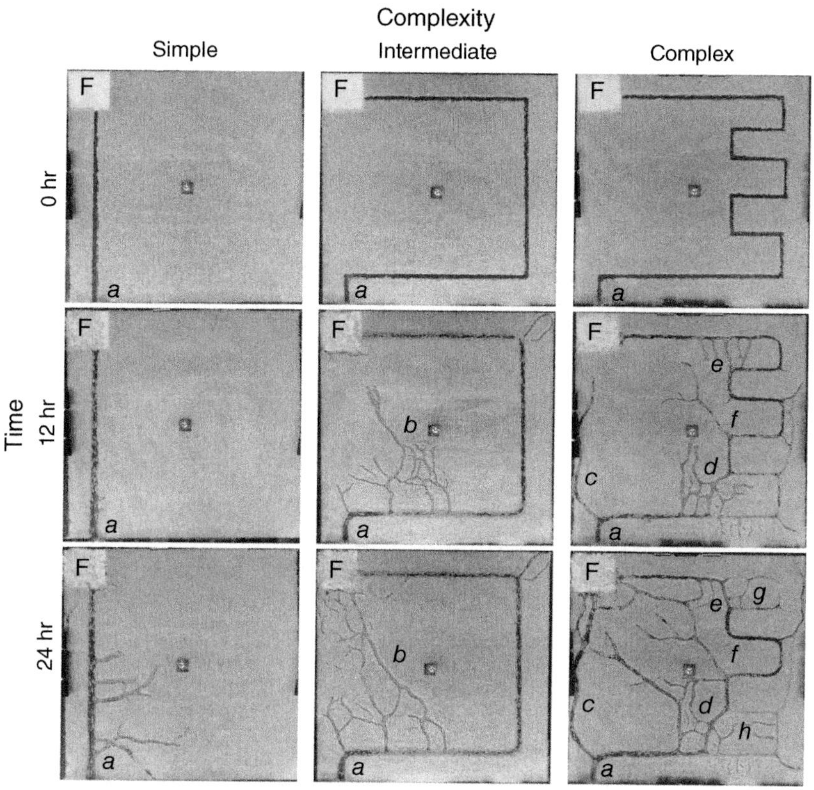

Fig. 5.9. When provided with preformed tunnels of three complexities, simple (straight), intermediate (four 90° turns), and complex (eleven 90° turns), *C. formosanus* reached food (F) immediately (0 h) after entering the arena from tunnel origin (*a*). Within 12 h, branch tunnels (*b, c,* and *f*) were formed and headed toward food in intermediate and complex arenas (12 h). By 24 h, these branches reached food directly (*b, c,* and f, 24 h). Many shortcuts were also created in the complex arena (*d*, and e, 12 h), and tunnels that were bypassed by the shortcuts were filled with sand (*g* and *h*, 24 h). (Reprinted with permission from Michael *et al*., 2023.)

reach the food source. Branches formed in "intermediate" arenas at 12 h continued to propagate and reached the food source by 24 h (Fig. 5.9*b*, 24 h). Similarly, branches heading toward food in the "complex" arena at 12 h also reached the food source by 24 h (Fig. 5.9*c*, *f*, 24 h). Tunnels that were bypassed by shortcuts (Fig. 5.9*d*, *e*) in complex arenas were abandoned and filled with sand (Fig. 5.9*g*, *h*, 24 h) removed from newly excavated tunnels. Su and Lee (2009) reported the same observation that unused tunnels were filled with sand removed from newly excavated tunnels.

Coptotermes formosanus excavated new tunnels heading to food sources following its discovery (Fig. 5.9*b*, *c*, *f*), suggesting that termites were able to geolocate newly discovered food via path integration (Bardunias and Su 2009a), which probably also enabled termites to create shortcuts observed in a complex arena (Fig. 5.9*d*, *e*). These new tunnels provided more direct paths (than preformed tunnels) to food sources and optimized food transportation efficiency.

5.4.3 Formation of *C. formosanus* foraging galleries

Information obtained from studies with planar arenas (GAV, path integration, GAV update) illustrates how a foraging gallery of *C. formosanus* shown in field excavation studies (Fig. 5.1B) may be formed. Initially, several tunnels radiate out of the central nest (Fig. 5.10A, solid circle) as described by Su *et al.* (2004) (see also Fig. 5.5A), and one of which (Fig. 5.10A, *a*) intercepts food source F1 (Fig. 5.10A). Following the discovery, termites construct a shortcut tunnel to reach food source F1 directly from the nest (Fig. 5.10B, dotted line *b*) (Michael *et al.*, 2023). At the same time, termites update their GAV origin (Bardunias and Su, 2010a) with more tunnels radiating out of F1 (Fig. 5.10B), and one of the branches (*c*) intercepts food source F2. Another tunnel (*d*) from the central nest also intercepts another food source F3. After intercepting F2, a shortcut tunnel (Fig. 5.10C, *e*) is excavated to form a direct path between F1 and F2, and more tunnels radiate out of F2 with one (*f*) intercepting F4. As with F2, more tunnels radiate out of F3, and one (*g*) intercepts F5. In the meantime, less-traveled tunnels and those that did not intercept food (Fig. 5.10C, gray lines, including tunnel *a*) are filled with soil and eventually disappear (Fig. 5.10D). With the same process, F5 and F6 are intercepted by tunnels *h* and *i* (Fig. 5.10D), and as unused tunnels are filled with soil and disappear, the main gallery is formed to connect food F1–F7 by tunnels *h*, *f*, *e*, *b*, *g*, and *i* (Fig. 5.10D).

5.4.4 Age polytheism in foraging

Du *et al.* (2016), who documented behaviors of 4-year-old colonies of *C. formosanus* near central nests in planar arenas, reported that older workers (W_{2+}) engaged in nest maintenance, sanitation, and caring for royal pairs. In contrast, young workers (W_1) performed most of the grooming activity. When *C. formosanus* colonies of different ages were compared for their ethograms, age polyethism was more extensive and elaborated in 4-year-old colonies than 10-month-old colonies (Du *et al.*, 2017a). Even in a small planar arena where the distance between the royal chamber and food was less than 25 cm, older workers were more frequently found at feeding sites than younger individuals (Du *et al.*, 2017b). In the laboratory study to investigate colony size and foraging distance relationship, where *C. formosanus* incipient colonies were introduced into extended arenas and allowed to forage for as long as they could, older workers were always found further away from the royal chamber (Su *et al.*, 2017a). Results of these studies confirmed age polyethism during the foraging of *C. formosanus* and that older workers were more involved in foraging activity, which agreed with the earlier finding of Crossland *et al.* (1997) with *R. fukiensis*. The ~20% workers (of a colony) that engaged in foraging activity were mainly older workers, but when these older foragers were removed (to simulate natural disturbances at foraging sites), *C. formosanus* colonies replaced them with the next-oldest workers to maintain the ~20% ratio (Lee *et al.*, 2022). The behavior plasticity suggested the importance of the forager proportion in *C. formosanus* colonies.

Aside from the instar-wise age polyethism in foraging, there appeared to be another morphologically linked, inter-molt polyethism. Feeding on hard substances such as wood inevitably wears down termite mandibles that have to be renewed after each molt (Roisin and Korb, 2011). This observation led to the hypothesis that newly molted and fully sclerotized workers with the sharpest mandibles are most likely the wood-feeding foragers (Lee *et al.*, unpublished data). To test this hypothesis, 3-year-old *C. formosanus* colonies in plastic containers (central nests) were connected to other containers provisioned with sand and wood (foraging sites) by using 3 m tubing to count and collect workers at the foraging sites. As with previous studies, larger numbers of older workers (W3 and W4) were found in the foraging sites and connecting tubes. When the mandible sharpness of W3 workers was measured by using the method of Šobotník *et al.* (2014), those collected from foraging sites were the sharpest, followed by those in the connecting tubes, and those in the central nest were the dullest. The results showed that age-polyethism in foraging is not limited to instar-wise age, and a worker may perform different tasks in foraging

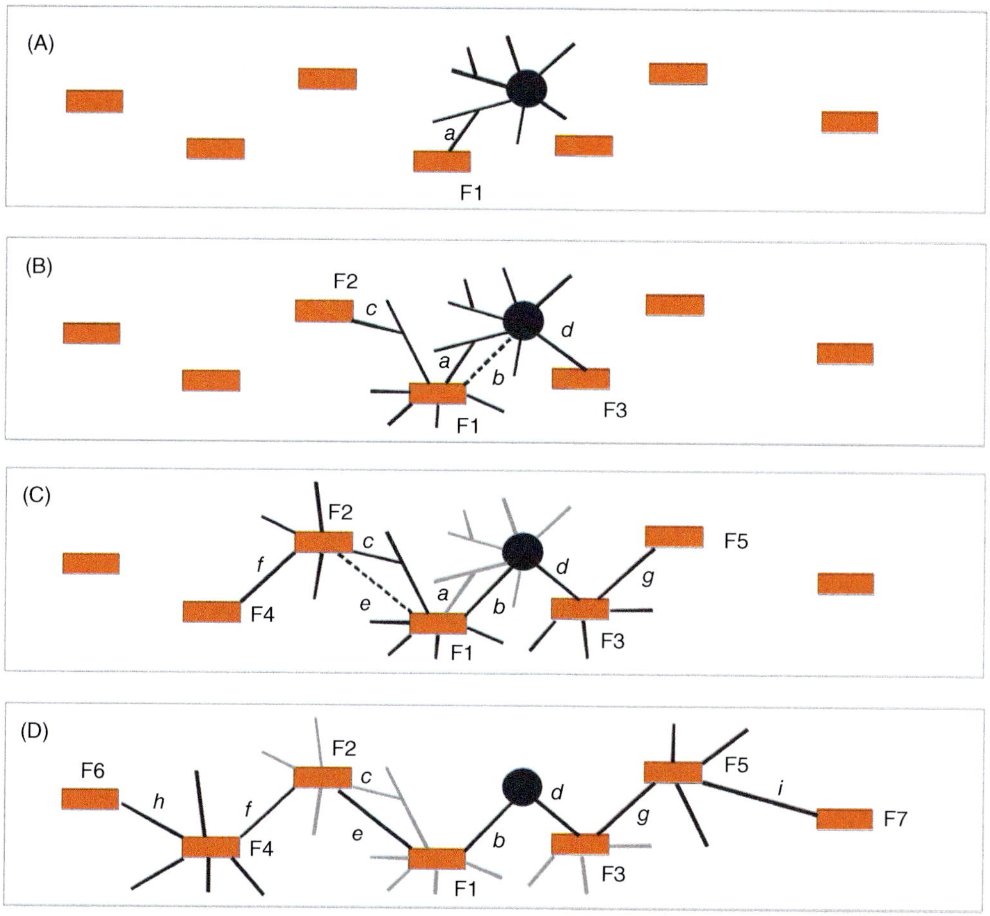

Fig. 5.10. (A) Formation of *C. formosanus* foraging gallery began with several tunnels radiating out of the central nest (solid circle) with one of which (*a*) intercepting food source F1 (brown rectangular). (B) Following the discovery, termites construct a shortcut tunnel to reach F1 directly from the nest (dotted line *b*). At the same time, termites update their GAV origin with more tunnels radiating out of F1, and one of the branches (*c*) intercepts food source F2. Another tunnel (*d*) from the central nest also intercepts another food source F3. (C) After intercepting F2, a shortcut tunnel (dotted line *e*) is excavated to form a direct path between F1 and F2, and more tunnels would radiate out of F2 with one (*f*) intercepting F4. As with F2, more tunnels also radiate out of F3, and one (*g*) intercepts F5. In the meantime, less-traveled tunnels and those that did not intercept food (gray lines, including tunnel *a*) are filled with soil. (D) These filled tunnels eventually disappear. With the same process, F5 and F6 are intercepted by tunnels *h* and *i*. As the unused tunnels are filled with soil and disappear, the main gallery is formed to connect food F1–F7 by tunnels *h*, *f*, *e*, *b*, *g*, and *i*.

depending on the age between two molts, i.e., the same instar.

Labor division in foraging subterranean termites such as *C. formosanus* is probably more complex than previously thought. In a laboratory study, Lee *et al.* (2021) tracked individually marked *C. formosanus* foragers and discovered that one cluster of foragers specialized in chewing wood, whereas the other cluster devoted most of their time in receiving and transporting wood fragments back to the central nest. Similar behaviors were reported from other social insects such as harvester termite, *Hodotermes mossambicus* (Hagen) (Leuthold *et al.*, 1976), and the "bucket brigade" of leaf-cutting ant, *Atta cephalotes* L.

(Hubbell *et al.*, 1980). Studies with *C. formosanus* suggested that the foraging polyethism of this lower termite species is extensive. A colony of *C. formosanus* may contain millions of termites that forage up to 100 m in search of food. It is unsurprising that a complex set of age polyethism in foraging is needed to maintain the organization of this magnitude.

References

Adams, E.S. and Levings, S.C. (1987) Territory size and population limits in mangrove termites. *Journal of Animal Ecology* 56, 1069–1081.

Arab, A., Costa-Leonardo, A.M., Casarin, F.E., Guaraldo, A.C. and Chaves, R.C. (2005) Foraging activity and demographic patterns of two termite species (Isoptera: Rhinotermitidae) living in urban landscapes in southern Brazil. *European Journal of Entomology* 102, 691–697.

Atkinson, T.H., Rust, M.K. and Smith, J.L. (1993) The Formosan subterranean termite, *Coptotermes formosanus* Shiraki (Isoptera: Rhinotermitidae) established in California. *The Pan-Pacific Entomologist* 69, 111–113.

Bardunias, P. and Su, N.-Y. (2009a) Dead reckoning in the tunnel propagation of the Formosan subterranean termite (Isoptera: Rhinotermitidae). *Annals of the Entomological Society of America* 102, 158–165.

Bardunias, P. and Su, N.-Y. (2009b) Opposing headings of excavating and depositing termites facilitate branch formation in the Formosan subterranean termite. *Animal Behavior* 78, 755–759.

Bardunias, P. and Su, N.-Y. (2010a) Change in tunnel heading in the Formosan subterranean termite (Isoptera: Rhinotermitidae) in response to movement through opened space. *Annals of the Entomological Society of America* 103, 449–454. doi: 10.1603/AN08116

Bardunias, P., and Su, N.-Y. (2010b) Queue size determines the width of tunnels in the Formosan subterranean termite (Isoptera: Rhinotermitidae). *Journal of Insect Behavior* 23, 189–204.

Bardunias, P., Su, N.-Y. and Yang, R.-L. (2010) Behavioral variation among tunnelers in the Formosan subterranean termite. *Journal of Asia-Pacific Entomology* 13, 45–49.

Begon, M. (1979) *Investigating Animal Abundance: Capture-Recapture for Biologists*. University Park Press, Baltimore, Maryland.

Benemann, J.R. (1973) Nitrogen fixation in termites. *Science* 181, 164–165. doi: 10.1126/science.181.4095.164

Bernard, S., Osbrink, W. and Su, N.-Y. (2017) Response of the Formosan subterranean termite to neighboring con-specific populations after baiting with noviflumuron. *Journal of Economic Entomology* 110, 575–583. doi: 10.1093/jee/tow317

Bess, H.A. (1970) Termites of Hawaii and the Oceanic islands. In: Krishna, K. and Weesner, F.M. (eds) *Biology of Termites, Vol.* 2. Academic Press, New York, pp. 449–476.

Breznak, J.A. (2000) Ecology of prokaryotic microbes in the guts of wood- and litter-feeding termites. In: Abe, T., Bignell, D.E. and Higashi, M. (eds) *Termites: Evolution, Sociality, Symbioses, Ecology*. Springer Netherlands, Dordrecht, pp. 209–231.

Breznak, J.A., Brill, W.J., Mertins, J.W. and Coppel, H.C. (1973) Nitrogen fixation in termites. *Nature* 244, 577–580.

Campora, C.E. and Grace, J.K. (2001) Tunnel orientation and search sequence of the Formosan subterranean termite (Isoptera: Rhinotermitidae). *Journal of Economic Entomology* 94, 1193–1199.

Cao, R. and Su, N.-Y. (2014) Tunneling and food transportation activity of four subterranean termite species (Isoptera: Rhinotermitidae) at various temperatures. *Annals of the Entomological Society of America* 107, 696–701. doi: 10.1603/AN13181

Cao, R. and Su, N.-Y. (2016) Temperature preferences of four subterranean termite species (Isoptera: Rhinotermitidae) and temperature-dependent survivorship and wood-consumption rate. *Annals of the Entomological Society of America* 109, 64–71. doi: 10.1093/aesa/sav095

Chiu, C.I., Yeh, H.T., Tsai, M.-J. and Li, H.-F. (2016) Naturalization and control of *Coptotermes gestroi* (Blattodea: Rhinotermitidae) in a Taiwanese forest. *Journal of Economic Entomology* 109, 1317–1325.

Clément, J.L. (1986) Open and closed societies in *Reticulitermes* termites (Isoptera, Rhinotermitidae): Geographic and seasonal variation. *Sociobiology* 11, 311–323.

Cornelius, M.L. and Osbrink, W.L.A. (2011) Effect of seasonal changes in soil temperature and moisture on wood consumption and foraging activity of Formosan subterranean termite (Isoptera: Rhinotermitidae). *Journal of Economic Entomology* 104, 1024–1030. doi: 10.1603/EC10332

Cornelius, M.L., Duplessis, L.M. and Osbrink, W.L.A. (2007) The impact of hurricane Katrina on the distribution of subterranean termite colonies (Isoptera: Rhinotermitidae) in City Park, New Orleans, Louisiana. *Sociobiology* 50, 311–335.

Cottone, C. and Su, N.-Y. (2017) Survival of Formosan subterranean termite colonies during periods of flooding. In: Dhang, P. (ed.) *Climate Change: Effects on Urban Pests*. CAB International, Boston, Massachusetts, pp. 65–79.

Cottone, C., Su, N.-Y., Scheffrahn, R. and Riegel, C. (2015) Survivorship of the Formosan subterranean termite (Isoptera: Rhinotermitidae) in a hypoxic environment. *Sociobiology* 62, 76–81. doi: 10.13102/sociobiology.v62il.76-81

Crossland, M.W.J., Lok, C.M., Wong, T.C., Shakarad, M. and Traniello, J.F.A. (1997) Division of labour in a lower termite: the majority of tasks are performed by older workers. *Animal Behavior* 54, 999–1012.

Darlington, J.P.E.C. (1984) A method for sampling the populations of large termite nests. *Annals of Applied Biology* 104, 427–436.

Delaplane, K.S. (1991) Foraging and feeding behaviors of the Formosan subterranean termite (Isoptera: Rhinotermitidae). *Sociobiology* 19, 101–114.

Du, H., Chouvenc, T., Osbrink, W.L.A. and Su, N.-Y. (2016) Social interactions in the central nest of *Coptotermes formosanus* juvenile colonies. *Insectes Sociaux* 63, 279–290. doi: 10.1007/s00040-016-0464-4

Du, H., Chouvenc, T. and Su, N.-Y. (2017a) Development of age polyethism with colony maturity in *Coptotermes formosanus* (Isoptera: Rhinotermitidae). *Environmental Entomology* 46, 311–318. doi: 10.1093/ee/nvw162

Du, H., Chouvenc, T., Osbrink, W.L.A. and Su, N.-Y. (2017b) Heterogeneous distribution of castes/instars and behaviors in the nest of *Coptotermes formosanus* Shiraki. *Insectes Sociaux* 64, 103–112. doi: 10.1007/s00040-016-0520-0

Emlen, J.M. (1966) The role of time and energy in food preference. *American Naturalist* 100, 611–617.

Erhorn, E.M. (1934) The termites of Hawaii, their economic significance and control, and the distribution of termites by commerce. In: Kofoid, C.A. (ed.) *Termites and Termite Control*. University of California Press, Berkeley, California, pp. 321–333.

Etthershank, G., Etthershank, A. and Whitford, W.G. (1980) Locations of food sources by subterranean termites. *Environmental Entomology* 9, 645–648.

Evans, T.A., Lenz, M. and Gleeson, P.V. (1998) Testing assumption of mark-recapture protocols for estimating population size using Australian mound-building subterranean termites. *Ecological Entomology* 23, 139–159.

Fei, H. and Henderson, G. (2004) Effects of temperature, directional aspects, light conditions, and termite species on subterranean termite activity (Isoptera: Rhinotermitidae). *Environmental Entomology* 33, 242–248

Forchler, B.T. and Townsend, M.L. (1996). Mark-release-recapture estimates of *Reticulitermes* spp. (Isoptera: Rhinotermitidae) colony foraging populations from Georgia, USA. *Environmental Entomology* 25, 952–962.

Grace, J.K., Abdallay, A. and Farr, K.R. (1989) Eastern subterranean termite (Isoptera: Rhinotermitidae) foraging territories and population in Toronto. *Canadian Entomologist* 121, 551–555. doi: 10.4039/Ent121551-7

Grace, J.K., Yamamoto, R.T. and Tamashiro, M. (1995) Relationship of individual worker mass and population decline in a Formosan subterranean termite colony (Isoptera: Rhinotermitidae). *Environmental Entomology* 24, 1258–1262.

Grace, J.K., Aihara-Sasaki, M.A. and Yates, J.R. (2004) Differences in tunneling behavior of *Coptotermes vasatator* and *C. formosanus* (Isoptera: Rhinotermitidae). *Sociobiology* 43, 153– 158.

Haagsma, K.A. and Rust, M.K. (1993) Two marking dyes useful for monitoring field populations of *Reticulitermes hesperus* (Isoptera: Rhinotermitidae). *Sociobiology* 23, 155–166.

Hapukotuwa, N.K. and Grace, J.K. (2012) Comparison of *C. formosanus* and *C. gestroi* (Blattodea: Rhinotermitidae) field sites and seasonal foraging activity in Hawaii. *Sociobiology* 59, 203–212.

Higa, S.Y. (1981) Flight, colony foundation, and development of the gonads of the primary reproductives of the Formosan subterranean termite, *Coptotermes formosanus* Shiraki. PhD Dissertation, University of Hawaii, Honolulu.

Higa, S.Y. and Tamashiro, M. (1983) Swarming of the Formosan subterranean termite, *Coptotermes formosanus* Shiraki, in Hawaii (Isoptera: Rhinotermitidae). *Proceedings of Hawaiian Entomological Society* 24, 233–238.

Howard, R.W., Jones, S.C., Mauldin, J.K. and Beal, R.H. (1982) Abundance, distribution, and colony size estimates for *Reticulitermes* spp. (Isoptera: Rhinotermitidae) in southern Mississippi. *Environmental Entomology* 11, 1290–1293.

Hubbell, S.P., Johnson, L.K., Stanislav, E. and Wilson, B. (1980) Foraging by bucket-brigade in leaf-cutting ants. *Biotropica* 12, 210–213.

Husseneder, C., Grace, J.K., Messenger, M.T., Vargo, E.L. and Su, N.-Y. (2003) Describing the spatial and social organization of Formosan subterranean termite colonies in Armstrong Park, New Orleans. *Sociobiology* 41, 61–65.

Ikehara, S. (1980) Termite fauna of Ryukyu archipelago. *ANIMA Magazine of Natural History* 91, 18–22 [in Japanese].

Jones, S.C. (1990) Colony size of desert subterranean termite *Heterotermes aureus* (Isoptera: Rhinotermitidae). *The Southwestern Naturalist* 35, 285–291.

Kakkar, G., Osbrink, W., Mullins, A. and Su, N.-Y. (2017) Molting site fidelity in workers of Formosan subterranean termites (Isoptera: Rhinotermitidae). *Journal of Economic Entomology* 110, 2512–2517. doi: 10.1093/jee/tox246

King, E.G. and Spink, W.T. (1969) Foraging galleries of the Formosan termite, *Coptotermes formosanus*, in Louisiana. *Annals of the Entomological Society of America* 62, 536–542.

Krebs, J.R., MacRoberts, M.H. and Cullen, J.M. (1972) Flocking and feeding in the great tit *Parus major*: an experimental study. *Ibis* 114, 507–530.

La Fage, J.P. and Nutting, W.L. (1978). Nutrient dynamics of termites. In Brian, M.V. (ed.) *Production Ecology of Ants and Termites*. Cambridge University Press, Cambridge, UK, pp. 165–232.

Lai, P.Y. (1977) Biology and ecology of the Formosan subterranean termite, *Coptotermes formosanus*, and its susceptibility to the entomogenous fungi, *Beauveria bassiana* and *Metarrhizium anisopliae*. PhD Dissertation, University of Hawaii, Honolulu.

Lai, P.Y., Tamashiro, M., Fujii, J.K., Yates J. and Su, N.-Y. (1983a) Sudan Red 7B, a dye marker for *Coptotermes formosanus*. *Proceedings of Hawaiian Entomological Society* 24, 277–282.

Lai, P.Y., Tamashiro, M., Yates J.R., Fujii, J.K., Su, N.-Y. *et al.* (1983b) Living plants in Hawaii attacked by *Coptotermes formosanus*. *Proceedings of Hawaiian Entomological Society* 24, 283–286.

Lee, C.-Y. (2002) Control of foraging colonies of subterranean termites, *Coptotermes travians* (Isoptera: Rhinotermitidae) in Malaysia using hexaflumuron baits. *Sociobiology* 39, 411–416.

Lee, S.-B., Chouvenc, T. and Su, N.-Y. (2021) Differential time allocation of foraging workers in the subterranean termite. *Frontiers in Zoology* 18, 61. doi: 10.1186/s12983-021-00446-5

Lee, S.-B., Chouvenc, T., Mizumoto, N., Mullins, A. and Su, N.-Y. (2022) Age-based spatial distribution of workers is resilient to worker loss in a subterranean termite. *Science Report* 12, 7837. doi: 10.1038/s41598-022-11512-1

Lee, S.-H. and Su, N.-Y. (2009) A simulation study of subterranean termites' territory formation. *Ecological Informatics* 4, 111–116.

Lee, S.-H., Bardunias, P. and Su, N.-Y. (2006) Food encounter rates of simulated termite tunnels with variable food size/distribution pattern and tunnel branch length. *Journal Theoretical Biology* 243, 493–500.

Lee, S.-H., Bardunias, P. and Su, N.-Y. (2007) Optimal length distribution of termite tunnel branches for efficient food search and resource transportation. *BioSystems* 90, 802–807.

Lee, S.-H., Bardunias, P. and Su, N.-Y. (2008) Two strategies for optimizing the food encounter rate of termite tunnels simulated by a lattice model. *Ecological Modeling* 213, 381–388.

Leong, K.L.H., Tamashiro, M., Yates, J.R. and Su, N.-Y. (1983) Microenvironmental factors regulating the flight of *Coptotermes formosanus* Shiraki in Hawaii. (Isoptera: Rhinotermitidae). *Proceedings of Hawaiian Entomological Society* 24, 287–291.

Leuthold, R.H., Bruinsma, O. and Van Huis, A. (1976) Optical and pheromonal orientation and memory for homing distance in the harvester termite *Hodotermes mossambicus* (Hagen). *Behavioral Ecology and Sociobiology* 1, 127–139.

Levings, S.C. and Adams, E.S. (1984) Intra- and interspecific territory in *Nasutitermes* (Isoptera: Terrnitidae) in a Panamanian mangrove forest. *Journal of Animal Ecology* 53, 705–714.

Li, H.-F. and Su, N.-Y. (2008) Sand displacement during tunnel excavation by the Formosan subterranean termite (Isoptera: Rhinotermitidae). *Annals of the Entomological Society of America* 101, 456–462.

MacArthur, R.H. and Pianka, E.R. (1966) On optimal use of a patchy environment. *American Naturalist* 100, 603–610.

Majid, A.H.A. and Ahmad, A.H. (2015) Define colony number of subterranean termites *Coptotermes gestroi* (Isoptera: Rhinotermitidae) in selected infested structures. *Sains Malaysia* 44, 211–216.

Matsumura, F., Coopel, H.C. and Tai, A. (1968) Isolation and identification of termite trail-following pheromone. *Nature* 219, 963–964.

Messenger, M.T. and Mullins, A. J. (2005) New flight distance record for *Coptotermes formosanus* (Isoptera: Rhinotermitidae). *Florida Entomologist* 88, 99–100.

Messenger, M.T. and Su, N.-Y. (2005a) Colony characteristics and seasonal activity of the Formosan subterranean termite (Isoptera: Rhinotermitidae) in Louis Armstrong Park, New Orleans, Louisiana. *Journal of Entomological Science* 40, 268–279.

Messenger, M.T. and Su, N.-Y. (2005b) Agonistic behavior between colonies of the Formosan subterranean termite from Louis Armstrong Park, New Orleans, Louisiana. *Sociobiology* 45, 331–345.

Messenger, M.T., Su, N.-Y., Husseneder, C. and Grace, J.K. (2005) Elimination and reinvasion studies with *Coptotermes formosanus* (Isoptera: Rhinotermitidae) in Louisiana. *Journal of Economic Entomology* 98, 916–920.

Michael, M., Chouvenc, T., Su, N.-Y. and Lee, S.-B. (2023) Finding shortcuts through collective tunnel excavations in a subterranean termite. *Behavioral Ecology* 34, 354–362. doi: 10.1093/beheco/arad007

Mizumoto, N., Bardunias, P. and Pratt, S.C. (2020) Complex relationship between tunneling patterns and individual behaviors in termites. *American Naturalist* 196, 555–565. doi: 10.1086/711020

Mullins, A. and Su, N.-Y. (2018) Parental nitrogen transfer and apparent absence of N_2 fixation during colony foundation in *Coptotermes formosanus* Shiraki. *Insects* 9, 37. doi: 10.3390

Mullins, A.J., Su, N.-Y. and Owens, C. (2011) Reinvasion and colony expansion of *Coptotermes formosanus* (Isoptera: Rhinotermitidae) after areawide elimination. *Journal of Economic Entomology* 104,1687–1697. doi: 10.1603/EC11036

Mullins, A. J., Messenger, M., Hochmair, H. H., Tonini, F., Su, N.-Y. *et al.* (2015) Dispersal flights of the Formosan subterranean termite (Isoptera: Rhinotermitidae). *Journal of Economic Entomology* 108, 702–719. doi: 10.1093/jee/tov022

Mullins, A., Chouvenc T. and Su, N.-Y. (2021) Soil organic matter is essential for colony growth in subterranean termites. *Scientific Report* 11, 21252. doi: 10.1038/s41598-021-00674-z

Mullins, A., Sheffhrahn, R.H. and Su, N.Y. (2022) Nitrogen inventories and nitrogenase expression rates of a drywood and a subterranean termite. *Annals of the Entomological Society of America* 115, 387–390. doi: 10.1093/aesa/saac014

Myles, T.G. (1996) Development and evaluation of transmissible coating for control of subterranean termites. *Sociobiology* 28, 373–457.

Nel, J.J.C. (1968) Aggressive behavior of the harvester termites *Hodotermes mossambicus* (Hagen) and *Trinervitermes trinervoides* (Sjöstedt). *Insectes Socaiux* 15, 145–156.

Noda, S., Ohkuma, M. and Kudo, T. (2002) Nitrogen fixation genes expressed in the symbiotic microbial community in the gut of the termite *Coptotermes formosanus*. *Microbes and Environment* 17, 139–143.

Owens, C.B., Su, N.-Y., Husseneder, C., Riegel, C. and Brown, K. (2012) Molecular genetic evidence of Formosan subterranean termite (Isoptera: Rhinotermitidae) colony survivorship after prolonged inundation. *Journal of Economic Entomology* 105, 518–522. doi: 10.1603/EC1150

Puche, H. and Su, N.-Y. (2001) Tunneling formation by *Reticulitermes flavipes* and *Coptotermes formosanus* (Isoptera: Rhinotermitidae) in response to wood in sand. *Journal of Economic Entomology* 94, 1398–1404.

Raina, A., Yong, L.P. and Galman, D. (2008) Molting in workers of the Formosan subterranean termite *Coptotermes formosanus*. *Journal of Insect Physiology* 54, 155–161.

Ratcliffe, F.N. and Greaves, T. (1940) The subterranean foraging galleries of *Coptotermes lacteus* (Frogatt). *Journal of the Council of Scientific and Industrial Research of Australia* 13, 150–161

Reinhard, J., Hertel, H. and Kaib, H. (1997) Systematic search for food in the subterranean termite *Reticulitermes santonensis* De Feytaud (Isoptera, Rhinotermitidae). *Insectes Sociaux* 44, 147–158.

Robson, S.K., Lesniak, M.G., Kothandapani, R.V. and Traniello J.F.A. (1995) Nonrandom search geometry in subterranean termites. *Naturwissenschaften* 82, 526–528.

Rohrmann, G.F. and Rossman, A.Y. (1980) Nutrient strategies of *Macrotermes ukuzii* (Isoptera: Termitidae). *Pedobiologia* 20, 61–73.

Roisin, Y. and Korb, J. (2011) Social organization and the status of workers in termites. In: Bignell, D.W., Roisin, Y. and Lo, N. (eds) *Biology of Termites: A Modern Synthesis*. Springer, Dordrecht, Netherlands, pp. 133–164.

Santos, M.N., Teixeira, M.J.F., Pereira M.B. and Menezes, M.B. (2010) Environmental factors influencing the foraging and feeding behavior of two termite species (Isoptera: Rhinotermitidae) in natural habitats. *Sociobiology* 55, 763–777.

Šobotník, J., Kutalová, K., Vytisková, B., Roisin, Y. and Bourguignon, T. (2014) Age-dependent changes in ultrastructure of the defensive glands of *Neocapritermes taracua* workers (Isoptera, Termitidae). *Arthropod Structure and Development* 43, 205–210.

Sornnuwat, Y., Tsunoda, K., Yoshimura, T., Takahashi, M. and Vongkaluang, C. (1996) Foraging populations of *Coptotermes gestroi* (Isoptera: Rhinotermitidae) in an urban area. *Journal of Economic Entomology* 89, 1485–1490. doi: 10.1093/jee/89.6.1485

Su, N.-Y. (1994) Field evaluation of a hexaflumuron bait for population suppression of subterranean termites (Isoptera: Rhinotermitidae). *Journal of Economic Entomology* 87, 389–397.

Su, N.-Y. (2005) Directional change in tunneling of subterranean termites (Isoptera: Rhinotermitidae) in response to decayed wood attractants. *Journal of Economic Entomology* 98, 471–475.

Su, N.-Y. and Haverty, M.I. (1991) Agonistic behavior among colonies of the Formosan subterranean termite, *Coptotermes formosanus* Shiraki (Isoptera: Rhinotermitidae), from Florida and Hawaii: Lack of correlation with cuticular hydrocarbon composition. *Journal of Insect Behavior* 4, 115–128.

Su, N.-Y. and La Fage, J.P. (1984) Differences in survival and feeding activity among colonies of the Formosan subterranean termite (Isoptera: Rhinotermitidae). *Zeitschrift für Angewandte Entomologie* 94, 134–138.

Su, N.-Y. and Lee, S.-H. (2008) Estimating the population size and colony boundary of subterranean termites by using the density functions of directionally-averaged capture probability. *Journal of Economic Entomology* 101, 592–604.

Su, N.-Y. and Lee, S.-H. (2009) Tunnel volume regulation and group size of subterranean termites (Isoptera: Rhinotermitidae). *Annals of the Entomological Society of America* 102, 1158–1164.

Su, N.-Y. and Lee, S.-B. (2023) A comparison of tunnel geometry between the Formosan subterranean termite and Asian subterranean termite (Blattodea: Rhinotermitidae). *Pest Management Science* 79(1), 3999–4003. doi: 10.1002/ps.7594

Su, N.-Y. and Puche, H. (2003) Tunneling activity of subterranean termites (Isoptera: Rhinotermitidae) in sand with moisture gradients. *Journal of Economic Entomology* 96, 88–93.

Su, N.-Y. and Scheffrahn, R.H. (1986) A method to access, trap, and monitor field populations of the Formosan subterranean termite (Isoptera: Rhinotermitidae) in the urban environment. *Sociobiology* 12, 299–304.

Su, N.-Y. and Scheffrahn, R.H. (1987) Alate production of a field colony of the Formosan subterranean termite (Isoptera: Rhinotermitidae). *Sociobiology* 13, 167–172.

Su, N.-Y. and Scheffrahn, R.H. (1988a) Foraging population and territory of the Formosan subterranean termite (Isoptera: Rhinotermitidae) in an urban environment. *Sociobiology* 14, 353–359.

Su, N.-Y. and Scheffrahn, R.H. (1988b) Intra- and interspecific competition of the Formosan and eastern subterranean termite: Evidence from field observations (Isoptera: Rhinotermitidae). *Sociobiology* 14, 157–164.

Su, N.-Y. and Scheffrahn, R.H. (1998) A review of subterranean termite control practices and prospects for integrated pest management programs. *Integrated Pest Management Reviews* 3, 1–13.

Su, N.-Y. and Tamashiro, M. (1986) Wood-consumption rate and survival of the Formosan subterranean termite (Isoptera: Rhinotermitidae) when fed one of six woods commercially used in Hawaii. *Proceedings of Hawaiian Entomological Society* 26, 109–113.

Su, N.-Y., Ban, P.M. and Scheffrahn, R.H. (1991) Evaluation of twelve dye markers for population studies of the eastern and Formosan subterranean termites (Isoptera: Rhinotermitidae). *Sociobiology* 19, 349–362.

Su, N.-Y., Scheffrahn, R.H. and Ban, P.M. (1995) Effects of sulfluramid-treated bait blocks on field colonies of the Formosan subterranean termite (Isoptera: Rhinotermitidae). *Journal of Economic Entomology* 88, 1343–1348.

Su, N.-Y., Scheffrahn, R.H. and Weissling, T. (1997) A new introduction of a subterranean termite, *Coptotermes havilandi* Holmgren (Isoptera: Rhinotermitidae) in Miami, Florida. *Florida Entomologist* 80, 408–411.

Su, N.-Y., Ban, P.M. and Scheffrahn, R.H. (2000) Control of *Coptotermes havilandi* (Isoptera: Rhinotermitidae) with hexaflumuron baits and a sensor incorporated into a monitoring-baiting program. *Journal of Economic Entomology* 93, 415–421. doi: 10.1603/0022-0493-93.2.415

Su, N.-Y., Stith, B.M., Puche, H. and Bardunias, P. (2004) Characterization of tunneling geometry of subterranean termites (Isoptera: Rhinotermitidae) by computer simulation. *Sociobiology* 44, 471–483.

Su, N.-Y., Guidry, E., Mullins, A.J. and Cottone, C. (2016) *Reinvasion dynamics of subterranean termites (Isoptera: Rhinotermitidae) following the elimination of all detectable colonies in a large area*. *Journal of Economic Entomology* 109, 809–814. doi: 10.1093/jee/tow018

Su, N.-Y., Osbrink, W., Kakkar, G., Mullins, A. and Chouvenc, T. (2017a) Foraging distance and population size of juvenile colonies of the Formosan subterranean termite (Isoptera: Rhinotermitidae) in laboratory extended arenas. *Journal of Economic Entomology* 110, 1728–1735. doi: 10.1093/jee/tox153

Su, N.-Y., Chouvenc, T. and Li, H.F. (2017b) Potential hybridization between two invasive termite species, *Coptotermes formosanus* and *C. gestroi* (Isoptera: Rhinotermitidae), and its biological and economic implications. *Insects* 8(1), 14. doi: 10.3390/insects8010014

Su, N.-Y., Mullins, A. and Chouvenc, T. (2023) Elimination of structural and tree infestations of the Asian subterranean termite, *Coptotermes gestroi* (Wasmann) (Blattodea: Rhinotermitidae) with noviflumuron baits in above-ground stations. *Journal of Economic Entomology* 116(3), 909–915. doi: 10.1093/jee/toad077

Sugio, K. (1995) Trunk trail foraging of the fungus-growing termite *Macrotermes carbonarius* (Hagen) in southeastern Thailand. *Tropics* 4, 211–222.

Tamashiro, M., Fujii J.K. and Lai P.Y. (1973) A simple method to observe, trap, and prepare large numbers of subterranean termites for laboratory and field experiments. *Environmental Entomology* 2, 721–22.

Tayasu, I., Inoue, T., Miller, L., Sugimoto, A., Takeichi, S. *et al.* (1998) Confirmation of soil-feeding termites (Isoptera; Termitiade; Termitinae) in Australia using stable isotope ratios. *Functional Ecology* 12, 536–542.

Thorne, B.L. (1982) Termite-termite interactions: Workers as an agonistic caste. *Psyche* 89, 133–150.

Thorne, B.L., Russek-Cohen, E., Forschler, B.T., Breisch, N.L. and Traniello, J.F.A. (1996) Evaluation of mark-recapture methods for estimating forager population size of subterranean termite (Isoptera: Rhinotermitidae) colonies. *Environmental Entomology* 25, 938 –951.

Tong, R.L., Aguilera-Olivares, D., Chouvenc, T. and Su, N.-Y. (2021) Nitrogen content of the exuviae of *Coptotermes gestroi* (Wasmann) (Blattodea: Rhinotermitidae). *Heliyon* 7, 4. doi: 10.1016/j.heliyon.2021.e06697

Tong, R.L., Choi, E.-K., Ugarelli, K., Chouvenc, T. and Su, N.-Y. (2023a) Trophic path of marked exuviae within colonies of *Coptotermes gestroi* (Blattodea: Rhinotermitidae). *Journal of Insect Science* 23(3), 1–5. doi: 10.1093/jisesa/iead007

Tong, R.L., Patel, J.S., Gordon, J.M., Lee, S.-B., Chouvenc, T. *et al.* (2023b) Exuviae recycling can enhance queen oviposition and colony growth in subterranean termites (Blattodea: Rhinotermitidae: Coptotermes). *Environmental Entomology* 52(2), 254–258. doi: 10.1093/ee/nvad009

Traniello, J.F.A. (1981) Enemy deterrence in the recruitment strategy of a termite: solider-organized foraging in *Nasutitermes costalis*. *Proceedings of National Academy of Science USA* 78, 1976–1979.

Traniello, J.F.A. and Leuthold, R.H. (2000) Behavior and ecology of foraging in termites. In: Abe, T., Bignell, D.E. and Higashi, M. (eds) *Termites: Evolution, Sociality, Symbioses, Ecology*. Kluwer Academic Publisher, Dordrecht, Netherlands, pp. 141–168.

Tseng, S.P., Boone, J., Boone, N., King, S., Taravati, S. *et al.* (2021) Genetic analysis of Formosan subterranean termite (Blattodea: Rhinotermitidae) populations in California. *Journal of Economic Entomology* 114, 1264–1269.

Tseng, S. P., Taravati, S., Choe, D. H., Rust, M. K. and Lee, C. Y. (2022) Genetic evidence for multiple invasions of *Coptotermes formosanus* (Blattodea: Rhinotermitidae) in California. *Journal of Economic Entomology* 115, 1251–1256.

Uchima, S.Y. and Grace, J.K. (2003) Characteristics of *Coptotermes vastator* (Isoptera: Rhinotermitidae) colonies on Oahu, Hawaii. *Sociobiology* 41, 281–288.

Umar, W.A.S.W. and Majid, A.H. (2020) Efficacy of minimum application of chlorfluazuron baiting to control urban subterranean termite populations of *Coptotermes gestroi* (Wasmann) (Blattodea: Rhinotermitidae). *Insects* 11, 569. doi:10.3390/insects1190569

Wood, T.G. (1978) Food and feeding habit of termites. In Brian, M.V. (ed.) *Production Ecology of Ants and Termites*. Cambridge University Press, Cambridge, UK, pp. 55–110.

Xing, L., Chouvenc, T. and Su, N.-Y. (2013) Molting process in the Formosan subterranean termite (Isoptera: Rhinotermitidae). *Annals of the Entomological Society of America* 106, 619–625. doi: 10.1603/AN13007

Yang, R.-L., Su, N.-Y. and Bardunias, P. (2009) Individual task load in tunnel excavation by the Formosan subterranean termite (Isoptera: Rhinotermitidae). *Annals of the Entomological Society of America* 102, 906–910.

Yeoh, B.-H. and Lee, C.-Y. (2007) Tunneling responses of the Asian subterranean termite, *Coptotermes gestroi* in termiticide-treated sand (Isoptera: Rhinotermitidae). *Sociobiology* 50, 457–468.

Zahavi, A. (1971) The social behavior of the White Wagtail *Motacilla alba alba* wintering in Israel. *Ibis* 113, 203–211.

Zheng, D., Hunt Jr, E.R. and Running, S.W. (1993) A daily soil temperature model based on air temperature and precipitation for continental applications. *Climate Research* 2, 183–191.

Zukowski, J. and Su, N.-Y. (2017) Survival of termites (Order: Isoptera) exposed to different levels of relative humidity (RH) and water availability and their RH preferences. *Florida Entomologist* 100, 532–538. doi: 10.1653/024.100.0307

6 Physiology of the Formosan subterranean termite, with special reference to wood degradation and metabolism

Gaku Tokuda[1], Shuji Itakura[2] and Nathan Lo[3]
[1]*Tropical Biosphere Research Center, University of the Ryukyus, Japan, tokuda@comb.u-ryukyu.ac.jp;* [2]*Graduate School of Agriculture, Kindai University, Japan, itakura@nara.kindai.ac.jp;* [3]*School of Life and Environmental Sciences, University of Sydney, Australia, nathan.lo@sydney.edu.au*

Abstract

It has been widely recognized that the Formosan subterranean termite, *Coptotermes formosanus* Shiraki, is a significant pest species damaging wooden structures, owing to its prominent capability for degradation of wood. The primary component of wood is a complex structure consisting of cellulose, hemicellulose, and lignin, termed lignocellulose. This chapter provides an overview of current knowledge on the digestive enzymes involved in lignocellulose degradation produced by *C. formosanus* and their microbial symbionts in the hindgut, as well as the mode of action by which these enzymes attack lignocellulose. We also summarize our understanding of catabolism of simple sugars released by hydrolysis of cellulose and hemicellulose in the tissues of *C. formosanus*, with reference to pyruvate metabolism and energy budget.

6.1 Introduction

The Formosan subterranean termite, *Coptotermes formosanus* Shiraki, has been recognized as a significant pest against wooden structures. The economic impact of subterranean termites, including *C. formosanus*, is estimated to be US$32 billion annually (Rust and Su, 2012). *Coptotermes formosanus* belongs to the family Rhinotermitidae. The genera *Coptotermes*, *Heterotermes*, and *Reticulitermes* form a clade within Rhinotermitidae which is the sister group to the most speciose termite family: the Termitidae or so-called higher termites, which harbor only a prokaryotic community in the hindgut (Bucek *et al.*, 2019). However, rhinotermitids are among the so-called lower termites, whose hindguts harbor the cellulolytic protists. These protists significantly contribute to the digestion of wood primarily composed of lignocellulose, a complex structure of cellulose, hemicellulose, and lignin. It has been assumed that digestive enzymes from the host and the symbiotic microbes act independently to degrade lignocellulose either in the midgut or in the hindgut (Nakashima *et al.*, 2002a,b; Arakawa *et al.*, 2009; Fujita *et al.*, 2010), while such a division of labor between the host and the hindgut symbionts is not limited to lignocellulose digestion. Indeed, other essential digestive enzymes,

 Biology and Management of the Formosan Subterranean Termite and Related Species (eds Nan-Yao Su and Chow-Yang Lee)
DOI: 10.1079/9781800621596.0006

such as proteases, are also secreted both in the midgut and in the hindgut protists, and their enzymatic properties are distinct from each other (Sethi *et al.*, 2011). In this chapter, we discuss the diversity of these enzymes from the host and the hindgut symbionts, and how they play a role in lignocellulose degradation, in light of previous findings from classical biochemistry to recent genomics and transcriptomics. We then describe how cellulosic and hemicellulosic hydrolysates are catabolized in the termite tissues.

6.2 Cellulose and hemicellulose digestion in *C. formosanus*

6.2.1 Cellulose digestion by endogenous digestive enzymes in *C. formosanus*

After the first discovery of a cellulase gene from a subterranean rhinotermitid species, *Reticulitermes speratus* (Kolbe) (Watanabe *et al.*, 1998), remarkable progress has been made in our understanding of the endogenous cellulolytic system in workers of lower termites. As we describe below, the genomes of termites that have been investigated to date encode only cellulases among various lignocellulolytic enzymes that cleave backbones of either cellulose, hemicellulose, or lignin. Therefore, this section focuses on the current knowledge of endogenous cellulases in workers of *C. formosanus* unless otherwise indicated.

Coptotermes formosanus secretes cellulases in the salivary glands and the midgut, which are endo-β-1,4-glucanases (EC 3.2.1.4) that randomly hydrolyze cellulose chains and β-glucosidases (EC 3.2.1.21) that hydrolyze cellobiose or cello-oligomers into glucose (Fig. 6.1). When carboxylmethylcellulose (CMC), which is hydrolyzed by endo-β-1,4-glucanases, was used as the substrate, the salivary glands

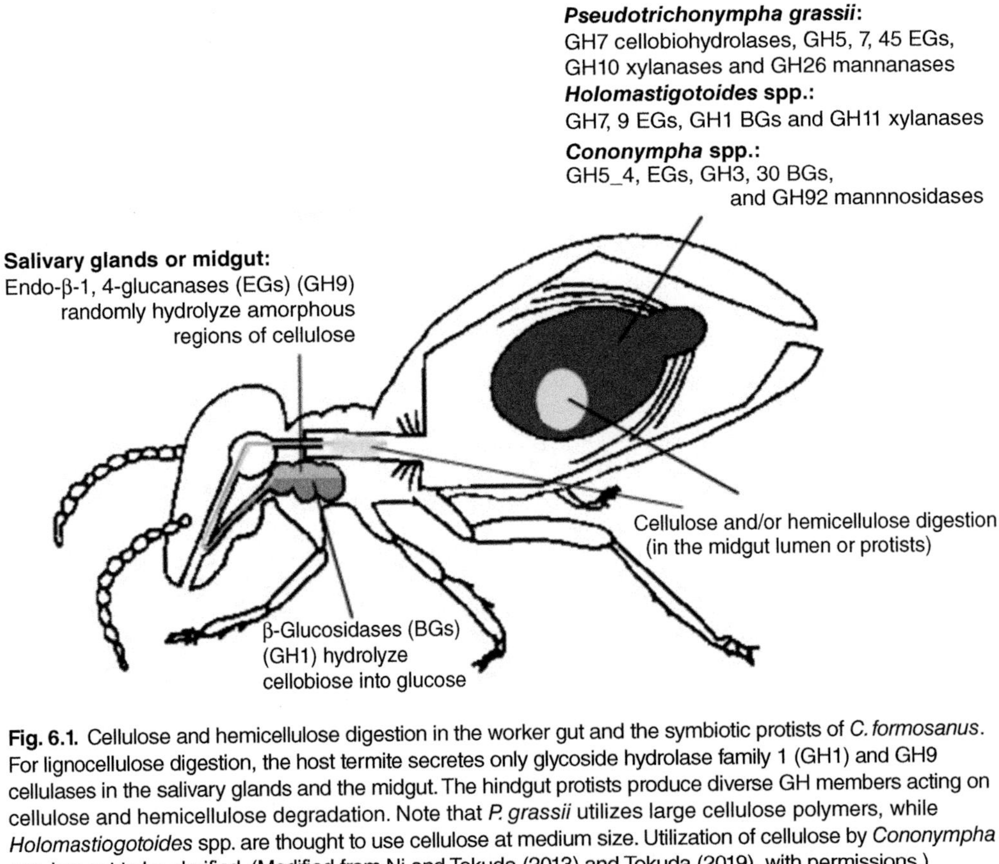

Fig. 6.1. Cellulose and hemicellulose digestion in the worker gut and the symbiotic protists of *C. formosanus*. For lignocellulose digestion, the host termite secretes only glycoside hydrolase family 1 (GH1) and GH9 cellulases in the salivary glands and the midgut. The hindgut protists produce diverse GH members acting on cellulose and hemicellulose degradation. Note that *P. grassii* utilizes large cellulose polymers, while *Holomastiogotoides* spp. are thought to use cellulose at medium size. Utilization of cellulose by *Cononympha* spp. has yet to be clarified. (Modified from Ni and Tokuda (2013) and Tokuda (2019), with permissions.)

always showed higher cellulase activity (35–75% of the total cellulase activity in the whole gut and the salivary glands) than the hindgut (10–38% of the total activity), which harbors the cellulolytic protists (Fig. 6.1) (Yoshimura *et al.*, 1992a; Itakura *et al.*, 1997; Nakashima *et al.*, 2002a; Tokuda *et al.*, 2004; Fujita *et al.*, 2010). The midgut, which is free of the symbiotic protists, also contains 15–18% of the total cellulase activity (Itakura *et al.*, 1997; Tokuda *et al.*, 2004; Fujita *et al.*, 2010). Such a distribution pattern of cellulase activity in the digestive system was dramatically changed when microcrystalline cellulose was used as substrate. Native cellulose consists primarily of crystalline regions with some amorphous structures. Amorphous cellulose can be attacked by endo-β-1,4-glucanases, but the crystalline structure protects internal cellulose fibers from penetration of water and enzyme molecules. Thus, efficient hydrolysis of such crystalline structure often requires cellobiohydrolases (EC 3.2.1.91 or EC 3.2.1.176), which act from the reducing or non-reducing ends of cellulose fibers and release cellobiose. Hydrolysis of native cellulose is achieved by the synergistic action between endo-β-1,4-glucanases and cellobiohydrolases (Lynd *et al.*, 2002). The hindgut of *C. formosanus* contains 36–87% of the total cellulase activity (that may depend on experimental conditions and the degree of cellulose crystallinity) against microcrystalline cellulose, while the salivary glands possess less than 25% of the total cellulase activity against crystalline cellulose (Yoshimura *et al.*, 1992a; Itakura *et al.*, 1997; Nakashima *et al.*, 2002a; Tokuda *et al.*, 2005). These studies indicate that endogenous cellulases have distinct enzymatic properties from those produced by the hindgut protists. See Ni and Tokuda (2013) for more details on the enzymatic properties of cellulases from termites.

Although endogenous endo-β-1,4-glucanases are assumed to act weakly against microcrystalline cellulose, one of them primarily released cellobiose from filter paper, similar to a catalytic reaction of cellobiohydrolases (Zhang *et al.*, 2009). Nakashima *et al.* (2002a) proposed that its activity may produce a certain amount of glucose (1.05 μmol/g termite/h) from microcrystalline cellulose, which compensates for half of the respiratory requirements (2.12 μmol/g termite/h) by the termites. This suggests that the activity of endogenous cellulases is insufficient for the termites to thrive on wood. This activity in the midgut is probably a result of synergistic action between endogenous endo-β-1,4-glucanase and β-glucosidase, which has been proposed in a closely related species, *Coptotermes gestroi* (Wasmann) (Franco Cairo *et al.*, 2013), while the net flow of carbon for energy metabolism should be accounted for by the efficiency of the downstream metabolic pathways (see section 6.5).

In an early study of the molecular biology of cellulases in *C. formosanus*, five similar cDNA sequences of endo-β-1,4-glucanases (i.e., *CfEG*s that are 99.5% identical to each other) were identified (Nakashima *et al.*, 2002a). Amino acid sequences of CfEGs are affiliated with glycoside hydrolase family 9 (GH9), widespread among bilaterian animals (Watanabe and Tokuda, 2010; Tokuda, 2019). An attempt of RNA interference (RNAi) targeting conserved regions of *CfEG*s, by either microinjection or oral ingestion, resulted in a 24–41% reduction in the expression levels of *CfEG*s, which caused an increase in mortality (12–23% vs 7–17% mortality without dsRNA) and a significant reduction in body weight of the treated termites (Wu *et al.*, 2019). These results suggest that the endogenous cellulases support the viability of the termites. Still, those from the gut symbionts might have compensated for the loss of the endogenous activity to some extent.

Multiple cDNA sequences of endogenous β-glucosidases (*CfGlu1A* to *CfGlu1E*, *CfBGIa*, *CfSNI1172*, and *CFSNI2059*) have also been cloned from *C. formosanus* (Liu *et al.*, 2017). Among them, the enzymatic properties of glucose-tolerant CfBGI, predominantly expressed in the salivary glands and later termed CfGluB, were initially reported (Zhang *et al.*, 2010, 2012a). Subsequently, the properties of CfGlu1C, which was also found in the salivary glands, were characterized (Feng *et al.*, 2015). CfGluB hydrolyzed cellobiose more efficiently than CfGluC, while CfGluC preferentially hydrolyzed sucrose. These enzymes did not show a synergistic effect on hydrolysis of cellobiose or sucrose, but they synergistically hydrolyzed lactose (Feng *et al.*, 2015). On the other hand, properties of the other β-glucosidases remain elusive.

These β-glucosidase sequences belong to GH1 but are phylogenetically divided into at least three distinct clusters (Liu *et al.*, 2017). One of these clusters consists of bona fide digestive β-glucosidases, which contain two β-glucosidases from *C. formosanus* such as CfBGa and CfGluB. A previous report indicated that RNAi effectively reduced the expression levels of target digestive β-glucosidases, but it increased in expressions of non-target β-glucosidases (Wu and Li, 2018). Thus, RNAi-treated individuals against β-glucosidases did not significantly increase in mortality compared with control individuals. On the other hand, β-glucosidases of other clusters either lack a signal peptide for secretion or show sex-specific expressions, suggesting that they may not be involved in cellulose digestion in the gut.

6.2.2 Genomic and transcriptomic perspectives on endogenous cellulases in *C. formosanus*

Metatranscriptomic analysis of the gut tissue and the hindgut protist community in *C. formosanus* commenced in 2012 (Xie *et al.*, 2012; Zhang *et al.*, 2012b), but an accurate determination of the source of mRNA (i.e., whether it is from the host, protists, or bacteria) has been challenging. Geng *et al.* (2018b) dissected the salivary glands and the gut into the foregut, the midgut, and the hindgut, and analyzed the transcriptomes in each segment. As a result, it was discovered that the salivary glands express a more comprehensive range of redox enzymes than the midgut. They were presumed to be involved in lignin modification during the passage from the foregut to the midgut. With regard to enzymes dominantly expressed in the salivary glands or the midgut, transcripts encoding GH13 (primarily acting on α-glucoside bonds including amylases), GH16 (that contains some different types of enzymes, but they are commonly found as β-1,3-glucanases and β-glucan-binding proteins in insects (da Silva Moraes *et al.*, 2014)), GH22 (mostly lysozymes), and GH23 (peptidoglycan lyases) in addition to GH1 and GH9 cellulases were detected. However, enzymes involved in the hydrolysis of hemicellulose backbones, such as xylan and (gluco)mannan, are lacking in the salivary glands and the midgut. The absence of endogenous hemicellulases in termites was further supported by the recent genome sequencing projects of termites, including *C. formosanus* (Itakura *et al.*, 2020). As enzymes involved in the hydrolysis of cellulose or hemicellulose backbones, the draft genome of *C. formosanus* contains only GH1 and GH9 members, consisting of at least nine and three genes, respectively.

Fig. 6.2 shows a phylogenetic tree of major GH9 members from insects. All genes of GH9 detected from the genome of *C. formosanus* are most closely related to those from *R. speratus* (Shigenobu *et al.*, 2022). However, no cellulase gene of *C. formosanus*, which should have been most closely associated with *RsEG1*, the first cellulase gene from termites (Watanabe *et al.*, 1998), was found, despite the presence of the corresponding cDNAs. This is probably due to the incompleteness of the draft genome, and we assume that the genome of *C. formosanus* contains four orthologs of GH9, as is the case of *R. speratus* (Shigenobu *et al.*, 2022). Insect GH9 members are divided into two large clusters, in which clade A mainly consists of *bona fide* digestive endo-β-1,4-glucanases (Fig. 6.2). Clade B, which contains a single gene from every termite species whose genomes have been sequenced, consists of GH9 homologs from diverse insects, including hymenopterans and dipterans that may not utilize cellulosic diets. In *R. speratus*, a gene in Clade B was weakly but uniformly expressed regardless of body parts, sex, or castes, while the expression level was slightly increased upon molting (Shigenobu *et al.*, 2022). Thus, it is reasonable to assume that the members of Clade B may have a role other than cellulose digestion in insects.

Alignment of amino acid sequences detected from *C. formosanus* and *R. speratus* indicates that amino acid residues involving a co-nucleophile (D) and a proton donor (E) are substituted in both GFG41057 (*C. formosanus*) and RS012684 (*R. speratus*), which clustered within Clade A-2 (Figs 6.2 and 6.3). These

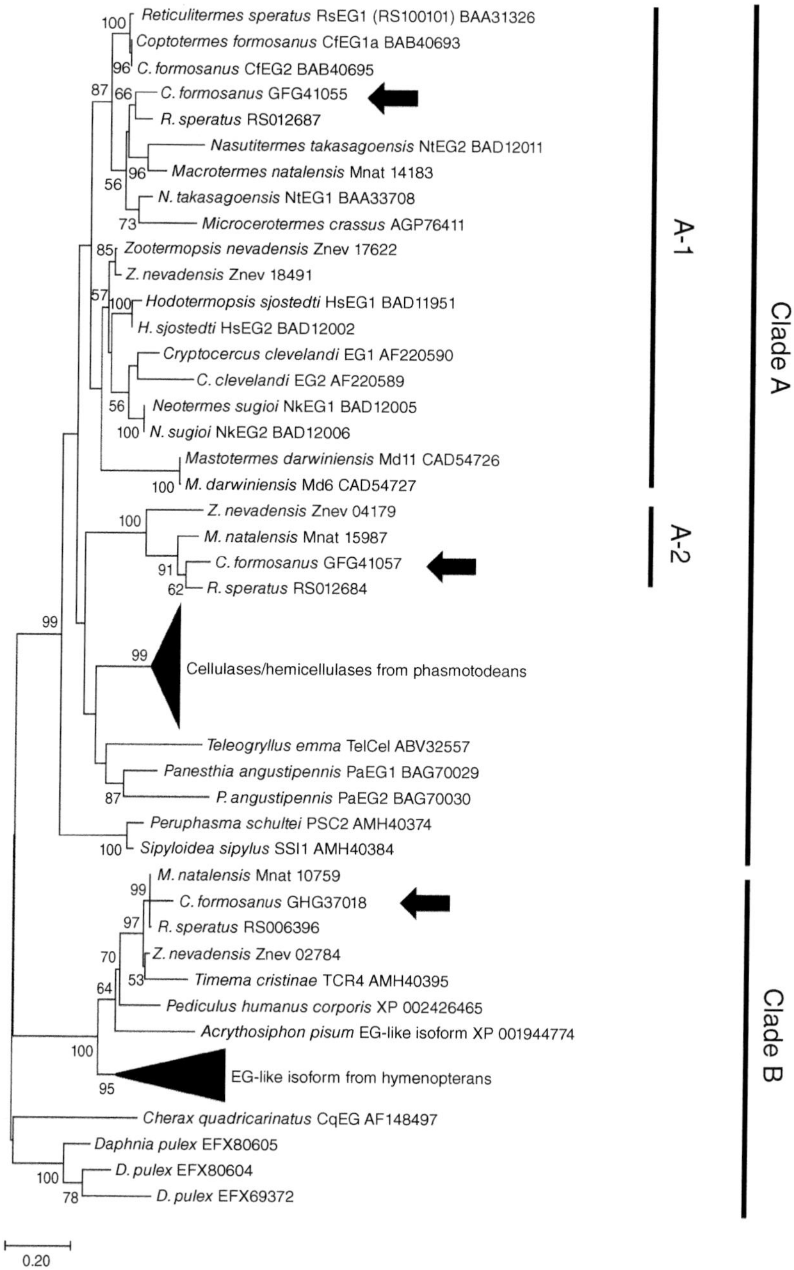

Fig. 6.2. Molecular phylogeny of endogenous GH9 members derived from insects, including *C. formosanus*. The evolutionary history was inferred using the Maximum Likelihood method based on amino acid sequences with the procedure previously mentioned by Shigenobu *et al.* (2022). Sequences derived from the genome of *C. formosanus* are denoted by arrows. Clades A and B correspond to those described in Fig. S8 by Shigenobu *et al.* (2022).

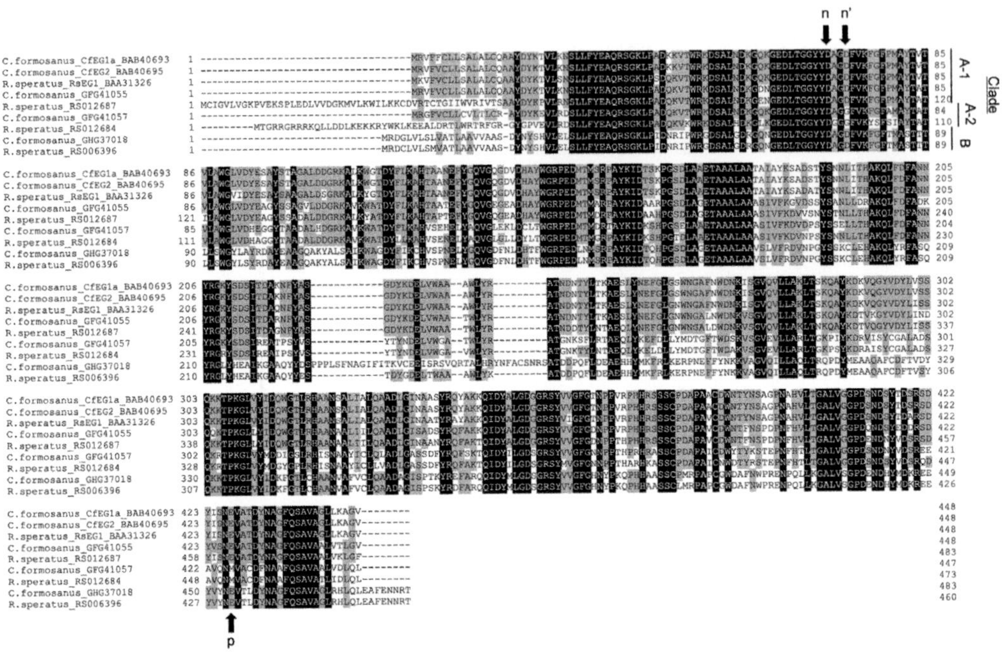

Fig. 6.3. Multiple alignment of amino acid sequences of endogenous GH9 members obtained from *C. formosanus* and *R. speratus*. Clades A-1, A-2 and B correspond to those described in Fig. 6.2. Symbols such as n, n', and p under the aligned sequences indicate nucleophile, co-nucleophile, and proton donor residues, respectively.

proteins are probably incapable of catalyzing the hydrolysis of cellulose chains. Nevertheless, a previous study indicated that the expression pattern of *RS012684* was very similar to *bona fide* digestive endo-β-1,4-glucanase *RsEG1* (Shigenobu *et al.*, 2022), suggesting that these genes may have the function of regulating cellulose hydrolysis with *bona fide* digestive enzymes.

The genome of *R. speratus* contains 16 GH1 paralogs, which were clustered into seven different clades showing unique expression patterns to each other (Shigenobu *et al.*, 2022). Although the draft genome of *C. formosanus* contains nine GH1 paralogs, it may increase in number after a more careful assessment. Since functions of GH1 members are diverse, including cellulose digestion, caste differentiation, and egg recognition pheromones (Shigenobu *et al.*, 2022), the precise roles of GH1 members in termites other than these functions are still awaiting clarification.

It has recently been recognized that diverse invertebrates possess genes encoding lytic polysaccharide monooxygenases (LPMOs) belonging to family15 of auxiliary activities (AA15) in their genomes. One such gene, found in the firebrat, has been demonstrated to oxidize and cleave cellulose chains, whereas another has been shown to oxidize and cleave the internal linkages of chitin (Sabbadin *et al.*, 2018). In *C. gestroi*, two members of AA15 have been found. These LPMOs were demonstrated to oxidize chitin, but they were unlikely to be involved in lignocellulose digestion (Franco Cairo *et al.*, 2021). Nevertheless, it is noteworthy that our preliminary searches indicated that the genome of *C. formosanus* contains three or more orthologs of AA15 LPMOs. The functions of these genes have yet to be elucidated.

6.3 Cellulose and hemicellulose degradation by the gut symbionts in *C. formosanus*

Until recently, it had been believed that the hindgut of *C. formosanus* harbored three protist species such as *Pseudotrichonympha grassii* Koidzumi (which is largest in body size), *Holomastgotoides hartmanii* Koidzumi (which is medium in size), and *Spirotrichonympha leidyi* Koidzumi (which is smallest in size) since the early taxonomic study by Koidzumi (1921). Most studies involving the symbiotic protists in *C. formosanus* were conducted on this premise. Still, these protists have been reclassified into three genera and five species (see Chapter 8, this volume). In this section, "*Holomastigotoides* spp." refers to a species originally described as "*Holomastigotoides hartmanii* (which was sometimes wrongly described as *H. mirabile*)," which might have included both *H. hartmanii* and *Holomastgotoides minor* Nishimura (Nishimura *et al.*, 2020). "*Cononympha* spp." refers to a species originally described as "*Spirotrichonympha leidy*," which might have consisted of two species such as *Cononympha leiydi* Koidzumi and *C. koidzumii* Jasso-Selles & Gile (Jasso-Selles *et al.*, 2020).

A pioneering study on the digestion of various saccharides in *C. formosanus* indicated that *P. grassii* cells were lost within 1 week of the termites feeding on non-cellulosic saccharides. At the same time, the numbers of *Holomastigotoides* spp. and *Cononympha* spp. were stable in termites feeding on non-cellulosic but hemicellulosic poly-, di-, and mono-saccharides for up to 4 weeks. However, these protist cells on non-cellulosic diets gradually decreased in number and were eventually lost over 8 weeks (Kanai *et al.*, 1982).

When termites were fed on cellulose that has a small degree of polymerization (DP < 27), the number of the largest protist, *P. grassii*, was decreased more rapidly than numbers of other protist species (Yoshimura *et al.*, 1993a). However, such a decrease in the number of *P. grassii* was not observed when the termites were fed on the cellulose of large DP (≈320). This suggests that *P. grassii*, which is distributed in the anterior part of the hindgut (Yoshimura *et al.*, 1992b), specifically depends on large cellulose polymers entering the hindgut but cannot utilize cellulose polymers that are small in size. After removal of the largest protist species (*P. grassii*) by administration of cellulose with small DP (≈17), the termites could not thrive on native cellulose any longer with a rapid decrease in the number of the other protist species (Yoshimura *et al.*, 1993b). Although polarizing and transmission electron microscopical observations indicated that the smallest protist cells (*Cononympha* spp.), which primarily distribute in the posterior part of the hindgut (Yoshimura *et al.*, 1992b), did not show uptake of cellulose particles (Yoshimura *et al.*, 1996), the protist species other than *P. grassii* may also depend on short polymers of cellulose. This is because administration of cellobiose to the termites for 30 days completely removed not only the large protists such as *P. grassii* and *Holomastigotoides* spp. but also most cells of *Cononympha* spp. from the hindgut (Tanaka *et al.*, 2006).

Using N-terminal amino acid sequences of the partially purified major cellulase components in the hindgut of *C. formosanus*, cDNAs encoding cellobiohydrolases and endo-β-1,4-glucanase were obtained from the symbiotic protists (Nakashima *et al.*, 2002b; Watanabe *et al.*, 2002). The largest protist *P. grassii* expressed genes encoding GH7 cellobiohydrolases and endo-β-1,4-glucanases, while *Holomastigotoides* spp. produced only GH7 endo-β-1,4-glucanases. No cellulase genes were detected from *Cononympha* spp. Moreover, Arakawa *et al.* (2009) suggested that 88% of the total xylanase activity was present in the hindgut, while 87% of total xylanase activity in the hindgut was derived from GH11 members expressed by *Holomastigotoides* spp.

Nishimura *et al.* (2020) confirmed many of these observations, through analysis of single-cell transcriptomics for each protist species in *C. formosanus*. This study demonstrated the strong expressions of GH7 cellobiohydrolases in *P. grassii* (Fig. 6.1). This species also produces xylanases and mannanases in addition to endo-β-1,4-glucanases, suggesting that *P. grassi* hydrolyzes backbones of cellulose and hemicellulose. *Holomastigotoides* spp. and *Cononympha* spp. also produce some endo-β-1,4-glucanases. However, only a trace of cellobiohydrolase expression was observed, suggesting that these species cannot efficiently hydrolyze the crystalline structure of native cellulose (Fig. 6.1).

Compared with *P. grassii* and *Holomastigitoides* spp., *Cononympha* spp. showed lower expressions of hemicellulase genes. Thus, *Cononympha* spp. may not be a specialist of hydrolyzing large polymers of cellulose or hemicellulose. On the other hand, a previous report indicated that a larger number of chitinase transcripts (GH18), which were assumed to have originated from fungi, were detected in the gut microbiome of *C. formosanus* (Zhang *et al.*, 2012b). Nishimura *et al.* (2020) demonstrated that *Cononympha* spp. have a complete pathway of chitin degradation, owing to the possession of chitinase genes that were acquired from fungi by horizontal gene transfer. Thus, *Cononympha* spp. are likely to obtain carbon and nitrogen through the digestion of chitin.

Regarding lignocellulolytic systems in the hindgut bacteria, recent studies employing single-cell genomics assumed that some ectosymbionts on oxymonadids are involved in cellulose and hemicellulose degradation (Yuki *et al.*, 2015; Treitl *et al.*, 2019; Utami *et al.*, 2019). However, this is probably not the case for *C. formosanus*, because *Coptotermes* termites harbor only parabasalids. Metagenomic analyses suggest that some free-living bacteria in their hindguts may produce cellulases and hemicellulases with a phosphotransferase system to yield their own energy (Do *et al.*, 2014; Gao *et al.*, 2016).

6.4 Lignin degradation in the gut of *C. formosanus*

Some studies suggest a possible involvement of endogenous AA members in lignin degradation in lower termites, including *C. gestroi* (see Franco Cairo *et al.*, 2016 and references therein). A series of studies by Ke *et al.* (2010, 2011a,b,c, 2012a,b, 2013), which analyzed the gut contents or feces using pyrolysis gas chromatography–mass spectrometry, indicated that at least some partial modifications of lignin occur during the passage in the gut in *C. formosanus*. On the other hand, a chemical analysis between a diet (red pine sapwood) and feces of *C. formosanus* indicated that 72–78% of ingested cellulose was hydrolyzed in the gut, while only 17–24% of lignin was digested (Itakura *et al.*, 1995). Both solid and solution-state NMR analyses in *C. formosanus* indicated that a large part of lignin polymers remains intact, while some were degraded only partially (Hyodo *et al.*, 1999; Tarmadi *et al.*, 2018). Although the possibility that salivary redox enzymes such as laccases, hemocyanins, and peroxidases modify lignin cannot be ruled out (Qiu *et al.*, 2015; Geng *et al.*, 2018a,b), these studies, regardless of analytical techniques, agreed that the abundant interunits in lignin such as ether-bonded β-O-4′ linkages were mostly unchanged, in contrast to considerable degradation of polysaccharides, suggesting a limited ability of lignin degradation or mineralization by termites or their gut symbionts. These results also indicate that lignin has little nutritional value to termites and their gut symbionts. Indeed, the administration of purified ball-milled lignin to *C. formosanus* individuals for 4 weeks had a detrimental effect on the termites (Tarmadi *et al.*, 2017a). Nevertheless, compared with the symbiotic protists in the termites feeding solely on polysaccharides such as cellulose or holocellulose (i.e., a complex of cellulose and hemicellulose), adding milled wood lignin to these polysaccharides significantly increased survival rates of the large protist species such as *P. grassii* and *Holomastigotoides* spp. after 8 weeks of feeding (Tarmadi *et al.*, 2017b). Although a mechanism by which lignin or lignin derivatives triggered this viable effect remains unclear, the result suggests that lignin is an indispensable component in wood to maintain the ability of efficient lignocellulose degradation by the symbiotic protists in *C. formosanus*.

6.5 Catabolism of cellulose and hemicellulose hydrolysates in *C. formosanus*

6.5.1 Pyruvate and acetate metabolism in termite tissues

Lower and higher termites possess all glycolytic pathway enzymes and a functional TCA cycle (O'Brien and Breznak, 1984; Slaytor *et al.*, 1997; Slaytor, 2000). Pyruvate, the final product of glycolysis, is produced directly from glucose via the glycolytic pathway and from other hexoses and pentoses via the pentose phosphate pathway linked to glycolysis (Slaytor *et al.*, 1997; Slaytor, 2000). In aerobic metabolism, pyruvate is converted to acetyl-CoA by the PDH (pyruvate dehydrogenase) complex. In termite tissue, excluding microorganisms in the gut, the PDH complex was maintained either in an inactive form (O'Brien and Breznak, 1984, Slaytor *et al.*, 1997) or in a moderately active state (Itakura *et al.*, 1999, 2003; Slaytor, 2000). The presence of an active PDH complex was demonstrated in the lower termite, *C. formosanus* (Itakura *et al.*, 2003), and the higher termite, *Nasutitermes walkeri* (Hill) (Slaytor, 2000). Mitochondrial preparations isolated from termite tissue of these termites can readily transport and oxidize pyruvate. Using an oxygen electrode, oxygen consumption by isolated mitochondria during pyruvate oxidation was determined. Pyruvate was oxidized at 8.3 nmol/termite/h and 17.6 nmol/termite/h by mitochondria isolated from *C. formosanus* (Itakura *et al.*, 2003) and *N. walkeri* (Slaytor, 2000), respectively. Mitochondria isolated in the presence of inhibitors of PDH complex interconversion, dichloroacetophenone, and fluphenazine to inhibit PDH kinase and PDH phosphatase, revealed that about 60% of the PDH complex was maintained in the active form in both *C. formosanus* and *N. walkeri* (Itakura *et al.*, 2003).

In tissues of the lower termites, *Reticulitermes flavipes* (Kollar) and *Coptotermes lacteus* (Froggatt), and the higher termite, *Nasutitermes exitiosus* (Hill), the PDH complex did not convert pyruvate to acetyl-CoA because the PDH complex remained inactive. Instead, acetyl-CoA synthetase converted acetate to acetyl-CoA (O'Brien and Breznak, 1984). Acetate produced by the hindgut microbiota was shown to be transported into termite tissue across the hindgut epithelium in *Mastotermes darwiniensis* Froggatt (Hogan *et al.*, 1985) and has been considered to be the source of acetyl-CoA for the TCA cycle in termite tissue based on the results that 7.2 mM acetate was in the hemolymph and that acetyl-CoA synthetase activity of 12 nmol/termite/h was detected in termite tissue of *N. walkeri* (Slaytor *et al.*, 1997). These results are the basis for the hypothesis that acetate synthesized by gut microbiota rather than pyruvate produced by glycolysis is a major energy and carbon source in *C. formosanus* (Yoshimura, 1995; Ohkuma *et al.*, 2015).

The amounts of activities of PDH complex in termite tissue, 8.3 nmol/termite/h in *C. formosanus* and 17.6 nmol/termite/h in *N. walkeri*, were insufficient considering the rates of respiration, i.e., O_2 consumption rates of 58–83 nmol/termite/h in *C. formosanus* (Wheeler *et al.*, 1996) and 86 nmol/termite/h in *N. walkeri* (Slaytor *et al.*, 1997). The activities of acetyl-CoA synthetase in termite tissue, 5.6 nmol/termite/h in *C. formosanus* (Itakura *et al.*, 1999) and 12 nmol/termite/h in *N. walkeri* (Slaytor *et al.*, 1997), which were also not sufficient compared with the rates of respiration described above, were weaker than the activities of PDH complex.

The amounts of acetyl-CoA synthetase activity were 55.5, 27.6, and 19.2 nmol/termite/h in *R. flavipes*, *C. lacteus*, and *N. exitiosus*, respectively (O'Brien and Breznak, 1984). In *R. flavipes*, the activity of acetyl-CoA synthetase seems to be enough to support an O_2 consumption rate of 33–46 nmol/termite/h (Wheeler *et al.*, 1996), whereas the O_2 consumption rates of the remaining two termite species are unknown, making it difficult to assess whether the activities of acetyl-CoA synthetase are sufficient for the rates of respiration.

The presence of malic enzyme and malate dehydrogenase (O'Brien and Breznak, 1984; Slaytor *et al.*, 1997) in termite tissue indicates that they may be necessary as anaplerotic enzymes for the conversion of pyruvate to oxalacetate (Fig. 6.4). Excess pyruvate not consumed by PDH complex could be converted to oxalacetate by the combined actions of the malic enzyme and malate dehydrogenase (O'Brien and Breznak, 1984; Slaytor *et al.*, 1997) and to lactate by lactate dehydrogenase (Itakura *et al.*, 1999), because pyruvate did not accumulate in termite tissue (Slaytor *et al.*, 1997). Oxalacetate flows into the TCA cycle, whereas lactate may be an important intermediate as mobile

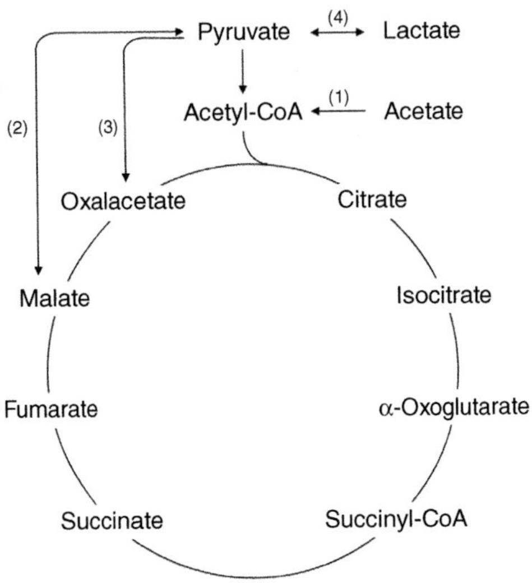

Fig. 6.4. Catabolism of pyruvate and acetate via TCA cycle in termite tissue of *C. formosanus*. (1) Acetyl-CoA synthetase, (2) malic enzyme, (3) pyruvate carboxylase or malic enzyme combined with malate dehydrogenase, and (4) lactate dehydrogenase.

fuel for aerobic metabolism (Gladden, 2004). It should be noted that pyruvate carboxylase, another anaplerotic enzyme that converts pyruvate to oxaloacetate, was slightly active (0.7 nmol/termite/h) in termite tissue of *N. walkeri* (Slaytor *et al.*, 1997), but inactive in tissues of *R. flavipes*, *C. lacteus*, and *N. exitiosus* (O'Brien and Breznak, 1984). In addition to these pyruvate metabolic pathways, the concept that pyruvate produced via glycolysis in termite tissue is transported into the hindgut and converted to acetyl-CoA by pyruvate-ferredoxin oxidoreductase of hindgut microbiota has been hypothesized, although not tested (Slaytor *et al.*, 1997). Under this scenario, it has been suggested that acetyl-CoA is converted to acetate in the hindgut to be transported across hindgut epithelium, such that the acetate is reconverted to acetyl-CoA by acetyl-CoA synthetase in termite tissue (Slaytor *et al.*, 1997).

6.5.2 Other energy metabolism in termite tissues

Mitochondrial fatty acid β-oxidation is the catabolic process by which fatty acid molecules are broken down in the mitochondria to generate acetyl-CoA that enters the TCA cycle. Fatty acid β-oxidation is an essential pathway for maintaining energy homeostasis (Houten and Wanders, 2010). The majority of long-chain fatty acids are stored as long-chain triglycerides. As glucose levels drop, triglycerides are hydrolyzed, and free fatty acids become available. Due to their negative charge, long-chain fatty acids cannot penetrate the mitochondrial membrane. Fatty acids are converted to fatty acyl-CoA ester and transported into the mitochondrial matrix via the carnitine shuttle (Knottnerus *et al.*, 2018). Total lipid contents of workers and nymphs of *C. formosanus* were 11.3% and 32.3%, respectively. The triglycerides contained in workers and nymphs were mainly composed of palmitic acid, stearic acid, oleic acid, and linoleic acid (Itakura *et al.*, 2006). These triglycerides have the potential to be converted to acetyl-CoA by β-oxidation and used as an energy source. However, since the oxidation rate of palmitoylcarnitine by isolated mitochondria was as slow as 2.8 nmol/termite/h in *C. formosanus* (Itakura *et al.*, 2003), mitochondrial fatty acid β-oxidation seems to play only an auxiliary role in the substrate metabolism via TCA cycle.

6.5.3 Genomic and transcriptomic perspectives on PDX complex in *C. formosanus*

Whole-genome analysis of *C. formosanus* has provided new insights into the PDH complex (Itakura *et al.*, 2020). As shown in Fig. 6.5, the PDH complex comprises three catalytic subunits, two regulatory subunits, and an E3 binding protein X component. All genes (i.e., those encoding three catalytic subunits, pyruvate dehydrogenase: E1 subunits α and β; dihydrolipoyl transacetylase: E2; dihydrolipoyl dehydrogenase: E3; and two regulatory subunits: PDH kinase and PDH phosphatase) are found in the genome of *C. formosanus*, except for the PDH protein X component that is required for anchoring E3 dimers to E2 core of PDH complex of eukaryotes and essential for a functional PDH complex (Patel and Roche, 1990). As shown in Table 6.1, two PDH E1 component subunits

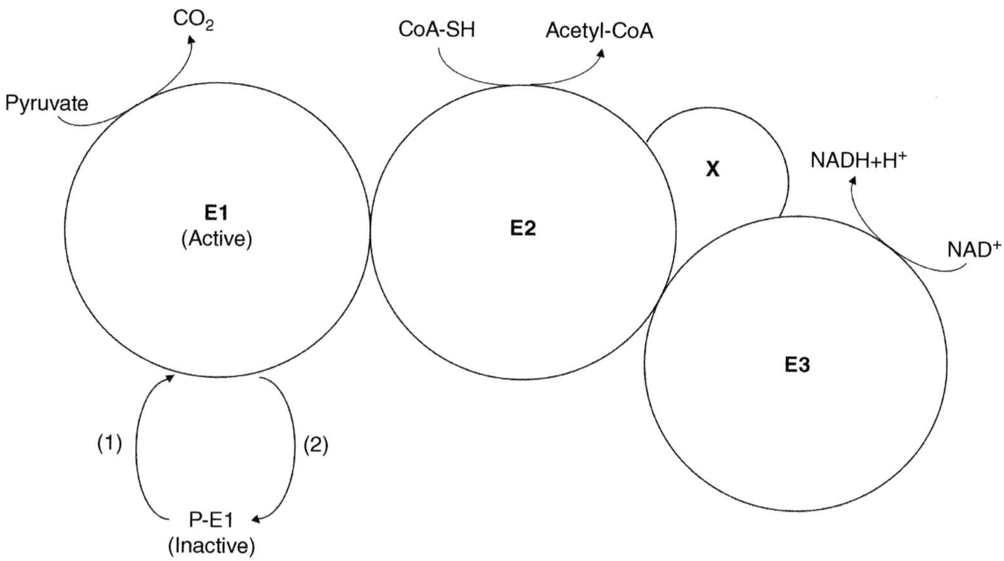

Fig. 6.5. Schematic representation of the reactions catalyzed by PDH complex, including its regulation by phosphorylation and dephosphorylation. E1: PDH, E2: dihydrolipoyl transacetylase, E3: dihydrolipoyl dehydrogenase, X: PDH protein X component, P-E1: phosphorylated PDH, (1) PDH phosphatase, and (2) PDH kinase.

Table 6.1. Expression of genes related to PDH complex of *C. formosanus*

Genes	Transcript expression level (TPM[a])		Accession No.
	Worker	Nymph	
Pyruvate dehydrogenase: E1 component subunit α	63.2 ± 19.2 19.6 ± 4.3	47.5 ± 1.8 21.4 ± 4.0	GFG37618 GFG35176
Pyruvate dehydrogenase: E1 component subunit ß	25.3 ± 3.0	22.2 ± 2.4	GFG40711
Pyruvate dehydrogenase complex dihydrolipoamide acetyltransferase: E2 component	21.0 ± 6.3 6.2 ± 0.6	36.4 ± 8.3 6.2 ± 2.0	GFG31755 GFG38465
Dihydrolipoyl dehydrogenase: E3 component	38.1 ± 2.1	42.1 ± 6.1	GFG32738
Pyruvate dehydrogenase [acetyl-transferring]-phosphatase 1	72.7 ± 26.2	37.3 ± 13.7	GFG31008
Pyruvate dehydrogenase [acetyl-transferring] kinase	3.6 ± 2.3	3.6 ± 1.4	GFG33726

[a]Transcripts per kilobase million

α (GFG37618, GFG35176), one PDH E1 component subunit β (GFG40711), two dihydrolipoamide acetyltransferase: E2 subunits (GFG31755, GFG38465), one dihydrolipoyl dehydrogenase: E3 subunit (GFG32738), one PDH [acetyl-transferring]-phosphatase 1 (GFG31008), and one PDH [acetyl-transferring] kinase (GFG33726) were found in the genome. However, no PDH protein X component was found in *C. formosanus*. The expression of these genes was supported by the TPM (transcripts per kilobase million) values (Table 6.1) (S. Itakura *et al.*, 2022, unpublished results). There was no significant difference in transcript expression levels of these genes between workers and nymphs.

Still, the expression level of PDH [acetyl-transferring]-phosphatase 1 in workers was about twice that of nymphs. PDH complex activity of nymphs has not yet been measured, but PDH activity of workers may be higher than that of nymphs, since PDH phosphatase is an enzyme that activates the PDH complex through dephosphorylation of PDH: E1 subunit α. As mentioned above, PDH complex activity is not sufficient to maintain the respiration of *C. formosanus* by itself. The absence of a PDH protein X component could cause the PDH complex's destabilization, because E3 is loosely anchored to the E2 core in the absence of the PDH protein X component. Insufficient PDH activity may have been detected by a standard spectrophotometric assay using termite tissue homogenate (Itakura *et al.*, 1999) because of destabilization of the PDH complex lacking the protein X component within disrupted mitochondria. In contrast, a moderate PDH activity was successfully determined by measuring the oxygen consumption rate using intact mitochondria (Itakura *et al.*, 2003). The PDH complex could be maintained in a stable state without dissociation.

All genes encoding proteins associated with the glycolytic pathway and TCA cycle, as well as xylulose kinase (responsible for the entry of xylose into the pentose phosphate pathway), are found in the genome of *C. formosanus* (Tables 6.2, 6.3). The oxoglutarate dehydrogenase complex (OGDC) is an enzyme complex analogous to the PDH complex and consists of three enzymes: α-oxoglutarate dehydrogenase: E1 component; dihydrolipoyl succinyltransferase: E2 component; and dihydrolipoyl dehydrogenase: E3 component. OGDC has the same subunit structure as the PDH complex, and the E3 component is shared in common between OGDC and PDH complex (Carothers *et al.*, 1989; Bhagavan and Ha, 2015). Succinate dehydrogenase (SDH) is also an enzyme complex. It consists of succinate dehydrogenase [ubiquinone] flavoprotein subunit, succinate dehydrogenase [ubiquinone] iron-sulfur subunit, succinate dehydrogenase cytochrome b560 subunit, and succinate dehydrogenase [ubiquinone] cytochrome b small subunit (Oyedotun and Lemire, 2004), plus succinate dehydrogenase assembly factors 2 and 3 as flavinator of succinate dehydrogenase (McNeil *et al.*, 2012). Unlike the lack of the protein X component in the PDH complex, transcript expression of all components of OGDC and SDH was observed. Transcript expression levels of hexokinase, glyceraldehyde-3-phosphate dehydrogenase, and enolase of the glycolytic pathway and malate dehydrogenase of the TCA cycle in nymphs were significantly higher than those in workers.

In contrast, expression levels of isocitrate dehydrogenase, succinate dehydrogenase [ubiquinone] cytochrome b small subunits, and succinate dehydrogenase assembly factor 2 of the TCA cycle in workers were significantly higher than those in nymphs (Tables 6.2, 6.3) (S. Itakura *et al.*, 2022, unpublished results). Among enzymes involved in pyruvate metabolism, transcript expression levels of lactate dehydrogenase in nymphs were significantly higher than that in workers, and expression level in nymphs was about ten times higher than that of workers (Table 6.3). Nymphs may actively convert pyruvate to lactate by lactate dehydrogenase, since lactate is an important intermediate for regeneration and mobile fuel for aerobic metabolism (Gladden, 2004). In *C. formosanus,* the amount of lactate dehydrogenase was 54.6 nmol/termite/h in the termite tissue of workers (Itakura *et al.*, 1999). On the other hand, transcript expression levels of pyruvate carboxylase that catalyzes the carboxylation of pyruvate to oxalacetate were about the same as those of acetyl-CoA synthetase. It suggests that pyruvate carboxylase may function as an anaplerotic enzyme involved in the metabolism of pyruvate via the TCA cycle.

In summary, it is hypothesized that glucose produced by cellulose catabolism is converted to pyruvate via the glycolytic pathway. Other hexoses and pentoses produced by the catabolism of hemicellulose are converted to pyruvate via the pentose phosphate pathway linked to glycolysis. A portion of the pyruvate is converted to acetyl-CoA by the PDH complex to be oxidized via the TCA cycle. In contrast, excess pyruvate could be converted to malate by malic acid enzymes or oxalacetate by pyruvate carboxylase to flow into the TCA cycle. In parallel with

Table 6.2. Expression of genes related to glycolysis and pentose phosphate pathway of *C. formosanus*

Genes	Transcript expression level (TPM[a])		Accession No.
	Worker	Nymph	
Hexokinase	22.8 ± 4.0**	47.4 ± 8.0**	GFG39358
	0.9 ± 1.2	0.6 ± 0.3	GFG37586
Phosphoglucose isomerase	87.1 ± 7.8	94.9 ± 0.6	GFG36723
Phosphofructokinase	12.3 ± 2.7	10.9 ± 1.9	GFG34551
Fructose-bisphosphate aldolase	134.3 ± 29.9	157.2 ± 35.3	GFG33513
Triosephosphate isomerase	98.8 ± 13.1	127.1 ± 14.2	AGM32485
Glyceraldehyde-3-phosphate dehydrogenase	1801.0 ± 19.9**	2597.7 ± 285.4**	GFG31790
Phosphoglycerate kinase	67.0 ± 2.2	57.9 ± 5.0	GFG38536
Phosphoglycerate mutase	203.9 ± 41.3	174.1 ± 3.4	GFG39130
Enolase	144.1 ± 13.9*	297.7 ± 56.6*	AGM32397
Pyruvate kinase	73.9 ± 11.7	87.1 ± 12.6	GFG33448
Pentose phosphate pathway			
Xylulose kinase	3.9 ± 3.1	7.0 ± 0.9	GFG31953

[a]Transcripts per kilobase million
**$P < 0.01$, *$P < 0.05$

Table 6.3 Expression of genes related to the tricarboxylic acid cycle of *C. formosanus*

Genes	Transcript expression level (TPM[a])		Accession No.
	Worker	Nymph	
Citrate synthase	28.6 ± 1.0	19.4 ± 4.6	GFG35845
Aconitase	9.3 ± 1.3	6.6 ± 1.7	GFG33245
Isocitrate dehydrogenase	87.4 ± 9.4**	59.4 ± 4.6**	GFG28298
	50.1 ± 23.2	41.5 ± 3.7	GFG37407
	11.4 ± 4.8	14.5 ± 1.3	GFG31514
α-Oxoglutarate dehydrogenase: E1 component	41.4 ± 1.8	33.6 ± 7.4	GFG31025
	0 ± 0	0 ± 0	GFG30105
	2.1 ± 0.9	2.1 ± 1.0	GFG30109
Dihydrolipoyl succinyltransferase: E2 component	13.7 ± 4.1	14.1 ± 2.7	GFG29709
Dihydrolipoyl dehydrogenase: E3 component	38.1 ± 2.1	42.1 ± 6.1	GFG32738
Succinate-CoA ligase	3.4 ± 1.9	3.4 ± 2.0	GFG33509
Succinate dehydrogenase [ubiquinone] flavoprotein subunit	91.2 ± 5.7	96.8 ± 15.6	GFG34166
Succinate dehydrogenase [ubiquinone] iron-sulfur subunit	82.5 ± 29.7	67.5 ± 5.9	GFG33962
Succinate dehydrogenase cytochrome b560 subunit	810.7 ± 208.1	974.6 ± 160.3	GFG28709
	512.5 ± 153.2	278.5 ± 14.0	GFG28711
Succinate dehydrogenase [ubiquinone] cytochrome b small subunit	1920.5 ± 463.9*	887.0 ± 66.1*	GFG29172
	631.6 ± 35.0*	388.6 ± 111.9*	GFG29171
Succinate dehydrogenase assembly factor 2	119.7 ± 10.0**	82.1 ± 3.8**	GFG38467
Succinate dehydrogenase assembly factor 3	20.1 ± 3.2	18.4 ± 13.3	GFG30715
Fumarase	50.7 ± 8.0	37.4 ± 7.6	GFG39019
Malate dehydrogenase	289.8 ± 28.3**	395.3 ± 22.6**	GFG35833

Continued

Table 6.3 Continued.

Genes	Transcript expression level (TPM[a])		Accession No.
	Worker	Nymph	
Anaplerosis			
Malic enzyme	120.8 ± 47.9	101.5 ± 7.7	GFG34131
	15.1 ± 3.3	16.1 ± 4.4	GFG38773
Pyruvate carboxylase	12.4 ± 3.7	8.5 ± 1.4	GFG37080
Pyruvate metabolism			
Acetyl-CoA synthetase	15.7 ± 2.4	12.1 ± 1.3	GFG31781
Lactate dehydrogenase	3.3 ± 2.9*	29.5 ± 15.9*	GFG32648

[a]Transcripts per kilobase million
$^{**}P < 0.01$, $^{*}P < 0.05$

these pathways, some of the pyruvates may be converted to lactate by lactate dehydrogenase. Since the amount of acetyl-CoA produced by the PDH complex is insufficient to maintain termite respiration, the acetyl-CoA produced by acetyl-CoA synthetase is supplied coordinately to the TCA cycle. Both PDH complex and acetyl-CoA synthetase are mitochondrial enzymes. Some regulatory mechanisms should work between these enzymes, but the details are currently unknown and are expected to be elucidated in the future.

6.6 Future perspectives

The present chapter summarizes current knowledge of the mechanism by which Formosan subterranean termites digest and metabolize lignocellulose, which have been unveiled by recent genomic and transcriptomic studies together with accumulation of classical biochemical studies of digestion and catabolism in termites. Many researchers have assumed that the termite–microbe symbiosis would be an ideal model to convert plant cell walls to biofuels and that the clarification of gene repertoires and their expressions involved in lignocellulose degradation from the host and the symbionts would be applicable to industrial use. However, to our knowledge, no one has successfully reproduced or reconstructed such an efficient digestive system *in vitro*. The functions of many genes derived from (meta)genomic and (meta)transcriptomic sequencings from the host and the symbionts have yet to be corroborated by biochemical evidence, and interactions between these gene products mostly remain a mystery. Therefore, our current knowledge might still be the tip of iceberg toward an elucidation of a total scheme of digestive physiology in termites. Our future efforts into a combination of gene sequencing and biochemical characterizations would greatly enhance our understanding of the true mechanism of lignocellulose digestion in termites.

References

Arakawa, G., Watanabe, H., Yamasaki, H., Maekawa, H. and Tokuda, G. (2009) Purification and molecular cloning of xylanases from the wood-feeding termite, *Coptotermes formosanus* Shiraki. *Bioscience, Biotechnology, and Biochemistry* 73, 710–718.

Bhagavan, N.V. and Ha, C.-E. (2015) Carbohydrate Metabolism I: Glycolysis and the Tricarboxylic Acid Cycle. In: Bhagavan, N.V. and Ha, C.-E. (eds) *Essentials of Medical Biochemistry* (2nd Edn). Academic Press, San Diego, California, pp. 165–185.

Bucek, A., Sobotník, J., He, S., Shi, M., McMahon, D.P. *et al.* (2019) Evolution of termite symbiosis informed by transcriptome-based phylogenies. *Current Biology* 29, 3728–3734.

Carothers, D.J., Pons, G. and Patel, M.S. (1989) Dihydrolipoamide dehydrogenase: Functional similarities and divergent evolution of the pyridine nucleotide-disulfide oxidoreductases. *Archives of Biochemistry and Biophysics* 268 409–425.

da Silva Moraes, C., Diaz-Albiter, H.M., do Valle Faria, M., Sant'Anna, M.R.V., Dillon, R.J. *et al.* (2014) Expression pattern of glycoside hydrolase genes in *Lutzomyia longipalpis* reveals key enzymes involved in larval digestion. *Frontiers in Physiology* 5, 276.

Do, T.H., Nguyen, T.T., Nguyen, H.N., Le, Q.G., Nguyen, C. *et al.* (2014) Mining biomass-degrading genes through Illumina-based *de novo* sequencing and metagenomic analysis of free-living bacteria in the gut of the lower termite *Coptotermes gestroi* harvested in Vietnam. *Journal of Bioscience and Bioengineering* 118, 665–671.

Feng, T., Liu, H., Xu, Q., Sun, J. and Shi, H. (2015) Identification and characterization of two endogenous β-glucosidases from the termite *Coptotermes formosanus*. *Applied Biochemistry and Biotechnology* 176, 2039–2052.

Franco Cairo, L.P.L., Oliveira, L.C., Uchima, C.A., Alvarez, T.M., Citadini, A.P.S. *et al.* (2013) Deciphering the synergism of endogenous glycoside hydrolase families 1 and 9 from *Coptotermes gestroi*. *Insect Biochemistry and Molecular Biology* 43, 970–981.

Franco Cairo, J.P.L., Carazzolle, M.F., Leonardo, F.C., Mofatto, L.S., Brenelli, L.B. *et al.* (2016) Expanding the knowledge on lignocellulolytic and redox enzymes of worker and soldier castes from the lower termite *Coptotermes gestroi*. *Frontiers in Microbiology* 7, Article 1518.

Franco Cairo, J.P.L., Cannella, D., Oliveira, L.C., Gonçalves, T.A., Rubio, M.V. *et al.* (2021) On the roles of AA15 lytic polysaccharide monooxygenases derived from the termite *Coptotermes gestroi*. *Journal of Inorganic Biochemistry* 216, 111316.

Fujita, A., Hojo, M., Aoyagi, T., Hayashi, Y., Arakawa, G. *et al.* (2010) Details of the digestive system in the midgut of *Coptotermes formosanus* Shiraki. *Journal of Wood Science* 56, 222–226.

Gao, G., Wang, A., Gong, B.-L., Li, Q.-Q., Liu, Y.-H. *et al.* (2016) A novel metagenome-derived gene cluster from termite hindgut: encoding phosphotransferase system components and high glucose tolerant glucosidase. *Enzyme and Microbial Technology* 84, 24–31.

Geng, A., Wu, J., Xie, R.-R., Li, X., Chang, F.-X. *et al.* (2018a) Characterization of a laccase from a wood-feeding termite, *Coptotermes formosanus*. *Insect Science* 25, 251–258.

Geng, A., Cheng, Y., Wang, Y., Zhu, D., Le, Y. *et al.* (2018b) Transcriptome analysis of the digestive system of a wood-feeding termite (*Coptotermes formosanus*) revealed a unique mechanism for effective biomass degradation. *Biotechnology for Biofuels* 11, 24.

Gladden, L.B. (2004) Lactate metabolism: a new paradigm for the third millennium. *Journal of Physiology* 558, 5–30.

Hogan, M.E., Slaytor, M. and O'Brien, R.W. (1985) Transport of volatile fatty acids across the hindgut of the cockroach *Panesthia cribrata* Saussure and the termite, *Mastotermes darwiniensis* Froggatt. *Journal of Insect Physiology* 31, 587–591.

Houten, S.M. and Wanders, R.J. (2010) A general introduction to the biochemistry of mitochondrial fatty acid β-oxidation. *Journal of Inherited Metabolic Disease* 33, 469–477.

Hyodo, F., Azuma, J.-I. and Abe, T. (1999) Estimation of effect of passage through the gut of a lower termite, *Coptotermes formosanus* Shiraki, on lignin by solid-state CP/MAS ^{13}C NMR. *Holzforschung* 53, 244–246.

Itakura, S., Ueshima, K., Tanaka, H. and Enoki, A. (1995) Degradation of wood components by subterranean termite, *Coptotermes formosanus* Shiraki. *Mokuzai Gakkaishi* 41, 580–586.

Itakura, S., Tanaka, H. and Enoki, A. (1997) Distribution of cellulases, glucose and related substances in the body of *Coptotermes formosanus*. *Material und Organismen* 31, 17–29.

Itakura, S., Tanaka, H. and Enoki, A. (1999) Occurrence and metabolic role of the pyruvate dehydrogenase complex in the lower termite *Coptotermes formosanus* (Shiraki). *Insect Biochemistry and Molecular Biology* 29, 625–633.

Itakura, S., Tanaka, H., Enoki, A., Chappell, D. J. and Slaytor, M. (2003) Pyruvate and acetate metabolism in termite mitochondria. *Journal of Insect Physiology* 49, 917–926.

Itakura, S., Okuda, J., Utagawa, K., Tanaka, H. and Enoki, A. (2006) Nutritional value of two subterranean termite species, *Coptotermes formosanus* Shiraki and *Reticulitermes speratus* (Kolbe) (Isoptera: Rhinotermitidae). *Japanese Journal of Environmental Entomology and Zoology* 17, 107–115.

Itakura, S., Yoshikawa, Y., Togami, Y. and Umezawa, K. (2020) Draft genome sequence of the termite, *Coptotermes formosanus*: Genetic insights into the pyruvate dehydrogenase complex of the termite. *Journal of Asia-Pacific Entomology* 23, 666–674.

Jasso-Selles, D.E., Martini, F.D., Velenovsky IV, J.F., Mee, E.D., Montoya, S.J. *et al.* (2020) The complete protist communities of *Coptotermes formosanus* and *Coptotermes gestroi*: morphological and molecular characterization of five new species. *Journal of Eukaryotic Microbiology* 67, 626–641.

Kanai, K., Azuma, J.-I. and Nishimoto, K. (1982) Studies on digestive system of termites. I. Digestion of carbohydrates by termite *Coptotermes formosanus* Shiraki. *Wood Research* 68, 47–57.

Ke, J., Sun, J.Z., Nguyen, H.D., Singh, D., Lee, K.C., Beyenal, H. and Chen, S. (2010) *In-situ* oxygen profiling and lignin modification in guts of wood-feeding termites. *Insect Science* 17, 277–290.

Ke, J., Laskar, D.D., Singh, D. and Chen, S. (2011a) *In situ* lignocellulosic unlocking mechanism for carbohydrate hydrolysis in termites: crucial lignin modification. *Biotechnology for Biofuels* 4,17.

Ke, J., Singh, D. and Chen, S. (2011b) Aromatic compound degradation by the wood-feeding termite *Coptotermes formosanus* (Shiraki). *International Biodeterioration & Biodegradation* 65, 744–756.

Ke, J., Singh, D., Yang, X. and Chen, S. (2011c) Thermal characterization of softwood lignin modification by termite *Coptotermes formosanus* (Shiraki). *Biomass and Bioenergy* 35, 3617–3626.

Ke, J., Laskar, D.D., Gao, D.F. and Chen, S. (2012a) Advanced biorefinery in lower termite-effect of combined pretreatment during the chewing process. *Biotechnology for Biofuels* 5, 11.

Ke, J., Singh, D. and Chen, S. (2012b) Metabolism of polycyclic aromatic hydrocarbons by the wood-feeding termite *Coptotermes formosanus* (Shiraki). *Journal of Agricultural and Food Chemistry* 60, 1788–1797.

Ke, J., Laskar, D.D. and Chen, S. (2013) Tetramethylammonium hydroxide (TMAH) thermochemolysis for probing *in situ* softwood lignin modification in each gut segment of the termite. *Journal of Agricultural and Food Chemistry* 61, 1299–1308.

Knottnerus, S.J.G., Bleeker, J.C., Wüst, R.C.I., Ferdinandusse, S., Ijlst, L. *et al.* (2018) Disorders of mitochondrial long-chain fatty acid oxidation and the carnitine shuttle. *Reviews in Endocrine and Metabolic Disorders* 19, 93–106.

Koidzumi, M. (1921) Studies on the intestinal protozoa found in the termites of Japan. *Parasitology* 13, 235–309.

Liu, X., Wu, W., Zhang, D. and Li, Z. (2017) Molecular cloning and characterization of a putative β-glucosidase from the Formosan subterranean termite (Isoptera: Rhinotermitidae). *Journal of Entomological Science* 52, 177–192.

Lynd, L.R., Weimer, P.J., van Zyl, W.H. and Pretorius, I.S. (2002) Microbial cellulose utilization: fundamentals and biotechnology. *Microbiology and Molecular Biology Reviews* 66, 506–577.

McNeil, M.B., Clulow, J.S., Wilf, N.M., Salmond, G.P.C. and Fineran, P.C. (2012) SdhE is a conserved protein required for flavinylation of succinate dehydrogenase in bacteria. *The Journal of Biological Chemistry* 287, 18418–18428.

Nakashima, K., Watanabe, H., Saitoh, H., Tokuda, G. and Azuma, J. (2002a) Dual cellulose digesting system of the wood-feeding termite, *Coptotermes formosanus* (Shiraki). *Insect Biochemistry and Molecular Biology* 32, 777–784.

Nakashima, K., Watanabe, H. and Azuma, J.I. (2002b) Cellulase genes from the parabasalian symbiont *Pseudotrichonympha grassii* in the hindgut of the wood-feeding termite *Coptotermes formosanus*. *Cellular and Molecular Life Sciences* 59, 1554–1560.

Ni, J. and Tokuda, G. (2013) Lignocellulose-degrading enzymes from termites and their symbiotic microbiota. *Biotechnology Advances* 31, 838–850.

Nishimura, Y., Otagiri, M., Yuki, M., Shimizu, M., Inoue, J.-I. *et al.* (2020) Division of functional roles for termite gut protists revealed by single-cell transcriptomes. *ISME Journal* 14, 2449–2460.

O'Brien, R.W. and Breznak, J.A. (1984) Enzymes of acetate and glucose metabolism in termites. *Insect Biochemistry* 14, 639–643.

Ohkuma, M., Noda, S., Hattori, S., Iida, T., Yuki, M. *et al.* (2015) Acetogenesis from H_2 plus CO_2 and nitrogen fixation by an endosymbiotic spirochete of a termite-gut cellulolytic protist. *Proceedings of the National Academy of Sciences of the USA* 112, 10224–10230.

Oyedotun, K.S. and Lemire, B.D. (2004) The quaternary structure of the *Saccharomyces cerevisiae* succinate dehydrogenase. Homology modeling, cofactor docking, and molecular dynamics simulation studies. *Journal of Biological Chemistry* 279, 9424–9431.

Patel, M.S. and Roche, T.E. (1990) Molecular biology and biochemistry of pyruvate dehydrogenase complexes. *FASEB Journal* 4, 3224–3233.

Qiu, H., Geng, A., Zhu, D., Le, Y., Wu, J. *et al.* (2015) Purification and characterization of a hemocyanin (Hemo1) with potential lignin-modification activities from the wood-feeding termite, *Coptotermes formosanus* Shiraki. *Applied Biochemistry and Biotechnology* 175, 687–697.

Rust, M.K. and Su, N.-Y. (2012) Managing social insects of urban importance. *Annual Review of Entomology* 57, 355–375.

Sabbadin, F., Hemsworth, G.R., Ciano, L., Henrissat, B., Dupree, P. *et al.* (2018) An ancient family of lytic polysaccharide monooxygenases with roles in arthropod development and biomass digestion. *Nature Communications* 9, 756.

Sethi, A., Xue, Q.-G., La Peyre, J.F., Delatte, J. and Husseneder, C. (2011) Dual origin of gut proteases in Formosan subterranean termites (*Coptotermes formosanus* Shiraki) (Isoptera: Rhinotermitidae). *Comparative Biochemistry and Physiology* 159A, 261–267.

Shigenobu, S., Hayashi, Y., Watanabe, D., Tokuda, G., Hojo, M.Y. *et al.* (2022) Genomic and transcriptomic analyses of the subterranean termite *Reticulitermes speratus*: gene duplication facilitates social evolution. *Proceedings of the National Academy of Sciences of the USA* 119, e2110361119.

Slaytor, M. (2000) Energy metabolism in the termite and its gut microbiota. In: Abe, T., Bignell, D.E. and Higashi, M. (eds) *Termites: Evolution, Sociality, Symbioses, Ecology*. Springer, Dordrecht, Netherlands, pp. 307–332.

Slaytor, M., Veivers, P.C. and Lo, N. (1997). Aerobic and anaerobic metabolism in the higher termite *Nasutitermes walkeri* (Hill). *Insect Biochemistry and Molecular Biology* 27, 291–303.

Tanaka, H., Aoyagi, H., Shina, S., Dodo, Y., Yoshimura, T. *et al.* (2006) Influence of the diet components on the symbiotic microorganisms community in hindgut of *Coptotermes formosanus* Shiraki. *Applied Microbiology and Biotechnology* 71, 907–917.

Tarmadi, D., Yoshimura, T., Tobimatsu, Y., Yamamura, M., Miyagawa, Y. *et al.* (2017a) The effects of various lignocelluloses and lignins on physiological responses of a lower termite, *Coptotermes formosanus*. *Journal of Wood Science* 63, 464–472.

Tarmadi, D., Yoshimura, T., Tobimatsu, Y., Yamamura, M. and Umezawa, T. (2017b) Effects of lignins as diet components on the physiological activities of a lower termite, *Coptotermes formosanus* Shiraki. *Journal of Insect Physiology* 103, 57–63.

Tarmadi, D., Tobimatsu, Y., Yamamura, M., Miyamoto, T., Miyagawa, Y. *et al.* (2018) NMR studies on lignocellulose deconstructions in the digestive system of the lower termite *Coptotermes formosanus* Shiraki. *Scientific Reports* 8, 1290.

Tokuda, G. (2019) Plant cell wall degradation in insects: Recent progress on endogenous enzymes revealed by multi-omics technologies. *Advances in Insect Physiology* 57, 97–136.

Tokuda, G., Lo, N., Watanabe, H., Arakawa, G., Matsumoto, T. *et al.* (2004) Major alteration of the expression site of endogenous cellulases in members of an apical termite lineage. *Molecular Ecology* 13, 3219–3228.

Tokuda, G., Lo, N. and Watanabe, H. (2005) Marked variations in patterns of cellulase activity against crystalline- versus carboxymethyl-cellulose in the digestive systems of diverse, wood-feeding termites. *Physiological Entomology* 30, 372–380.

Treitli, S.C., Kolisko, M., Husnik, F., Keeling, P.J. and Hampl, V. (2019) Revealing the metabolic capacity of *Streblomastix strix* and its bacterial symbionts using single-cell metagenomics. *Proceedings of the National Academy of Sciences of the USA* 116, 19675–19684.

Utami, Y.D., Kuwahara, H., Igai, K., Murakami, T., Sugaya, K. *et al.* (2019) Genome analyses of uncultured TG2/ZB3 bacteria in 'Margulisbacteria' specifically attached to ectosymbiotic spirochetes of protists in the termite gut. *ISME Journal* 13, 455–467.

Watanabe, H. and Tokuda, G. (2010) Cellulolytic systems in insects. *Annual Review of Entomology* 55, 609–632.

Watanabe, H., Noda, H., Tokuda, G. and Lo, N. (1998) A cellulase gene of termite origin. *Nature* 394, 330–331.

Watanabe, H., Nakashima, K., Saito, H. and Slaytor, M. (2002) New endo-β-1,4-glucanases from the parabasalian symbionts, *Pseudotrichonympha grassii* and *Holomastigotoides mirabile* of *Coptotermes* termites. *Cellular and Molecular Life Sciences* 59, 1983–1992.

Wheeler, G.S., Tokoro, M., Scheffrahn, R.H. and Su, N.Y. (1996) Comparative respiration and methane production rates in Nearctic termites. *Journal of Insect Physiology* 42, 799–806.

Wu, W. and Li, Z. (2018) dsRNA injection successfully inhibited two endogenous β-glucosidases in *Coptotermes formosanus* (Isoptera: Rhinotermitidae). *Journal of Economic Entomology* 111, 860–867.

Wu, W., Gu, D., Yan, S. and Li, Z. (2019) RNA interference of endoglucanases in the formosan subterranean termite *Coptotermes formosanus* shiraki (Blattodea: Rhinotermitidae) by dsRNA injection or ingestion. *Journal of Insect Physiology* 112, 15–22.

Xie, L., Zhang, L., Zhong, Y., Liu, N., Long, Y., Wang, S., Zhou, X., Zhou, Z., Huang, Y. and Wang, Q. (2012) Profiling the metatranscriptome of the protistan community in *Coptotermes formosanus* with emphasis on the lignocellulytic system. *Genomics* 99, 246–255.

Yoshimura, T. (1995) Contribution of the protozoan fauna to nutritional physiology of the lower termite, *Coptotermes formosanus* Shiraki (Isoptera: Rhinotermitidae). *Wood Research* 82, 68–129.

Yoshimura, T., Tsunoda, K. and Takahashi, M. (1992a) Distribution of the symbiotic protozoa in the hindgut of *Coptotermes formosanus* Shiraki (Isoptera: Rhinotermitidae). *Japanese Journal of Environmental Entomology and Zoology* 4, 115–120.

Yoshimura, T., Watanabe, T., Tsunoda, K. and Takahashi, M. (1992b) Distribution of the cellulolytic activities in the lower termite, *Coptotermes formosanus* Shiraki (Isoptera: Rhinotermitidae). *Material und Organismen* 27, 273–284.

Yoshimura, T., Azuma, J., Tsunoda, K. and Takahashi, M. (1993a) Cellulose metabolism of the symbiotic protozoa in termite, *Coptotermes formosanus* Shiraki (Isoptera: Rhinotermitidae) I. Effect of degree of polymerization of cellulose. *Mokuzai Gakkaishi* 39, 221–226.

Yoshimura, T., Azuma, J., Tsunoda, K. and Takahashi, M. (1993b) Cellulose metabolism of the symbiotic protozoa in termite, *Coptotermes formosanus* Shiraki (Isoptera: Rhinotermitidae) II. Selective defaunation of protozoa and its effect on cellulose metabolism. *Mokuzai Gakkaishi* 39, 227–230.

Yoshimura, T., Fujino, T., Itoh, T., Tsunoda, K. and Takahashi, M. (1996) Ingestion and decomposition of wood and cellulose by the protozoa in the hindgut of *Coptotermes formosanus* Shiraki (Isoptera: Rhinotermitidae) as evidenced by polarizing and transmission electron microscopy. *Holzforschung* 50, 99–104.

Yuki, M., Kuwahara, H., Shintani, M., Izawa, K., Sato, T. *et al.* (2015) Dominant ectosymbiotic bacteria of cellulolytic protists in the termite gut also have the potential to digest lignocellulose. *Environmental Microbiology* 17, 4942–4953.

Zhang, D., Lax, A.R., Raina, A.K. and Bland, J.M. (2009) Differential cellulolytic activity of native-form and C-terminal tagged-form cellulase derived from *Coptotermes formosanus* and expressed in *E. coli*. *Insect Biochemistry and Molecular Biology* 39, 516–522.

Zhang, D., Lax, A.R., Bland, J.M., Yu, J., Fedorova, N. *et al.* (2010) Hydrolysis of filter-paper cellulose to glucose by two recombinant endogenous glycosyl hydrolases of *Coptotermes formosanus*. *Insect Science* 17, 245–252.

Zhang, D., Allen, A.B. and Lax, A.R. (2012a) Functional analyses of the digestive β-glucosidase of Formosan subterranean termites (*Coptotermes formosanus*). *Journal of Insect Physiology* 58, 205–210.

Zhang, D., Lax, A.R., Henrissat, B., Coutinho, P., Katiya, N. *et al.* (2012b) Carbohydrate-active enzymes revealed in *Coptotermes formosanus* (Isoptera: Rhinotermitidae) transcriptome. *Insect Molecular Biology* 21, 235–245.

7 Chemical ecology

Qian Sun
Department of Entomology, Louisiana State University, Baton Rouge, LA, 70803, USA, QSun@agcenter.lsu.edu

Abstract

The Formosan subterranean termites, *Coptotermes formosanus* Shiraki, live in colonies with a division of labor among castes. Their social activities are organized through sophisticated interactions among colony members, and colony function relies on detecting environmental risks and resources. *Coptotermes formosanus* termites produce and perceive a variety of chemicals for social organization and defense, highlighting the importance of chemical communication in their ecological adaptation. The sex and trail pheromones, which share the same major component, have been characterized in this species. Many glandular and cuticular compounds have been investigated, but biological activities are not fully determined. Additionally, several semiochemicals associated with pathogens and food resources have been identified to regulate termite behavior in disease resistance and foraging. This chapter reviews the semiochemicals, chemically mediated behaviors, and chemosensory mechanisms in *C. formosanus* and discusses outstanding questions to be addressed in the future.

7.1 Introduction

The interaction of insects with the chemical environment is critical for their ecological success. In social insects, colonies operate under environmental inputs in a decentralized manner. Individuals respond to local cues through interactions with colony members, leading to the emergence of colony-level behavior (Gordon, 2016). Chemical communication is an integral component of the social organization (Blum, 1974). Termites maintain a division of labor through a caste system consisting of reproductives (queens and kings), workers, and soldiers (Bignell *et al.*, 2011). In the Formosan subterranean termite, *Coptotermes formosanus* Shiraki, colonies are founded by a pair of dispersal alates, which construct an underground nest and rear the first cohort of the brood as the primary queen and king. Newly developed workers take over brood care, food collection, and hygienic activities. Soldiers defend the colony chemically and mechanically from competitors and predators. Subterranean termites, except for the dispersal alates, do not develop functional compound eyes and rely heavily on chemical and tactile cues to organize social behavior.

The behavioral and chemical ecology of *C. formosanus* has been increasingly studied since its introduction to the continental U.S. in the mid-20th century. Various compounds have been examined for their chemical identity

DOI: 10.1079/9781800621596.0007

and effects on behavior and physiology. Four exocrine glands, including the labial, frontal, tergal, and sternal glands, as well as cuticular compounds, have been investigated in *C. formosanus*, revealing various chemicals and some with confirmed biological activities. Other chemicals produced in carton nests, entomopathogenic fungi, and decayed wood have been investigated for their roles in disease resistance and foraging behavior. Termite chemical ecology has been previously reviewed (Sillam-Dussès, 2010; Šobotník *et al.*, 2010; Bordereau and Pasteels, 2011; Bagnères and Hanus, 2015; Mitaka and Akino, 2021). This chapter summarizes the current knowledge of chemical ecology in *C. formosanus* and discusses future research directions.

7.2 Semiochemicals and chemically mediated behaviors

7.2.1 Mating and reproduction

Mature termite colonies produce winged reproductive individuals (i.e., alates) that engage in seasonal dispersal flights known as swarms. Dispersal flights of *C. formosanus* are highly synchronized. Colonies release massive amounts of alates on the same days around sunset during late spring to early summer (Su and Scheffrahn, 1987; Henderson and Delaplane, 1994; Chouvenc *et al.*, 2017). After flying away from the natal colony, alates land, shed their wings (i.e., dealation), exhibit a courtship behavior, and form tandem-running pairs before constructing a new nest chamber and mating (Raina *et al.*, 2003). The courtship is mediated by a sex pheromone, (3*Z*,6*Z*,8*E*)-dodeca-3,6,8-trien-1-ol, produced in the tergal glands located at the 9th and 10th dorsal abdominal segments of females (Chouvenc *et al.*, 2020). This pheromone is commonly found in many species of Rhinotermitidae and Termitidae (Sillam-Dussès, 2010; Bordereau and Pasteels, 2011). As males approach females, they antennate the female's dorsolateral region of their terminal abdominal segments before following in tandem, and then use their antennae and mouthparts to maintain contact with the female's posterior abdomen while she leads the search for a nesting site (Fig. 7.1) (Raina *et al.*, 2003). It was shown that males took a longer time to locate a female and failed to establish tandem pairs when their antennae were ablated, but the removal of maxillary or labial palps did not affect the behavior (Raina *et al.*, 2003). This observation suggests that antennae are the primary sensory organs for female recognition, presumably through their sex pheromone and tactile cues. In *Coptotermes gestroi* (Wasmann), if the tandem was accidentally separated, the male and the female displayed asymmetrical behaviors, where the female paused while

Fig. 7.1. A male dealate of *C. formosanus* follows a female as the pair engages in tandem running. (Photo courtesy of Thomas Chouvenc.)

the male increased mobility to increase the odds of reconnecting (Mizumoto *et al.*, 2020). Such a behavior has yet to be confirmed in *C. formosanus*. While the tergal glands are absent in *C. formosanus* males, alates of both sexes have a sternal gland that produces a trail pheromone chemically identical to the female sex pheromone (Bland *et al.*, 2007). It remains to be determined if the trail pheromone is released during tandem behavior.

Female alates of *C. gestroi* produce the same sex pheromone as *C. formosanus*, and the two species have partially overlapping swarm seasons in their invasive ranges (e.g., south Florida), which can trigger heterospecific mating resulting in hybrid colonies (Chouvenc *et al.*, 2015, 2020). When given the choice of tandem running partners, males of both species prefer *C. formosanus* females, likely because they have larger tergal glands that produce and store more sex pheromone than *C. gestroi* females (Chouvenc *et al.*, 2020). Nevertheless, *C. gestroi* males form stable conspecific pairs due to their capacity to recognize the smaller amount of sex pheromone produced by the females (Mizumoto *et al.*, 2021).

In many termite species, female dealates display a calling behavior by raising the abdomen and releasing a sex pheromone that attracts males before forming a tandem pair (Costa-Leonardo and Haifig, 2014). The calling behavior is ephemeral, therefore only rarely documented in *Coptotermes* (Chouvenc *et al.*, 2020). Calling is expected to be coupled with the use of a long-range sex pheromone, but the *C. formosanus* sex pheromone is effective at short distances or upon contact (Raina *et al.*, 2003; Chouvenc *et al.*, 2020). Long-range attraction is probably unnecessary for this species as mates are frequently encountered at gregarious sites after massive swarms. While (3*Z*,6*Z*,8*E*)-dodeca-3,6,8-trien-1-ol is a volatile compound, it is converted to a contact pheromone at least partially due to trilinolein. This viscous, co-secreted tergal gland substance drastically reduces its volatility (Bland *et al.*, 2004; Chouvenc *et al.*, 2020). Trilinolein has been identified as a female-specific compound and is known to be transferred to males, potentially as a nuptial gift (Park *et al.*, 2004; Bland *et al.*, 2004), but this hypothesis remains to be tested.

Alates of *C. formosanus* sometimes aggregate on large trees after dispersal flights, but the chemical basis of this aggregation has not been identified. It is unknown if alates can detect semiochemicals from host trees or release an aggregation pheromone to attract other alates. Additionally, *C. formosanus* is a monogamous species with the queen and king mating repeatedly during their lifetime, but no known sex-mating pheromone exists for eliciting copulation (Bordereau and Pasteels, 2011).

After colony foundation, pheromones released by the reproductives likely mediate many aspects of the behavior and physiology of non-reproductive individuals (Matsuura, 2012). In *Reticulitermes speratus* (Kolbe), female neotenic reproductives and eggs emit a queen pheromone consisting of *n*-butyl-*n*-butyrate and 2-methyl-1-butanol, which attracts workers, inhibits differentiation of additional neotenics, regulates egg production among existing queens, and exhibits antifungal activities (Matsuura *et al.*, 2010; Yamamoto and Matsuura, 2011; Matsuura, 2012). In *Reticulitermes flavipes* (Kollar), a queen and king recognition pheromone, *n*-C21, has been demonstrated to induce antennation and vibratory shaking in workers (Funaro *et al.*, 2018). Pre-molt foraging workers of *C. formosanus* move close to the central nest, where the queen, king, and eggs are located, to complete their molting process (Kakkar *et al.*, 2017). The molting site fidelity is possibly mediated by chemical signals released by the royal pair or eggs. In addition, nymphs of *C. formosanus* can differentiate into neotenic reproductives upon the death of primary reproductives (Vargo *et al.*, 2003; Raina *et al.*, 2004), suggesting the existence of queen and king pheromones that regulate nestmate fertility. So far, reproductive pheromones have not been characterized in *Coptotermes*.

7.2.2 Foraging and feeding

Foraging in *C. formosanus* is a complex collective behavior mediated by various social and environmental factors (see Chapter 5 of this volume). Collective foraging is a convergent trait among termites and ants. As in social hymenopterans, foraging in termites is regulated through extensive interactions among nestmates and trail pheromones (Traniello and Leuthold, 2000; Czaczkes *et al.*, 2015). As multiple-site nesters, *C. formosanus* colonies feed on wood pieces connected by underground tunnels. The royal chamber is typically located in or near one piece of wood. Workers forage outward by tunneling through the soil to search for new food sources, usually accompanied by soldiers. When a new food source is discovered, more workers and soldiers are recruited to colonize the piece of wood. Workers and soldiers are densely distributed in foraging tunnels and food sources during this process. Chemical cues, alone or integrated with tactile cues, are likely involved in regulating all aspects of foraging, including food exploration, recruitment, aggregation, and feeding.

Trail pheromones are one of the most studied types of pheromones in termites (see reviews by Bordereau and Pasteels (2011) and Mitaka and Akino (2021)). These pheromones can be deposited by foraging individuals during food exploration and after food finding, and they elicit trail-following behavior in other individuals. The chemical components of trail pheromones are usually conserved within a genus or family (Mitaka and Akino, 2021). In *C. formosanus*, (3*Z*,6*Z*,8*E*)-dodeca-3,6,8-trien-1-ol is the major component of trail pheromone produced in the sternal gland (5th sternite) of workers and alates (Tokoro *et al.*, 1989; Bland *et al.*, 2007). This pheromone is chemically identical to the sex pheromone released from the tergal and sternal glands of alates in this species (Chouvenc *et al.*, 2020), and it is shared by at least 12 species in Rhinotermitidae as a trail pheromone (Mitaka and Akino, 2021). An isomer of this compound, (*Z*,*E*,*E*)-dodecatrien-1-ol, was reported as a minor component and proposed to confer species-specificity in *C. formosanus*; however, behavioral assays did not support this hypothesis (Tokoro *et al.*, 1994). Dodecatrienol is partially volatile and presumably reinforced by foraging individuals for the trail to remain active. Trail longevity may be increased through integration with fecal substances, as is shown in Termitidae (Souto and Kitayama, 2000). Trail-following in *C. formosanus* can also be elicited by non-pheromonal chemicals, such as naphthalene in their carton nest (Chen *et al.*, 1998b) and 2-phenoxyethanol from a ballpoint pen ink (Chen *et al.*, 1998c), though the sensory mechanisms are unknown. Many questions remain regarding the use of trail pheromone in *C. formosanus*. It is unknown whether the pheromone is reinforced according to food quality or quantity, whether it is released only after food finding or during exploratory tunneling, and if other chemicals synergize with or complement the activity of the major component.

The foraging activity of *C. formosanus* can be influenced by semiochemicals in the soil, such as the attractants from decayed wood. Su (2005) showed that *C. formosanus* and *R. flavipes* workers directed tunnel excavation toward the extracts of wood decayed by the brown rot fungus, *Gloeophyllum trabeum* (pers. ex Fr.) Murr, but the attractive range was relatively short in the sand (less than 18 cm). Interestingly, (3*Z*,6*Z*,8*E*)-dodeca-3,6,8-trien-1-ol has also been isolated from the wood extracts that have an attractant effect (Smythe *et al.*, 1967; Matsumura *et al.*, 1968; Grace, 1991; Rust *et al.*, 1996). How or if food-associated semiochemicals are transmitted below ground to affect termite foraging is not understood. The effects are potentially dependent upon many variables, including soil type, climatic condition (e.g., temperature and moisture), and property of the chemicals (e.g., water solubility and volatility).

Workers of many termite species, including *C. formosanus*, produce hydroquinone from the labial glands (Reinhard *et al.*, 2002). Hydroquinone has been suggested as a general phagostimulant pheromone in diverse families and species; however, its activity as a phagostimulant in *C. formosanus* is not supported (Raina *et al.*, 2005). The social organization of feeding and food colonization may involve pheromones composed of multiple components. In *R. speratus*, workers at food sources deposit a multicomponent aggregation pheromone consisting of six compounds – an aromatic compound, two hydrocarbons, two fatty acids, and cholesterol. This pheromone attracts foraging workers and arrests them at the odor source (Mitaka *et al.*, 2020). Given the similarity in their foraging and feeding ecology, *C. formosanus* workers may also produce an aggregation pheromone with the same function, which awaits further characterization.

7.2.3 Cuticular hydrocarbons and nestmate recognition

Cuticular hydrocarbons (CHCs) are the primary components of a waxy layer on the epicuticle of insects (Howard and Blomquist, 2005; Chung and Carroll, 2015). CHCs form a hydrophobic barrier that prevents transpirational water loss, and they have also evolved as signaling molecules in many species for the recognition of species, nestmate, caste, and reproductive status (Martin and Drijfhout, 2009; Van Oystaeyen *et al.*, 2014; Bagnères and Hanus, 2015). Insect CHCs are synthesized in the oenocytes following a conserved biosynthetic pathway and then transported to the cuticular surface (Chung and Carroll, 2015). The CHC profile of an insect consists of a blend of long-chain hydrocarbons, which vary in chain length as well as the number and position of branches and double bonds (Howard and Blomquist, 2005; Chung and Carroll, 2015). Longer-chain linear alkanes generally have higher melting temperatures and increased waterproofing capacity. In comparison, methyl-branched and unsaturated hydrocarbons have lower melting temperatures but allow a greater diversity of information for chemical communication (Gibbs and Pomonis, 1995; Chung and Carroll, 2015).

Using gas chromatography–mass spectrometry, Haverty *et al.* (1996) identified 52 individual or isomeric mixtures of hydrocarbons in *C. formosanus*. These CHCs included linear and methyl-branched alkanes with main chain lengths ranging from 25 to 43 carbons, and no alkenes were detected. The shorter-chain compounds (C25–C30) appear to be the most abundant and detectable in other studies (Haverty *et al.*, 1990; Bland *et al.*, 2001; Chouvenc and Su, 2017; Chouvenc *et al.*, 2020). The CHC composition of *C. formosanus* varies quantitatively with geographic population, colony, caste, and season, but no qualitative changes have been reported (Haverty *et al.*, 1990, 1996). Four introduced populations of *C. formosanus* collected from Honolulu (Hawai'i), Hallandale (Florida), New Orleans (Louisiana), and Lake Charles (Louisiana) were distinguishable based on the quantitative differences in CHC profiles using canonical discriminant analysis (Haverty *et al.*, 1990). Based on an investigation of seven colonies collected from four locations in Oahu, Hawai'i, Haverty *et al.* (1996) showed that each colony carried a blend of CHCs distinguishable according to the relative abundances; in addition, workers and soldiers were separated from each other and from nymphs and alates based on the quantitative composition of CHCs, but nymph and alate CHCs were not significantly different.

In *R. flavipes*, *n*-C21 has been identified as a queen and king recognition pheromone (Funaro *et al.*, 2018), but no caste-specific hydrocarbon has been found in *C. formosanus*. Several hydrocarbon peaks, each representing an individual compound or a co-eluted isomeric mixture, showed distinct seasonal changes depending on the caste. During swarm season (May to July with Hawai'i as the study site), two peaks, 13-; 15-meC39 + 15,17-dimeC39 and 13,17-; 15,19-dimeC39, decreased in their proportions to total CHCs in workers, while two other peaks, 9-; 11-; 13-meC25 and *n*-C27, reduced and increased in relative proportion, respectively, in soldiers (Haverty *et al.*, 1996). The changes in CHCs are possibly associated with variations in colony demography, as the developmental status and age structures of different castes greatly vary with season (Chouvenc *et al.*, 2022). CHC profiles also vary quantitatively with colony age, as in *C. gestroi* (Gordon *et al.*, 2020). So far, the behavioral effects of CHCs and the influence of CHC plasticity on caste recognition have not been investigated in *C. formosanus*.

Social insect colonies typically have well-defined territorial boundaries maintained through a recognition system. The discrimination of colony-specific odors mediates aggressive behavior between colonies, and CHCs have been considered a critical mechanism of nestmate recognition (Bagnères and Hanus, 2015). However, there is a lack of evidence supporting the role of CHCs in intercolonial aggression within *C. formosanus*. Research on introduced and native populations of *C. formosanus* showed that intercolonial interactions were greatly variable: some colony pairs were not aggressive while others were highly aggressive to each other (Su and Haverty, 1991; Shelton and Grace, 1997; Cornelius and Osbrink, 2003; Messenger and Su, 2005; He *et al.*, 2005, 2006; Pan *et al.*, 2006). Based on an investigation of Florida and Hawai'i colonies, differences in CHC composition were not correlated with intercolonial aggression in *C. formosanus* (Su and Haverty, 1991). As CHCs are highly plastic in response to intrinsic (e.g., individual and colony developmental status) and exogenous factors (e.g., diet and seasonal changes) (Haverty *et al.*, 1996; Liang and Silverman, 2000; Gordon *et al.*, 2020), these conditions should be considered in the analysis of CHCs and behavioral responses. Chouvenc and Su (2017) found that, under the same developmental status and environmental condition, CHCs were quantitatively different among colonies of *C. formosanus*, *C. gestroi*, and their hybrids based on south Florida populations, but the detectable CHCs were not correlated with aggressive interactions between species or mating combinations.

Genetic distance was positively correlated with aggression levels between *C. formosanus* colonies from a Louisiana population (Husseneder *et al.*, 2005), but this correlation was not found in a Hawai'i population (Husseneder and Grace, 2001). Environmental factors, such as diet, can affect intercolonial aggression through the modulation of termite chemical profiles. Florane *et al.* (2004) found that aggression between *C. formosanus* colonies diminished when they were switched from a different to the same diet, along with a shift in volatile compounds emitted by workers. However, changes in laboratory diet were not found to influence aggression in a different study (Cornelius and Osbrink, 2009). This discrepancy in previous studies suggests that interactions between *C. formosanus* colonies are complex, and nestmate recognition is likely mediated by multiple chemical cues due to differences in genetic factors and environmental conditions. So far, the recognition cues and mechanisms of intercolonial aggression remain unclear, and future investigations are required to examine each potential factor under well-controlled experimental conditions.

7.2.4 Chemical defense

Termites employ various strategies to defend against predators, mainly ants, and compete with intra- and interspecific colonies for nesting and foraging resources (Prestwich, 1984). The soldier caste is morphologically specialized for defense and has evolved in most termite species (Tian and Zhou, 2014). Soldiers usually constitute about 10% of the colony population in *C. formosanus*, but the proportion varies considerably depending on social and environmental conditions (Haverty, 1977; Delaplane *et al.*, 1991; Su and La Fage, 1986).

When a colony is disturbed, the mobility of workers and soldiers is differentially altered through alarm communication (Wang *et al.*, 2016). Soldiers rush out and aggregate at the opening of the foraging gallery, while workers retreat and then re-construct the gallery under the guard of soldiers. This alarm behavior is conserved in termites, and similar alarm communication is present in the sub-social cockroach *Cryptocercus* (Sillam-Dussès *et al.*, 2023). *Coptotermes formosanus* soldiers employ both mechanical and chemical defenses in the forms of mandibular attacks and the secretion of a milky fluid from their fontanelles, i.e., frontal gland opening (Fig. 7.2). The fontanelle is located above the mandibles and allows the secretion to pool into a droplet that douses enemies as the soldiers bite them (Noirot, 1969; Mill, 1983). The frontal gland is relatively large, extending from the fontanelle into the abdomen, comprising up to 36% of the soldier's body weight (Waller and La Fage, 1987). The defensive secretion of

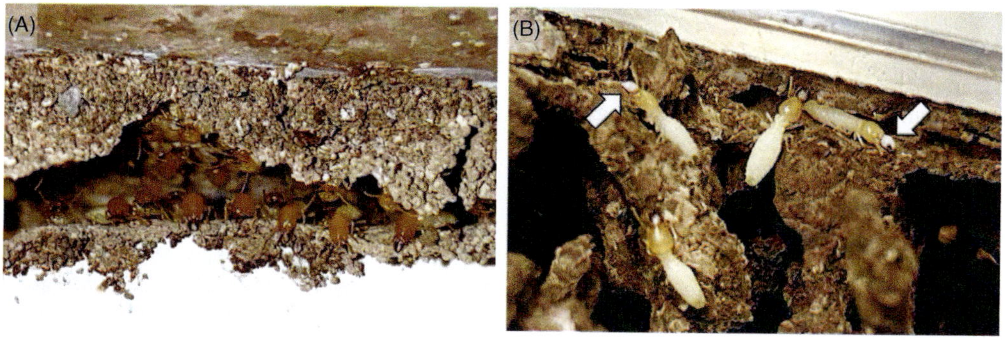

Fig. 7.2. Defensive behavior of *C. formosanus* soldiers. (A) A large number of soldiers aggregate at the opening of a disturbed mud tube during swarm season. (B) Soldiers secrete defensive fluids (arrows point to white droplets). (Photo courtesy of Joseph McCarthy, 2022.)

C. formosanus is a white, sticky, and glue-like mixture, including fatty acids, mucopolysaccharides, hydrocarbons, ceramides, aromatic compounds, ketones, alcohols, aldehydes, and esters (Zhang *et al.*, 2006; Ohta *et al.*, 2007). Lignoceric and hexacosanoic acids have been identified as the major components of the secretion in three colonies collected from an introduced Louisiana population (Chen *et al.*, 1999). The two fatty acids may be highly variable in quantity across populations, as they were not detected in three different colonies from a native *C. formosanus* population (Zhang *et al.*, 2006).

The function of frontal gland secretions in *C. formosanus* is not fully determined. These chemicals do not appear to be repellent or toxic to ants (Mill, 1983; Waller and La Fage, 1987; Mao *et al.*, 2005). Nevertheless, it has been suggested to contribute to the unpalatability of soldiers (Waller and La Fage, 1987), and the adhesive fluid may clog the spiracles and suffocate ants (Mill, 1983). In other species of Rhinotermitidae and Termitidae, the defensive secretions induce various physiological and behavioral responses (Šobotník *et al.*, 2010; Mitaka and Akino, 2021). In *Nasutitermes*, frontal gland secretions can trigger alarm responses in workers and soldiers (Eisner *et al.*, 1976), and they contain compounds that influence soldier differentiation within the colony (Lefeuve and Bordereau, 1984). Mao *et al.* (2005) showed that the crude extract of frontal gland secretion in *C. formosanus* did not influence soldier differentiation, but some of its fatty acid components stimulated soldier formation.

Additionally, *C. formosanus* defensive secretion contains lysozymes (Hardt *et al.*, 2003) and Kazal-type serine protease inhibitors (Negulescu *et al.*, 2015). Although the function has not been investigated, they may protect the colony against bacterial or fungal infection (Negulescu *et al.*, 2015). The defensive secretion of *C. formosanus* likely has other unidentified components, and the chemical complexity suggests that it is responsible for additional physiological and behavioral effects on both termites and the organisms they interact with.

7.2.5 Disease resistance

Subterranean termites live in highly crowded colonies and forage in microbe-rich environments. While an increased risk of epizootic events in social colonies has been suggested (Schmid-Hempel, 1995), termites have evolved sophisticated disease resistance mechanisms to limit the spread of pathogens and enhance social immunity, thus challenging the application of biological control toward pest termite species (Chouvenc *et al.*, 2011; Rosengaus *et al.*, 2011). Chemical defense and hygienic behavior are among their prominent disease resistance strategies.

Workers of *C. formosanus* fill the voids of feeding sites using chewed wood particles mixed with saliva and fecal materials, forming a sponge-like carton nest, and they coat the inner surfaces of foraging galleries with a fecal envelope (King and Spink, 1969; Chouvenc *et al.*, 2013). The nest materials are incorporated with endogenous termite secretions and support the growth of actinobacteria recruited from the surrounding soil, creating an environment poorly conducive to entomopathogens (Chouvenc *et al.*, 2013, 2018). Eleven antimicrobial metabolites were identified from the actinobacteria; in particular, bafilomycin C1 elicited a strong chemical response from a common soil saprophyte *Trichoderma harzianum* Rifai, highlighting chemical interactions among microbes within a termite nest (Mevers *et al.*, 2017). In addition, Chen *et al.* (1998b) found that the carton nest contained naphthalene, a volatile fumigating the nest as an antimicrobial agent and a repellent to ants, while *C. formosanus* was highly tolerant to this volatile. The origin of naphthalene is unknown but has been suggested to be associated with the processing of wood or soil that termites ingest (Chen *et al.*, 1998a). Hussain *et al.* (2017) detected naphthalene as one of the volatile organic compounds emitted by alates, but not from workers, soldiers, or nymphs, though this observation remains to be explained. In the same study, diverse volatile compounds were identified in *C. formosanus* with a caste-specific pattern, including fatty acids, aldehydes, aromatics, flavonoids, sesquiterpenes, alkanes, ketones, esters, alcohols, and nitrogen- or sulfur-containing compounds, many potentially contributing to antifungal activities in the colony.

Termites perform allogrooming behavior to remove fungal conidia from the cuticle of nestmates, and the ingested conidia are eliminated through alimentary tracts (Yanagawa and Shimizu, 2007). *Coptotermes formosanus* workers can perceive fungal odors using antennae and distinguish the odors from three entomopathogenic fungi: *Beauveria brongniartii* (Sacc.) Petch, *Isaria fumosorosea* (Wize), and *Metarhizium anisopliae* (Metchnikoff) Sorokin (Yanagawa *et al.*, 2009b, 2010). Some fungal odors have been identified, including 3-methyl-1-butanol, 3-octanone, 1-octen-3-ol, and 3-octanol, with termites showing an aversion to 3-octanone and a preference for 3-octanol depending on the concentration (Yanagawa *et al.*, 2011, 2012). In addition, Chen *et al.* (2022) found that ergosterol was released by *C. formosanus* termites exposed to *M. anisopliae* and showed a dose-dependent attractant effect, potentially facilitating grooming and removal of fungal conidia. Ergosterol has also been found as a feeding stimulant (Henderson *et al.*, 1999; Cornelius, 2003), and its natural occurrence is usually associated with fungal communities in the soil (Stahl and Parkin, 1996).

Termites dispose of dead colony members through cannibalism and burial as a prophylactic strategy to maintain colony hygiene (Sun *et al.*, 2018). In *R. flavipes*, this behavior is mediated by a dynamic change of chemical cues from workers after death, with 3-octanone and 3-octanol emitted as early death cues that attract live workers to retrieve and consume the dead, and fatty acids (mainly oleic acid) as late death cues that elicit corpse burial (Sun *et al.*, 2017). While cannibalism also frequently occurs in *C. formosanus*, 3-octanone and 3-octanol were not detected from dead workers in this species (Q. Sun, unpublished results). Workers of *C. formosanus* cannibalize nestmates killed by fungal pathogens and efficiently bury most of them when mortality is high (Chouvenc and Su, 2012). Oleic acid is a common decomposition product that allows death recognition in many arthropods (Sun *et al.*, 2018). It is possibly one of the chemical stimuli for burial behavior in *C. formosanus*.

7.3 Sensory mechanisms of chemical communication

7.3.1 Sensory structures

In insects, antennae are the primary sensory organs for perception of the chemical environment. Termite antennae play important roles in chemical-mediated behaviors, such as nestmate recognition (Huang, 2012), mating (Raina *et al.*, 2003), and pathogen detection (Yanagawa *et al.*, 2009b). As in other termites, *C. formosanus* has moniliform antennae, and each antenna is divided into three basic regions, including a scape, a pedicel, and a segmented flagellum with varying numbers of flagellomeres (Fig. 7.3A). The proliferation of antennal segments occurs at the proximal end of the flagellum upon molting, increasing the number of flagellomeres as the individual develops (Fuller, 1920). In *C. formosanus*, the total number of antennal segments (including scape and pedicel) varies from nine in the first instar larvae to 21 in alates (Chouvenc and Su, 2014). The antennae of *C. formosanus* are mainly chemosensory organs but also sensitive to mechanical stimuli, humidity, and temperature. Yanagawa *et al.* (2009a) and Castillo *et al.* (2021) found nine morphological types of antennal sensilla in *C. formosanus*, which were categorized as basiconicum, campaniformium, capitulum, chaeticum I, chaeticum II, chaeticum III, marginal, trichodeum I, and trichodeum I (Fig. 7.3B), while different nomenclatures were used by Deng *et al.* (2006) and Fu *et al.* (2020). Among them, basiconicum, chaeticum II, trichodeum I, and trichodeum II are putative chemoreceptor sensilla, which constitute more than 90% of the total number of antennal sensilla. The abundance of chemosensilla increases toward the distal end of the antennae, suggesting that the distal flagellomeres are more sensitive to chemical stimuli. Interestingly, basiconicum sensilla are found exclusively on the ventral side of antennae in *C. formosanus*, and the spatial-specific distribution is similar to previous observations in several ant species (Renthal *et al.*, 2003; Nakanishi *et al.*, 2009; McKenzie *et al.*, 2016).

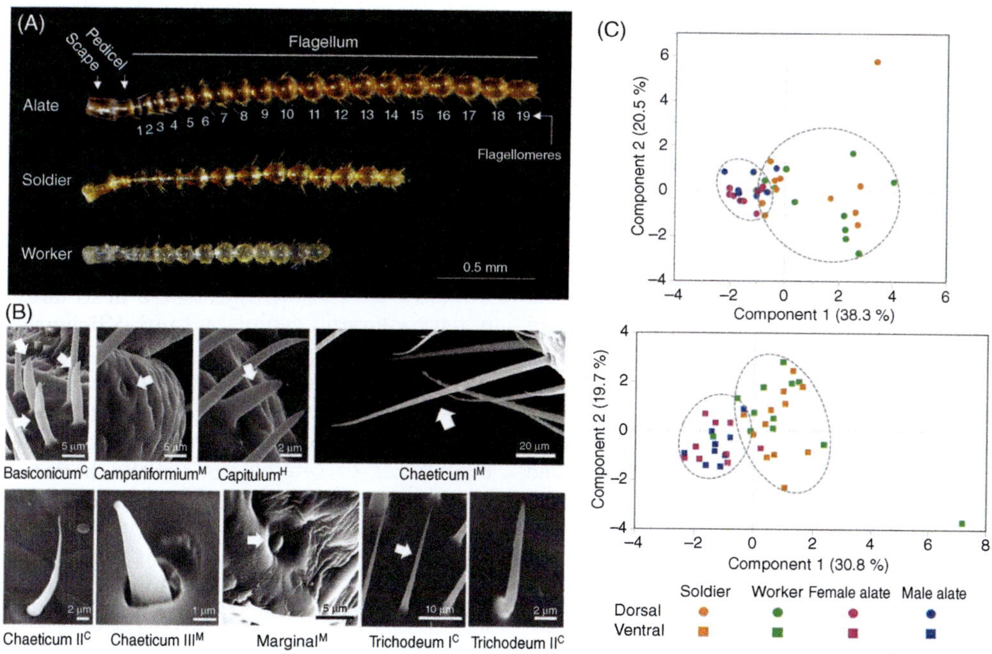

Fig. 7.3. Antennal morphology and antennal sensilla in *Coptotermes formosanus*. (A) Antennae of three castes. The antenna is divided into scape, pedicel, and flagellum, composed of flagellomeres numbered from the proximal to the distal end, as exemplified in the alate. (B) Morphological types of sensilla and functional categories. C: putative chemosensory sensilla; H: putative hygro/thermoreceptive sensilla; M: putative mechanosensory sensilla. (C) Principal component analysis of dorsal and ventral sensilla in different castes. Each gray ellipse encircles samples in the reproductive (female and male alate) or non-reproductive (soldier and worker) caste. Reused from Castillo *et al.* (2021) (no permission required).

Basiconicum sensilla are responsive to CHCs for nestmate recognition in ants (Ozaki *et al.*, 2005). Although their function has not been determined in termites, the ventral-specific distribution in *C. formosanus* is consistent with the observations that termites often touch their nestmates using the ventral surfaces of antennae, which suggests that basiconicum sensilla may play a similar role in social communication in termites.

The antennal morphology and organization of sensilla exhibit caste-biased patterns in *C. formosanus* (Castillo *et al.*, 2021). Dispersal alates, the newly emerged adults, have longer and more heavily sclerotized antennae than workers and soldiers that are considered immature individuals (Fig. 7.3A). Each alate antenna carries more than 3,000 sensilla, twice as abundant as those in workers and soldiers. While alates, workers, and soldiers possess the same types of sensilla, the quantitative composition diverges between the reproductive and non-reproductive castes based on principal component analysis. However, no significant differences are found between female and male alates or between worker and soldier castes (Fig. 7.3C). The quantitative variation in antennal sensilla indicates differential sensitivity to environmental stimuli between alates and non-reproductive individuals. Alates are the only individuals exposed to the open environment and have an expanded behavioral repertoire. In addition to vision, enhanced chemosensation also facilitates their behavior. The increased number of antennal sensilla is correlated with their developmental maturity. It may allow alates to perceive a greater range of environmental cues needed for dispersal, mate-seeking, nest construction, and brood care. It is worth noting that, in many termite species, both female and male alates crop their antennae after colony foundation (Nalepa *et al.*, 2011; Chiu *et al.*, 2018). On average, *C. gestroi* and *Coptotermes lacteus* (Froggatt) dealates lose 7.3 and 5.1 flagellomeres, respectively (Nalepa *et al.*, 2011). The loss of distal antennal segments has also been observed in the primary reproductives of *C. formosanus*. A "reduced sensitivity" hypothesis has been proposed, which suggests that reproductive individuals reduce peripheral sensory inputs upon the behavioral transition from dispersal flight to reproduction when their task is highly specialized and sensing external environmental cues is no longer necessary (Nalepa *et al.*, 2011). Antennal cropping may also reflect a transition from chemical to tactile communication between the reproductive pair once they are inside the royal chamber.

7.3.2 Molecular and neural bases

Insect chemosensory organs express various chemosensory genes, encoding odorant-binding proteins (OBPs), chemosensory proteins (CSPs), and diverse families of odorant receptors (ORs), ionotropic receptors (IRs), and gustatory receptors (GRs) (Hansson and Stensmyr, 2011). Olfactory sensilla are located primarily on the antennae and characterized by a multiporous surface that allows volatiles from the environment to enter the sensillar lymph. Dendrites of olfactory sensory neurons (OSNs) extend to the sensilla, and ORs are expressed on the dendritic membranes. Through activities of OSNs, odorant signals are transduced into electrical impulses, which are integrated into the antennal lobes and may further project to higher brain centers, such as mushroom bodies, to trigger a behavioral response (McGuire *et al.*, 2001; Hansson and Stensmyr, 2011). ORs have undergone rapid evolution in insects, facilitating their adaptation to a wide range of ecological niches (Hansson and Stensmyr, 2011; Robertson, 2019). Compared with solitary species, social insects have expanded repertoires of olfactory receptors that accommodate the increased need for social communication. Different from ants and bees with larger OR profiles (Zhou *et al.*, 2015; McKenzie *et al.*, 2016), termites evolved greater numbers of IRs (Harrison *et al.*, 2018), but their functions remain undetermined.

While each species possesses a distinct OR repertoire, most species express only one odorant receptor co-receptor (Orco), which is highly conserved across insects and required for the function of all ORs (Hansson and Stensmyr, 2011). While ORs confer odorant specificity, Orco and OR subunits form heterotetramers that compose the odorant-gated ion channels. The Orco ortholog in *C. formosanus* is highly conserved in sequence and structure compared with other insects (Castillo *et al.*, 2022). Orco is expressed in multiple developmental stages, including eggs, larvae, workers, pre-soldiers, nymphs, and alates. While antennae are not developed in eggs, expression of Orco during the early developmental stage is possibly required for the proper development of olfactory processing neurons in the brain, which is the case in ants (Trible *et al.*, 2017; Yan *et al.*, 2017). The antennal expression of Orco is nearly twice as high in alates as in workers and soldiers. This pattern is consistent with the number of antennal sensilla, highlighting differential olfactory sensitivity between the dispersal alates and the non-reproductive individuals. While the full profile and functions of chemosensory genes in *C. formosanus* are still under investigation, some chemoreceptors are found differentially expressed among castes, which may underlie the division of labor in chemically mediated tasks (Q. Sun, unpublished results).

7.4 Conclusions

The Formosan subterranean termites possess a suite of exocrine glands similar to other termite species (Prestwich, 1983; Costa-Leonardo and Haifig, 2010). Many glandular and cuticular components and some biological activities have been examined in this species (Table 7.1). While (3*Z*,6*Z*,8*E*)-dodeca-3,6,8-trien-1-ol has been characterized for its pheromonal activities in sex pairing and trail following, the functions of other compounds are not well understood. The use of the same compound in mating and foraging, although from different glandular sources, reflects pheromonal parsimony in *C. formosanus* as widely observed in other termite species (Bordereau and Pasteels, 2011). Some other semiochemicals associated with the *C. formosanus* habitat, such as naphthalene, fungal odors, and decayed wood extracts, have been investigated, revealing complex interactions of termites with their chemical environment. However, our current knowledge of chemical ecology in *C. formosanus* is still limited. The chemical modulation of several major social activities, including reproduction, caste differentiation, and nestmate recognition, remains to be explored. Biological functions or synthetic origins are often undetermined for many of the identified compounds associated with *C. formosanus*.

Additionally, understanding termite chemical ecology will benefit from investigating

Table 7.1. Major glandular and cuticular compounds in *C. formosanus*.

Sources	Identified compounds	Castes	Functions	References
Labial glands	Hydroquinone	Worker	Undetermined	Reinhard *et al.* (2002); Raina *et al.* (2005)
Frontal gland	Fatty acids, mucopolysaccharides, hydrocarbons, ceramides, and other minor components	Soldier	Defense, possibly through soldier unpalatability and ant asphyxiation	Waller and La Fage (1987); Chen *et al.* (1999); Zhang *et al.* (2006); Ohta *et al.* (2007)
Tergal glands	(3*Z*,6*Z*,8*E*)-dodeca-3,6,8-trien-1-ol	Female alate	Sex pheromone, eliciting tandem pairing	Chouvenc *et al.* (2020)
Sternal gland	(3*Z*,6*Z*,8*E*)-dodeca-3,6,8-trien-1-ol	Worker, alate	Trail pheromone, eliciting trail following in foraging	Tokoro *et al.* (1989); Bland *et al.* (2007); Chouvenc *et al.* (2020)
Cuticular surface	Linear and methyl-branched alkanes	All castes	Undetermined, possibly desiccation resistance and nestmate recognition	Haverty *et al.* (1990, 1996)

trophallactic fluid and fecal composition, as they are widely used in social communication (LeBoeuf *et al.*, 2016). While chemical communication is essential for termites, they also perceive tactile cues, and alates possess compound eyes to sense visual stimuli. It remains to be examined how different sensory modalities complement or synergize in social activities. In addition, the investigation of molecular and neural mechanisms underlying chemosensation in termites is still in its infancy. Future interdisciplinary research integrating analytical chemistry, molecular biology, neurophysiology, and behavioral ecology will provide insights into the mechanisms and adaptation of *C. formosanus* to the chemical environment.

References

Bagnères, A.-G. and Hanus, R. (2015) Communication and social regulation in termites. In: Aquiloni L. and Tricarico E. (eds) *Social Recognition in Invertebrates*. Springer International Publishing, Heidelberg, pp. 193–248.

Bignell, D.E., Roisin, Y. and Lo, N. (2011) *Biology of Termites: A Modern Synthesis*, 2nd edn. Springer, Dordrecht, Netherlands.

Bland, J.M., Osbrink, W.L., Cornelius, M.L., Lax, A.R. and Vigo, C.B. (2001) Solid-phase microextraction for the detection of termite cuticular hydrocarbons. *Journal of Chromatography A* 932, 119–127.

Bland, J.M., Park, Y.I., Raina, A.K., Dickens, J.C. and Hollister, B. (2004) Trilinolein identified as a sex-specific component of tergal glands in alates of *Coptotermes formosanus*. *Journal of Chemical Ecology* 30, 835–849.

Bland, J.M., Raina, A.K., Carpita, A. and Dickens, J.C. (2007) Comparative analysis of the sex/trail pheromone, 3, 6, 8-dodecatrien-1-ol, from three species of subterranean termites (Isoptera: Rhinotermitidae). *Sociobiology* 50, 535–551.

Blum, M. (1974) Pheromonal bases of social manifestations in insects. In: Birch M.C. (ed.) *Pheromones*. North-Holland, Amsterdam, pp. 190–199.

Bordereau, C. and Pasteels, J.M. (2011) Pheromones and chemical ecology of dispersal and foraging in termites. In: Bignell, D.E., Roisin, Y. and Lo, N. (eds) *Biology of Termites: a Modern Synthesis*. Springer, Dordrecht, The Netherlands, pp. 279–320.

Castillo, P., Le, N. and Sun, Q. (2021) Comparative antennal morphometry and sensilla organization in the reproductive and non-reproductive castes of the Formosan subterranean termite. *Insects* 12, 576.

Castillo, P., Husseneder, C. and Sun, Q. (2022) Molecular characterization and expression variation of the odorant receptor co-receptor in the Formosan subterranean termite. *PLoS ONE* 17, e0267841.

Chen, J., Henderson, G., Grimm, C.C., Lloyd, S.W. and Laine, R.A. (1998a) Naphthalene in Formosan subterranean termite carton nests. *Journal of Agricultural and Food Chemistry* 46, 2337–2339.

Chen, J., Henderson, G., Grimm, C.C., Lloyd, S.W. and Laine, R.A. (1998b) Termites fumigate their nests with naphthalene. *Nature* 392, 558–559.

Chen, J., Henderson, G. and Laine, R.A. (1998c) Isolation and identification of 2-phenoxyethanol from a ballpoint pen ink as a trail-following substance of *Coptotermes formosanus* Shiraki and *Reticulitermes* sp. *Journal of Entomological Science* 33, 97–105.

Chen, J., Henderson, G. and Laine, R.A. (1999) Lignoceric acid and hexacosanoic acid: major components of soldier frontal gland secretions of the Formosan subterranean termite (*Coptotermes formosanus*). *Journal of Chemical Ecology* 25, 817–824.

Chen, Y., Zhao, C., Zeng, W., Wu, W., Zhang, S. *et al.* (2022) The effect of ergosterol on the allogrooming behavior of termites in response to the entomopathogenic fungus *Metarhizium anisopliae*. *Insect Science* 30(1), 185–196.

Chiu, C.-I., Neoh, K.-B. and Li, H.-F. (2018) Colony-founding success of pleometrosis in a fungus-growing termite *Odontotermes formosanus*. *Behavioral Ecology and Sociobiology* 72, 1–9.

Chouvenc, T. and Su, N.-Y. (2012) When subterranean termites challenge the rules of fungal epizootics. *PLoS ONE* 7, e34484.

Chouvenc, T. and Su, N.Y. (2014) Colony age-dependent pathway in caste development of *Coptotermes formosanus* Shiraki. *Insectes Sociaux* 61, 171–182.

Chouvenc, T. and Su, N.-Y. (2017) Testing the role of cuticular hydrocarbons on intercolonial agonism in two subterranean termite species (*Coptotermes*) and their hybrids. *Insectes Sociaux* 64, 347–355.

Chouvenc, T., Su, N.-Y. and Grace, J.K. (2011) Fifty years of attempted biological control of termites – analysis of a failure. *Biological Control* 59, 69–82.

Chouvenc, T., Efstathion, C.A., Elliott, M.L. and Su, N.-Y. (2013) Extended disease resistance emerging from the faecal nest of a subterranean termite. *Proceedings of the Royal Society B: Biological Sciences* 280, 20131885.

Chouvenc, T., Helmick, E.E. and Su, N.-Y. (2015) Hybridization of two major termite invaders as a consequence of human activity. *PLoS ONE* 10, e0120745.

Chouvenc, T., Scheffrahn, R.H., Mullins, A.J. and Su, N.-Y. (2017) Flight phenology of two *Coptotermes* species (Isoptera: Rhinotermitidae) in southeastern Florida. *Journal of Economic Entomology* 110, 1693–1704.

Chouvenc, T., Elliott, M.L., Šobotník, J., Efstathion, C.A. and Su, N.-Y. (2018) The termite fecal nest: a framework for the opportunistic acquisition of beneficial soil *Streptomyces* (Actinomycetales: Streptomycetaceae). *Environmental Entomology* 47, 1431–1439.

Chouvenc, T., Sillam-Dussès, D. and Robert, A. (2020) Courtship behavior confusion in two subterranean termite species that evolved in allopatry (Blattodea, Rhinotermitidae, *Coptotermes*). *Journal of Chemical Ecology* 46, 461–474.

Chouvenc, T., Ban, P.M. and Su, N.-Y. (2022) Life and death of termite colonies, a decades-long age demography perspective. *Frontiers in Ecology and Evolution* 10, 911042.

Chung, H. and Carroll, S.B. (2015) Wax, sex and the origin of species: Dual roles of insect cuticular hydrocarbons in adaptation and mating. *BioEssays* 37, 822–830.

Cornelius, M.L. (2003) Evaluation of semiochemicals as feeding stimulants. *Sociobiology* 41, 583–591.

Cornelius, M.L. and Osbrink, W.L.A. (2003) Agonistic interactions between colonies of the Formosan subterranean termite (Isoptera: Rhinotermitidae) in New Orleans, Louisiana. *Environmental Entomology* 32, 1002–1009.

Cornelius, M.L. and Osbrink, W.L.A. (2009) Bioassay design and length of time in the laboratory affect intercolonial interactions of the Formosan subterranean termite (Isoptera, Rhinotermitidae). *Insectes Sociaux* 56, 203–211.

Costa-Leonardo, A.M. and Haifig, I. (2010) Pheromones and exocrine glands in *Isoptera*. In: Litwack, G. (ed.) *Vitamins and Hormones: Pheromones*. Elsevier Academic Press, San Diego, California, pp. 521–549.

Costa-Leonardo, A.M. and Haifig, I. (2014) Termite communication during different behavioral activities. In: Witzany, G. (ed.) *Biocommunication of Animals*. Springer Netherlands, Dordrecht, pp. 161–190.

Czaczkes, T.J., Grüter, C. and Ratnieks, F.L.W. (2015) Trail pheromones: an integrative view of their role in social insect colony organization. *Annual Review of Entomology* 60, 581–599.

Delaplane, K.S., Saxton, A.M. and La Fage, J.P. (1991) Foraging phenology of the Formosan subterranean termite (Isoptera: Rhinotermitidae) in Louisiana. *American Midland Naturalist* 125, 222–230.

Deng, T., Mo, J., He, H., Pan, C. and Cheng, J. (2006) Differences of morphology of antennae between soldiers and workers in *Coptotermes formosanus* (Isoptera: Rhinotermitidae). *Sociobiology* 48, 689–700.

Eisner, T., Kriston, I. and Aneshansley, D.J. (1976) Defensive behavior of a termite (*Nasutitermes exitiosus*). *Behavioral Ecology and Sociobiology* 1, 83–125.

Florane, C.B., Bland, J.M., Husseneder, C. and Raina, A.K. (2004) Diet-mediated inter-colonial aggression in the Formosan subterranean termite *Coptotermes formosanus*. *Journal of Chemical Ecology* 30, 2559–2574.

Fu, B.-X., Rong, N.-H., Hong, J., Zhu, Z.-R., Mo, J.-C. *et al.* (2020) Comparative study with scanning electron microscopy on the antennal sensilla of two main castes of *Coptotermes formosanus* Shiraki (Blattaria: Rhinotermitidae). *Micron* 129, 102777.

Fuller, C. (1920) Studies on the post-embryonic development of the antennae of termites. *Annals of the Natal Museum* 4, 235–295.

Funaro, C.F., Böröczky, K., Vargo, E.L. and Schal, C. (2018) Identification of a queen and king recognition pheromone in the subterranean termite *Reticulitermes flavipes*. *Proceedings of the National Academy of Sciences* 115, 3888–3893.

Gibbs, A. and Pomonis, J.G. (1995) Physical properties of insect cuticular hydrocarbons: The effects of chain length, methyl-branching and unsaturation. *Comparative Biochemistry and Physiology Part B: Biochemistry and Molecular Biology* 112, 243–249.

Gordon, D.M. (2016) From division of labor to the collective behavior of social insects. *Behavioral Ecology and Sociobiology* 70, 1101–1108.

Gordon, J.M., Šobotník, J. and Chouvenc, T. (2020) Colony-age-dependent variation in cuticular hydrocarbon profiles in subterranean termite colonies. *Ecology and Evolution* 10, 10095–10104.

Grace, J.K. (1991) Semiochemical mediation and manipulation of *Reticulitermes* behavior (Isoptera: Rhinotermitidae). *Sociobiology* 19, 147–162.

Hansson, B.S. and Stensmyr, M.C. (2011) Evolution of insect olfaction. *Neuron* 72, 698–711.

Hardt, M., Guo, Y., Henderson, G. and Laine, R.A. (2003) Zymogram with Remazol brilliant blue-labeled *Micrococcus lysodeikticus* cells for the detection of lysozymes: example of a new lysozyme activity in Formosan termite defense secretions. *Analytical Biochemistry* 312, 73–76.

Harrison, M.C., Jongepier, E., Robertson, H.M., Arning, N., Bitard-Feildel, T. *et al.* (2018) Hemimetabolous genomes reveal molecular basis of termite eusociality. *Nature Ecology and Evolution* 2, 557–566.

Haverty, M.I. (1977) The proportion of soldiers in termite colonies: a list and a bibliography. *Sociobiology* 2, 199–216.

Haverty, M.I., Grace, J.K., Nelson, L.J. and Yamamoto, R.T. (1996) Intercaste, intercolony, and temporal variation in cuticular hydrocarbons of *Copotermes formosanus* shiraki (Isoptera: Rhinotermitidae). *Journal of Chemical Ecology* 22, 1813–1834.

Haverty, M.I., Nelson, L.J. and Page, M. (1990) Cuticular hydrocarbons of four populations of *Coptotermes formosanus* Shiraki in the United States. *Journal of Chemical Ecology* 16, 1635–1647.

He, H., Mo, J., Pan, C. and Zhang, S. (2005) Intercolony agonism and protein variation based on environmental cues in *Coptotermes formosanus* (Isoptera: Rhinotermitidae). *Sociobiology* 46, 615–625.

He, H., Mo, J., Teng, L., Pan, C., Zhang, C. and Cheng, J.A. (2006) No influence of exocrine glands on nestmate discrimination in *Coptotermes formosanus* (Isoptera: Rhinotermitidae). *Sociobiology* 47, 253–264

Henderson, G. and Delaplane, K. (1994) Formosan subterranean termite swarming behavior and alate sex-ratio (Isoptera: Rhinotermitidae). *Insectes Sociaux* 41, 19–28.

Henderson, G., Chen, J. and Laine, R.A. (1999) Compositions and methods for detecting and killing termites. United States Patent US5874097.

Howard, R.W. and Blomquist, G.J. (2005) Ecological, behavioral, and biochemical aspects of insect hydrocarbons. *Annual Review of Entomology* 50, 371–393.

Huang, Q. (2012) Aggressive behavior and the role of antennal sensillae in the termite *Reticulitermes chinensis* (Isoptera: Rhinotermitidae). *Sociobiology* 59, 1239–1251.

Hussain, A., Wen, S.-Y. and Tian, M.-Y. (2017) Exploring the caste-specific multi-layer defense mechanism of Formosan subterranean termites, *Coptotermes formosanus* Shiraki. *International Journal of Molecular Sciences* 18, 2694.

Husseneder, C. and Grace, J.K. (2001) Evaluation of DNA fingerprinting, aggression tests, and morphometry as tools for colony delineation of the Formosan subterranean termite. *Journal of Insect Behavior* 14, 173–186.

Husseneder, C., Messenger, M.T., Su, N.-Y., Grace, J.K. and Vargo, E.L. (2005) Colony social organization and population genetic structure of an introduced population of Formosan subterranean termite from New Orleans, Louisiana. *Journal of Economic Entomology* 98, 1421–1434.

Kakkar, G., Osbrink, W., Mullins, A. and Su, N.-Y. (2017) Molting site fidelity in workers of Formosan subterranean termites (Isoptera: Rhinotermitidae). *Journal of Economic Entomology* 110, 2512–2517.

King, E.G.J. and Spink, W.T. (1969) Foraging galleries of the Formosan subterranean termite, *Coptotermes formosanus*, in Louisiana. *Annals of the Entomological Society of America* 62, 536–542.

LeBoeuf, A.C., Waridel, P., Brent, C.S., Gonçalves, A.N., Menin, L. *et al.* (2016) Oral transfer of chemical cues, growth proteins and hormones in social insects. *eLife* 5, e20375.

Lefeuve, P. and Bordereau, C. (1984) Soldier formation regulated by a primer pheromone from the soldier frontal gland in a higher termite, *Nasutitermes lujae*. *Proceedings of the National Academy of Sciences* 81, 7665–7668.

Liang, D. and Silverman, J. (2000) "You are what you eat": Diet modifies cuticular hydrocarbons and nestmate recognition in the Argentine ant, *Linepithema humile*. *Naturwissenschaften* 87, 412–416.

Mao, L., Henderson, G. and Laine, R.A. (2005) Formosan subterranean termite (Isoptera: Rhinotermitidae) frontal gland secretion and their fatty acid constituent activity on termites and red imported fire ants (Hymenoptera: Formicidae). *Sociobiology* 46, 141–154.

Martin, S. and Drijfhout, F. (2009) A review of ant cuticular hydrocarbons. *Journal of Chemical Ecology* 35, 1151–1161.

Matsumura, F., Coppel, H.C. and Tai, A. (1968) Isolation and identification of termite trail-following pheromone. *Nature* 219, 963–964.

Matsuura, K. (2012) Multifunctional queen pheromone and maintenance of reproductive harmony in termite colonies. *Journal of Chemical Ecology* 38, 746–754.

Matsuura, K., Himuro, C., Yokoi, T., Yamamoto, Y., Vargo, E.L. *et al.* (2010) Identification of a pheromone regulating caste differentiation in termites. *Proceedings of the National Academy of Sciences* 107, 12963–12968.

McGuire, S.E., Le, P.T. and Davis, R.L. (2001) The role of *Drosophila* mushroom body signaling in olfactory memory. *Science* 293, 1330–1333.

McKenzie, S.K., Fetter-Pruneda, I., Ruta, V. and Kronauer, D.J.C. (2016) Transcriptomics and neuroanatomy of the clonal raider ant implicate an expanded clade of odorant receptors in chemical communication. *Proceedings of the National Academy of Sciences* 113, 14091–14096.

Messenger, M.T. and Su, N.-Y. (2005) Agonistic behavior between colonies of the Formosan subterranean termite (Isoptera: Rhinotermitidae) from Louis Armstrong Park, New Orleans, Louisiana. *Sociobiology* 45, 331–346.

Mevers, E., Chouvenc, T., Su, N.-Y. and Clardy, J. (2017) Chemical interaction among termite-associated microbes. *Journal of Chemical Ecology* 43, 1078–1085.

Mill, A.E. (1983) Behavioural and toxic effects of termite defensive secretions on ants. *Physiological Entomology* 8, 413–418.

Mitaka, Y. and Akino, T. (2021) A review of termite pheromones: multifaceted, context-dependent, and rational chemical communications. *Frontiers in Ecology and Evolution* 8, 595614.

Mitaka, Y., Matsuyama, S., Mizumoto, N., Matsuura, K. and Akino, T. (2020) Chemical identification of an aggregation pheromone in the termite *Reticulitermes speratus*. *Scientific Reports* 10, 7424.

Mizumoto, N., Rizo, A., Pratt, S.C. and Chouvenc, T. (2020) Termite males enhance mating encounters by changing speed according to density. *Journal of Animal Ecology* 89, 2542–2552.

Mizumoto, N., Lee, S.-B., Valentini, G., Chouvenc, T. and Pratt, S.C. (2021) Coordination of movement via complementary interactions of leaders and followers in termite mating pairs. *Proceedings of the Royal Society B: Biological Sciences* 288, 20210998.

Nakanishi, A., Nishino, H., Watanabe, H., Yokohari, F. and Nishikawa, M. (2009) Sex-specific antennal sensory system in the ant *Camponotus japonicus*: structure and distribution of sensilla on the flagellum. *Cell and Tissue Research* 338, 79–97.

Nalepa, C.A., Evans, T.A. and Lenz, M. (2011) Antennal cropping during colony foundation in termites. *ZooKeys* 148, 185–196.

Negulescu, H., Guo, Y., Garner, T.P., Goodwin, O.Y., Henderson, G. *et al.* (2015) A Kazal-type serine protease inhibitor from the defense gland secretion of the subterranean termite *Coptotermes formosanus* Shiraki. *PLoS ONE* 10, e0125376.

Noirot, C.H. (1969) Glands and secretion. In: Krishna, K. and Weesner, F.M. (eds) *Biology of Termites*. Academic Press, New York, pp. 89–123.

Ohta, M., Matsuura, F., Henderson, G. and Laine, R.A. (2007) Novel free ceramides as components of the soldier defense gland of the Formosan subterranean termite (*Coptotermes formosanus*). *Journal of Lipid Research* 48, 656–664.

Ozaki, M., Wada-Katsumata, A., Fujikawa, K., Iwasaki, M., Yokohari, F. *et al.* (2005) Ant nestmate and non-nestmate discrimination by a chemosensory sensillum. *Science* 309, 311–314.

Pan, C., Mo, J. and Cheng, M. (2006) Influence of diet and soil on inter-colonial aggression of *Coptotermes formosanus* (Isoptera: Rhinotermitidae). *Sociobiology* 48, 841–848.

Park, Y.I., Bland, J.M. and Raina, A.K. (2004) Factors affecting post-flight behavior in primary reproductives of the Formosan subterranean termite, *Coptotermes formosanus* (Isoptera: Rhinotermitidae). *Journal of Insect Physiology* 50, 539–546.

Prestwich, G.D. (1983) Chemical systematics of termite exocrine secretions. *Annual Review of Ecology and Systematics* 14, 287–311.

Prestwich, G.D. (1984) Defense mechanisms of termites. *Annual Review of Entomology* 29, 201–232.

Raina, A.K., Bland, J.M., Dickens, J.C., Park, Y.I. and Hollister, B. (2003) Premating behavior of dealates of the Formosan subterranean termite and evidence for the presence of a contact sex pheromone. *Journal of Insect Behavior* 16, 233–245.

Raina, A., Osbrink, W. and Park, Y.I. (2004) Nymphs of the Formosan subterranean termite (Isoptera: Rhinotermitidae): aspects of formation and transformation. *Annals of the Entomological Society of America* 97, 757–764.

Raina, A.K., Bland, J.M. and Osbrink, W. (2005) Hydroquinone is not a phagostimulant for the formosan subterranean termite. *Journal of Chemical Ecology* 31, 509–517.

Reinhard, J., Lacey, M.J., Ibarra, F., Schroeder, F.C., Kaib, M. *et al.* (2002) Hydroquinone: a general phagostimulating pheromone in termites. *Journal of Chemical Ecology* 28, 1–14.

Renthal, R., Velasquez, D., Olmos, D., Hampton, J. and Wergin, W.P. (2003) Structure and distribution of antennal sensilla of the red imported fire ant. *Micron* 34, 405–413.

Robertson, H.M. (2019) Molecular evolution of the major arthropod chemoreceptor gene families. *Annual Review of Entomology* 64, 227–242.

Rosengaus, R., Traniello, J.A. and Bulmer, M. (2011) Ecology, behavior and evolution of disease resistance in termites. In: Bignell, D.E., Roisin, Y. and Lo, N. (eds) *Biology of Termites: a Modern Synthesis*. Springer, Dordrecht, The Netherlands, pp. 165–191.

Rust, M.K., Haagsma, K. and Nyugen, J. (1996) Enhancing foraging of western subterranean termites (Isoptera: Rhinotermitidae) in arid environments. *Sociobiology* 28, 275–286.

Schmid-Hempel, P. (1995) Parasites and social insects. *Apidologie* 26, 255–271.

Shelton, T.G. and Grace, J.K. (1997) Suggestion of an environmental influence on intercolony agonism of Formosan subterranean termites (Isoptera: Rhinotermitidae). *Environmental Entomology* 26, 632–637.

Sillam-Dussès, D. (2010) *Trail Pheromones and Sex Pheromones in Termites*, Nova Science Publishers/ Novinka, New York.

Sillam-Dussès, D., Jandák, V., Stiblik, P., Delattre, O., Chouvenc, T. *et al.* (2023) Alarm communication predates eusociality in termites. *Communications Biology* 6, 83.

Smythe, R.V., Coppel, H.C., Lipton, S.H. and Strong, F.M. (1967) Chemical studies of attractants associated with *Reticulitermes flavipes* and *R. virginicus*. *Journal of Economic Entomology* 60, 228–233.

Šobotník, J., Jirošová, A. and Hanus, R. (2010) Chemical warfare in termites. *Journal of Insect Physiology* 56, 1012–1021.

Souto, L. and Kitayama, K. (2000) *Constrictotermes cyphergaster* (Isoptera: Termitidae: Nasutitermitinae) maintain foraging trails for a longer period by means of faecal droplets. *Sociobiology* 35, 367–372.

Stahl, P.D. and Parkin, T.B. (1996) Relationship of soil ergosterol concentration and fungal biomass. *Soil Biology and Biochemistry* 28, 847–855.

Su, N.-Y. (2005) Directional change in tunneling of subterranean termites (Isoptera: Rhinotermitidae) in response to decayed wood attractants. *Journal of Economic Entomology* 98, 471–475.

Su, N.-Y. and Haverty, M.I. (1991) Agonistic behavior among colonies of the Formosan subterranean termite, *Coptotermes formosanus* Shiraki (Isoptera: Rhinotermitidae), from Florida and Hawaii: Lack of correlation with cuticular hydrocarbon composition. *Journal of Insect Behavior* 4, 115–128.

Su, N.-Y. and La Fage, J.P. (1986) Effects of starvation on survival and maintenance of soldier proportion in laboratory groups of the Formosan subterranean termite, *Coptotermes formosanus* (Isoptera: Rhinotermitidae). *Annals of the Entomological Society of America* 79, 312–316.

Su, N.-Y. and Scheffrahn, R. (1987) Alate production of a field colony of the Formosan subterranean termite (Isoptera: Rhinotermitidae). *Sociobiology* 13, 167–172.

Sun, Q., Haynes, K.F. and Zhou, X. (2017) Dynamic changes in death cues modulate risks and rewards of corpse management in a social insect. *Functional Ecology* 31, 697–706.

Sun, Q., Haynes, K.F. and Zhou, X. (2018) Managing the risks and rewards of death in eusocial insects. *Philosophical Transactions of the Royal Society B: Biological Sciences* 373, 20170258.

Tian, L. and Zhou, X. (2014) The soldiers in societies: defense, regulation, and evolution. *International Journal of Biological Sciences* 10, 296–308.

Tokoro, M., Takahashi, M., Tsunoda, K. and Yamaoka, R. (1989) Isolation and primary structure of trail pheromone of the termite, *Coptotermes formosanus* Shiraki (Isoptera: Rhinotermitidae). *Wood Research: Bulletin of the Wood Research Institute Kyoto University* 76, 29–38.

Tokoro, M., Takahashi, M. and Yamaoka, R. (1994) (*Z*, *E*, *E*)-Dodecatrien-1-ol: a minor component of trail pheromone of termite, *Coptotermes formosanus* Shiraki. *Journal of Chemical Ecology* 20, 199–215.

Traniello, J.F.A. and Leuthold, R.H. (2000) Behavior and ecology of foraging in termites. In: Abe, T., Bignell, D.E. and Higashi, M. (eds) *Termites: Evolution, Sociality, Symbioses, Ecology*. Springer, Dordrecht, The Netherlands, pp. 141–168.

Trible, W., Olivos-Cisneros, L., McKenzie, S.K., Saragosti, J., Chang, N.-C. *et al.* (2017) Orco mutagenesis causes loss of antennal lobe glomeruli and impaired social behavior in ants. *Cell* 170, 727–735.

Van Oystaeyen, A., Oliveira, R.C., Holman, L., Van Zweden, J.S., Romero, C. *et al.* (2014) Conserved class of queen pheromones stops social insect workers from reproducing. *Science* 343, 287–290.

Vargo, E.L., Husseneder, C. and Grace, J.K. (2003) Colony and population genetic structure of the Formosan subterranean termite, *Coptotermes formosanus*, in Japan. *Molecular Ecology* 12, 2599–2608.

Waller, D.A. and La Fage, J.P. (1987) Unpalatability as a passive defense of *Coptotermes formosanus* Shiraki soldiers against ant predation. *Journal of Applied Entomology* 103, 148–153.
Wang, C., Henderson, G., Gautam, B.K., Chen, J. and Bhatta, D. (2016) Panic escape polyethism in worker and soldier *Coptotermes formosanus* (Isoptera: Rhinotermitidae). *Insect Science* 23, 305–312.
Yamamoto, Y. and Matsuura, K. (2011) Queen pheromone regulates egg production in a termite. *Biology Letters* 7, 727–729.
Yan, H., Opachaloemphan, C., Mancini, G., Yang, H., Gallitto, M. *et al.* (2017) An engineered orco mutation produces aberrant social behavior and defective neural development in ants. *Cell* 170, 736–747, e9.
Yanagawa, A. and Shimizu, S. (2007) Resistance of the termite, *Coptotermes formosanus* Shiraki to *Metarhizium anisopliae* due to grooming. *BioControl* 52, 75–85.
Yanagawa, A., Shimizu, S., Noma, K., Nishikawa, M., Kazumasa, O. *et al.* (2009a) Classification and distribution of antennal sensilla of the termite *Coptotermes formosanus* (Isoptera: Rhinotermitidae). *Sociobiology* 54, 327–349.
Yanagawa, A., Yokohari, F. and Shimizu, S. (2009b) The role of antennae in removing entomopathogenic fungi from cuticle of the termite, *Coptotermes formosanus*. *Journal of Insect Science* 9, 1–9.
Yanagawa, A., Yokohari, F. and Shimizu, S. (2010) Influence of fungal odor on grooming behavior of the termite, *Coptotermes formosanus*. *Journal of Insect Science* 10, 141.
Yanagawa, A., Fujiwara-Tsujii, N., Akino, T., Yoshimura, T., Yanagawa, T. *et al.* (2011) Musty odor of entomopathogens enhances disease-prevention behaviors in the termite *Coptotermes formosanus*. *Journal of Invertebrate Pathology* 108, 1–6.
Yanagawa, A., Fujiwara-Tsujii, N., Akino, T., Yoshimura, T., Yanagawa, T. *et al.* (2012) Odor aversion and pathogen-removal efficiency in grooming behavior of the termite *Coptotermes formosanus*. *PLoS One* 7, e47412.
Zhang, S., Mo, J., Teng, L., Cheng, M. and Cheng, J.A. (2006) Inter-colonial variation in the compositions of the frontal gland secretion of *Coptotermes formosanus* (Isoptera: Rhinotermitidae). *Sociobiology* 47, 553–561.
Zhou, X., Rokas, A., Berger, S.L., Liebig, J., Ray, A. *et al.* (2015) Chemoreceptor evolution in Hymenoptera and its implications for the evolution of eusociality. *Genome Biology and Evolution* 7, 2407–2416.

8 Symbiosis and microbiome in termite guts: a unique quadripartite system

Claudia Husseneder
Department of Entomology, Louisiana State University, USA, CHusseneder@agcenter.lsu.edu

Abstract
A prominent feature of the biology of lower termites is the multilayered Russian doll-like symbiosis with gut microbes featuring "bugs in bugs in bugs." Lower termites harbor a diverse community of microorganisms in their guts spanning the three domains of life: Eukaryota (Protists), Archaea, and Bacteria. Most termite gut microorganisms belong to obligate functional groups contributing to energy and nutrient production and benefit the termite host directly or indirectly as ecto- and endosymbionts of the protists. Advances in -omics technologies have recently enabled researchers to shed light on spatial and phylogenetic distribution of the gut microbiota, their co-diversification with various hosts, and to make predictions about physiological functions and trophic interactions even though most gut symbionts remain uncultured. Although still fragmentary, this knowledge makes termite gut microbes excellent models for investigating the complex interactions in gut ecology and function but also makes these microbes tools and targets for controlling invasive termite pest species.

8.1 Introduction

Termites evolved from wood-consuming cockroach-like ancestors and feed exclusively on plant material consisting of lignocellulose (Brune, 2014). Efficient digestion and augmentation of lignocellulose diets, which lack vitamins and components necessary for the synthesis of protein and fat, require obligate symbiosis with a complex community of microorganisms in the hindguts of termite workers (Brune, 2014). This symbiosis forms the collaborative dual cellulose digestion system (Nakashima *et al.*, 2002a; Zhou *et al.*, 2007; Zeng *et al.*, 2016) unique to termites where endogenous termite lignocellulolytic enzymes work in concert with symbiont cellulases to ensure adequate termite nutrition (reviewed in Scharf and Peterson, 2021; see also Chapter 6, this volume). This symbiosis is obligate, as termites are unable to survive without their symbionts (Cleveland, 1923; Breznak and Brune, 1994; Sethi *et al.*, 2014; Peterson *et al.*, 2015). In the so-called lower termites, gut protists provide the main cellulolytic and xylanolytic enzymes (Peterson and Scharf, 2016a) supplementing endogenous termite cellulases, while in higher termites (family Termitidae) cellulolytic symbionts consist of fungus gardens (Macrotermitinae) and gut bacteria (Brune, 2014).

Subterranean termites of the genera *Coptotermes* and *Reticulitermes* belong to the lower termite family Rhinotermitidae and harbor taxonomically and functionally diverse gut

 Biology and Management of the Formosan Subterranean Termite and Related Species (eds Nan-Yao Su and Chow-Yang Lee)
DOI: 10.1079/9781800621596.0008

microbiota consisting of all three domains of life: the protists (Eukaryota), archaea, and bacteria (Prokaryota) (Cleveland, 1923; Breznak, 2000; Brune and Dietrich, 2015; Benjamino and Graf, 2016). In addition, termite guts are "hot spots" for viruses, including bacteriophages (Tikhe and Husseneder, 2018). This quadripartite relationship makes subterranean termites unique models to explore the evolution and intricacies of a multi-layered obligate symbiosis. This includes gaining knowledge of interdependencies among symbionts and termites, mutual diversification and adaptation, stability, and resilience, as well as the role of symbionts in maintaining the ecological balance of the gut. Since the health, survival, and fitness of termite colonies is dependent upon the symbiont community acting as providers of nutrition and energy as well as environmental buffer and pathogen defense (Chouvenc *et al.*, 2013; Rosengaus *et al.*, 2014), gut microbiomes can be employed as tools and targets for novel termite control methods (Zhou *et al.*, 2008; Sethi *et al.*, 2014; Scharf, 2015; Husseneder *et al.*, 2016).

Symbiont inventories of subterranean termite guts derived from culture-dependent and genetic marker techniques and their symbiotic roles, including carbohydrate metabolism, nitrogen fixation and recycling, amino acid and vitamin production, hydrogen capture as well as methano- and acetogenesis for energy production, have been previously reviewed (Breznak and Brune, 1994; Breznak, 2000; Brune, 2013, 2014; Ohkuma, 2003; König *et al.*, 2006; Husseneder, 2010). However, knowledge about symbiont species, functions, ecology, interaction among symbionts and with the host, host-symbiont co-evolution, and complementary roles continues to advance significantly with recent developments in modern -omics technologies, including genome, metagenome and transcriptome analyses (Scharf and Peterson, 2021; Arora *et al.*, 2022).

The large body of literature on termite gut symbiosis dating back over 150 years (Leidy, 1881) is certainly beyond the scope of this chapter. To achieve a balance between breadth and depth, this review will focus on general principles of termite gut symbiosis with emphasis on examples from *Coptotermes* and related rhinotermitid species.

8.2 Evolution of symbiotic microbiota in termites

The wood-feeding ancestors of termites separated from their sister group of cockroaches in the late Jurassic 150–170 million years ago (Bourguignon *et al.*, 2015). Termite gut symbionts from 20 million years ago have been found preserved in amber (Wier *et al.*, 2002) and have been studied for almost 150 years (Leidy, 1881). The early history of symbiont discovery in lower termites was thoroughly reviewed by Brune (2013) and Scharf and Peterson (2021). The first symbionts were likely ingested with plant material by xylophagous gregarious cockroaches similar to the extant wood-consuming cockroach *Cryptocercus*, which is considered to be the closest living relative to the ancestor of termites (Lo *et al.*, 2000). At the beginning of the termite host–symbiont relationship, symbionts were horizontally transferred via coprophagy and/or trophallaxis within social groups (Nalepa *et al.*, 2001). After symbionts established themselves in the gut, mutualistic dependencies developed with symbionts providing nutrients and energy to the host (Brune and Ohkuma, 2011; Scharf *et al.*, 2011) in exchange for a stable food-rich habitat. Increasing interdependencies facilitated termite evolution and transition to eusociality resulting in a trophic and brood care shift: symbionts became obligate and vertically transmitted to the emerging worker caste, which assumed the role of feeding the colony, while parental care shifted to alloparental care by the worker caste (Nalepa *et al.*, 1994; Nalepa, 2020). Predominantly vertically transmitted termite core microbiomes started to co-diversify with the termite host, preserving largely species-specific communities of core taxa while diet and gut microenvironment kept shaping relative abundance (Dietrich *et al.*, 2014; Abdul Rahman *et al.*, 2015). The extent to which the symbiont community is shaped by termite phylogeny (Tai *et al.*, 2015) is dependent on whether microbes are eukaryotic protists (section 8.3, this chapter), prokaryotic archaea (section 8.4, this chapter), bacteria (section 8.5, this chapter) or bacteriophages (section 8.7, this chapter).

8.3 Eukaryote symbionts: protists

While the so-called "higher" termites (family Termitidae) secondarily replaced the protists of their cockroach-like ancestors with cellulolytic fungus gardens (subfamily Macrotermitinae, Aanen *et al.*, 2002) and symbiotic bacteria (in all other Termitidae: Brune, 2014), the lower termites, which include subterranean termites (family Rhinotermitidae), still retain cellulolytic protists similar to those acquired from ancestral wood-feeding cockroaches (Ohkuma *et al.*, 2009). Termite gut protists are flagellated and belong to two groups of the Excavata supergroup that are unique to termites: the parabasalids (phylum Parabasalia) and oxymonads (class Oxymonadea, phylum Preaxostyla). Four of the five oxymonad families are found exclusively in termites and wood-feeding cockroaches, in contrast to the Parabasalia, which are not unique to termites (Brugerolle and Lee, 2000). Subterranean termite guts are dominated by protists, which can number up to 100,000 and account for up to 60–90% of hindgut weight of individual workers (Brune, 2011, 2014).

The protists complement the termites' endogenous cellulolytic enzymes with their own powerful cellulases that are different from the termite glycosyl hydrolases families as discovered through meta- and single-species genome and transcriptome projects in *Coptotermes formosanus* Shiraki (Nakashima *et al.*, 2002b; Inoue *et al.*, 2005; Hongoh *et al.*, 2008a; Xie *et al.*, 2012), *Coptotermes lacteus* (Froggatt) (Watanabe *et al.*, 2002), *Reticulitermes speratus* (Kolbe) (Ohtoko *et al.*, 2000; Todaka *et al.*, 2007; Hongoh *et al.*, 2008b), and *Reticulitermes flavipes* (Kollar) (Zhou *et al.*, 2007; Tartar *et al.*, 2009; see also Chapter 6, this volume). Protists perform the majority of fermentation of lignocellulose to acetate and short-chain fatty acids as energy source for the host (Hungate, 1939; Brune, 2014) and produce hydrogen in the process (Inoue *et al.*, 2005, 2007). Protists species have specific niches and functions in the gut, forming a lignocellulose "disassembly line" as shown in *C. formosanus* workers: large protists engulf wood particles via phagocytosis and break them down in digestive vacuoles, while smaller protists downstream perform further breakdown of the metabolites and/or consume bacteria (Lai *et al.*, 1983; Yoshimura, 1992; Yoshimura, 1995; Inoue *et al.*, 2000; Xie *et al.*, 2012).

In addition to cellulose digestion, protists produce proteases supplementing the proteases of the termite host for digestion of bacteria to retrieve nitrogenous compounds. For example, Sethi *et al.* (2011) described a dual protease system in the guts of *C. formosanus* workers where protists showed similar levels of trypsin-like and subtilisin/chymotrypsin-like activities as midgut extracts; however, protist proteases were clearly distinct from the endogenous proteases of the termite gut tissue, having different molecular mass.

Generally, the gut protist community exhibits a tight phylogenetic association with their termite hosts, caused by vertical transmission and co-diversification (Tai *et al.*, 2015). For example, *Reticulitermes* is a sister clade to other rhinotermitids like *Coptotermes* and *Heterotermes*. This distinction is reflected in the protist community with oxymonads occurring in *Reticulitermes* but not in other rhinotermitids, along with differences in parabasalid communities (Kitade and Matsumoto, 1998). In particular, Trichonymphea protists, which were present in the wood-feeding roach ancestor of termites, diversified with evolving termite lineages and gave rise to the *Trichonympha* spp. in *Reticulitermes* which are related to those in the Archotermopsidae. All other rhinotermitids, including *Coptotermes*, harbor *Pseudotrichonympha* spp. Moreover, Spirotrichonymphea protists originated alongside the Rhinotermitidae and radiated to their greatest diversity in this family. However, *Reticulitermes* spp. do not have the typical rhinotermitid Spirotrichonymphea, but related Spirotrichonymphea-like protists (Tai *et al.*, 2015).

Most termite gut protists are unique and restricted to their host termite species or closely related congeneric species. There are only a few protist groups that are shared across distant termite taxa. For example, a rare member of the Trichomonadea was found in *Coptotermes testaceus* (Linnaeus), a rhinotermitid species, as well as in the Stolotermitidae family (Tai *et al.*, 2015).

Over 500 protists have been described from wood-feeding roaches, including termites, using microscopy, targeted fluorescent probes, real-time qPCR and 18S rRNA gene sequencing, but few have been isolated and cultured (Yamin, 1979; Ohkuma and Brune, 2010; Inagaki *et al.*, 2020; Jasso-Selles *et al.*, 2020; Song *et al.*, 2021). The numbers of protist species are generally specific to each termite species and range from a few to a dozen but vary even among closely related species, as shown within the genus *Reticulitermes* (Yamin, 1979; Lewis and Forschler, 2004; Song *et al.*, 2021) and *Coptotermes* (Jasso-Selles *et al.*, 2020). Relative abundance of protist populations can vary within subterranean termite species depending on caste, diet, and environment (Yamin, 1979; Kitade and Matsumoto, 1998; Kitade *et al.*, 2013). Recent advances in DNA sequencing suggest a higher protists diversity than originally described (Koidzumi, 1921; Su *et al.*, 2016; Jasso-Selles *et al.*, 2020). Workers of *C. formosanus* harbor three genera of flagellates in their hindgut. Originally these were morphologically described as three species: *Pseudotrichonympha grassii* (Trichonymphida), *Holomastigotoides hartmanni*, and *Cononympha (Spirotrichonympha) leidyi* (Spirotrichonymphida) (Koidzumi, 1921; Lai *et al.*, 1983; Kitade *et al.*, 2013). Recent studies employing molecular methods and classical microscopy described an additional *Holomastigotoides* (*H. minor*) and *Cononympha* (*Con. koidzumii*) species (Nishimura *et al.*, 2020; Jasso-Selles *et al.*, 2020), totaling five species for *C. formosanus* (Fig. 8.1). Workers of the closely related invasive Asian subterranean termite, *Coptotermes gestroi* (Wasmann), which shared common ancestry with the *C. formosanus* approximately 18 million years ago (Bourgouignon *et al.*, 2018), contain a similar community of protists with one *Pseudotrichonympha*, and two *Holomastigotoides* and *Cononympha* species each (del Campo *et al.*, 2017; Jasso-Salles *et al.*, 2020). All *Coptotermes* spp. have *Pseudotrichonympha* and *Holomastigotoides* species (Koidzumi, 1921; Yamin, 1979); however, each protist species is unique to its termite host. *Pseudotrichonympha* species are exclusively found in the majority of Rhinotermitidae, except for *Reticulitermes* spp., which have more diversity of protists than *Coptotermes* spp. but lack *Pseudotrichonympha* (Yamin, 1979; Saldarriaga *et al.*, 2011). *Pseudotrichonympha* phylogeny is congruent with their termite hosts indicating a tight co-evolution of these termite specific obligate protists lineages (Yamin, 1979; Noda *et al.*, 2007; Saldarriaga *et al.*, 2011).

While the ancestral protists in the wood-feeding cockroach survive through the host's molting cycle by protecting themselves through encystment (Cleveland, 1934; Cleveland, 1959), most, if not all, termite flagellates die off when hormonal changes trigger ecdysis, and recently molted termites must be reinoculated via anal trophallaxis from nest mates (Nalepa, 2017). Guts of juveniles of *C. formosanus* lost all protists 6 days before the actual molt, likely triggered by rising juvenile hormone titers (Raina *et al.*, 2008). Juvenile hormone also caused downregulation of gene expression and numbers of protists in *R. flavipes* (Sen *et al.*, 2013). In rare cases, such as in a *Pseudotrichonympha* species of *Coptotermes acinaciformis* (Froggatt), encystment might be temporarily initiated but the cycle is aborted, and protists die prior to molting (Cleveland, 1965). Interestingly, there is some evidence that large wood-feeding flagellates persist through the final imaginal molt to adult alate, which is governed by a different hormonal network than the juvenile molts (Nalepa, 2017). Although most gut content is voided prior to the dispersal flight to reduce

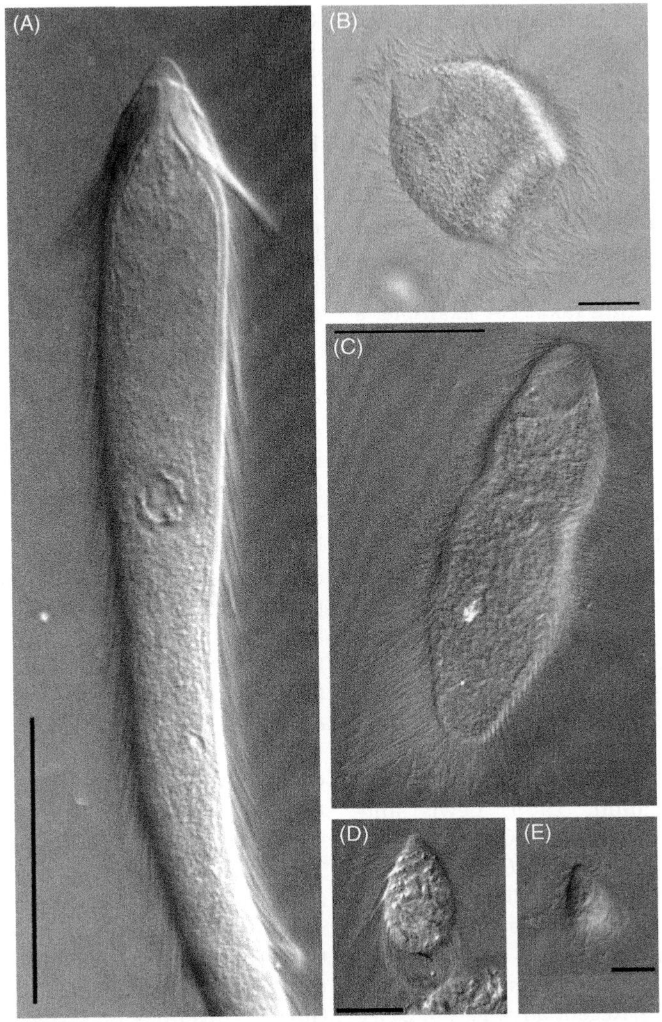

Fig. 8.1. Protist symbionts of *Coptotermes formosanus* (live), differential interference contrast light microscopy. (A) *Pseudotrichonympha grassii*, scale bar = 100 µm. (B) *Holomastigotoides minor*, scale bar = 20 µm. (C) *Holomastigotoides hartmanni*, scale bar = 50 µm. (D) *Cononympha leidyi*, scale bar = 20 µm. (E) *Cononympha koidzumii*, scale bar = 10 µm. (Photo courtesy of G. Gile, Arizona State University.)

weight, alates of a few termite species, such as *Reticulitermes lucifugus* (Rossi), retain their protists through the imaginal molt (Grasse and Noirot, 1945; Grasse, 1952). The mechanism for surviving the molt is unknown and likely differs from the sexual encystment cycle that protists of the wood-feeding cockroach undergo, since cysts are rarely found in termites (Nalepa, 2017).

Nalepa (2017) proposed an intriguing hypothesis that termites and protists coordinate their life cycles. Protists might be functionally asexual when living as captured symbionts of the colonies' sterile castes; and during juvenile molts the flagellates die, due to incomplete sexual encystment cycles, and must be replaced via trophallaxis from colony mates. In contrast, imaginal molts might trigger a rudimentary but functional sexual cycle similar to the process observed in ancestral cockroach symbionts, which benefits dispersing alates and reproductive development of the colony founders. Recent research in Rhinotermitidae is thought to lend support to this hypothesis, although surprisingly little is known how protists reproduce and regulate their population numbers in termite guts. Subterranean termite alates have fewer protists compared with other castes (Lai *et al.*, 1983; Lewis and Forschler, 2004; Shimada *et al.*, 2013). Losses during the molting process might be mitigated via reinoculation of alates by colony mates (Nutting, 1969; Michaud *et al.*, 2020); however, it is possible that alates retain a sufficient population of protists through their final molt, as early observational studies suggest that alates of some species do not receive nutrition after their final molt until the new colony is established (Heath, 1903).

Protist numbers in *C. formosanus*, *C. gestroi* and *R. speratus* royal pairs increase sharply immediately after colony foundation (Shimada *et al.*, 2013; Inagaki *et al.*, 2020; Velenovsky *et al.*, 2021; Chen *et al.*, 2022b). Protists reproduction is fueled by wood-consumption of the founder pair but the mode of reproduction (asexual and/or sexual) is not known (Shimada and Maekawa, 2010). The spike in numbers also occurs in same-sex pairs in *R. speratus* (Inagaki *et al.*, 2020); thus, mating of the royal pair is not the sole trigger of the increase in protist reproduction. Protist numbers peak between 35 and 60 days, when first workers emerge, and decline sharply thereafter when king and queen transfer the symbiont community to the first brood of workers via parental anal trophallaxis (Shimada *et al.*, 2013; Velenovsky *et al.*, 2021). This and subsequent worker generations will assume the task of foraging, digestion and providing nutrition to the royal pair and the brood. When the transition from biparental to alloparental care is complete, king and queen irreversibly become nutritionally dependent on the worker caste and eventually lose most, if not all, their protists (Shimada *et al.*, 2013). Males of *R. speratus* have higher protist numbers than females, indicating that kings may donate protists as proteinaceous nutrients to queens and offspring (Shimada *et al.*, 2013). Queens of *C. formosanus* and *C. gestroi* lose protists faster than kings (Velenovsky *et al.*, 2021), presumably because they donate more hindgut fluid to the emerging workers than males (Brossette *et al.*, 2019), and their developing ovaries demand reduction in hindgut size (Shimada *et al.*, 2013). All protists in a termite colony are descendants of the "starter package" of the founder pair, which also includes bacterial symbionts of the protists. Many protists harbor endo-and ectosymbiotic bacteria and archaea that perform vital services for protists and the termite host (section 8.6, this chapter).

8.4 Prokaryote symbionts: archaea

Archaea and bacteria are both prokaryote symbionts in termite guts, but archaea are less diverse and abundant than bacteria and account for only around 0.1–10% of bacteria in subterranean termite guts (Ohkuma *et al.*, 1995, 1999; Leadbetter and Breznak, 1996; Shinzato *et al.*, 1999; Brauman *et al.*, 2001; Hongoh, 2010). The 16 families of Archaea in termite guts (Arora *et al.*, 2022) are mostly represented by different lineages of strictly anaerobic methanogens. Methanogens were reported in guts of many higher and lower termites with different diets ranging from wood- to soil- and fungus-feeding species (reviewed by Purdy, 2007). Methanogenic archaea use two different processes to generate methane: hydrogenotrophic and acetoclastic methanogenesis, but there is no proof of the latter in termite guts (Brune, 2010; Abdul Rahman *et al.*, 2015). Termite methanogens generate methane from consuming H_2 and CO_2 (Ebert and Brune, 1997).

Methanogenic Archaea have been detected in *Coptotermes* and *Reticulitermes* by culture-dependent (Leadbetter and Breznak, 1996) and culture-independent techniques (Ohkuma *et al.*, 1995, 1999; Ohkuma and Kudo, 1999; Shinzato *et al.*, 1999, 2001; Tokura *et al.*, 2000; Hara *et al.*, 2004). High-throughput amplicon sequencing of the gut microbiota of *R. flavipes* (Boucias *et al.*, 2013) and several other species (Abdul Rahman *et al.*, 2015) showed that methanogens in lower termites are almost exclusively represented by the genus *Methanobrevibacter*, but also revealed rare lineages previously found only in higher termites and other insects. Leadbetter and Breznak (1996) were the first to culture and quantify *Methanobrevibacter* species from *R. flavipes* workers, which possess around a million methanogens (5% of the prokaryotes) per gut. *Coptotermes* species tend to have higher numbers and multiple representatives of methanogens (Abdul Rahman *et al.*, 2015), including methanogenic endosymbionts in *C. formosanus* protists (Inoue *et al.*, 2008) (section 8.6, this chapter).

Different species of *Methanobrevibacter* can coexist in the same termite, but they prefer separate microhabitats. Most lineages are attached to the gut wall or filamentous bacteria colonizing the gut wall (Leadbetter and Breznak, 1996). Their existence near the gut periphery, where oxygen seeps into the gut, has long been considered as enigmatic, because methanogenesis is sensitive to oxygen and methanogens are considered to be strict anaerobes (Ebert and Brune, 1997; Brune and Friedrich, 2000; Brune, 2010). Several protective mechanisms have been suggested to ensure survival of methanogens in the microoxic periphery of the termite gut, including the ability to reduce molecular oxygen (Brune, 2010). Other methanogens are associated with protists as ecto- and endosymbionts (Tokura *et al.*, 2000; Hara *et al.*, 2004; Hongoh and Ohkuma, 2018). For example, methanogenic endosymbionts were detected using methanogen-specific fluorescence in small protists such as *Spirotrichonympha leidyi* in *C. formosanus* but not in *Holomastogotoides hartmanni* and *Pseudotrichonympha grassii* (Tsunoda *et al.*, 1993; Inoue *et al.*, 2008). Single *S. leidyi* cells can contain 80 cells of a *Methanobrevibacter* species, making methanogens account for 0.3% of the prokaryotes in *C. formosanus* (Inoue *et al.*, 2008). The *Methanobrevibacter* species of *S. leidyi* in *C. formosanus* forms a monophyletic group with methanogens from *R. speratus* protists *Dinenympha parva* (Oxymonadid) and *Microjoenia* sp. (Spirotrichonymphida) (Tokura *et al.*, 2000; Inoue *et al.*, 2008).

Methanogens in the termite gut likely benefit from the interspecies transfer of H_2. Fermentation of sugars by cellulolytic protists produces ample amounts of hydrogen and methanogens metabolize the excess H_2 accumulated by anaerobic fermentation, thereby promoting lignocellulose digestion by protists (Odelson and Breznak,1985; Shinzato *et al.*, 1999; Tokura *et al.*, 2000; Hara *et al.*, 2004; Hongoh and Ohkuma, 2018). However, acetogenic microbes in wood-feeding lower termites predominate over methanogens, with which they compete for H_2 access (Breznak and Switzer, 1986; Pester and Brune, 2007). Nevertheless, both H_2 consuming processes can coexist in the termite gut because of the spatial separation of acetogens and methanogens in different gut microenvironments, which reduces direct competition for H_2 (Brune and Friedrich, 2000). Consequently, termites are a substantial source of methane release (Brune 2010).

8.5 Prokaryote symbionts: bacteria

8.5.1 Taxa composition

Bacteria represent the majority (> 90%) of termite gut prokaryotes with 10^5–10^8 bacteria cells per gut, which contains approximately one cubic millimeter of gut fluid. While many bacteria species have been cultured and described from *Coptotermes* and *Reticulitermes* species (reviewed in König *et al.*, 2006; Husseneder, 2010; Brune, 2013), the vast uncultured bacterial diversity was described recently by 16S rRNA gene sequencing. Thousands of bacteria strains defined by their unique 16S rRNA gene sequences from up to 29 phylum-level clusters have been detected across termite families (Hongoh, 2010; Su *et al.*, 2016; Arora *et al.*, 2022) with several hundred to almost 5,000 strains per termite species, depending on the detection method. Many of these bacteria are uncultured, undescribed, unique to termites, and are not known to live outside the termite gut. The nomenclature can be confusing, because different terms are used across the literature. Gene sequence-based strains, often called amplicon sequence variants (ASVs), can represent different species or genetic variants within the same species. Phylotypes, which are sometimes called ribotypes when based on rRNA gene sequences, are sequences that are binned into groups based on their similarity to a reference database. Operational taxonomical units (OTUs) are clustered based on sequence similarities to each other and do not require a reference database; however, OTUs can be taxonomically tagged via comparison to reference sequences. The higher the sequence similarity cut-off level for clustering and taxonomic assignment (typically ≥ 97%), the higher is the stringency, which means better resolution and reliability of identification, but also increasing proportions of bacteria that will not make the threshold for taxonomical assignment.

Over 20 bacteria phyla have been recognized from the guts of lower termites (Hongoh *et al.*, 2005; Ohkuma, 2008; Hongoh, 2010; Boucias *et al.*, 2013; Dietrich *et al.*, 2014). At least 11 core phyla and ten core genera of bacterial symbionts are present across all lower wood-feeding termites. The relative abundance of core phyla and genera differs between *Reticulitermes* and *Coptotermes* species. In *R. flavipes* members, the most dominant phyla (> 90%; Boucias *et al.*, 2013) are the Spirochaetes (mainly Treponema), the Elusimicrobia formerly known as Termite Group 1, which contains the 'Endomicrobia' as a candidate class (Stingl *et al.*, 2005; Tai *et al.*, 2015; Su *et al.*, 2016), the Firmicutes (mainly Clostridiales and Lactobacillales), Bacteroidetes, Proteobacteria, and Fibrobacteres. Minor phyla with low relative abundance include Verrucomicrobia, Actinobacteria, Tenericutes, and others. Methodological differences and PCR bias notwithstanding, similar phyla composition has been recorded in *R. flavipes* specimens by 454 pyrosequencing of 19 phyla with 581 total OTUs (Boucias *et al.*, 2013) and 69 core OTUs (Benjamino and Graf, 2016) when clustered at the 97% sequence similarity level. Similar phyla were also detected in *R. flaviceps* members with 454 pyrosequencing (Su *et al.*, 2016), albeit with some variation in abundance due to species difference and selection of different variable regions of the 16S rRNA gene. These deep sequencing studies added considerable diversity to earlier inventories of a subset of these phyla obtained by clone-based methods with comparatively limited throughput in *R. flavipes* (Yang *et al.*, 2005; Fisher *et al.*, 2007; Huang *et al.*, 2013) and *R. speratus* specimens (Ohkuma and Kudo, 1996; Kudo *et al.*, 1998; Hongoh *et al.*, 2003). Most dominant core taxa in *R. flavipes* members on the family level and below include the spirochaete *Treponema* (42%), 'Endomicrobia' (19%) and 'Azobacteroides' (3%) (Benjamino and Graf, 2016), which was similar to the bacterial inventory of *R. flaviceps* termites (Su *et al.*, 2016).

The main difference between *Reticulitermes* and *Coptotermes* gut bacteria composition is that the latter has greater abundance of Bacteroidetes than Spirochaetes. Clone-based 16S rRNA gene amplicon sequencing of *C. formosanus* workers identified at least 213 different bacteria strains from 13 phyla (with Bacteroidetes, Firmicutes, and Spirochaetes being the most dominant) defined by DNA sequence variability (Shinzato *et al.*, 2005; Hongoh *et al.*, 2008a; Husseneder *et al.*, 2010a). The dominant bacterium species in *Coptotermes* was a Bacteroides species later identified as *Candidatus* 'Azobacteroides pseudotrichonymphae', an endosymbiont of protists (Hongoh *et al.*, 2008a) (section 8.6, this chapter). *Candidatus* 'A. pseudotrichonymphae' showed low sequence similarity to the '*Azobacteroides* sp.' core bacterium described in *R. flavipes* and *R. flaviceps* (Benjamino and Graf, 2016; Su *et al.*, 2016; Chen *et al.*, 2022a), which is expected, since *Reticulitermes* lack *Pseudotrichonympha* protists (Lewis and Forschler, 2004; Song *et al.*, 2021). The second most dominant bacterium species was an uncultured *Clostridium* (Firmicutes). The third most abundant bacterium species has since been characterized as the novel termite-specific lactic acid-producing *Pilibacter termitis* (Lactobacillales), an obligate abundant symbiont in *C. formosanus* workers and alates (Chen and Husseneder, unpublished) involved in sugar fermentation and regulation of pH and gut homoeostasis (Higashiguchi *et al.*, 2006). *Pilibacter* species also encode adenylyl-sulfate reductase, a gene involved in the sulfate reduction pathway (Arora *et al.*, 2022). The bacterial diversity in *C. formosanus* guts reported in these clone-based studies was most likely underestimated, since next-generation metagenome sequencing of a closely related species, *C. gestroi*, reported 1,460 taxa of free-living bacteria along with a plethora of prokaryotic genes involved in lignocellulose digestion (Do *et al.*, 2014). Since the protist-associated symbionts, which belong mostly to the Bacteroidales, were not included in the *C. gestroi* study, the Spirochaetales and Lactobacillales were more abundant than the Bacteroidales, followed by Clostridiales, Enterobacteriales, Pseudomonades, Synergistales, Desulfovibrionales, Xanthomonadales, Burkholderiales, Bacillales, and Actinobacteria (Do *et al.*, 2014). Most abundant genera in *C. gestroi* were *Treponema*, *Lactococcus*, *Pseudomonas*, *Enterobacter*, *Dysgonomonas*, *Clostridium*, and *Bacteroides*. The presence of the main taxa in *C. gestroi* was generally consistent with the earlier culture-dependent (Husseneder *et al.*, 2009) and culture-independent studies in *C. formosanus* (Shinzato *et al.*, 2005; Husseneder *et al.*, 2010a) and *Coptotermes niger* Snyder (Dietrich *et al.*, 2014).

8.5.2 Roles of bacteria in subterranean termite guts

While protists of the lower termites are responsible for the bulk of lignocellulose digestion, bacteria play important supporting roles (Brune, 2013). These include metabolic functions for nutrient acquisition to supplement the termites' wood diet in the form of nitrogen fixing and recycling for amino acid synthesis, vitamin and co-factor synthesis, and aromatic hydrocarbon polymer (lignin and xylan) degradation. Furthermore, gut bacteria are involved in detoxification via nitrite reduction, sulfide oxidation, and selenate reduction as well as pathogen defense via chitin degradation (Brune and Friedrich, 2000; Inoue *et al.*, 2000; Ohkuma, 2003; Su *et al.*, 2016; Zeng *et al.*, 2021; Arora *et al.*, 2022). Bacteria also serve as a sink for the large amounts of hydrogen produced by fermentation (acetogenesis) (Hongoh *et al.*, 2008a,b), capture oxygen to protect the anoxic gut microbes including the protists, and assist in protist motility (Wenzel *et al.*, 2003; Hongoh *et al.*, 2007b). A comprehensive metagenomic analysis of the prokaryote community across representatives of the global termite diversity including subterranean termites lists taxa and genes encoding enzymes in important metabolic pathways, such as carbohydrate metabolism, reductive acetogenesis, methanogenesis, sulfate reduction, nitrogen recycling and fixation. Genes involved in essential nutritional functions were largely conserved and were likely present in the common ancestor of termites (Arora *et al.*, 2022).

The dominant phyla play a variety of roles in subterranean termites that complement each other and benefit the protists and the termite host overall. Spirochaetes are the most abundant and diverse bacteria in *Reticulitermes* and to a slightly lesser degree in *Coptotermes* termites (Lilburn *et al.*, 1999; Hongoh *et al.*, 2003; Husseneder *et al.*, 2010a; Boucias *et al.*, 2013; Do *et al.*, 2014). For example, using Sanger sequencing of 16S rRNA clones, 67 different treponemal phylotypes with 21 new species of *Treponema* were found in *R. flavipes* and similar novel *Treponema* diversity was revealed in *R. speratus* and *C. formosanus* specimens (Lilburn *et al.*, 1999). Since then, several next-generation sequencing studies have confirmed that *Treponema* species represent the most phylogenetically diverse group within the spirochaetes (Brune *et al.*, 2022) and are present in every termite species (Arora *et al.*, 2022). Some *Treponema* species are involved in major functions of gut lignocellulose hydrolysis, fermentation, acetogenesis, and N_2 fixation (Leadbetter and Breznak, 1996; Leadbetter *et al.*, 1999; Breznak, 2000; Graber and Breznak, 2004). Interestingly, the full pathway for acetogenesis was not detected in any of the *Treponema* genomes assembled from the metagenomes of *Coptotermes* and other termite species, but many genes involved in acetogenesis were detected (Arora *et al.*, 2022). Although acetogenesis was confirmed in cultured *Treponema* (Graber and Breznak, 2004), the abundance of acetogenic *Treponema* populations might be lower and their diversity higher than previously thought (Arora *et al.*, 2022). Similarly, *Treponema* species were not a major source of N_2-fixing genes in *Coptotermes* specimens (Arora *et al.*, 2022), since this function is performed by Bacteroidales in *Coptotermes* species. Multiple spirochaetes including *Treponema* are protist symbionts in *Reticulitermes* and *Coptotermes* species and some benefit protists by providing mobility as ectosymbionts and serve as H_2 sink to increase protist fermentation (Iida *et al.*, 2000; Inoue *et al.*, 2008) (section 8.6, this chapter).

Bacteroidetes are the most dominant bacteria and represent the majority of ecto- and endosymbionts of protists in *C. formosanus* guts (Noda *et al.*, 2005, 2007, 2009) (section 8.6, this chapter), but also are present in considerable proportions in *Reticulitermes* species (Boucias *et al.*, 2013; Benjamino and Graf, 2016; Su *et al.*, 2016). Bacteroidetes have genes encoding enzymes for a variety of glycoside hydrolase families, making them significant contributors of cellulose and hemicellulose degradation (Arora *et al.*, 2022). Bacteroidetes ectosymbionts of protists are likely responsible for the bulk of bacterial cellulolytic activity in lower termites in contrast to higher termites where Fibrobacteres, Spirochaetes and Firmicutes were identified as the main cellulolytic symbionts (Arora *et al.*, 2022). Bacteroidetes are also involved in nitrogen recycling and fixation. The predominant nitrogen-fixing bacterium species in lower termites was an undescribed member of the Bacteroidetes; however, this bacterium was mostly absent from *Coptotermes* and *Heterotermes*, which harbor the protist symbiont *Ca.* 'Azobacteroides' as the main nitrogen fixing species (Arora *et al.*, 2022). The genome of the *Ca.* 'Azobacteroides pseudotrichonymphae' symbiont from *C. formosanus* has been fully sequenced and genes involved in nitrogen and carbohydrate metabolism have been confirmed (Hongoh *et al.*, 2008a) (section 8.6, this chapter). Bacteroidetes also have many genes for reductive acetogenesis in *Reticulitermes*, but only a few of such genes in *Coptotermes* individuals (Arora *et al.*, 2022). Species of *Tannerella*, a dominant Bacteroidetes genus in *C. formosanus* and *Reticulitermes* members, harbor genes for amino acid synthesis (Su *et al.*, 2016; Arora *et al.*, 2022).

Elusimicrobia represent mostly uncultured obligate endosymbionts of lower termite gut protists (Ikeda-Ohtsubo *et al.*, 2007; Geissinger *et al.*, 2009). They may account for up to 20% of the bacteria community in *Reticulitermes* species (Hongoh *et al.*, 2003; Yang *et al.*, 2005; Boucias *et al.*, 2013) but represent only 1% of the bacteria community in *C. formosanus* (Husseneder *et al.*, 2010a), where *Ca.* 'A. pseudotrichonympha' fills the functional niche as main protist endosymbiont (Noda *et al.*, 2005; Hongoh *et al.*, 2008a). 'Endomicrobia' are likely involved in N_2-fixation, and vitamin and amino acid synthesis (Hongoh *et al.*, 2008b; Ikeda Ohtsubo *et al.*, 2009; Zheng *et al.*, 2016); however, there is still much to be explored about their unique physiology and function in symbiosis.

Although not among the most dominant in relative abundance, the phylum Firmicutes is highly diverse in termite guts (Hongoh *et al.*, 2003). Firmicutes are mainly represented by the orders Clostridiales and Lactobacillales in *Reticulitermes*, and *Coptotermes* members Clostridiales are abundant and fall mostly into termite-specific lineages (Hongoh *et al.*, 2003). This order contains several known cellulose- and hemicellulose-digesting as well as N_2-fixating species (Yang *et al.*, 2005; Boucias *et al.*, 2013). Lactic acid bacteria were the premier source of prokaryote cellulolytic enzymes, followed by Enterobacteriales, Bacillales and Actinobacteria (Boucias *et al.*, 2013).

8.5.3 Intraspecific variation of the bacteria community

Variation in bacteria communities of the same rhinotermitid species is relatively low, as the microbiota is largely consistent across colonies, even if they are from different geographic regions (Boucias *et al.*, 2013; Husseneder *et al.*, 2010a). Composition of dominant bacteria taxa in terms of bacteria richness, diversity, and relative abundance of phyla was similar in *C. formosanus* worker guts from their native (China) and introduced ranges (U.S., Japan) suggesting only minimal environmental effects on the core bacteria across geographic regions and invasion history (Husseneder *et al.*, 2010a; Shinzato *et al.*, 2005). Similarly, 88% of bacterial OTUs (Boucias *et al.*, 2013), including the core taxa (Benjamino and Graf, 2016), in *R. flavipes* specimens were consistent among colonies. The consistency in intraspecific microbiota across termite colonies and geographic separation supports the close association of core bacteria groups with the termite host.

Minor inter-colony differentiation in microbiota can be caused by environmental and/or genetic differences at the colony level, as well as stochastic founder effects during colony initiation (Vargo and Husseneder, 2009); new colonies in subterranean termites are founded by a single pair of king and queen which bring a "starter package" of microbiota for inoculation of the first worker brood. These differences in microbiota cannot be eradicated by rearing termite colonies on the same diet (Sapkota and Scharf, 2022). However, dietary differences can cause some adaptive shifts in the microbiota. Short-term treatment with different diets showed small shifts in bacterial communities, with a complex wood diet resulting in slightly higher diversity than a simple cellulose diet; however, variability induced among colony members by diet was smaller than inter-colonial differences (Boucias *et al.*, 2013). The diet effect detected in *R. flavipes* was smaller than was found in *Coptotermes*. Lab rearing of *C. formosanus* workers on different cellulose sources caused rapid shifts in the counts of cultured gut bacteria of several morphotypes (Husseneder *et al.*, 2009). Although there were only small differences in the proportions of these morphotypes among field-collected workers and those from colonies reared in the lab on a wood diet, there were significant shifts when termites were fed a filter paper diet. Noticeably, the changes did not affect dominant obligate gut bacteria, such as *Pilibacter termitis* (Higashiguchi *et al.*, 2006), but were caused by facultative anaerobic generalists in the Enterobacteraceae and Lactobacillales. Similarly, Tanaka *et al.* (2006) reported 40% change of bacterial diversity for *C. formosanus* workers fed on wood vs cellulose in a month-long assay. Their study reported high termite mortality, which might have confounded the treatment effect on microbiota since termites under nutritional stress can digest their own gut microbiota to access protein and vitamins (Martin and Kukor, 1984). The low impact of diet changes on microbiota might be due to the fact that the termite host pre-digests and detoxifies complex lignocellulose material, providing the bacteria with ready-to-use simple carbohydrates regardless of the complexity of the original input material (Peterson and Scharf, 2016a). Not surprisingly, differences in bacteria concentrations and relative proportions among castes in the same colony were detected due to different functions of workers, soldiers and reproductives reflected in their gut morphology. The reduced number of protists in alates and soldiers impacted the abundance of gut bacteria that are associated with protists (Benjamino and Graf, 2016; section 8.6, this chapter).

8.5.4 Co-diversification

While studies agree on the strict phylogenetic association of protists and termite host, the results concerning the degree to which bacteria communities are structured by host phylogeny are more varied. Obligately symbiotic protists are stable across individuals and populations within a termite species, but the presence of many bacteria might be defined by their membership in a functional group rather than taxonomic placement when distantly related termite host taxa were concerned (Tai *et al.*, 2015).

Most bacteria in subterranean termites belonged to termite-specific bacteria lineages and were, thus, likely inherited from a termite ancestor. For example, Hongoh *et al.* (2003) found at least 35 phylotypes in guts of *C. formosanus* workers that were from termite-specific lineages. These bacteria are most likely adapted to the special conditions in the termite gut and are permanent vertically transmitted residents instead of transients recently acquired from the environment (Hongoh *et al.*, 2003, 2005; Shinzato *et al.*, 2005; Husseneder *et al.*, 2010a).

While most bacteria were specific to their termite host, the unique occurrence of a symbiont is not necessarily evidence of co-diversification without additional information about the distribution and evolution of related bacteria across host species (Tai *et al.*, 2015). Since increasing numbers of termite microbiomes have been subjected to deep sequencing, a considerable fraction of core bacteria has been found to be shared among closely related termite species. Clades of bacteria and termites do not necessarily coincide on higher taxonomic levels, probably due to multiple independent acquisitions and transfer among host species (Ohkuma *et al.*, 2007; Noda *et al.*, 2009; Hongoh, 2010; Tai *et al.*, 2015; Bourguignon *et al.*, 2018). However, co-cladogenesis of host and certain bacterial symbionts is frequently observed in closely related termite species and genera, suggesting a tight coevolutionary relationship (Noda *et al.*, 2007; Bourguignon *et al.*, 2018).

The degree of co-speciation varies among bacteria phyla depending on their lifestyle (free-living vs symbiont of protist) and whether their function is general or essential with impact on termite fitness. For example, the highly diverse Firmicutes are not well correlated with host phylogeny, probably because this phylum contains transient bacteria with general or non-essential functions that are easily taken up from the environment and transferred among hosts (Tai *et al.*, 2015). In contrast, bacteria groups that contain many endosymbionts of protists (section 8.6, this chapter), such as Bacteroides (Noda *et al.*, 2005, 2006), 'Endomicrobia' (Ikeda-Ohtsubo *et al.*, 2007; Ohkuma *et al.*, 2007), Spirochaetes (Noda *et al.*, 2003) and Synergistes (Hongoh *et al.*, 2007a) were more likely to be structured by host phylogeny, even though some members of these groups are free-living (Tai *et al.*, 2015; Mikaelyan *et al.*, 2017). For example, strict co-specification has been documented for *Endomicrobium* bacteria and *Trichonympha* protists in *Reticulitermes* species (Ikeda-Ohtsubo and Brune, 2009; Hongoh *et al.*, 2005) and 'Azobacteroides' bacteria and *Pseudotrichonympha* protists in *Coptotermes* (Noda *et al.*, 2005, 2007). In the latter example co-diversification was tripartite including the host termite with congruent phylogenetic topologies of *Coptotermes* species and closely related subterranean termite genera (Noda *et al.*, 2005, 2007; Chen *et al.*, 2022a).

8.6 Bacterial ecto- and endosymbionts of protists

Some termite gut bacteria move freely in the gut lumen, others form biofilms on the gut wall or are attached to the surface, cytoplasm or nucleus of protists (Brune and Stingl, 2006; Noda *et al.*, 2006, 2009, 2018; Ikeda-Ohtsubo and Brune, 2009; Brune, 2014; Sato *et al.*, 2014; Mikaelyan *et al.*, 2017; Hongoh and Ohkuma, 2018). Most prokaryotes (85%) in the guts of lower termites are associated with protists (Ohkuma, 2008). Most of these bacterial symbionts of protists belong to five bacteria phyla: the Bacteroidetes (Noda *et al.*, 2005), the Spirochaetes (Iida *et al.*, 2000), the Elusimicrobia with the candidate class 'Endomicrobia' (Ohkuma and Kudo, 1996; Stingl *et al.*, 2005), the Proteobacteria, and Actinobacteria (Strassert *et al.*, 2012). In addition, Methanobacteriales of the Archaea are also associated with protists (section 8.4, this chapter). These symbiotic bacterial lineages are associated with flagellated protists as ecto- (surface) or endosymbionts (nucleus, cytoplasm).

The most common ectosymbionts are from the order Bacteroidales (phylum Bacteroidetes) and two lineages of the extremely diverse *Treponema* (phylum Spirochaetes, Brune *et al.*, 2022); both phyla can coexist on the same protist cell (Hongoh *et al.*,

2007b). Across lower termites, Noda *et al.* (2009) reported 31 Bacteroidales taxa as ectosymbionts from 17 protist genera in ten families, showing multiple independent acquisitions by different flagellates. However, within a host genus the ectosymbionts co-speciated with their protist host (Noda *et al.*, 2009). Single-cell genome sequencing revealed that Bacteroidales ectosymbiotic bacteria such as *Candidatus* 'Symbiothrix dinenymphae' can collaborate with flagellates in cellulose digestion of wood-feeding termites (Yuki *et al.*, 2015). *Candidatus* 'Symbiothrix dinenymphae' is attached to the cell surface of the cellulolytic gut protists *Dinenympha* spp. in *Reticulitermes*. The genome contains genes involved in amino acid, cofactor and nucleotide biosynthetic pathways. In addition, the genome encodes for various glycoside hydrolases for lignocellulose digestion. *Candidatus* 'S. dinenymphae' bacteria likely contribute to the carbon and energy household of their host termite by supplementing the protists' fermentation of lignocellulose into acetate and, possibly, function as a pretreatment to ease the breakdown of crystalline cellulose for the host protist (Yuki *et al.*, 2015).

Some ectosymbiontic bacteria, like *Treponema* spirochaetes, provide motility to flagellates (Wenzel *et al.*, 2003); however, most *Treponema* ectosymbionts add other benefits, like nitrogen fixing and sugar fermentation (Graber and Breznak, 2004). Ectosymbiotic *Treponema* also might increase fermentation of their protist host by performing hydrogen-consuming reductive acetogenesis and participating in inter-species H_2-transfer similar to the methanogenic endosymbionts (section 8.4, this chapter). Detection via fluorescent probes and sequencing of 16S clone libraries found three distinct ectosymbiotic *Treponema* sharing the surface of about 5% of *H. hartmanni* in *C. formosanus*. The low rate of association indicates that this relationship is not obligate for the protist in *Coptotermes* (Inoue *et al.*, 2008).

Notably, some *Treponema* spirochaetes attached to protists have their own ectosymbiotic bacteria, adding a fourth level to the tripartite relationship. *Candidatus* 'Termititenax' of the candidate phylum 'Margulisbacteria' specifically attach to ectosymbiotic spirochetes of oxymonad gut protists and were detected in at least 40 termite and cockroach species, including *C. formosanus* and different *Reticulitermes* species (Utami *et al.*, 2019). Genome analyses showed that the bacteria are likely forming a mutual symbiosis with their partner *Treponema*: the bacteria have genes that enable them to hydrolyze and ferment cellulose and cellobiose to acetate and ethanol, producing H_2 and CO_2 in the process, while the *Treponema* spirochaetes contain genes for reductive acetogenesis using H_2 and CO_2 (Utami *et al.*, 2019).

Other ectosymbionts besides Bacteroidales and spirochaetes include *Ca.* 'Desulfovibrio trichonymphae', which resides almost completely embedded in *Trichonympha* cytoplasm of *Reticulitermes* but remains connected to the outside by a small pore (Sato *et al.*, 2009; Kuwahara *et al.*, 2017). Its genome contains pathways for synthesis of amino acids and cofactors, and for oxidizing H_2 by sulfate and/or fumarate, while many genes otherwise common in desulfovibrios are non-functional pseudogenes or lacking. Thus, *Ca.* 'D. trichonymphae' is an integral part of an elaborate interdependent relationship in *Reticulitermes* where the protist *T. agilis* provides habitat and metabolites to its acetogenic endosymbiont *Ca.* 'Endomicrobium trichonymphae' (Hongoh *et al.*, 2008b) and *Ca.* 'D. trichonymphae' takes up the acetate, H_2 and CO_2 generated by the protist and its endosymbiont as energy and carbon source (Brune, 2014). The protist/endosymbiont system can increase fermentation due to the removal of H_2 and receives various amino acids and cofactors from the ectosymbiont (Kuwahara *et al.*, 2017).

In addition to ectosymbionts, termite gut protists harbor endosymbiotic bacteria (Stingl *et al.*, 2005; Hongoh and Ohkuma, 2018). One of the best-studied protists/bacteria endosymbiotic systems exists in *Reticulitermes*, where 'Endomicrobia' endosymbionts can be found in various species of protists (Stephens and Gage, 2020). The 'Endomicrobia' endosymbiont *Ca.* 'Endomicrobium pyrsonymphae' was described in the *R. flavipes* protist (*Pyrsonymphae veterans*) (Stingl *et al.*, 2005). *Trichonympha agilis* cells in *R. speratus* and *R. flavipes* always contain the cytoplasmic endosymbiont *Ca.* 'E. trichonymphae' (Hongoh *et al.*, 2003; Stingl *et al.*, 2005; Ohkuma *et al.*, 2007), which is vertically transmitted and co-diversified for 40 million years with its host (Ikeda-Ohtsubo and Brune, 2009; Zheng *et al.*, 2016). *Trichonympha agilis* also provides habitat and metabolites to an ectosymbiotic prokaryotic associate, *Ca.* 'Desulfovibrio trichonymphae' as described above (Sato *et al.*, 2009; Kuwahara *et al.*, 2017). In addition, *T. agilis* harbor Verrucomicrobia species in their nucleus (Sato *et al.*, 2014).

Although predominantly described from *Reticulitermes* species, protist cells can harbor up to a thousand 'Endomicrobia' cells in a variety of lower termite species (Stingl *et al.*, 2005; Ikeda-Ohtsubo *et al.*, 2007; Ohkuma *et al.*, 2007). 'Endomicrobia'-like bacteria were also detected in *C. formosanus* specimens by 16S PCR (Stingl *et al.*, 2005) and next-generation sequencing (Chen and Husseneder, unpublished) but not described in further detail. 'Endomicrobia' reflect the diversity of their protist hosts and were detected in at least 24 Parabasalia and Oxymonada protists (Ikeda-Ohtsubo *et al.*, 2007). 'Endomicrobia' show strict co-speciation with *Trichonympha* (Ikeda-Ohtsubo and Brune, 2009); however, the presence in both parabasalids and oxymonads suggests multiple acquisitions and horizontal transfer occurred between flagellate phyla cohabitating in the termite gut (Ikeda-Ohtsubo *et al.*, 2007; Ohkuma *et al.*, 2007; Mikaelyan *et al.*, 2017).

An endosymbiotic system with similar importance exists in *Coptotermes* species, but based on a partnership with three distinct lineages of Bacteroidales species, which is one of the most predominant orders in subterranean termite guts (Noda *et al.*, 2005). In contrast to the 'Endomicrobia', Bacteroidales are only found in parabasalid protists of the genus *Pseudotrichonympha* (Noda *et al.*, 2007), which is almost exclusively found in Rhinotermitidae (Yamin, 1979; Kitade and Matsumoto, 1998). The Bacteroidales endosymbionts are the dominant bacteria in most rhinotermitid guts, except *Reticulitermes* (Song *et al.*, 2021). Related Bacteroidales accounted for 96% of the prokaryote community in the gut of *Prorhinotermes simplex*, 82% in *Heterotermes tenuis*, 94% in *C. testaceus*, 65% in *C. curvignatus*, 89% in another *Coptotermes* sp. and 43%–70% in *C. formosanus* specimens (Noda *et al.*, 2005; Hongoh *et al.*, 2008a; Husseneder *et al.*, 2010a; King *et al.*, 2014).

The majority of the Bacteroidales in *C. formosanus* are represented by a single genus, *Candidatus* 'Azobacteroides', which has close relatives in many lower termites (Abdul Rahman *et al.*, 2015). *Candidatus* 'Azobacteroides' endosymbionts co-speciated with their *Pseudotrichonympha* protist hosts and co-diversified with their subterranean termite hosts (Noda *et al.*, 2007; Hongoh, 2010), indicative of the importance of the tripartite relationship. Single *Pseudotrichonympha* cells in *C. formosanus* workers can harbor up to 100,000 endosymbiotic *Ca.* 'Azobacteroides pseudotrichonymphae' cells (formerly known as phylotype CfPt1-2) (Noda *et al.*, 2005). The total population of *Ca.* 'Azobacteroides pseudotrichonymphae' accounts for up to 70% of the prokaryotic community in *C. formosanus* workers (Shinzato *et al.*, 2005; Hongoh *et al.*, 2008a; Husseneder *et al.*, 2010a).

Candidatus 'Azobacteroides pseudotrichonymphae' contains genes for fermentation, carbohydrate metabolism, and nitrogen acquisition (Hongoh *et al.*, 2008a). This endosymbiont imports and utilizes monosaccharides from lignocellulose (including glucose, xylose) as carbon and energy sources. This species is capable of N_2 fixation, but also recycles nitrogen waste products of the host protist and provides amino acids and co-factors for the host (summarized in Hongoh, 2010). *Candidatus* 'Azobacteroides pseudotrichonymphae' showed strong uptake hydrogenase activity, suggesting that together with other H_2-consuming prokaryotes like spirochaetes and methanogens it also may contribute to H_2 elimination and inter-species H_2-transfer from H_2-producing protists (Inoue *et al.*, 2007; Hongoh *et al.*, 2008a).

The close relationship of protists with their archaea and bacteria ecto- and endosymbionts in the gut of lower termites, regardless of whether the symbiosis is permanently integrated or transient with frequent host switches, leads to interesting research questions concerning the extent of lateral gene transfer among symbiotic partners and how this impacts their molecular and physiological evolution (Sato *et al.*, 2014). One mechanism hypothesized to facilitate gene transfer is via bacteriophages, which has become an important emerging research field in the past decade.

8.7 Bacteriophages

Bacteriophages add a fourth level to the tripartite termite–protist–bacteria relationship. While bacteriophages are not considered living organisms and are certainly not symbionts in the traditional sense, they play important regulatory functions in the gut. Bacteriophages are viruses that infect bacteria. Most phages either follow a lytic life cycle where the phage lyses the bacteria immediately after infection, or a temperate/lysogenic life cycle where the phage genome integrates into the bacterial chromosome genome to replicate with the host (Hobbs and Abedon, 2016; Howard-Varona *et al.*, 2017). Phages drive bacterial genome evolution and diversification by horizontal gene transfer and selection pressure (Salmond and Fineran, 2015). Phage–bacteria interactions shape the ecology of a microbial community and thus affect its host. Since the subterranean termite gut is a "hot spot" for bacterial diversity, it is also a good model for research into bacteriophages and their role in gut ecology.

The first studies on termite bacteriophages were conducted in higher termites (Termitidae) using metagenome sequences and found that phage-related genes were ubiquitous in termite guts (Ottesen *et al.*, 2006; Tadmor *et al.*, 2011; Kerr *et al.*, 2018). *Coptotermes formosanus* was the first termite species of which individual phages were isolated, sequenced, and characterized. The first phage isolated was CVT22, which infects *Citrobacter* (Tikhe *et al.*, 2015), followed by two *Enterobacter* phages in the guts of *C. formosanus* workers (Tikhe *et al.*, 2018a,b), including a temperate phage of the Podoviridae family and a phage from the Myoviridae family whose genome showed strong similarity to lytic phages but also contained nonfunctional lysogeny-related pseudogenes.

Metavirome sequencing revealed that guts of *C. formosansus* workers contained a diversity of previously unknown bacteriophages determined by gene matches to virus sequences in databases and the assembly of the genomes of dozens of novel phages in this termite species (Tikhe and Husseneder, 2018). Fully assembled genomes included those of ten Circoviridae (small ssDNA viruses) and 12 Microviridae (ssDNA viruses with small genomes). The majority of Microviridae in the termite gut belonged to a novel subfamily with the proposed name of Sukshmavirinae (Sukshma = Sanskrit word for "small"). Microviruses were subsequently confirmed in *C. formosanus* by Schmidlin *et al.* (2019). The majority of genes (90%) and seven assembled genomes belonged to Caudovirales (dsDNA tailed phages) with Siphoviridae-related bacteriophages being the most diverse (483 different bacteriophages), followed by Myoviridae (335) and Podoviridae (142) as reported by Tikhe and Husseneder (2018).

Among the full genomes assembled from *C. formosanus* guts were two phages from the crAss-like phage group known to be dominant in the human gut (Tikhe and Husseneder, 2018; Yutin *et al.*, 2018; Chen *et al.*, 2022a). These crAss-like phages showed similarity to a phage genome that was coincidentally detected during genome analysis of *Ca.* 'Azobacteroides pseudotrichonymphae', which is an obligate endosymbiotic bacterium of the protist *Pseudotrichonympha* sp. in the termite *Prorhinotermes japonicus* (Pramono *et al.*, 2017; section 8.6, this chapter).

Many predicted genes in the termite gut metavirome matched to phage sequences found to be integrated in bacterial chromosomes (prophages), suggesting that the gut virome is dominated by temperate phages. Previous studies showed that bacteria endosymbionts of *Endomicrobium* protists have functional CRISPR systems, which indicates phage infection, despite being intracellular symbionts of protists (Izawa *et al.*, 2016). However, the termite gut crAss-like phages do not integrate into bacterial chromosomes but exist in the form of extrachromosomal elements in their 'Azobacteroides' host and thus are not following a classical temperate/lysogenic life cycle (Pramono *et al.*, 2017; Chen *et al.*, 2022a). Variants of the *C. formosanus* crAss-like phages were confirmed in several *Coptotermes* species across different geographic regions and are present in both worker and reproductive castes. Thus, they are considered part of the core virome in this genus. In contrast to their co-diversifying bacteria, protist and termite host species, phages diversification was incongruent with host phylogeny (Chen *et al.*, 2022a).

With the advent of high-throughput sequencing techniques that do not require culture, phage research has garnered tremendous interest. This is one field where termites are at the forefront, because they are the hosts of obligate gut protists that contain bacteria endosymbionts with known phages. This quadripartite relationship makes termites a good model to understand how multi-level interactions of gut microbes benefit gut function in the individual worker and overall colony nutrition. Understanding termite gut symbiosis also opens avenues for use in industrial applications and in termite control by disrupting gut function.

8.8 Symbionts as biotechnological tools and targets for termite control

The synergistic dual digestive system in the termite gut, also dubbed the "smallest bioreactor in the world" (Brune, 2014), can be exploited in various ways, including the design of termite-derived biotechnology (Scharf, 2015). Assembling inventories of symbiont species, genes and genomes of subterranean termite species promises the continued discovery of novel and highly efficient enzymes and metabolic processes. These materials could create technologies for tapping into sustainable plant-based energy sources (biogas and biofuel), breaking down recalcitrant material (lignin and plastics), detoxification, bioremediation, and soil fertilization via nitrogen fixation (Scharf and Tartar, 2008). Termite-modelled biotechnology approaches using the termite/symbiont digestomes as resource for biomass and biomaterial industrial applications were reviewed in Scharf (2015).

A different termite-modeled biotechnology concerns the use of gut microbes as tools and targets for novel termite control (Zhou *et al.*, 2008; Sethi *et al.*, 2014; Husseneder *et al.*, 2016). The Rhinotermitidae and especially the genus *Coptotermes* contains the largest number of economically important and invasive pest species among the termites (Vargo and Husseneder, 2009; Evans *et al.*, 2013). There is a need to develop fast, cost-effective, and environmentally friendly biotechnologies to reduce the dependence on chemical pesticides for termite control. Conventional biological control using entomopathogens remains largely unsuccessful for termite management under field conditions, due the termites' hygienic behavior, use of antimicrobial substances in their nest, and their own immune responses (Chouvenc *et al.*, 2011).

There is increasing evidence for a collaborative physiology of termites and their symbionts in terms of pathogen defense. Protists and bacteria are likely contributing to pathogen defense via upregulation of specific antimicrobial and antifungal genes in the termite gut or termite nest structures. For example, fungal challenge leads to upregulation of bacterial amidohydrolase genes and other protist and bacteria genes with yet-to-be-determined function in the microbiome of *Reticulitermes* specimens (Peterson and Scharf, 2016b). Other protist-expressed immune genes include beta-1, 3-glucanase that prevents fungus spore germination and is conserved in wood-feeding roaches and lower termites (Bulmer *et al.*, 2009; Rosengaus *et al.*, 2014). A less specific but still effective defense against pathogens is provided by proteases and cellulases of termites and symbionts, which contribute to both digestion and destruction of pathogens and fungal spores (Peterson and Scharf, 2016b). In addition to the pathogen defense in the termite gut, termite nest material is laced with antimicrobial compounds. Hundreds of different bacteria species were identified from the nest material of *Coptotermes* colonies; some of them, like the Actino- and Acidobacteria, might be linked to pathogen defense (Oberpaul *et al.*, 2020; Soukup *et al.*, 2021). In *C. formosanus* colonies, Actinobacteria of the genus *Streptomyces* that produce antimicrobial compounds are recruited from the environment and incorporated as external symbionts in nest material as defense against entomopathogenic fungi (Chouvenc *et al.*, 2013, 2018).

The integrated immune defense by termite and symbionts might explain why most studies failed to identify entomopathogens in healthy termite guts (Fisher *et al.*, 2007; Shinzato *et al.*, 2005; Husseneder *et al.*, 2010a) and why most promising pathogens identified in lab studies failed to control termites under field conditions (Chouvenc *et al.*, 2011). Consequently, there is a need to design fortified biological control methods that knock out or circumvent the collaborative immune defense mounted by the termites' endogenous immune system and their symbiotic gut microbiota (Veivers *et al.*, 1982; Rosengaus *et al.*, 1999; Sen *et al.*, 2015).

One experimental approach to increase pathogen efficiency involved the suppression of the termites' immune system by pharmacological compounds or by silencing immune genes that produce Gram-negative bacteria-binding proteins and termicin/defensin antimicrobial peptides via RNA interference (Bulmer *et al.*, 2009; Hamilton and Bulmer, 2012). However, the functional redundancy in the combined termite/symbiont pathogen defense might limit practical applications of these methods. Similar limitations apply for the use of chemical inhibitors and RNA interference-based silencing of protist or termite cellulases (Zhou *et al.*, 2008; Liu *et al.*, 2017; Wu *et al.*, 2019; Scharf, 2015).

A completely different approach from immune suppression is the use of paratransgenesis, where genetically engineered microbial symbionts act as "Trojan horses" that can bypass the immune system to deliver and express genes in the gut of *C. formosanus* workers. Proof of concept that recombinant microbes can be stably introduced into termite lab colonies by feeding, express their payload in the termite gut, and are transferred among colony members was achieved by initially engineering a lab strain of *Escherichia coli* and subsequently an *Enterobacter cloacae* strain isolated from the termite gut to express green fluorescent protein (GFP) (Husseneder *et al.*, 2005; Husseneder and Grace, 2005).

Several natural and synthetic lytic peptides were tested for their efficiency to kill the cellulose-digesting protists *in vitro* and via microinjection into the hindguts of *C. formosanus* workers (Husseneder and Collier, 2009; Husseneder *et al.*, 2010b). To increase target specificity, a ligand was developed via phage display that, when coupled to the lytic peptide Hecate, binds Hecate specifically to the membranes of the termite gut protists, but not to gut tissue or bacterial symbionts. The attached ligand increased not only target specificity but also efficiency: ligand–Hecate killed the protists in the guts of *C. formosanus* workers faster than Hecate alone (Husseneder *et al.*, 2010b).

Sethi *et al.* (2014) engineered *Kluyveromyces lactis* yeast to express the ligand–Hecate gene and secrete an active form of the ligand–lytic peptide into the termite gut. The yeast was incorporated into cellulose bait, which *C. formosanus* workers readily consumed. Workers transferred the genetically engineered yeast among nestmates in the laboratory. The expressed ligand–lytic peptide killed the cellulolytic protists in the gut of workers and termite lab colonies died of starvation within 3 weeks. A similar yeast-based paratransgenesis system was developed for using natural lytic peptides, like the bee venom melittin, to kill gut protists and termite colonies (Husseneder *et al.*, 2016).

To ensure that the genetically engineered microbes do not escape into the environment, bacteria species only known to exist in the termite gut were tested to assess their usefulness as "Trojan horses" (Tikhe *et al.*, 2016, 2017). *Trabulsiella odontotermitis*, a termite-specific bacterium isolated from the guts of *C. formosanus* workers, was genetically engineered to express either GFP from a plasmid or the antibiotic Kanamycin from the chromosome and fed to termite lab colonies (Tikhe *et al.*, 2016). Fluorescence and antibiotic resistance confirmed that bacteria were ingested and horizontally transferred, colonized the hindgut wall, and expressed the foreign genes for several weeks. This *Trabulsiella* strain tolerated higher concentrations of ligand–Hecate than were required to kill the gut protist and thus was considered a potential "Trojan horse" for lytic peptide expression (Tikhe *et al.*, 2017). Although a pilot study succeeded in engineering *T. trabulsiella* with a plasmid containing the ligand–Hecate gene, no lytic peptide expression or effects on protists could be detected, suggesting that the small peptide may have been degraded by proteases in the bacteria's intracellular environment. When peptide size was increased by adding GFP to the ligand–Hecate construct, *T. odontotermitis* was able to express functional ligand–Hecate from a plasmid; however, several attempts to express the gene construct from the chromosome of this non-model bacteria species failed. Since plasmids can be laterally transferred to other bacteria species, only termite-specific bacteria strains that can stably carry and express lytic peptide in their chromosome should be considered a viable "Trojan horse" (Tikhe *et al.*, 2017). Recently developed gene editing methods are expected to make it easier to genetically engineer non-model organisms like termite-specific bacteria which might boost interest in paratransgenic delivery systems for pest control (Riley and Guss, 2021).

A completely unexplored biocontrol method of termites could involve the use of bacteriophages (see section 8.7). Bacteriophages have immense potential as a biotechnology tool that can be exploited for termite control. Bacteriophages have been used to benefit human health via phage therapy as an alternative to antibiotics. They also have been utilized as biocontrol agents for pathogens in plants and aquaculture (Reardon, 2014). Moreover, the use of bacteriophages as a tool for molecular biology and genetic manipulation is well-established (Pines *et al.*, 2015). Bacteriophages can be utilized in two ways as termite biocontrol agents: (1) a phage therapy approach whereby a cocktail of lytic bacteriophages is used to infect and kill essential hindgut bacteria in the termite workers and thus deprive the colony of nutrition; and (2) as a means of directly expressing toxin genes from the phage genome during the phage lytic cycle, thereby providing a mechanism for the delivery and spread of toxins throughout the termite colony similar to previously developed delivery systems using paratransgenesis. An excellent example of a similar type of situation is found in nature, as toxins encoded by a bacterial symbiont phage protect aphids from parasitoid wasps when the phage genome is expressed (Moran *et al.*, 2005; Oliver *et al.*, 2009). Exposing termites to an engineered bacteriophage that is specific for an obligate hindgut symbiont has the potential to be an environmentally safe method of biocontrol. The product would be a significant improvement over current baits containing insect growth regulators, because phages in baiting systems will likely become self-replicating and self-sustaining in the termite colony, while being rather specific in the range of their host bacteria.

8.9 Conclusion

Research in termite symbiosis has come a long way since emerging from the field of parasitology over a century ago. Great advances have been made in recent decades which can be attributed to the increasing accessibility and affordability of -omics technologies; however, studies have barely started uncovering and understanding the basic and applied aspects of the "termite holobiont" and its trophic network (Margulis, 1991). The ultimate goal would be to assemble a highly detailed holistic organizational chart of the patterns and processes of symbiotic interactions that includes cross-feeding of and interspecific competition for metabolites, nutrients, and energy in termite individuals and superorganisms (colonies), where one microbe's metabolic waste might be another's nutritional treasure.

As the field advances fueled by the rapid pace of an ever-evolving suite of sequencing technologies, key areas offer future challenges and opportunities. These include detailed quantitative and functional analysis of uncultured protists, including their endo- and ectosymbionts. In addition, there is still great need to culture protists, bacteria and archaea to study their physiology and physical and nutritional integration with each other and the termite host. While nutritional collaboration within the intestinal tract has drawn a lot of interest, there is much more to discover concerning the symbionts' roles in host physiology, development, and behavior. Inclusion of biotic and abiotic stressors as metadata or in experimental challenges will lead to better understanding of adaptability and resilience of the termite holobiont which also will generate hypotheses pertaining to the roles that the virome might be playing in the termite gut. Overall, the nested quadripartite termite gut microbiome continues to be one of the most fascinating models for basic and applied research into animal symbioses.

References

Aanen, D.K., Eggleton, P., Rouland-Lefevre, C., Guldberg-Frøslev, T., Rosendahl, S. *et al.* (2002) The evolution of fungus-growing termites and their mutualistic fungal symbionts. *Proceedings of the National Academy of Sciences* 99, 14887–14892.

Abdul Rahman, N., Parks, D.H., Willner, D.L., Engelbrektson, A.L., Goffredi, S.K. *et al.* (2015) A molecular survey of Australian and North American termite genera indicates that vertical inheritance is the primary force shaping termite gut microbiomes. *Microbiome* 3, 1–16.

Arora, J., Kinjo, Y., Šobotník, J., Buček, A., Clitheroe, C. *et al.* (2022) The functional evolution of termite gut microbiota. *Microbiome* 10, 1–22.

Benjamino, J. and Graf, J. (2016) Characterization of the core and caste-specific microbiota in the termite, *Reticulitermes flavipes. Frontiers in Microbiology* 7, 171.

Boucias, D.G., Cai, Y., Sun, Y., Lietze, V.U., Sen, R. *et al.* (2013) The hindgut lumen prokaryotic microbiota of the termite *Reticulitermes flavipes* and its responses to dietary lignocellulose composition. *Molecular Ecology* 22, 1836–1853.

Bourguignon, T., Lo, N., Cameron, S.L., Šobotník, J., Hayashi, Y. *et al.* (2015) The evolutionary history of termites as inferred from 66 mitochondrial genomes. *Molecular Biology and Evolution* 32, 406–421.

Bourguignon, T., Lo, N., Dietrich, C., Šobotník, J., Sidek, S. *et al.* (2018) Rampant host switching shaped the termite gut microbiome. *Current Biology* 28, 649–654 (e642).

Brauman, A., Doré, J., Eggleton, P., Bignell, D., Breznak, J.A. *et al.* (2001) Molecular phylogenetic profiling of prokaryotic communities in guts of termites with different feeding habits. *FEMS Microbiology Ecology* 35, 27–36.

Breznak, J.A. (2000) Ecology of prokaryotic microbes in the guts of wood- and litter-feeding termites. In: Abe, T., Bignell, D.E. and Higashi, M. (eds) *Termites: Evolution, sociality, symbioses, ecology*. Springer Netherlands, Dordrecht, pp. 209–231.

Breznak, J.A. and Brune, A. (1994) Role of microorganisms in the digestion of lignocellulose by termites. *Annual Review of Entomology* 39, 453–487.

Breznak, J.A. and Switzer, J.M. (1986) Acetate synthesis from H2 plus CO2 by termite gut microbes. *Applied and Environmental Microbiology* 52, 623–630.

Brossette, L., Meunier, J., Dupont, S., Bagnères, A.G. and Lucas, C. (2019) Unbalanced biparental care during colony foundation in two subterranean termites. *Ecology and Evolution* 9, 192–200.

Brugerolle, G.and Lee, J.J. (2000) Order Oxymonadida. In: Lee, J.J., Leedale, G.F. and Bradbury, P. (eds) *Illustrated Guide to the Protozoa*. Society of Protozoologists, Lawrence, Kansas, pp. 1186–1195.

Brune, A. (2010) Methanogens in the digestive tract of termites. In: Hackstein, J.H.P. (ed.) *(Endo)symbiotic Methanogenic Archaea, Microbiology Monograph*. Springer Berlin Heidelberg, Berlin, pp. 81–100.

Brune, A. (2011) Microbial symbioses in the digestive tract of lower termites. In: Rosenberg, E. and Gophna, U. (eds) *Beneficial Microorganisms in Multicellular Life Forms*. Springer Berlin Heidelberg, Berlin, pp. 3–25.

Brune, A. (2013) Symbiotic associations between termites and prokaryotes. In: Rosenberg, E., Delong, E.F., Lory, S., Stackebrandt, E. and Thompson, F. (eds) *The Prokaryotes*. Springer Berlin Heidelberg, Berlin, pp. 545–577.

Brune, A. (2014) Symbiotic digestion of lignocellulose in termite guts. *Nature Reviews Microbiology* 12, 168–180.

Brune, A. and Dietrich, C. (2015) The gut microbiota of termites: digesting the diversity in the light of ecology and evolution. *Annual Review of Microbiology* 69, 145–166.

Brune, A. and Friedrich, M. (2000) Microecology of the termite gut: structure and function on a microscale. *Current Opinion in Microbiology* 3, 263–269.

Brune, A. and Ohkuma, M. (2011) Role of the termite gut microbiota in symbiotic digestion. In: Bignell, D.E., Roisin, Y. and Lo, N. (eds) *Biology of Termites: A Modern Synthesis*. Springer Netherlands, Dordrecht, pp. 439–475.

Brune, A. and Stingl, U. (2006) Prokaryotic symbionts of termite gut flagellates: phylogenetic and metabolic implications of a tripartite symbiosis. In: Overmann, J. (ed.) *Molecular Basis of Symbiosis*. Springer Berlin Heidelberg, Berlin, pp. 39–60.

Brune, A., Song, Y., Oren, A. and Paster, B.J. (2022) A new family for 'termite gut treponemes': description of Breznakiellaceae fam. nov., Gracilinema caldarium gen. nov., comb. nov., Leadbettera azotonutricia gen. nov., comb. nov., Helmutkoenigia isoptericolens gen. nov., comb. nov., and Zuelzera stenostrepta gen. nov., comb. nov., and proposal of Rectinemataceae fam. nov. *International Journal of Systematic and Evolutionary Microbiology* 72, 005439.

Bulmer, M.S., Bachelet, I., Raman, R., Rosengaus, R.B. and Sasisekharan, R. (2009) Targeting an antimicrobial effector function in insect immunity as a pest control strategy. *Proceedings of the National Academy of Sciences USA* 106,12652–12657. doi: 10.1073/pnas.0904063106

Chen, J., Gissendanner, C.R., Tikhe, C.V., Li, H.-F., Sun, Q. *et al.* (2022a) Genomics and geographic diversity of bacteriophages associated with endosymbionts in the guts of workers and alates of *Coptotermes* species (Blattodea: Rhinotermitidae). *Frontiers in Ecology and Evolution* 507.

Chen, J., Setia, G., Sun, Q. and Husseneder, C. (2022b) Dynamics of gut symbionts in termite kings and queens during early colony development. *Louisiana Agriculture* 65, 28–29.

Chouvenc, T., Su, N.-Y. and Grace, J.K. (2011) Fifty years of attempted biological control of termites–analysis of a failure. *Biological Control* 59, 69–82.

Chouvenc, T., Efstathion, C.A., Elliott, M.L. and Su, N.-Y. (2013) Extended disease resistance emerging from the faecal nest of a subterranean termite. *Proceedings of the Royal Society B: Biological Sciences* 280, 20131885.

Chouvenc, T., Elliott, M.L., Šobotník, J., Efstathion, C.A. and Su, N.-Y. (2018) The termite fecal nest: a framework for the opportunistic acquisition of beneficial soil Streptomyces (Actinomycetales: Streptomycetaceae). *Environmental Entomology* 47, 1431–1439.

Cleveland, L.R. (1923) Symbiosis between termites and their intestinal protozoa. *Proceedings of the National Academy of Sciences* 9, 424–428.

Cleveland, L.R. (1934) The wood feeding roach *Cryptocercus*, its protozoa, and the symbiosis between protozoa and roach. *Memoirs of the American Academy of Arts and Sciences* 17, 185–382.

Cleveland, L.R. (1959) Sex induced with ecdysone. *Proceedings of the National Academy of Sciences* 45, 747–753.

Cleveland, L.R. (1965) Fertilization in Trichonympha from termites. *Archiv für Protistenkunde* 108, 1–5.

Del Campo, J., James, E.R., Hirakawa, Y., Fiorito, R., Kolisko, M. *et al.* (2017) *Pseudotrichonympha leei*, *Pseudotrichonympha lifesoni*, and *Pseudotrichonympha pearti*, new species of parabasalian flagellates and the description of a rotating subcellular structure. *Scientific Reports* 7, 1–8.

Dietrich, C., Köhler, T. and Brune, A. (2014) The cockroach origin of the termite gut microbiota: patterns in bacterial community structure reflect major evolutionary events. *Applied and Environmental Microbiology* 80, 2261–2269.

Do, T.H., Nguyen, T.T., Nguyen, T.N., Le, Q.G., Nguyen, C. *et al.* (2014) Mining biomass-degrading genes through illumina-based de novo sequencing and metagenomic analysis of free-living bacteria in the gut of the lower termite *Coptotermes gestroi* harvested in Vietnam. *Journal of Bioscience and Bioengineering* 118, 665–671.

Ebert, A. and Brune, A. (1997) Hydrogen concentration profiles at the oxic-anoxic interface: a microsensor study of the hindgut of the wood-feeding lower termite *Reticulitermes flavipes* (Kollar). *Applied and Environmental Microbiology* 63, 4039–4046.

Evans, T.A., Forschler, B.T. and Grace, J.K. (2013) Biology of invasive termites: a worldwide review. *Annual Review of Entomology* 58, 455–474.

Fisher, M., Miller, D., Brewster, C., Husseneder, C. and Dickerman, A. (2007) Diversity of gut bacteria of *Reticulitermes flavipes* as examined by 16S rRNA gene sequencing and amplified rDNA restriction analysis. *Current Microbiology* 55, 254–259.

Geissinger, O., Herlemann, D.P., Mörschel, E., Maier, U.G., and Brune, A. (2009) The ultramicrobacterium "Elusimicrobium minutum" gen. nov., sp. nov., the first cultivated representative of the termite group 1 phylum. *Applied and Environmental Microbiology* 75, 2831–2840. doi: 10.1128/AEM.02697-08

Graber, J.R. and Breznak, J.A. (2004) Physiology and nutrition of *Treponema primitia*, an H2/CO2-acetogenic spirochete from termite hindguts. *Applied and Environmental Microbiology* 70, 1307–1314.

Grassé, P. (1952) Role des flagellés symbiotiques chez les blattes et les termites. *Tijdschrift voor Entomologie* 95, 70–80.

Grassé, P. and Noirot, C. (1945) La transmission des flagellés symbiotiques et les aliments des termites. *Bulletin Biologique de la France et de la Belgique* 79, 273–297.

Hamilton, C. and Bulmer, M.S. (2012) Molecular antifungal defenses in subterranean termites: RNA interference reveals in vivo roles of termicins and GNBPs against a naturally encountered pathogen. *Developmental & Comparative Immunology* 36, 372–377.

Hara, K., Shinzato, N., Oshima, T. and Yamagishi, A. (2004) Endosymbiotic methanobrevibacter species living in symbiotic protists of the termite *Reticulitermes speratus* detected by fluorescent in situ hybridization. *Microbes and Environments* 19, 120–127.

Heath, H. (1903) The habits of California termites. *The Biological Bulletin* 4, 47–63.

Higashiguchi, D.T., Husseneder, C., Grace, J.K. and Berestecky, J.M. (2006) *Pilibacter termitis* gen. nov., sp. nov., a lactic acid bacterium from the hindgut of the Formosan subterranean termite (*Coptotermes formosanus*). *International Journal of Systematic and Evolutionary Microbiology* 56, 15–20.

Hobbs, Z. and Abedon, S.T. (2016) Diversity of phage infection types and associated terminology: the problem with 'lytic or lysogenic'. *FEMS Microbiology Letters* 363.

Hongoh, Y. (2010) Diversity and genomes of uncultured microbial symbionts in the termite gut. *Bioscience, Biotechnology, and Biochemistry* 74, 1145–1151.

Hongoh, Y. and Ohkuma, M. (2018) Termite gut flagellates and their methanogenic and eubacterial symbionts. In: Hackstein, J.H.P. (ed.) *(Endo)symbiotic Methanogenic Archaea*. Microbiology Monographs, Vol. 19, 2nd edn (1st edn 2010). Springer Verlag, Berlin Heidelberg, pp. 55–80.

Hongoh, Y., Ohkuma, M. and Kudo, T. (2003) Molecular analysis of bacterial microbiota in the gut of the termite *Reticulitermes speratus* (Isoptera; Rhinotermitidae). *FEMS Microbiology Ecology* 44, 231–242.

Hongoh, Y., Deevong, P., Inoue, T., Moriya, S., Trakulnaleamsai, S. *et al.* (2005) Intra- and interspecific comparisons of bacterial diversity and community structure support coevolution of gut microbiota and termite host. *Applied and Environmental Microbiology* 71, 6590–6599.

Hongoh, Y., Sato, T., Dolan, M.F., Noda, S., Ui, S. *et al.* (2007a) The motility symbiont of the termite gut flagellate *Caduceia versatilis* is a member of the "Synergistes" group. *Applied and Environmental Microbiology* 73, 6270–6276.

Hongoh, Y., Sato, T., Noda, S., Ui, S., Kudo, T. *et al.* (2007b) Candidatus symbiothrix dinenymphae: bristle-like bacteroidales ectosymbionts of termite gut protists. *Environmental Microbiology* 9, 2631–2635.

Hongoh, Y., Sharma, V.K., Prakash, T., Noda, S., Toh, H. *et al.* (2008a) Genome of an endosymbiont coupling N2 fixation to cellulolysis within protist cells in termite gut. *Science* 322, 1108–1109.

Hongoh, Y., Sharma, V.K., Prakash, T., Noda, S., Taylor, T.D. *et al.* (2008b) Complete genome of the uncultured termite group 1 bacteria in a single host protist cell. *Proceedings of the National Academy of Sciences* 105, 5555–5560.

Howard-Varona, C., Hargreaves, K.R., Abedon, S.T. and Sullivan, M.B. (2017) Lysogeny in nature: mechanisms, impact and ecology of temperate phages. *The ISME Journal* 11, 1511–1520.

Huang, X.-F., Bakker, M.G., Judd, T.M., Reardon, K.F. and Vivanco, J.M. (2013) Variations in diversity and richness of gut bacterial communities of termites (*Reticulitermes flavipes*) fed with grassy and woody plant substrates. *Microbial Ecology* 65, 531–536.

Hungate, R.E. (1939) Experiments on the nutrition of Zootermopsis. III. The anaerobic carbohydrate dissimilation by the intestinal protozoa. *Ecology* 20, 230–245.

Husseneder, C. (2010) Symbiosis in subterranean termites: a review of insights from molecular studies. *Environmental Entomology* 39, 378–388.

Husseneder, C. and Collier, R.E. (2009) Paratransgenesis for termite control. In: Bourtzis, K. and Miller, M.A. (eds) *Insect Symbiosis*. CRC Press, Boca Raton, Florida, pp. 361–376.

Husseneder, C. and Grace, J.K. (2005) Genetically engineered termite gut bacteria (*Enterobacter cloacae*) deliver and spread foreign genes in termite colonies. *Applied Microbiology and Biotechnology* 68, 360–367.

Husseneder, C., Grace, J.K. and Oishi, D.E. (2005) Use of genetically engineered *Escherichia coli* to monitor ingestion, loss, and transfer of bacteria in termites. *Current Microbiology* 50, 119–123.

Husseneder, C., Berestecky, J.M. and Grace, J.K. (2009) Changes in composition of culturable bacteria community in the gut of the formosan subterranean termite depending on rearing conditions of the host. *Annals of the Entomological Society of America* 102, 498–507.

Husseneder, C., Ho, H.Y. and Blackwell, M. (2010a) Comparison of the bacterial symbiont composition of the Formosan subterranean termite from its native and introduced range. *The Open Microbiology Journal* 4, 53–66.

Husseneder, C., Sethi, A., Foil, L. and Delatte, J. (2010b) Testing protozoacidal activity of ligand-lytic peptides against termite gut protozoa in vitro (protozoa culture) and in vivo (microinjection into termite hindgut). *Journal of Visualized Experiments* 46 e2190.

Husseneder, C., Donaldson, J.R. and Foil, L.D. (2016) Genetically engineered yeast expressing a lytic peptide from bee venom (Melittin) kills symbiotic protozoa in the gut of Formosan subterranean termites. *PLoS ONE* 11, e0151675.

Iida, T., Ohkuma, M., Ohtoko, K. and Kudo, T. (2000) Symbiotic spirochetes in the termite hindgut: phylogenetic identification of ectosymbiotic spirochetes of oxymonad protists. *FEMS Microbiology Ecology* 34, 17–26.

Ikeda-Ohtsubo, W., Desai, M., Stingl, U. and Brune, A. (2007) Phylogenetic diversity of 'Endomicrobia' and their specific affiliation with termite gut flagellates. *Microbiology* 153, 3458–3465.

Ikeda-Ohtsubo, W. and Brune, A. (2009) Cospeciation of termite gut flagellates and their bacterial endosymbionts: Trichonympha species and 'Candidatus Endomicrobium trichonymphae'. *Molecular Ecology* 18, 332–342.

Inagaki, T., Yanagihara, S., Fuchikawa, T. and Matsuura, K. (2020) Gut microbial pulse provides nutrition for parental provisioning in incipient termite colonies. *Behavioral Ecology and Sociobiology* 74, 64.

Inoue, T., Kitade, O., Yoshimura, T. and Yamaoka, I. (2000) Symbiotic associations with protists. In: Abe, T., Bignell, D.E. and Higashi, M. (eds) *Termites: Evolution, Sociality, Symbioses, Ecology*. Springer, Dordrecht, Netherlands, pp. 275–288.

Inoue, T., Moriya, S., Ohkuma, M. and Kudo, T. (2005) Molecular cloning and characterization of a cellulase gene from a symbiotic protist of the lower termite, *Coptotermes formosanus*. *Gene* 349, 67–75.

Inoue, J.-I., Saita, K., Kudo, T., Ui, S. and Ohkuma, M. (2007) Hydrogen production by termite gut protists: characterization of iron hydrogenases of parabasalian symbionts of the termite *Coptotermes formosanus*. *Eukaryotic Cell* 6, 1925–1932.

Inoue, J.-I., Noda, S., Hongoh, Y., Ui, S. and Ohkuma, M. (2008) Identification of endosymbiotic methanogen and ectosymbiotic spirochetes of gut protists of the termite *Coptotermes formosanus*. *Microbes and Environments* 23, 94–97.

Izawa, K., Kuwahara, H., Kihara, K., Yuki, M., Lo, N. *et al.* (2016) Comparison of intracellular "Ca. endomicrobium trichonymphae" genomovars illuminates the requirement and decay of defense systems against foreign DNA. *Genome Biology and Evolution* 8, 3099–3107.

Jasso-Selles, D.E., De Martini, F., Velenovsky IV, J.F., Mee, E.D., Montoya, S.J. *et al.* (2020) The complete protist symbiont communities of *Coptotermes formosanus* and *Coptotermes gestroi*: morphological and molecular characterization of five new species. *Journal of Eukaryotic Microbiology* 67, 626–641.

Kerr, M., Rosario, K., Baker, C.C.M. and Breitbart, M. (2018) Discovery of four novel circular single-stranded DNA viruses in fungus-farming termites. *Genome Announcements*, 6, e00318–00318.

King, J.H.P., Mahadi, N.M., Bong, C.F.J., Ong, K.H. and Hassan, O. (2014) Bacterial microbiome of *Coptotermes curvignathus* (Isoptera: Rhinotermitidae) reflects the coevolution of species and dietary pattern. *Insect Science* 21, 584–596.

Kitade, O. and Matsumoto, T. (1998) Characteristics of the symbiotic flagellate composition within the termite family Rhinotermitidae (Isoptera). *Symbiosis* 25, 271–278.

Kitade, O., Hayashi, Y. and Noda, S. (2013) Symbiotic protist communities in the termite *Coptotermes formosanus* in Japan and a comparison of community structures between workers and soldiers. *Japanese Journal of Protozoology* 46, 21–29.

Koidzumi, M. (1921) Studies on the intestinal protozoa found in the termites of Japan. *Parasitology* 13, 235–309.

König, H., Fröhlich, J. and Hertel, H. (2006) Diversity and lignocellulolytic activities of cultured microorganisms. In: König, H. and Varma, A. (eds) *Intestinal Microorganisms of Termites and Other Invertebrates*. Springer Berlin Heidelberg, Berlin, pp. 271–301.

Kudo, T., Ohkuma, M., Moriya, S., Noda, S. and Ohtoko, K. (1998) Molecular phylogenetic identification of the intestinal anaerobic microbial community in the hindgut of the termite, *Reticulitermes speratus*, without cultivation. *Extremophiles* 2, 155–161.

Kuwahara, H., Yuki, M., Izawa, K., Ohkuma, M. and Hongoh, Y. (2017) Genome of 'Ca. Desulfovibrio trichonymphae', an H2-oxidizing bacterium in a tripartite symbiotic system within a protist cell in the termite gut. *The ISME Journal* 11, 766–776.

Lai, P.Y., Tamashiro, M. and Fujii, J.K. (1983) Abundance and distribution of the three species of symbiotic protozoa in the hindgut of *Coptotermes formosanus* (Isoptera: Rhinotermitidae). *Proceedings of the Hawaiian Entomological Society* 24, 271–276.

Leadbetter, J.R. and Breznak, J.A. (1996) Physiological ecology of methanobrevibacter cuticularis sp. nov. and methanobrevibacter curvatus sp. nov., isolated from the hindgut of the termite *Reticulitermes flavipes*. *Applied and Environmental Microbiology* 62, 3620–3631.

Leadbetter, J.R., Schmidt, T.M., Graber, J.R. and Breznak, J.A. (1999) Acetogenesis from H2 plus CO2 by spirochetes from termite guts. *Science* 283, 686–689.

Leidy, J. (1881) Parasites of the termites. *Journal of The Academy of Natural Sciences of Philadelphia* 8, 425–450.

Lewis, J. and Forschler, B. (2004) Protist communities from four castes and three species of *Reticulitermes* (Isoptera: Rhinotermitidae). *Annals of the Entomological Society of America* 97, 1242–1251.

Lilburn, T., Schmidt, T. and Breznak, J. (1999) Phylogenetic diversity of termite gut spirochaetes. *Environmental Microbiology* 1, 331–345.

Liu, X.J., Xie, L., Liu, N., Zhan, S., Zhou, X.G. and Wang, Q. (2017) RNA interference unveils the importance of *Pseudotrichonympha grassii* cellobiohydrolase, a protozoan exoglucanase, in termite cellulose degradation. *Insect Molecular Biology* 26, 233–242.

Lo, N., Tokuda, G., Watanabe, H., Rose, H., Slaytor, M., *et al.* (2000) Evidence from multiple gene sequences indicates that termites evolved from wood-feeding cockroaches. *Current Biology* 10, 801–804.

Margulis, L. (1991) Symbiosis as a source of evolutionary innovation: speciation and morphogenesis. In: Cambridge, MA MLFR Marulis, L. and Fester, R.(eds) *Symbiogenesis and Symbionticism*. MIT Press, Cambridge, Massachusetts, pp. 1–14.

Margulis, L. (1993) *Symbiosis in Cell Evolution: Microbial Communities in the Archean and Proterozoic Eons*. Freeman, New York.

Martin, M.M. and Kukor, J.J. (1984) Role of mycophagy and bacteriophagy invertebrate nutrition. In: Klug, M.J. and Reddy, C.A. (eds) *Current Perspectives in Microbial Ecology*. ASM Press, Washington, DC, pp. 257–263.

Michaud, C., Hervé, V., Dupont, S., Dubreuil, G., Bézier, A.M. *et al.* (2020) Efficient but occasionally imperfect vertical transmission of gut mutualistic protists in a wood-feeding termite. *Molecular Ecology* 29, 308–324.

Mikaelyan, A., Thompson, C.L., Meuser, K., Zheng, H., Rani, P. *et al.* (2017) High-resolution phylogenetic analysis of endomicrobia reveals multiple acquisitions of endosymbiotic lineages by termite gut flagellates. *Environmental Microbiology Reports* 9, 477–483.

Moran, N.A., Degnan, P.H., Santos, S.R., Dunbar, H.E. and Ochman, H. (2005) The players in a mutualistic symbiosis: insects, bacteria, viruses, and virulence genes. *Proceedings of the National Academy of Sciences* 102, 16919–16926.

Nakashima, K., Watanabe, H., Saitoh, H., Tokuda, G. and Azuma, J.-I. (2002a) Dual cellulose-digesting system of the wood-feeding termite, *Coptotermes formosanus* (Shiraki). *Insect Biochemistry and Molecular Biology* 32, 777–784.

Nakashima, K., Watanabe, H. and Azuma, J.-I. (2002b) Cellulase genes from the parabasalian symbiont *Pseudotrichonympha grassii* in the hindgut of the wood-feeding termite *Coptotermes formosanus*. *Cellular and Molecular Life Sciences* 59, 1554–1560.

Nalepa, C.A. (2017) What kills the hindgut flagellates of lower termites during the host molting cycle? *Microorganisms* 5, 82.

Nalepa, C.A. (2020) Origin of mutualism between termites and flagellated gut protists: transition from horizontal to vertical transmission. *Frontiers in Ecology and Evolution* 8, 14.

Nalepa, C., Hunt, J. and Nalepa, C. (1994) Nourishment and the evolution of termite sociality. In: Hunt, J.H. and Nalepa, C.A. (eds) *Nourishment and Evolution in Insect Societies*. Westview Press, Boulder, Colorado, pp. 57–104.

Nalepa, C., Bignell, D. and Bandi, C. (2001) Detritivory, coprophagy, and the evolution of digestive mutualisms in dictyoptera. *Insectes Sociaux* 48, 194–201.

Nishimura, Y., Otagiri, M., Yuki, M., Shimizu, M., Inoue, J.-I. *et al.* (2020) Division of functional roles for termite gut protists revealed by single-cell transcriptomes. *The ISME Journal*, 14, 2449–2460.

Noda, S., Ohkuma, M., Yamada, A., Hongoh, Y. and Kudo, T. (2003) Phylogenetic position and in situ identification of ectosymbiotic spirochetes on protists in the termite gut. *Appl Anrion Microbiol.* 2003 Jan; 69(1) 625–633. doi: 10.1128/AEM.69.1.625-633.2003

Noda, S., Iida, T., Kitade, O., Nakajima, H., Kudo, T. *et al.* (2005) Endosymbiotic bacteroidales bacteria of the flagellated protist *Pseudotrichonympha grassii* in the gut of the termite *Coptotermes formosanus*. *Applied and Environmental Microbiology* 71, 8811–8817.

Noda, S., Kawai, M., Nakajima, H., Kudo, T. and Ohkuma, M. (2006) Identification and in situ detection of two lineages of Bacteroidales ectosymbionts associated with a termite gut protist, Oxymonas sp. *Microbes and Environments* 21, 16–22.

Noda, S., Kitade, O., Inoue, T., Kawai, M., Kanuka, M. *et al.* (2007) Cospeciation in the triplex symbiosis of termite gut protists (*Pseudotrichonympha spp.*), their hosts, and their bacterial endosymbionts. *Molecular Ecology* 16, 1257–1266.

Noda, S., Hongoh, Y., Sato, T. and Ohkuma, M. (2009) Complex coevolutionary history of symbiotic Bacteroidales bacteria of various protists in the gut of termites. *BMC Evolutionary Biology* 9, 1–12.

Noda, S., Shimizu, D., Yuki, M., Kitade, O. and Ohkuma, M. (2018) Host-symbiont cospeciation of termite-gut cellulolytic protists of the genera teranympha and eucomonympha and their treponema endosymbionts. *Microbes and Environments* 33(1), 26–33.

Nutting, W.L. (1969) Flight and colony foundation. In: Krishna, K. and Weesner, F.M. (eds) *Biology of Termites*. Academic Press, New York, pp. 233–282.

Oberpaul, M., Zumkeller, C. M., Culver, T., Spohn, M., Mihajlovic, S. *et al.* (2020) High-throughput cultivation for the selective isolation of acidobacteria from termite nests. *Frontiers in Microbiology* 11, 597628.

Odelson, D.A. and Breznak, J.A. (1985) Nutrition and growth characteristics of *Trichomitopsis termopsidis*, a cellulolytic protozoan from termites. *Applied and Environmental Microbiology* 49, 614–621.

Ohkuma, M. (2003) Termite symbiotic systems: efficient bio-recycling of lignocellulose. *Applied Microbiology and Biotechnology* 61, 1–9.

Ohkuma, M. (2008) Symbioses of flagellates and prokaryotes in the gut of lower termites. *Trends in Microbiology* 16, 345–352.

Ohkuma, M. and Brune, A. (2010) Diversity, structure, and evolution of the termite gut microbial community. In: Bignell, D.E., Roisin, Y. and Lo, N. (eds) *Biology of Termites: A Modern Synthesis*. Springer, New York, pp. 413–438.

Ohkuma, M. and Kudo, T. (1996) Phylogenetic diversity of the intestinal bacterial community in the termite *Reticulitermes speratus*. *Applied and Environmental Microbiology* 62, 461–468.

Ohkuma, M., Noda, S., Horikoshi, K. and Kudo, T. (1995) Phylogeny of symbiotic methanogens in the gut of the termite *Reticulitermes speratus*. *FEMS Microbiology Letters* 134, 45–50.

Ohkuma, M., Noda, S. and Kudo, T. (1999) Phylogenetic relationships of symbiotic methanogens in diverse termites. *FEMS Microbiology Letters* 171, 147–153.

Ohkuma, M., Sato, T., Noda, S., Ui, S., Kudo, T. *et al.* (2007) The candidate phylum 'termite group 1'of bacteria: phylogenetic diversity, distribution, and endosymbiont members of various gut flagellated protists. *FEMS Microbiology Ecology* 60, 467–476.

Ohkuma, M., Noda, S., Hongoh, Y., Nalepa, C.A. and Inoue, T. (2009) Inheritance and diversification of symbiotic trichonymphid flagellates from a common ancestor of termites and the cockroach Cryptocercus. *Proceedings of the Royal Society B: Biological Sciences* 276, 239–245.

Ohtoko, K., Ohkuma, M., Moriya, S., Inoue, T., Usami, R. *et al.* (2000) Diverse genes of cellulase homologues of glycosyl hydrolase family 45 from the symbiotic protists in the hindgut of the termite *Reticulitermes speratus*. *Extremophiles* 4, 343–349.

Oliver, K.M., Degnan, P.H., Hunter, M.S. and Moran, N.A. (2009) Bacteriophages encode factors required for protection in a symbiotic mutualism. *Science* 325, 992–994.

Ottesen, E.A., Hong, J.W., Quake, S.R. and Leadbetter, J.R. (2006). Microfluidic digital PCR enables multigene analysis of individual environmental bacteria. *Science* 314, 1464–1467.

Pester, M. and Brune, A. (2007) Hydrogen is the central free intermediate during lignocellulose degradation by termite gut symbionts. *The ISME Journal* 1, 551–565.

Peterson, B.F. and Scharf, M.E. (2016a) Lower termite associations with microbes: synergy, protection, and interplay. *Frontiers in Microbiology* 7, 422.

Peterson, B.F. and Scharf, M.E. (2016b) Metatranscriptome analysis reveals bacterial symbiont contributions to lower termite physiology and potential immune functions. *BMC Genomics* 17, 1–12.

Peterson, B.F., Stewart, H.L. and Scharf, M.E. (2015) Quantification of symbiotic contributions to lower termite lignocellulose digestion using antimicrobial treatments. *Insect Biochemistry and Molecular Biology* 59, 80–88.

Pines, G., Freed, E.F., Winkler, J.D. and Gill, R.T. (2015) Bacterial recombineering: genome engineering via phage-based homologous recombination. *ACS Synthetic Biology* 4, 1176–1185.

Pramono, A.K., Kuwahara, H., Itoh, T., Toyoda, A., Yamada, A. *et al.* (2017) Discovery and complete genome sequence of a bacteriophage from an obligate intracellular symbiont of a cellulolytic protist in the termite gut. *Microbes and Environments* 32(2), 1121–117.

Purdy, K.J. (2007) The distribution and diversity of euryarchaeota in termite guts. *Advances in Applied Microbiology* 62, 63–80.

Raina, A., Park, Y.I. and Gelman, D. (2008) Molting in workers of the Formosan subterranean termite *Coptotermes formosanus*. *Journal of Insect Physiology* 54, 155–161.

Reardon, S. (2014) Phage therapy gets revitalized: the rise of antibiotic resistance rekindles interest in a century-old virus treatment. *Nature* 510, 15–17.

Riley, L.A. and Guss, A.M. (2021) Approaches to genetic tool development for rapid domestication of non-model microorganisms. *Biotechnology for Biofuels* 14, 30.

Rosengaus, R.B., Jordan, C., Lefebvre, M.L. and Traniello, J.F.A. (1999) Pathogen alarm behavior in a termite: a new form of communication in social insects. *Naturwissenschaften* 86, 544–548.

Rosengaus, R.B., Schultheis, K.F., Yalonetskaya, A., Bulmer, M.S., DuComb, W.S. *et al.* (2014) Symbiont-derived β-1, 3-glucanases in a social insect: mutualism beyond nutrition. *Frontiers in Microbiology* 5, 607.

Saldarriaga, J.F., Gile, G.H., James, E.R., Horak, A., Scheffrahn, R.H. *et al.* (2011) Morphology and molecular phylogeny of *Pseudotrichonympha hertwigi* and *Pseudotrichonympha paulistana* (Trichonymphea, parabasalia) from neotropical rhinotermitids. *Journal of Eukaryotic Microbiology* 58, 487–496.

Salmond, G.P. and Fineran, P.C. (2015) A century of the phage: past, present and future. *Nature Reviews Microbiology* 13 777–786.

Sapkota R. and Scharf, M.E. (2022) Intercolony Comparisons of Gut Microbiome Composition From Lab Reared Eastern Subterranean Termites (Blattodea: Rhinotermitidae). *Journal of Insect Science* 22(2), 9.

Sato, T., Hongoh, Y., Noda, S., Hattori, S., Ui, S. *et al.* (2009) Candidatus *Desulfovibrio trichonymphae*, a novel intracellular symbiont of the flagellate *Trichonympha agilis* in termite gut. *Environmental Microbiology* 11, 1007–1015.

Sato, T., Kuwahara, H., Fujita, K., Noda, S., Kihara, K. *et al.* (2014) Intranuclear verrucomicrobial symbionts and evidence of lateral gene transfer to the host protist in the termite gut. *The ISME Journal* 8, 1008–1019.

Scharf, M.E. (2015) Termites as targets and models for biotechnology. *Annual Review of Entomology* 60, 77–102.

Scharf, M.E. and Peterson, B.F. (2021) A century of synergy in termite symbiosis research: linking the past with new genomic insights. *Annual Review of Entomology* 66, 23–43.

Scharf, M.E. and Tartar, A. (2008) Termite digestomes as sources for novel lignocellulases. *Biofuels, Bioproducts and Biorefining* 2, 540–552.

Scharf, M.E., Karl, Z.J., Sethi, A. and Boucias, D.G. (2011) Multiple levels of synergistic collaboration in termite lignocellulose digestion. *PLoS ONE* 6, e21709.

Schmidlin, K., Kraberger, S., Fontenele, R.S., De Martini, F., Chouvenc, T., Gile, G.H. and Varsani, A. (2019) Genome sequences of microviruses associated with *Coptotermes formosanus.Microbiology Resource Announcements* 8(16), e00185–19. doi: 10.1128/MRA.00185-19

Sen, R., Raychoudhury, R., Cai, Y., Sun, Y., Lietze, V.-U. *et al.* (2013) Differential impacts of juvenile hormone, soldier head extract and alternate caste phenotypes on host and symbiont transcriptome composition in the gut of the termite *Reticulitermes flavipes*. *BMC Genomics* 14, 1–18.

Sen, R., Raychoudhury, R., Cai, Y., Sun, Y., Lietze, V.-U. *et al.* (2015) Molecular signatures of nicotinoid-pathogen synergy in the termite gut. *PLoS ONE* 10(4), e0123391.

Sethi, A., Xue, Q.-G., La Peyre, J.F., Delatte, J. and Husseneder, C. (2011) Dual origin of gut proteases in Formosan subterranean termites (*Coptotermes formosanus* Shiraki) (Isoptera: Rhinotermitidae). *Comparative Biochemistry and Physiology Part A: Molecular & Integrative Physiology* 159, 261–267.

Sethi, A., Delatte, J., Foil, L. and Husseneder, C. (2014) Protozoacidal trojan-horse: use of a ligand-lytic peptide for selective destruction of symbiotic protozoa within termite guts. *PLoS ONE* 9, e106199.

Shimada, K. and Maekawa, K. (2010) Changes in endogenous cellulase gene expression levels and reproductive characteristics of primary and secondary reproductives with colony development of the termite *Reticulitermes speratus* (Isoptera: Rhinotermitidae). *Journal of Insect Physiology* 56, 1118–1124.

Shimada, K., Lo, N., Kitade, O., Wakui, A. and Maekawa, K. (2013) Cellulolytic protist numbers rise and fall dramatically in termite queens and kings during colony foundation. *Eukaryotic Cell* 12, 545–550.

Shinzato, N., Matsumoto, T., Yamaoka, I., Oshima, T. and Yamagishi, A. (1999) Phylogenetic diversity of symbiotic methanogens living in the hindgut of the lower termite *Reticulitermes speratus* analyzed by PCR and in situ hybridization. *Applied and Environmental Microbiology* 65, 837–840.

Shinzato, N., Matsumoto, T., Yamaoka, I., Oshima, T. and Yamagishi, A. (2001) Methanogenic symbionts and the locality of their host lower termites. *Microbes and Environments* 16, 43–47.

Shinzato, N., Muramatsu, M., Matsui, T. and Watanabe, Y. (2005) Molecular phylogenetic diversity of the bacterial community in the gut of the termite *Coptotermes formosanus*. *Bioscience, Biotechnology, and Biochemistry* 69, 1145–1155.

Song, Y.Q., Zhang, D., Chen, W., Dang, X.X. and Yang, H. (2021) Phylogenetic identification of symbiotic protists of five chinese *Reticulitermes* species indicates a cospeciation of gut microfauna with host termites. *Journal of Eukaryotic Microbiology* 68, e12862.

Soukup, P., Větrovský, T., Stiblik, P., Votýpková, K., Chakraborty, A. *et al.* (2021) Termites are associated with external species-specific bacterial communities. *Applied and Environmental Microbiology* 87, e02042–02020.

Stephens, M.E. and Gage, D.J. (2020) Single-cell amplicon sequencing reveals community structures and transmission trends of protist-associated bacteria in a termite host. *PLoS ONE* 15, e0233065.

Stingl, U., Radek, R., Yang, H. and Brune, A. (2005) "Endomicrobia": cytoplasmic symbionts of termite gut protozoa form a separate phylum of prokaryotes. *Applied and Environmental Microbiology* 71, 1473–1479.

Strassert, J.F., Köhler, T., Wienemann, T.H., Ikeda-Ohtsubo, W., Faivre, N. *et al.* (2012) 'Candidatus Ancillula trichonymphae', a novel lineage of endosymbiotic Actinobacteria in termite gut flagellates of the genus *Trichonympha*. *Environmental Microbiology* 14, 3259–3270.

Su, L., Yang, L., Huang, S., Su, X., Li, Y. *et al.* (2016) Comparative gut microbiomes of four species representing the higher and the lower termites. *Journal of Insect Science* 16, 1–9.

Tadmor, A.D., Ottesen, E.A., Leadbetter, J.R. and Phillips, R. (2011) Probing individual environmental bacteria for viruses by using microfluidic digital PCR. *Science* 333, 58–62.

Tai, V., James, E.R., Nalepa, C.A., Scheffrahn, R.H., Perlman, S.J. *et al.* (2015) The role of host phylogeny varies in shaping microbial diversity in the hindguts of lower termites. *Applied and Environmental Microbiology* 81, 1059–1070.

Tanaka, H., Aoyagi, H., Shina, S., Dodo, Y., Yoshimura, T. *et al.* (2006) Influence of the diet components on the symbiotic microorganisms community in hindgut of *Coptotermes formosanus* Shiraki. *Applied Microbiology and Biotechnology* 71, 907–917.

Tartar, A., Wheeler, M.M., Zhou, X., Coy, M.R., Boucias, D.G. *et al.* (2009) Parallel metatranscriptome analyses of host and symbiont gene expression in the gut of the termite *Reticulitermes flavipes*. *Biotechnology for Biofuels* 2, 1–19.

Tikhe, C.V. and Husseneder, C. (2018) Metavirome sequencing of the termite gut reveals the presence of an unexplored bacteriophage community. *Frontiers in Microbiology* 8, 2548.

Tikhe, C.V., Martin, T.M., Gissendanner, C.R. and Husseneder, C. (2015) Complete genome sequence of citrobacter phage CVT22 isolated from the gut of the Formosan subterranean termite, *Coptotermes formosanus* Shiraki. *Genome Announcements* 3, e00408–00415.

Tikhe, C.V., Martin, T.M., Howells, A., Delatte, J. and Husseneder, C. (2016) Assessment of genetically engineered *Trabulsiella odontotermitis* as a 'trojan horse' for paratransgenesis in termites. *BMC Microbiology* 16, 1–11.

Tikhe, C.V., Sethi, A., Delatte, J. and Husseneder, C. (2017) Isolation and assessment of gut bacteria from the Formosan subterranean termite, *Coptotermes formosanus* (Isoptera: Rhinotermitidae), for paratransgenesis research and application. *Insect Science* 24, 93–102.

Tikhe, C.V., Gissendanner, C.R. and Husseneder, C. (2018a) Whole-genome sequence of the novel temperate enterobacter bacteriophage tyrion, isolated from the gut of the Formosan subterranean termite. *Genome Announcements* 6, e00839-00817.

Tikhe, C.V., Gissendanner, C.R. and Husseneder, C. (2018b) Whole-genome sequence of the novel enterobacter bacteriophage arya with an integrase pseudogene, isolated from the gut of the Formosan subterranean termite. *Genome Announcements* 6, e00838–00817.

Todaka, N., Moriya, S., Saita, K., Hondo, T., Kiuchi, I. *et al.* (2007) Environmental cDNA analysis of the genes involved in lignocellulose digestion in the symbiotic protist community of *Reticulitermes speratus*. *FEMS Microbiology Ecology* 59, 592–599.

Tokura, M., Ohkuma, M. and Kudo, T. (2000) Molecular phylogeny of methanogens associated with flagellated protists in the gut and with the gut epithelium of termites. *FEMS Microbiology Ecology* 33, 233–240.

Tsunoda, K., Ohmura, W., Yoshimura, T. and Tokoro, M. (1993) Methane emission by the termite, *Coptotermes formosanus* Shiraki (Isoptera: Rhinotermitidae) I. Effect of termite caste,population size and volume of test containers. *Wood Research* 79, 34–40.

Utami, Y.D., Kuwahara, H., Igai, K., Murakami, T., Sugaya, K. *et al.* (2019) Genome analyses of uncultured TG2/ZB3 bacteria in 'Margulisbacteria' specifically attached to ectosymbiotic spirochetes of protists in the termite gut. *The ISME Journal* 13, 455–467.

Vargo, E. L. and Husseneder, C. (2009) Biology of subterranean termites: insights from molecular studies of *Reticulitermes* and *Coptotermes*. *Annual Review of Entomology* 54, 379–403.

Veivers, P., O'Brien, R. and Slaytor, M. (1982) Role of bacteria in maintaining the redox potential in the hindgut of termites and preventing entry of foreign bacteria. *Journal of Insect Physiology* 28, 947–951.

Velenovsky, J., Gile, G., Su, N.-Y. and Chouvenc, T. (2021) Dynamic protozoan abundance of *Coptotermes* kings and queens during the transition from biparental to alloparental care. *Insectes Sociaux* 68, 33–40.

Watanabe, H., Nakashima, K., Saito, H. and Slaytor, M. (2002) New endo-β-1, 4-glucanases from the parabasalian symbionts, *Pseudotrichonympha grassii* and *Holomastigotoides mirabile* of *Coptotermes* termites. *Cellular and Molecular Life Sciences* 59, 1983–1992.

Wenzel, M., Radek, R., Brugerolle, G. and König, H. (2003) Identification of the ectosymbiotic bacteria of *Mixotricha paradoxa* involved in movement symbiosis. *European Journal of Protistology* 39, 11–23.

Wier, A., Dolan, M., Grimaldi, D., Guerrero, R., Wagensberg, J. *et al.* (2002) Spirochete and protist symbionts of a termite (*Mastotermes electrodominicus*) in Miocene amber. *Proceedings of the National Academy of Sciences* 99, 1410–1413.

Wu, W., Gu, D., Yan, S., and Li, Z. (2019) RNA interference of endoglucanases in the Formosan subterranean termite *Coptotermes formosanus* Shiraki (Blattodea: Rhinotermitidae) by dsRNA injection or ingestion. *Journal of Insect Physiology* 112, 15–22. doi: 10.1016/j.jinsphys.2018.11.007

Xie, L., Zhang, L., Zhong, Y., Liu, N., Long, Y. *et al.* (2012). Profiling the metatranscriptome of the protistan community in *Coptotermes formosanus* with emphasis on the lignocellulolytic system. *Genomics* 99, 246–255.

Yamin, M.A. (1979) Flagellates of the orders Trichomonadida Kirby, Oxymonadida Grassé, and Hypermastigida Grassi and Foà reported from lower termites (Isoptera families Mastotermitidae, Kalotermitidae, Hodotermitidae, Termopsidae, Rhinotermitidae, and Serritermitidae) and from the wood-feeding roach *Cryptocercus* (Dictyoptera: Cryptocercidae). *Sociobiology* 4, 5–119.

Yang, H., Schmitt-Wagner, D., Stingl, U. and Brune, A. (2005) Niche heterogeneity determines bacterial community structure in the termite gut (*Reticulitermes santonensis*). *Environmental Microbiology* 7, 916–932.

Yoshimura, T. (1992) Distribution of the symbiotic protozoa in the hindgut of *Coptotermes formosanus* Shiraki (Isoptera; Rhinotermitidae). *Japanese Journal of Environmental Entomology and Zoology* 4, 115–120.

Yoshimura, T. (1995) Contribution of the protozoan fauna to nutritional physiology of the lower termite, *Coptotermes formosanus* Shiraki (Isoptera: Rhinotermitidae). *Wood Research* 82, 68–129.

Yuki, M., Kuwahara, H., Shintani, M., Izawa, K., Sato, T. *et al.* (2015) Dominant ectosymbiotic bacteria of cellulolytic protists in the termite gut also have the potential to digest lignocellulose. *Environmental Microbiology* 17, 4942–4953.

Yutin, N., Makarova, K.S., Gussow, A.B., Krupovic, M., Segall, A. *et al.* (2018) Discovery of an expansive bacteriophage family that includes the most abundant viruses from the human gut. *Nature Microbiology* 3, 38–46.

Zeng, W., Li, Z., Liu, B., Li, Q. and Zhong, J. (2016) Disruption of *Coptotermes formosanus* (Isoptera: Rhinotermitidae) gut microbiota reproduction and its consequences for the cellulolytic system. *Journal of Entomological Science* 51, 138–150.

Zeng, W., Liu, B., Wu, W., Zhang, S., Chen, Y., *et al.* (2021) Collaborative response of the host and symbiotic lignocellulytic system to non-lethal toxic stress in *Coptotermes formosanus* Skiraki. *Insects* 12, 510.

Zheng, H., Dietrich, C., Radek, R. and Brune, A. (2016) Endomicrobium proavitum, the first isolate of endomicrobia class. Nov.(phylum Elusimicrobia)–an ultramicrobacterium with an unusual cell cycle that fixes nitrogen with a group IV nitrogenase. *Environmental Microbiology* 18, 191–204.

Zhou, X., Smith, J.A., Oi, F. M., Koehler, P.G., Bennett, G.W. *et al.* (2007) Correlation of cellulase gene expression and cellulolytic activity throughout the gut of the termite *Reticulitermes flavipes*. *Gene* 395, 29–39.

Zhou, X., Wheeler, M.M., Oi, F.M. and Scharf, M.E. (2008) RNA interference in the termite *Reticulitermes flavipes* through ingestion of double-stranded RNA. *Insect Biochemistry and Molecular Biology* 38, 805–815.

9 Dispersal and genetic structure of colonies and populations of *Coptotermes formosanus* and *Coptotermes gestroi*

Edward L. Vargo
Department of Entomology, Texas A&M University, USA, ed.vargo@tamu.edu

Abstract
The application of molecular markers has greatly increased our knowledge of the dispersal, colony breeding structure, population structure, and routes of invasion of *Coptotermes formosanus* Shiraki and its congener *Coptotermes gestroi* (Wasmann). There is evidence of partner selection during tandem pair formation in *C. formosanus*. Colonies of both species begin as simple families headed by a monogamous pair of reproductives and may later become extended families headed by numerous inbreeding neotenics. There is a large variation among populations in the numbers of simple and extended families, presumably reflecting differences in age structure. Some genetic structure exists within populations in the native ranges of these species. There have been several attempts to retrace their introduction histories, but the conclusions from such studies are hampered by low genetic diversity in the mitochondrial genome. A recent study using genome-wide nuclear markers shows the power of this approach to unravel the complex spread of these species by humans.

9.1 Introduction

The Formosan subterranean termite, *Coptotermes formosanus* Shiraki, is one of the worst invasive species in the world. Native to southern China and Taiwan, it has been introduced to many parts of the world, where it causes significant economic damage (Evans *et al.*, 2013). The worst affected areas are Hawai'i, the U.S. mainland, and Japan. It has recently become established in Israel (Scheffrahn *et al.*, 2020), and continues to threaten nearly all subtropical and some tropical areas worldwide. The ability of this species to disperse and spread to new areas depends on both human-mediated transport and its own dispersal ability. Obviously, the introduction of *C. formosanus* into new areas, often located several thousand kilometers across an ocean, can only occur through human transport. Once established in an area, it can spread locally and increase in abundance through mating flights and the establishment of new colonies. Knowing its dispersal range can help predict how quickly it will spread in an area once it has become established.

The breeding structure of colonies, that is, the number and types of reproductives present in colonies, is a fundamental aspect of the biology of social insects, including termites (Vargo and Husseneder, 2011; Vargo, 2019). In addition, changes in the colony breeding structure in some social insects, such as ants, are associated with their invasion success (Eyer and Vargo, 2021). Thus, understanding the breeding structure of

 Biology and Management of the Formosan Subterranean Termite and Related Species (eds Nan-Yao Su and Chow-Yang Lee)
DOI: 10.1079/9781800621596.0009

invasive social insects in their native and introduced ranges may provide insights into their invasion success. However, the nests of termites, especially subterranean species, such as *C. formosanus*, are often cryptic and hard to access. Hence, determining the colony breeding structure through direct censusing of colonies is impractical. Thanks to the application of genetic markers, we have seen remarkable advances in our understanding of termite colony breeding structure over the past two decades (Vargo and Husseneder, 2011; Vargo, 2019).

Genetic markers provide insights into the structure of populations and routes of invasion. The genetic structure of populations can give us valuable information on the dispersal ability of alates and the extent to which gene flow occurs within a given area. The more structured the populations are, that is, the more they are genetically differentiated from one another, the greater our ability to identify potential source populations by matching their genetic composition to that of invasive populations. However, it is important to keep in mind that *C. formosanus* is spread by human activity. If colonies are frequently moved by people in the native range, then signatures of population structure that may have arisen over time through natural processes will be reduced, making it difficult, if not impossible, to identify potential source populations. This appears to be the case in the native range of the eastern subterranean termite, *Reticulitermes flavipes* (Kollar), in which extensive homogenization of the native population has likely occurred, hindering our ability to identify populations that may have served as the source of invasive populations in France, Chile, Canada, and elsewhere (Eyer *et al.*, 2021).

Genetic markers provide especially powerful tools to retrace the invasion history of a species, including the identification of source populations, the routes of invasion, the number of independent introductions, and pathways of secondary spread (Cristescu, 2015). There have been several studies using mostly mitochondrial DNA markers to identify potential source populations and invasion routes of *C. formosanus*. These studies have met with mixed success, depending on sampling intensity and the types of genetic markers used.

This chapter reviews what is known about the dispersal, colony genetic structure, population genetic structure, and invasion history of *C. formosanus*. Much of what we know is inferred from genetic data, but we also have other forms of empirical data, including historical records of the invasion history, to help inform inferences based on genetic analyses. These studies have advanced our understanding of the biology of *C. formosanus* and how it has spread around the world. To add to our understanding of *C. formosanus*, the chapter will also review the literature on its congener, *Coptotermes gestroi* (Wasmann), a native to Southeast Asia that is even more widespread and destructive than *C. formosanus* (Evans *et al.*, 2013). *Coptotermes gestroi* mainly inhabits tropical areas and has been spreading in south Florida (Su *et al.*, 1997; Scheffrahn and Su, 2005; Chouvenc *et al.*, 2015). Studies of these two species give us a more complete picture of the biology of this important genus and its success as a major group of invasive species.

9.2 Colony and population genetics

9.2.1 Alate dispersal and colony foundation

Termite colonies, including those of *Coptotermes*, are generally founded by monogamous pairs of primary reproductives who disperse from their natal colonies during mating flights (Nutting, 1969). Termite reproductives mate for life, and both males and females engage in extensive parental care in rearing their first offspring (Rosengaus and Traniello, 1991; Shellman-Reeve, 1999, 2001). Therefore, males and females are expected to be selective in choosing partners, since the physiological and genetic characteristics of their mates can have important consequences on colony founding success (Shellman-Reeve, 1999, 2001; Chouvenc, 2019; Eyer and Vargo, 2022). Males and females may be expected to avoid pairing with siblings to prevent inbreeding, but this does not appear to be the case in *C. formosanus*. In New Orleans, Lousiana, where swarm aggregations consist of alates from on average 13 different colonies (Husseneder *et al.*, 2006), there is no evidence of sibling avoidance (Husseneder and Simms, 2008), even though inbred incipient colonies raised by sibling reproductive pairs have lower long-term productivity than outbred colonies headed by unrelated reproductives (Fei and Henderson, 2003). The lack of discrimination against siblings during tandem pair formation may be more general among subterranean termites. Males and females of the eastern subterranean termite *R. flavipes* were equally likely to pair with siblings and non-siblings in natural mating aggregations (DeHeer and Vargo, 2006). However, primary reproductives heading mature colonies of *R. flavipes* are unrelated, suggesting colonies formed by sibling pairs are unlikely to survive to maturity (DeHeer and Vargo, 2006). These conclusions are supported by experimental studies on *R. flavipes* showing that the offspring of inbred colonies are more susceptible to fungal pathogens than the offspring of outbred colonies (Eyer and Vargo, 2022). The inherent costs of pairing with siblings during colony foundation may be offset by strong pressure on reproductives to rapidly find a mate and begin colony foundation to avoid predation and/or desiccation (Eyer and Vargo, 2022).

Although there is no evidence that *C. formosanus* reproductives exercise kin avoidance in choosing a partner, studies show that both weight (a measure of general body size and quantity of stored body reserves) and genetic diversity affect partner formation. In naturally occurring mating aggregations in *C. formosanus*, Husseneder and Sims (2008) found that males paired with females were larger in size than those that had no partner and that females with a partner were genetically more heterozygous (indicating greater genetic diversity) than females that remained single. These results are consistent with partner selection in which males favor more genetically diverse females and females favor larger males; however, it is possible that there is competition among individuals leading to the observed patterns in partner composition. Laboratory studies of *C. gestroi* show that larger males and females are more likely to succeed in colony founding (Chouvenc, 2019), indicating that large body size is an advantage in tandem pairs in *Coptotermes* termites. Further studies are needed to see if there is active partner selection or mate competition occurring during tandem pair formation in *C. formosanus* and *C. gestroi*.

Differences among colonies may play a role in inbreeding avoidance and in the colony foundation success of tandem pairs. In a genetic analysis of swarm aggregations of *C. formosanus* in New Orleans, Louisiana, Husseneder *et al.* (2006) found that males and females were genetically differentiated, indicating that they came from different colonies. Moreover, these authors showed that male alates were more inbred than female alates, suggesting that colonies headed by pairs of primary reproductives produce mainly females, whereas extended family colonies headed by neotenic reproductives produce mostly males. Husseneder *et al.* (2006) suggested that such sex-biased investment in colonies based on age and relatively long dispersal distances of 500–1,000 m (Mullins *et al.*, 2015) may facilitate inbreeding avoidance during tandem pair formation.

The microbial load carried by alates on their cuticle from their natal colonies may be another source of colony-specific differences affecting colony foundation success. A study of the subterranean termite, *R. flavipes*, showed large variation among colonies in the microbial loads (bacteria and fungi) on the cuticles of alates and workers, and that alates from colonies with high microbial loads were less likely to succeed in colony founding whether paired with a sibling or with an alate from another colony (Eyer and Vargo, 2022). It is possible that microbial loads on the cuticle of alates may influence colony founding success in *C. formosanus*.

Husseneder *et al.* (2010) reported that workers from extended family colonies (headed by inbreeding neotenics) had higher bacterial loads on their cuticle than those from simple family colonies (headed by a single king and queen). The microbial loads in *C. formosanus* alates in relation to the colony breeding structure and its possible effects on colony founding success could be a fruitful area of future research.

A recent important development in colony founding of *Coptotermes* termites is the formation of hybrid colonies between *C. formosanus* and *C. gestroi* (see Chapter 17 of this volume). In areas of south Florida, where the two species are sympatric, they have overlapping mating flights and reproductives naturally form hybrid pairs – both males of *C. formosanus* and females of *C. gestroi* and vice versa (Chouvenc *et al.*, 2015). Surprisingly, hybrid incipient colonies grow twice as quickly in the laboratory as incipient colonies of either parental species (Chouvenc *et al.*, 2015). Established hybrid colonies have recently been found in Taiwan (Chen, 2022), and this is a major concern for the pest management industry in south Florida and other areas where the ranges of these invasive species overlap (Patel *et al.*, 2023), such as Hawai'i (Woodrow *et al.*, 2001) and Taiwan (Li *et al.*, 2009).

9.2.2 Colony development and breeding structure

Like the colonies of all social insects, the genetic structure of colonies of *Coptotermes* and other termites is determined by the number and types of reproductives in the colony and the degree of relatedness among them (Ross, 2001; Vargo and Husseneder, 2009; Vargo and Husseneder, 2011). In termites, there are generally three types of colonies as defined by their breeding system (Thorne *et al.*, 1999; Vargo, 2003b; Vargo and Husseneder, 2009). **Simple families** are headed by a monogamous pair of reproductives, usually the founding pair of primaries. **Extended families** contain multiple inbreeding reproductives descended from the initial founding pair. These reproductives are generally neotenics but can be adultoid in some higher termites, e.g., *Nasutitermes corniger* (Thorne, 1982). **Mixed families** are headed by multiple unrelated reproductives, most likely through colony fusion (DeHeer and Vargo, 2004, 2008), but possibly through pleometrosis, as occurs in some higher termites, e.g., *Macrotermes michaelensis* (Darlington, 1985).

The colony breeding structure in *C. formosanus* has been well-characterized using genetic markers. Finding the nests of *Coptotermes* species and then collecting the reproductives can be difficult and very labor-intensive. Therefore, determining colony breeding structures by censusing colonies is impractical. The use of genetic markers is a powerful way to infer the breeding structure of colonies. Microsatellite markers are generally the marker of choice for studies of social insect breeding systems, owing to their high variability, biparental inheritance, and reliability of genotyping. Microsatellites consist of short tandem repeats of DNA that are inherited from both the mother and the father. Because they mutate at a rapid rate, microsatellites generate numerous alleles that differ in their size due to variation in the number of repeat units. Microsatellite loci are amplified using PCR, and allele sizes are scored based on the size of the resulting PCR amplicon. By genotyping nestmate workers, the colony breeding structure can be inferred, and by genotyping a number of colonies within a population, the degree of inbreeding within colonies can be determined (Thorne *et al.*, 1999; Husseneder *et al.*, 2003). The specific composition of colonies within a population depends on a combination of colony age (younger colonies are simple families), the lifespan of colonies, and whether colonies fuse. Populations with a younger age structure will have simpler families, whereas populations with an older age structure will have older colonies headed by neotenics.

There have been a number of studies on the colony breeding structure of *C. formosanus* in both the native and invasive ranges (Vargo *et al.*, 2003a,b, 2006a; Husseneder *et al.*, 2005, 2007, 2008, 2012; Aluko and Husseneder, 2007; Owens *et al.*, 2012, Tseng *et al.*, 2022). These results are summarized in Husseneder *et al.* (2012) and in Fig. 9.1. In total, over 200 colonies from three native populations and ten introduced populations, including populations in Japan (n = 2), Hawai'i (n = 4), and the U.S. mainland (n = 4) have been studied. Only simple and extended families have been reported and there is large variation among populations, ranging from 90% (out of ten colonies) simple families in Fukue, Japan to 100% (of 12 colonies) extended families in Guangzhou City, China. For the most part, native populations had more extended family colonies (range = 86–100% in three populations) than introduced populations (range = 10–86% in ten populations); in fact, seven of ten introduced populations had mostly simple families (Fig. 9.1), but more native populations need to be sampled before any robust conclusions can be drawn regarding the breeding structure in China.

The level of inbreeding in extended family colonies can be used to infer the number of neotenics actively reproducing (Thorne *et al.*, 1999; Vargo, 2003b). Genetic studies suggest that extended colonies in most populations of both the native and introduced ranges generally have only a few neotenic pairs present (in the order of five or fewer). Notably, colonies in some populations show an excess of heterozygotes among nestmate workers, suggesting that these colonies may have spatially separate reproductive centers located in different nests, each with its own set of neotenics and with limited mating between reproductive centers (Husseneder *et al.*, 2012). In their study of colony breeding structure and worker size (head width and body weight), Husseneder *et al.* (2008) found that extended family colonies with higher levels of inbreeding had smaller workers, suggesting that colonies with more neotenics produce smaller workers than those with fewer neotenics. However, these authors found no differences in worker size between simple family colonies and extended family colonies, nor was there a relationship between worker size and degree of heterozygosity of

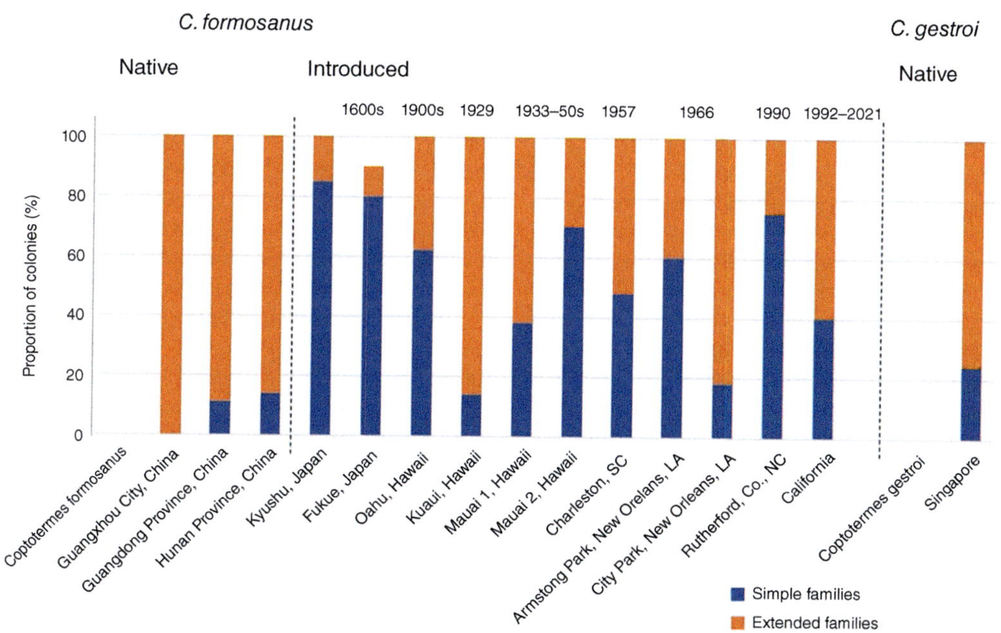

Fig. 9.1. Comparison of family structure of colonies in native and introduced populations of *C. formosanus* and *C. gestroi*. Data are based on studies by Vargo *et al.* (2003a,b, 2006a), Husseneder *et al.* (2005, 2008, 2012), Aluko and Husseneder (2007), Owens *et al.* (2012) and Tseng *et al.* (2022) for *C. formosanus*, and by Zhang and Evans (2017) for *C. gestroi*.

individuals within colonies. Clearly breeding structure can affect a number of factors, such as sex ratio of alates produced, degree of heterozygosity of individuals, including alates, microbial load of individuals, and worker body size. In the long term, inbreeding within extended family colonies can lead to weakened individual and social immunity, contributing to the demise of the colony (Eyer and Vargo, 2022). The effects of colony breeding structure, especially as they relate to levels of inbreeding, are likely to be more pervasive and certainly merit more detailed investigation.

The colony breeding structure of invasive populations of *C. formosanus* stand in contrast to those of *R. flavipes*. There are no reports of mixed family colonies in *C. formosanus*, which are common in invasive populations of *R. flavipes*, and even in some native populations (Vargo and Husseneder, 2011). Mixed family colonies are common in the introduced range of *R. flavipes*, where it forms exclusively extended family and mixed family colonies in France and Chile (Dronnet *et al.*, 2005; Perdereau *et al.*, 2010; Perdereau *et al.*, 2015), possibly due to a breakdown in colony recognition cues following a genetic bottleneck. As discussed below, the invasive population of *C. formosanus* in the U.S. did not undergo a genetic bottleneck (Eyer *et al.*, 2021). Lack of aggression among colonies may be associated with colony fusion in subterranean termites (Matsuura and Nishida, 2001; Fisher *et al.*, 2004). Aggression among *C. formosanus* colonies is variable in the native (Vargo and Husseneder, personal observation) and introduced ranges (Husseneder and Grace, 2001). Despite intercolony aggression levels sometimes being quite low (Su and Haverty, 1991; Husseneder and Grace, 2001; Cornelius and Osbrink, 2003) and population density quite high (Messenger *et al.*, 2005), colony integrity is typically maintained such that colony fusion is rare or non-existent in *C. formosanus*.

The breeding structure of *C. gestroi* colonies has received little attention. In the only genetic study of this species conducted so far, Zhang and Evans (2017) found that seven of 29 (24%) colonies in Singapore, where this species is native, were simple families, and 22 (76%) were extended families (Fig. 9.1). The extended families were only slightly inbred, suggesting that they were headed by relatively few neotenics. Clearly, more work needs to be done on *C. gestroi* to better understand its colony breeding structure in both the native and introduced ranges.

9.2.3 Population genetic structure

The genetic structure of populations can inform us about many aspects of a species' biology and history. For example, it can tell us about the dispersal abilities of a species, and whether dispersal is through natural means or through human transport, the amount of gene flow that occurs among different geographic areas, the level of genetic diversity within a population, and whether populations have undergone genetic bottlenecks in the recent past. Also, our ability to identify source locations of invasive populations depends on the existence of distinct genetic signatures of potential source populations. If there is little to no genetic differentiation among potential source populations, then it is difficult or impossible to trace the origin of an invasive population by matching its genetic signature to a source location. Much of what we know regarding the population structure of *C. formosanus* and *C. gestroi* is in the context of their invasion biology. This will be covered below under section 9.3 ("Invasion biology and global spread"). Here, the focus is on work that has been done in native populations.

We have some information about the genetic structure of *C. formosanus* populations. Husseneder *et al.* (2012) examined the population structure of three native populations (Guangzhou City, greater Guangdong Province, and Hunan Province) and introduced populations in Hawai'i, the U.S. mainland, and Japan. There was similar genetic differentiation among the native populations (mean F_{ST} = 0.09) as there was among the introduced populations in Hawai'i and the U.S. mainland (range F_{ST} = 0.04 to 0.18), showing a close connection among the native and introduced populations. Populations in Kyushu and Nagasaki, Japan, separated by some 60 km of water, showed moderate levels of differentiation (F_{ST} = 0.12) (Vargo *et al.*, 2003a).

In a more comprehensive study of native populations using single nucleotide polymorphisms, Blumenfeld *et al.* (2021) found three distinct genetic groups in China that clustered geographically: a western China cluster, an eastern China cluster, and a Hong Kong region cluster (Fig. 9.2). Interestingly, all three of these regions seemed to have contributed to the introduced populations in Hawai'i and the U.S. mainland.

Isolation by distance analysis comparing the genetic similarity of colonies in relation to distance has been conducted in a number of native and introduced populations. Strong isolation by distance indicates high population viscosity due to either short-range mating flights and/or frequent colony reproduction by budding. Weak but significant isolation by distance has only been reported in native populations in Guangdong Province, China (R^2 = 0.18, over a distance of 200 km), Hunan Province, China (R^2 = 0.19, over a distance of 150 km), and in introduced populations on Maui, Hawai'i (R^2 = 0.16, over a distance of 11 km) and Charleston, South Carolina (R^2 = 0.09), but no significant relationship in other populations in Maui, Kauai and Oahu (Hawai'i), Japan, New Orleans

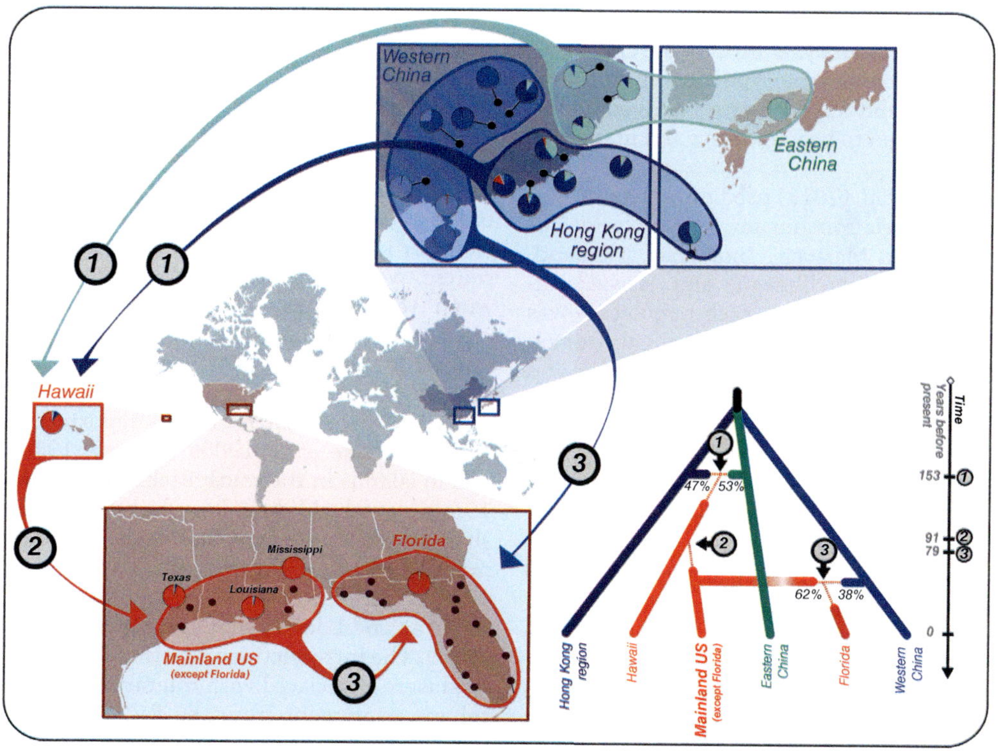

Fig. 9.2. Inferred spread of invasive populations of *C. formosanus* to Hawai'i and the U.S. mainland based on Approximate Bayesian Computation (ABC) analysis of single nucleotide polymorphism data. Native populations grouped into three genetic clusters. The lower right depicts the most likely invasion history scenario for *C. formosanus* out of the various scenarios tested through ABC. The meeting of two colored lines, such as at points 1 and 3, indicates admixture, with the percentage originating from each lineage indicated. Thin dotted lines represent bottleneck events. Data from Blumenfeld *et al.* (2021).

(Louisiana), or Rutherford County (North Carolina) (Husseneder *et al.*, 2012; Vargo *et al.*, 2006a). These results suggest only weak population viscosity in some areas, indicating that there is widespread gene flow and that colony reproduction occurs primarily by dispersal flights rather than by budding.

There are three studies of the genetic structure of *C. gestroi* in its native range. Cheng *et al.* (2014) studied two mitochondrial DNA markers on samples collected from different habitats in peninsular Malaysia. Analysis of the 16S rRNA gene yielded clusters separating into eastern, western, and southern populations, irrespective of habitat type (beach forest, mangrove swamp, plantation, urban or semi-urban). Cheng *et al.* (2014) also analyzed the cytochrome c oxidase subunit I (COI) gene and found that some haplotypes were habitat specific. Phylogenetic analysis of COI provided evidence of a basal haplotype inhabiting a wide range of habitats, with more derived haplotypes limited to beach forests and mangrove swamps.

Yeap *et al.* (2011) used microsatellites to study the relationship among populations of *C. gestroi* in Malaysia, Indonesia, Thailand, the Philippines, Taiwan, and Hawai'i, although in many cases, each population was represented by only a few individuals (for example, a single individual for Hawai'i and two individuals each for Indonesia and Thailand). Peninsular Malaysia was well sampled in this study (n = 53), but there was no evidence of genetic structure among different populations in Malaysia. These results suggested the occurrence of one large, genetically diverse population in Malaysia and Singapore, with Thailand, Taiwan, and Indonesia forming distinct genetic clusters, while the Philippines was genetically diverse. These authors found significant isolation by distance within Taiwan, across Malaysia, and across the Asian populations sampled as a group. The authors concluded that there was frequent human-mediated dispersal of *C. gestroi*. If so, their results indicate that most movement occurs on a small scale, with transport of colonies more likely to occur among areas located relatively close to each other and long-range dispersal being rare.

Using microsatellite markers, Zhang and Evans (2017) found no evidence of genetic differentiation between colonies in urban areas and those in forested areas in Singapore. These authors concluded that the *C. gestroi* colonies occurring in Singapore form one large panmictic population. They further proposed that *C. gestroi* is the first native species demonstrated to be an "urban exploiter." However, there are many native species of insects that thrive in urban environments, for example many native ants and species of *Reticulitermes*, where they are also urban pests, so the fact that *C. gestroi* can flourish in urban areas where it is native is not unique.

9.3 Invasion biology and global spread

It is widely believed that *C. formosanus* is native to southeastern China and Taiwan, based on the associated termitophilous beetles (Kistner, 1985; Liang *et al.*, 2020) and the elevated genetic diversity of *C. formosanus* in this region (Li *et al.*, 2009). Of course, we have good historical records on when this species became established in areas outside China and Taiwan. The first known introduction of *C. formosanus* outside its native range occurred prior to 1600 in Japan (Mori, 1987). It was reported in Hawai'i in the early 1900s (Swezey, 1914) and in islands in the south Pacific in the late 1950s, namely, Midway Island, and the Marshall Islands. It was recorded on the U.S. mainland in Charleston, South Carolina, in 1957 (Chambers *et al.*, 1988) and was found in Texas and Louisiana in the 1960s (Spink, 1967). The introductions into the U.S. mainland were associated with military ports receiving equipment from the Pacific theater after World War II, including Hawai'i (Spink, 1967). It now occurs throughout the Gulf Coast states in the U.S. and infestations in Georgia, Tennessee, North Carolina, Virginia, and California (Woodson *et al.*, 2001; Vargo *et al.*, 2006a; Tseng *et al.*, 2021; Hottel, 2022; Tseng *et al.*, 2022) (also see Fig. 2.3). It was also reported in South Africa in the

1960s, where it appears to have been eradicated (Evans *et al.*, 2013; Scheffrahn, 2013).

Retracing the invasion history of *C. formosanus* and *C. gestroi* has proven challenging. The ability to draw robust conclusions regarding potential source populations and routes of spread depends on several factors. First, one needs intensive sampling to ensure that genetic variation within each sampled population has been fully captured. As mentioned earlier, most studies of *C. formosanus* and *C. gestroi* have very small sample sizes, well below the levels required for accurately characterizing a population's genetic signature. For many studies, the most we can say is that the results are suggestive. Second, there needs to be sufficient variation in the genetic markers used to allow robust inferences. If there is too little variation, then there will not be enough signal to distinguish individuals and populations from each other. Also, there needs to be sufficient genetic differentiation within the native range. If the native range is one large, panmictic population, then there is no way to conclusively tie an introduced population of interest to a region within the native range. Finally, it is well established that bridgehead effects are important in the spread of many invasive species (Bertelsmeier and Keller, 2018; Bertelsmeier *et al.*, 2018), including termites (Blumenfeld and Vargo, 2020). When dealing with globally invasive species, such as *C. formosanus* and *C. gestroi*, one must be aware that introduced populations are often the source of other introduced populations, so even if one can link an introduced population to a possible source location in the native range it may be several introduction events removed. This is where historical records can help. Linking the time of introduction to trade routes and historic events can provide important connections among regions. In this regard, the relatively good records for the first appearance of *C. formosanus* and *C. gestroi* in new areas can help confirm conclusions drawn from genetic data.

Several genetic studies have investigated the relationships among native and introduced populations of *C. formosanus*. Analysis of mitochondrial DNA sequences show surprisingly low levels of genetic variation, even in native populations (Broughton and Grace, 1994; Austin *et al.*, 2006; Fang *et al.*, 2008; Li *et al.*, 2009). For example, based on sequencing a 667 bp region containing parts of the COI gene, the tRNA-Leu, and a section of the COII gene, Austin *et al.* (2006) found only two mtDNA haplotypes in 23 samples from China, Japan, and the U.S. mainland. Similarly, Li *et al.* (2009) analyzed 1406 base pairs from parts of the COII gene, 16S rRNA, and 12S rRNA, in samples from China, Taiwan, Japan, Hawai'i, and the U.S. mainland, and found five haplotypes with a maximum difference of only 10–11 base pairs (0.7–0.8% divergence). Jenkins *et al.* (2002) reported eight mtDNA lineages in their analysis of the COII gene in samples from China, Hawai'i, and the U.S. mainland, but Austin *et al.* (2006) concluded that many of the haplotypes identified by these authors were likely scoring errors.

Based on their findings of two mtDNA haplotypes in Hawai'i and the U.S. mainland, Austin *et al.* (2006) concluded that there were at least two introductions to the continental U.S. and Hawai'i. However, given the limited genetic variation in the mitochondrial genome, as described above, mtDNA markers are of restricted utility in identifying potential source populations. Using nuclear microsatellite markers, Husseneder *et al.* (2012) investigated the genetic relationships among three native populations in China, two introduced populations in Japan, four introduced populations in Hawai'i, and four introduced populations in the U.S. mainland. Genetic similarities, as determined by pairwise F_{ST}s, suggested that the introduced populations in Hawai'i originated from populations in Guangdong Province, and then the Hawai'ian populations served as a source for the introductions into the U.S. mainland. However, these authors urged caution in this conclusion, due to the limited sampling of native Chinese populations.

More recently, Blumenfeld *et al.* (2021) conducted an extensive study of 359 samples from the native and introduced ranges using single nucleotide polymorphisms (SNPs) generated by double digest restriction assisted DNA sequencing (ddRADseq). The advantage of this approach is that it assays markers throughout the nuclear genome, providing a powerful means to detect genetic variation (Elshire *et al.*, 2011; Peterson *et al.*, 2012). Using some 22,000 SNPs and a statistical inference approach based on approximate Bayesian computation (Beaumont *et al.*, 2002), the authors inferred a complex invasion history into the U.S. as depicted in Fig. 9.2. The evidence suggests that there were initially two separate introductions into Hawai'i, one from an eastern China population and one from the Hong Kong region. Once in Hawai'i, termites from the two introductions interbred, leading to admixture. The admixed Hawai'ian population then served as a bridgehead for a single introduction into the southeastern U.S. There appears to have been a separate introduction from western China into Florida which then interbred with the existing introduced population from Hawai'i. The estimated dates for the introduction events comport well with historical records (Fig. 9.2).

A notable finding in the study by Blumenfeld *et al.* (2021) was the overall high genetic diversity in the introduced populations compared with native populations. Their genome-wide SNP data showed that introduced populations, analyzed either separately or as a group, were as genetically diverse as the native populations. These results stand in contrast to a previous study using eight microsatellites that reported some 46% lower genetic diversity in introduced populations (Husseneder *et al.*, 2012) than native populations. The difference between the two studies is most likely due to the much larger number of loci with genome wide coverage (22,000 versus eight) in the former study. The relatively high genetic diversity in the introduced populations is attributed to multiple introductions from different source populations following by admixture in the introduced populations in the U.S. Consistent with this hypothesis, the level of genetic diversity in the introduced population in Japan, in which there is no evidence of admixture, was lowest.

The origin of the recent colonies discovered in southern California (Tseng *et al.*, 2021; Tseng *et al.*, 2022) remains to be determined. There are two mtDNA haplotypes among the five colonies discovered, indicating that there were at least two introductions to the area (Tseng *et al.*, 2022). Both haplotypes are present in the southeastern U.S. and Hawai'i, so it is not possible to exclude either of these locations, *or* the native range in China, as potential sources of these infestations.

Coptotermes gestroi is native to Southeast Asia. It is one of the most widespread invasive termites in the world (Evans *et al.*, 2013). It has become established in Mauritius and Réunion Island to the west and in Taiwan to the north. It is also found in Guam, Hawai'i, Mexico, Brazil, Paraguay, throughout the West Indian Islands in the Caribbean, Florida, and limited occurrences in Cleveland, Ohio, and Australia (Su and Scheffrahn, 1998; Jenkins *et al.*, 2007; Evans *et al.*, 2013; Scheffrahn *et al.*, 2015). Studies of the invasion biology of *C. gestroi* have so far been limited to mtDNA analysis. Similar to *C. formosanus*, there is limited mtDNA diversity in *C. gestroi*, which has hindered our ability to retrace its introduction history. Jenkins *et al.* (2007) compared samples collected in Ohio, U.S., Puerto Rico, Florida, and Brisbane, Australia to samples from two native areas (Singapore and Malaysia) using sequence data from two mitochondrial genes (16S rRNA and COII) and part of the nuclear internal transcribed spacer (ITS) region. These authors found that the sequences of the sole sample from Ohio were most closely related to samples from Singapore, whereas the samples from Florida, Australia, and Puerto Rico were most closely related to samples from Malaysia. However, more extensive studies of mitochondrial DNA variation show limited genetic differentiation between Singapore and Malaysia (Cheng *et al.*, 2014), indicating that the ability to distinguish between Singapore and Malaysia as potential sources is limited. In addition, because *C. gestroi* is so widespread, the possibility of secondary introductions cannot be ruled out, so it is impossible to say if the samples came directly from native populations or were the result of secondary introductions from bridgeheads in other areas where it is invasive.

A study by Martins *et al.* (2010) analyzed 23 samples from Brazil plus an additional 12 samples from native and introduced populations from around the world. In all the samples they studied, they found variation at only two sites out of a total of 706 base pairs of the COII gene, resulting in four haplotypes that varied at most at two positions. These authors concluded that the populations in Brazil may have originated from Malaysia and that the populations in Cleveland, Ohio, and Miami, Florida, were the result of separate introductions, possibly from Malaysia. Like the other studies mentioned above, the sample sizes are very small and geographically limited, so any conclusions regarding source populations from this study should be treated as tentative.

Li *et al.* (2012) examined COII and 16S mtDNA sequences from a limited number of samples from Hainan Island, China, and Taiwan, which they compared with a small number of samples from other Asian locations that had been studied previously. These authors concluded that *C. gestroi* on Hainan Island and Taiwan had been introduced from the Philippines. However, these conclusions are based on only two samples from Hainan Island and three samples from the Philippines, the latter samples all coming from a single location (Yeap *et al.*, 2007). Clearly more detailed sampling is required before firm conclusions can be drawn.

The most thorough genetic study of *C. gestroi* to date was conducted by Cheng *et al.* (2014), who sequenced 50 samples from Peninsular Malaysia at two mitochondrial genes, the 16S rRNA and COI genes. In their analysis, they also included a number of samples available in Genbank from around the world. Studies of the 16S rRNA gene showed ten haplotypes, with five haplotypes in Malaysia, four of which also occurred in Singapore (n = 7 samples), two occurred in Indonesia (n = 2 samples), and one in Thailand (n = 4 samples). The most common haplotype among all these countries is also present in invasive colonies collected from Puerto Rico, Miami, Florida (n = 1 sample for each location), Key West, Florida (n = 2 samples), and from colonies on yachts in Catania, Italy (Ghesini *et al.*, 2011), and Brisbane, Australia, so the origin of these introduced colonies is still uncertain. A common haplotype shared by Malaysia, Singapore, and Indonesia was found in Cleveland, Ohio (n = 1 sample), so again pinpointing the origin of the introduced colony is not possible. In contrast, Taiwan (n = 7 samples) had only one haplotype, which was unique, and the Philippines (n = 9 samples) had two haplotypes, neither of which were found anywhere else in the native range. Notably, the only haplotype reported in Hawai'i (n = 4 samples) is shared in common with the Philippines, suggesting a Philippine origin for the Hawai'ian population. Samples from Antigua and Barbuda and from Grand Turk, Turks and Caicos Islands in the Caribbean, each had unique haplotypes, with only one and two base pairs that were different from the most common haplotype in Malaysia and Thailand. While the above results are suggestive of potential sources of origin, at least in some cases, it should be noted that sample sizes are small in both the native and introduced ranges, and differences between the 16S rRNA haplotypes were minimal, with only a single base pair separating many of them. Thus, these inferences should be taken with caution. These same authors also analyzed the COI gene on the 50 samples they collected from Malaysia. Despite much more variation in this gene, yielding 14 haplotypes, there were very few sequences available for samples outside of Malaysia, and all of them from China, so inferences regarding invasion history with this gene were not possible.

The low mtDNA sequence variation in both *C. formosanus* and *C. gestroi* stands in sharp contrast to studies of the eastern subterranean termite, *R. flavipes*. For example, Vargo and Carlson (2006) found that of 16 sympatric colonies studied from a single site, there were 12 haplotypes, with a mean sequence divergence of 1.5%. In the study by Cheng *et al.* (2014) mentioned above, out of 50 samples in Malaysia, *C. gestroi* showed only 0.3% sequence divergence on average in the COI gene. Similarly in their study of *C. formosanus*, Li *et al.* (2009) found seven haplotypes out of eight samples from Taiwan (n = 3), China (n = 1), Japan (n = 2), Hawai'i (n = 1), and Florida (n = 1), and there was only 0.3% nucleotide divergence out of

1406 positions that included the COII, 12S rRNA, and 16S rRNA gene sequences. Thus, for the data we currently have, it seems as though mtDNA sequence variation is an order of magnitude lower in *C. formosanus* and *C. gestroi* than in *R. flavipes*. In contrast to the apparent low variability in the mitochondrial genome, variation at nuclear microsatellite markers is more substantial, with about seven alleles per locus in native populations of *C. formosanus* ($n = 8$ loci; Husseneder *et al.*, 2012) and ten alleles per locus in Malaysian populations of *C. gestroi* ($n = 10$ loci; Yeap *et al.*, 2011). These levels of variability are close to the reported 6.0–12.6 alleles per locus for native populations of *R. flavipes* ($n = 8$ loci) (Vargo, 2003a,b; Vargo *et al.*, 2006b, 2013). The reasons for the low mitochondrial DNA diversity in *C. formosanus* and *C. gestroi* are unknown but could be due to selective sweeps, possibly related to *Wolbachia* or other maternally inherited bacteria. *Wolbachia* has been reported in some species of *Coptotermes* (Lo and Evans, 2007; Salunke *et al.*, 2010), although there are no records of it being present in *C. formosanus* or *C. gestroi*. Whatever the reason, the low mitochondrial DNA variation in these species is striking and has hindered our ability to trace the spread of invasive populations. Greater emphasis should be placed on using genome-wide nuclear markers such as SNPs for studies on routes of invasion.

9.4 Conclusions and future directions

Our knowledge of the dispersal, colony breeding structure, population structure, and routes of invasion of *C. formosanus* and *C. gestroi* has greatly improved in the past couple of decades, largely due to the application of genetic markers. There have been a number of studies on tandem pair formation in *C. formosanus* showing that individuals that have larger body size and greater genetic diversity are more successful at forming tandem pairs. Future studies should focus on whether there is active partner selection, and the costs and benefits of choosing a partner. We know very little about alate dispersal and tandem pair formation in *C. gestroi*. Colony breeding structure of both *C. formosanus* and *C. gestroi* is typical of most subterranean termites, with younger colonies forming simple families and older colonies forming extended families. There is no evidence of mixed family colonies, suggesting that colony fusion in these species is rare or non-existent. There is large variation in the proportions of simple families and extended families among populations of *C. formosanus* (only a single population of *C. gestroi* has been studied), presumably due to differences in the demographic structure of populations. Understanding the factors underlying the different age structures of populations will be an important area for future research, especially understanding how long primary reproductives can live and the costs of inbreeding once neotenics begin reproducing. Long-term studies of colonies, such as that conducted by Chouvenc *et al.* (2022), combined with genetic data to determine changes in colony breeding structure will be particularly insightful (see Chapter 4 of this volume).

Both *C. formosanus* and *C. gestroi* show limited mtDNA sequence variation. Thus, use of mtDNA markers by themselves to study population structure and invasion routes is of limited utility, and the conclusions based on these markers should be viewed with an abundance of caution. Much more detailed information can be gleaned from genome-wide nuclear markers, such as SNPs. Studies to date show genetic structure in the native ranges of both species. There is likely sufficient signal in the nuclear genome to trace the invasion history of these species, which, at least in some cases, may be complex, involving bridgehead effects and multiple introductions with admixture (Blumenfeld *et al.*, 2021).

Acknowledgments

Writing of this chapter was supported by the Texas A&M Urban Entomology Endowment and AgriLife Research.

References

Aluko, G.A. and Husseneder, C. (2007) Colony dynamics of the Formosan subterranean termite in a frequently disturbed urban landscape. *Journal of Economic Entomology* 100, 1037–1046.

Austin, J.W., Szalanski, A.L., Scheffrahn, R.H., Messenger, M.T., McKern, J.A. *et al.* (2006) Genetic evidence for two introductions of the Formosan subterranean termite, *Coptotermes formosanus* (Isoptera: Rhinotermitidae), to the United States. *Florida Entomologist* 89, 183–193.

Beaumont, M.A., Zhang, W.Y. and Balding, D.J. (2002) Approximate Bayesian Computation in population genetics. *Genetics* 162, 2025–2035.

Bertelsmeier, C. and Keller, L. (2018) Bridgehead effects and role of adaptive evolution in invasive populations. *Trends in Ecology & Evolution* 33, 527–534.

Bertelsmeier, C., Ollier, S., Liebhold, A.M., Brockerhoff, E.G., Ward, D. *et al.* (2018) Recurrent bridgehead effects accelerate global alien ant spread. *Proceedings of the National Academy of Sciences of the United States of America* 115, 5486–5491.

Blumenfeld, A.J. and Vargo, E.L. (2020) Geography, opportunity and bridgeheads facilitate termite invasions to the United States. *Biological Invasions* 22, 3269–3282.

Blumenfeld, A.J., Eyer, P.-A., Husseneder, C., Mo, J., Johnson, L.N.L. *et al.* (2021) Bridgehead effect and multiple introductions shape the global invasion history of a termite. *Communications Biology* 4, 196.

Broughton, R.E. and Grace, J.K. (1994) Lack of mitochondrial-DNA variation in an introduced population of the Formosan subterranean termite (Isoptera, Rhinotermitidae). *Sociobiology* 24, 121–126.

Chambers, D.M., Zungoli, P.A. and Hill, H.S. Jr (1988) Distribution and habitats of the Formosan subterranean termite (Isoptera: Rhinotermitidae) in South Carolina. *Journal of Economic Entomology* 81, 1611–1619.

Chen, G.Y. (2022) Hybridization between Formosan subterranean termite and Asian subterranean termite in Taiwan. MSc dissertation, National Chung Hsing University, Taichung, Taiwan.

Cheng, S., Thinagaran, D., Mohanna, S.Z.M. and Noh, N.A.M. (2014) Haplotype-habitat associations of *Coptotermes gestroi* (Termitoidae: Rhinotermitidae) from mitochondrial DNA genes. *Environmental Entomology* 43, 1105–1116.

Chouvenc, T. (2019) The relative importance of queen and king initial weights in termite colony foundation success. *Insectes Sociaux* 66, 177–184.

Chouvenc, T., Ban, P.M. and Su, N.-Y. (2022) Life and death of termite colonies, a decades-long age demography perspective. *Frontiers in Ecology and Evolution* 10, 911042.

Chouvenc, T., Su, N.Y. and Helmick, E.E. (2015) Hybridization of two major termite invaders as a consequence of human activity. *PLoS ONE* 10, e0120745.

Cornelius, M.L. and Osbrink, W.L.A. (2003) Agonistic interactions between colonies of the Formosan subterranean termite (Isoptera: Rhinotermitidae) in New Orleans, Louisiana. *Environmental Entomology* 32, 1002–1009.

Cristescu, M.E. (2015) Genetic reconstructions of invasion history. *Molecular Ecology* 24, 2212–2225.

Darlington, J.P.E.C. (1985) Multiple primary reproductives in the termite *Macrotermes michaleseni* (Sjöstedt). In: Watson, J.A.L., Okot-Kotber, B.M. and Noirot, C. (eds) *Caste Differentiation in Social Insects*. Pergamon Press, Oxford, UK, pp. 187–200.

DeHeer, C.J. and Vargo, E.L. (2004) Colony genetic organization and colony fusion in the termite *Reticulitermes flavipes* as revealed by foraging patterns over time and space. *Molecular Ecology* 13, 431–441.

DeHeer, C.J. and Vargo, E.L. (2006) An indirect test of inbreeding depression in the termites *Reticulitermes flavipes* and *Reticulitermes virginicus*. *Behavioral Ecology and Sociobiology* 59, 753–761.

DeHeer, C.J. and Vargo, E.L. (2008) Strong mitochondrial DNA similarity but low relatedness at microsatellite loci among families within fused colonies of the termite *Reticulitermes flavipes*. *Insectes Sociaux* 55, 190–199.

Dronnet, S., Chapuisat, M., Vargo, E.L., Louhou, C. and Bagnères, A.-G. (2005) Genetic analysis of the breeding system of an invasive subterranean termite, *Reticulitermes santonensis*, in urban and natural habitats. *Molecular Ecology* 14, 1311–1320.

Elshire, R.J., Glaubitz, J.C., Sun, Q., Poland, J.A., Kawamoto, K. *et al.* (2011) A robust, simple genotyping-by-sequencing (GBS) approach for high diversity species. *PLoS ONE* 6, e19379.

Evans, T.A., Forschler, B.T. and Grace, J.K. (2013) Biology of invasive termites: A worldwide review. *Annual Review of Entomology* 58, 45--474.

Eyer, P.-A. and Vargo, E.L. (2021) Breeding structure and invasiveness in social insects. *Current Opinion in Insect Science* 46, 24–30.

Eyer, P.A. and Vargo, E.L. (2022) Short and long-term costs of inbreeding in the lifelong-partnership in a termite. *Communications Biology* 5, 389.

Eyer, P.-A., Blumenfeld, A.J., Johnson, L.N.L., Perdereau, E., Shults, P. *et al.* (2021) Extensive human-mediated jump dispersal within and across the native and introduced ranges of the invasive termite *Reticulitermes flavipes*. *Molecular Ecology* 30, 3948–3964.

Fang, R., Huang, L. and Zhong, J.H. (2008) Surprising low levels of genetic diversity of Formosan subterranean termites in south China as revealed by the COII gene (Isoptera: Rhinotermitidae). *Sociobiology* 51, 1–20.

Fei, H.X. and Henderson, G. (2003) Comparative study of incipient colony development in the Formosan subterranean termite, *Coptotermes formosanus* Shiraki (Isoptera, Rhinotermitidae). *Insectes Sociaux* 50, 226–233.

Fisher, M.L., Gold, R.E., Vargo, E.L. and Cognato, A.I. (2004) Behavioral and genetic analysis of colony fusion in *Reticulitermes flavipes* (Isoptera: Rhinotermitidae). *Sociobiology* 44, 565–576.

Ghesini, S., Puglia, G., and Marini, M. (2011). First report of *Coptotermes gestroi* in Italy and Europe. *Bulletin of Insectology* 64, 53–54.

Hottel, B. (2022) *Formosan termites expand northward into Virginia*. Pest Management Professional, Cleveland, Ohio. Available at: https://www.mypmp.net/2022/10/20/formosan-termites-expand-northward-into-virginia/ (accessed 6 September 2023)

Husseneder, C. and Grace, J.K. (2001) Evaluation of DNA fingerprinting, aggression tests and morphometry as tools for colony delineation of the Formosan subterranean termite. *Journal of Insect Behavior* 14, 173–186.

Husseneder, C. and Simms, D.M. (2008) Size and heterozygosity influence partner selection in the Formosan subterranean termite. *Behavioral Ecology* 19, 764–773.

Husseneder, C., Vargo, E.L. and Grace, J.K. (2003) Molecular genetic methods: New approaches to termite biology. In: Goodell, B., Schultz, T.P. and Nicholas, D.D. (eds) *Wood Deterioration and Preservation: Advances in our Changing World*. American Chemical Society Symposium Series 845. American Chemical Society, Washington, DC, pp. 358–370.

Husseneder, C., Messenger, M.T., Su, N.-Y., Grace, J.K. and Vargo, E.L. (2005) Colony social organization and population genetic structure of an introduced population of Formosan subterranean termite from New Orleans, Louisiana. *Journal of Economic Entomology* 98, 1421–1434.

Husseneder, C., Simms, D.M. and Ring, D.R. (2006) Genetic diversity and genotypic differentiation between the sexes in swarm aggregations decrease inbreeding in the Formosan subterranean termite. *Insectes Sociaux* 53, 212–219.

Husseneder, C., Simms, D.M. and Riegel, C. (2007) Evaluation of treatment success and patterns of reinfestation of the Formosan subterranean termite (Isoptera: Rhinotermitidae). *Journal of Economic Entomology* 100, 1370–1380.

Husseneder, C., Powell, J.E., Grace, J.K., Vargo, E.L. and Matsuura, K. (2008) Worker size in the Formosan subterranean termite in relation to colony breeding structure as inferred from molecular markers. *Environmental Entomolology* 37, 400–408.

Husseneder, C., Simms, D.M., Aluko, G.K. and Delatte, J. (2010) Colony breeding system influences cuticular bacterial load of Formosan subterranean termite (Isoptera: Rhinotermitidae) workers. *Environmental Entomology* 39, 1715–1723.

Husseneder, C., Simms, D., Delatte, J., Wang, C., Grace, J.K. *et al.* (2012) Genetic diversity and colony breeding structure in native and introduced ranges of the Formosan subterranean termite, *Coptotermes formosanus*. *Biological Invasions* 14, 419–437.

Jenkins, T.M., Dean, R.E. and Forschler, B.T. (2002) DNA technology, interstate commerce, and the likely origin of Formosan subterranean termite (Isoptera: Rhinotermitidae) infestation in Atlanta, Georgia. *Journal of Economic Entomology* 95, 381–389.

Jenkins, T.M., Jones, S.C., Lee, C.Y., Forschler, B.T., Chen, Z.B. *et al.* (2007) Phylogeography illuminates maternal origins of exotic *Coptotermes gestroi* (Isoptera: Rhinotermitidae). *Molecular Phylogenetics and Evolution* 42, 612–621.

Kistner, D.H. (1985) A new genus and species of termitophilous Aleocharinae from mainland China associated with *Coptotermes formosanus* and its zoogeographic significance (Coleoptera: Staphylinidae). *Sociobiology* 10, 93–104.

Li, H.F., Ye, W.M., Su, N.Y. and Kanzaki, N. (2009) Phylogeography of *Coptotermes gestroi* and *Coptotermes formosanus* (Isoptera: Rhinotermitidae) in Taiwan. *Annals of the Entomological Society of America* 102, 684–693.

Li, Z., Liu, B., Zeng, W. and Zhong, J. (2012) Genetic variation in *Coptotermes gestroi* (Isoptera: Rhinotermitidae) populations from Hainan, Taiwan, and the Philippines. *Journal of Agricultural and Urban Entomology* 28, 57–65.

Liang, W.-R., Maruyama, M., Kanao, T., Iwata, R. and Li, H.-F. (2020) Discovery of termitophilous rove beetles associated with Formosan subterranean termite *Coptotermes formosanus* in Taiwan, with the first larval description for the tribe Termitohospitini (Coleoptera: Staphylinidae). *Acta Entomologica Musei Nationalis Pragae* 60, 77–87.

Lo, N. and Evans, T.A. (2007) Phylogenetic diversity of the intracellular symbiont *Wolbachia* in termites. *Molecular Phylogenetics and Evolution* 44, 461–466.

Martins, C., Fontes, L.R., Bueno, O.C. and Martins, V.G. (2010) *Coptotermes gestroi* (Isoptera: Rhinotermitidae) in Brazil: Possible origins inferred by mitochondrial cytochrome oxidase II gene sequences. *Genome* 53, 651–657.

Matsuura, K. and Nishida, T. (2001) Colony fusion in a termite: what makes the society "open"? *Insectes Sociaux* 48, 378–383.

Messenger, M.T., Su, N.-Y., Husseneder, C. and Grace, J.K. (2005) Elimination and reinvasion studies with *Coptotermes formosanus* (Isoptera: Rhinotermitidae) in Louisiana. *Journal of Economic Entomology* 98, 916–929.

Mori, H. (1987) The Formosan subterranean termite in Japan: Its distribution, damage, and current and potential control measures. In: Tamashiro, M. and Su, N.-Y. (eds) *Biology and Control of the Formosan Subterranean Termite.* Research Extension Series 083. University of Hawaii, Honolulu, Hawaii, pp. 23–26.

Mullins, A.J., Messenger, M.T., Hochmair, H.H., Tonini, F., Su, N.Y. *et al.* (2015) Dispersal flights of the Formosan subterranean termite (Isoptera: Rhinotermitidae). *Journal of Economic Entomology* 108, 707–719.

Nutting, W.L. (1969) Flight and colony foundation. In: Krishna, K. and Weesner, F.M. (eds) *Biology of Termites,* Vol. 1. Academic Press, New York, pp. 233–282.

Owens, C.B., Su, N.Y., Husseneder, C., Riegel, C. and Brown, K.S. (2012) Molecular genetic evidence of Formosan subterranean termite (Isoptera: Rhinotermitidae) colony survivorship after prolonged inundation. *Journal of Economic Entomology* 105, 518–522.

Patel, J.S., Lee, S.B., Chouvenc, T. and Su, N.-Y. (2023). Equivalent colony growth of hybrids of two invasive *Coptotermes* species can threaten urban areas. *Journal of Economic Entomology* 116, 538–545.

Perdereau, E., Bagnères, A.G., Dupont, S. and Dedeine, F. (2010) High occurrence of colony fusion in a European population of the American termite *Reticulitermes flavipes*. *Insectes Sociaux* 57, 393–402.

Perdereau, E., Bagnères, A.G., Vargo, E.L., Baudouin, G., Xu, Y. *et al.* (2015) Relationship between invasion success and colony breeding structure in a subterranean termite. *Molecular Ecology* 24, 2125–2142.

Peterson, B.K., Weber, J.N., Kay, E.H., Fisher, H.S. and Hoekstra, H.E. (2012) Double digest RADseq: An inexpensive method for *de novo* SNP discovery and genotyping in model and non-model species. *PLoS ONE* 7, e37135.

Rosengaus, R.B. and Traniello, J.F.A. (1991) Biparental care in incipient colonies of the dampwood termite *Zootermopsis angusticollis* Hagen (Isoptera: Termopsidae). *Journal of Insect Behavior* 4, 633–647.

Ross, K.G. (2001) Molecular ecology of social behaviour: Analyses of breeding systems and genetic structure. *Molecular Ecology* 10, 265–284.

Salunke, B.K., Salunkhe, R.C., Dhotre, D.P., Khandagale, A.B., Walujkar, S.A. *et al.* (2010) Diversity of *Wolbachia* in *Odontotermes* spp. (Termitidae) and *Coptotermes heimi* (Rhinotermitidae) using the multigene approach. *FEMS Microbiology Letters* 307(1), 55–64.

Scheffrahn, R.H. (2013) Overview and current status of non-native termites (Isoptera) in Florida. *Florida Entomologist* 96, 781–788.

Scheffrahn, R.H. and Su, N.-Y. (2005) Distribution of the termite genus *Coptotermes* (Isoptera: Rhinotermitidae) in Florida. *Florida Entomologist* 88, 201–203.

Scheffrahn, R.H., Carrijo, T.F., Křeček, J., Su, N.-Y., Szalanski, A.L. *et al.* (2015) A single endemic and three exotic species of the termite genus *Coptotermes* (Isoptera, Rhinotermitidae) in the New World. *Arthropod Systematics & Phylogeny* 73, 333–348.

Scheffrahn, R.H., Bahder, B.W. and Lu, T. (2020) *Coptotermes formosanus* (Blattodea: Rhinotermitidae) established in Israel and world distribution of a major termite pest. *Check List* 16, 1537–1543.

Shellman-Reeve, J.S. (1999) Courting strategies and conflicts in a monogamous, biparental termite. *Proceedings of the Royal Society of London. Series B: Biological Sciences* 266, 137–144.

Shellman-Reeve, J.S. (2001) Genetic relatedness and partner preference in a monogamous wood-dwelling termite. *Animal Behaviour* 61, 869–876.

Spink, W.T. (1967) The Formosan subterranean termite in Louisiana. *Louisiana State University Circular* 89, 1–12.

Su, N.-Y. and Haverty, M.I. (1991). Agonistic behavior among colonies of the Formosan subterranean termite, *Coptotermes formosanus* Shiraki (Isoptera: Rhinotermitidae), from Florida and Hawaii: Lack of correlation with cuticular hydrocarbon composition. *Journal of Insect Behavior* 4, 115–128.

Su, N.-Y. and Scheffrahn, R.H. (1998) *Coptotermes vastator* Light (Isoptera: Rhinotermitidae) in Guam. *Proceedings of the Hawaiian Entomological Society* 33, 13–18.

Su, N.-Y., Scheffrahn, R.H. and Weissling, T. (1997) A new introduction of a subterranean termite, *Coptotermes havilandi* Holmgren (Isoptera: Rhinotermitidae) in Miami, Florida. *Florida Entomologist* 80, 408–411.

Swezey, O.H. (1914) Notes and exhbitions. *Proceeding of the Hawaiian Entomological Society* 3, 27.

Thorne, B.L. (1982) Polygyny in termites: Multiple primary queens in colonies of *Nasutitermes corniger* (Motschuls) (Isoptera: Termitidae). *Insectes Sociaux* 29, 102–117.

Thorne, B.L., Traniello, J.F.A., Adams, E.S. and Bulmer, M. (1999) Reproductive dynamics and colony structure of subterranean termites of the genus *Reticulitermes* (Isoptera Rhinotermitidae): a review of the evidence from behavioral, ecological and genetic studies. *Ethology, Ecology and Evolution* 11, 149–169.

Tseng, S.-P., Boone, J., Boone, L., King, N., Taravati, S. *et al.* (2021) Genetic analysis of Formosan subterranean termite (Blattodea: Rhinotermitidae) populations in California. *Journal of Economic Entomology* 114, 1264–1269.

Tseng, S.P., Taravati, S., Choe, D.H., Rust, M.K. and Lee, C.Y. (2022) Genetic evidence for multiple invasions of *Coptotermes formosanus* (Blattodea: Rhinotermitidae) in California. *Journal of Economic Entomology* 115, 1251–1256.

Vargo, E.L. (2003a) Genetic structure of *Reticulitermes flavipes* and *R. virginicus* (Isoptera: Rhinotermitidae) colonies in an urban habitat and tracking of colonies following treatment with hexaflumuron bait. *Environmental Entomology* 32, 1271–1282.

Vargo, E.L. (2003b) Hierarchical analysis of colony and population genetic structure in the eastern subterranean termite, *Reticulitermes flavipes*, using two classes of molecular markers. *Evolution* 57, 2805–2818.

Vargo, E.L. (2019) Diversity of termite breeding systems. *Insects* 10, 52.

Vargo, E.L. and Carlson, J.C. (2006) Comparative study of breeding systems of sympatric subterranean termites (*Reticulitermes flavipes* and *R. hageni*) in central North Carolina using two classes of molecular genetic markers. *Environmental Entomology* 35, 173–187.

Vargo, E.L. and Husseneder, C. (2009) Biology of subterranean termites: Insights from molecular studies of *Reticulitermes* and *Coptotermes*. *Annual Review of Entomology* 54, 379–403.

Vargo, E.L. and Husseneder, C. (2011) Genetic structure of termite colonies and populations. In Bignell, D.E., Roison, Y. and Lo, N. (eds) *Biology of Termites: A Modern Synthesis*. Springer Science+Business Media, Dordrecht, Heidelberg, London, New York, pp. 321–347.

Vargo, E.L., Husseneder, C. and Grace, J.K. (2003a) Colony and population genetic structure of the Formosan subterranean termite, *Coptotermes formosanus*, in Japan. *Molecular Ecology* 12, 2599–2608.

Vargo, E.L., Husseneder, C., Grace, J.K., Henderson, G. and Ring, D. (2003b) Colony and population genetic structure of Formosan subterranean termites from Hawaii and Louisiana. *Sociobiology* 41, 67–69.

Vargo, E.L., Husseneder, C., Woodson, D., Waldvogel, M.G. and Grace, J.K. (2006a) Genetic analysis of colony and population genetic structure of three introduced populations of the Formosan subterranean termite (Isoptera: Rhinotermitidae) in the continental United States. *Environmental Entomology* 35, 151–166.

Vargo, E.L., Juba, T.R. and DeHeer, C.J. (2006b) Relative abundance and comparative breeding structure of subterranean termite colonies (*Reticulitermes flavipes*, *R. hageni*, *R. virginicus*, and *Coptotermes formosanus*) in a South Carolina lowcountry site as revealed by molecular markers. *Annals of the Entomological Society of America* 99, 1101–1109.

Vargo, E.L., Leniaud, L., Swoboda, L.E., Diamond, S.E., Weiser, M.D. *et al.* (2013) Clinal variation in colony breeding structure and level of inbreeding in the subterranean termites *Reticulitermes flavipes* and *R.grassei*. *Molecular Ecology* 22, 1447–1462.

Woodrow, R.J., Grace, J.K. and Higa, S.Y. (2001) Occurrence of *Coptotermes vastator* (Isoptera: Rhinotermitidae) on the island of Oahu, Hawaii. *Sociobiology* 38, 667–673.

Woodson, W.D., Wiltz, B.A. and Lax, A.R. (2001) Current distribution of the Formosan subterranean termite (Isoptera: Rhinotermitidae) in the United States. *Sociobiology* 37, 661–671.

Yeap, B.-K., Othman, A.S., Lee, V.S. and Lee, C.-Y. (2007) Genetic relationship between *Coptotermes gestroi* and *Coptotermes vastator* (Isoptera: Rhinotermitidae). *Journal of Economic Entomology* 100, 467–474.

Yeap, B.-K., Othman, A.S. and Lee, C.-Y. (2011) Genetic analysis of population structure of *Coptotermes gestroi* (Isoptera: Rhinotermitidae) in native and introduced populations. *Environmental Entomology* 40, 470–476.

Zhang, M.P. and Evans, T.A. (2017) Determining urban exploiter status of a termite using genetic analysis. *Urban Ecosystems* 20, 535–545.

10 Inspection and monitoring

Faith M. Oi
Entomology and Nematology Department, University of Florida, foi@ufl.edu

Abstract
Inspection and monitoring are related to surveillance and detection. They are distinct processes and are multi-layered. Surveillance and monitoring should be tied to early-response, rapid detection protocols designed to prevent the spread of the invasive Formosan and Asian subterranean termites. Inspection and monitoring are critical components in devising treatment strategies and evaluating treatment effectiveness. The key to effective inspection and monitoring is a well-trained inspector who is called to do inspections for different purposes, on widely varying construction types, and often under challenging conditions. Inspectors must have a firm understanding of termite biology, behavior, and identification. During an inspection, they must know what to look for, including evidence, damage, live termites, and identifying conducive conditions that can support termite populations. This chapter covers these aspects of inspections and monitoring as well as how inspections may vary due to changes in the International Building Code and building construction materials.

10.1 Introduction

Inspection and monitoring are two distinct processes. Inspection is defined as "a careful and critical examination" (Merriam-Webster, 2023). Monitoring and surveillance are sometimes used interchangeably in pest management. They are related, but in this chapter, surveillance is defined as the continual process to identify undetected cases of invasive termites, while monitoring is an ongoing evaluation designed to provide feedback after the problem is identified. Inspection, surveillance, and monitoring should be systematic and tied to correct identification, early detection, and rapid local and areawide response protocols. Regrettably, early detection and early response are not common for termites. Inspection, surveillance, and monitoring are multi-layered and critical to slowing the spread of invasive species such as the Formosan subterranean termite.

The keys to thorough inspections and effective monitoring for subterranean termites are well-trained inspectors and an informed public. Subterranean termites infest all manner of structures in the built environment which is defined as "man-made or modified structures that provide people with living, working, and recreational spaces" (USEPA, 2023). Inspectors can increase their inspection effectiveness by synthesizing their working knowledge of termite biology

DOI: 10.1079/9781800621596.0010

and behavior with the unique information about any given parcel of the built environment. The inspection should place emphasis on conditions observed at the time of the site visit that would provide for termite survival, with the goal of identifying termite entry points and other evidence of termite activity.

Subterranean termites have been described as "cryptobiotic" (Ebeling, 1975), meaning they live in concealment, making inspections challenging; thus, inspectors must be curious enough to investigate how termites gain access to the structure. After training thousands of pest management professionals, the author has heard, "I see so much more now that I know what I am looking for!" This chapter will cover what to look for during inspections, how to do a general inspection, inspection types, and monitoring and surveillance. The chapter is intended to provide guidance on inspections and monitoring. It is not intended to supplant regional and local regulations, which vary widely.

10.2 What to look for

The goal of termite inspections in and around structures is to find: (1) evidence (i.e., mud or shelter tubes, wings, dead termites); (2) damage; and/or (3) live termites. Damage is evidence of termite activity, but not all evidence of termite activity results in damage. The consequences of discovering damage versus other evidence can be quite different; hence, the distinction.

A species-level determination is the ultimate result of an inspection because the inspector now has the advantage of access to vast bodies of research that can inform the next steps in control, elimination or management, and future surveillance or prevention. The best way to confirm the identification of species is by morphological examination of a good-quality sample that includes workers, soldiers, and alates (i.e., swarmers with wings). Realistically, an inspector is more likely to find evidence of termite activity than live or dead termite specimens. Thus, inspectors must recognize evidence of a termite infestation and distinguish between evidence of subterranean and non-subterranean termites, because treatment recommendations will be different. Signs of previous treatment (Fig. 10.1) should alert an inspector to the possibility of previous termite activity. It is also important to recognize conditions conducive to termite survival for the different groups of termites.

10.2.1 Mud tubes

Mud tubes alone are not enough to provide a genus or species identification, but they are characteristic of subterranean termites as a group (Fig. 10.2) and some higher termites,

Fig. 10.1. Drill holes for injecting soil termiticides are evidence of previous treatment.

Fig. 10.2. Formosan subterranean termite mud tube by-passing a makeshift "termite shield."

such as *Nasutitermes corniger* (Motschulsky). Termites can use a variety of substrates to build tubes, but tubes constructed with soil are the most common. Mud tubes are said to be comprised of soil and "bits of wood, frass (partially digested wood)," and to be "coated with a gluey substance" (Pickens, 1934, p. 166). The "gluey substance" that holds mud tubes together is what differentiates subterranean termite mud tubes from tubes built by the big-headed ant, *Pheidole megacephala* (Fabricius), another invasive species that may be found in similar areas. Mud tubes built by *P. megacephala* crumble easily when touched. Inspectors must be able to distinguish between subterranean termite tubes and those constructed by big-headed ants if using mud tubes as part of termite identification.

Mud tubes were described by Light (1934, p. 147) as "covered runways," which inferred that their function was to facilitate the movement of termites from one location to another. Ebeling (1968) also ascribed various functions to mud tubes, calling them shelter tubes, utility tubes, exploratory or migratory tubes, suspended or dropped tubes that are directed down to soil from a wooden member, and swarm tubes. Ebeling (1968, p. 15) provided descriptions for each type of tube based on his observations. For example, exploratory tubes were "not as strong" as utility tubes (i.e., working tubes that were runways for termites to travel from wood to soil); drop tubes were "lighter colored because they contained more wood fiber." In fact, there are few studies that have tested the composition of mud tubes or their purposes. However, one study measured 18 chemical elements recovered from termite mud tubes as part of a study to demonstrate the impact of subterranean termites in nutrient recycling (Myer and Forschler, 2019). Despite numerous descriptions, any mud tube is cause for closer examination by a well-trained termite inspector.

10.2.2 Carton

The formation of carton by *Coptotermes* species is distinctive. Carton is "material cemented by saliva or excrement" (Emerson, 1938, p. 261). Carton can fill voids such as the space between wooden studs used in framing, chimneys, and even under membrane roofs (Fig. 10.3). The challenge for the termite inspector is that most voids in the built environment where carton may be found are concealed, allowing infestations to persist, unnoticed, until there is a swarm indoors or there is substantial activity resulting in damage to interior cladding or structural members. One of the earliest descriptions of the carton nest of the Formosan subterranean termite, *Coptotermes formosanus* Shiraki, was given by Oshima (1919, p. 333): "The nest consists of a mixture of abdominal excreta and clay or sand, pasted together with a special secretion of the salivary glands. Sometimes it is rigid and compact and seems like a piece of rock." Du *et al.* (2016) provided a more detailed description of second instar workers incorporating fecal material into the carton nest via two methods: (1) smearing a liquid substance on the nest surface with mouthparts; or (2) collecting fecal material through proctodeal trophallaxis and directly applying it to the nest.

10.2.3 Damage

Damage is defined as the chewing of wood fibers by termites. It is important to be able

Fig. 10.3. Carton of a probable aerial infestation of Formosan subterranean termites in a membrane roof. The termites chewed through the membrane of this flat roof, allowing rainwater to seep through the Styrofoam with ample moisture for survival.

to distinguish subterranean termite damage from that done by other wood-destroying organisms, because treatment recommendations can be vastly different. For example, subterranean termites preferentially consume the softer spring wood, and thus follow the grain of the wood (Fig. 10.4), while drywood termites feed across the wood grain (Fig. 10.5). An inspector also may observe the presence of soil within damaged wood that is brought in by subterranean termites (Fig. 10.4), while drywood termite galleries are smooth and without soil because they lack a ground connection (Fig. 10.5). Any genus of termite will eventually cause loss of structural integrity to a piece of lumber through their feeding activity, but *Coptotermes* are particularly voracious, sometimes leaving only splinters of wood behind (Fig. 10.6).

Fig. 10.4. Formosan subterranean termites feeding along the grain of the wood. Note soil and copious numbers of soldiers.

10.2.4 Wings, dead termites, and termite parts

One of the most obvious signs of a termite infestation is when a swarm occurs inside a building or home. In the case of *Coptotermes* termites, alates numbering in the hundreds to thousands can be diagnostic. The natural reaction of occupants is to clean up after a swarm event, thereby removing alates. The inspector is then left with only a few whole termites and wings or pieces of termites for identification purposes. The

Fig. 10.5. (A) Drywood termites create smooth galleries and there is no evidence of soil, because there is no ground connection. (B) Note six-sided pellets.

Fig. 10.6. Splinters remaining due to Formosan subterranean termite feeding. (Photo: J. Maurer.)

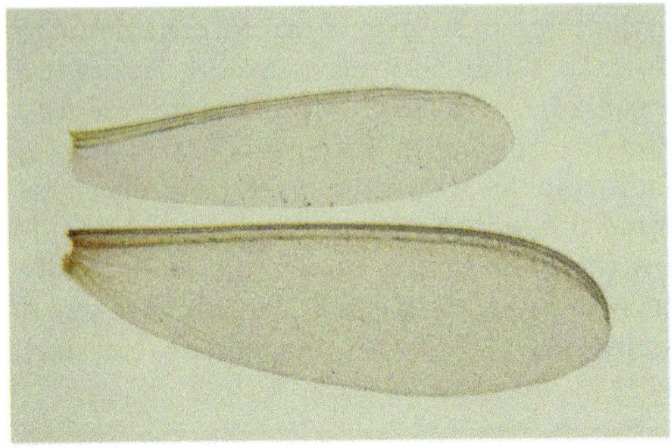

Fig. 10.7. *C. formosanus* wing (bottom): prominent costal margin, sclerotized radial sector vein, and dense setae. Compared with *Incisitermes snyderi* wing (top): costal margin, sclerotized radius and radial sector veins with an unsclerotized medial vein and no setae on wing. (Photo: UF, L. Buss.)

quality of field-collected samples of dead termites often is not pristine. Specimens may be desiccated and brittle after being recovered from light fixtures, windowsills, or countertops. However, securing specimens of any quality is better than nothing and adds to pieces of the puzzle.

One of the most accurate ways to identify *Coptotermes* species is by collecting several intact alates and examining the wings. A difficult task in doing termite identifications with wings is to get the wing to lie flat so that the costal field is completely visible. It may be necessary to rehydrate (i.e., relax) the wing. As is true for all subterranean termites, *Coptotermes* wings have a prominent costal margin and sclerotized radial sector vein (giving the appearance of two "veins" without cross-veins toward to end of the wing) and densely populated with setae on the wing membrane (Fig. 10.7). Drywood termite alates such as *Incisitermes snyderi* (Light) may resemble *Coptotermes* alates at a glance, but drywood termite wings are distinctly different upon careful examination. Drywood termite wings have no setae and contain a costal margin, sclerotized radius and radial sector veins (giving the appearance of 3 "veins" with cross-veins at the end of the wing) with an unsclerotized medial vein (Fig. 10.7) (Scheffrahn and Su, 1994). Depending on lighting, it may be possible to see a prismatic sheen (Fig. 10.8).

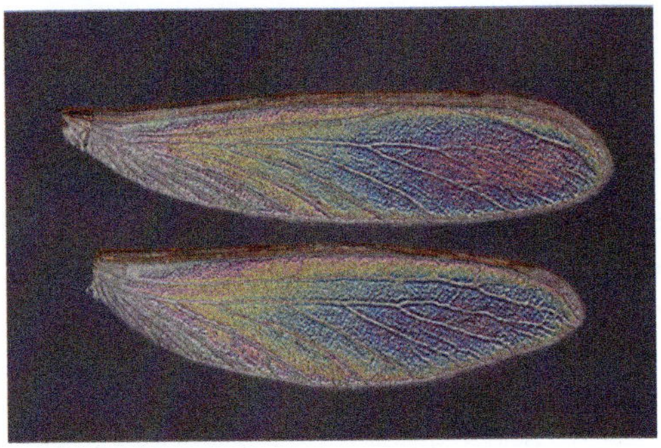

Fig. 10.8. Prismatic sheen on *Cryptotermes brevis* wing. (Photo: UF, L. Buss.)

It is not uncommon for a field sample to contain only one or two dealates, or just parts of swarmers and no wings. It still may be possible to see the difference between a *Coptotermes* and drywood dealate (Fig 10.9). The wing scale on *Coptotermes* will have more setae, much like the wing membrane, whereas the drywood termite wing scale will have significantly fewer setae. Without a side-by-side comparison this character can be hard to see. The other character that can be used is the pronotum. *Coptotermes* swarmers have a more rounded pronotum, while those of drywood termites tend to be more rectangular.

10.2.5 Live termites

Finding live infestations of *Coptotermes* lends itself to helpful field diagnostic characters. *Coptotermes* and *I. snyderi* swarm

Fig. 10.9. Wing scales of Formosan subterranean termite (right) have more setae than those of drywood (*Incisitermes snyderi*) alates (left). Formosan subterranean termite pronotum is more rounded while those of drywoods are more rectangular. (Photo: UF, L. Buss.)

around the same time of year, at dusk, and have honey-colored bodies that are about the same size, which may lead to misidentifications without careful examination. In observing infested wood that is probed during an inspection, it is easy to see live *Coptotermes* soldiers congregate at the opening (Fig. 10.4). Soldiers often produce a white, latex-based defensive secretion from the fontanelle (Prestwich, 1979). The pronotum of a subterranean termite soldier is narrower than the width of its head and body. Subterranean termite mandibles have a smooth appearance. What distinguishes a *Coptotermes* soldier from other species is its sclerotized, teardrop-shaped head (Fig. 10.10). It is a myth that the crossing of mandibles is an identifying character of *Coptotermes* termites.

Fig. 10.10. Formosan subterranean termite soldier with pronotum narrower than the width of the head and body, teardrop-shaped head, and smooth mandibles.

Fig. 10.11. Multiple conducive conditions: faux rock wall covering and pine bark mulch covering the inspection space along the concrete slab combined with irrigation installed too close to the wall. There is enough moisture for weed growth.

10.2.6 Signs of prior treatment

Signs of prior treatment warrant further investigation by the inspector. While this section may be repetitive, it is intended to provide a general checklist of what to look for depending on area of the structure being inspected.
Exterior, look for:

- Evidence of trenching adjacent to foundation walls.
- Termite baiting systems (Fig. 10.11).
- Drill holes through slabs (Fig. 10.1), concrete block, and along expansion joints or stress cracks found in garage and adjacent slab areas.

Interior, look for:

- Treatment notices that may be required by state or local regulations can be posted on water heaters and electrical boxes (Fig. 10.12).

- Drill holes near plumbing.
- Above ground termite bait stations.
- Dye that is evidence of a wood-surface treatment on exposed wood in the garage or unfinished rooms (Fig. 10.13).

A termite bait was wisely installed in a termite conducive area (Fig 10.11).
In attics, look for:

- Treatment notices at or near access opening (Fig. 10.12).
- Drill holes related to wood-destroying organism (WDO) treatments.
- Dye on exposed wood rafters, joists, or sheathing (Fig 10.13).

In substructures, look for:

- Treatment notices on or near crawl space access door (Fig. 10.12).
- Drill holes in block foundation walls and/or block piers, and in exposed wood related to WDO treatments.
- Evidence of trenching adjacent to foundation walls, piers, or pier footings.
- Dye on exposed wood sills, floor joists or flooring (Fig 10.13).
- Moisture (= vapor) barriers.
- Termite baiting (Fig 10.11).

(Modified from WDO Inspections Committee, 2007, p. 11)

10.2.7 Conducive conditions

Conducive conditions are anything in the built environment that provides conditions required by termites – particularly food and water – that enables the establishment and maintenance of an infestation. Wood-to-ground contact, cellulosic debris, stucco, and "wood walls" with soil built up against the wall (i.e., earth banking), landscaping that maintains moisture near the foundation, have been known as conducive conditions for almost a century (Brown *et al.*, 1934; Horner *et al.*, 1934). Conducive conditions can be related to architectural design and construction, or human behavior such as stacking firewood against a house. Conducive conditions related to construction tend to be less subject to change. For example, wall coverings installed below the soil line are less likely to be

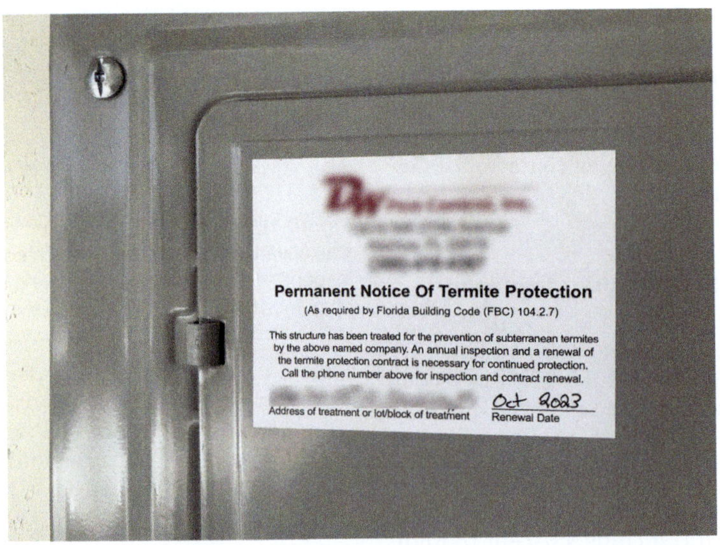

Fig. 10.12. Treatment notice that is tied to regulatory requirements for the state of Florida is evidence of previous treatment.

Fig. 10.13. Dye as evidence of previous treatment on wood and sheathing.

modified over time compared with landscaping that may change seasonally (Fig. 10.11).

Current building and landscape architecture rarely considers termite-proofing during design and construction, despite the existence of termite-proofing recommendations (Geiger and Cox, 2012), and despite building codes that also address termite protection (Oi, 2022). In some countries, such as the U.S., conducive conditions can void termite treatment warranties. Thus, it is important for inspectors to know where and what to look for.

Conducive conditions fall into the following overlapping categories: (1) wood-to-ground contact; (2) excessive moisture in and around structures; (3) landscaping practices; (4) polyurethane spray foam and rigid board insulation; and (5) limited visibility and accessibility.

Wood-to-ground contact

Wood-to-ground contact has been recognized as a conducive condition for over a century. Examples include form boards (Fig. 10.14) and grade stakes left in place, wooden piers under crawl spaces, wooden stairs placed directly on soil, firewood piled against a structure. If termites are found emerging from the middle of the floor, it may be because a wooden grade stake was not removed after the concrete was poured. In practice, any building construction material that is installed below the soil grade could create conducive conditions. Common building construction materials installed below grade include insulation, wall coverings such as stucco, brick, or exterior insulating finishing systems (EIFSs).

Excessive moisture

The category of "moisture" as a conducive condition is large. Obvious causes of excessive moisture around a structure include water flowing toward a building, improperly installed downspouts, torrential rains, dripping outdoor water pipes and heating, ventilation and air conditioning (HVAC) condensate that causes soil erosion. Soil erosion displaces liquid termiticide treatments or can displace bait monitors, leaving structures with little or no protection.

Excessive moisture is exploited by subterranean termites in the development of

Fig. 10.14. Concrete forms are often made of wood. Form boards should be removed after the concrete cures, to avoid becoming a food source and conducive condition for subterranean termites.

aerial colonies. Aerial colonies do not have a ground connection. *Coptotermes* termites seem to have a greater propensity to create aerial colonies compared with *Reticulitermes* species. Most subterranean termite inspectors are trained to focus on building construction elements that are close to the ground (i.e., baseboard, outside foundations walls). Inspectors should be reminded to "look up" when inspecting structures in areas known to have *Coptotermes* species.

Roof and plumbing leaks are common sources of moisture that contribute to aerial colony survival. Light and Pickens (1934) made an early reference to aerial colonies in association with infestations on ships because there was no connection to the soil. Su and Scheffrahn (1987) reported that of the 14 buildings inspected in south Florida, five buildings (~36%) contained aerial infestations and three buildings (21%) had both ground and aerial infestations. Aerially infested structures were designed as flat roofs with pressed wood insulation. Flat roofs notoriously collect pools of water. Termites easily chew through roofing material, allowing water to provide ample moisture for termite survival (Fig. 10.3).

Landscaping practices

Landscaping practices contribute significantly to creating conducive conditions around structures. Planting trees and shrubs too close to buildings will allow roots to grow and create guidelines for termites to follow to the outside foundation wall. Subterranean termites preferentially follow guidelines (Pitts-Singer and Forschler, 2000) and preformed tunnels that can mimic tree roots (Pitts-Singer and Forschler, 2000; Lee *et al.*, 2008). *Coptotermes* also exploit trees and their branches as an avenue to enter structures (Fig. 10.15); thus, trees can be excellent monitors for *Coptotermes* species.

Mulching is often used as part of a landscape plan because it helps retain moisture, moderates soil temperatures, and prevents soil from splashing against walls. However, mulch piled high enough so that the inspection space is obscured creates a conducive condition by enhancing the ability of subterranean termites to enter structures undetected behind wall coverings (Fig. 10.11).

Subterranean termites do consume some mulches, although mulch from melaleuca

Fig. 10.15. Formosan subterranean termites entered this structure via branches resting on the roof. When the branches were pruned away from the structure, the infestation ceased.

and cypress heartwood were least preferred and termite survival was less than 15% (Duryea *et al.*, 1999). Long *et al.* (2001) tested the ability of *Reticulitermes virginicus* (Banks) to find monitors containing cardboard rolls under organic mulches (shredded hardwood bark, miniature pine bark nuggets, shredded eucalyptus bark) and an inorganic ground cover (pea gravel, Fig. 10.16) in the field. While there was no significant difference in the ability of subterranean termites to find those monitors, regardless of mulch or ground cover type, significantly more cardboard was consumed from monitors under pea gravel. In a separate laboratory study, the survival of termites fed on hardwood, eucalyptus, or pine bark mulches was significantly lower than that of termites fed on white birch as a control, suggesting that even though subterranean termites feed on mulch it may not provide adequate nutrition for prolonged survival (Long *et al.*, 2001).

Other landscaping practices that are considered to be conducive conditions include, but are not limited to, allowing any plant material to grow against outside foundation walls, planter boxes built against the outside foundation wall, misdirected irrigation heads that spray water onto walls, and "green roofs" or rooftop gardens. If there is enough moisture for plants to grow, it will be enough to sustain subterranean termite survival. Planter boxes built against the outside foundation pose a particular threat, because the visual inspection space between the ground and foundation is obscured (Fig. 10.17). When plants in the box are watered, moisture remains against the outside foundation wall, and if the plants are edible (herbs, tomatoes, etc.) it cannot be treated with liquid termiticide.

Insulation

Insulation is critically important for decreasing global energy consumption and is

Fig. 10.16. Subterranean termites consumed more cardboard from monitors under inorganic mulch depicted here. They may also pose challenges to perimeter treatments with liquid soil termiticides and termite bait installation.

codified in various sections of the International Building Code, International Energy Conservation Code (IECC), the American Society of Heating, Refrigerating and Air-Conditioning Engineers, and in local codes. Rodriguez-Soria *et al.* (2014) reviewed international regulations on thermal insulation requirements and examined the factors that contribute to energy loss through the building envelope. Although important to energy conservation, insulation can be a conducive condition by creating hidden access points, encouraging the accumulation of moisture if installed incorrectly, and providing thermal regulation so that termites might survive outside their natural range. Thus, it is important for the inspector to know where insulation may be installed in the built environment.

Insulation can be installed as part of the building thermal envelope. The IECC defines the building thermal envelope as "(t)he basement walls, exterior walls, floors, ceiling, roofs, and any other building element assemblies that enclose conditioned space or provide a boundary between conditioned space and exempt or unconditioned space." (IECC, 2021, Section R202).

Kalhor and Emaminejad (2020) described the types of insulation used in the built environment in the context of energy codes and the properties that should be considered in selection, including moisture resistance. Moisture resistance is an important consideration of insulation wall assemblies. If installed incorrectly, condensation can occur within walls due to the temperature differential between the inward-facing and outward-facing wall elements, creating a conducive condition.

Rigid board insulation and other wall coverings installed below grade around the

Fig. 10.17. Planter boxes obscure inspection spaces and provide moisture for termite survival.

perimeter of a structure will hamper an inspector's ability to observe signs of subterranean termites before entry. Before building codes recognized that insulation installed below grade created a conducive condition, Smith and Zungoli (1995) surveyed 225 pest control companies and found that 12% were sued because termite damage was concealed by rigid foam board. Building codes were eventually modified but numerous homes and buildings in the U.S. continue to be built with insulation installed below grade, due to an exception in the International Building Code. The International Building Code (2021) recognized that insulation could create conducive conditions, but allowed its installation below grade with "approved methods" of protection against termites. It did not specify which "approved methods" must be met or under what authority or performance standards (Section 2603.8, Protection against termites):

> "In areas where the probability of termite infestation is very heavy in accordance with Figure 2603.8, extruded and expanded polystyrene, polyisocyanurate and other foam plastics shall not be installed on the exterior face or under interior or exterior foundation walls or slab foundations located below grade. The clearance between foam plastics installed above grade and exposed earth shall be no less than 6 inches (152 mm).
> Exceptions:
> 1. Buildings where structural members of walls, floors ceilings and roofs are entirely of noncombustible materials or preservative-treated wood.
> 2. An approved method of protecting the foam plastic and structure from subterranean termite damage is provided.
> 3. On the interior side of basement walls."

The "Exceptions" still leave significant room for conducive conditions to persist. Oi (2022) reviewed the performance standards for termite control products in the U.S. Products claiming structural protection may only need to produce two years of efficacy, depending on use pattern. Preservative-treated lumber can still be damaged by subterranean termites, particularly the Formosan and the Asian subterranean termites, and approved methods of termite control can fail. A possible solution to insulation as a conducive condition is the incorporation of termiticides into insulation. Some commercially available products are marketed, but the extent to which they are used is not publicly documented. Rigid foam board insulation has been experimentally impregnated with various termiticides and tested against *C. formosanus* (Su *et al.*, 2003; Yoshimura *et al.*, 2003). Pyrethroid termiticides were more protective compared with borates.

Structures are not at risk of infestation if subterranean termites are not in an area, regardless of the conducive conditions present; hence, baiting is effective in mitigating conducive conditions that are related to building construction, including foam insultation. There are numerous publications demonstrating that termite baits using chitin synthesis inhibitors such as hexaflumuron, noviflumuron, novaluron, chlorfluazuron, or bistrifluron, can eliminate specifically *Coptotermes* from an area (Su, 1994; Grace *et al.*, 1996; Su and Scheffrahn, 1996; Lee, 2002; Peters and Fitzgerald, 2003; Cabrera and Thoms, 2006; Evans, 2010; Osbrink *et al.*, 2011; Su *et al.*, 2016a, 2018) (see Chapter 11 of this volume).

Limited accessibility

Even with technological advances in termite detection, areas that are inaccessible can become locations of conducive conditions. An accessible area is defined as "an area of a structure to which the person performing the inspection can either gain physical entry in order to identify and/or determine the presence of wood-destroying organisms" (WDO Inspection Committee, 2007, p. 4). Limited accessibility can be caused by planned and unplanned elements of the built environment itself and human behavior (i.e., clutter and debris). An inaccessible area includes, but is not limited to, wall coverings including stucco exteriors (particularly at corners near downspouts), floor coverings, insulation or items that are not moved due to their size (e.g., furniture, stored articles) or safety concerns (e.g., equipment). Seismic upgrades installed to minimize damage from earthquakes may use plywood boards that are attached to subarea support posts and obstruct visual inspections (V. Lewis, 2023, personal communication).

10.3 Inspections procedures

Some local governments or professional associations adhere to formal processes for termite inspections, depending on the type of inspection requested (see Section 10.4, this chapter, "Types of inspections"). The following outline is used with permission from the Florida Wood-Destroying Organisms (WDO) Inspection Committee chairperson. The original document was written to satisfy what is minimally acceptable in Florida (WDO Inspection Committee, 2007). However, this guidance is updated here to accommodate the behavior of *Coptotermes* species and is more encompassing than an inspection for the purposes of a real-estate transaction. It is important to check with local regulatory authorities on inspection requirements specific to your area.

Consistency is critically important to the inspection process. Experienced inspectors do inspections in a clockwise or counterclockwise direction, but not both. They are prepared to climb and crawl, looking for termites in 3 dimensions and do not simply focus on lower construction elements such as concrete slabs or outside foundation walls, particularly with *Coptotermes* termites that can build aerial colonies. One of the challenges of doing inspections where the Formosan and Asian subterranean termites are emerging invasive pests is educating inspectors that they must change their inspection process to include areas above ground level. Experienced inspectors methodically search for evidence, damage, live termites, and conducive conditions as described in Section 10.2 of this chapter, accurately describing infestation locations, and recording observations in a legally defensible manner (i.e., following regulatory requirements). Inspectors may employ "sounding and probing" to areas that are "accessible" to determine the extent of an infestation. In instances where an area is not accessible, inspectors are encouraged to do a "visual" assessment, which usually requires flashlights to sufficiently illuminate the inspection area, or binoculars that extend an inspector's line of sight. Please consult Appendix A of this chapter for definitions used and Appendix B for tools and equipment used.

10.3.1 Exterior of the structure

"WDO inspectors must familiarize themselves with the structure(s) to be inspected.

The inspection of each structure shall include all sides of the building exterior" (WDO Inspections Committee, 2007, p.7). Familiarizing oneself with a structure can be daunting. Structure has been defined as "(a) ny building, regardless of design or type of material and used in its construction, whether public or private, vacant or occupied, and adjacent outside areas" (Georgia Structural Pest Control Commission's Honey Bee Control and Removal workgroup, 2023, p. 33). Examples of structures include but are not limited to simple storage sheds, mansions, commercial buildings, or historic buildings. Building construction materials also vary widely depending on geographic region and consumer preference. Not all construction materials are consumed by termites for nutritional purposes, but they can be damaged or used as a guideline to enter a structure. Structural elements that subterranean termites exploit to enter structures such as studs, pipes, and conduits are often neither "visual nor accessible," increasing the complexity of inspections for subterranean termites (Fig. 10.18). Termite inspectors who understand the built environment are at a distinct advantage compared with inspectors who do not.

Creating a graph while inspecting the exterior of the structure will aid the inspector in identifying critical areas. Someone other than the person who drew the graph should be able to find points of infestation and recognize landmarks on the property. To graph a structure manually, an inspector will need a clipboard, graph paper, pen or pencil, ruler, measuring wheel, and/ or tape measure to recreate the footprint of the structure. Measuring the width of a wall at a doorway can help deduce whether walls are made of concrete masonry units, wooden studs, or other material (Fig. 10.19). Wall coverings will prevent the inspector from directly seeing these building construction elements. For example, concrete masonry units are 8 inches wide; thus, the depth at the doorway would be greater than 8 inches.

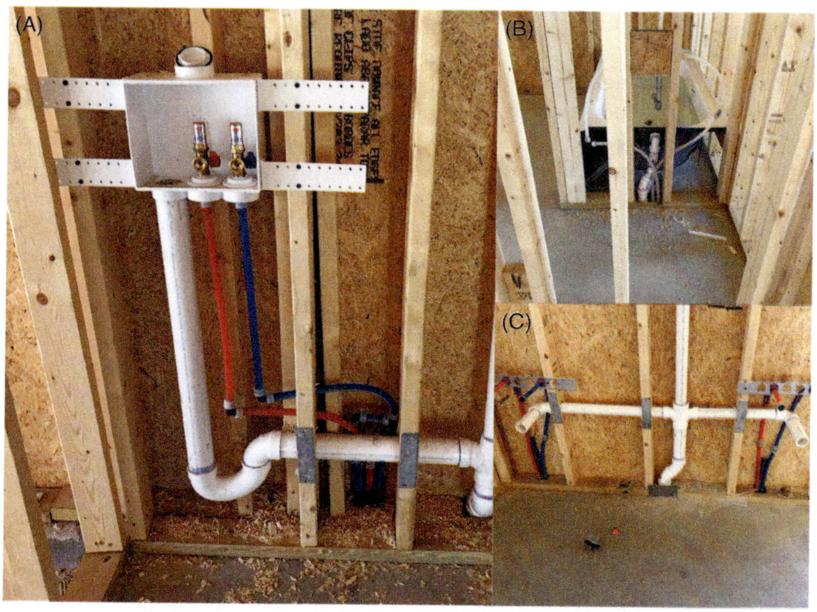

Fig. 10.18. Critical areas for termite entry include pipes and conduits and can be along both exterior and interior walls. They breach the soil and provide a guideline for termites to follow into a structure. They may be moved during the construction process, disturbing soil termiticide treatments. Pipes and conduits also may not be visible or accessible in a completed building and can conceal termite entry. (A) Clothes washer water connection; (B) bathtub water and drain connections; (C) sink water connections.

Fig. 10.19. This home is built with wooden framing using 2 × 4 studs; thus, the width at the doorway is roughly 6 inches (15.2 cm), including the wall covering.

An experienced inspector will learn as much as possible about the structure, potentially leading to deeper investigations into areas where termites may enter (for example, concrete masonry units are hollow and can provide subterranean termites hidden access to structures).

After the footprint of the structure is graphed, walk around the building, and add the landscaping elements and conducive conditions. Landscaping elements include shrubs, trees, and plant beds that may be bounded by railroad ties. Railroad ties for landscaping have been implicated in the spread of Formosan subterranean termites and should be noted (Forschler *et al.* 2001). Potted plants and personal items have been connected to the first infestation of Formosan subterranean termites in La Mesa, California (Atkinson *et al.*, 1993; Haagsma *et al.*, 1995; Rust *et al.*, 1998).

Trees should always be included during inspections for *Coptotermes* termites. They have been found in trees (Greaves, 1962; Kalshoven, 1963; King and Spink, 1969; Osbrink *et al.*, 1999; Chouvenc and Foley, 2018) and 47 different living plants (Lai *et al.*, 1983). Trees can be used as monitors in subsequent inspections as measures of change in activity either as a result of treatment or natural changes in termite distribution. If mulch obscures the inspection space along slabs, pull it away, if permissible, to do a more thorough inspection. Some hedges close to the perimeter of the structure contain thorns and can be a hazard to the inspectors. Exposure to thorns can be mitigated by placing a tarp over the thorny plants, allowing the inspector (and applicator) to pass between the outside foundation wall and hedge.

Many pest control companies use management software that contains reporting functions with tools to create maps of service areas. These tools incorporate digital pictures and inspection reports, including those from bait systems such as Sentricon®, the National Pest Management Association (NPMA)-33 or state equivalent. They are mobile-device accessible and can sync to company cloud storage while the technician is in the field.

In areas where *Coptotermes* termites are found, it is advisable to inspect above the first floor of structures. While ladders should be part of an inspection toolkit, there are varying levels of risk associated with the height one may climb to confirm an infestation. Professional inspectors should use ladders that are compliant with the regulatory requirements for their area, and they should be trained in proper ladder use. Ladder fall data specific to the pest control industry are difficult to obtain and are usually associated with fumigators who must climb to place tarps on roofs. However, a query of the National Trauma Data Bank for "ladder falls," returned 168,227 patient results from 2007 to 2017 (Liasidis *et al.*, 2022). The median patient age was 56 years old. Work-related falls comprised ~20% of cases. While the number of work-related ladder falls is relatively small, the severity of injuries is great: broken bones, including skull and spinal fractures, and in a small number of cases, death. These findings agree with anecdotal evidence from at least one pest control insurance carrier. For context, there are more slip-and-fall accidents than ladder falls in the industry, as technicians may be distracted by tasks such as mapping a property or texting while walking.

To minimize the risk of using a ladder, use binoculars, drones, and infrared cameras to extend the range of what is "visual" and identify areas of interest. A camera with macro lens can also extend the range of what is "visual" and document points of infestation at the time of inspection. Digital pictures can provide invaluable documentation of the condition of the property at the time of inspection.

In summary, exterior areas to inspect visually for signs of termite activity and include in a report should involve but are not limited to:

- Wood to ground (soil) contact, planter boxes, windows, door frames, shutters, and window wells (i.e., a barrier that surrounds a basement window to prevent moisture damage, to provide an additional emergency exit, and to provide additional natural light).
- Siding, particularly if installed below grade (i.e., below the soil line).
- Lattice work and skirting around mobile homes and crawl spaces.
- Fences, decks, balconies, particularly at the point of attachment to structures.
- Fascia and soffits.
- Chimneys.
- Porches

(Modified from WDO Inspections Committee, 2007, pp. 7–8).

Finally, it is not uncommon for the exterior of structures to be made inaccessible by attached sheds, decks, fences, landscape plantings, mulch and/or ground cover, stored items, and other debris. Obstructed and inaccessible areas should be documented on reports for future reference.

10.3.2 Interior of the structure

> "Evidence of and damage caused by WDOs in living areas is often hidden by furniture, floor coverings and interior finish and trim. Inspectors need to concentrate their efforts in those areas most likely to present visible evidence of WDOs ..." (WDO Inspections Committee, 2007, p. 9).

A summary of areas that most likely contain visible evidence of WDOs in an interior inspection include the following:

- Light fixtures that can be a collection point for termite wings and swarmers.
- Doors and windows. Door frames are subject to subterranean termite infestations because of their proximity to the ground and to moisture (Fig. 10.20). Window frames are also subject to water intrusion, particularly if flashing is installed improperly. Look for evidence of termite swarmers around windows and in spider webs around windows and doors.
- Baseboards and trim. Emphasis should be placed on the inspection of all accessible baseboards and trim, due to the propensity for infestation to be visible in these areas. Probing and sounding are recommended to confirm the presence of WDOs.

"Probing" is invasive and done to discover the extent of observed WDO activity. Sounding and probing should be done with an adjustable trekking pole that can be equipped with rubber tips to minimize marring surfaces or cause unnecessary damage (Fig. 10.20). "Sounding" is a non-invasive method of inspection. An inspector taps an area with a tool such as the handle of a screwdriver. Experienced inspectors can hear the difference between sound wood and wood that is hollow or otherwise compromised, indicating the possibility of termite damage. Other tools used in sounding include the rubber grip end of a golf putter, walk or trekking poles (Fig. 10.20).

- Interior walls. "Inspect wall surfaces for waviness, blistering, slightly raised areas and peeling" (WDO Inspections Committee, 2007, p. 9). Rolling a flashlight parallel to the wall can help reveal imperfections by casting a shadow beyond the raised areas (Fig. 10.21). Some experienced inspectors brush their hands over a wall to feel imperfections. Termites can consume the paper between the gypsum and the paint, causing crumbling.
- Cabinets. Inspect accessible areas of the kitchen, bathroom, and other cabinets; particularly those that contain sinks and pipes. Also check built-in bookcases and alcoves.
- Wood flooring. "Inspect accessible wood flooring for WDO evidence and damage. Special attention should be given to flooring adjacent to entry

Fig 10.20. (A) The result of probing to discover the extent of termite damage in a door frame. (B) Sounding and probing base board using a trekking pole as part of a WDO inspection. (C) Close up of trekking pole tip.

Fig 10.21. Rolling flashlight parallel to a wall can magnify imperfections.

areas, plumbing penetrations, HVAC (heating, venting, and air conditioning) registers or vents, areas of obvious buckling in the floor, fireplaces, and staircases. Wood floors at ground level may indicate the presence of a substructure crawl space" (WDO Inspections Committee, 2007, p. 10).

- Access panels. "Open and/or remove access panels and inspect the accessible area for evidence or activity of WDOs and/or damage such as termite wings, mud tubes and galleries. Access panels may include areas that are not specifically related to plumbing" (WDO Inspections Committee, 2007, p. 9).

"Interior areas are commonly obstructed by furniture, wall coverings/hangings, floor coverings (rugs, mats), blinds and drapery (cornice and valance), stored items (boxes, containers, clothing, etc.), plants, or other household items" (WDO Inspections Committee, 2007, p. 9). Obstructed areas should be clearly documented.

10.3.3 Substructure inspections

Substructure areas are also known as crawl spaces and basements. There are instances when substructure spaces are low enough to the ground that it is almost impossible to access without risk to the inspector. Substructure areas become increasingly inaccessible because they contain ductwork, conduits, and data cables. Search for mud tubes along the sill plates, around piers, pipes, floor joists, and any vertical elements to the subfloor. Debris is often found in crawl spaces, and it should be removed and noted in written reports.

Some homes also may have termite physical barriers made of stainless-steel mesh, sand, or crushed granite (particle size barriers) in substructure areas. Disturbance of these physical barriers during an inspection may compromise their effectiveness, thus effectively rendering them inaccessible (V. Lewis, personal communication).

Substructures can pose safety concerns, including (but not limited to) unprotected electrical lines, exposed nails from flooring, venomous arthropods, people using the crawl space for shelter in areas with high homeless populations, snakes and other wildlife, broken glass, sewage, mold spores and other respiratory hazards. Consider personal protection equipment such as hard hats, bump caps (some of which can be inserted into company-logoed baseball caps), or rock-climbing hard hats with chin straps to protect the skull. Consider using a head lamp to keep hands free. Experienced inspectors will do most of the substructure inspection on their backs in order to observe the flooring above.

Subterranean termites also may access structures through the basement. Check the sill plates, cracks in the foundation, any pipes or conduits, and look for evidence of swarmers if there are windows. "(T)hose areas that cannot be physically entered into yet can be observed by use of a lighting device should be inspected visually to the extent possible. Investigation should be made to ascertain if an interior entryway exists for crawl space areas that do not have an exterior entryway" (WDO Inspections Committee, 2007, p. 10).

10.3.4 Attic inspections

Attics should be inspected if there is an opening to the attic and if the truss webbing is sufficient to pass through. Truss tails (i.e., rafter tails) can provide ready evidence of termite infestation but it may be difficult to see from the inside of the attic. Inspecting truss tails from the exterior of the structure may reveal termite activity. Spray foam insulation and increased duct work associated with HVAC systems can make attic spaces difficult to inspect (Fig. 10.22). These conditions should be documented for future reference.

True attic access openings have a removable panel that can include pull-down ladders while others require the use of a portable ladder. In commercial buildings, some ladders are "fixed." Dropped ceiling panels are not an attic access opening. Climbing ladders increases risk to the inspector and should be considered when setting the scope of the inspection.

Heat is another hazard of attic inspections. Inspectors should be trained to recognize the symptoms of heat stress that can also lead to other on-the-job injuries. For example, sweaty palms and dizziness can cause loss of grip while climbing and result in a fall. Factors that contribute to heat stress

Fig. 10.22. Spray foam insulation installed as part of the thermal envelope. When combined with duct work integrated within truss webs, the combined elements can make some parts of an attic difficult to inspect.

include the physical condition of the inspector (e.g., weight, general health) and required personal protective equipment that can increase core body temperature. Some companies have protocols to ensure worker safety, such as calling the office every 15 minutes or scheduling inspections to be done during cooler morning hours.

10.4 Types of inspections

Depending on geographic regions and regulatory requirements, inspections can be put into four categories: (1) post-construction inspections to develop treatment plans; (2) annual renewal inspections; (3) inspections for the purposes of a real-estate transaction; and (4) forensic inspections. Inspections are impacted by purpose of the inspection, and regional surveillance and monitoring efforts (see Section 10.5). The keys to the depth of an inspection are time spent, and the knowledge of the inspector. The time dedicated to forensic inspections is almost unlimited. The knowledge required of someone doing the first three categories of inspection can be as little as informal on-the-job training without regulatory certification or regulatory certification that the WDO inspector has received training in biology, behavior, and identification, which is affirmed by affidavit but not testing (Florida Statutes (FS) 482.091(9)). In Florida, "training" is not specified and could range from one extreme of passively sitting through webinars and learning from a poorly trained senior technician to the other extreme of attending university-level programs. In contrast, an inspector retained as an expert witness for litigation will likely have accrued many years of field experience, is credentialed by the state, and may have a college degree (bachelors, masters, and/or PhD).

10.4.1 Post-construction inspections

Post-construction treatments are usually done as a result of the property owner discovering an initial termite infestation, then contacting a pest control company for treatment. The property owner is not likely trained in termite identification, behavior, and treatment methods; thus, it is imperative that an inspection is done by someone who is well-trained. Post-construction inspections have revealed the Formosan and the Asian subterranean termites in new geographic areas. A common scenario is that a property owner or pest control company reports an infestation that is "different" than anything they have seen before. Understanding that the infestation is "different" implies that there is a recognition of what is "normal" for the area. In the absence of any coordinated regional or worldwide surveillance for invasive termites, these reports are vitally important (see Section 10.5 of this chapter).

Company salespeople are often tasked with doing inspections as part of developing a quote for treatment, so it is important that they also are trained; however, these inspections can be simply to confirm that there is a termite infestation. Some states, like Florida, allow Formosan subterranean termites to be excluded from contract warranties (Florida Administrative Code 5E-14.105(2)(e)), which would require a salesperson be able to do a field identification to species level before a treatment contract is signed. However, it is often the applicator who discovers the extent of the infestation and must confirm the species. Relying on the inspection of anyone who is not well-trained will almost inevitably lead to a cascade of unfortunate events such as re-treatment, litigation, and in the worst case, the complete demolition and replacement of the structure, particularly in areas where Formosan and Asian subterranean termites are found.

10.4.2 Annual renewal inspections

An annual renewal contract usually includes an inspection to detect termites before significant damage occurs and can be offered to a consumer after treatment of new construction (i.e., "pre-treatment") is completed or post-construction. Re-treatment only or re-treatment and replacement warranties may accompany annual renewal contracts. As the name implies, these inspections have historically been done at least once a year, but consumers, particularly in areas of Formosan and Asian subterranean termites, should be aware of changes in contract warranties and inspection practices.

Inspecting only the exterior of a structure during an annual renewal inspection is the lowest common denominator for inspections. Not all annual renewal inspections include the interior of the structure. Sometimes interiors are not inspected because the inspector and customer cannot find a mutually acceptable time for inspector access. It is more concerning that some companies forgo interior inspections because they believe that termite baits will always be successful in eliminating subterranean termites from an area, and therefore, the structure is no longer at risk. Subterranean termite baits have a tremendous success rate, but this rationale ignores the possibility of aerial colonies forming, bait failure due to improper use, changes in landscape that may disrupt station placement, and a host of other reasons that may allow subterranean termites to enter a structure regardless of treatment methods used. Conversely, some companies may do interior inspections if liquid termiticides were used because they believe that the only way to know if a liquid treatment is failing is to inspect for termites around critical areas such as pipes. Regardless, there is no research to support either justification. It is advisable to do interior and exterior inspections regardless of treatment.

10.4.3 Inspections for the purposes of a real-estate transaction

Inspections for the purposes of a real-estate transaction are associated with significantly more regulations. For example, property transactions that are guaranteed by the US Department of Housing and Urban Development (HUD) or US Department of Veterans Affairs must be inspected for wood-destroying insects, including termites, and the National Pest Management Association (NPMA) Form-33 must be used to report any wood-destroying insect or wood-destroying fungi activity (HUD, 2019, pp. 179, 357, 362). Many conventional mortgage lenders also require the use of the NPMA-33 form or a state equivalent.

In regions such as Florida where the risk of termite damage to property is extremely high, regulatory bodies may impose additional requirements. For example, the Florida Department of Agriculture and Consumer Services requires the use of form 13645 for inspections done as part of real-estate transactions. Additionally, Formosan subterranean termite infestations have risen to a level of recognition that they are specifically referenced in Florida Statute 482.242 (1)(c)(1) (Preemption). Simply stated: a local government can enact ordinances that "require, for multicomplex dwellings in excess of 10 units, annual termite inspections for termite activity or damage, including Formosan termites, which must be performed by a person licensed under Florida Statute 482" (C. Frazier, 2023, personal communication).

Termite inspections for real-estate transactions and termite damage rank #1 and #2 in general liability insurance claims in the pest control industry (A. Fugler, 2023, personal communication). Termite damage is not covered by homeowners' insurance. If a property owner discovers termite damage during the course of renovating a property, the inspector and company are often named in lawsuits that seek to recover damage costs. In the most serious cases, termite damage can result in structures being declared "catastrophic losses" and the pest control company is likely responsible for the cost to replace the structure.

Fig. 10.23. Measuring the angle of every drill hole to ensure treatment was done properly during a forensic investigation.

10.4.4 Forensic inspections

Forensic inspections are by far the most detailed. The best expert witnesses are the ones who outdo all others in inspections. They have the greatest working knowledge of termite biology, behavior, control, regulations, and the law. Inspectors who are called as experts in a lawsuit will have an almost unlimited amount of time to complete their investigation compared with inspections done to sell a post-construction treatment or even for the purposes of a real-estate transaction. They can inspect every aspect of the structure and may have the advantage of walls being opened to reveal the extent of the termite damage. They may not limit their inspections to what is simply "visual and accessible," they may probe, sound, and use detection devices such as moisture meters, boroscopes, radar and thermal sensor devices to determine likely termite entry points into the structure. If the treatment is in question, some expert witnesses use protractors to measure the angle of every drill hole to evaluate if termiticide application was appropriate for the foundation type (Fig. 10.23). Forensic inspections are accompanied by meticulously written reports that are legally defensible, using pictures that may number in the hundreds, and will be used to present a story of how termites entered a structure and caused damage.

Expert witnesses may have access to documents that relate to a particular case and can include graphs, treatment records (i.e., work orders and service tickets), possibly technician training records, and will work with other experts (structural engineers, regulatory officials) to determine whether previous inspections and treatments were negligent or not. During the discovery phase of a lawsuit, an expert witness can compare their inspection report to the one done by the company being sued. Every missed piece of infestation evidence, conducive condition, and flawed communication that is revealed will be shared with both plaintiff and defense counsel. Settlements and judgments can be in the millions of US dollars. Consequently, companies that engage in inspections and treatments for termites should assign their best-trained and most experienced employees rather than assigning employees who simply meet a minimum standard.

10.5 Surveillance and monitoring

10.5.1 Surveillance

Formosan subterranean termites are on the top 100 global invasive species list (GISD, 2023), yet there is no coordinated effort to surveille for and monitor the spread of this highly destructive termite. Given the potentially devastating nature of this cryptobiotic social insect, the ease with which it is transported (sometimes in plant material that is inspected for other pests by regulatory agencies), and its economic importance, on the surface it is simply bewildering that there is not a single regulatory agency in the U.S. tasked with stemming its spread, let alone global interagency cooperation. Not unexpectedly, upon further examination, federal agencies and laws are complex. In the U.S., several federal and state agencies list “invasive species” as a priority, but in an analysis of federal invasive species programs, efforts were described as “fractured and incomplete” (Burgos-Rodriguez and Burgiel, 2020, p. 131).

The National Invasive Species Council (NISC) within the US Department of Interior specifically lists Formosan subterranean termites as a species of concern because it poses risks to homes and infrastructure. However, the primary role of NISC is to coordinate work plans with other federal agencies in “priority setting” and ironically, to “remove institutional and policy barriers.” There are no NISC initiatives to surveille for or monitor the spread of Formosan subterranean termites, partially due to its internal institutional criteria for decision making. Decision-making criteria to prioritize work plans mainly include: (1) having two or more federal council member agencies (e.g., the US Department of Commerce and the Department of Agriculture) commit to a work plan; and (2) the plan must be feasible “given the capacity of council staff coupled with human or financial resources contributed by engaged agencies” (NISC, 2023, p. 6). There are no U.S. federal or state agencies specifically looking for invasive termites or other pests that damage structures as part of their mission. Thus, the path to federally coordinated efforts to monitor the introduction and spread of Formosan subterranean termites will be difficult, because invasive termites simply have not been elevated to a high enough priority.

Additionally, in the U.S., personal rights are described as an impediment to invasive species surveillance and monitoring: Government authorities can inspect federally owned lands, but not private property (Burgos-Rodriguez and Burgiel, 2020). Thus, hypothetically, if a federally coordinated surveillance program existed, it would not prevent Formosan subterranean termites being transported in potted plants via private vehicles or infested private boats. In short, structural pests such as the Formosan subterranean termite fall between the bureaucratic cracks of federal and state agencies tasked with invasive species response and management. Invasive termites are not considered economically important pests of agriculture, including forestry. They are not a threat to human health or the environment. However, termites are economically important pests to property owners, causing billions of dollars of damage globally (Rust and Su, 2012; Nagro, 2015). Thus, surveillance is left to pest control industry professionals and property owners who suffer the consequences of wood-destroying organisms that silently damage the places where they live, work, and play.

The US Department of Agriculture does not actively inspect for termites; however, it has intercepted termites for ~100 years. Blumenfeld and Vargo (2020) identified 906 non-native termite interceptions from 29 different states. States with the highest number of interceptions are Florida (232), New York (104), California (95), Louisiana (82), and Texas (70). Interceptions included invasive species that are now established in the U.S., such as *C. formosanus* and *C. gestroi*, as well as species that have the potential to establish, such as *Nasutitermes ephratae* (Holmgren) and *N. nigiceps* (Haldeman). The need for effective control methods is underscored by the absence of an active surveillance system and proof of interceptions. It is not “if” but “when” invasive termite species will establish and re-establish in areas outside their native range.

While federally coordinated surveillance programs are non-existent for invasive termites in the U.S., it is not impossible to collect data on the distribution of Formosan and Asian subterranean termites. Invasive termites are most frequently found by property owners. Citizen science-based surveillance is not systematic, but when data are collected over time and distributions analyzed, a compelling story emerges on the spread of Formosan, Asian and other termites over time (Chouvenc *et al.*, 2022). In turn, informed property owners seek monitoring and control methods appropriate to the species in their area.

10.5.2 Monitoring

Monitoring (i.e., evaluation, over time, designed to provide feedback after the problem is identified) for invasive termites outside of research projects is rare. Monitors can be as simple as wooden stakes pounded into the soil, rolls of toilet paper, or bucket traps that contain wooden blocks or cardboard (Su and Scheffrahn, 1986). Monitors are checked over time to provide information on colony distribution or treatment effectiveness. Termite monitoring is labor intensive.

The concept of remote monitoring for termites was pioneered over 20 years ago when Su *et al.* (2000) demonstrated that electronic sensors could be used in monitors for termite baits in lieu of manually checking bait stations. Su (2001) further tested sensors that were connected to dataloggers and telephone lines. Data were downloaded every 4 days. Sensor longevity was only ~4 months. It was improved upon (Su, 2002), and over 20 years ago the concept of remote monitoring was visionary. Su (2011) identified cost as an impediment to remote termite monitoring and speculated that cost may be reduced in time. The explosion of remote monitoring for rodents has proven this true, but remote monitoring for any pest will suffer from poor performance if monitors are installed without knowledge of biology and behavior. In fact, bait monitors were infested twice as often and more quickly when placed in areas of conducive conditions (Jones, 2003). Technology has improved and cost for remote monitoring has decreased. Remote monitoring for subterranean termites has been deployed at the Forbidden City in China (Zhang *et al.*, 2022), but requires more development and has not been widely accepted in the commercial sector yet.

Monitoring Formosan subterranean termite colony movement can also be evaluated using genetic methods to inform surveillance. Jenkins *et al.* (2002) were the first to use mitochondrial DNA to document the probable spread of Formosan subterranean termites from New Orleans to Atlanta because they shared the same maternal lineage. The following studies sought to answer the question of whether invasive termite populations have been eliminated from an area or if newly discovered infestations are the same or different from initial infestations. In 2022, Tseng *et al.* used genetic markers to determine that, in addition to the 1992 La Mesa introduction, Formosan subterranean termites found in Canyon Lake and Rancho Santa Fe/Highland Park were distinctly different, indicating multiple introductions into California. Monitoring using genetic data can also provide important feedback on the effectiveness of termite management (Husseneder *et al.*, 2007; Mullins *et al.*, 2011), particularly in measuring the progress of areawide management programs designed to control the spread of Formosan subterranean termites (Rust *et al.*, 1998; Guillot *et al.*, 2010; Su *et al.*, 2016b) (see Chapter 15 of this volume) and other invasive termites (Thorne *et al.*, 2021).

Non-research monitoring is usually associated with baiting systems. Termite baiting is increasing in popularity because of the challenges of applying liquid termiticides correctly, including substantially more applicator training needed compared with installing bait stations every 10–20 feet. There was an attempt to commercialize remote monitoring for subterranean termite baits (Sentricon® ESP™, Dow AgroSciences, now Corteva Agriscience), but it is no longer available (see Chapter 11 of this volume). There are some commercially available monitors without the active ingredients that can be effectively used in integrated pest management approaches to invasive termite management. For example, TM-1 Termite Monitor (B&G Equipment) consists of a wooden stake and plastic housing that contains cellulosic material.

10.6 Closing

Inspections and monitoring are related to surveillance and detection. These practices are multi-layered and will change as building construction practices and materials changes. The 2021 International Building Code allows construction as tall as 18 stories using mass timber and other non-combustible material which will impact inspections and termite detection. Research needs for the mass timber industry include evaluating termite resistance of pressure-treated glulam and cross-laminated timbers, investigating the feasibility of landscape-level termite management to protect mass timber construction, and updating termite distribution maps (Zelinka *et al.*, 2019). Zelinka *et al.* (2019, p. 21) also recognized that "research is needed to better understand the performance of mass timber construction in southern high moisture climates" and that there is "no guarantee that termites or other wood-destroying organisms" will not attack cross-laminated timbers. Wang *et al.* (2018) specifically mentioned Formosan subterranean termites as a risk to mass timber construction. However, while research is being published on the termite resistance of mass timber, it does not appear to account for invasive species such as the Formosan and Asian subterranean termite, the immense size of their colonies, and their global spread (França *et al.*, 2022).

Inspections and the effectiveness of detection technologies will also change as building construction material and practices change. Understanding the limits and capabilities of those technologies is paramount in appropriate implementation for termite inspections (Katz and Forschler, 2023). For example, in a demonstration project to test the efficacy of detection devices with and without spray foam, Forschler (2022) demonstrated that even with experienced inspectors, the number of areas identified as having evidence of termites varied using a combination of visual inspection, and device aided inspections using moisture meters, infra-red cameras, microwave devices (Terma-Trac®), and thermometers. After spray foam was applied to areas identified with active termites, only the microwave device detected movement in one area.

There are numerous studies related to developing detection technologies for termites (Oliver-Villanueva and Abián-Pérez, 2013; Himmi *et al.*, 2017) but few are commercialized. Green *et al.* (2018) provided a review of acoustic emission detection, electrical and light sensors, radar and microwave detection, remote moisture monitoring, and gas emission detection with the primary goal of subterranean termite detection. Lewis *et al.* (2009) tested an acoustic emission device, borescope, infra-red camera, microwave, and X-ray device on drywood termites. They concluded that there was no signal device that could reliably detect drywood termites, and that the best option is a trained inspector.

Monitoring and detection technology will continue to advance. The global distribution of invasive termites will continue to change, but the biggest challenge to inspections and monitoring will continue to be developing and maintaining a well-trained work force who can recognize the invasive species such as Formosan and Asian subterranean termites, then develop a strategy for management and control.

Appendix A. Definitions

List modified from WDO Inspections Committee (2007, pp. 4–5, except where otherwise referenced).

1. Attic Access – "a visible and unobstructed opening, scuttle or ladder installed specifically for physical entry into attic area of a structure."
2. Accessible Area – "an area of a structure to which the person performing the inspection can either gain physical entry in order to identify and/or determine the presence of wood-destroying organisms."
3. Access Panel – "a visible and unobstructed removable cover, plate or panel installed specifically for the purposes of conducting a visual inspection of the interior of a wall or void."
4. Built Environment – "man-made or modified structures that provide people with living, working, and recreational spaces" (USEPA, 2023).
5. Carton – "material cemented by saliva or excrement" (Emerson, 1938, p. 261)
6. Conducive Conditions – anything in the built environment that provides conditions termites require – particularly food and water – that enables the establishment and maintenance of an infestation.

7. Damage (caused by WDOs) – “any degradation or deterioration of wood, cellulosic, or non-cellulosic components of a structure through the activity of WDO. Damage is not limited to wooden structural members. Examples of damage include insect consumption of cellulose materials (galleries and mining), or WDO created holes (adult emergence holes, kick out holes, swarm holes) in any structural material. Additionally, the deterioration of any cellulose material by the presence of wood decay fungi creates damage. Areas of a structure that may have been obscured by paint or patching materials but that can still be identified by the inspector as WDO damage, must be reported as damage.”
8. Evidence – “visible signs of wood-destroying organisms other than damage or live activity the presence of which indicates possible infestation. Examples of evidence include termite wings, dead insects, drywood termite fecal pellets, powder post beetle frass, old house borer frass, subterranean mud shelter tubes and carton material. Areas of a structure that may have been obscured by paint or patching materials but that can still be identified by the inspector as WDO evidence, must be reported as evidence.”
9. Inaccessible Area – “any area of a structure which is not accessible including, but not necessarily limited to those areas that are enclosed or concealed by wall coverings, floor coverings, furniture, equipment, stored articles, insulation or any portion of the structure in which inspection would necessitate removing or defacing any part of the structure, or cannot be inspected due to the area’s height, width, or safety concerns.”
10. Live – “the detection and observation of living immature or adult wood-destroying insects and active wood decay fungi.”
11. Monitoring – the ongoing evaluation designed to provide feedback after the problem is identified.
12. Multi-Unit Structure – “A structure, consisting of one or more stories, composed of functionally distinct units, such as stores, offices, apartments, town houses, and condominiums, in which units share a common wall of wood or metal studs or single masonry walls, including fire walls.”
13. Normal Means – inspection of visible and accessible areas performed following the guidance in this document.
14. Probing – “the act of penetrating through the surface of a suspected area to determine the type of WDO present.”
15. Sounding – “the non-invasive act of tapping on suspected areas of the structure with a tool in order to detect possible evidence of a WDO that may not be visible.”
16. Structure – “(a)ny building, regardless of design or type of material and used in its construction, whether public or private, vacant or occupied, and adjacent outside areas” (Georgia Structural Pest Control Commission’s Honey Bee Control and Removal workgroup, 2023; p. 33). Structures include buildings with a roof built for habitation or storage, including attached decks, pergolas, pavilions, gazebos and may include the portion of any fence touching or adjoining to a structure, on the designated property. Definitions vary depending on region and purpose. Please follow local regulations.
17. Substructure – “a shallow space below the living/occupied areas of at least a partially basementless structure normally enclosed by the foundation walls.”
18. Substructure Access – “a visible and unobstructed removable cover or opening that provides for physical entry beneath a structure.”
19. Surveillance – the continual process to identify undetected cases of invasive termites.
20. Tools – see Tools and Equipment section for the list of required tools.
21. Treatment – “is any method or device or the application of any substance routinely performed by a Pest Control Company for the purposes of preventing, controlling, eradicating, mitigating, diminishing, or curtailing a WDO in, on, or under a structure.”
22. Visible – Can be seen by the person performing the inspection at the time of the inspection without the need for visual enhancement other than illumination.

Appendix B. Tools and Equipment

(Modified from WDO Inspections Committee, 2007, pp. 4–5)

1. Flashlights or portable lights sufficient to illuminate areas which need to be illuminated during inspection.

2. Ladder sufficient for inspector to safely reach up to 8 feet to access interior attic openings.
3. Slotted head and Phillips head screw drivers.
4. Hand lens or magnifying device with sufficient power to distinguish and identify wings, pellets, frass, etc., of a WDO.
5. Measuring device or tape measure.
6. Binoculars, and possibly crawl bots and drones.
7. Camera to digitally document inspection.

Also consider employing, when appropriate, a moisture meter, boroscope, canines (Brooks *et al.*, 2003), infra-red cameras, microwave technology (Taravati, 2018) or acoustic emission devices (Mankin *et al.*, 2002).

References

Atkinson, T., Rust, M. and Smith, J. (1993) The Formosan subterranean termite, *Coptotermes formosanus* Shiraki (Isoptera, Rhinotermitiddae), established in California. *Pan-Pacific Entomologist* 69, 111–113.

Blumenfeld, A. and Vargo, E. (2020) Geography, opportunity and bridgeheads facilitate termite invasions to the United States. *Biological Invasions* 22, 3269–3282.

Brooks, S., Oi, F. and Koehler, P. (2003) Ability of canine termite detectors to locate live termites and discriminate them from non-termite material. *Journal of Economic Entomology* 96, 1259–1266.

Brown, A.A., Herms, W.B., Horner, A.C., Kelly, J.W., Kofoid, C.A. *et al.* (1934) General Recommendations for the Control of Termite Damage. In: Kofoid, C.A. (ed.) *Termites and Termite Control*. University of California Press, Berkeley, California, pp. 579–591.

Burgos-Rodriguez, J. and Burgiel, S. (2020) Federal legal authorities for the early detection of and rapid response to invasive species. *Biological Invasions* 22, 129–146.

Cabrera, B. and Thoms, E. (2006) Versatility of baits containing noviflumuron for control of structural infestations of Formosan subterranean termites (Isoptera: Rhinotermitidae). *Florida Entomologist* 89, 20–31.

Chouvenc, T. and Foley, J. (2018) *Coptotermes gestroi* (Wasmann) (Blattodea [Isoptera]: Rhinotermitidae), a threat to the southeastern Florida urban tree canopy. *Florida Entomologist* 101, 79–90.

Chouvenc, T., Scheffrahn, R.H. and Buss, L. (2022) Termite species distribution in Florida and UF termite identification services. (ENY-2079), 6 pp. Available at: https://edis.ifas.ufl.edu/publication/IN1360 (accessed 25 February 2023)

Du, H., Chouvenc, T., Osbrink, W. and Su, N.-Y. (2016) Social interactions in the central nest of *Coptotermes formosanus* juvenile colonies. *Insectes Sociaux* 63, 279–290.

Duryea, M.L., Huffman, J.B., English, R.J. and Osbrink, W. (1999) Will subterranean termites consume landscape mulches? *Journal of Arboriculture* 25, 143–150.

Ebeling, W. (1968) Identification, habits, and distribution, termites. In: Ebeling, W. (ed.) *Identification, Biology, and Control of Termites Attacking Buildings*. University of California, Division of Agricultural Sciences, Berkeley, California, pp. 11-–26.

Ebeling, W. (1975) Wood-destroying insects and fungi. In: Ebeling. W. (ed.) *Urban Entomology*. University of California, Division of Agricultural Sciences, Richmond, California, pp. 128–216.

Emerson, A.E. (1938) Termite nests – a study of the phylogeny of behavior. *Ecological Monographs* 8, 247–284.

Evans, T.A. (2010) Rapid elimination of field colonies of subterranean termite (Isoptera: Rhinotermitidae) using bistrifluron solid bait pellets. *Journal of Economic Entomology* 103, 423–432.

Forschler, B. (2022) Demonstration project reporting detection of subterranean termite (Blattodea: Rhinotermitidae) infestation and spray polyurethane foam. *Journal of Entomological Science* 57, 138–144.

Forschler, B., Harron, J. and Jenkins, T. (2001) Case histories involving the Formosan subterranean termite in Atlanta, Georgia USA (Isoptera: Rhinotermitidae). *Sociobiology* 37, 311–322.

França, T., Stokes, C. and Tang, J. (2022) Development of a modified standard termite test for mass timber products. *Wood and Fiber Science* 54, 24–34.

Geiger, C., and Cox, C. (2012) Pest prevention by design: Authoritative guidelines for designing pests out of structures. 89 pp. Available at: https://sfenvironment.org/sites/default/files/fliers/files/final_ppbd_guidelines_12-5-12.pdf (accessed 8 May 2023)

Georgia Structural Pest Control Commission's Honey Bee Control and Removal Workgroup (2023) Honey bee control and removal: A Georgia pest control certification manual. 36 pp. Available at: https://agr.ga.gov/sites/default/files/documents/pest-control/HBR-Manual.pdf (accessed 3 May 2023)

GISD (Global Invasive Species Database) (2023) *100 of the World's Worst Invasive Alien Species. A selection from the Global Invasive Species Database.* Invasive Species Specialist Group (ISSG) of Species Survival Commission (SSC) of the World Conservation Union (IUCN), 12 pp. Available at: http://www.iucngisd.org/gisd/100_worst.php (accessed 25 February, 2023)

Grace, J., Tome, C., Shelton, T., Oshiro, R. and Yates, J. (1996) Baiting studies and considerations with *Coptotermes formosanus* (Isoptera: Rhinotermitidae) in Hawaii. *Sociobiology* 28, 511–520.

Greaves, T. (1962) Studies of foraging galleries and the invasion of living trees by *Coptotermes acinaciformis* and *C. brunneus* (Isoptera). *Australian Journal of Zoology* 10, 630–651.

Green, F., Boardman, C., and Munson, R. (2018) *Remote detection of termite activity in wooden bridge structures*. General Technical Report FPL-GTR-261. United States Department of Agriculture, Forest Service, Forest Products Laboratory, Madison, Wisconsin, 20 pp.

Guillot, F., Ring, D., Lax, A., Morgan, A., Brown *et al.* (2010) Area-wide management of the Formosan subterranean termite, *Coptotermes formosanus* Shiraki (Isoptera: Rhinotermitidae), in the New Orleans French Quarter. *Sociobiology* 55, 311–338.

Haagsma, K., Rust, M.K., Reierson, D.A., Atkinson, T.H. and Kellum, D. (1995) Formosan subterranean termite established in California. *California Agriculture* 49, 30–33.

Himmi, S.K., Yoshimura, T., Yanase, Y., Torigoe, T., Akada, M. *et al.* (2017) Volume visualization of hidden gallery system of drywood termite using computed tomography: A new approach on monitoring of termite infestation. In: McLellan, B. (ed.) *Sustainable Future for Human Security*. Springer, Singapore. doi: 10.1007/978-981-10-5430-3_6

Horner, A.C., Bowe, E.E., Putnam, W. and Chase, E.G. (1934) Buildings. In: Kofoid, C.A. (ed.) *Termites and Termite Control*. University of California Press, Berkeley, California, pp. 599–642.

HUD (2019) *FHA Single Family Housing Policy Handbook*. US Department of Housing and Urban Development Available at: https://www.hud.gov/sites/dfiles/OCHCO/documents/4000.1hsgh.pdf (accessed 25 February 2023)

Husseneder, C., Simms, D. and Riegel, C. (2007) Evaluation of treatment success and patterns of reinfestation of the Formosan subterranean termite (Isoptera: Rhinotermitidae). *Journal of Economic Entomology* 100, 1370–1380.

IECC (International Energy Conservation Code) (2021) Section R202. General Definitions. Available at: https://codes.iccsafe.org/content/IECC2021P2/chapter-2-ce-definitions (accessed 25 February 2023)

International Building Code (2021) Protection against termites. Section 2603.8. Available at: https://codes.iccsafe.org/content/IBC2021P2/chapter-26-plastic (accessed 25 February 2023)

Jenkins, T., Dean, R. and Forschler, B. (2002) DNA technology, interstate commerce, and the likely origin of Formosan subterranean termite (Isoptera: Rhinotermitidae) infestation in Atlanta, Georgia. *Journal of Economic Entomology* 95, 381–389.

Jones, S. (2003) Targeted versus standard bait placement affects subterranean termite (Isoptera Rhinotermitidae) infestation rates. *Journal of Economic Entomology* 96, 1520–1525.

Kalhor, K. and Emaminejad, N. (2020) Qualitative and quantitative optimization of thermal insulation materials: Insights from the market and energy codes. *Journal of Building Engineering* 30, 101275. doi: 10.1016/j.jobe.2020.101275

Kalshoven, L.G.E. (1963) *Coptotermes curvignathus* causing the death of trees in Indonesia and Malaya. *Entomologische Berichten* 23, 90–100.

Katz, O. and Forschler, B. (2023) *Finding termite infestations with today's tools*. Pest Control Technology, February. Available at: https://pctonline.com/article/finding-termite-infestations-with-todays-tools/ (accessed 8 May 2023)

King, E. and Spink, W. (1969) Foraging galleries of Formosan subterranean termite, *Coptotermes formosanus*, in Louisiana. *Annals of the Entomological Society of America* 62, 536–542.

Lai, P.-Y., Tamashiro, M., Yates, J.R., Su, N.-Y., Fujii, J.K. *et al.* (1983) Living plants in Hawaii attacked by *Coptotermes formosanus*. *Proceedings of the Hawaiian Entomological Society* 24, 283–286.

Lee, C.-Y. (2002) Control of foraging colonies of subterranean termites, *Coptotermes travians* (Isoptera: Rhinotermitidae) in Malaysia using hexaflumuron baits. *Sociobiology* 39, 411–416.

Lee, S., Yang, R. and Su, N.-Y. (2008) Tunneling response of termites to a pre-formed tunnel. *Behavioural Processes* 79, 192–194.

Lewis, V., Moore, S., Yang, L., Leighton, Tabuchi, R. *et al.* (2009) Laboratory and field evaluation of detection devices for drywood termites, *Incisitermes minor*. In: Final Report. Research proposal to the California Structural Pest Control Board for Structural Pest Control Research #084-3635-6, pp. 16–33. Available at: https://www.pestboard.ca.gov/howdoi/research/ucbfinal.pdf (accessed 8 May 2023)

Liasidis, P., Jakob, D., Schellenberg, M., Matsushima, K., Lam *et al.* (2022) Falls from ladders: Injury patterns and outcomes. *Journal of Trauma and Acute Care Surgery* 93, 323–331.

Light, S.F. (1934) Habit and habit types of termites and their economic significance. In: Kofoid, C.A. (ed.) *Termites and Termite Control*. Johnson Reprint Corporations, New York: University of California Press, pp. 136–149.

Light, S.F. and Pickens, A.L. (1934) American subterranean termites, their classification and distribution. In: Kofoid, C.A. (ed) *Termites and Termite Control*, Johnson Reprint Corporations, New York: University of California Press, pp. 150–156.

Long, C., Thorne, B., Breisch, N. and Douglass, L. (2001) Effect of organic and inorganic landscape mulches on subterranean termite (Isoptera: Rhinotermitidae) foraging activity. *Environmental Entomology* 30, 832–836.

Mankin, R., Osbrink, W.L., Oi, F.M. and Anderson, B.L. (2002). Acoustic detection of termite infestations in urban trees. *Journal of Economic Entomology* 95, 981–988.

Merriam-Webster (2023) Inspection, 2023. Available at: https://www.merriam-webster.com/dictionary/inspection (accessed 25 February 2023)

Mullins, A., Su, N.-Y. and Owens, C. (2011) Reinvasion and colony expansion of *Coptotermes formosanus* (Isoptera: Rhinotermitidae) after areawide elimination. *Journal of Economic Entomology* 104, 1687–1697.

Myer, A. and Forschler, B. (2019) Evidence for the role of subterranean termites (*Reticulitermes* spp.) in temperate forest soil nutrient cycling. *Ecosystems* 22, 602–618.

Nagro, A. (2015) A-maze-ing journey. *Pest Control Technology* (February 2015). Available at: https://www.pctonline.com/article/pct0215-annual-termite-damage-quest/ (accessed 25 February 2023)

NISC (2023) National Invasive Species Council Annual Work Plan FY 2023. Available at: https://doi.gov/sites/doi.gov/files/final-nisc-fy2023-wp-10-20-2022.pdf (accessed 25 February 2023)

Oi, F. (2022) A review of the evolution of termite control: A continuum of alternatives to termiticides in the United States with emphasis on efficacy testing requirements for product registration. *Insects* 13, 50. doi: 10.3390/insects13010050

Oliver-Villanueva, J.V. and Abián-Pérez, M.A. (2013) Advancing wireless sensors for termite detection in wood constructions. *Wood Science Technology* 47, 269–280.

Osbrink, W.L., Woodson, W.D., and Lax, A.R. (1999) Population of Formosan subterranean termite, *Coptotermes formosanus* (Isoptera:Rhinotermitidae), established in living urban trees in New Orleans, Louisiana, USA. *3rd International Conference on Urban Pests*. Prague, Czech Republic: Graficke zavody hronov. pp. 19–22.

Osbrink, W., Cornelius, M. and Lax, A. (2011) Areawide field study on effect of three chitin synthesis inhibitor baits on populations of *Coptotermes formosanus* and *Reticulitermes flavipes* (Isoptera: Rhinotermitidae). *Journal of Economic Entomology* 104, 1009–1017.

Oshima, M. (1919) Formosan termites and methods of preventing their damage. *Philippine Journal of Science* 15, 319–383.

Peters, B.C. and Fitzgerald, C.J. (2003) Field evaluation of the bait toxicant chlorfluazuron in eliminating *Coptotermes acinacifromis* (Frogatt) (Isoptera: Rhinotermitidae). *Journal of Economic Entomology* 96, 1828–1831.

Pickens, A.L. (1934) The biology and economic significance of the western subterranean termite, *Reticulitermes hesperus*. In: Kofoid, C.A. (ed.) *Termites and Termite Control*. University of California Press, Berkeley, California, pp. 157–183.

Pitts-Singer, T. and Forschler, B. (2000) Influence of guidelines and passageways on tunneling behavior of *Reticulitermes flavipes* (Kollar) and *R. virginicus* (Banks) (Isoptera: Rhinotermitidae). *Journal of Insect Behavior* 13, 273–290.

Prestwich, G. (1979) Chemical defense by termite soldiers. *Journal of Chemical Ecology* 5, 459–480.

Rodriguez-Soria, B., Dominguez-Hernandez, J., Perez-Bella, J. and del Coz-Diaz, J. (2014) Review of international regulations governing the thermal insulation requirements of residential buildings and the harmonization of envelope energy loss. *Renewable and Sustainable Energy Reviews* 34, 78–90.

Rust, M. and Su, N.-Y. (2012) Managing social insects of urban importance. *Annual Review of Entomology* 57, 355–375.

Rust, M., Reierson, D., Paine, E., Kellum, D. and Haagsma, K. (1998) Ravenous Formosan subterranean termites persist in California. *Hilgardia* 52, 34–37.

Scheffrahn, R., and Su, N.-Y. (1994) Keys to soldier and winged adult termites (Isoptera) of Florida. *Florida Entomologist* 77, 460–474.

Smith, B. and Zungoli, P. (1995) Rigid board insulation in South Carolina – its impact on damage, inspection and control of termites (Isoptera, Rhinotermitidae). *Florida Entomologist* 78, 507–515.

Su, N.- Y. (1994) Evaluation of a hexaflumuron bait for population suppression of subterranean termites (Isoptera: Rhinotermitidae). *Journal of Economic Entomology* 87, 389–397.
Su, N.-Y. (2001) A computerized system for remote monitoring of subterranean termites near structures. *Journal of Economic Entomology* 94, 1518–1525.
Su, N.-Y. (2002) Dimensionally stable sensors for a continuous monitoring program to detect subterranean termite (Isoptera: Rhinotermitidae) activity. *Journal of Economic Entomology* 95, 975–980.
Su, N.-Y. (2011) Technological needs for sustainable termite management. *Sociobiology* 58, 229–239.
Su, N.-Y. and Scheffrahn, R. (1986) A method to access, trap, and monitor field populations of the Formosan subterranean termite (Isoptera, Rhinotermitidae) in the urban environment. *Sociobiology* 12, 299–304.
Su, N.-Y. and Scheffrahn, R.H. (1987) Current status of the Formosan subterranean termite in Florida. In: Tamashiro, M. and Su, N.-Y. (eds) *Biology and Control of the Formosan Subterranean Termite*. University of Hawai'i, Honolulu, pp. 27–31.
Su, N.-Y. and Scheffrahn, R. (1996) Fate of subterranean termite colonies (Isoptera) after bait applications – An update and review. *Sociobiology* 27, 253–275.
Su, N.-Y., Ban, P. and Scheffrahn, R. (2000) Control of *Coptotermes havilandi* (Isoptera: Rhinotermitidae) with hexaflumuron baits and a sensor incorporated into a monitoring and baiting program. *Journal of Economic Entomology* 93, 415–421.
Su, N.-Y., Ban, P. and Scheffrahn, R. (2003) Resistance of insecticide-treated foam board insulation against the eastern subterranean termite and the Formosan subterranean termite (Isoptera: Rhinotermitidae). *Journal of Economic Entomology* 96, 1526–1529.
Su, N.-Y., Guidry, E. and Cottone, C. (2016a) Sustainable management of subterranean termite populations (Isoptera: Rhinotermitidae) in Armstrong Park, New Orleans, with durable baits. *Journal of Economic Entomology* 109, 1326–1332.
Su, N.-Y., Guidry, E., Mullins, A. and Cotonne, C. (2016b) Reinvasion dynamics of subterranean termites (Isoptera: Rhinotermitidae) following the elimination of all detectable colonies in a large area. *Journal of Economic Entomology* 109, 809–814.
Su, N.-Y., Yokum, B., Mullins, A., Bordes, E., Cottone, C. and Riegel, C. (2018) Field evaluations of fluid baits against colonies of the Formosan subterranean termite (Blattodea: Rhinotermitidae). *Journal of Economic Entomology* 111, 1806–1812.
Taravati, S. (2018) Evaluation of low-energy microwaves technology (Termatrac) for detecting western drywood termite in a simulated drywall system. *Journal of Economic Entomology* 111, 1323–1329.
Thorne, B., Alspach, S., Tenn, K. and Clark, M. (2021) Actions toward containment, control, and eradication of invasive conehead termites (Blattodea: Isoptera: Termitidae). *Journal of Integrated Pest Management* 12, 1–16.
Tseng, S.-P., Taravati, S., Choe, D.-H., Rust, M.K. and Lee, C.-Y. (2022) Genetic evidence for multiple invasions of *Coptotermes formosanus* (Blattodea: Rhinotermitidae) in California. *Journal of Economic Entomology* 115, 1251–1256.
USEPA (2023) *What is the built environment?* US Environmental Protection Agency. Available at: https://www.epa.gov/smm/basic-information-about-built-environment#builtenviron (accessed 25 February 2023)
Wang, J., Stirling, R., Morris, P., Taylor, A., Lloyd *et al.* (2018) Durability of mass timber structures: A review of the biological risks. *Wood and Fiber Science* 50, 110–127.
WDO Inspections Committee (2007) *Baseline practices for performing 13645 WDO inspections, 2007*. Available at: https://www.fdacs.gov/content/download/3137/file/industry_baseline_final_10-07.pdf (accessed 25 February 2023)
Yoshimura, T., Imamura, Y. and Takahashi, M. (2003) Attacks on foam insulation materials by *Coptotermes formosanus* Shiraki and *Reticulitermes speratus* (Kolbe). *Japanese Journal of Environmental Entomology and Zoology* 14, 213–221.
Zelinka, S.L., Williamson, T., Martinson, K.L. and Ritter, M.A. (2019) *Research needs assessment for the mass timber industry, 2nd North American Mass Timber Research Needs Workshop*. General Technical Report FPL-GRT-273. United States Department of Agriculture, Forest Service, Forest Products Laboratory, Madison, Wisconsin, 41 pp.
Zhang, G., Gu, A., Zhang, D. and Shen, J. (2022) Risk assessment and monitoring of termites in the Forbidden City under global warming. *Studies in Conservation* 67, 334–340.

11 Management using baits

Nan-Yao Su
Entomology and Nematology Department, Ft. Lauderdale Research and Education Center, University of Florida, nysu@ufl.edu

Abstract
Unlike soil insecticide treatments that exclude soilborne termites from structures, the aim of bait application is to eliminate the source of infestation, i.e., termite colony(s). The first bait product using the chitin synthesis inhibitor (CSI), hexaflumuron, was commercialized in 1995, and all current commercial baits use CSIs as the active ingredient. These CSIs are slow-acting and non-repellent and their lethal times (i.e., times until death) are dose-independent. Moreover, due to termites' molting site fidelity, CSI-affected termites return to the central nest and die there. To escape the accumulating corpses, the royal pair relocate to other locations, but only to be surrounded by another wave of corpses again. The cycle repeats itself several times until most workers die of molting inhibitory effects, resulting in colony elimination when all remaining dependent castes (queen, king, soldiers) starve to death.

11.1 Introduction

Before the widespread use of soil termiticide barriers that began in the 1950s, the goal of termite control had been to stop termite damage by eliminating the source of infestation, i.e., termite colony(s). Arsenic dust was the popular choice for termite control in the early 20th century. The statement of Randall and Doody (1934a) that "a small quantity of arsenic dust placed in the galleries of the termite colony would ... kill the insects even in the most remote chambers" (p. 463) is indicative of the then commonly expected outcome of termite control. In the early 20th century, a variety of insecticides were applied in soil surrounding a house with claims that the soil treatments would "eradicate" subterranean termites from the soil (Randall and Doody, 1934b). After conducting a series of experiments, however, Randall and Doody (1934b) concluded that ground treatments "can at best serve only as barriers between sources of infestation and the structures to be protected" (p. 503) and that "ground treatments may kill local termite group infesting a particular location, but it is quite hopeless to expect thus [*sic*] to kill the entire colony system or to make the ground completely free of subterranean termites" (p. 504). Despite their findings, the termite control industry continued to treat soil with termiticides, and when the persistent chlorinated hydrocarbon pesticides became available, they were extensively used by the industry. In the 1950s, one of the chlorinated hydrocarbons, chlordane, became the pesticide of choice, and the earlier goal of eliminating the source of infestation was replaced by the use of

 Biology and Management of the Formosan Subterranean Termite and Related Species (eds Nan-Yao Su and Chow-Yang Lee)
DOI: 10.1079/9781800621596.0011

termiticide barriers to exclude termites from a structure. Notwithstanding the dominance of soil termiticide in the termite control industry, research on measures to eliminate the source of infestation (i.e., termite baits) continued and the first commercial bait product, Sentricon® (DowElanco), was commercialized in 1995. This chapter reviews progress in termite bait research and discusses current and future trends in the development of bait technologies.

11.2 Early bait research

As early as the 1930s, arsenic was incorporated in sugar-based baits, but they did not control termite populations according to field trials (Randall and Doody, 1934a). Even in the 1960s when the soil termiticide was the industry standard for subterranean termite control, research on baits was pursued to "eliminate a termite population in the ground" (Ostaff and Gray, 1975, p. 1321) as many believed baits were more efficacious and inexpensive than termiticide barriers (Esenther and Gray, 1968). Being slow-acting was recognized as one property required for a bait toxicant, as the bait has to be transferred among nestmates, and mirex (dechlorane) was deemed a promising candidate (Esenther and Gray, 1968). Another property that was considered essential was the attractiveness of a bait, and mirex was typically impregnated in wood blocks decayed with brown rot fungi, *Gloeophyllum trabeum* (Pers. ex Fr.) to attract termites in soil (Esenther *et al.*, 1961; Esenther and Beal, 1974). Beard (1974), who observed that shelter tubes of the eastern subterranean termite, *Reticulitermes flavipes* (Kollar), did not intercept a decayed wood block placed as close as 1 cm, questioned the need for an attractant in bait. The experiment of Beard (1974), however, tested the airborne attraction of decayed wood, which is unrealistic because the bait blocks are typically buried in soil. If soil is moist or sufficiently wet, attractants such as 3*Z*,6*Z*,8*E*-dodecatrienol (Matsumura *et al.*, 1968) may permeate through soil via water to direct termite tunnels toward the source at the distance of 12–18 cm (Su, 2005a). In the end, however, incorporation of attractants in baits may be impractical, because baits typically stay in soil for an extended period, during which the chemical attractants may degrade or become too diluted (through water permeation) before the arrival of termites.

In the 1960s–1970s, a series of field trials with decayed wood blocks impregnated with mirex was carried out to "eliminate" populations of *R. flavipes* in the ground, but none claimed that baited termite populations were eliminated (Esenther and Gray, 1968; Beard, 1974; Esenther and Beal, 1974, 1978; Ostaff and Gray, 1975). These studies typically placed mirex-bait blocks alternately with untreated blocks in treated plots, whereas only untreated blocks were placed in control plots. The test periods ranged from 1 to 3.5 years. Their results showed the overall decline of termite activity in treated plots (both treated and untreated blocks) with unchanged activity levels in control plots. Because termite activity declined in treated plots, they claimed the "suppression" of termite activity by mirex baits. These studies failed to claim elimination of *R. flavipes* populations because of lingering termite activities even after years of baiting, and because "there is no way to estimate the initial size of termite colonies" (Esenther and Beal, 1978, p. 607). Their evaluation protocols were hampered by the lack of proper tools to identify termite colonies in the test plots. Without knowing the identity of the termite colony infesting the wood blocks, it was impossible to know if the post-baiting termites in the wood blocks were the surviving individuals of baited colonies, or those of other colonies. Beard (1974), who tested mirex blocks in 20 field plots, also stressed the importance of recognizing termite territories, and the need for improved detection methods. Coincidentally, two sampling techniques for subterranean termites had been independently published in the previous year by La Fage *et al.* (1973) and Tamashiro *et al.* (1973).

11.3 Monitoring stations and *C. formosanus* colony

Tamashiro *et al.* (1973) used the "trap stake" method of Erhorn (1934) to first detect activity of the Formosan subterranean termite, *Coptotermes formosanus* Shiraki, in soil by driving wooden stakes in the ground. An infested stake was replaced by a bundle of wood and covered by a steel can on the ground. Termites in the on-the-ground monitoring station represent a subset of the underground termite population, and the station serves as a window to study the activity of the cryptic subterranean termites. The monitoring station of Tamashiro *et al.* (1973) has been modified and used by many in the study of subterranean termites (Su and Scheffrahn, 1986; Grace *et al.*, 1989). To conceal the monitoring stations in a populated urban environment, for example, Su and Scheffrahn (1986) used a PVC pipe to create an in-ground station to monitor *C. formosanus* activity.

11.3.1 Dye markers and a subterranean termite colony

Another key development in the study of subterranean termites is the use of the dye marker, Sudan Red 7B, to identify interconnection of monitoring stations (Lai, 1977). After establishing a cluster of monitoring stations (Tamashiro *et al.*, 1973), termites from one station were collected, marked, and released to the original station. Results of the mark–recapture study showed that foraging galleries of *C. formosanus* may extend up to 110 m (Lai, 1977). Stations that contained marked termites were deemed to be occupied by the same colony and the area encompassing interconnected stations was considered the colony territory. In this context, a colony is defined as "a group of termites sharing interconnected foraging sites" (Su and Scheffrahn, 1998, p. 3).

Because the underground nesting environment of *C. formosanus* colonies are relatively warm and humid and are suitable for growth of fungi, Lai (1977) hypothesized that application of entomopathogenous fungi, *Beauveria bassiana* (Balsamo) Vuillemin and *Metarrhizium anisopliae* (Metchnikoff) Sorokin, may trigger an epizootic to eliminate a colony. After confirming the interconnection of a group of monitoring stations with the dye marker, Lai (1977) released fungi-contaminated termites in one station and observed termite activities in all interconnected stations. The result showed that termites did not return to the release station, but remained active in interconnected and untreated stations, which indicated that termites were repelled by the fungi. The application of these entomopathogenous fungi did not trigger an epizootic but the protocol used by Lai (1977) for field trials with these biological control agents formed the basis for field evaluations of candidate baits.

11.3.2 Foraging site fidelity of *C. formosanus*

During the early stage of bait development for *C. formosanus* in the 1980s, it was unclear if or how bait toxicants may be transferred among nestmates through trophallaxis. Because a *C. formosanus* colony is large (millions of termites per colony that forage up to 100 m; see Chapter 5 of this volume), the toxicant may be too diluted if the majority of nestmates have to receive the toxicant from a small number of foragers that directly feed on baits. There was a concern that "if the foragers have a site preference and feed only at one or just a few sites, then only those foragers that frequent the treated area will be affected" (Su *et al.*, 1984, p. 1466). Whereas bait might affect the entire colony if "termites had no foraging site preferences, and selected the foraging sites at random, then with time most of the foragers would visit the treated sites. This could eliminate the entire colony ..." (Su *et al.*, 1984, p. 1466). It was under this concern that Su *et al.* (1984) investigated the feeding site fidelity to determine if a toxicant could be delivered by direct feeding to affect the entire colony.

Before the field study, a laboratory study confirmed that trophallaxis transfer of the dye Sudan Red 7B was minimum, and only those directly fed on dyed papers were visibly marked (Su *et al.*, 1983). Dyed papers were then placed in a cluster of stations to emulate bait placement, and termites from other interconnected stations were collected to count the proportions of marked termites. The results showed that, within 3 weeks, 70–100% of foragers of three tested colonies were visibly marked, and there appeared no foraging site fidelity, because the dyed termites would never have reached 100% if some foragers preferred sites with no dyed paper.

The laboratory dye study of Su *et al.* (1983) also showed that the amount of dye in termites could be measured by the absorbance values with a spectrophotometer, and was correlated with the amount of dye paper consumption. When the absorbance values of field-collected termites were examined, they displayed a normal distribution with a single peak. The results led Su *et al.* (1984) to conclude that *C. formosanus* workers appeared to select among the established feeding locations at random. This finding indicated the possibility of colony elimination by bait applications because "foraging sites were selected at random, all the foragers in a given colony would visit a specific foraging site, given sufficient time. It is theoretically possible, therefore, to introduce a suitable control agent in a single foraging site and eliminate the entire colony" (Su *et al.*, 1984, p. 1466).

The absence of foraging site preference of *C. formosanus* was later challenged by Evans (2002), who claimed that movement data of mound-building termites *Coptotermes lacteus* (Froggatt) and *Nasutitermes exitiosus* (Hill) suggested that these termites preferentially foraged at specific sites. A re-examination of data presented by Evans (2002), however, indicated that his data could be explained by a random site selection model instead of otherwise (Su and Scherer, 2003). A simulation program that was constructed to re-analyze the data of Su *et al.* (1984) also suggested that the most plausible explanation of the single peak absorbance distribution was due to the random selection of foraging sites (Su and Scherer, 2003). The study of Su *et al.* (1984) was initiated because the trophallactic transfer of bait toxicants was unknown in the 1980s. Trophallactic transfer of toxicants such as hexaflumuron and other commercially used CSIs has since been confirmed (Sheets *et al.*, 2000; Lewis and Forschler, 2017; Gordon *et al.*, 2022) and numerous field studies have demonstrated eliminations of subterranean termite colonies by bait applications (Su, 2003). In the context of developing termite bait technologies, the argument regarding foraging site fidelity is no longer relevant (Su, 2019)

11.4 Bait toxicant and evaluation

Mirex was widely used in early termite bait studies because it is "slow-acting" and could be spread among colony members before the onset of death (Esenther and Gray, 1968). Su *et al.* (1982) used behavioral responses of *C. formosanus* to classify insecticides into three types: Type I includes repellents such as pyrethroids; Type II includes acute toxicants such as chlorpyrifos; and Type III insecticides include slow-acting non-repellents such as hydramethylnon. It was concluded that Type I and Type II insecticides should be used for soil barrier treatments, and only Type III should be used as a bait toxicant (Su *et al.*, 1982). These two properties would be the first of many to be identified as the requirements for a bait toxicant for elimination of subterranean termite colonies.

11.4.1 Laboratory evaluation of bait toxicants

Because baits are typically incorporated in feeding medium, "no feeding deterrence" was later considered more appropriate than being "non-repellent." Another question for discovery of bait toxicants was the time-trend in mortality of a toxicant. In the screening of candidate toxicants for imported fire ants, *Solenopsis invicta* Buren, the delayed action was defined as < 15% at 1 day and > 89% at 14 days post-treatment (Stringer *et al.*, 1964). Esenther and Gray (1968) used similar criteria for selecting mirex as the bait toxicant. In a laboratory study, Su *et al.* (1987) used Probit analysis to estimate the time required to achieve 90% mortality of *C. formosanus* 14 days post-exposure, and defined ELT_{90} (effective lethal time 90%) as the "amount of time required to cause 90% mortality by a fixed concentration of a toxicant that killed at least 90% of termites in 14 days" (p. 8). Their results showed that lethal times of tested insecticides were concentration-dependent, i.e., at higher concentrations, even those considered to be "slow-acting" could cause a fast kill. Slow-acting insecticides, however, are recognizable by their ELT_{90} values that span a broad time range, whereas acute toxicants tend to exhibit a narrow range of ELT_{90} values. It should be noted that the term "lethal time" that has been used in termite bait studies implies "time until the onset of death after exposure," instead of toxicant "exposure time" as in the discipline of invertebrate pathology.

Toxicant lethality and feeding deterrence are inter-related and concentration-dependent (Su and Lees, 2009). Once a candidate slow-acting AI (active ingredient) is identified, a laboratory choice test is needed to determine the minimum concentration that may cause significant mortality (α: mortality threshold concentration) and the minimum concentration that may cause feeding deterrence (β: deterrence threshold concentration). Both α and β of an AI are species specific and may differ substantially depending on feeding medium. If $\alpha < \beta$, then the AI may have a potential to be a successful bait toxicant, and the better if α is substantially smaller than β, or $\alpha << \beta$. On the other hand, if $\alpha > \beta$, the AI may deter termite feeding without causing significant mortality and should be excluded from further testing.

11.4.2 Field evaluation protocol

The use of dye markers to establish interconnection of monitoring stations (Tamashiro *et al.*, 1973) as proposed by Lai (1977) is probably the single most important development for the field evaluation of baits that are intended to eliminate the target colonies of subterranean termites. After releasing marked termites from a station, any stations containing marked termites are considered interconnected and belong to the same colony. The area encompassing interconnected stations is considered the territory of a subterranean termite colony. When a bait is applied in a station, its efficacy against the colony could be objectively assessed by measuring the change of termite activities (wood consumption rate or number of termites) in untreated stations that are interconnected with the treated station within the territory (Su and Scheffrahn, 1998). This information on the colony affinity of untreated stations or feeding blocks was missing in early bait toxicant studies of the 1960s to 1970s, rendering their inability to determine if a baited colony was eliminated (Esenther and Gray, 1968; Beard, 1974; Esenther and Beal, 1974, 1978; Ostaff and Gray, 1975). In recent years, DNA fingerprinting instead of dye markers has been used to identify colony affinity of subterranean termites (Husseneder and Grace, 2001). DNA fingerprinting technique is faster than dye markers and can be used to determine if termites that reappear in the foraging territories of supposedly "eliminated" colonies originated from neighboring colonies, or if they are survivors of baited colonies. Most field evaluation studies with baits since the 1990s adopted the protocol of first establishing colony affinity with dye markers or DNA fingerprinting, applying baits in one or a few of the stations, and measuring the change in termite activities in untreated and interconnected stations. The exception has been some evaluation studies with mound-building termites that rely on direct excavations (Peters and Fitzgerald, 2003).

11.4.3 Metabolic inhibitor: laboratory evaluation

Hydramethylnon was identified as a bait toxicant due to its slow-acting and no-feeding deterrent property. When hydramethylnon (impregnated in paper towels) was field tested against two colonies of *C. formosanus*, the activity of one colony was partially suppressed, but the other colony was unaffected (Su, 1982). Matsuoka *et al.* (1998), however, reported a successful elimination of an artificially planted colony of *C. formosanus* with hydramethylnon baits.

Other post-mirex metabolic inhibitors were identified from laboratory choice tests. In a choice test with wood slices (Su and Scheffrahn, 1988) impregnated with a biocide, A-9248 (dihaloalkyl arylsulfone), mortality threshold (α) and deterrence threshold (β) were determined at 400 and 8,000 ppm, respectively, for *C. formosanus* (Table 11.1). Because $\alpha < \beta$ (and $\beta/\alpha = 20$), A-9248 was considered a good candidate for further field evaluation. In another laboratory choice study, sulfluramid-impregnated wood blocks were tested against *C. formosanus* and *R. flavipes* (Su and Scheffrahn, 1991), and the results showed that α and β of sulfluramid were 4 and 10 ppm ($\beta/\alpha = 2.5$) for *C. formosanus*, and 18 and 30 ppm ($\beta/\alpha = 1.7$) for *R. flavipes*, respectively (Table 11.1). For a comparison, mirex-impregnated wood blocks were also tested. The results showed that α and β of mirex were 10 and 90 ppm for *C. formosanus* ($\beta/\alpha = 9$), and 9 and 15 for *R. flavipes* ($\beta/\alpha = 1.7$), respectively. With the largest β/α ratio of 20 for *C. formosanus*, A-9248 appeared to be the most promising bait toxicant, followed by mirex ($\beta/\alpha = 9$), but mirex had not been field-tested against *C. formosanus* before it was banned. Due to the generally low β/α ratio for *C. formosanus* (2.5) and *R. flavipes* (1.7), sulfluramid was considered the least promising bait toxicant. The low β/α ratio of mirex for *R. flavipes* (1.7) may explain its failures to eliminate colonies of this species despite being extensively tested in the field (Esenther and Gray, 1968; Esenther and Beal, 1974, 1978; Beard, 1974; Ostaff and Gray, 1975).

11.4.4 Metabolic inhibitor: field evaluation

Due to the favorable β/α ratio, A-9248 was selected for field evaluation against *C. formosanus* colonies (Su *et al.*, 1991). Composite wooden blocks (three each of treated and untreated wood boards) were placed in some of the in-ground stations of three *C. formosanus* colonies, and termite activities were monitored from untreated and interconnected stations every 30–40 days (Su *et al.*, 1991). After one year of baiting, activities of three baited *C. formosanus* colonies declined, but were not eliminated. The same protocol was used for field evaluation of sulfluramid against another three colonies of *C. formosanus* (Su *et al.*, 1995a). When composite bait blocks were placed in in-ground

Table 11.1. Minimum AI concentrations (ppm) that caused significant mortality (α: mortality threshold concentration) or feeding deterrence (β: deterrence threshold concentration) in laboratory choice tests with *C. formosanus* or *R. flavipes* fed on candidate AIs impregnated in different feeding medium

		C. formosanus		*R. flavipes*		
AI	Feeding medium	α	β	α	β	References
A-9248	Wood slice	400	8,000	–	–	Su and Scheffrahn (1988)
Sulfluramid[a]	Wood block	4	10	18	30	Su and Scheffrahn (1991)
Mirex	Wood block	10	90	9	15	
Diflubenzuron	Wood block	1,000	2	7.8	31.3	Su and Scheffrahn (1993)
Hexaflumuron	Wood block	15.6	125	2	62.5	
Lufenuron	Sawdust	8,000	2,000	800	100	Su and Scheffrahn (1996)
Hexaflumuron	Sawdust	125	20,000	31.3	8,000	

[a]Learned avoidance

stations, wooden boards treated with non-deterrence concentrations (Su and Scheffrahn, 1991) were initially fed upon by termites, but were later avoided by two colonies. After sulfluramid concentrations were reduced, termite feeding resumed for a few months before the feeding deterrence was again detected. Because reduced concentrations only resulted in the reoccurrence of feeding deterrence, Su *et al.* (1995a) suggested that some *C. formosanus* that ingested sublethal doses of sulfluramid probably "learned" to avoid this AI. After 12-month baiting, sulfluramid reduced activities of three *C. formosanus* colonies, but could not eliminate them.

Field trial results of A-9248, sulfluramid, and hydromethylnon against *C. formosanus* resembled those of mirex against *R. flavipes* colonies, i.e., following months of baiting, termite colonies were suppressed but not eliminated. These results led Su *et al.* (1995a) to conclude that, because "lethal time" of these metabolic inhibitors is dose-dependent, termites that consumed higher doses of these AIs were probably killed too quickly to satisfy the slow-acting criteria, and dose-independent "lethal time" had to be one of the requirements of a candidate bait-toxicant in addition to being allow-acting and non-deterrent. In this context, lethal time to produce 100% mortality (lethal doses) of fast-acting AIs (Type II) (Su *et al.*, 1982), is also dose-independent; i.e., regardless of doses ingested by termites, as long as the doses produce 100% mortality, the time required for an AI to fully express its effect is the same (Fig. 11.1A, T_h), which is usually within hours post-exposure (Su and Lees, 2009). At sublethal doses of fast-acting AIs (Fig. 11.1A, T_{mh} and T_{ml}), however, the time required to reach the final mortalities may increase (Smith and Rust, 1990). On the other hand, the time required for slow-acting metabolic inhibitors (Type III AIs) to fully express their effects on termites may be prolonged if mid-low doses are ingested (Fig. 11.1B, T_{ml}). Their lethal times are reduced when mid-high doses are ingested (Fig. 11.1B, T_{mh}), and may be similar to fast-acting AIs at high doses (Fig. 11.1B, T_h). At a low dose (sublethal dose), the final mortality may not reach 100% (Fig. 11.1B).

11.4.5 Chitin synthesis inhibitor (CSI): laboratory evaluation

Another class of potential bait toxicant that may meet the requirement of dose-independent lethal time is the chitin synthesis inhibitors (CSIs). The relationship between lethal time and dose of a CSI is substantially different from those of fast-acting AIs or slow-acting metabolic inhibitors. Because CSIs mainly affect termite molting, and the timing of ecdysis is independent of doses, as long as a lethal dose is ingested, the high (T_h), medium-high (T_{mh}) or medium low (T_{ml}) doses would have produced more or less the similar lethal time (Fig. 11.1C).

Early laboratory studies (Doppelreiter and Korioth, 1981) demonstrated the ecdysis inhibitory effects of diflubenzuron against *Heterotermes indicola* (Wasmann) and *R. flavipes*. Although lethal time of diflubenzuron is theoretically dose-independent, results of a field trial with this CSI on *Microcerotermes* species were inconclusive (Faragalla *et al.*, 1985). The unfavorable field results suggested that other factors such as mortality (α) and deterrence threshold (β) need to be considered before carrying out a field trial.

In a laboratory study, Su and Scheffrahn (1993) compared effects of wooden blocks impregnated with diflubenzuron or hexaflumuron against *C. formosanus* and *R. flavipes*. Three weeks after feeding on one of the CSIs, some termite workers displayed symptoms of molting inhibitory effects as shown in Fig. 11.2. The results showed that diflubenzuron-impregnated blocks yielded significant mortality of *C. formosanus* at > 1,000 ppm (α = 1,000 ppm) but termite feeding was deterred at a low concentration of 2 ppm (β = 2 ppm), i.e., $\alpha > \beta$ for diflubenzuron against *C. formosanus*, and this CSI was deemed ineffective for this species (Table 11.1). For *R. flavipes*, α = 7.8 ppm and β = 31.3 ppm, and with β/α = 4.0, diflubenzuron was considered a bait AI candidate for *R. flavipes* (Su and Scheffrahn, 1993). With hexaflumuron-impregnated blocks, α and β for *C. formosanus* were 15.6 and 125 ppm, and 2 and 62.5 ppm, respectively, for *R. flavipes* (Table 11.1). With $\alpha < \beta$ for both

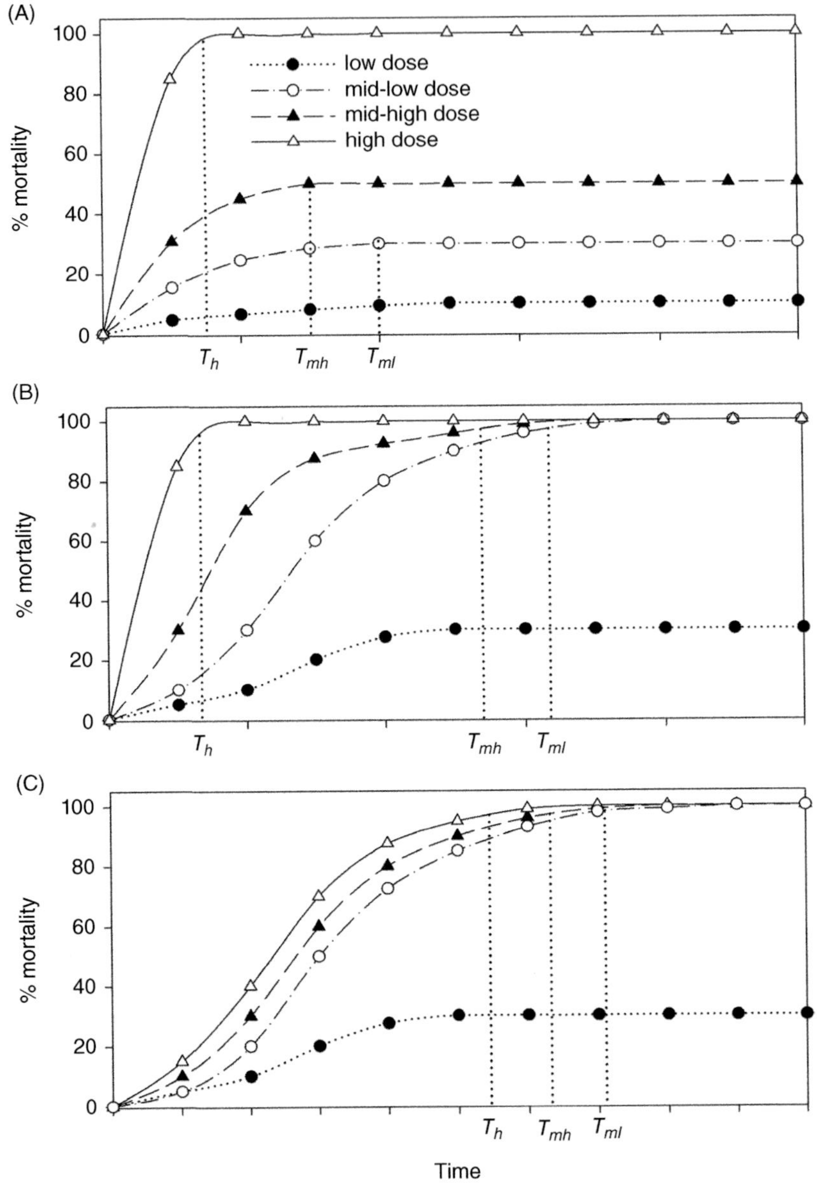

Fig. 11.1. (A) The relationship between lethal time and dose of fast-acting toxicants, (B) slow-acting metabolic inhibitors, and (C) chitin synthesis inhibitors (CSIs) such as hexaflumuron or noviflumuron. Lethal time is defined as the time for an active ingredient (AI) to fully express its effects against test insects. At the high doses, both fast-acting AIs and slow-acting metabolic inhibitors cause substantial mortalities quickly (T_h, A and B). The reduced doses of a fast-acting AI produce lower final mortalities, and time required to reach the final mortality tends to increase as shown in T_{mh} and T_{ml} (A). When the doses are reduced (T_{mh} and T_{ml}) for slow-acting metabolic inhibitors, the final mortalities remain high and are reached at protracted time frames (B). CSIs mainly affect termite molting, and the timing of ecdysis is independent of doses. As long as a lethal dose is ingested, the high, medium-high or medium-low doses would have produced more or less the similar lethal time (C). (Modified with permission from Su and Lees, 2009.)

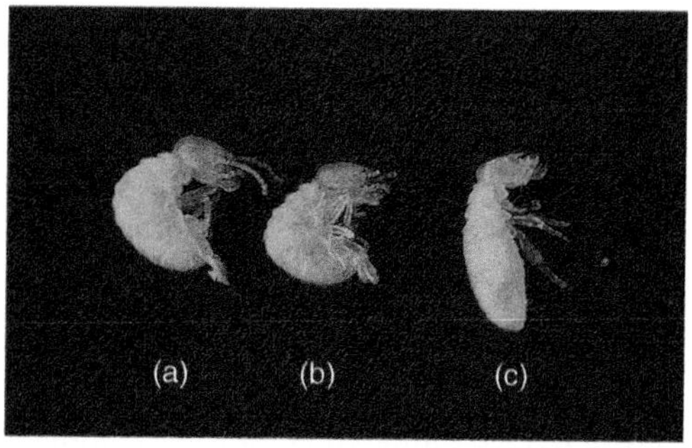

Fig. 11.2. Symptoms of molting inhibition exhibited by workers of *C. formosanus* when exposed to one of the two chitin synthesis inhibitors, hexaflumuron or diflubenzuron. Some affected workers were unable to shed exuviae fully, resulting in a "jackknife" pose (a), whereas others had their antennae (b) or appendages (c) cannibalized by the nestmates. (Reprinted with permission from Su and Scheffrahn, 1993.)

species, hexaflumuron was considered a promising bait toxicant for these two economically important termite pests (Su and Scheffrahn, 1993). When the feeding medium was changed from wooden block to sawdust in another laboratory test (Su and Scheffrahn, 1996), feeding deterrence thresholds (β) of hexaflumuron were markedly improved from 125 to 20,000 ppm and 62.5 to 8,000 ppm for *C. formosanus* and *R. flavipes*, respectively (Table 11.1). This may be due to the uneven distribution of AIs in impregnated wood blocks. Wood blocks are typically pressure-impregnated in acetone or alcohol solution of an AI and left to air dry before use. Because the solvent evaporates from the wood surface, higher AI concentrations tend to accumulate there, which deter termite feeding when they encounter the treated blocks. On the other hand, sawdust was usually mixed with a solvent solution of an AI and constantly stirred during the drying process, which would yield a feeding medium with a relatively homogeneous AI concentration. Even with the sawdust feeding medium, however, mortality threshold of lufenuron (α) was higher than its deterrence threshold (β), or α > β for *C. formosanus* and *R. flavipes*, suggesting that lufenuron was ineffective as a bait toxicant against either species (Table 11.1).

11.4.6 Chitin synthesis inhibitor (CSI): field evaluation

With the favorable β/α ratio of 160 and 255 for *C. formosanus* and *R. flavipes*, respectively (Table 11.1), hexaflumuron-impregnated sawdust was selected for field trials against three colonies each of these two species in 1991 (Su, 1994). Activities of these colonies (wood consumption rate, g wood/day/station) were monitored from in-ground monitoring stations 5–24 months before bait application, and their foraging territories were confirmed by using the mark–release protocol of Lai (1977). Wooden survey stakes were driven into the ground of foraging territories of target colonies, and infested stakes were replaced with bait tubes partially filled with sawdust baits impregnated with hexaflumuron. Coined "self-recruitment" procedure, termites extracted from infested stakes were placed in the head space of a bait tube to recruit nestmates after they fed on baits and exited the tube. After 2–4 months of baiting, activities of three baited *R. flavipes* colonies ceased. It took 5–9 months of bait applications to eliminate two colonies of *C. formosanus*, and activity of the third *C. formosanus* declined to near zero after 3 months of baiting but the low activity lingered on for another 12 months before a complete cessation. Following the results of Su (1994), a total of 34 independent field studies with hexaflumuron were carried out between 1995 and 2003 that demonstrated elimination of 158 of 165 (96%) baited colonies of 13 pest species of subterranean termites in the U.S. and seven other countries (Su, 2003). Most of these field studies followed the evaluation protocol of Lai (1977).

11.5 Commercial baits and CSIs

The positive field evaluation results with hexaflumuron led to the commercialization of the first bait product for subterranean termite control, Sentricon® system (Dow-Elanco), in 1995. Before long, other bait products that used sulfluramid (FirstLine®, FMC Corp.), hydramethylnon (Subterfuge®, BASF Corp.), and diflubenzuron (Labyrinth®, Ensystex, Inc.) were also commercialized, but baits containing metabolic inhibitors such as sulfluramid and hydramethylnon have since been discontinued.

Unlike metabolic inhibitors, increasing numbers of CSIs have been laboratory- and field-tested over the years, and currently all commercial bait products contain CSIs. One of the post-hexaflumuron CSIs, noviflumuron, was found to be two- to threefold more efficacious than hexaflumuron against *R. flavipes*, according to a laboratory study (Karr *et al.*, 2004). Field evaluation with noviflumuron baits demonstrated elimination of four structural infestations of *C. formosanus* (Cabrera and Thoms, 2006). Between 2004 and 2006, Eger *et al.* (2012) conducted a large-scale field evaluation study in the U.S. and showed that applications of noviflumuron baits eliminated a total of 122 colonies of six species of subterranean termites (number of colonies), including *C. formosanus* (5), *Heterotermes aureus* (Snyder) (4), *R. flavipes* (72), *Reticulitermes hageni* Banks (22), *Reticulitermes hesperus* Banks (9), and *Reticulitermes viginicus* (Banks) (10). Area-wide IPM projects with hexaflumuron and noviflumuron baits were successfully carried out in several regions (Chapter 15 of this volume). Noviflumuron (Recruit® HD termite bait) is the AI of the current Sentricon® system, whereas hexaflumuron is the AI of Shatter® termite bait for the Hex Pro® system (Corteva Agriscience) and Terminate® (Spectracide).

Notwithstanding its unfavorable laboratory results (Su and Scheffrahn, 1996), lufenuron baits were successfully field-tested against six *R. hesperus* populations in California (Haverty *et al.*, 2010), and one aerial population of *R. flavipes* in Oklahoma (Bowen and Kard, 2012). However, lufenuron baits have not been commercialized. Diflubenzuron has been used in the commercial bait Labyrinth® of the Exterra® System, but a field test showed that it "did not have any observable effect on populations of either *C. formosanus* or *R. flavipes*" (Osbrink *et al.*, 2011, p. 1011).

Another CSI bait that is available in the U.S. is Trelona® (Advance® Termite Bait System, BASF Corp.). A field study in Texas showed that novaluron, the AI of Trelona®, eliminated nine and two structural infestations of *Reticulitermes* species and *C. formosanus*, respectively (Keefer *et al.*, 2015). Using 4-year-old incipient colonies connected to extended (15 m) foraging arenas, Chouvenc (2021) demonstrated that novaluron baits eliminated laboratory colonies of *C. gestroi* in 45 days. Shults *et al.* (2021) also reported elimination of 15 colonies of *Reticulitermes* species with novaluron baits in a large area.

Osbrink *et al.* (2011) showed the elimination of *C. formosanus* populations by chlorfluazuron baits in New Orleans, but this CSI is not currently available in the U.S. Outside of the U.S., baits containing chlorfluzuron (Requiem™, Exterra® System, Ensystex Australia Pty Ltd.) or bistrifluron (Exterm®, Sumitomo Chemical), have been extensively tested. Field evaluation studies showed that chlorfluzuron baits were successfully applied to eliminate colonies of the mound-building termite, *Coptotermes acinaciformis* (Froggatt), in Australia (Peters and Fitzgerald, 2003), *Macrotermes gilvus* (Hagen) (Dhang, 2011), and *C. gestroi* (Rojo and Acda, 2019) in the Philippines, *Coptotermes curvignathus* Holmgren (Sukartana *et al.*, 2009), *M. gilvus* (Lee *et al.*, 2014), and *C. gestroi* (Umar and Majid, 2020) in Malaysia, and *Heterotermes indicola* Wasmann in Pakistan (Manzoor and Pervez, 2016). Studies also demonstrated that bistrifluron baits were used to eliminate colonies of *C. acinaciformis* (Evans, 2010), *Nasutitermes exitiosus* (Hill) (Webb and McClintock, 2015), *Coptotermes lacteus* (Froggatt) (Webb, 2017) in Australia, and *Gobitermes sulphureus* (Haviland) in Malaysia (Neoh *et al.*, 2011).

11.5.1 Molting site fidelity and CSI baits

It is apparent from the above that only CSI baits consistently eliminated colonies of subterranean termite, including mound-builders. It should be noted, however, that not all CSIs possess such capability, and some CSIs are probably more species-specific than others. Their dose-independent lethal time may account for successful colony eliminations by these CSIs, but it may not explain everything. For example, Su (2005b) kept large *C. formosanus* groups collected from field foraging populations in extended foraging arenas (50 m), and all termites progressively died throughout the network of arenas after feeding on a CSI bait, regardless of the distance from bait placement. In many field-evaluation studies, however, individuals with symptoms of molting inhibitory effects (Fig. 11.2) have rarely, if any, been observed from monitoring stations or bait stations, i.e., foraging sites. Moreover, *C. formosanus* workers are known to undergo a 6-day fasting period before molting but field-collected individuals do not molt in the laboratory until day 11 after being collected (Raina *et al.*, 2008), i.e., if some termites undergoing a pre-molt fasting are found in foraging groups, they should have molted within 6 days after being collected to the laboratory.

These observations suggested that *C. formosanus* workers do not molt at foraging sites such as monitoring stations or bait stations, which prompted Kakkar *et al.* (2017) to hypothesize that *C. formosanus* leave the foraging sites before the onset of ecdysis and move to specific site(s) to go through the fasting and molting process. To test the hypothesis, Kakkar *et al.* (2017) re-examined termite specimens (in 95% EOH) from the foraging distance study of Su *et al.* (2017) and discovered that pre-molt, molting, and post-molt workers were found in the main nest in the vicinity of royal pairs, eggs, and brood (Fig. 11.3). The finding suggested that *C. formosanus* workers returned to the main nest to molt. In a follow-up study, 7-year-old incipient colonies of *C. formosanus* were placed in extended foraging arenas composed of six planar arenas connected with tubes to form a 25 m foraging distance (Kakkar *et al.*, 2018). Noviflumuron baits were introduced into treatment chambers located at the farthest end from the central nest arenas where royal pairs, eggs, and brood were found. Following the consumption of noviflumuron baits, pre-molt workers gathered near the eggs and royal pairs and died there when they failed to complete the ecdysis (Fig. 11.4). As dead workers accumulated in the central nest, royal pairs migrated to another planar arena, and eggs and brood were also transported near them. As time passed, however, pre-molt workers that ingested lethal doses of

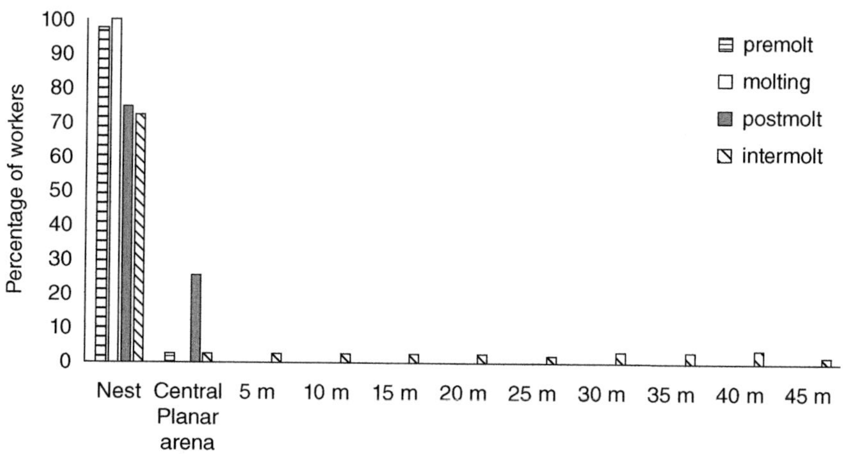

Fig. 11.3. A re-examination of data of Su *et al.* (2017) revealed that pre-molt (identified by the mottled head capsule), molting and post-molt (identified by the pale and less sclerotized mandibles) were found exclusively near the central nest in the vicinity of royal pairs, eggs, and broods. (Reprinted with permission from Kakkar *et al.*, 2017.)

Fig. 11.4. After ingesting lethal doses of noviflumuron, workers return to the central nest and die of chitin synthesis inhibitory effects (arrows) near royal pair and eggs. (Reprinted with permission from Kakkar *et al.*, 2018.)

noviflumuron again gathered in the planar arena with royal pairs and eggs, and again died there, which promoted the migration of royal pairs, eggs, and brood to other planar arenas. During the 60-day test period, the position of the central nest changed six to eight times until the entire colony collapsed (Kakkar *et al.*, 2018).

Results of Kakkar *et al.* (2018) illustrate a sequence of events that might take place following applications of CSI baits to a field colony of subterranean termites. After intercepting bait stations (Fig. 11.5A), foraging termites consume baits and spread CSIs to other individuals via trophallaxis throughout the colony, even with minimum feeding on baits (< 3 days of feeding) by foragers that accounts for < 5% of the colony population (Chouvenc and Su, 2017; Gordon *et al.*, 2022). Because the lethal time of a CSI is dose-independent, even those that have acquired lethal doses continue to function normally. Individuals that are about to molt then return to the central nest (Fig. 11.3) and die of molting inhibitory effects in the vicinity of royal pairs and eggs (Fig. 11.4). The queen is then exposed to CSI through trophallaxis and rapidly becomes unable to lay viable eggs (Chouvenc and Lee, 2021; Tong, 2021). To escape the accumulating corpses, the royal pair abandon the corpse-filled central nest and move to other locations. Eggs and brood are also transported to the corpse-free new central nest. A new group of pre-molt workers, however, eventually gather at the new central nest, and die there due to molting inhibitory effects (Xing *et al.*, 2014), which prompts the royal pair to migrate to another site, but only to be surrounded by corpses again. The cycle repeats itself several times until the majority of workers die of molting inhibitory effects, and the queen, king, and soldiers progressively die from starvation (Chouvenc, 2018), ultimately resulting in colony elimination (Fig. 11.5B). As concluded by Kakkar *et al.* (2018), the molting site fidelity of termites explains why only CSI baits were able to achieve colony elimination, and why all current commercial bait products contain CSIs as the AIs. In comparison, other bait products using non-CSIs inevitably lead to the accumulation of dead individuals near treatment sites, resulting in emergence of secondary repellency, which often prevents colony elimination (Su, 2005b; Chouvenc, 2018).

11.5.2 Monitoring–baiting program, in-ground (IG), and above-ground (AG) bait stations

The earlier protocol for Sentricon® System employed a monitoring–baiting procedure (Su *et al.*, 1995b) similar to that described in the field evaluation study of Su (1994). Wooden monitors are first placed in the IG (in-ground) station to detect the presence of termites. A termite control technician has to be on-site to open the stations manually and inspect the wood monitors. When termites are found in the IG stations, infested wooden monitors are replaced with tubes containing baits (Fig. 11.5A). The routine on-site monitoring is required monthly to quarterly, depending on termite species, season, and previous incidents of infestation. The monitoring–baiting procedure was necessary because baits were too fragile to be kept in soil for a prolonged period. The routine on-site visit, however, was labor intensive and expensive to operate. To save the expenses associated with routine monitoring, the current Sentricon® System no longer employs the monitoring–baiting procedure, but most other commercial bait products do.

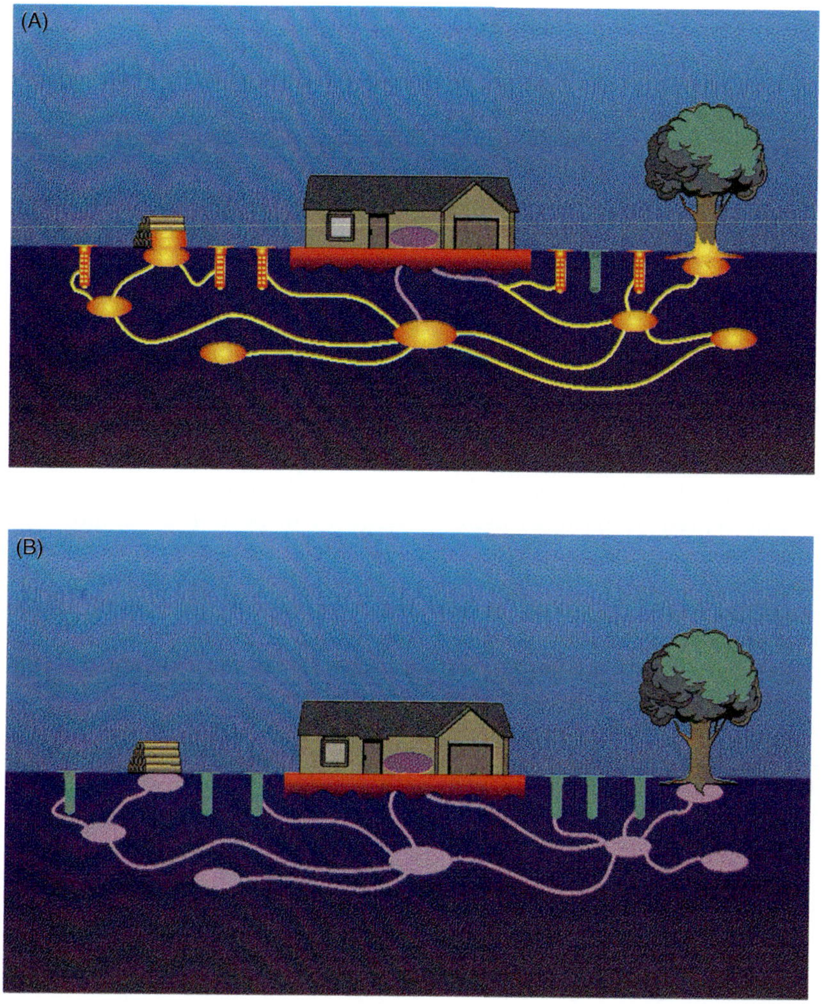

Fig. 11.5. (A) After intercepting bait stations, foraging termites consume baits and spread chitin synthesis inhibitors to other individuals via trophallaxis. Pre-molting individuals return to the central nest and die of molting inhibitory effects in the vicinity of royal pairs and eggs. To escape the accumulating corpses, the royal pair relocate to other locations, but only to be surrounded by another wave of corpses again. (B) The cycle repeats itself several times until the majority of workers die of molting inhibitory effects, resulting in colony elimination.

In addition to IG (in-ground) stations, AG (above-ground) bait stations were offered for the Sentricon® System. Live termite infestations in a home or trees are often noticed by homeowners, and if not tampered with, provide an opportunity for immediate delivery of baits to the colony. The first AG bait station comprised a re-closable plastic bag containing moistened bait (Su *et al.*, 1997). A hole was cut from one side of the plastic bag and adhered on an active infestation of *C. formosanus* to provide termite access to baits. The bait bag was further covered by a piece of close-cell polyethylene foam for protection. A field study with AG stations showed that, of the five *C. formosanus* colonies that received the AG bait stations, four were eliminated in 2–9 months (Su *et al.*, 1997), and IG bait stations were used to eliminate the lingering populations of the last colony. Two commercial prototype AG stations, soft-style (bait in a flexible plastic bag) and hard-style (bait in a plastic box), were also evaluated against colonies of *R. flavipes* and *C. gestroi* (Su *et al.*, 2001). Recruit® AG FlexPack™ (Corteva Agriscinece) offers an AG station consisting of a flexible package containing the bait matrix with two cover options, a hard cover or a flexible cover. Field studies with the hard-style commercial AG system demonstrated elimination of all detectable activities of subterranean termites (Weissling and Thoms, 1999; Lee, 2002). Beside Sentricon® System, the Exterra® System also offers AG bait stations.

If properly installed, baits in AG stations are readily consumed by termites (especially if baits are kept moist), resulting in faster colony elimination than IG stations that have to be intercepted by termites before baits are consumed. Live infestations in a home, however, are often sprayed with over-the-counter insecticides or otherwise tampered with by homeowners, and the opportunity to employ AG stations is lost before the arrival of termite control professionals. Moreover, when installed in visible locations in a home, AG stations are unsightly and are often tampered with or removed by homeowners, rendering them useless. If used indoors, the homeowners have to be at home to provide access to termite control professionals for a routine service, which is inconvenient for both parties.

Despite these challenges, an AG station is an important tool that can yield a faster colony elimination and, for species such as *C. gestroi* in some regions, may be the only choice to eliminate colonies. For bait evaluation studies against *C. gestroi* colonies in Malaysia (Lee, 2002; Sajap *et al.*, 2005) and the Philippines (Rojo and Acda, 2019), for example, only AG stations were used to deliver baits. In other studies in Malaysia (Lee, 2002; Umar and Majid, 2020) and Little Cayman (Su *et al.*, 2000), both IG and AG baits were used to eliminate *C. gestroi* colonies. Su *et al.* (2023) reported that in southeastern Florida, none of IG bait stations was intercepted by *C. gestroi*, but all baited colonies were eliminated with the use of AG bait stations only. The exact reason for the inability of *C. gestroi* populations in some regions to intercept survey stakes, IG monitoring, or bait stations is unknown, but they readily consumed AG baits, resulting in colony eliminations. For pest control professions serving the areas with *C. gestroi* populations, a priority use of AG bait stations is recommended, at least until the issues with IG stations for this species are resolved.

11.5.3 Durable bait

The routine on-site monitoring of in-ground stations was needed primarily because cellulosic bait matrix degrades quickly in the ground, but it is also the most labor-intensive and costly element of the monitoring–baiting procedure. For some termite species, a frequent inspection of monitoring stations may also result in station avoidance (Swoboda *et al.*, 2004). With aggressive species such as *C. formosanus*, on the other hand, infrequent inspection may result in the total consumption of wood monitors and station abandonment.

To avoid these problems associated with the monitoring phase of early baiting system, a durable bait was developed to withstand the hostile soil environment (Su, 2007). Cellulosic baits (impregnated with hexaflumuron) hermetically enveloped with closed-cell polyethylene sheets were field-tested against colonies of *C. formosanus* and *R. flavipes*. With a few exceptions of infiltration by plant roots that compromised the polyethylene envelopes, all other hermetically sealed baits withstood the elements for over 12 months. In the presence of termites, the closed-cell polyethylene sheet was readily tunneled through, and baits were consumed by termites. After 1–3 placements of durable baits, three colonies each of *C. formosanus* and *R. flavipes* were eliminated (Su, 2007).

The first durable bait, Recruit® HD (High Density) with Always Active Technology of Sentricon® System, became commercially available in 2010. The introduction of Recruit® HD removed the monitoring phase of the monitoring–baiting procedure of the early version of the Sentricon® System. Field studies showed that Recruit® HD baits were preferred over wooden monitors by *C. formosanus*, *H. aureus*, *R. flavipes*, *R. malletei*

Clément, *R. hageni*, and *R. hesperus* (Eger *et al.*, 2011), and no significant degradation of its active ingredient, noviflumuron, was detected after 5 years of aging in the field (Eger *et al.*, 2014). Current labels for Recruit® HD prescribe an on-site monitoring at least once a year, and HD baits are replaced only when more than one-third of the bait has been consumed. Because of the less frequent on-site monitoring, more termite control firms adopted Recruit® HD with Always Active™ Technology, and the uses of Sentricon® System for subterranean termite control increased twofold between 2010 and 2015. Recruit® HD provides a viable tool for a sustainable area-wide termite management program, as the durable baits can eliminate invading populations of subterranean termites and maintain a termite-free area (Su *et al.*, 2016) (Chapter 15 of this volume). Durable baits such as Recruit® HD may also be employed in a large agricultural field where the monitoring of bait stations is impractical.

11.6 Future development

Numerous changes and improvements have been made since the launching of the first commercial bait in 1995 (Su, 2019). The bait matrices of the Sentricon® System alone, for example, have been changed from the original wood flour (Recruit®) (i.e., sawdust) to Laminated Textured Cellulose (LTC) with hexaflumuron (Recruit® II) in 1996, LTC with noviflumuron (Recruit® III) in 2003, a compressed briquette cellulose formulation (Recruit® IV) in 2005, and polymer/cellulose mixture of current Recruit® HD in 2010. Technologies do not stand still and have to keep up with society's demands as well as the competition. The following are some examples of potential future innovations in termite bait technology.

11.6.1 Fluid bait

To overcome difficulties with applications of AG stations mentioned above, a fluid bait was developed that can be injected into active infestation of termites for a faster elimination of termite colonies (Su, 2015). The fluid bait comprised a mixture of α-cellulose, fine-ground Summon® disc (FMC corp.), and methylcel solution to form a paste-like texture. Summon® is made of compressed cellulosic powder and is known to increase feeding by *C. formosanus*, according to a field study by Cornelius and Lax (2005). The paste-like fluid baits were packed in caulking tubes and injected into active termite galleries through drill holes. A field study using fluid bait with hexaflumuron showed that *C. formosanus* colonies in buildings and trees were eliminated in 4–22 months (Su *et al.*, 2018). The longest elimination time of 22 months was probably due to the presence of multiple colonies that invaded territories of previously baited and eliminated colonies.

A large quantity of fluid bait can be injected into active foraging galleries or voids in wood, and a single visit to a site may provide sufficient baits to eliminate the target colony(s). If used in conjunction with termite detectors such as acoustic or motion detectors, multiple active loci may be identified to apply more fluid baits throughout a house. Fluid baits are injected through small drill holes that can be patched up and become less conspicuous than an AG station. Moreover, fluid baits can be injected into a hole of any surface or position that may not accommodate the installation of an AG station. These advantages over the current AG bait station should encourage more bait applications into active infestation of termites, which should achieve a faster colony elimination.

11.6.2 Remote monitoring

To alleviate labor expenses of the routine monitoring phase of the early monitoring–baiting program, an electronic sensor was developed to detect termite feeding in IG stations (Su *et al.*, 2000). The sensor consisted of an IG wooden monitor on which a conductive circuit was painted, and termite feeding was detected through circuit breakage.

The sensors were further incorporated in a datalogger that routinely recorded the circuit integrity, and a remote computer was programmed to communicate with the datalogger by telephone lines for data download (Su, 2001). These studies proved the concept of transforming termite feeding activity into electronic signals that could be transmitted through telecommunication. Due to cost issues, the computerized remote system did not become a product, but the sensor technology was further refined and commercialized as Sentricon® System with ESP™ (Electronic Sensing and Protection). The commercial sensor comprised a polymer-enforced paper painted with a carbon-based circuit and was attached to an RFID (Radio Frequency Identification) chip that was activated by a hand-held interrogator. With the ESP™ technology, a pest control professional can detect sensor circuit breakage without opening an IG station. The ESP™ technology was used primarily in large commercial sites with many IG stations to save labor and time needed to manually open and inspect the stations.

With the introduction of durable baits such as Recruit® HD, a routine on-site inspection of IG wooden monitor is no longer necessary, but the incorporation of ESP™ in IG stations still reduced labor expenses for on-site monitoring even with the currently label-prescribed annual inspection. Moreover, the current Sentricon® label dictates that Recruit® HD baits have to be replaced when more than one-third of the bait has been consumed. With the availability of inexpensive internet-based communication technologies, it is now possible to design an automated system to alert the host computer when the event of substantial bait consumption occurs. Such a system will further reduce the expense of on-site service and overall cost of baiting systems.

11.6.3 Colony elimination time

One shortcoming of baiting systems over the soil insecticide treatment is the relatively lengthy time it takes to achieve the final goal, i.e., colony elimination, notwithstanding that soil treatments do not eliminate termite colony(s). For IG baiting systems, the time required to eliminate a colony can be divided into three segments: (1) interception time; (2) bait toxicant acquisition time; and (3) lethal time (Su, 2019).

Interception time refers to the time required for subterranean termites to encounter IG bait station(s) in soil, and depends on termite population pressure, number and size of bait stations, their placement, and bait attractiveness and palatability. Employment of larger stations in larger quantities may improve interception time, but it will also increase operation costs. Hence, the size and number of IG stations to be installed have to be a compromise between economic concern and efficacy. Use of attractants may lure termites in from a distance, but considering their limited effective distance (Su, 2005a) and short residual activity in soil, their practical use is questionable. However, use of phagostimulants such as Summon® discs is known to increase the interception rates of IG stations (Cornelius and Lax, 2005). As with other subterranean termites, *C. formosanus* is known to prefer moist conditions, and excavates more tunnels in sand with higher moisture content (Su and Puche, 2003). It has been recommended to place IG stations in moist soil, or add water to keep the stations moist, but except for testimonials, data are scare to show to what extent such practices may improve interception time.

Bait toxicant acquisition time is the time required for termites to acquire lethal doses of toxicants and depends on the availability of competing food source and bait palatability. For this, it is recommended to remove other food sources at bait sites, but again, no data are available to show how effective such practice is. Baits with phagostimulants will probably increase the rate of bait intake and reduce toxicant acquisition time. Field studies showed that Recruit® HD is preferred over wooden stakes (Eger *et al.*, 2011), and may help to improve the toxicant acquisition time.

Lethal time refers to the time required for a toxicant to kill termites after ingesting lethal doses. Because the lethal time of a CSI is dose-independent, as long as a CSI is used in a bait, it was assumed lethal time cannot be reduced. However, a recent discovery that termites return to the central nest to molt (Kakkar *et al.*, 2017), which accounts for the success of CSI baits in colony elimination, Kakkar *et al.* (2018), suggested that the use of ecdysteroids may achieve the same efficacy of CSI baits with reduced lethal time. Ecdysone is known to trigger insect molting, but the excessive amounts of ecdysone and its derivative, 20-hydroxyecdysone (20E), that persist in the blood stream prevents insects from completing the ecdysis, resulting in “hyperecdysonism” (Williams, 1968). The symptoms of hyperecdysonism in termites are similar to those of CSI effects and the outcome is fatal (Su and Monteagudo, 2017). Ingestion of lethal doses of 20E immediately triggered the onset of ecdysis, followed by hyperecdysonism and mortality in 12 days. Because the 12-day lethal time of ecdysteroids is shorter than that of a CSI (45 days), baits with ecdysteroids alone or in conjunction with CSIs may reduce the colony elimination time by more than a month.

The above three technologies are but some of the innovations that may enter the termite bait market in the future. Baits are safer and more sustainable than soil insecticides (Chapter 15, this volume). Introduction of these innovations will hopefully make bait technologies more acceptable to the termite control industry.

References

Beard, R.L. (1974) Termite biology and bait-block method of control. *Connecticut Agricultural Experiment Station Bulletin* 748.

Bowen, C.J. and Kard, B. (2012) Termite aerial colony elimination using lufenuron bait (Isoptera: Rhinotermitidae). *Journal of Kansas Entomological Society* 85, 273–284.

Cabrera, B.J. and Thoms, E. (2006) Versatility of baits containing noviflumuron for control of structural infestations of Formosan subterranean termites (Isoptera: Rhinotermitidae). *Florida Entomologist* 89, 20–31.

Chouvenc, T. (2018) Comparative impact of chitin synthesis inhibitor baits and non-repellent liquid termiticides on subterranean termite colonies over foraging distances: colony elimination versus localized termite exclusion. *Journal of Economic Entomology* 111, 2317–2328.

Chouvenc, T. (2021) Subterranean termite (*Coptotermes gestroi* (Blattodea: Rhinotermitidae)) colony elimination through exposure to a novaluron CSI bait formulation in laboratory. *Journal of Economic Entomology* 114, 1249–1255.

Chouvenc, T. and Lee, S.B. (2021) Queen egg laying and egg hatching abilities are hindered in subterranean termite colonies when exposed to a chitin synthesis inhibitor bait formulation. *Journal of Economic Entomology* 114, 2466–2472.

Chouvenc, T. and Su, N.Y. (2017) Subterranean termites feeding on CSI baits for a short duration still results in colony elimination. *Journal of Economic Entomology* 110, 2534–2538.

Cornelius, M.L. and Lax, A.R. (2005) Effect of Summon preferred food source on feeding, tunneling, and bait station discovery by the Formosan subterranean termite (Isoptera: Rhinotermitidae). *Journal of Economic Entomology* 98, 502–508.

Dhang, P. (2011) A preliminary study on elimination of colonies of the mound building termite *Macrotermes gilvus* (Hagen) using a chlorfluazuron termite bait in the Philippines. *Insects* 2, 486–490.

Doppelreiter, V.H. and Korioth, M. (1981) Entwicklungshemmung durch Diflubenzuron bei den Bodentermiten *Heterotermes indicola* und *Reticulitermes flavipes*. *Zeitschrift für Angewandte Entomologie* 91, 131–137.

Eger, J.E., Lees, M.D., Fisher, M.L., Messenger, M.T. and Tolley, M.P. (2011) Discovery and consumption of a novel durable bait matrix by subterranean termites (Isoptera: Rhinotermitidae). *Sociobiology* 58, 651-666.

Eger, J.E., Lees, M.D., Neese, P.A., Atkinson, T.H., Thoms, E.M. *et al.* (2012) Elimination of subterranean termite (Isoptera: Rhinotermitidae) colonies using a refined cellulose bait matrix containing noviflumuron when monitored and replenished quarterly. *Journal of Economic Entomology* 105, 533–539.

Eger, J.E., Hamm, R.L., DeMark, J.J., Chin-Heady, E., Tolley, M.P. *et al.* (2014) Durability of a novel durable bait for control of subterranean termites (Isoptera: Rhinotermitidae): Results of five-year field aging studies. *Journal of Economic Entomology* 107, 1201–1205.

Erhorn, E.M. (1934) The termites of Hawaii, their economic significance and control, and the distribution of termites by commerce. In: Kofoid, C.A. (ed.) *Termites and Termite Control*. 2nd Ed. University of California Press, Berkeley, California, pp. 321–333.

Esenther, G.R. and Beal, R.H. (1974) Attractant-mirex bait suppresses activity of *Reticulitermes* spp. *Journal of Economic Entomology* 67, 85–88

Esenther, G.R. and Beal, R.H. (1978) Insecticidal baits on field plot perimeters suppress *Reticulitermes*. *Journal of Economic Entomology* 71, 604–607

Esenther, G.R. and Gray D.E. (1968) Subterranean termite studies in southern Ontario. *Canadian Entomologist* 100, 827–834.

Esenther, G.R., Allen, T.C., Casida, J.E. and Shenefelt, R.D. (1961) Termite attractant from fungus-infected wood. *Science* 134, 50.

Evans, T.A. (2002) Tunnel specificity and forager movement in subterranean termites (Isoptera: Rhinotermitidae and Termitidae). *Bulletin of Entomological Research* 92, 193–201.

Evans, T.A. (2010) Rapid elimination of field colonies of subterranean termite (Isoptera: Rhinotermitidae) using bistrifluron solid bait pellets. *Journal of Economic Entomology* 103, 423–432.

Faragalla, A.A., Badawi, A.I. and Dabbour, A.I. (1985) Field evaluation of the effects of the juvenile hormone analogous (JHA's) and diflubenzuron (Dimilin) on termites of the genus *Microcerotermes* (Isoptera: Termitidae) in the central region of Saudi Arabia. *Sociobiology* 11, 29–37.

Gordon, J.M., Velenovsky, J.F. and Chouvenc, T. (2022) Subterranean termite colony elimination can be achieved even when only a small proportion of foragers feed upon a CSI bait. *Journal of Pest Science* 95, 1207–1216.

Grace, J.K., Abdallay, A. and Farr, K.R. (1989) Eastern subterranean termite (Isoptera: Rhinotermitidae) foraging territories and populations in Toronto. *Canadian Entomologist* 121, 551–556.

Haverty, M.I., Tabuchi, R.L., Vargo, E.L., Cox, D.L., Nelson, L.J. *et al.* (2010) Response of *Reticulitermes hesperus* (Isoptera: Rhinotermitidae) colonies to baiting with lufenuron in northern California. *Journal of Economic Entomology* 103, 770–780.

Husseneder, C. and Grace, J.K. (2001) Evaluation of DNA fingerprinting, aggression tests and morphometry as tools for colony identification of the Formosan subterranean termite. *Journal of Insect Behavior* 14, 173–186.

Kakkar, G., Osbrink, W. Mullins, A. and Su, N.Y. (2017) Molting site fidelity in workers of Formosan subterranean termites (Isoptera: Rhinotermitidae). *Journal of Economic Entomology* 110, 2512–2517. doi: 10.1093/jee/tox246

Kakkar, G., Osbrink, W. and Su, N.Y. (2018) Molting site fidelity accounts for colony elimination of the Formosan subterranean termites (Isoptera: Rhinotermitidae) by chitin synthesis inhibitor baits. *Science Reports* 8, 1259. doi:10.1038/s41598-018-19603-8

Karr, L.L., Sheets, J.L., King, J.E. and Dripps, J.E. (2004) Laboratory performance and pharmacokinetics of the benzoylphenylurea noviflumuron in eastern subterranean termites (Isoptera: Rhinotermitidae). *Journal of Economic Entomology* 97, 593–600.

Keefer, T.C., Puckett, R.T., Brown, K. and Gold, R. (2015) Field trials with 0.5% novaluron insecticide applied as a bait to control subterranean termites (*Reticuliterms* sp. and *C. formosanus* [Isoptera: Rhinotermitidae]) on structures. *Journal of Economic Entomology* 108, 2407–2413.

La Fage, J.P., Nutting, W.L. and Haverty, M.I. (1973) Desert subterranean termites: A method for studying foraging behavior. *Environmental Entomology* 2, 954–956.

Lai, P.-Y. (1977) Biology and ecology of the Formosan subterranean termite, *Coptotermes formosanus*, and its susceptibility to the entomogenous fungi, *Beauveria bassiana* and *Metarrhizium anisopliae*. PhD dissertation, University of Hawaii, Honolulu.

Lee, C.-Y. (2002) Control of foraging colonies of subterranean termites, *Coptotermes travians* (Isoptera: Rhinotermitidae) in Malaysia using hexaflumuron baits. *Sociobiology* 39, 411–416.

Lee, C.-C., Neoh, K.-B. and Lee, C.-Y. (2014) Colony size affects the efficacy of bait containing chlorfluazuron against the fungus-growing termite *Macrotermes gilvus* (Blattodea: Rhinotermitidae). *Journal of Economic Entomology* 107, 2154–2162.

Lewis, J.L. and Forschler, B.T. (2017) Transfer of five commercial termite bait formulations containing benzoylphenyl urea chitin synthesis inhibitors within groups of the subterranean termite *Reticulitermes flavipes* (Blattodea: Rhinotermitidae). *International Journal of Pest Management* 63, 224–233.

Manzoor, F. and Pervez, M. (2016) Evaluation of chlorfluazuron against subterranean termites *Heterotermes indicola* (Isoptera: Rhinotermitidae) in Pakistan. *Journal of Economic Entomology* 109, 2264–2271.

Matsumura, F., Coopel, H.C. and Tai, A. (1968) Isolation and identification of termite trail-following pheromone. *Nature* 219, 963–964.

Matsuoka, H., Tsunoda, K., Yoshimura, T. and Takahashi, M. (1998) Elimination of colonies of *Coptotermes formosanus* Shiraki (Isoptera: Rhinotermitidae) and *Reticulitermes speratus* (Kolbe) (Isoptera: Rhinotermitidae) by bait system. *Wood Research* 85, 84–86.

Neoh, K.-B., Jalaludin, N.A. and Lee, C.-Y. (2011) Elimination of field colonies of mound-building termite *Globitermes sulphureus* (Isoptera: Termitidae) by bistrifluron bait. *Journal of Economic Entomology* 104, 607–613.

Osbrink, W.L.A., Cornelius, M.L. and Lax, A.R. (2011) Areawide field study of three chitin synthesis inhibitor baits on populations fo *Coptotermes formosanus* and *Reticulitermes flavipes* (Isoptera: Rhinotermitidae). *Journal of Economic Entomology* 104, 1009–1017.

Ostaff, D. and Gray, D.E. (1975) Termite (Isoptera) suppression with toxic baits. *Canadian Entomologist* 107, 1321–1325.

Peters, B.C. and Fitzgerald, C.J. (2003) Field evaluation of the bait toxicant chlorfluazuron in eliminating *Coptotermes acinacifromis* (Frogatt) (Isoptera: Rhinotermitidae). *Journal of Economic Entomology* 96, 1828–1831.

Raina, A., Yong, I.P. and Galman, D. (2008) Molting in workers of the Formosan subterranean termite *Coptotermes formosanus. Journal of Insect Physiology* 54, 155–161.

Randall M. and Doody T.C. (1934a) Poison dusts. I. Treatments with poisonous dusts. In: Kofoid, C.A. (ed.) *Termites and Termite Control.* 2nd Ed. University of California Press, Berkeley, California, pp. 463–476.

Randall M. and Doody T.C. (1934b) Ground treatment. In: Kofoid, C.A. (ed.) *Termites and Termite Control.* 2nd Ed. University of California Press, Berkeley, California, pp. 502–513.

Rojo, M.J.A. and Acda, M.N. (2019) Elimination of the Asian subterranean termite *Coptotermes gestroi* (Blattodea: Rhinotermitidae) using above-ground chlorfluazuron baits. *Philippine Agricultural Scientist* 102, 93–96.

Sajap, A.S., Lee, L.C., Ouimette, D. and Jaafar, A.M. (2005) Field evaluation of noviflumuron for controlling Asian subterranean termite, *Coptotermes gestroi* (Isoptera: Rhinotermitidae). In: Lee, C.-H. and Robinson, W.H. (eds) *Proceedings of the 5th International Conference on Urban Pests*, P&Y Design Network, Singapore, pp. 239–241.

Sheets, J.L., Karr, L.L. and Dripps, J.E. (2000) Kinetics of uptake, clearance, transfer, and metabolism of hexaflumuron by eastern subterranean termites (Isoptera: Rhinotermitidae). *Journal of Economic Entomology* 93, 871–877.

Shults, P., Richardson, S., Eyer, P.-A., Chura, M., Barreda, H. *et al.* (2021) Area-wide elimination of subterranean termite colonies using a novaluron bait. Insects 12, 192. doi: 10.3390/insects12030192

Smith, J.L. and Rust, M.K. (1990) Tunneling response and mortality of the western subterranean termite (Isoptera: Rhinotermitidae) to soil treated with termiticides. *Journal of Economic Entomology* 83, 1395–1401.

Stringer, C.E. Jr, Lofgren, C.S. and Bartlett, F.J. (1964) Imported fire ant toxic bait study: evaluation of toxicants. *Journal of Economic Entomology* 57, 941–945.

Su, N.-Y. (1982) An ethological approach to the remedial control of the Formosan subterranean termite, *Coptotermes formosanus* Shiraki. PhD dissertation, University of Hawaii, Honolulu.

Su, N.-Y. (1994) Field evaluation of a hexaflumuron bait for population suppression of subterranean termites (Isoptera: Rhinotermitidae). *Journal of Economic Entomology* 87, 389–397.

Su, N.-Y. (2001) A computerized system for remote monitoring of subterranean termites near structures. *Journal of Economic Entomology* 94, 1518–1525.

Su, N.-Y. (2003) Baits as a tool for population control of the Formosan subterranean termite. *Sociobiology* 41, 177–192.

Su, N.-Y. (2005a) Directional change in tunneling of subterranean termites (Isoptera: Rhinotermitidae) in response to decayed wood attractants. *Journal of Economic Entomology* 98, 471–475.

Su, N.Y. (2005b) Response of the Formosan subterranean termites (Isoptera: Rhinotermitidae) to baits or nonrepellent termiticides in extended foraging arenas. *Journal of Economic Entomology* 98, 2143–2152.

Su, N.-Y. (2007) Hermetically sealed baits for subterranean termites (Isoptera: Rhinotermitidae). *Journal of Economic Entomology* 100, 475–482.

Su, N.-Y. (2015) A fluid-bait for remedial control of subterranean termites. *Journal of Economic Entomology* 108, 274–276.

Su, N.-Y. (2019) Development of baits for management of subterranean termites. *Annual Review of Entomology* 64, 115–130. doi: 10.1146/annurev-ento-011118112429

Su, N.-Y. and Lees, M. (2009) Biological activities of a bait toxicant for population management of subterranean termites. In: Perterson, C. and Stout, D. (eds) *Household, Structural and Residential Pest Management*. Oxford University Press, New York, pp. 87–96.

Su, N.-Y. and Monteagudo, E.J. (2017) Hyperecdysonism in the Formosan subterranean termite and eastern subterranean termite (Isoptera: Rhinotermitidae). *Journal of Economic Entomology* 110, 1736–1739. doi: 10.1093/jee/tox178

Su, N.-Y. and Puche, H. (2003) Tunneling activity of subterranean termites (Isoptera: Rhinotermitidae) in sand with moisture gradients. *Journal of Economic Entomology* 96, 88–93.

Su, N.-Y. and Scheffrahn, R.H. (1986) A method to access, trap, and monitor field populations of the Formosan subterranean termite (Isoptera: Rhinotermitidae) in the urban environment. *Sociobiology* 12, 299–304.

Su, N.-Y. and Scheffrahn, R.H. (1988) Toxicity and feeding deterrency of a dihaloalkyl arylsulfone biocide, A-9248 against the Formosan subterranean termite (Isoptera: Rhinotermitidae). *Journal of Economic Entomology* 81, 850–854.

Su, N.-Y. and Scheffrahn, R.H. (1991) Laboratory evaluation of two slow-acting toxicants against Formosan and eastern subterranean termites. (Isoptera: Rhinotermitidae). *Journal of Economic Entomology* 84, 170–175.

Su, N.-Y. and Scheffrahn, R.H. (1993) Laboratory evaluation of two chitin synthesis inhibitors, hexaflumuron and diflubenzuron, as bait toxicants against Formosan and eastern subterranean termites (Isoptera: Rhinotermitidae). *Journal of Economic Entomology* 86, 1453–1457.

Su, N.-Y. and Scheffrahn, R.H. (1996) Comparative effects of two chitin synthesis inhibitors, hexaflumuron and lufenuron, in a bait matrix against subterranean termites (Isoptera: Rhinotermitidae). *Journal of Economic Entomology* 89, 1156–1160.

Su, N.-Y. and Scheffrahn, R.H. (1998) A review of subterranean termite control practices and prospects for integrated pest management programs. *Integrated Pest Management Review* 3, 1–13.

Su, N.-Y. and Scherer, P. (2003) Feeding site selection by workers of the Formosan subterranean termite (Isoptera: Rhinotermitidae) – A re-analysis of field data from a mark-recapture study. *Bulletin of Entomological Research* 93, 471–477.

Su, N.-Y., Tamashiro, M., Yates, J.R. and Haverty, M.I. (1982) Effects of behavior on the evaluation of insecticides for prevention of or remedial control of the Formosan subterranean termite. *Journal of Economic Entomology* 75, 188–193.

Su, N.-Y., Tamashiro, M., Yates, J.R., Lai, P.Y. and Haverty, M.I. (1983) A dye, Sudan Red 7B, as a marking material for foraging studies with the Formosan subterranean termite. *Sociobiology* 8, 91–97.

Su, N.-Y., Tamashiro, M., Yates, J.R. and Haverty, M.I. (1984) Foraging behavior of the Formosan subterranean termite (Isoptera: Rhinotermitidae). *Environmental Entomology* 13, 1466–1470.

Su, N.-Y., Tamashiro, M. and Haverty, M.I. (1987) Characterization of slow-acting insecticides for the remedial control of the Formosan subterranean termite (Isoptera: Rhinotermitidae). *Journal of Economic Entomology* 80, 1–4.

Su, N.-Y., Ban, P.M. and Scheffrahn, R.H. (1991) Suppression of foraging populations of the Formosan subterranean termite (Isoptera: Rhinotermitidae) by field applications of a slow-acting toxicant bait. *Journal of Economic Entomology* 84, 1525–1531.

Su, N.-Y., Scheffrahn, R.H. and Ban, P.M. (1995a) Effects of sulfluramid-treated bait blocks on field colonies of the Formosan subterranean termite (Isoptera: Rhinotermitidae). *Journal of Economic Entomology* 88, 1343–1348.

Su, N.-Y., Thoms, E.M., Ban, P.M. and Scheffrahn, R.H. (1995b) Monitoring/baiting station to detect and eliminate foraging populations of subterranean termites (Isoptera: Rhinotermitidae) near structures. *Journal of Economic Entomology* 88, 932–936.

Su, N.-Y., Ban, P.M. and Scheffrahn, R.H. (1997) Remedial baiting with hexaflumuron in above-ground stations to control structure-infesting populations of the Formosan subterranean termite (Isoptera: Rhinotermitidae). *Journal of Economic Entomology* 90, 809–817.

Su, N.-Y., Ban, P.M. and Scheffrahn, R.H. (2000) Control of *Coptotermes havilandi* (Isoptera: Rhinotermitidae) with hexaflumuron baits and a sensor incorporated into a monitoring-baiting program. *Journal of Economic Entomology* 93, 415–421.

Su, N.-Y., Ban, P.M. and Scheffrahn, R.H. (2001) Control of subterranean termites (Isoptera: Rhinotermitidae) using commercial prototype aboveground stations and hexaflumuron baits. *Sociobiology* 37, 111–120

Su, N.-Y., Guidry, E. and Cotonne, C. (2016) Sustainable management of subterranean termite populations (Isoptera: Rhinotermitidae) in Armstrong Park, New Orleans, with durable baits. *Journal of Economic Entomology* 109, 1326–1332. doi: 10.1093/jee/tow051

Su, N.-Y., Osbrink,W., Kakkar, G., Mullins, A. and Chouvenc, T. (2017) Foraging distance and population size of juvenile colonies of the Formosan subterranean termite (Isoptera: Rhinotermitidae) in laboratory extended arenas. *Journal of Economic Entomology* 110, 1728–1735. doi: 10.1093/jee/tox153

Su, N.-Y., Yokum, B., Mullins, A.J., Bordes, E., Cotonne, C. *et al.* (2018) Field evaluations of fluid baits against colonies of the Formosan subterranean termite (Blattodea: Rhinotermitidae). *Journal of Economic Entomology* 111, 1806–1812. doi: 10.1093/jee/toy153

Su, N.-Y., Mullins, A. and Chouvenc, T. (2023) Elimination of structural and tree infestations of the Asian subterranean termite, *Coptotermes gestroi* (Wasmann) (Blattodea: Rhinotermitidae) with noviflumuron baits in above-ground stations. *Journal of Economic Entomology* 116(3), 909–915. doi: 10.1093/jee/toad077

Sukartana, P., Sumarni, G. and Broadbent, S. (2009) Evaluation of chlorfluazuron in controlling the subterranean termite *Coptotermes curvignathus* (Isoptera: Rhinotermitidae) in Indonesia. *Journal of Tropical Forest Science* 21, 13–18.

Swoboda, L.E., Miller, D.M., Fell, R.J. and Mullins, D.E. (2004) The effects of nutrient compounds (sugars and amino-acids) on bait consumption by *Reticulitermes* spp. (Isoptera: Rhinotermitidae). *Sociobiology* 44, 547–563.

Tamashiro, M., Fujii. J.K. and Lai, P.Y. (1973) A simple method to observe, trap, and prepare large numbers of subterranean termites for laboratory and field experiments. *Environmental Entomology* 2, 721–722.

Tong, R. (2021) Nitrogen conservation in subterranean termites. PhD dissertation, University of Florida, Gainesville, Florida.

Umar, W.A.S.W. and Majid, A.H.A. (2020) Efficacy of minimum application of chlorfluazuron baiting to control urban subterranean termite populations of *Coptotermes gestroi* (Wasmann) (Blattodea: Rhinotermitidae). *Insects* 11, 569. doi: 10.3390/insects11090569

Webb, G. (2017) Elimination of *Coptotermes lacteus* (Froggatt) (Blattodea: Rhinotermitidae) colonies using bistrifluron bait applied through in-ground bait station surrounding mounds. *Insects* 8, 98. doi: 10.3390/insects8030098

Webb, G. and McClintock, C. (2015) Elimination of the mound-building termite, *Nasutitermes exitiosus* (Isoptera: Rhinotermitidae) in south-eastern Australia using bistrifluron bait. *Journal of Economic Entomology* 108, 2702–2710.

Weissling, T.J. and Thoms, E.M. (1999) Use of acoustic emission detector for locating Formosan subterranean termite (Isoptera: Rhinotermitidae) feeding activity when installing and inspecting aboveground termite bait stations containing hexaflumuron. *Florida Entomologist* 82, 60–71.

Williams, C.M. (1968) Ecdysone and ecdysone-analogues: Their assay and action on diapausing pupae of the Cynthia silkworm. *Biological Bulletin* 134, 344–355.

Xing, L., Chouvenc, T. and Su, N.Y. (2014). Behavioral and histological changes in the Formosan subterranean termite (Isoptera: Rhinotermitidae) induced by the chitin synthesis inhibitor noviflumuron. *Journal of Economic Entomology* 107, 741–747.

12 Management of subterranean termites using liquid termiticides

Chow-Yang Lee[1] and Kok-Boon Neoh[2]

[1]*Department of Entomology, University of California, Riverside, chowyang.lee@ucr.edu;*
[2]*Department of Entomology, National Chung Hsing University, Taiwan, neohkokboon@nchu.edu.tw*

Abstract

Liquid termiticides remain the most popular option for subterranean termite management worldwide. This chapter describes the historical and present perspectives on liquid termiticide use against subterranean termites, particularly the Formosan subterranean termite (FST). We also discuss the types of termiticides (repellent and non-repellent) and the factors affecting their performance and bioavailability in soil. The challenges on the use of repellent and non-repellent termiticides and important research directions that warrant further investigations are also discussed.

12.1 Introduction

The Formosan subterranean termite (FST), *Coptotermes formosanus* Shiraki, is one of the most destructive urban pests in the world (Su, 1990; Evans *et al.*, 2013; Cuthbert *et al.*, 2021). It is also the only termite species listed under the 100 of the World's Worst Invasive Alien Species (Lowe *et al.*, 2000). The damages due to *C. formosanus* infestation amounted to > US$1 billion annually in the U.S. (Su, 2002; Pimentel *et al.*, 2005). In a random survey carried out on 5000 homeowners in the regions in the U.S. where the FST is a known or potential problem, 30% responded that termites had damaged the home they presently lived in (Vlosky *et al.*, 2009).

Since the 1960s, many classes of insecticides have been evaluated and used against the FST. The control of FST has relied heavily on the use of liquid termiticides. Soil treatment using liquid termiticides, either to exclude termites from buildings and structures or as an agent to suppress termite activity, has been widely accepted as a method for managing subterranean termites (Hu, 2011; Potter, 2011). Besides liquid termiticides, baiting (see Chapter 11 of this volume) is an effective option for subterranean termite control. However, liquid termiticide remains popular, although its usage has decreased over the past several years. While it is not possible to subdivide according to the termite groups and species they are used for, the sole use of liquid termiticides in termite control services decreased from 52% in 2020 (n = 396 respondents) to 42% in 2021 (n = 129), to 33% in 2022 (n = 243) in the U.S. On the other hand, the use of both liquid termiticide and baits in termite control services increased from 32% in 2020 to 38% in 2021 and 62% in 2022 (PCT, 2021, 2022, 2023).

 Biology and Management of the Formosan Subterranean Termite and Related Species (eds Nan-Yao Su and Chow-Yang Lee)
DOI: 10.1079/9781800621596.0012

In this chapter, we discuss the use of liquid termiticides against the FST. For more specific details on using wood preservatives against termites and other botanical insecticides, readers can refer to Chapters 13 and 14 of this volume. For general reviews of termiticides and other control options against subterranean termites, please refer to Hu (2011), Potter (2011), Rust (2014), and Oi (2022).

12.2 Application of liquid termiticides

Liquid termiticides are used for treatments at pre- and post-construction. Pre-construction treatment allows termiticides to be evenly applied to the soil and fill, and a polyethylene sheet barrier is usually laid before the building foundation is poured, creating an uninterrupted barrier between the soil and the wooden components of the structure. The termiticide is usually applied to the soil and fill around the perimeter and underneath the foundation (Hu, 2011; Potter, 2011).

Post-construction treatments are applied to structures that have already been built when a termite infestation is found. Occasionally, some building owners may be proactive and opt for soil treatment every several years. They are more labor-intensive and harder to achieve a comprehensive barrier than those that are treated during pre-construction. At post-construction, treatments include using liquid termiticides, termite baits, wood treatments, dusting, etc. When treating with liquid termiticides, they are applied around the perimeter of the building and underneath the foundation. Post-construction treatments require knowledge of building construction, especially on the potential termite entry points. To treat soil along foundations, trenching and rodding techniques are used. For treatment under concrete slabs, termiticides are applied by drilling and injecting through the slab or horizontally treating through the foundation from the exterior (Potter, 2011).

When treating with liquid termiticide, ensuring that it is delivered to potential termite entry points is crucial. However, achieving consistency in coverage and deposition of the liquid treatment is challenging when treating under the slab. To overcome this issue, termiticide is applied as a foam, enabling it to disperse laterally in space under the slabs. Because foam is less dense than the liquid, it will dispense more uniformly, and the bubbles will give the chemical more time to penetrate hard-to-reach areas. Commercial foaming equipment and surfactants are normally used to generate termiticide foam. The foam will spread once injected into the concrete slab, filling up untreated gaps, enveloping any termite entryways, and transporting the termiticides longer distances. The foam's surfactants promote termiticide penetration, resulting in a homogeneous and continuous barrier. Many studies have shown that foam can reach gaps that liquid termiticide cannot and is more effective than liquid termiticide treatment alone. In addition, foam can be injected into cracks, expansion joints, plumbing gaps, wall voids, etc. (Thomas and Robinson, 1991; Barlow and Robinson, 1993; Robinson and Barlow, 1993; Thoms *et al.*, 1993; Potter, 2011). Ready-to-use termiticide in foam in pressurized cans is also available.

12.3 Historical and present perspectives of liquid termiticide usages

Liquid termiticides have been applied to the soil to treat subterranean termites for more than 120 years (Kofoid, 1934). This practice was known as soil poisoning (USDA, 1942). Because of environmental concerns, the availability of novel chemistries, advances in insect toxicology, and regulatory changes, the choice of termiticides and their use have significantly changed in the past (Rust, 2014). Over the past 70 years, there have been at least ten classes of synthetic organic insecticides (chlorinated hydrocarbons, carbamates, diamide, neonicotinoid, organophosphates, oxadiazine, phenyl-pyrazole, pyrrole, pyrethroids, spinosyn, and others) that have been tested both in the laboratory and in the field as liquid termiticides against the FST (Table 12.1).

The earliest compounds tested and/or used against subterranean termites were arsenic trioxide, sodium arsenate, diphenylamine, orthodichlorobenzene, trichlorobenzene, DDT, coal-tar creosote, and pentachlorophenol (Randall and Doody, 1934; Snyder, 1935; Hockenyos, 1939; Kowal and George, 1948; Hetrick, 1950; Mori, 1987). Ehrhorn (1934) described the control of FST in Hawai'i by first attracting them to planted pine wood stakes before the stakes were removed and liquid fumigants such as carbon bisulfide or calcium cyanide were poured into the holes and the holes were covered.

Chlorinated hydrocarbons (cyclodienes) – chlordane and heptachlor were the first synthetic organic insecticides marketed for termite control (Potter, 2011). Cyclodienes are GABA-gated chloride channel antagonists (Yu, 2015). They were inexpensive and had the longest residuals (Bess *et al.*, 1966; Beal, 1971; Beal and Smith; 1971). In field tests, Kard *et al.* (1989) demonstrated that chlordane-treated soils were active for more than 35 years. In another long-term field study in Hawai'i, soil treated with 1% chlordane weathered in exposed field plots for 28 years still effectively killed 54–83% of *C. formosanus* workers (Grace *et al.*, 1993). However, its persistence in the environment and EPA risk assessment (EPA, 1987, 2000) led to its termination of usage in 1988.

Even before chlordane usage was terminated, repellent compounds from the class of organophosphates had been used and became popular. A survey in 1985 indicated that chlorpyrifos (Dursban® TC) was used by 61.4% of the pest management professionals interviewed, and 28.6% reported that it was their principal termiticide. The usage increased to 63.6% the following year, and 33.8% used it as their principal termiticide (Mix, 1987). Another survey of pest management professionals (PMPs) in 1991 revealed chlorpyrifos (Dursban® TC) as the most commonly used termiticide (65%), followed by 12% cypermethrin (Demon® TC), 10.6% permethrin (Dragnet® and Torpedo®), 10% isofenphos (Pryfon®), and 1.8% fenvalerate (Tribute®) (Mix, 1991).

Organophosphates (e.g., chlorpyrifos, fenitrothion, isofenphos, etc.) affect the regular function of acetylcholine at post-synapse by disrupting the function of acetylcholinesterase, prolonging the binding of acetylcholine at its postsynaptic receptor (Yu, 2015). Among chlorpyrifos, fenitrothion, and isofenphos, the latter was the least effective in preventing FST penetration in simulated field studies (Su *et al.*, 1999). Laboratory observation found that FSTs that entered the chlorpyrifos-treated zone were killed quickly, but active termites continued to enter the treated area for 48–96 h. Dead termites piled up in the tunnels, and eventually, the tunnels were sealed, and the termites avoided the treated area in response to dead termites (Su *et al.*, 1982). In laboratory studies on FST, tunneling distance in treated soils was affected by treatment age (Su *et al.*, 1993), termite number (Jones, 1990), and soil type (Osbrink and Lax, 2002).

Pyrethroids (e.g., bifenthrin, cypermethrin, deltamethrin, fenvalerate, lambda-cyhalothrin, permethrin, etc.) showed longer persistence in the soil than organophosphates (Su *et al.*, 1999). They interfere with the normal nerve function of the axon, targeting the sodium channel by altering the rate of sodium gate closure, resulting in an increase in sodium ion influx into the cell. The repetitive stimulation of nerves induces excitatory paralysis and death in termites (Yu, 2015). Unlike organophosphates, pyrethroids are highly toxic and repellent, and the FST avoids tunneling into the treated zone, limiting mortality (Su *et al.*, 1982, 1993).

Oi (2022) pointed out that termites breach through the treatment due to their ability to find the untreated gaps in soil, which could be due to the treatment concentration used (Forschler, 1994), treatment thickness (Su *et al.*, 1995), termite population density (Jones, 1990), and degradation of termiticides (Gold *et al.*, 1996a). On their tests using pyrethroids at 100, 500 and 1000 ppm, Kuriachan and Gold (1998) showed that *R. flavipes* could locate and breach

Table 12.1. Past and present liquid termiticides evaluated against *C. formosanus* in laboratory and field conditions.

Insecticide class	Termiticide(s)	Evaluation method	Concentration[1]	Findings	Reference
Chlorinated hydrocarbon	aldrin, chlordane, dieldrin, heptachlor	Tube test	25–500 ppm	*C. formosanus* penetrated through 50 mm of treated soil of all insecticides	Beal and Smith (1971)
	chlordane, dieldrin	Insecticide added to the concrete	0.5–4%	Termites died when placed directly on dieldrin-treated concrete, but chlordane-treated ones caused little or no mortality	Beal (1971)
	aldrin, dieldrin, chlordane, DDT, heptachlor, lindane	Exposure in the field for 7 years and evaluated in the lab using Petri dish and penetration assays		Toxicity at 7th year ranked in the following order: dieldrin > aldrin > heptachlor > chlordane > DDT > lindane.	Bess and Hylin (1970)
	aldrin, dieldrin, chlordane, DDT, heptachlor, lindane	Mixed with 3 soil materials (crushed coral, sandy loam and clay) and tested at 2, 4, 5 and 6 years	0.075–8% (varies between insecticides)	Aldrin and dieldrin-treated soils are still effective at 6th year. Lindane lost much of its toxicity, while DDT, chlordane and heptachlor-treated soils showed reduced toxicity	Bess *et al.* (1966)
	aldrin, chlordane, DDT, dieldrin and heptachlor	Mixed with 3 soil materials (crushed coral, sandy loam and clay) and tested at 17, 20, 24, 28 and 33 years using penetration bioassay method	0.075–8% (varies between insecticides)	Termites tunneled significantly less in soils treated with 0.5% aldrin (in coral and clay), 0.3% dieldrin (in clay), 1.0% heptachlor (in sandy loam), and 2.0% DDT (in coral) at 28 years. No insecticides prevented termite penetration at 33 years. All except DDT (8.0%) caused high termite mortality up to 28 years after application. Only 0.5% aldrin, 0.3% dieldrin, and 1.0% heptachlor caused significant mortality at 33 years.	Grace *et al.* (1993)
	chlordane	Petri dish and tube tests	500, 1000, 5000 ppm	Termites that entered treated area were killed quickly, but active termites continued to enter treated area for 2–4 days. Eventually, the tunnels were sealed, and termites avoided treated area.	Su *et al.* (1982)

Continued

Table 12.1. Continued.

Insecticide class	Termiticide(s)	Evaluation method	Concentration[1]	Findings	Reference
	chlordane	Tunneling assay through treated soil at different population densities (100, 1000 termites)	500 ppm	100 termites: cumulative tunneling distance = 34 mm, 81% mortality; 1000 termites: cumulative tunneling distance 55.2 mm, 100% mortality	Jones (1990)
	chlordane	Topical assay, glass tube tunneling assay (50 mm)		LD_{50} = 45.34 μg/g insect, termites penetrated 100% at 10 ppm, the lowest concentration to stop penetration was 80–100 ppm.	Su and Scheffrahn (1990)
	chlordane, methoxyclor	Treated filter paper exposure on worker and soldier termites of 4 colonies	630.65 μg/cm^2 (chlordane), 5260 μg/cm^2 (methoxychlor)	chlordane LT_{50} = 10.2–44.1 min (workers), 115.7–236.2 min (soldiers); methoxychlor LT_{50} = 153.8–223.3 min (workers), 97.1–137.4 min (soldiers)	Osbrink and Lax (2001)
	chlordane	Topical assays of termites being held individually or in group		LD_{50} = 59.3 μg/g insect (individual), 45.5 μg/g insect (group)	Valles and Woodson (2002)
	chlordane, methoxychlor	Glass tube tunneling assay (150 mm) through 3 treated substrates (sand, soil and clay) up to 30 days	5 ppm	Chlordane: mortality at 30 days (100% sand, 53.3–70.8% soil, 58.3–74.2% clay), % penetration at 30 days (65–74.5% sand, 100% soil, clay); methoxychlor: mortality at 30 days (80–100% sand, 83.3–86.7% soil, 75.8–83.3% clay), % penetration at 30 days (86–87% sand, 100% soil, clay)	Osbrink and Lax (2002)
Carbamate	bendiocarb	Treated filter paper exposure on worker and soldier termites of 4 colonies	2.63 μg/cm^2	LT_{50} = 198–46.7 min (workers), 30.3–68.1 min (soldiers)	Osbrink and Lax (2001)
	bendiocarb	Glass tube tunneling assay (150 mm) through 3 treated substrates (sand, soil and clay) up to 30 days	5 ppm	Mortality at 30 days (42.5–100% sand, 80–83.3% soil, 44.2–88.3% clay), % penetration at 30 days (16.5–24.5% sand, 90–100% soil, 100% clay).	Osbrink and Lax (2002)

	carbaryl	Petri dish and tube tests	500, 1000, 5000 ppm	Termites that entered treated area were killed quickly, but active termites continued to enter treated area for 2–4 d. Eventually, the tunnels were sealed, and termites avoided treated area.	Su *et al.* (1982)
	fenobucarb	Filter paper assay at 2 h and 21 days exposure	0.00157–1.57 μg/cm^2 (= 0.0001–1.0 mg/ filter paper)	EC_{50} = > 0.01 mg/filter paper (21 days), 0.094 mg/filter paper (2 h).	Kubota *et al.* (2008)
	fenobucarb	Topical assay, glass tube tunneling assay	0.0626–1 μg/termite, 20–1000 ppm fenobucarb in treated soil	LD_{50} (mouthpart) = 0.064–0.11 μg/ termite, LD_{50} (tergite) = 0.29–0.62 μg/termite; termites tunneled completely (50 mm) in soil treated with 20 mg/kg even at 1 d, termites only tunneled <10 mm in soil treated with ≥ 200 mg/kg up to 7 d.	Kubota *et al.* (2007)
Diamide	chlorantraniliprole	Termites exposed in Petri dish with treated soil/sand at different exposure times, and then transferred to a clean dish with healthy termites on moistened filter paper	25, 50 ppm	Delayed mortality (when compared to fipronil) in sand; in soil, chlorantraniliprole did not cause higher mortality to either donor or recipient termites.	Gautam and Henderson (2011)
	chlorantraniliprole	Topical assay, exposure on treated sand	0.4–2.026 ng/termite (topical), 0.16–1.28 ppm (sand)	LD_{50} = 1.32 ng/termite (7 days), 0.72 ng/termite (14 d); LC_{50} = 0.47 ppm (7 days)	Mao *et al.* (2011)
	chlorantraniliprole	Feeding cessation assay, horizontal transfer assays, and field treatment	N/S	Chlorantraniliprole caused feeding cessation in treated *C. formosanus*. Mortality of recipient termites ranged from 15–90% in horizontal transfer experiment. Field trials of chlorantraniliprole revealed 27.3% of the structures have infestations of termites that required re-treatment.	Puckett *et al.* (2012)

Continued

Table 12.1. Continued.

Insecticide class	Termiticide(s)	Evaluation method	Concentration[1]	Findings	Reference
	chlorantraniliprole	JWPA[2] tunneling assay and confined test	125, 250 ppm	Termites tunneled 3–26 mm and died within 3 days. In the confined test, termites died within 48–72 h.	Fujitomo *et al.* (2012)
	chlorantraniliprole	Lab-simulated evaluation of in-ground treatment on above-ground termite survival	1, 10 ppm	40% survival for 10 ppm, 1 ppm did not cause significant mortality compared to control.	Henderson *et al.* (2016)
	chlorantraniliprole	Flooding simulation on treated substrates (sand, soil)	1, 10, 25 ppm	Toxicity of chlorantraniliprole for 10 and 25 ppm was not affected by flooding irrespective of substrate type, except at the lowest concentration (1 ppm) on sand.	Sapkota *et al.* (2020)
	cyantraniliprole	Topical assay, exposure on treated sand	0.64–2.5 ng/termite (topical), 0.16–1.28 ppm (sand)	LD_{50} = 1.50 ng/termite (7 days), 0.95 ng/termite (14 days); LC_{50} = 0.45 ppm (7 days)	Mao *et al.* (2011)
Neonicotinoid	clothianidin, dinotefuran, imidacloprid, thiamethoxam	JWPA[2] tunneling assay and confined test	Clothianidin, dinotefuran: 25, 125 ppm; imidacloprid: 25, 37.5 ppm; thiamethoxam: 50, 100 ppm	Termites tunneled 10–33 mm and died within 3–4 days. In the confined test, termites died within 24–48 h.	Fujimoto *et al.* (2012)
	acetamiprid	Topical assay, treated sandy soil exposure, glass tube tunneling assay in loam soil	0.0384–120 ppm (soil exposure), 0.5–8 ppm (tunneling assay)	LD_{50} = 5.8 µg/termite (120 h), 13.8 µg/termite (72 h), 119 µg/termite (24 h); LC_{50} = 1.05 ppm (72 h), 2.12 ppm (72 h), 3.87 ppm (24 h); 0.4–8.3 cm tunneling distance (0.5–8 ppm).	Mo *et al.* (2005)

cycloxaprid	Continuous and brief (2 h) exposure assays in Petri dish, choice and non-choice tunneling assays, horizontal transfer assay	0.1, 1, 10, 100 ppm in soil and sand	Continuous exposure to 100 ppm caused 100% mortality; brief exposure to 100 ppm killed all termites at 21 days; 0.1 ppm reduced wood consumption; length and area of tunnels in sand treated with 10 and 100 ppm were lower than that of untreated sand; donor termites exposed to soil or sand containing 10 ppm cycloxaprid caused significantly mortality in the recipient termites	Zhang *et al.* (2022)
imidacloprid	Field treatment of trees and periodic sampling of independent monitors	Injection of 0.1% foam	Latent mortality in termites after treatment due to imidacloprid intoxication. However, 6–15 months after treatment, termite populations recovered.	Osbrink and Lax (2003)
imidacloprid	Horizontal transfer experiment: donor termites were exposed to treated sand in Petri dish and then mixed with healthy termites at ratio of 5:95 (donors: recipients)	1, 10, 100 ppm	Mean recipient mortality ranged from 3.9% (1 ppm) to 86.7% (100 ppm). Results of recipient mortality were not consistent with the duration of interaction and concentrations used.	Shelton and Grace (2003)
imidacloprid	Soil treatment around and under building and sampling of termites in 30 independent monitors	0.05%	No termites showed latent mortality after imidacloprid treatment, the soil treatment did not reduce *C. formosanus* populations adjacent to the treatment.	Osbrink *et al.* (2005)
imidacloprid	Lab-simulated evaluation of in-ground treatment on above-ground termite survival	1, 10 ppm	60–70% mortality for the concentrations tested	Henderson *et al.* (2016)
imidacloprid	Topical assay, exposure on treated sand	0.4–2.4 ng/termite (topical), 0.1–0.8 ppm (sand)	LD_{50} = 0.74 ng/termite (7 days), 0.52 ng/termite (14 days); LC_{50} = 0.24 ppm (7 days)	Mao *et al.* (2011)
imidacloprid	Field evaluation – trees were drilled and foamed with imidacloprid	0.5%	No significant difference in termite activity between treated and untreated trees	Osbrink and Cornelius (2013)

Continued

Table 12.1. Continued.

Insecticide class	Termiticide(s)	Evaluation method	Concentration[1]	Findings	Reference
	imidacloprid	Flooding simulation on treated substrates (sand, soil)	1, 10, 25 ppm	For all concentrations, termite mortality was reduced after flooding. Imidacloprid-treated homes may require a retreatment after a flood.	Sapkota *et al.* (2020)
	imidacloprid	Exposure in Petri dish	2.5–40 mg/l	LC_{50} = 17.386 mg/l	Zeng *et al.* (2020)
	thiamethoxam	Treated filter paper exposure in Petri dishes	0.5–500 ppm	Mortality: 45.78–99.56% (1 day), 61.33–100% (4 days), 63.33–100% (7 days), 66.56–100% (15 days)	Remmen and Su (2005a)
	thiamethoxam	Glass tube tunneling assay method	0.1–800 ppm	LC_{50} = 0.32 ppm (1 week exposure), ≥8 ppm thiamethoxam prevented termites from tunneling through 50 mm	Remmen and Su (2005b)
	thiamethoxam	Extended foraging arena of 50 m using tygon tube up to 10 weeks	135.8 ppm (wt/wt)	Mortality: 48.4% (workers), 20.4% (soldiers). Horizontal transfer of lethal effects of the toxicant was shorter than 5 m.	Su (2005)
Organophosphate	chlorpyrifos, diazinon	Petri dish and tube tests	500, 1000, 5000 ppm	Termites that entered treated area were killed quickly, but active termites continued to enter treated area for 2–4 days. Eventually, the tunnels were sealed, and termites avoided treated area.	Su *et al.* (1982)
	chlorpyrifos, fenitrothion, isofenphos	Termiticide applied to simulate commercial pre-construction treatment, and treated sand was sampled at 3 h and 1 year post-treatment and evaluated using glass tube tunneling assay	0.75, 1% (chlorpyrifos), 1.6% (fenitrothion), 0.75% (isofenphos)	3 h: tunneling distance = 0.4–2.1 cm, mortality = 69.8–88.3%; 1 year: tunneling distance = 1.4–5 cm, mortality = 5.2–94%	Su *et al.* (1993)
	chlorpyrifos, fenitrothion, isofenphos	Field degradation studies using miniature slabs, followed by termite penetration assay	0.75, 1% (chlorpyrifos), 1.6% (fenitrothion), 0.75% (isofenphos)	Isofenphos was least effective in preventing penetration.	Su *et al.* (1999)

	chlorpyrifos	Topical assays on normal and defaunated termites		LD_{50} (95% FL) = 2.33 µg/g (2.21–2.45), 3.18 (3.05–3.31) for normal ones, while LD_{50} (95% FL) = 2.22 µg/g (2.10–2.34), 2.83 (2.69–2.96)	Khoo and Sherman (1979)
	chlorpyrifos	Tunneling assay through treated soil at different population densities (100, 1000 termites)	500 ppm	100 termites: cumulative tunneling distance = 1.6 mm, 89% mortality; 1000 termites: cumulative tunneling distance 11.2 mm, 57% mortality	Jones (1990)
	chlorpyrifos	Topical assay, glass tube tunneling assay (50 mm)		LD_{50} = 3.39 µg/g insect. Effective concentration in preventing penetration = 40–100 ppm.	Su and Scheffrahn (1990)
	chlorpyrifos	Glass tube tunneling assay through different thickness of treated sand (0.15–5 cm)	1, 10, 100, 500 ppm (wt/wt)	1 ppm: ~4 cm penetrated, 10 ppm: 2 cm	Su *et al.* (1995)
	chlorpyrifos	Treated filter paper exposure on worker and soldier termites of 4 colonies	526.13 µg/cm^2	LT_{50} = 11.8–35.9 min (workers), 31.6–51.2 min (soldiers)	Osbrink amd Lax (2001)
	chlorpyrifos	Glass tube tunneling assay (150 mm) through 3 treated substrates (sand, soil and clay) up to 30 days	5 ppm	Mortality at 30 days (100% sand, 45–99.2% soil, 10.0–86.7% clay), % penetration at 30 days (8–30% sand, 100% soil, clay)	Osbrink and Lax (2002)
	chlorpyrifos	Topical assays of termites being held individually or in group		LD_{50} = 5.61 µg/g insect (individual), 4.0 µg/g insect (group)	Valles and Woodson (2002)
	chlorpyrifos	Exposure in Petri dish	2.5–40 mg/l	LC_{50} = 1.081 mg/l	Zeng *et al.* (2020)
Oxadiazine	indoxacarb	Topically treated termites (donors) are mixed with healthy termites (recipients) at different ratios and post-exposure times	10, 20, 50, 100, 200 ng/termite	200 ng killed 100% donors and recipients in 20 d; 100 ng killed 68–100%, while ≤50 ng killed ≤60% in 24 days. At 24 d, greater recipient mortalities were observed at ratios ≥1:4 at doses ≥20 ng, but only at a ratio 1:1 at dose 10 ng. Recipient mortality increased significantly as post-exposure time increased.	Hu *et al.* (2005)

Continued

Table 12.1. Continued.

Insecticide class	Termiticide(s)	Evaluation method	Concentration[1]	Findings	Reference
	indoxacarb	Penetration tube bioassay with different thickness of treated zones (20, 50 mm)	1, 10, 50, 100 ppm	Concentrations of ≥10 ppm of both thicknesses caused mortality 30–100%, but mortality caused by 1 ppm did not differ from the control. Termites penetrated through all indoxacarb-treated soil for both thicknesses.	Hu (2005)
	indoxacarb	Topical assay, exposure on treated sand	14–28.8 ng/termite (topical), 0.16–1.28 ppm (sand)	LD_{50} = 17.16 ng/termite (7 days), 14.97 ng/termite (14 days); LC_{50} = 0.79 ppm (7 days)	Mao *et al.* (2011)
	indoxacarb	Treated filter paper in Petri dishes and kept at 16°, 22° and 28°C	50, 75, 100 ppm	Toxicity increased with increasing temperatures	Quarcoo *et al.* (2019)
Phenylpyrazole	fipronil	Treated filter paper exposure on worker and soldier termites of 4 colonies	630.65 μg/cm², 630.65 μg/cm² + 526.13 oil	Fipronil only: LT_{50} = 164.2–248.7 min (workers), 366.3 min (soldiers); fipronil + oil: LT_{50} = 182.7–300.0 min (workers), 229.7 min (soldiers)	Osbrink amd Lax (2001)
	fipronil	Glass tube tunneling assay (150 mm) through 3 treated substrates (sand, soil and clay) up to 30 days	5 ppm	Mortality at 30 days (100% in sand, soil, clay), % penetration at 30 days (28–29.5% sand, 27–30% soil, 47.5–55% clay).	Osbrink and Lax (2002)
	fipronil	Topical assay, horizontal transfer experiment, repellency experiment	0.625–40 ng/insect (topical), 2.5 ng/insect (horizontal transfer), 0.01, 0.0625, 0.125% (repellency)	LD_{50} at 72 h = 1.33–1.39 ng/insect (worker), 1.53–1.80 ng/insect (soldier); horizontal transfer: 5–150% treated soldier caused 22–97.9% mortality in untreated workers; <0.063% no repellency, 0.125% repellent.	Ibrahim *et al.* (2003)
	fipronil	Horizontal transfer experiment: donor termites were exposed to treated sand in Petri dish and then mixed with healthy termites at ratio of 5:95 (donors: recipients)	1, 10, 100 ppm	Mean recipient mortality ranged from 5.6% (1 ppm) to 28.1% (100 ppm). Results of recipient mortality were not consistent with the duration of interaction and concentrations used.	Shelton and Grace (2003)

fipronil	Extended foraging arena of 50 m using tygon tube up to 10 weeks	80.8 ppm (wt/wt)	Mortality: 43% (workers), 27.6% (soldiers). Horizontal transfer of lethal effects of the toxicant was ≤5 m.	Su (2005)
fipronil	JWPA[2] tunneling assay and confined test	45, 75 ppm	Termites tunneled 11–31 mm and died within 2 days. In the confined test, termites died within 24 h.	Fujimoto *et al.* (2012)
fipronil	Termites exposed in Petri dish with treated soil/sand at different exposure times, and then transferred to a clean dish with healthy termites on moistened filter paper	25, 50, 60 ppm	Fipronil showed less delayed mortality (compared to chlorantraniliprole); greater number of donor termites died in the soil at 14 days after treatment, while higher death of recipients occurred at 21 days at 60 ppm exposure.	Gautam and Henderson (2011)
fipronil	Topical assay, exposure on treated sand	0.1–1.6 ng/termite (topical), 0.0025–0.02 ppm (sand)	LD_{50} = 0.20 ng/termite (7 days), 0.14 ng/termite (14 d); LC_{50} = 0.009 ppm (7 days)	Mao *et al.* (2011)
fipronil	Glass tube bioassay, horizontal transfer assay, and field evaluation	0.06%, 0.12%	Tube assay: tunneled distance at 72 h =10 mm (0.06%), 16 mm (0.12%), 100% mortality at 72 h; horizontal transfer: mortality 100% at 96 h (donor:recipient = 1:19), 100% at 72 h (donor:recipient – 10:10); 1 infested house treated showed no termite infestation at 36 months	Keefer *et al.* (2012)
fipronil	Horizontal transfer studies of different group sizes (10, 25 and 50 termites) at donor: recipient ratio of 1:4 using Petri dish method	5, 10 ppm	5 ppm: LT_{50} = 51.3–196.6 h, 44.7–70.7 h, 37.7–121.5 h (10, 25, 50 termites, respectively); 10 ppm: 32.5–47.7 h, 30–35.7 h, 25.9–35.6 h (10, 25, 50 termites, respectively)	Wang *et al.* (2013)
fipronil	Lab-simulated evaluation of in-ground treatment on above-ground termite survival	1, 10 ppm	>95 mortality for both 1 and 10 ppm. Fast acting and more toxic at the concentrations tested than chlorantraniliprole and imidacloprid.	Henderson *et al.* (2016)

Continued

Table 12.1. Continued.

Insecticide class	Termiticide(s)	Evaluation method	Concentration[1]	Findings	Reference
	fipronil	Penetration tube bioassay with different thickness of treated zones (20, 50 mm)	1, 10, 50, 100 ppm	All concentrations of both thicknesses caused mortality 100% at 28 days. Termites penetrated through fipronil-treated soil (all concentrations) at 20 mm thickness, but only 1 and 10 ppm at 50 mm thickness were penetrated.	Hu (2005)
	fipronil	Field evaluation of spot treatment on 4 colonies	0.0125% applied into one monitoring station	All colonies survived the treatment, and only 1 colony showed reduction. The affected colony was treated ~1 m from its primary nest.	Osbrink *et al.* (2014)
	fipronil	Specially designed test arena with a center container connected to 5 peripheral containers with each further connected to 10 more peripheral containers using Tygon tubings	60 ppm to one peripheral container, or 60 ppm to the one of the 10 second peripheral container	~80% mortality at 96 h, and ~90% at 9 days	Gautam *et al.* (2014)
	fipronil	Treated filter paper in Petri dishes and kept at 16°, 22° and 28°C	0.5, 1, 5 ppm	Toxicity increased with increasing temperatures	Quarcoo *et al.* (2019)
	fipronil	Field treatment on 2 colonies using discontinuous soil treatment		One of the 2 colonies were suppressed and possibly eliminated.	Itakura *et al.* (2021)
	fipronil	Treated filter paper exposure in Petri dishes	1–10 ppm	Mortality: 13.67–99.33% (1 day), 63–100% (4 days), 77.78–100% (7 days), 81.56–100% (15 days)	Remmen and Su (2005a)
	fipronil	Glass tube tunneling assay method	0.1–800 ppm	LC_{50} = 0.11 ppm (1 week exposure), ≥1 ppm fipronil prevented termites from tunneling through 50 mm	Remmen and Su (2005b)
	fipronil	Flooding simulation on treated substrates (sand, soil)	1, 10, 25 ppm	Toxicity of fipronil was not affected by flooding irrespective of substrate type, except at the lowest concentration (1 ppm).	Sapkota *et al.* (2020)

Pyrrole	chlorfenapyr	JWPA[2] tunneling assay and confined test	250, 2000 ppm	Termites tunneled 4–42 mm, and died within 2 days. In the confined test (250 ppm), termites died within 6–8 h.	Fujimoto *et al.* (2012)
	chlorfenapyr	Topical assay, exposure on treated sand	8–19 ng/termite (topical), 0.02–0.16 ppm (sand)	LD_{50} = 9.92 ng/termite (7 days), 5.70 ng/termite (14 days); LC_{50} = 0.052 ppm (7 days)	Mao *et al.* (2011)
Pyrethroid	Bifenthrin, cyhalothrin, cypermethrin, deltamethrin, fenvalerate, esfenvalerate, permethrin, silaneophane, tralomethrin	Topical assay, glass tube tunneling assay (50 mm)		LD_{50} (in µg/g insect) = 0.73 (bifenthrin), 0.29 (cyhalothrin), 0.64 (cypermethrin), 0.12 (deltamethrin), 2.14 (fenvalerate), 0.56 (esfenvalerate), 2.03 (permethrin), 5.24 (silaneophane), 0.14 (tralomethrin); effective concentration in preventing penetration: 1 ppm (permethrin), 6–12 ppm (tralomethrin, bifenthrin, cypermethrin, deltamethrin), 8–80 ppm (esfenvalerate), 40–80 ppm (fenvalerate), 40–100 ppm (silaneophane)	Su and Scheffrahn (1990)
	Bifenthrin, cypermethrin, lambdacyhalothrin, permethrin	Termiticide applied to simulate commercial preconstruction treatment, and treated sand was sampled at 3 h and 1 year post-treatment and evaluated using glass tube tunneling assay	0.031% (bifenthrin), 0.25, 0.3, 0.5% (cypermethrin), 0.125, 0.25% (lambdacyhalothrin), 0.5% (permethrin)	3 h: tunneling distance = 0–0.4 cm, mortality = 7.5–17.9%; 1 y: tunneling distance = 0–2 cm, mortality = 6.3–12.1%	Su *et al.* (1993)
	Bifenthrin, cypermethrin, lambdacyhalothrin, permethrin	Field degradation studies using miniature slabs, followed by termite penetration assay	0.031% (bifenthrin), 0.25, 0.5% (cypermethrin, 0.125, 0.25% (lambdacyhalothrin), 0.5% (permethrin)	Permethrin has the longest half-life (21.9 months) and most persistent.	Su *et al.* (1999)

Continued

Table 12.1. Continued.

Insecticide class	Termiticide(s)	Evaluation method	Concentration[1]	Findings	Reference
	Bifenthrin, etofenprox, permethrin, silafluofen	JWPA[2] tunneling assay and confined test	Permethrin: 25, 50 ppm; bifenthrin: 6.25, 12.5 ppm; etofenprox: 100, 200 ppm, silafluofen: 200, 375 ppm	Termites tunneled 1–27 mm, and died within 2 days. In the confined test, termites died within 24–48 h.	Fujimoto *et al.* (2012)
	bifenthrin	Topical assay, exposure on treated sand	0.9–5.0 ng/termite (topical), 0.01–0.08 ppm (sand)	LD_{50} = 3.68 ng/termite (7 days), 2.98 ng/termite (14 days); LC_{50} = 0.07 ppm (7 days)	Mao *et al.* (2011)
	bifenthrin	Flooding simulation on treated substrates (sand, soil)	1, 10, 25 ppm	Toxicity of bifenthrin was not affected by flooding irrespective of substrate type, except at the lowest concentration (1 ppm).	Sapkota *et al.* (2020)
	bifenthrin	Exposure in Petri dish	2.5–40 mg/l	LC_{50} = 3.873 mg/l	Zeng *et al.* (2020)
	cypermethrin	Topical assays of termites being held individually or in group		LD_{50} = 1.13 µg/g insect (individual), 0.74 µg/g insect (group)	Valles and Woodson (2002)
	cypermethrin, permethrin, deltamethrin	Treated filter paper exposure on worker and soldier termites of 4 colonies	5.26 µg/cm² (cypermethrin, permethrin) 0.053 µg/cm² (deltamethrin)	Cypermethrin: LT_{50} = 11.4–34.1 min (workers), 20.8–108.2 min (soldiers); permethrin: LT_{50} = 23.6–83.5 min (workers), 47.4–96.5 min (soldiers); deltamethrin: LT_{50} = 21.0–250.0 min (workers), 31.8–191.2 min (soldiers)	Osbrink and Lax (2001)
	deltamethrin, permethrin	Glass tube tunneling assay (150 mm) through 3 treated substrates (sand, soil and clay) up to 30 days	5 ppm	Deltamethrin: mortality at 30 days (10–83.3% sand, 60–83.3% soil, 13.3–58.3% clay), % penetration at 30 days (14–43% sand, 57.5–100% soil, 90–100% clay); permethrin: mortality at 30 days (93.3–96.7% sand, 51.7–100% soil, 35.8–55.8% clay), % penetration at 30 days (20–34.5% sand, 100% soil, 90–100% clay)	Osbrink and Lax (2002)

	etofenprox	Topical assay, glass tube tunneling assay through 4 cm of treated coral or silica sand (vertical penetration), sandwich assay (horizontal penetration)	Tunneling (1, 10, 100, 1000 ppm)	LD_{50} = 4.8–6.6 µg/g termite; vertical penetration: 1 ppm (100%), 10 ppm (75–87.5%), 100 ppm (4.2–49.2%), 1000 ppm (1.7–14.2%); horizontal penetration: 1 ppm (79.2–100%), 10 ppm (87.5–100%), 100 ppm (2.8–21.7%), 1000 ppm (1.4–14.2%).	Tamashiro *et al.* (1996)
	fenvalerate, permethrin, pyrethrin	Petri dish and tube tests	500, 1000, 5000 ppm	At high concentration (5000 ppm), termites sealed the tunnels to the treated area within 24 h, while at lower concentrations (500, 1000 ppm), termites did not enter the treated area even though the tunnels were not sealed.	Su *et al.* (1982)
	permethrin	Tunneling assay through treated soil at different population densities (100, 1000 termites)	500 ppm	100 termites: cumulative tunneling distance = 0.2 mm, 8% mortality; 1000 termites: cumulative tunneling distance 12.1 mm, 6% mortality	Jones (1990)
	permethrin	Filter paper assay at 2 h and 21 days exposure	0.00157–1.57 µg/cm^2 (= 0.0001–1.0 mg/filter paper)	EC_{50} = 0.030 mg/filter paper (21 days), <0.2 mg/filter paper (2 h).	Kubota *et al.* (2008)
	silafluofen	Petri dish (forced exposure) and tunneling (indirect exposure) assays	1–1000 ppm (Petri dish) 1000–7500 ppm (tunneling)	In forced exposure assays, 100 ppm killed all termites in 24 h. In tunneling assays, 500 ppm prevented complete termite penetration through 7.5 cm tunneling arena, while ≥ 1000 ppm reduced tunneling distance to ≤ 1 cm.	Grace *et al.* (1992)
Spinosyn	spinosad	Horizontal transfer using Petri dish method	1, 25, 50 ppm on sand and soil	Spinosad was better transferred in the sand than in soil. At 25 and 50 ppm in soil, high mortalities of donors and recipients were observed at 7 days.	Bhatta and Henderson (2016)

Continued

Table 12.1. Continued.

Insecticide class	Termiticide(s)	Evaluation method	Concentration[1]	Findings	Reference
	spinosad, spinetoram	Non-choice, two-choice and multi-choice assays using Petri dish method	1, 25, 50 ppm on sand, soil and filter paper	For 25 and 50 ppm, >85% mortality in the sand in non-choice assay after 1 day, and 100% mortality at 7 days in two-choice treatment. For 1 ppm, 80–100% mortality at 14 days for all non-choice treatments.	Bhatta *et al.* (2016)
Others	eugenol, citral, citronellal, geraniol, limonene, nerol, piperidine, pyrrolidine, n-undecane	Toxicity in sand, tunneling, fumigant and feeding deterrence assays		Eugenol was most effective as termiticides and fumigant against *C. formosanus.* It also prevented tunneling of treated sand barrier for at least 5 days. Eugenol was not effective as a feeding deterrent, except at high concentration (100,000 ppm)	Cornelius *et al.* (1997)
	sulfluramid	Topical assay, Feeding assay	0.025–0.35 g/L; (topical assay) 1–1000 ppm (feeding assay)	LD_{50} = 9.04 μg/g insect (at 8 days), 8.16 μg/g insect (at 10 days), 7.78 μg/g insect (at 14 days); 100% mortality for 10 ppm (13 days), 100 ppm (5 days) and 1000 ppm (3 days)	Grace *et al.* (2000)
	menadione	Survivor, feeding and tunneling distance in treated sand assay (choice and non-choice)	1–600 ppm	Non-choice: Survivorship and feeding reduced significantly in all treatments, 10 ppm showed significant reduction in tunneling activity, while 60–600 ppm reduced tunneling activity by > 90%. Choice: Survivorship reduced significantly in 10 and 60 ppm treatments, feeding and tunneling activity were reduced significantly in treatments ≥6 ppm.	Ngo *et al.* (2021)

[1]N/S = not stated, [2]JWPA = Japan Wood Protection Association

untreated gaps at a minimum size of 3–4 cm. Unlike cyclodienes, which provide significant mortality and persist longer in the soil, pyrethroids do not kill many of the test insects (Johnston *et al.*, 1971; Su *et al.*, 1982; Grace *et al.*, 1993). Despite their longer soil persistence than the organophosphates, PMPs often experienced control failures and callbacks with pyrethroid treatments (Gold *et al.*, 1996b).

As a result, the industry and academia supported termite research from 1990 to the early 2000s to develop new termiticides and termite control strategies that were both effective and had a lower impact on the environment. Many advances in termite biology and treatment innovations were made during this period, amongst which were the introduction of termite bait and the non-repellent termiticides such as imidacloprid, fipronil, chlorfenapyr, indoxacarb, and chlorantraniliprole.

Imidacloprid was the first non-repellent neonicotinoid termiticide introduced into the market in the early 1990s. Neonicotinoids cause nerve hyperstimulation and death in insects by mimicking acetylcholine and binding to acetylcholine receptors at post-synapse nerve cells (Yu, 2015). Imidacloprid has a delayed action and causes feeding inhibition (Ramakrishnan *et al.*, 2000) and enhanced susceptibility to infection by entomopathogenic fungi *Metarhizium anisopliae* (Ramakrishnan *et al.*, 1999) in termites. Laboratory studies on FST revealed horizontal transmission of imidacloprid from intoxicated (donor termites) to healthy individuals (recipient termites) (Table 12.1). However, the recipient mortalities were not consistent with the duration of interaction and imidacloprid concentrations used (Shelton and Grace, 2003).

Parman and Vargo (2010) used microsatellite markers to track the fate of the *Reticulitermes* colonies after treatment with imidacloprid exterior perimeter/interior spot treatment. Termite samples were collected at selected time intervals after treatment for up to 2 years. Of all treated colonies, 75% disappeared after 90 days and were not found again. Potter and Hillery (2002) used exterior-targeted imidacloprid treatment (trenching and rodding only) on four structures that were infested by *R. flavipes* and monitored the wood consumption of termites in in-ground monitoring stations. The termite activities in the monitoring stations in three structures were less affected by the treatment, although the visible termite activity involving structural components eventually diminished. Osbrink and Lax (2003) evaluated the field performance of imidacloprid against FST by injecting 0.1% imidacloprid in foam into infested trees and periodically sampled independent monitoring stations. They reported latent mortality of termites collected from the stations after the treatment. However, 6–15 months after the treatment, the FST populations recovered. Another similar study evaluated 0.5% imidacloprid in foam and found no significant difference in FST activity through periodic acoustical monitoring between treated and untreated trees. Soil treatment using 0.05% imidacloprid around and under buildings infested with FST showed no latent mortality in termites sampled from 30 monitoring stations. The soil treatment did not reduce FST termite populations (Osbrink *et al.*, 2005).

Fipronil, a phenylpyrazole compound, was introduced in the late 1990s as a liquid termiticide. It blocks both GABA-gated and glutamate-gated chloride channels (Yu, 2015). It is highly toxic to FST, with topical toxicity (LD_{50}) of 0.14 ng/termite, compared with imidacloprid (LD_{50} = 0.52 ng/termite) and bifenthrin (LD_{50} = 2.98 ng/termite) at 14 days post-treatment (Mao *et al.*, 2011). An earlier study by Ibrahim *et al.* (2003) reported fipronil topical toxicity of 1.33–1.39 ng/termite at 72 h. The discrepancy in toxicity between these two studies could be due to the differences in post-treatment time when the mortality data were collected (72 h vs 14 days), termite colonies used, and experimental handling methods.

Optimal concentrations applied to soil substrate are required to ensure fipronil is non-repellent and provides delayed mortality to termites (Table 12.1). Exposure to fipronil-treated soil at 1 and 5 ppm treated soil concentrations for 1 h resulted in 100% mortality within 5 and 7 days, respectively, in *Reticulitermes hesperus* Banks; while 0.5 ppm treated soil resulted in 78% mortality at 7 days. Although delayed mortality was observed in all test concentrations, the tunneling distances of termites at the respective concentrations were significantly reduced to 10% relative to the control (Saran and Rust, 2007).

Fipronil did not repel *C. formosanus* even at concentrations as high as 64 ppm in glass tube assay. Fipronil in the sand at 1 ppm reduced the penetration distance of *C. formosanus* to approximately 50% compared with the control treatment. Mortality at 1 ppm was high, > 80% at 7 days post-treatment (Remmen and Su, 2005a). Ibrahim *et al.* (2003) used three-chambered transparent plastic containers in which one of the lateral chambers served as a treated chamber, and the other was untreated. Termites were placed in the middle chamber. The authors demonstrated that, at 48 h, significantly fewer termites (3% of workers) were found in the treated chamber (introducing chamber) in fipronil treatments at the highest test concentrations (0.125%) when compared with non-treated chambers. At 0.063% or less, fipronil showed no repellency, while 0.125% was repellent to the worker termites.

Only a few studies have demonstrated the performance of fipronil in suppressing or eliminating subterranean termite populations (Potter and Hillery, 2002; Vargo and Parman, 2012; Keefer *et al.*, 2012). Using exterior-targeted fipronil treatment, Potter and Hillery (2002) treated eight structures infested with *R. flavipes* and found a substantial and rapid reduction in termite foraging activity. They found that 93% (n = 45) of the in-ground monitoring stations within 0.3–4 m became inactive after 3 months, with dead and decaying termites in some of them, reinforcing the fact that the impact of the treatment is short-range. Keefer *et al.* (2012) treated fipronil to termite-infested structures (11 with *R. flavipes* and one with *C. formosanus*), and no termites were detected throughout the 36 months post-treatment. Itakura *et al.* (2021) demonstrated that FST and *Reticulitermes speratus* (Kolbe) colonies moved 28 and 6 m, respectively, for up to 3 weeks before their activities ceased.

Given that fipronil is highly toxic to termites even at low concentrations (1 ppm and 5 ppm in soil), the protective effects of a soil treatment probably have a relatively short range (< 5 m) (Su, 2005; Ripa *et al.*, 2007; Saran and Rust, 2007; Henderson *et al.*, 2016). In southern Arizona, Baker and Miguelena (2020) evaluated the effects of fipronil-treated soil against the desert subterranean termite, *Heterotermes aureus* (Snyder), and found that the areas adjacent to where the plots were treated with fipronil at the recommended rate experienced significant reductions in termite activity, but no reduction in termite activity was found in areas located further away from the soil treatment (> 1.7 m). Their results concurred with the study by Ripa *et al.* (2007) in Chile, which reported that the number of active monitoring stations with *R. flavipes* only declined within 2 m of the fipronil-treated zone.

Indoxacarb, an oxadiazine, is a pro-insecticide that, before the metabolite can kill the insects, must first undergo biotransformation into an N-decarbomethoxylated metabolite (DCJW) by an esterase/amidase (Yu, 2015). Because of this, it is relatively slow-acting. Unlike pyrethroids which cause uncontrolled sodium ion influx in sodium channels, the sodium channels in axons are blocked by indoxacarb, which stops the nerve impulses from firing. When tested on FST, indoxacarb toxicity (LD_{50} = 15 ng/termite at 14 days post-treatment (Mao *et al.*, 2011)) increased with increasing temperatures (Quarcoo *et al.*, 2019). Hu *et al.* (2005) investigated the horizontal transfer of indoxacarb using FST and found that recipient termite mortality increased significantly with increased concentration, donor number, and post-exposure time.

Chlorantraniliprole and cyantraniliprole are classified under the class diamide. Chlorantraniliprole demonstrated delayed toxicity at its lowest label rate in soils (50 ppm) (Neoh *et al.*, 2012). They stimulate calcium release from the sarcoplasmic reticulum, resulting in impaired regulation of muscle contraction, which affects skeletal and cardiac muscle and causes rapid cessation of feeding and immobility (Yu, 2015). Unlike the termiticides described previously, chlorantraniliprole caused *C. formosanus* and *C. gestroi* to cease their ability to feed and subsequently starve (Neoh *et al.*, 2012; Puckett *et al.*, 2012), as evident from the lower wood consumption in chlorantraniliprole-treated termites (Yeoh and Lee, 2007). The affected termites received intensive reciprocal food exchange and frequent attention from untreated nestmates, leading to the greater horizontal transfer of chlorantraniliprole (Neoh *et al.*, 2012, 2014). Jones *et al.* (2017) demonstrated that chlorantraniliprole severely impacted or possibly eliminated *R. flavipes* in 12 infested structures in Texas, North Carolina and Ohio, USA, with the use of molecular identification of termite colonies and intensive monitoring of infested structures and surrounding landscape. Another field trial on FST found that during the first 24 months post-treatment, 27.3% of the chlorantraniliprole-treated structures had FST infestations that required re-treatment (Puckett *et al.*, 2012). However, no FSTs were found at 30- and 36-month post-treatment inspections.

Another nonrepellent termiticide is chlorfenapyr, which is classified as a pyrrole. It must first undergo oxidative removal of the N-ethoxymethyl group before the metabolite disrupts the production of ATP by uncoupling oxidative phosphorylation in the mitochondria (Yu, 2015). Unlike fipronil, imidacloprid, and chlorantraniliprole, the compound is less toxic to FST (LD_{50} = 5.7 ng/termite at 14 days) (Mao *et al.*, 2011). Burkart *et al.* (2002) reported that termite mortality after exposure to chlorfenapyr can occur between 2 and 14 days, depending on the concentration, soil type, and exposure period. Rust and Saran (2006) found that *R. hesperus* exposed to the compound showed impaired movements within 4 h after exposure, reducing their ability to return to the nest to transfer the toxicants to the nestmates. The success of chlorfenapyr as a termiticide is likely related to its non-repellent and delayed toxicity characteristics and not horizontal transfer.

12.4 Types of termiticides

Based on termiticide characteristics and insecticidal activity, we adapted the description by Su *et al.* (1982) and divided them into four types (Table 12.2). Type I are repellent compounds, and termites will avoid contacting them, resulting in no or very low mortality. Type II are sometimes regarded as less repellent compounds, usually fast-acting, and the lethal outcome of the exposure can be observed within hours to days (Smith and Rust, 1990). The repellent properties likely come from the accumulation of carcasses of the killed termites that contact the toxicants (secondary repellency effect). Type III are non-repellent compounds at lower doses and are slower-acting than Types I and II. Lastly, Type IV are non-repellent chitin synthesis inhibitors and widely used in termite baits.

The lethal time (time required to onset of death after exposure, or survival time) of Types II and III are dose-dependent. Type IV are dose-independent (please refer to Chapter 11, sections 11.4.5 and 11.4.6 for detailed definitions of dose-dependent and dose-independent). Types I and II termiticides are more suitable for preventive treatment, while Types III and IV termiticides are often used for remedial treatment. Economics is another reason for using Type I and II termiticides for preventive treatment. This type of treatment requires a large volume of termiticides, and inexpensive generic formulations of Types I and II termiticides are available. There are some efforts now to use termite baits (containing Type IV termiticides) for preventive treatment, but it will depend on the termite species found in the locations and regions and the termite pressure, as termite baits are ineffective against non-rhinotermitids, especially the fungus-growing termites (Lee *et al.*, 2007, 2014).

Although termiticides are often classified based on whether they are repellent or non-repellent, their properties may differ depending on the concentrations required. Using the glass tube method, Yeoh and Lee (2007) demonstrated that *C. gestroi* were not repelled and showed 100% cumulative tunneling distance when 1 ppm of Types I and III termiticides (bifenthrin, chlorfenapyr, chlorantraniliprole, and indoxacarb) were used; however, when the concentration was increased to 100 ppm, only 0–29.8% cumulative tunneling distances were registered.

Table 12.2. Classification of termiticides (type I–IV) based on termite behavioral responses[a]

Type	I	II	III	IV
Lethal time	Not applicable	Dose-dependent	Dose-dependent	Dose-independent
Characteristics	Highly repellent, termite avoid treated area, sealed off tunnels, and with little termite mortality	Less repellent, fast-acting. Regardless of doses, toxicant effect is fully expressed within hours to days. Termites may breach into treated area, be killed, and tunnels to the treated area sealed off. Dead termites may pile in tunnels, resulting in secondary repellent effect	Non-repellent, fast-acting at higher doses and slow-acting at lower doses. Termites contacted treated area, killed, mortality increased with time	Non-repellent, slow-acting, termites readily feed on the toxicant, mortality related to molting
Classes	Pyrethroids	Cyclodienes, carbamates, organophosphates	Phenylpyrazole, neonicotinoid, oxadiazine, diamide, pyrrole	Chitin synthesis inhibitor
Examples	Bifenthrin, cypermethrin, lambdacyhalothrin, permethrin,	Chlordane, heptachlor, chlorpyrifos, fenitrothion, isofenphos	Fipronil, imidacloprid, thiamethoxam, indoxacarb, chlorantraniliprole, cyantraniliprole, chlorfenapyr	Hexaflumuron, bistrifluron, chlorfluazuron, noviflumuron, novaluron
Toxicant transfer to other colony members	No	No	Limited[b]	Yes
Colony suppression/ elimination	No	No	Limited	Yes
Formulation	Liquid termiticide	Liquid termiticide	Liquid termiticide	Bait
Suitable treatment(s)	Preventive	Preventive	Remedial	Preventive, remedial treatment

[a]Adapted after Su *et al.* (1982) with modification.

[b]Only shown for fipronil – distance from treatment zone: <5 m (Su, 2005), 28 m (Itakura *et al.*, 2021)

This implies that any termiticide can be repellent at high concentrations but non-repellent at low concentrations. As concentration increases, efficacy and repellency may increase. Higher concentrations may prevent tunneling by repellency and/or kill in the galleries. Hence, PMPs must use the manufacturer's recommended concentration to achieve the desired termiticidal activity.

12.4.1 Repellent termiticides

Conventional termite control involves creating a termiticide barrier in the soil that will repel termites from building structures. Repellent barriers must protect the structure by preventing the termites from penetrating through treated soil for at least 5 years (EPA, 1998). Repellent termiticides must be applied at a concentration that termites will readily detect so that they avoid breaching the treated zone (Table 12.2). With termiticide treatment, termites either completely avoid the treated zone (Type I termiticides), or some may enter the treated area and be killed (Type II). Smith and Rust (1990) showed that termites tunneled closest to soil treated with chlordane (Type II) but never approached the soil treated with bifenthrin or cypermethrin (Type I). The tunnels with dead termites leading to the treated zone may subsequently be sealed by the healthy termites. Su *et al.* (1982) defined this avoidance behavior that is due to carcasses and decomposed bodies as necrophobic behavior. Because repellent insecticides are typically highly toxic and fast-acting, they are ideal for preventive treatments (Su *et al.*, 1982) at the pre-construction level.

When using repellent termiticides such as pyrethroids, thoroughly applying a termiticide barrier is crucial for long-term structural protection. However, despite the best efforts, termites are often able to breach repellent termiticide barriers due to the presence of untreated gaps. These gaps can result from post-application activities by homeowners (e.g., floor renovation, new piping system, etc.), soil erosion, burrowing vertebrates, and other physical soil disturbances (Hu, 2011). Besides, the application of termiticides below the label rate is also a common factor that leads to breaches of repellent barriers (Mampe, 1994).

12.4.2 Non-repellent termiticides

In soil treated with non-repellent termiticides (Type III), the compounds are usually undetectable by termites when applied at the manufacturer's recommended label rates. Thus, termites are exposed to the chemicals when they forage through the treated areas. Unlike repellent termiticides, these slow-acting termiticides allow the termites to acquire a lethal dose, permitting some transfer to healthy nestmates within the colony via grooming, thereby potentially leading to colony decline or suppression (Potter, 1999). However, because lethal times of type III termiticides are dose-dependent, if termites repeatedly enter or continuously tunnel through the treated zone they may die quickly in or near the treatment. Similar to Type II termiticides, this could result in the accumulation of a large number of termite carcasses near the treated zone that may result in a secondary repellent effect (Potter and Hillery, 2002; Su, 2005; Henderson *et al.*, 2016; Chouvenc, 2018).

Non-repellent termiticides exhibited delayed toxicity that enabled horizontal transfer of the toxicants from treated termites to healthy ones (Ibrahim *et al.*, 2003; Kard, 2003; Shelton and Grace, 2003; Hu *et al.*, 2005, 2006; Tomalski and Vargo, 2005; Shelton *et al.*, 2006; Song and Hu, 2006; Tsunoda, 2006; Yeoh and Lee, 2007; Spomer *et al.*, 2008; Bagneres *et al.*, 2009). The intoxicated insects showed induced behaviors such as ataxia, moribundity, proctodeal (anus to mouth) or stomodeal (mouth to mouth) fluid release, reduced mobility, impaired tunneling, impaired ability to repair mounds, and reduced or impaired feeding (Thorne and Breisch, 2001; Hu *et al.*, 2006; Quarcoo *et al.*, 2010; Neoh *et al.*, 2012; Puckett *et al.*, 2012).

However, Hu *et al.* (2006) argued that most of these studies used small groups of termites in small arenas. Besides, most of these studies were conducted in laboratory conditions in the presence of light, air movement, and human disturbance, which do not reflect termites' natural habitats. Using a simulated laboratory experiment on 12 *R. flavipes* colonies aged 4–5 years old (6,000–8,000 termites), Hu *et al.* (2006) investigated the effects of localized fipronil treatments (1 ppm) by recording the dynamics of termite-toxicant and nestmate interactions for 1.5 months in complete darkness. They were able to demonstrate: (1) horizontal transfer of fipronil from poisoned to healthy termites; (2) that dying termites always returned to the nest sites rather than dying at the treated zone; and (3) that dying termites received intensive grooming and care from nestmates, hence increasing horizontal transfer of the toxicant. All treated colonies were either eliminated or had few remaining survivors. The findings of Hu *et al.* (2006), however, contradict the laboratory findings of Su (2005), Chouvenc (2018), and field studies of Ripa *et al.* (2007). Chouvenc (2018) evaluated fipronil on whole laboratory colonies of *C. gestroi* in extended foraging arenas (12 m distance) and found that all termites within 1.5 m away died from the treated zone within 2 weeks. This accumulation of carcasses resulted in secondary repellency effects, and the remaining colony members avoided the treated zone.

Laboratory studies have shown that non-repellent liquid termiticides could be horizontally transferred from treated individuals to healthy ones when the concentration on treated soils is low. It remains unclear whether the colony-level effect observed in field studies was due to their non-repellent properties, which enabled termites to tunnel through the treated zone, hence resulting in termite mortality, or a combination of both direct contact and horizontal transfer. Judging from the toxicity of some non-repellent termiticides, such as fipronil and imidacloprid, it could likely be that these toxicants affected the colony more through direct contact with the treated zone than via horizontal transfer. Since non-repellent termiticides are dose-dependent toxicants, the effects on the colony could also be influenced by the termite numbers, frequency of contact with the toxicant, and speed of action. This explains why the published studies found that non-repellent termiticides were only effective against termite species with smaller colonies closer to the treatment zone. For the larger colony size FST with the propensity to forage far from their nest, it is unlikely that the non-repellent termiticides will be able to impact at colony level.

One of the major challenges of liquid termiticides is the differences in insecticide susceptibility between termite species (Table 12.3). Topical toxicity results have shown that insecticide susceptibility between species could vary multiple-fold. For example, the LD_{50} of bifenthrin on *C. formosanus* is 6.8-fold and 9.2-fold higher than that of *R. flavipes* and *R. hesperus*, respectively. Even within the same species, there could be major differences in insecticide susceptibility. Osbrink and Lax (2001) found in their studies on multiple colonies of *C. formosanus* that one colony was 16-fold more tolerant than another to deltamethrin. Other studies reported that imidacloprid and chlorfenapyr are 14.2-fold and 3-fold more toxic to *C. formosanus* than to *R. hesperus*, respectively (Rust and Saran, 2008; Mao *et al.*, 2011). As manufacturers typically only recommend one label rate, termiticide performance may be inconsistent between different termite species (and different colonies), especially in the case of non-repellent termiticides in which it is pivotal for the toxicant to be horizontally transferred to nestmates. The more susceptible species or colony will be killed at a faster rate before it can interact with its conspecifics. It is obvious that "one size does not fit all."

Despite claims by chemical manufacturers that non-repellent termiticides could suppress or even eliminate termite colonies, only a limited number of field studies have demonstrated large-scale transfer of the toxicant beyond the boundaries of the small termite colonies, especially on the FST. The claim of colony elimination by non-repellent termiticides remains an elusive idea until there is sufficient field evidence to substantiate it.

Table 12.3. Topical toxicity (LD_{50} in ng/mg insect) of termiticides on workers of *C. formosanus* and other important species at 7 days post-treatment (unless stated otherwise).

Insecticide	Class[1]	LD_{50} (in ng/insect)				Reference(s)
		C. formosanus	*C. gestroi*	*R. flavipes*	*R. hesperus*	
acetamiprid	NNC	–	–	–	0.02	Rust and Saran (2008)
bifenthrin	PY	3.68, 0.21[3]	0.13[2]	0.54, 0.03[3]	0.40	Rust and Saran (2008), Mao *et al*. (2011), Neoh *et al*. (2012)
chlorantraniliprole	DA	1.32	41.92[2]	0.98–5.66	–	Mao *et al*. (2011), Neoh et al. (2012), Saran *et al.* (2014)
chlordane	CYD	12.25[3]	–	8.38[3]	0.138	Su and Scheffrahn (1990), Smith and Rust (1992)
chlorfenapyr	PYR	9.92	10.01[2]	11.90	29.98	Rust and Saran (2006), Mao *et al.* (2011), Neoh *et al*. (2012)
chlorpyrifos	OP	0.97[3]	–	0.76[3]	14.01	Su and Scheffrahn (1990), Rust and Saran (2006)
cyantraniliprole	DA	1.50	–	1.55	–	Mao *et al*. (2011)
cyhalothrin	PY	0.08[3]	–	0.02[3]	–	Su and Scheffrahn (1990)
cypermethrin	PY	0.19[3]	–	0.06[3]	2.81	Su and Scheffrahn (1990), Rust and Saran (2006)
deltamethrin	PY	0.03[3]	–	0.004[3]	–	Su and Scheffrahn (1990)
esfenvalerate	PY	0.17[3]	–	0.04[3]	–	Su and Scheffrahn (1990)
fenobucarb	CARB	0.29–0.62	–	–	–	Kubota *et al.* (2007)
fenvalerate	PY	0.65[3]	–	0.17[3]	–	Su and Scheffrahn (1990)
fipronil	PP	0.20, 1.33–1.39[3]	0.87[2]	0.12–0.27	0.14–0.16	Ibrahim *et al.* (2003), Rust and Saran (2006), Saran and Rust (2007), Mao *et al.* (2011), Neoh *et al*. (2012), Saran *et al*. (2014)
imidacloprid	NNC	0.74	–	1.02–119.43	10.57	Rust and Saran (2008), Mao *et al.* (2011), Saran *et al.* (2014)
indoxacarb	OX	17.16	4.04[2]	13.88	–	Mao *et al*. (2011), Neoh *et al.* (2012)
permethrin	PY	0.60[3]	–	0.27[3]	–	Su and Scheffrahn (1990)
silaneophane	PY	1.50[3]	–	0.39[3]	–	Su and Scheffrahn (1990)
thiamethoxam	NNC	–	–	–	1.35	Rust and Saran (2008)
tralomethrin	PY	0.03[3]	–	0.02[3]	–	Su and Scheffrahn (1990)

[1]CARB, carbamate; CYD, cyclodiene; DA, diamide; NNC, neonicotinoid; OP, organophosphate; OX, oxadiazine; PP, phenylpyrazole; PYR, pyrrole; PY, pyrethroid
[2]24 h post-treatment
[3]72 h post-treatment

12.5 Factors influencing termiticide performance and bioavailability

Liquid termiticide persistence and degradation in soil are affected by a variety of variables, such as termiticide physical properties, soil type, soil moisture content, soil organic content, soil pH, soil temperature, microbial degradation, photodegradation, and rate of the first application (Forschler and Townsend, 1996; Racke *et al.*, 1996; Henderson *et al.*, 1998; Su and Scheffrahn, 1998; Osbrink and Lax, 2002; Noller *et al.*, 2006; Saran and Kamble, 2008; Mahapatra *et al.*, 2017; Sapkota *et al.*, 2020).

12.5.1 Termiticide physical properties

Several important physical properties of termiticides govern their bioavailability in the soil and performance against subterranean termites: (1) water solubility (Smith and Rust, 1992); (2) vapor pressure (Smith and Rust, 1991); (3) octanol–water partition coefficient (log K_{OW}); and (4) organic carbon–water partition coefficient (log K_{OC}) (Table 12.4).

Termiticides with high water solubility leach more than those with lower water solubility. Vapor pressure is the tendency of a chemical to volatilize; termiticides with low vapor pressures are less likely to volatilize, whereas those with high vapor pressures tend to volatilize easily. The log K_{OW} refers to the lipophilicity of the compound. The lower the value, the more soluble is the termiticide in the water. Log K_{OC}, also known as soil adsorption coefficient, is a measurement of the mobility of the compound in the soil. Compounds with high log K_{OC} will be strongly adsorbed onto the soil and organic matter and do not readily move throughout the soil. On the other hand, those with low K_{OC} have poor adsorption and are highly mobile.

Types I and II termiticides (pyrethroids, cyclodienes, and organophosphates) generally have higher vapor pressure than the Type III termiticides (Table 12.4), which also likely leads to their repellent properties. Among all the presently used termiticides, chlorantraniliprole has the lowest vapor pressure, followed by imidacloprid. Most pyrethroids (except cypermethrin) have very low water solubility and high log K_{OW} and log K_{OC}, implying that they are strongly absorbed into the soil and do not readily leach. Among the non-repellent termiticides (Type III), chlorfenapyr shows the lowest water solubility and the highest log K_{OW} and log K_{OC}, while imidacloprid has the lowest log K_{OW} and is highly soluble in water. Based on the physical properties of termiticides, imidacloprid will likely pollute groundwater resources, especially in regions with heavy rain patterns and soil with low organic matter.

12.5.2 Soil properties

Termiticide movement through the soil is affected by the physical properties of the termiticides, the presence of water, biota, and organic matter, and the solvent used in the formulation. Soil properties, pH, and the presence of organic matter greatly influence the longevity and effectiveness of termiticides. Sandy soils with low organic content and wet conditions increase the movement of the termiticide in the soil, while it would be otherwise for those with high clay and organic contents (Noller *et al.*, 2006).

Forschler and Townsend (1996) found that soil type significantly affected the efficacy of six termiticide formulations when tested with four soil varieties. All tested termiticide concentrations (including chlorpyrifos, fenvalerate, cypermethrin, and permethrin) were at least seven times lower in sandy soils than in sandy loam and sandy clay loam. Gold *et al.* (1996b) discovered that acidic soils with low clay and organic content were the most effective at retaining termiticide residues over time. In contrast, alkaline soils with high clay content and organic compositions greater than 1% were the least effective.

Smith and Rust (1993) investigated how cellulose and clay in sand affect termiticide treatments. They found that when cellulose concentrations in sand were increased, it reduced the mortality of *R. hesperus* in chlordane and chlorpyrifos treatments but increased mortality caused by cypermethrin.

Table 12.4. Physical properties of selected termiticides.

Termiticide type	Termiticide	Vapor pressure at 25°C (mm Hg or mPa)	Water solubility (mg/l) (at 20° or 25°C)	Log K_{OW} (Octanol-water partition coefficient)	Log K_{OC} (organic carbon-water partition coefficient)	References
I	bifenthrin	1.8×10^{-7}	0.001 (20°C)	6.00	5.4	Baskaran *et al.* (1999), Oros and Werner (2005), Gold *et al.* (1996a)
I	cypermethrin	1.1×10^{-7}	< 9 (20°C)	6.6	5.26–5.76	Oekotoxzentrum (2019)
I	fenvalerate	2.8×10^{-7}	0.0024 (22°C)	6.20	3.0–4.08	PubChem (2023a)
I	permethrin	2.0×10^{-8}	0.006 (20°C)	6.50	3.91–5.74	Gold *et al.* (1996)
II	chlordane	1.0×10^{-5}	0.056 (25°C)	6.16	3.49–4.64	PubChem (2023b)
II	chlorpyrifos	1.87×10^{-5}	1.40 (25°C)	4.96	3.9	Gold *et al.* (1996a), PubChem (2023c)
III	chlorfenapyr	7.4×10^{-8}	0.12 (25°C)	4.83	4.08	PubChem (2023d)
III	chlorantraniliprole	1.2×10^{-14}	1.023 (20°C)	2.90	2.6	Neoh *et al.* (2012), Vela *et al.* (2017)
III	fipronil	2.8×10^{-9}	1.9–2.4 (20°C)	3.5	3.0	Gunasekara *et al.* (2007)
III	imidacloprid	3×10^{-12}	610 (20°C)	0.57	1.8	Neoh *et al.* (2012), Aliste *et al.* (2021)

Clay concentration increased with toxicity of cypermethrin, but the effective repellency of chlorpyrifos decreased. On the other hand, the clay presence reduced chlordane binding to the sand particles, increasing the movement of chlordane.

Substrate composition affects termiticide treatments. Before treating a structure for subterranean termites, pest management professionals may need to assess the soil composition and choose the appropriate termiticide.

12.5.3 Environmental conditions and changing climate

Flooding of soil could affect the performance of termiticides through leaching and hydrolysis of termiticides (Bobe *et al.*, 1998; Zheng and Liu, 1999; Bentley *et al.*, 2010; Sapkota *et al.*, 2020). As discussed in section 12.4.2, water can play a role in the movement of the termiticides, especially for termiticides that are highly water-soluble and with low K_{OW}. The organic content in the soil during the flooding also affects the bioavailability of the termiticide. Shuai *et al.* (2012) reported that the concentration of fipronil in the leachate from three Hawai'ian soils taken from different sources (Makiki, Wahiawa, and Waialua) correlated inversely with soil organic carbon content. Based on Table 12.4, compounds with high K_{OC} (e.g., bifenthrin, permethrin, chlorfenapyr) in soil with high organic content will be less affected in the event of a flood, compared with those with low K_{OC} (e.g., imidacloprid).

Using a flooding simulation experiment in the laboratory, Sapkota *et al.* (2020) investigated the performance of four termiticides (fipronil, imidacloprid, chlorantraniliprole, and bifenthrin) at different concentrations (1, 10, and 25 ppm) treated in both sand (low organic content) and soil (high organic content) against field-collected colonies of FST. Flooding did not affect the toxicity of bifenthrin and fipronil, irrespective of soil type (except for those treated with 1 ppm). Chlorantraniliprole toxicity was low at 1 ppm in the flooded soil and sand, while the mortality of the FST was reduced for all concentrations of imidacloprid-treated sand after the flood. Flooding did not affect the mortality of the FST in soil treated with chlorantraniliprole, but the soil organic matter likely reduced its bioavailability to termites even at label rates, as demonstrated earlier by Gautam and Henderson (2011).

Soil moisture level affects the degradation rate of termiticides, as shown on fipronil, bifenthrin (Mohapatra and Ahuja, 2009, 2010), and imidacloprid (Baskaran *et al.*, 1999; Mahapatra *et al.*, 2017). Besides, termiticides also degraded faster under direct sunlight than when kept under dark conditions (Mahapatra *et al.*, 2017), under warmer soil temperatures (Ou *et al.*, 1982) as well due to enhanced microbial activity (Ou, 2000; Kamble, 2006). However, information on the role of soil microbiota in termiticide degradation remains scarce and warrants investigation.

With the impact of climate change, especially with irregular rainfall patterns and warmer soil temperatures, the rate of termiticide degradation will likely be accelerated. Warmer and moist soil conditions contribute toward greater desorption and greater volatilization losses. Heavy rainfalls increase leaching, especially termiticides with high water solubility and lower K_{OW}. This could affect soil termiticide performance and increase the prevalence of subterranean termite infestations in the long term.

12.6 Conclusion and future directions

Despite the many weaknesses and potentially long-term environmental impact, liquid termiticides remain relevant in termite management. Although there has been increasing effort to incorporate baiting in termite management, as evident from the recent surveys (see section 12.1), baits are only effective against rhinotermitids. Even within rhinotermitids, some species are more challenging to bait, such as *R. hesperus* in southern California (Haagsma and Rust, 2005) and *R. speratus* in Japan (Tsunoda *et al.*, 1998, 1999; Wang *et al.*, 2000; Tsunoda, 2005). In regions where termite pressures could come from multiple termite genera, such as Southeast Asia (Lee *et al.*, 2007), the use of liquid termiticide is ideal for preventive treatment at pre-construction level. At post-construction, the choice of remedial treatment can be decided based on the termite species involved or a combination of both baiting and liquid termiticide treatment.

Whether non-repellent termiticides can eliminate large field termite colonies of the FST remains unresolved. Except for one utilizing microsatellite genotyping and intensive monitoring of termite activity in monitoring stations that demonstrated the elimination of a colony of FST (Itakura *et al.*, 2021), the other studies observed either the absence/presence of termite infestation after the treatment (Keefer *et al.*, 2012; Puckett *et al.*, 2012) or the activity of termites (Osbrink and Lax, 2003; Osbrink *et al.*, 2005); little is known on the colony-level impact of non-repellent termiticides on the FST. As discussed in section 12.4.2, lethal times of non-repellent termiticides are dose-dependent and could act quickly at higher doses, resulting in the accumulation of dead carcasses in the treated zone. This would lead to secondary repellent effect and avoidance of the termites from the treated zone. Hence, the absence of termite activity does not imply that the termites have been eliminated. To resolve this question, more large-scale area-wide studies with nonrepellent termiticides on FST infestations utilizing genetic markers and intensive monitoring should be carried out.

It is known that high moisture and temperature increase the degradation of termiticides, as evident with the shorter longevity of termiticides in the tropics compared with those in the temperate region. In the changing climate with irregular rain patterns and increasing soil temperature, the performance of termiticides will likely be affected. These factors may also influence the abundance and diversity of soil microbiota. However, how soil microbiota affects the degradation of termiticides remains unknown. As the climate warms, termite activity and abundance will also increase (Zanne *et al.*, 2022). Besides, in regions with drought due to irregular rain patterns, it could potentially lead to the development of thicker cuticular hydrocarbons (CHCs) in termites to reduce desiccation, as has been shown in the FST (Richardson, 2022). As the insects adapt, increased CHCs production and thicker epicuticle may lead to increased tolerance to termiticides due to reduced insecticide penetration. In addition, overexpression of some cytochrome P450 genes involved in insecticide detoxification has been linked to increased CHC production (Balabanidou *et al.*, 2016; Chen *et al.*, 2020). How these factors interact and affect the future of termite management remains unknown and warrant further investigation.

References

Aliste, M., Perez-Lucas, G., Garrido, I., Fenoll, J., and Navarro, S. (2021) Mobility of insecticide residues and main intermediates in a clay loam soil, and impact of leachate components on their photocatalytic degradation. *Chemosphere* 274, 129965.

Bagnères, A.G., Pichon, A., Hope, J., Davis, R. and Clément, J.L. (2009) Contact versus feeding intoxication by fipronil in *Reticulitermes* termites (Isoptera: Rhinotermitidae): laboratory evaluation of toxicity, uptake, clearance, and transfer among individuals. *Journal of Economic Entomology* 102, 347–356.

Baker, P.B. and Miguelena, J.G. (2020) Field distance effects of fipronil and chlorfenapyr as soil termiticides against the desert subterranean termite, *Heterotermes aureus* (Blattodea: Rhinotermitidae). *Sociobiology* 67, 94–105.

Balabanidou, V., Kampouraki, A., MacLean, M., Blomquist, G.J., Tittiger, C. *et al.* (2016) Cytochrome P450 associated with insecticide resistance catalyzes cuticular hydrocarbon production in *Anopheles gambiae*. *Proceedings of the National Academy of Science* 113, 9268–9273.

Barlow, R.A. and Robinson, W.H. (1993) Using foam effectively. *Pest Control Technology* 21 (11), 39, 42–44, 46.

Baskaran, S., Kookana, R.S. and Naidu, R. (1999) Degradation of bifenthrin, chlorpyrifos and imidacloprid in soil and bedding materials at termiticidal application rates. *Pesticide Science* 55, 1222–1228.

Beal, R.H. (1971) Concrete impregnated with chlordane and dieldrin not tubed over by subterranean termites. *Journal of Economic Entomology* 64, 1289–1291.

Beal, R.H. and Smith, V.K. (1971) Relative susceptibilities of *Coptotermes formosanus*, *Reticulitermes flavipes*, and *R. virginicus* to soil insecticides. *Journal of Economic Entomology* 64, 472–475.

Bentley, K.S., Fletcher, J.L. and Woodward, M.D. (2010) Chlorantraniliprole: an insecticide of the anthranilic diamide class. In: Krieger R. (ed.) *Hayes' Handbook of Pesticide Toxicology*. Academic Press, London, pp. 2231–2242.

Bess, H.A. and Hylin, J.W. (1970) Persistence of termiticides in Hawaiian soils. *Journal of Economic Entomology* 63, 633–638.

Bess, H.A., Ota, A.K. and Kawanishi, C. (1966) Persistence of soil insecticides for control of subterranean termites. *Journal of Economic Entomology* 59, 911–915.

Bhatta, D. and Henderson, G. (2016) Horizontal transfer of spinosad in *Coptotermes formosanus* (Isoptera: Rhinotermitidae). *Journal of Economic Entomology* 109, 1813–1818.

Bhatta, D., Henderson, G. and Gautam, B.K. (2016) Toxicity and nonrepellency of Spinosad and spinetoram on Formosan subterranean termites (Isoptera: Rhinotermitidae). *Journal of Economic Entomology* 109, 1341–1349.

Bobé, A., Meallier, P., Cooper, J.F. and Coste, C.M. (1998) Kinetics and mechanisms of abiotic degradation of fipronil (hydrolysis and photolysis). *Journal of Agricultural and Food Chemistry* 46, 2834–2839.

Burkart, S.E., Klein, C., Akagawa, H. and Ikeda, Y. (2002) Performance of non-repellent termiticide, chlorfenapyr, in France, Japan, and the USA. In: Jones, S.C., Zhai, J., and Robinson, W.H. (eds) *Proceedings of the 4th International Conference on Urban Pests*, p. 494. https://www.icup.org.uk/media/sx0jakan/icup300.pdf

Chen, N., Pei, X.J., Li, S., Fan, Y.L. and Liu, T.X. (2020) Involvement of integument-rich CYP4G19 in hydrocarbon biosynthesis and cuticular penetration resistance in *Blattella germanica*. *Pest Management Science* 76, 215–226.

Chouvenc, T. (2018) Comparative impact of chitin synthesis inhibitor baits and nonrepellent liquid termiticides on subterranean termite colonies over foraging distances: colony elimination versus localized termite exclusion. *Journal of Economic Entomology* 111, 2317–2328.

Cornelius, M.L., Grace, J.K. and Yates, J.R. III (1997) Toxicity of monoterpenoid and other natural products to the Formosan subterranean termite (Isoptera: Rhinotermitidae). *Journal of Economic Entomology* 90, 320–325.

Cuthbert, R.N., Diagne, C., Haubrock, P.J., Turbelin, A.J. and Courchamp, F. (2021) Are the '100 of the world's worst' invasive species also the costliest? *Biological Invasions* 24, 1895–1904.

EPA (1987) OPP-60011; FRL-3286-8. Chlordane and heptachlor termiticides: cancellation order. *Federal Registration* 52, 42145–42149.

EPA (1998) *Product Performance Test Guidelines, OPPTS 810.3600* Structural Treatments. EPA-712-C-98-424.

EPA (2000) *Chlordane*. US Environmental Protection Agency, Washington, DC. Available at: https://www.epa.gov/sites/default/ files/2016-09/documents/chlordane.pdf (accessed 9 September 2023)

Erhorn, E.M. (1934) The termites of Hawaii, their economic significance and control, and the distribution of termites by commerce. In: Kofoid, C.A. (ed.) *Termites and Termite Control*. University of California Press, Berkeley, California, pp. 321–333.

Evans, T.A., Forschler, B.T., and Grace, J.K. (2013) Biology of invasive termites: a worldwide review. *Annual Review of Entomology* 58, 455–474.

Forschler, B.T. (1994) Survivorship and tunneling activity of *Retuculitermes flavipes* (Kollar) (Isoptera: Rhinotermitidae) in response to termiticide soil barriers with and without gaps of untreated soil. *Journal of Entomological Science* 29, 43–54.

Forschler, B.T. and Townsend, M.L. (1996) Mortality of eastern subterranean termites (Isoptera: Rhinotermitidae) exposed to four soils treated with termiticides. *Journal of Economic Entomology* 89, 678–681.

Fujimoto, I., Enoki, A., Itakura, S. and Tanaka, H. (2012) Comparison of the efficacy of eleven soil termiticides on the Formosan subterranean termite (*Coptotermes formosanus*). *Sociobiology* 59, 149–164.

Gautam, B.K. and Henderson, G. (2011) Effect of soil type and exposure duration on mortality and transfer of chlorantraniliprole and fipronil on Formosan subterranean termites (Isoptera: Rhinotermitidae). *Journal of Economic Entomology* 104, 2025–2030.

Gautam, B.K., Henderson, G. and Wang, C. (2014) Localized treatments using commercial dust and liquid formulations of fipronil against *Coptotermes formosanus* (Isoptera: Rhinotermitidae) in the laboratory. *Insect Science* 21, 174–180.

Gold, R.E., Howell, H.N., Pawson, B.M., Wright, M.S. and Lutz, J.C. (1996a) Evaluation of termiticides residues and bioavailability from five soil types and locations in Texas. In: Wildey, K.B. (ed.) *Proceedings of the 2nd International Conference on Insect Pests in the Urban Environment*. pp. 467–484. https://www.icup.org.uk/media/4eepradm/icup764.pdf

Gold, R.E., Howell, H.N., Pawson, B.M., Wright, M.S. and Lutz, J.C. (1996b) Persistence and bioavailability of termiticides to subterranean termites (Isoptera: Rhinotermitidae) from five soil types and locations in Texas. *Sociobiology* 28, 337–363.

Grace, J.K., Yamamoto, R.T. and Ebesu, R.H. (1992) Laboratory evaluation of the novel soil insecticide silafluofen against *Coptotermes formosanus* Shiraki (Isoptera: Rhinotermitidae). *Journal of Applied Entomology* 113, 466–471.

Grace, J.K., Yates, J.R., Tamashiro, M. and Yamamoto, R.T. (1993) Persistence of organochlorine insecticides for Formosan subterranean termite (Isoptera: Rhinotermitidae) control in Hawaii. *Journal of Economic Entomology* 86, 761–766.

Grace, J.K., Yamamoto, R.T. and Tome, C.H.M. (2000) Toxicity of sulfluramid to *Coptotermes formosanus* (Isoptera: Rhinotermitidae). *Sociobiology* 35, 457–466.

Gunasekara, A.S., Truong, T., Goh, K.S., Spurlock, F. and Tjeerdema, R.S. (2007) Environmental fate and toxicology of fipronil. *Journal of Pesticide Science* 32, 189–199.

Haagsma, K.A. and Rust, M.K. (2005) Effect of hexaflumuron on mortality of the Western subterranean termite (Isoptera: Rhinotermitidae) during and following exposure and movement of hexaflumuron in termite groups. *Pest Management Science* 61, 517–531.

Henderson, G., Walthall, O., Wiltz, B.A., Rivera-Montroy, V., Ganaway, D.R. and Selim, H.M. (1998) Analysis of soil properties in relation to termiticide performance in Louisiana. In: *Proceeding of the National Conference on Urban Entomology*, San Diego, California, pp. 65–75.

Henderson, G., Gautam, B.K. and Wang, C. (2016) Impact of ground-applied termiticides on the above-ground foraging behavior of the Formosan subterranean termite. *Insects* 7, 43. doi:10.3390/insects7030043

Hetrick, L.A. (1950) The toxicity of some organic insecticides to the eastern subterranean termite. *Journal of Economic Entomology* 43, 57–59.

Hockenyos, G. (1939) Laboratory evaluation of soil poisons used in termite control. *Journal of Economic Entomology* 32, 147–149.

Hu, X.P. (2005) Evaluation of efficacy and nonrepellency of indoxacarb and fipronil treated soil at various concentrations and thicknesses against two subterranean termites (Isoptera: Rhinotermitidae). *Journal of Economic Entomology* 98, 509–517.

Hu, X.P. (2011) Liquid termiticides: their role in subterranean termite management. In: Dhang, P. (ed.) *Urban Pest Management: An Environmental Perspective*. CABI, Wallingford, UK, pp. 114–132.

Hu, X.P., Song, D. and Scherer, C. (2005) Transfer of indoxacarb among workers of *Coptotermes formosanus* Shiraki (Isoptera: Rhinotermitidae): effects of dose, donor/ recipient ratio, and post-exposure time. *Pest Management Science* 61, 1209–1214.

Hu, X.P., Song, D. and Presley, W. (2006) Horizontal and vertical transfer of fipronil within functional colonies. In: Sutphin, T., Cartwright, B. and Houseman, R. (eds) *Proceedings of 2006 National Conference on Urban Entomology*, pp. 39–44.

Ibrahim, S.A., Henderson, G. and Fei, H. (2003) Toxicity, repellency, and horizontal transmission of fipronil in the Formosan subterranean termite (Isoptera: Rhinotermitidae). *Journal of Economic Entomology* 96, 461–467.

Itakura, S., Ohdake, J., Takino, T. and Umezawa, K. (2021) Colony suppression and possible colony elimination of the subterranean termites *Coptotermes formosanus* and *Reticulitermes speratus* by discontinuous soil treatment using a diluent of fipronil suspension concentrate. *Insects* 12, 334. doi: 10.3390/insects12040334

Johnston, H.R., Smith, V.K. and Beal, R.H. (1971) Chemicals for subterranean termite control: Results of long-term tests. *Journal of Economic Entomology* 64, 745–748.

Jones, S.C. (1990) Effects of population density on tunneling by Formosan subterranean termite (Isoptera: Rhinotermitidae) through treated soil. *Journal of Economic Entomology* 83, 875–878.

Jones, S.C., Vargo, E.L., Keefer, T.C., Labadie, P., Scherer, C.W. *et al.* (2017) Efficacy of chlorantraniliprole in controlling structural infestations of the Eastern subterranean termite in the USA. *Insects* 8, 92. doi:10.3390/insects8030092

Kamble, S.T. (2006) *Fate of insecticides used for termite control in soil*. NebGuide G1260. University of Nebraska, Lincoln, Nebraska. Available at: https://iranarze.ir/wp-content/uploads/2016/07/4348-english.pdf (accessed 9 September 2023)

Kard, B.M. (2003) Integrated pest management of subterranean termites (Isoptera). *Journal of Entomological Science* 38, 200–224.

Kard, B.M., Mauldin, J.K. and Jones, S.C. (1989) Evaluation of soil termiticides for the control of subterranean termites (Isoptera: Rhinotermitidae). *Sociobiology* 15, 285–297.

Keefer, T.C., Puckett, R.T. and Gold, R.E. (2012) Performance of Termidor HE High-Efficiency Termiticide co-pack with HE Technology against Eastern subterranean termites *Reticulitermes flavipes* and Formosan subterranean termite *Coptotermes formosanus* in laboratory trials and field applications. *Sociobiology* 59, 1547–1562.

Khoo, B.K. and Sherman, M. (1979) Toxicity of chlorpyrifos to normal and defaunated Formosan subterranean termites. *Journal of Economic Entomology* 72, 298–304.

Kofoid, C.A. (1934) *Termites and Termite Control*. University of California Press, Berkeley, California, 795 pp.

Kowal, R.J. and St George, R.A. (1948) Preliminary results of termite soil-poisoning tests. *Journal of Economic Entomology* 41, 112–113.

Kubota, S., Shono, Y., Matsunaga, T. and Tsunoda, K. (2007) Response of the subterranean termite *Coptotermes formosanus* (Isoptera: Rhinotermitidae) to soil treated with microencapsulated fenobucarb. *Pest Management Science* 63, 1224–1229.

Kubota, S., Shono, Y., Mito, N. and Tsunoda, K. (2008) Termiticidal efficacies of fenobucarb and permethrin against Japanese subterranean termites *Coptotermes formosanus* and *Reticulitermes speratus* (Isoptera: Rhinotermitidae). *Japanese Journal of Environmental Entomology and Zoology* 19, 31–37.

Kuriachan, I. and Gold, R.E. (1998) Evaluation of the ability of *Reticulitermes flavipes*, a subterranean termite, to differentiate between termiticide treated and untreated soil in laboratory tests. *Sociobiology* 32, 151–168.

Lee, C.C., Neoh, K.B. and Lee, C.Y. (2014) Colony size affects the efficacy of bait containing chitin synthesis inhibitor against the fungus-growing termite. *Journal of Economic Entomology* 107, 2154–2162.

Lee, C.Y., Vongkaluang, C. and Lenz, M. (2007) Challenges to subterranean termite management in multi-genera faunas in Southeast Asia and Australia. *Sociobiology* 50, 213–221.

Lowe, S., Browne, M., Boudjelas, S. and De Poorter, M. (2000) *100 of the World's Worst Invasive Alien Species. A selection from the Global Invasive Species Database*. Invasive Species Specialist Group (ISSG) of IUCN's Species Survival Commission (SSC), Auckland, New Zealand, 12 pp. Available at: http://www.iucngisd.org/gisd/pdf/100English.pdf (accessed 6 Sep 2023)

Mahapatra, B., Adak, T., Patil, N.K.B., Pandi, G.G.P., Gowda, G.B. *et al.* (2017) Effect of abiotic factors on degradation of imidacloprid. *Bulletin of Environmental Contamination and Toxicology* 99, 475–480.

Mampe, C.D. (1994) Reducing termite treatment. *Pest Control* 62, 4.

Mao, L., Henderson, G. and Scherer, C.W. (2011) Toxicity of seven termiticides on the Formosan and Eastern subterranean termites. *Journal of Economic Entomology* 104, 1002–1008.

Mix, J. (1987) Termiticide market stabilizes. *Pest Control* 55, 22–3.

Mix, J. (1991) Dursban TC tops termiticide survey. *Pest Control* 59, 22.

Mo, J., Pan, C., Zhang, S., Chen, C., He, H. and Cheng, J. (2005) Toxicity of acetamiprid to workers of *Reticulitermes flaviceps* (Isoptera: Rhnotermitidae), *Coptotermes formosanus* (Rhinotermitidae) and *Odontotermes formosanus* (Isoptera: Termitidae). *Journal of Pesticide Science* 30, 187–191.

Mohapatra, S. and Ahuja, A.K. (2009) Effect of moisture and soil type on the degradation of bifenthrin in soil. *Pesticide Research Journal* 21, 191–194.

Mohapatra, S. and Ahuja, A. K. (2010) Behaviour of fipronil in soil under different moisture levels. *Indian Journal of Agricultural Science* 80, 658–661.

Mori, H. (1987) The Formosan subterranean termite in Japan: Its distribution, damage, and current and potential control measures. In: Tamashiro, M. and Su, N.Y. (eds) *Biology and Control of the Formosan Subterranean Termite*. Research Extension Series 083. Hawaii Institute of Tropical Agriculture and Human Resources, Honolulu, Hawaii, pp. 23–30.

Neoh, K.B., Hu, J., Yeoh, B.H. and Lee, C.Y. (2012) Toxicity and horizontal transfer of chlorantraniliprole against the Asian subterranean termite *Coptotermes gestroi* (Wasmann): effects of donor: recipient ratio, exposure duration and soil type. *Pest Management Science* 68, 749–756.

Neoh, K.B., Yeoh, B.H. and Lee, C.Y. (2014) Mortality patterns in *Coptotermes gestroi* (Blattodea: Rhinotermitidae) following horizontal transfer of nonrepellent and repellent insecticides: effects of donor: recipient ratio and exposure time. *Journal of Economic Entomology* 107, 1563–1572.

Ngo, K., Castillo, P., Laine, R.A. and Sun, Q. (2021) Effects of menadione on survival, feeding, and tunneling activity of Formosan subterranean termite. *Insects* 12, 1109. doi: 10.3390/ insects12121109

Noller, B., Gilbert, D., Sadler, R., Truss, R., Zalucki, M. *et al.* (2006) Urban Australian cities under termite attack. In: *Second State of Australian Cities National Conference* 2005. 18 pp. https://www.researchgate.net/publication/43461396_Urban_Australian_cities_under_termite_attack

Oekotoxzentrum (2019) SQC (EQS_{sed}) – proposal from the Ecotox Center for cypermethrin. https://www.centreecotox.ch/media/195582/0_cypermethrin_may2022.pdf

Oi, F. (2022) A review of the evolution of termite control: a continuum of alternatives to termiticides in the United States with emphasis on efficacy testing requirements for product registration. *Insects* 13, 50. doi: 10.3390/insects13010050

Oros, D.R. and Werner, I. (2005) *Pyrethroid insecticides: an analysis of use patterns, distributions, potential toxicity and fate in the Sacramento-San Joaquin Delta and Central Valley*. White Paper for the Interagency Ecological Program. SFEI Contribution 415. San Francisco Estuary Institute, Oakland, California.

Osbrink, W. and Cornelius, M. (2013) Acoustical evaluation of trees for *Coptotermes formosanus* Shiraki (Isoptera: Rhinotermitidae) treated with imidacloprid and noviflumuron in historic Jackson Square, New Orleans, Louisiana. *Sociobiology* 60, 77–95.

Osbrink, W.L.A. and Lax, A.R. (2001) Insecticide susceptibility in *Coptotermes formosanus* and *Reticulitermes virginicus* (Isoptera: Rhinotermitidae). *Journal of Economic Entomology* 94, 1217–1228.

Osbrink, W.L.A. and Lax, A.R. (2002) Effect of tolerance to insecticides on substrate penetration by Formosan subterranean termites (Isoptera: Rhinotermitidae). *Journal of Economic Entomology* 95, 989–1000.

Osbrink, W.L.A. and Lax, A.R. (2003) Effect of imidacloprid tree treatments on the occurrence of Formosan subterranean termites, *Coptotermes formosanus* Shiraki (Isoptera: Rhinotermitidae). *Journal of Economic Entomology* 96, 117–125.

Osbrink, W.L.A., Cornelius, M.L. and Lax, A.R. (2005) Effect of imidacloprid soil treatments on occurrence of Formosan subterranean termites (Isoptera: Rhinotermitidae) in independent monitors. *Journal of Economic Entomology* 98, 2160–2168.

Osbrink, W.L.A., Cornelius, M.L., Showler, A.T. and Pound, J.M. (2014) Effects of a fipronil spot treatment on field colonies of *Coptotermes formosanus* (Isoptera: Rhinotermitidae). *Journal of Economic Entomology* 107, 727–740.

Ou, L.T. (2000) Pesticide biodegradation. In: *Encyclopedia of Microbiology*, 2nd edn, Volume 3. Academic Press, San Diego, p. 594.

Ou, L.T., Gancarz, D.H., Wheeler, W.B., Road, P.S.C. and Davison, J.M. (1982) Influence of soil temperature and soil moisture on degradation and metabolism of carbofuran in soils. *Journal of Environmental Quality* 11, 293–298.

Parman, V. and Vargo, E.L. (2010) Colony-level effects of imidacloprid in subterranean termites (Isoptera: Rhinotermitidae). *Journal of Economic Entomology* 103, 791–798.

PCT (2021) *2021 State of the Termite Market Report*. Pest Control Technology, Valley View, Ohio. http://giecdn.azurewebsites.net/fileuploads/document/2021/02/08/2021%20State%20of%20the%20Termite%20Market%20Report.pdf

PCT (2022) 2022 State of the Termite Market Report. Pest Control Technology, Valley View, Ohio. https://cdn.gie.net/fileuploads/document/2022/02/14/2022%20state%20of%20the%20termite%20market%20report.pdf

PCT (2023) 2023 State of the Termite Market Report. Pest Control Technology, Valley View, Ohio. https://cdn.gie.net/fileuploads/document/2023/03/27/2023_state_termite_market_report.pdf

Pimentel, D., Zuniga, R. and Morrison, D. (2005) Update on the environmental and economic costs associated with alien-invasive species in the United States. *Ecological Economics* 52, 273–288.

Potter, M.F. (1999) The changing face of termite control. II. *Pest Control Technology* 27, 33–23, 36, 38–39, 42, 90.

Potter, M.F. (2011) Termites. In: Moreland, D. (ed.) *Handbook of Pest Control*. GIE Media Inc, Cleveland, Ohio, pp. 293–441.

Potter, M.F. and Hillery, A.E. (2002) Exterior-targeted liquid termiticides: an alternative approach to managing subterranean termites (Isoptera: Rhinotermitidae) in buildings. *Sociobiology* 39, 373–405.

PubChem (2023a) *Fenvalerate*. National Center for Biotechnology Information, Bethesda, Maryland. Available at: https://pubchem.ncbi.nlm.nih.gove/compound/3347 (accessed 9 September 2023)

PubChem (2023b) *Chlordane*. National Center for Biotechnology Information, Bethesda, Maryland. Available at: https://pubchem.ncbi.nlm.nih.gov/compound/5993 (accessed 9 September 2023)

PubChem (2023c) *Chlorpyrifos*. National Center for Biotechnology Information, Bethesda, Maryland. Available at: https://pubchem.ncbi.nlm.nih.gov/compound/2730 (accessed 9 September 2023)

PubChem (2023d) Chlorfenapyr. National Center for Biotechnology Information, Bethesda, Maryland. Available at: https://pubchem.ncbi.nlm.nih.gov/compound/91778 (accessed 9 September 2023)

Puckett, R.T., Keefer, T.C. and Gold, R.E. (2012) Performance of Altriset™ (chlorantraniliprole) termiticide against Formosan subterranean termites, *Coptotermes formosanus* Shiraki, in laboratory feeding cessation and collateral transfer trials, and field applications. *Sociobiology* 59, 1427–1445

Quarcoo, F., Appel, A. and Hu, X.P. (2010) Effects of indoxacarb concentration and exposure time on the onset of abnormal behaviors, moribundity, and death in the eastern subterranean termite (Isoptera: Rhinotermitidae). *Journal of Economic Entomology* 103, 762–769.

Quarcoo, F.Y., Hu, X.P. and Appel, A.G. (2019) Temperature-mediated variations in behavior and mortality caused by non-repellent insecticides in subterranean termites (Blattodea: Rhinotermitidae). *Insects* 10, 37. doi: 10.3390/insects10020037

Racke, K.D., Steele, K.P., Yoder, R.N., Dick, W.A. and Avidov, E. (1996) Factors affecting the hydrolytic degradation of chlorpyrifos in soil. *Journal of Agricultural and Food Chemistry* 44, 1582–1592.

Ramakrishnan, R., Suiter, D.R., Nakatsu, C.H., Humber, R.A. and Bennett, G.W. (1999) Imidacloprid-enhanced *Reticulitermes flavipes* (Kollar) (Isoptera: Rhinotermitidae) susceptibility to the entomopathogen *Metarhizium anisopliae* (Metsch.) Sorokin. *Journal of Economic Entomology* 92, 1125–1132.

Ramakrishnan, R., Suiter, D.R., Nakatsu, C.H. and Bennett, G.W. (2000) Feeding inhibition and mortality in *Reticulitermes flavipes* (Isoptera: Rhinotermitidae) after exposure to imidacloprid-treated soils. *Journal of Economic Entomology* 93, 422–428.

Randall, M. and Doody, T.C. (1934) Poison dusts. In: Kofoid, C.A. (ed.) *Termites and termite control.* University of California Press, Berkeley, California, pp. 463–479.

Remmen, L.N. and Su, N.-Y. (2005a) Time trends in mortality for thiamethoxam and fipronil against Formosan subterranean termites and Eastern subterranean termites (Isoptera: Rhinotermitidae). *Journal of Economic Entomology* 98, 911–915.

Remmen, L.N. and Su, N.-Y. (2005b) Tunneling and mortality of Eastern and Formosan subterranean termites (Isoptera: Rhinotermitidae) in sand treated with thiamethoxam or fipronil. *Journal of Economic Entomology* 98, 906–910.

Richardson, S.J. (2022) Soil moisture mediated behavioral and physiological alterations of *Coptotermes formosanus* and *Reticulitermes flavipes* (Blattodea: Rhinotermitidae). MSc thesis. Louisiana State University, Baton Rouge, Louisiana, 60 pp.

Ripa, R., Luppichini, P., Su, N.-Y. and Rust, M.K. (2007) Field evaluation of potential control strategies against the invasive Eastern subterranean termite (Isoptera: Rhinotermitidae) in Chile. *Journal of Economic Entomology* 100, 1391–1399.

Robinson, W.H. and Barlow, R.A. (1993) Evaluating the foamers. *Pest Control Technology* 22(4), 46, 48–49, 50, 52.

Rust, M.K. (2014) Management strategies for subterranean termites. In: Dhang, P. (ed.) *Urban Insect Pests: Sustainable Management Strategies.* CABI, Wallingford, UK, pp. 114–129.

Rust, M.K. and Saran, R.K. (2006) Toxicity, repellency, and transfer of chlorfenapyr against western subterranean termites (Isoptera: Rhinotermitidae). *Journal of Economic Entomology* 99, 864–872.

Rust, M.K. and Saran, R.K. (2008) Toxicity, repellency, and effects of acetamiprid on Western subterranean termite (Isoptera: Rhinotermitidae). *Journal of Economic Entomology* 101, 1360–1366.

Sapkota, R., Stout, M.J., and Henderson, G. (2020) Residual effects of termiticides on mortality of Formosan subterranean termite (Isoptera: Rhinotermitidae) on substrates subjected to flooding. *Journal of Economic Entomology* 113, 367–374.

Saran, R.K. and Kamble, S.T. (2008) Concentration-dependent degradation of three termiticides in soil under laboratory conditions and their bioavailability to eastern subterranean termites (Isoptera: Rhinotermitidae). *Journal of Economic Entomology* 101, 1373–1383.

Saran, R.K. and Rust, M.K. (2007) Toxicity, uptake, and transfer efficiency of fipronil in Western subterranean termite (Isoptera: Rhinotermitidae). *Journal of Economic Entomology* 100, 495–508.

Saran, R.K., Ziegler, M., Kudlie, S., Harrison, D., Leva, D.M. *et al.* (2014) Behavioral effects and tunneling responses of Eastern subterranean termites (Isoptera: Rhinotermitidae) exposed to chlorantraniliprole treated soils. *Journal of Economic Entomology* 107, 1878–1889.

Shelton, T.G. and Grace, J.K. (2003) Effects of exposure duration on transfer of nonrepellent termiticides among workers of *Coptotermes formosanus* Shiraki (Isoptera: Rhinotermitidae). *Journal of Economic Entomology* 96, 456–460.

Shelton, T.G., Mulrooney, J E. and Wagner, T.L. (2006) Transfer of chlorfenapyr among workers of *Reticulitermes flavipes* (Isoptera: Rhinotermitidae) in the laboratory. *Journal of Economic Entomology* 99, 886–892.

Shuai, X., Chen, J. and Ray, C. (2012) Adsorption, transport and degradation of fipronil termiticide in three Hawaii soils. *Pest Management Science* 68, 731–739.

Smith, J.L. and Rust, M.K. (1990) Tunneling response and mortality of the western subterranean termite (Isoptera: Rhinotermitidae) to soil treated with termiticides. *Journal of Economic Entomology* 83, 1395–1401.

Smith, J.L. and Rust, M.K. (1991) Vapor activity of insecticides used for subterranean termite (Isoptera: Rhinotermitidae) control. *Journal of Economic Entomology* 84, 181–184.

Smith, J.L. and Rust, M.K. (1992) Activity and water-induced movement of termiticides in soil. *Journal of Economic Entomology* 85, 430–434.

Smith, J.L. and Rust, M.K. (1993) Cellulose and clay in sand affects termiticide treatments. *Journal of Economic Entomology* 86, 53–60.

Song, D. and Hu, X.P. (2006) Effects of dose, donor–recipient interaction time and ratio on fipronil transmission among the Formosan subterranean termite nestmates (Isoptera: Rhinotermitidae). *Sociobiology* 48, 237–246.

Spomer, N.A., Kamble, S.T., Warriner, R.A. and Davis, R.W. (2008) Influence of temperature on rate of uptake and subsequent horizontal transfer of [14C] fipronil by Eastern subterranean termites (Isoptera: Rhinotermitidae). *Journal of Economic Entomology* 101, 902–908.

Su, N.-Y. (1990) Economically important termites in the United States and their control. *Sociobiology* 17, 77–94.

Su, N.-Y. (2002) Novel technologies for subterranean termite control. *Sociobiology* 40, 95–102.

Su, N.-Y. (2005) Response of the Formosan subterranean termites (Isoptera: Rhinotermitidae) to baits and nonrepellent termiticides in extended foraging arenas. *Journal of Economic Entomology* 98, 2143–2152.

Su, N.Y. and Scheffrahn, R.H. (1990) Comparison of eleven soil termiticides against the Formosan subterranean termite and Eastern subterranean termite (Isoptera: Rhinotermitidae). *Journal of Economic Entomology* 83, 1918–1924.

Su, N.-Y. and Scheffrahn, R.H. (1998) A review of subterranean termite control practices and prospects for integrated pest management programs. *Integrated Pest Management Reviews* 3, 1–13.

Su, N.Y., Tamashiro, M., Yates, J.R., and Haverty, M.I. (1982) Effect of behavior on the evaluation of insecticides for prevention of or remedial control of the Formosan subterranean termite. *Journal of Economic Entomology* 75, 188–193.

Su, N.Y., Scheffrahn, R.H. and Ban, P.M. (1993) Barrier efficacy of pyrethroid and organophosphate formulations against subterranean termites (Isoptera: Rhinotermitidae). *Journal of Economic Entomology* 86, 772–776.

Su, N.Y., Wheeler, G.S., and Scheffrahn, R.H. (1995) Subterranean termite (Isoptera: Rhinotermitidae) penetration into sand treated at various thicknesses with termiticides. *Journal of Economic Entomology* 88, 1690–1694.

Su, N.Y., Ban, P.M. and Scheffrahn, R.H. (1999) Longevity and efficacy of pyrethroid and organophosphate termiticides in field degradation studies using miniature slabs. *Journal of Economic Entomology* 92, 890–898.

Snyder, T.E. (1935) Our enemy the termite. Comstock Publishing, Ithaca, New York, 196 pp.

Tamashiro, M., Grace, J.K. and Yamamoto, R.T. (1996) Insecticidal and behavioral activity of etofenprox against the Formosan subterranean termite (Isoptera: Rhinotermitidae). *Sociobiology* 28, 53–66.

Thomas, C.R. and Robinson, W.H. (1991) Foam formulations for subslab injection of termiticides. *Pest Control Technology* 19 (11), 38, 42–44.

Thoms, E.S., Nolen, S. and Bodalbhai, L. (1993) Termiticide foams fill slab treatment gaps. *Pest Control* 61(2), 47–51.

Thorne, B.L. and Breisch, N.L. (2001) Effects of sublethal exposure to imidacloprid on subsequent behavior of subterranean termite *Reticulitermes virginicus* (Isoptera: Rhinotermitidae). *Journal of Economic Entomology* 94, 492–498.

Tomalski, M. and Vargo, E.L. (2005) Acquisition, transfer and metabolism of [14C] imidacloprid among workers of the subterranean termite, *Reticulitermes flavipes* (Isoptera: Rhinotermitidae). In: Lee, C.Y. and Robinson, W.H. (eds) *Proceedings of the 5th International Conference on Urban Pests*, pp. 243–249. https://www.icup.org.uk/media/mmcmqe0q/icup038.pdf

Tsunoda, K. (2005) Improved management of termites to protect Japanese homes. In: Lee, C.Y. and Robinson, W.H. *Proceedings of the 5th International Conference on Urban Pests*. pp. 33–37. https://www.icup.org.uk/media/jzpllcve/icup005.pdf

Tsunoda, K. (2006) Transfer of fipronil, a nonrepellent termiticide, from exposed workers of *Coptotermes formosanus* (Isoptera: Rhinotermitidae) to unexposed workers. *Sociobiology* 47, 563–575.

Tsunoda, K., Matsuoka, H. and Yoshimura, T. (1998) Colony elimination of *Reticulitermes speratus* (Isoptera: Rhinotermitidae) by bait application and the effect on foraging territory. *Journal of Economic Entomology* 91, 1383–1386.

Tsunoda, K., Matsuoka, H., Yoshimura, T. and Tokoro, M. (1999) Foraging populations and territories of *Reticulitermes speratus* (Isoptera: Rhinotermitidae). *Journal of Economic Entomology* 92, 604–609.

USDA (1942) *Preventing damage to buildings by subterranean termites and their control*. Farmers' Bulletin 1911. US Department of Agriculture, Washington, DC.

Valles, S.M. and Woodson, W.D. (2002) Group effects on insecticide toxicity in workers of the Formosan subterranean termite, *Coptotermes formosanus* Shiraki. *Pest Management Science* 58, 769–774.

Vargo, E.L. and Parman, V. (2012) Effect of fipronil on subterranean termite colonies (Isoptera: Rhinotermitidae) in the field. *Journal of Economic Entomology* 105, 523–532.

Vela, N., Pérez-Lucas, G., Navarro, M. J., Garrido, I., Fenoll, J. and Navarro, S. (2017) Evaluation of the leaching potential of anthranilamide insecticides through the soil. *Bulletin of Environmental Contamination and Toxicology* 99, 465–469.

Vlosky, R.P., Shupe, T. and Wu, Q. (2009) Perceptions and use of termite resistant treated wood products. Part I: The perspective of homeowners in Formosan subterranean termite infected states. *Drvna Industrija* 60, 135–144.

Wang, C., Henderson, G., Chen, X. and Gautam, B.K. (2013) Horizontal transfer of fipronil is enhanced with increased group size in *Coptotermes formosanus* (Isoptera: Rhinotermitidae). *Journal of Economic Entomology* 106, 2630–2632.

Wang, J., Ohtani, Y. and Sameshima, K. (2000) Study on basic biology of *Reticulitermes speratus* (Kolbe) with both field and laboratory trials in Kochi. *Proceedings of the 3rd International Wood Science Symposium*, pp. 167–172.

Yeoh, B.H. and Lee, C.Y. (2007) Tunneling responses of the Asian subterranean termite, *Coptotermes gestroi* in termiticide-treated sand (Isoptera: Rhinotermitidae). *Sociobiology* 50, 457–468.

Yu, S.J. (2015) *The Toxicology and Biochemistry of Insecticides*. CRC Press, New York, 276 pp.

Zanne, A.E., Flores-Moreno, H., Powell, J.R., Cornwell, W.K., Dalling, J.W. *et al.* (2022) Termite sensitivity to temperature affects global wood decay rates. *Science* 377, 1440–1444.

Zeng, X., Wang, S., Xu, P., Zhang, L, Guo, Y. and Yang, J. (2020) A comparative study on toxicity of pyrethrin and three conventional pesticides against *Coptotermes formosanus* Shiraki. *Plant Diseases and Pests* 11, 22–25.

Zhang, L., Zhang, J., Wang, C., He, Y., Wen, X. *et al.* (2022) Toxicological, behavioral, and horizontal transfer effects of cycloxaprid against Formosan subterranean termites (Blattodea: Rhinotermitidae). *Journal of Economic Entomology* 115, 1240–1250.

Zheng, W. and Liu, W. (1999) Kinetics and mechanism of the hydrolysis of imidacloprid. *Pesticide Science* 55, 482–485.

13 Prevention of damage to building materials using wood preservatives

Wakako Ohmura[1] and Koichi Yamamoto[2]

[1]Laboratory of Innovative Humano-habitability, Research Institute for Sustainable Humanosphere, Kyoto University, Japan, ohmura.wakako.7h@kyoto-u.ac.jp; [2]Japan Wood Protection Association, koichi3859mt@gmail.com

Abstract

Wood and cellulosic materials can be pre- and post-treated with preservatives with fungicidal and termiticidal properties. This chapter reviews the methods applied to protect wood structures and the types of wood preservatives used, focusing on protecting various materials used in buildings (such as wood, insulation, and cables). We also discuss the modes of action of the protective measures, especially against Formosan subterranean termites, and the methods used to evaluate the performance of these wood preservatives.

13.1 Introduction

Termites are an important group of wood-destroying insects. They feed on wood fibers, causing a drastic reduction in wood strength, that could lead to wood damage and reduced structural integrity (Mori *et al.*, 2019). Due to the cryptic nature of termites, structural damage to buildings by termites is often not noticed until a later stage. If the damage occurs at the critical structural load-bearing areas (e.g., columns and joints), the risk of building collapse increases. Moreover, termite-infested buildings are more prone to collapse in a natural disaster such as a hurricane (typhoon) or earthquake (Doi *et al.*, 1998).

Compounds such as chlorinated hydrocarbons and organophosphates that were used to prevent termite attacks in the past are banned or prohibited due to their persistence in the environment and/or their impact on human health (Reinprecht, 2016; Kirker and Lebow, 2021; Khademibami and Bobadilha, 2022). Because of this, termiticides and control approaches with reduced impact on the environment and human health are preferred, especially those that are target-specific (Su and Schefferahn, 1998, 2002; Yoshimura, 2001; Gill and Garg, 2014).

Two approaches can be used to prevent termites from infesting buildings and structures: (1) protecting building materials from termite attacks; and (2) preventing termite entry into the building or structure. This includes baits that are used for both pre- and post-construction treatments to prevent termite infestation. In this chapter, we discuss the former approach. Wood and lignocellulose materials can be treated with fungicides and termiticides. Surface treatment or impregnated treatment of construction materials can be done to protect the material. In addition, since insulation is commonly used in buildings in temperate regions,

 Biology and Management of the Formosan Subterranean Termite and Related Species (eds Nan-Yao Su and Chow-Yang Lee)
DOI: 10.1079/9781800621596.0013

termiticides can be applied to insulation materials to help prevent termite damage (Kramer, 1993; Guyette, 1994; Smith and Zungoli, 1995a,b; Zungoli, 1999; Mankowski and Grace, 2004). Termites may also damage electrical cables, and treatments of cable sheathings with termiticides provide protection (Ruddel, 1985; Lynch, 2004; Lenz *et al.*, 2012).

13.2 Protection of wood by treatment

Subterranean termites enter buildings primarily through the subfloor. Once in the building, the subterranean termites secure moisture and proceed to damage each component of the building (Fig. 13.1). The strength of the building may deteriorate before the damage by termites is even recognized, and a termite-damaged building may collapse when it is subjected to sudden force such as earthquake or storms (Fig. 13.2A, B).

Preservative treatment is a means to directly prevent termite damage to the wood. Wood treatment with wood preservatives can be divided into two main types: pressure-impregnated and surface treatments. In pressure-impregnated treatment, the preservative is impregnated deep inside the wood. In contrast, in the case of surface treatment, only a small amount of the preservative is absorbed by the wood. In both treatments, the amount of chemical penetration varies greatly, depending on the species, size, and location (i.e., sapwood vs heartwood) of the wood to be treated.

In addition to these preservative-treated woods, some chemically modified woods, such as acetylated woods and wood–plastic composites (WPCs), are used as construction materials/components, and exterior materials with termite-resistant properties have been verified (Rust, 2014).

13.2.1 Pressure-impregnation method

The pressure-impregnation process is applied at a dedicated large-scale facility. Different mechanical pressure values are used to infuse a preservative into wood or woody materials. Before the treatment, it is sometimes necessary to first incise the wood, where minute scratches are made from the wood surface to a depth of 10 mm to ensure

Fig. 13.1. Wood materials damaged by *C. formosanus*. (Photo courtesy of T. Mori.)

Fig. 13.2. Termite damage in a building that became apparent after an earthquake. Termite damage may accelerate building destruction by earthquakes.

the prescribed degree of preservative penetration. It is also important that the wood has been dried to below the fiber saturation point (approximately 30%) of moisture content before pressure treatment, to obtain the appropriate amount of preservative retention (Reinprecht, 2016).

The process of pressure impregnation is completed as follows: (1) the wood is brought into the treatment cylinder, and the treatment liquid is introduced; (2) a vacuum is created in the cylinder to exhaust the air in the wood; (3) the pressure operation is then continued for a certain period under a specified pressure; (4) when the impregnation volume reaches the desired quantity, the pressure is returned to the atmospheric pressure value; and (5) the treating liquid is drained from the cylinder, and the wood is then removed from the cylinder. After inspecting the impregnation volume and other parameters, a curing period is used to ensure that the preservative becomes fixed in the wood.

In the wet type of pressure treatment, the production of the treated commodity is completed after the subsequent drying period. In dry pressure treatment, the treatment solution is discharged from the cylinder in the above-described step (5), and the organic solvent that was used to dissolve the active ingredient is also recovered from the treated wood during the post-vacuum phase in the same cylinder so that a dry treated product is produced. Compared with the wet treatment process, dry treatment does not require the post-treatment drying process and is less likely to cause wood deformation due to drying.

13.2.2 Surface treatment

Surface treatment is generally an on-site preservation treatment in which a chemical solution is applied to the wood surface by a simple method such as coating or spraying. Some wood and woody products are surface-treated in a factory. In general, the target of this treatment is building components that are relatively close to or in contact with the ground and pose a risk of fungal decay and termite infestation.

Dip-diffusion treatment with water-soluble preservatives such as borates may be considered a special type of surface treatment. In this case, lumber is dipped or soaked in a tank or pool of the preservative solution, then removed and wrapped with a tarp to allow the preservative to diffuse into the wood for several weeks before the tarp is removed and the wood dried for use.

When wood preservatives are applied to wood surfaces by a brush, roller, or spray, the process should be done at least twice to obtain sufficient absorption. The second application is made after the first coat of preservatives has been fully absorbed (Esaki *et al.*, 2020). An inter-application interval of ≥ 20 h is required for water-soluble agents and ≥ 5 h for oil-soluble agents. If a relatively large surface area needs to be treated, spraying is widely used. Special incising is also applied, and oil-soluble wood preservatives are sprayed at high pressure under atmospheric conditions to allow the preservative to penetrate the incision depth without using a pressure cylinder (Shigeyama, 2018).

Even with pressure-treated materials, there are cases where untreated areas of the wood appear on cracked or pre-cut surfaces, and there is a risk of decay or termite damage in these areas. It is desirable to treat such areas carefully with surface-treatment preservatives, using a brush to reduce the risk of damage.

13.2.3 Prior knowledge regarding wood characteristics in preservatives treatment

Liquid permeability into wood varies greatly depending on the tree species, portion (i.e., sapwood or heartwood), and orientation (radial, tangential, or longitudinal) (Hayashi and Nishimoto, 1965; Reinprecht, 2016; AWPA, 2022). Even with the same tree species, liquid permeability varies greatly depending on the tree's growing environment, such as the location, soil type, and sunlight conditions. Liquid permeability is generally higher in sapwood and lower in heartwood, although there can be significant individual differences even within the same species. This is because the heartwood has a high closure rate of pits involved in liquid flow; and the cell lumina are also covered with heartwood substances, which prevents liquid flow between cells. In the case of oaks, chestnuts, and other broad-leaf tree species, even sapwood has low liquid permeability, due to the closure of the vessels by tyloses (Zabel and Morrell, 2020).

The liquid permeability of anisotropic wood varies greatly depending on its direction, being greatest in the fiber direction, followed in order by the radial direction and tangential direction. Latewood has higher liquid permeability than earlywood, especially in conifers, where the distinction between early- and latewood is more pronounced (Esaki *et al.*, 2020). Latewood may have thicker pits than earlywood, resulting in less pit occlusion (Petty and Puritch, 1970; Usta and Hale, 2006). The cross-end section of wood, which crosses the fiber direction, has higher liquid permeability compared with the other sections. Still, it is also more susceptible to decay and termite damage.

Even in sapwood, cedar wood dried at higher temperatures tends to have lower liquid permeability (Sakagami *et al.*, 2016a,b). According to Sakagami *et al.* (2016a,b), the highest absorption of cedar sapwood by pressure impregnation of copper azole (CuAz) preservative was approximately 400 kg/m^3 when dried at 20°C, but the absorption decreased with increasing drying temperature and was < 100 kg/m^3 when the wood was dried at 100°C or 120°C. The extent of penetration in wood dried at 20°C was > 70% at 10 cm from the cross end of the wood, whereas the extent in wood dried at 100°C or 120°C was minimal, i.e., only at 6 cm from the end. Thus, the higher the drying temperature, the less is the liquid permeability of the cedar wood.

13.3 Termite damage to insulation and cables and its prevention

Insulation is used to improve the energy efficiency of buildings and has also become an essential building material for soundproofing. When insulation is used, especially when the insulation is in contact with ground, termites can penetrate the insulation from soil. From there, they could enter the interior of the structure and damage the building components (Fig. 13.3). Fig. 13.3 shows a situation in which termites entered through the inside of the foundation and insulated walls and caused damage to the upper part of the house. The only remedy for such infestation is to destroy the walls and remove the damaged insulation or to drill through the concrete foundation and walls to inject insecticides into the inner insulation to kill the termites within. When termites attack through the insulation, the damage may not be visible from the outside,

Fig. 13.3. Insulation damage by termites. Termites may enter buildings through or behind insulation.

and the damage detection may be delayed. Foam insulation presents another challenge. The insulation is easy to install but if termite damage occurs to the insulation itself or housing components covered by the insulation, the removal of the irregularly shaped foam insulation to control the termites is extremely time-consuming.

Termite damage to insulation materials was first reported in the 1990s, when insulation materials became more widely used in homes (Kramer, 1993; Guyette, 1994; Smith and Zungoli, 1995a,b; Zungoli, 1999). Products in which chemicals are mixed into the insulation itself have been evaluated (Su *et al.*, 2003; Mankowski and Grace, 2004; Williams and Bergstrom, 2005), and stainless-steel mesh has been used to cover the ground surface of the insulation and gaps between the foundation and insulation, to prevent termite infestation without the use of chemicals (Grace *et al.*, 1996).

Termite damage to power cables has also been reported (Ruddel, 1985; Lynch, 2004). To address this problem, approaches to damage prevention have included the addition of chemicals to cable materials and using materials with increased hardness (Lenz *et al.*, 2012; Zhang *et al.*, 2016).

13.4 Type and safety of chemicals used in wood treatment

The preservatives used to prevent various types of wood biodeterioration are usually formulated to have both fungicidal and insecticidal properties. The risk of termite damage is exceptionally high in areas with high termite pressure in subtropical to temperate zones (Bignell and Eggleton, 2000; Junqueira and Florencio, 2018). When the wood is used for outdoor and underfloor areas of buildings, it is essential to treat the wood with preservatives to stop termite feeding damage. The risk of termite damage to wood tends to decrease with latitude and altitude, while the risk of decay increases (Zanne *et al.*, 2022).

Wood and woody products that are pressure-treated with preservatives are expected to maintain high efficacy for an extended period in harsh environments. These products thus require deep penetration into the wood and long-term fixation. For infrastructure products such as utility poles, railway sleepers, and civil engineering materials, inorganic or oil-based agents are often the main ingredients, considering weather resistance, e.g., resistance to degradation by ultraviolet rays and elution by rainwater.

Chromated copper arsenate (CCA) preservative, patented and first manufactured in 1938 (Freeman *et al.*, 2003), dominated the global wood preservation market in the second half of the 20th century, due to its superior preservative efficacy, retention in wood, and economic efficiency (Mohajerani *et al.*, 2018). However, with environmental awareness in the 1990s, the production and the use of CCA has been greatly limited, beginning with Switzerland and Austria (Morais *et al.*, 2021). In the U.S., CCA-treated products were voluntarily discontinued in virtually all residential applications at the end of 2003 and are no longer used in most residential settings. The U.S. Environmental Protection Agency (EPA) has classified CCA-treated products as restricted-use products, allowing only certified contractors to use them in infrastructure products such as utility poles and railway sleepers (EPA, 2022a,b).

In Hawai'i, concern over the poor penetration of CCA in lumber used locally for construction led to the adoption of the alternative preservative ammoniacal copper zinc arsenate (ACZA), in which the ammonia in the formulation enables greater penetration; and subsequently to a widespread use of the diffusible preservative disodium octaborate tetrahydrate (DOT) (Grace, 2002).

Like CCA, creosote has been widely used worldwide as a wood preservative for applications that require high durability, such as utility poles, railway sleepers, and civil engineering materials. Being oil-based, creosote penetrates well into many kinds of wood, and post-treatment penetration is also observed (Kilmer *et al.*, 1998; Wacker, 2003). Creosote maintains high fungicidal and termiticidal performance for an extended period. However, besides its distinctive odor, it contains carcinogenic substances such as benzo[a]pyrene; hence, creosote is currently only limited for outdoor use. In Japan, the quality standard for the use of creosote in wood preservation was revised in 2013 by the Japanese Standards Association (JSA) to an extremely low concentration of 10 mg/kg each of benzo[a]pyrenes (JIS, 2013).

In contrast, wood preservatives for surface treatment are mainly formulations in which the main ingredient is an organic agent for both fungicidal and termiticidal efficacy. The most common types of surface-treatment formulations are water-based (such as emulsions and flowable formulations), due to concerns about volatile organic compounds (VOCs) and other factors. However, the penetration of these formulations is lower than that of oil-based formulations.

Inorganic borate compounds, particularly disodium octaborate tetrahydrate, are also used for surface treatment, either in a simple aqueous solution or in combination with a glycol to create a higher concentration treatment solution (Grace and Yamamoto, 1994).

13.5 Evaluation methods of wood preservatives

The termiticidal efficacy of a preservative depends in part on the climatic conditions (hours of sunlight, rainfall, temperature, humidity), the physical and chemical conditions of the soil, the presence or absence of termite infestation, the termite species, and the period of termite activity. The standards for the methods of evaluating the termiticidal performance of wood preservatives differ from country to country and region to region, depending on the major termite pest species, wood species, architectural styles, and termite control methods (Ahmed *et al.*, 2004; Mulrooney *et al.*, 2007; Reinprecht, 2016; Hassan and Morell, 2021). Therefore, it is imperative to conduct efficacy tests in each country or region where the wood preservative is used.

13.5.1 Actual and predicted service life of wood from field tests

Two ubiquitous outdoor wood products, utility poles and railway sleepers, have been gradually replaced by concrete products in many parts of the world, partly due to the lesser service life of the wood versions. When wood is used outdoors, preservatives are essential to extend its service life, and information on the service life of wood products is indispensable for producers and users. Although there are differences among national standards in the methods of determining the degree of damage, the field stake test is the simplest and most effective test for comparing the service life of treated wood and wooden materials for use in soil contact. Long-term data have accumulated from stake tests in various countries, including those of preservatives no longer in mainstream use.

In the American Wood Protection Association (AWPA) standard, regional differences within the U.S. in degradation environments when the wood is used in ground contact conditions are mapped as Deterioration Zones, ranging from 1 = Low to 5 = Severe (Kirker *et al.*, 2017). The US Department of Agriculture (USDA) Forest Products Laboratory (FPL) has regularly published stake test results for both nearly obsolete preservatives and recently developed preservatives, incorporating untreated southern pine wood as a control (Woodward *et al.*, 2011). According to those data, the service life of untreated southern pine was 0.7 years in the Panama Canal zone (Zone 5), and termites were the only cause of deterioration. In contrast, in Madison, Wisconsin, in the U.S. (Zone 2), the service life was 5.7 years, and 100% of the deterioration was caused by decay, with no damage caused by termites. Deterioration of the untreated control wood was thus ~8 times faster at the Panama

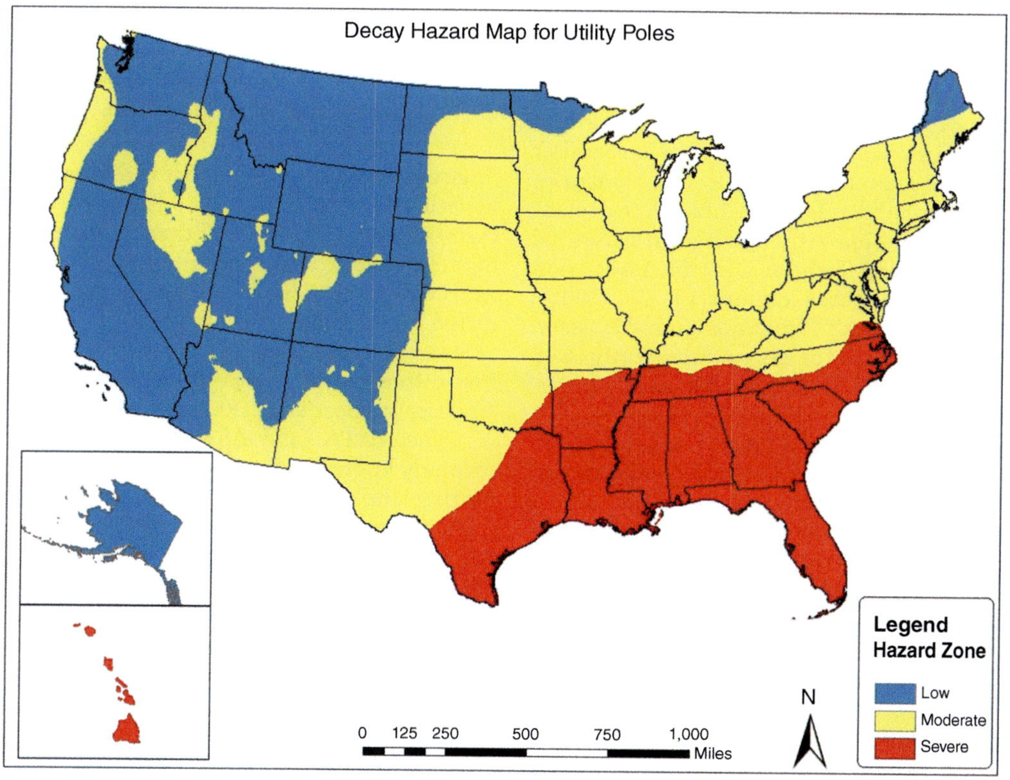

Fig. 13.4. In some modified environments, such as banks along irrigation canals or irrigated residential or agricultural lands, a higher degree of protection might be needed than would be required in the local natural environment. It must also be recognized that within individual regions, certain natural environments such as river valleys or coastlines may present greater potential for wood deterioration than the region as a whole. (Reprinted with permission from the American Wood Protection Association, Inc. This fungal decay hazard map was published in the 2022 AWPA Book of Standards in AWPA Standard U1, Commodity Specification D – Utility Poles.)

Canal compared with Wisconsin. The service life at the other three sites, Bogalusa, Louisiana (Zone 5), Jacksonville, Florida (Zone 5), and the Harrison Experimental Forest, Mississippi (Zone 5), was 2.8–2.9 years. At these three test sites, both decay and termites were the causes of deterioration, but only in Harrison were the termites responsible for most of the deterioration (70%), while damage was due to decay only and not termites in Bogalusa and Jacksonville. The AWPA standard Deterioration Zones Map has recently been updated and the zones have been reclassified as Low, Moderate, and Severe (AWPA, 2022) (Fig. 13.4).

Japan's two most damaging termite species are the subterranean termites *Coptotermes formosanus* Shiraki and *Reticulitermes speratus* (Kolbe). *Coptotermes formosanus* is distributed at lower latitudes along Japan's coast. *Reticulitermes speratus* is distributed throughout Japan except in mountainous areas and the eastern part of the island of Hokkaido. Based on the distribution areas of these two species, the application categories of preservative treatment and soil treatment at construction sites have been classified by the Japan Termite Control Association (JTCA) into three regions (JTCA, 2005): Region I for both *C. formosanus* and *R. speratus*, Region II for *R. speratus* only, and Region III where *R. speratus* are only partially present.

The Japan Wood Protection Association (JWPA)'s homepage (http://www.mokuzaihozon.org/info/chousa/yagaikui/) shows the results of stake tests in various regions of Japan. Field tests for termites in Japan have been conducted mainly in Region I using cedar sapwood as a substrate in both stake tests of treated materials impregnated with a preservative and non-ground non-exposure test of preservative surface-treated materials. The service life of cedar sapwood is 2.5 years in Kagoshima (Region I), 2.8 years in Tsukuba (Region II), and ~3.5 years in Asahikawa (Region III). However, in the case of Asahikawa, data are not for cedar but for todomatsu (*Abies sachalinensis*) sapwood, because the region is too cold for cedar trees. In terms of deterioration factors, even in Kagoshima (an area inhabited by *C. formosanus*), the progression of deterioration caused by termites alone is limited. In both Kagoshima and Tsukuba, termite damage progresses in combination with decay.

To increase the usefulness of the performance data of wood preservatives in field tests, a research group from the International Research Group on Wood Protection obtained data on six tree species in both ground-contact and above-ground conditions at 36 sites worldwide. The results showed no significant correlations between decay rates and the Scheffer climate index (SCI) under the ground-contact condition. Differences in decay rates were primarily associated with differences in moisture content under above-ground conditions (Stirling *et al.*, 2016).

13.5.2 Effect of different environmental conditions on degradation

Environmental conditions that include a low moisture content where wood is used will carry a lower risk of decay than those with high moisture content, resulting in a longer expected service life. Comparing the predicted service life of roofed and unroofed wooden bridges showed a greater than threefold difference in service life (144 years vs 43 years, respectively) (Reinprecht, 2016).

The AWPA (and ASTM International, formerly known as the American Society for Testing and Materials (ASTM)), the European Norm (European Standards) (EN), and the Australian Standards (AS) each refer to the exposure conditions that may be involved in the Use category, Use class, and Hazard class, respectively; and the former two classes, in particular, have established test methods that take into account the differences in the penetration of preservatives into wood species in each category/class. The Japan Agricultural Standards (JAS) specifies only the types of preservatives that can be used and the amount of retention according to the environment of uses from K1 to K5 (e.g., JAS 1083, Sawn timber). Note that "K" stands for "kankyo" (environment), and K1–K5 are assigned to group different environments. While the Japan Industrial Standards (JIS) K 1571 standard describes fungi and termite tests in both laboratory and field settings, it does not provide a termite field test method for pressure-treated preservatives used for wood products that are applied above-ground in buildings (JIS, 2010).

Comparative studies of termite damage levels in preservative-treated wood have been jointly conducted in Waimanalo (Hawai'i, USA), Kagoshima (Japan), and Kincardine (Canada) (Tsunoda *et al.*, 2002; Morris *et al.*, 2011, 2014). The test method used in this experiment was a covered above-ground test, suitable for the diffusible preservative disodium octaborate tetrahydrate (DOT) used in building wood components that are not in direct contact with soil or rainwater. The test method is a prototype of the AWPA's E21-06 standard (AWPA, 2022), which assumes that wood products may be used in above-ground environments, which is equivalent to Use Categories 1 and 2 in service. The termite activity is much more severe in Waimanalo, and the untreated wood degraded about 10 times faster than in Kagoshima. Termite attack was negligible in Kincardine, which is located at the northern limit of the distribution of *R. flavipes*. Various treated wood materials suffered little degradation in Kincardine and Kagoshima. In contrast, at Waimanalo, it was observed that even wood specimens that were surface-treated with CCA and high concentrations of DOT (3% boric acid equivalent (BAE)) were damaged by termites. In contrast, the wood specimens with very high preservative retention levels showed no termite damage.

We have been evaluating the durability of various preservative-treated wood samples under exposed ground-contact and unexposed above-ground conditions at a termite test site in Kagoshima, Japan, annually since 2005 (Ohmura *et al.*, 2019). The test methods we have been using for the exposed ground-contact conditions are in accordance with the JIS K 1571 standard (JIS, 2010), and the test methods for the unexposed above-ground conditions follow the AWPA standard E21-06 (AWPA, 2022). The test specimens were collected, and their moisture contents were examined in May 2010 and May 2015. The average moisture content of the exposed ground-contact samples was approximately 60%, and that of the unexposed above-ground specimens was approximately 20%. The termite damage and decay to the untreated wood were significant in the exposed ground-contact tests. In the preservative-treated wood, minor damage by *C. formosanus* was observed within 3 years of exposure, and degradation by decay fungi began to occur within 5 years of exposure. The damage by *R. speratus* tended to happen with the onset and progression of decay.

We also observed that although termites continued to form shelter tubes on the surface of treated specimens under the unexposed above-ground conditions, the damage by termites was minimal. The damage rate to untreated wood under the exposed above-ground conditions at 5 years of exposure was 40% by termites and 77% by decay (note that damage by termites and fungi may occur simultaneously in the same test specimens). The damage rate for untreated wood under the unexposed above-ground conditions was 76% by termites and nearly 0% by fungi. Notably, termite attacks occurred at a higher rate (76%) in the unexposed above-ground condition compared with that (40%) in the exposed ground-contact condition, probably due to the preference of termites for the moisture content of wood placed in an air-dry condition (20% moisture content) rather than in a wet condition (60% moisture content) (Nakayama *et al.*, 2004).

13.5.3 Test standards for evaluating the efficacy of preservatives and the need for standardization efforts

Hassan and Morell (2021) compared test standards and other information from around the world to evaluate the effectiveness of the preservative treatment of wood against termites, and they organized the results into those obtained in field tests and those obtained in laboratory tests. They concluded that combining similar standards from different countries into a unified standard is feasible by consolidating data on the same termite species into a broader range of typical termite species. Based on the commonalities and problems described in the present review, we hope that further consideration will be given to developing unified international standards.

Laboratory tests have the advantage of obtaining the results relatively quickly. Accelerated degradation, which is a combination of leaching and drying (heating), is often conducted before fungal and termite tests, simulating dissolution and volatilization of the preservative that may occur in service. A laboratory decay test of wood specimens conducted after subjecting samples to 9, 18, and 27 months of outdoor weathering revealed a significant decrease in decay resistance in some wood species (Keržič and Humar, 2022). Disadvantages of laboratory tests include small specimen sizes, short test periods, and the difficulty of reproducing the actual use conditions. A set of field tests should thus also be conducted to evaluate the performance of wood products under conditions closer to actual use.

The ability to predict the service life of wood preservatives is important. Models have been developed to predict the service life of wood structures and members. Factors relevant to prediction include weather conditions, structure dimensions, design details, and the materials' moisture absorption and desorption capacity (Brischke *et al.*, 2019). The biodeterioration factors in such models are often specific to decay more than to termites. Termite damage is highly temperature-dependent (Zanne *et al.*, 2022) and this needs to be reflected in service life models, especially in temperate regions where decay and termite damage could occur simultaneously.

13.6 Conclusion

It is impossible in practice to preserve all wood and plastic materials used in buildings to protect them from termite attack. However, effective prevention of termite damage to buildings can be achieved by preventing subterranean termites from entering the building, with the use of soil termiticides, baits, and physical barriers, among other techniques. When these methods are combined with the wood preservative treatments described in this chapter, it is possible to ensure buildings as safe living spaces for a longer period.

References

Ahmed, B.M., French, J.R. and Vinden, P. (2004) Review of remedial and preventative methods to protect timber in service from attack by subterranean termites in Australia. *Sociobiology* 44, 297–312.

AWPA (2022) *Book of Standards*. American Wood Protection Association, Birmingham, Alabama. [Includes standards on preservatives, treatments, methods of analysis, and inspection.]

Bignell, D.E. and Eggleton, P. (2000) Termites in ecosystems. In: Abe, T., Bignell, D.E. and Higashi, M. (eds) *Termites: Evolution, Sociality, Symbioses, Ecology*. Springer, Dordrecht, Netherlands, pp. 363–387.

Brischke, C., Niklewski, J., Humar, M. and Alfredsen, G. (2019) Service life planning of wooden structures: Mathematical prediction models versus professional experience. Document No. IRG/WP 19-20663. International Research Group on Wood Protection, Stockholm.

Doi, T., Kitamoto, H. and Han, L.H. (1998) Damage to wooden houses due to the 1995 great Hansin-Awaji earthquake disaster. *11th Earthquake Conference*, 856.

EPA (2022a) Overview of wood preservative chemicals. https://www.epa.gov/ingredients-used-pesticide-products/overview-wood-preservative-chemicals

EPA (2022b) Chromated Arsenicals (CCA). https://www.epa.gov/ingredients-used-pesticide-products/chromated-arsenicals-cca.

Esaki, T., Niimi, K., Doi, T. and Momohara, I. (2020) Pest control and construction. In: License Certification Committee (eds) *Fundamentals of Termite and Decay Control Construction*. The Japan Termite Control Association, Shinjuku, Japan, pp. 163–225.

Freeman, M.H., Shupe, T.F., Vlosky, R.P. and Barnes, H.M. (2003) Past, present, and future of the wood preservation industry. *Forest Products Journal* 53, 8–15.

Gill, H.K. and Garg, H. (2014) Pesticide: environmental impacts and management strategies. In: Soloneski, S. (ed.) *Pesticides – Toxic Aspects*. InTech Open, London, pp. 187–230.

Grace, J.K. (2002) Hawaiian experience with treated building components. In: *Enhancing the Durability of Lumber & Engineered Wood Products*. Forest Products Society, Madison, Wisconsin, pp. 305–309.

Grace, J.K., and Yamamoto R.T. (1994) Simulation of remedial borate treatments intended to reduce attack on Douglas-fir lumber by the Formosan subterranean termite (Isoptera: Rhinotermitidae). *Journal of Economic Entomology* 87, 1547–1554.

Grace, J.K., Yates, J.R., Tome, C.H. and Oshiro, R.J. (1996) Termite resistant construction: use of a stainless steel mesh to exclude *Coptotermes formosanus* (Isoptera: Rhinotermitidae). *Sociobiology* 28, 365–372.

Guyette, J.E. (1994) Termites targeting foam insulation. *Pest Control* 62, 49, 50, 52.

Hassan, B. and Morrell, J.J. (2021) Termite testing methods: A global review. *Journal of Testing and Evaluation*, doi: 10.1520/JTE20200455.

Hayashi, S. and Nishimoto, K. (1965) Studies on the water permeability of hardwoods. *Wood Research* 35, 33–43.

JIS (2010) JIS K 1571: 2010 *Wood preservatives – Performance requirements and their test methods for determining effectiveness.* Japanese Standards Association, Tokyo.

JIS (2013) JIS K 1570: 2013 *Wood preservatives*. Japanese Standards Association, Tokyo

JTCA (2005) *Standard specification for termite control.* The Japan Termite Control Association, Tokyo.

Junqueira, L.K. and Florencio, D.F. (2018) Termite damage in agriculture areas and implanted forests: An ecological approach. In: Khan, M.A. and Ahmad, W. (eds) *Termites and Sustainable Management*. Springer, Cham, Switzerland, pp. 37–50.

Keržič, E. and Humar, M. (2022) Natural durability and moisture dynamics of commercially important European wood species after weathering. Document no. IRG/WP 22-10992. International Research Group on Wood Protection, Stockholm.

Khademibami, L. and Bobadilha, G.S. (2022) Recent developments studies on wood protection research in academia: a review. *Frontiers in Forests and Global Change* 5, 793177. *doi:* 10.3389/ffgc.2022.793177

Kilmer, R.W., Blankenhorn, P.R., Labosky, P. and Janowiak, J.J. (1998) Laminating creosote-treated hardwoods. *Wood and Fiber Science* 30, 175–84.

Kirker, G.T. and Lebow, S.T. (2021) Wood preservatives. In: Forest Products Laboratory (ed.) *Wood Handbook – Wood as an Engineering Material*. United States Department of Agriculture Forest Service, Madison, Wisconsin, pp. 15.1–15.26.

Kirker, G.T., Bishell, A.B. and Hickey, W.J. (2017) AWPA Biodeterioration Hazard Map Revisited. *Proceedings One Hundred Thirteenth Annual Meeting of the American Wood Protection Association*, pp. 90–96.

Kramer, R.D. (1993) *Foam insulation and termites*. Service letter ESPC 055075, Number 1284. National Pest Control Association, Dunn Loring, Virginia, 3 pp.

Lenz, M., Creffield, J.W., Evans, T.A., Kard, B., Vongkaluang, C., Sornnuwat, Y., Lee, C.-Y., Yoshimura, T. and Tsunoda, K. (2012) Resistance of polyamide and polyethylene cable sheathings to termites in Australia, Thailand, USA, Malaysia and Japan: a comparison of four field assessment methods. *International Biodeterioration & Biodegradation* 66, 53–62.

Lynch, S. (2004) Termite corrosion. *Journal of Failure Analysis and Prevention* 4, 6.

Mankowski, M. and Grace, J.K. (2004) Response of the Formosan subterranean termite (*Coptotermes formosanus*) to cellulose insulation treated with boric acid in choice and no-choice tests. Document no. IRG/WP 04-10532. International Research Group on Wood Preservation, Stockholm.

Mohajerani, A., Vajna, J. and Ellcock, R. (2018) Chromated copper arsenate timber: A review of products, leachate studies and recycling. *Journal of Cleaner Production* 179, 292–307.

Morais, S., Fonseca, H.M., Oliveira, S.M., Oliveira, H., Gupta, V.K., Sharma, B. and de Lourdes Pereira, M. (2021) Environmental and health hazards of chromated copper arsenate-treated wood: A review. *International Journal of Environmental Research and Public Health* 18, 5518. doi: 10.3390/ijerph18115518

Mori, T., Otsbo, Y., Inoue, R., Ikeda, M., Kambara, K. and Ohmura, W. (2019) Experimental study on compressive strength performance of cross laminated timber with biodeterioration damage. Document no. IRG/WP 19-40883. International Research Group on Wood Protection, Stockholm.

Morris, P.I., Grace, J.K. and Tsunoda, K. (2011) Ten year field tests of borate-treated wood against subterranean termites. Document no. IRG/WP/11-30557. International Research Group on Wood Protection, Stockholm.

Morris, P.I., Grace, J.K., Yoshimura, T. and Tsunoda, K. (2014) An international termite field test of wood treated with insecticides in a buffered amine oxide carrier. *Forest Products Journal* 64, 156–160.

Mulrooney, J.E., Wagner, T.L., Shelton, T.G, Peterson, C.J. and Gerard, P.D. (2007) Historical review of termite activity at Forest Service termiticide test sites from 1971 to 2004. *Journal of Economic Entomology* 100, 488–494.

Nakayama, T., Yoshimura, T. and Imamura, Y. (2004) The optimum temperature-humidity combination for the feeding activities of Japanese subterranean termites. *Journal of Wood Science* 50, 530–534.

Ohmura, W., Kambara, K., Ito, Y., Matsunaga, H., Momohara, I. and Yoshimura, T. (2019) Evaluation of long-term durability of preservative-treated wood under different usage environments. *Abstracts of Annual Meeting of Japan Wood Society* 69, N15-P-14.

Petty, J.A. and Puritch, G.S. (1970) The effects of drying on the structures and permeability of the woods of *Abies grandis*. *Wood Science and Technology* 4, 140–154.

Reinprecht, L. (2016) *Wood Deterioration, Protection and Maintenance*. John Wiley & Sons, Chichester, UK, 337 pp.

Ruddell, H.J. (1985) Development of an insect resistant telecommunication cable for the Australian environment. *Telecommunication Journal of Australia* 35, 59e73.

Rust, M. (2014) Management strategies for subterranean termites. In: Dhang, P. (ed.) *Urban Insect Pests: Sustainable Management Strategies*. CABI Publishing, Wallingford, UK, pp. 114–129.

Sakagami, H., Tokunaga, A., Fujimoto, N., Koga, S., Kobayashi, I. and Momohara, I. (2016a) Effects of drying temperature for *Cryptomeria japonica* on the permeability of wood preservative. I: The permeability of dried logs. *BioResources* 11, 4781–4793.

Sakagami, H., Tokunaga, A., Fujimoto, N., Koga, S., Kobayashi, I. and Momohara, I. (2016b) Effects of drying temperature for *Cryptomeria japonica* on the permeability of wood preservative. II: The permeability of dried, split log pieces. *BioResources* 11, 6896–6908.

Shigeyama, T. (2018) Deeper penetration treatment method. *Mokuzai Hozon (Wood Protection)* 44, 140–141.

Smith, B.C. and Zungoli, P.A. (1995a) Rigid board insulation in South Carolina: Its impact on damage, inspection, and control of termites (Isoptera: Rhinotermitidae). *Florida Entomologist* 78, 507–515.

Smith, B.C. and Zungoli, P.A. (1995b) Termites in rigid board insulation, Part 2: Structural field inspections. *Pest Control Technology* 23, 56, 65, 66.

Stirling, R., Alfredsen, G., Brischke, C., De Windt, I., Francis, L.P. *et al.* (2016) Global survey on durability variation – on the effect of the reference species. Document no. IRG/WP 16-20573. International Research Group on Wood Protection, Stockholm.

Su, N.-Y. and Schefferahn, R.H. (1998) A review of subterranean termite control practices and prospects for integrated pest management programmes. *Integrated Pest Management Reviews* 3, 1–13.

Su, N.-Y. and Schefferahn R.H. (2002) Termites as pests of buildings, In: Abe, T., Bignell, D.E. and Higashi, M. (eds) *Termites: Evolution, Sociality, Symbioses, Ecology*. Kluwer Academic Publishers, Dordrecht, Netherlands, pp. 437–453.

Su, N.-Y., Ban, P., and Scheffrahn, R.H. (2003) Resistance of insecticide-treated foam board insulation against the eastern subterranean termite and the Formosan subterranean termite (Isoptera: Rhinotermitidae). *Journal of Economic Entomology* 96, 1526–1529.

Tsunoda, K., Grace, J.K., Byrne, T. and Morris, P.I. (2002) Effectiveness of disodium octaborate tetrahydrate (Tim-Bor®) in controlling subterranean termite attack and decay of house sill plate. *Mokuzai gakkaishi* 48, 107–114.

Usta, I. and Hale, M.D. (2006) Comparison of the bordered pits of two species of spruce (Pinaceae) in a green and kiln-dried condition and their effects on fluid flow in the stem wood in relation to wood preservation. *Forestry* 79, 467–475.

Wacker, J.P. (2003) *Creosote Retention Levels of Timber Highway Bridge Superstructures in Michigan's Lower Peninsula*. United States Department of Agriculture, Forest Service, Forest Products Laboratory. National Wood in Transportation Information Center, Iowa State University, Ames, Iowa.

Williams, L.H. and Bergstrom, T.B. (2005) Boron-treated expanded polystyrene insulation resists native subterranean termite damage after 3-year field exposure. *Forest Products Journal* 55, 56–60.

Woodward, B.M., Hatfield, C.A. and Lebow, S.T. (2011) Comparison of Wood Preservatives in Stake Tests 2011, Progress Report. Research Note FPL–RN–02. US Department of Agriculture, Forest Service, Forest Products Laboratory.

Yoshimura, T. (2001) Present status and future trends of termite controlling strategies in Japan. In: Imamura, Y. (ed.) *High Performance Utilization of Wood for Outdoor Uses*. Wood Research Institute, Kyoto University, Kyoto, Japan, pp. 131–144.

Zabel, R.A. and Morell, P.A. (2020) *Wood Microbiology – Decay and Its Prevention*, 2nd edn. Academic Press.
Zanne, A.E., Flores-Moreno, H., Powell, J.R., Cornwell, W.K., Dalling, J.W. *et al.* (2022) Termite sensitivity to temperature affects global wood decay rates. *Science* 377, 1440–1444.
Zhang, L., Cai, W., Chen, W.Y., Zhang, L., Hu, K. and Guan, Y.Q. (2016) Synthesis of AzPhchitosan-bifenthrin-PVC to protect cables against termites. *Carbohydrate Polymers* 139, 50–60.
Zungoli, P.A. (1999) Termite infestations in rigid board insulation: problems and solutions. In: Robinson, W. M., Rettich, F. and Rambo, G.W. (eds) *Proceedings of the 3rd International Conference on Urban Pests*, pp. 19–22.

14 Alternative and experimental management methods

J. Kenneth Grace

Department of Plant and Environmental Protection Sciences, College of Tropical Agriculture and Human Resources, University of Hawai'i at Mānoa, kennethg@hawaii.edu

Abstract

Alternative and experimental approaches to subterranean termite management can be broadly characterized as the use of natural products, physical barriers, and biological control. Methods currently available for *Coptotermes* Wasmann prevention are the use of naturally durable woods in building construction in lieu of preservative-treated wood and installation of physical barriers to prevent termite penetration rather than soil insecticide treatments. A particle barrier of crushed and screened basalt was the first physical barrier developed for *Coptotermes formosanus* Shiraki management. Particle barriers, stainless-steel mesh, and insecticide-treated plastic membranes are now available as management options in several regions; and insecticide-free plastic membranes acting purely as physical barriers are now of interest. Worldwide, a great deal of research effort has been devoted to identification of plant allelochemicals that deter termites, and evaluation of suspected entomopathogens as possible biological control agents; but with little to no commercial application to date. Greater emphasis on understanding biological mechanisms, particularly with relevance to termite field biology and ecology, and on overcoming commercial production issues are necessary for these areas of research to potentially produce useful alternative methods of termite management.

14.1 Introduction

There has been, and continues to be, a great deal of research interest in the development of alternatives to the use of synthetic pesticides and wood preservatives for termite management. The successful commercialization of baiting technology (see Chapter 11 of this volume), a termite control method thought to be something of an elusive dream only a few decades ago (La Fage, 1984), is an excellent example of a novel approach that minimizes pesticide use and non-target exposure. Baiting can be relatively expensive, however, and may not be readily available commercially in all regions where subterranean termites are a problem. Thus, interest has continued in developing additional effective and environmentally benign methods for termite prevention and control.

In addition to environmental concerns, cost, and availability issues with current commercial termite management methods in some regions of the world, the relatively low costs and infrastructure needs of conducting laboratory research on termites with locally found/produced materials have also been factors promoting research on novel controls. In the tropical and subtropical regions that are home to termite species

 Biology and Management of the Formosan Subterranean Termite and Related Species (eds Nan-Yao Su and Chow-Yang Lee)
DOI: 10.1079/9781800621596.0014

in the genus *Coptotermes* Wasmann, termites may be readily collected for laboratory experiments and the local environment is often a rich source of plants with possible insecticidal properties, locally occurring stone aggregates, and possible pathogenic microbes. Thus, research on a significant pest problem (termites) can be conducted even when only limited funding and facilities are available. This helps to explain, for example, why numerous laboratory studies on termite biological control continue to be published despite several critical reviews of this research area and the long odds of success with *Coptotermes* species (Culliney and Grace, 2000; Chouvenc *et al.*, 2011).

Alternative and experimental approaches to *Coptotermes* management can be broadly characterized as the use of natural products (including durable building materials), physical barriers, and biological control. A few researchers have also explored, at least in the laboratory, more unusual methods such as electromagnetic radiation (Diba *et al.*, 2013; Yanagawa *et al.*, 2020) and alternative fumigants (Delate *et al.*, 1995). Although such approaches will likely never be generally applicable in *Coptotermes* control, they might prove useful in the future for treatment of above-ground colonies, termite-infested boats, or infested materials in transit.

14.2 Natural products

There is a great deal of literature available on the durability of different tree species against termite attack and the heartwood extractives considered responsible for that durability. Like preservative-treated wood, naturally durable timbers are often incorporated into termite-resistant building construction. A great deal of effort has also gone into identification of the chemical compounds responsible for natural durability, and of bioactive compounds from plants reputed to have medicinal or insecticidal properties, although this has not yet led to practical termite management products.

14.2.1 Naturally durable building materials

The use of naturally durable timbers offers an alternative to preservative-treated wood for termite prevention. For example, redwood, *Sequoia sempervirens* (D. Don) Endl., heartwood has long been known to be more resistant to termites and decay than other softwoods commonly used in North American construction (Su and Tamashiro, 1986), although damage to redwood from *Coptotermes formosanus* Shiraki has been observed in the field (Su and Tamashiro, 1986) and the degree of durability may also differ depending upon which *Coptotermes* species is present (Hapukotuwa and Grace, 2011). Yellow cypress, also known as Alaska cedar (*Chamaecyparis nootkatensis* (D. Don) Spach), heartwood was found to be more durable than redwood (Grace and Yamamoto, 1994), presumably because of the presence of nootkatone, a sesquiterpene ketone also found in grapefruit and in vetiver grass (Zhu *et al.*, 2001b). These latter authors suggested that vetiver grass itself planted around buildings might form an effective barrier to foraging subterranean termites. Incorporation of termite-repellent plant materials in landscape mulches has also been proposed (Ding and Hu, 2010). Although minimizing termite activity in the landscape may well be beneficial, such approaches have not been tested in the field with respect to building protection.

In Hawai'i, the imported tropical hardwoods *Shorea laevis* Ridl. (known as bangkirai in Indonesia and balau in Malaysia) and *Intsia palembanica* Miq. (merbau) were found to be resistant to attack by *C. formosanus* (Grace and Tome, 2005). Ngee *et al.* (2004) reported that these woods were also avoided by *Coptotermes gestroi* (Wasmann), the other *Coptotermes* species found in Hawai'i. Although the cost of wood importation is not insubstantial, these two woods have been approved by local building authorities as options to preservative-treated wood (required throughout the entire structure in Hawai'i) and are used in high-end custom home construction (Prefabworld Bali International, 2014; Teak Bali, 2019).

Heartwood extractives, both as biocides (Carter and de Camargo, 1983) and possibly also as antioxidants (Ragon *et al.*, 2008), are considered the primary contributors to wood durability, although hardness imparted by high silica content may play a role in some tropical hardwoods (Nobre and Nunes, 2007). It is possible that antioxidants play a larger role in termite avoidance than has generally been thought to be the case (Ragon *et al.*, 2008). For example, Stirling *et al.* (2015) found no association between termite resistance and measured extractive content of western red cedar (*Thuja plicata* Donn ex D. Don) heartwood but did find a correlation between the degree of red coloration, linked to antioxidant polyphenolic polymers, and durability against *C. formosanus* attack. Termite avoidance of antioxidants was also demonstrated by Grace *et al.* (1992), who found that small-molecular-weight iron chelators (siderophores) produced by fungi inhibited feeding by *C. formosanus*.

Researchers have identified several concerns with the use of durable tree species in construction. Most importantly, the compounds conveying durability are found only in the heartwood and sometimes the bark (Delate and Grace, 1995; Morris *et al.*, 1999). In comparison, the outer sapwood of almost all trees is susceptible to termite attack. Although yellow cypress heartwood is quite durable, the sapwood is readily attacked (Fig. 14.1). The same is true of Japanese cypress or hinoki, *Chamaecyparis obtusa* (Siebold and Zucc.) Endl., a durable timber commonly used in Japan (Ngee *et al.*, 2004) (Fig. 14.1). Thus, it is essential to specify heartwood lumber when purchasing durable woods, and to minimize any sapwood content in that lumber.

Extractive content may differ depending upon tree age, size, and growing conditions at particular sites, but the relative importance and relationships among these factors are poorly understood. Different termite species may also have different responses to the same wood. Teak, *Tectona grandis* L. f., from Burma and Laos was found to be more resistant to *C. formosanus* attack in laboratory tests than teak of Malaysian origin (Grace and Yamamoto, 1994; Grace *et al.*, 1998), while both Burmese and Malaysian teak proved extremely durable in a field test against *Coptotermes curvignathus* (Holmgren) (Wong *et al.*, 1998). In laboratory tests with Malaysian teak, Ngee *et al.* (2004) recorded greater feeding by *C. curvignathus* than by *C. gestroi*. These differences in tree characteristics and termite responses make it dangerous to overgeneralize about tree species durability.

Throughout the world, there is interest in the use of agricultural fibers in construction elements, either in combination with wood or as a replacement (Youngquist *et al.*, 1996). Although enhanced resistance to wood-destroying insects is sometimes mentioned by proponents as a benefit of fibers such as hemp, this has not been supported by research with *C. formosanus* (Grace, 1996, 2005a, 2005b).

14.2.2 Allelochemicals

Wood extractives and other plant phytochemicals as possible termite control agents have been the subjects of a great deal of research activity (Nobre and Nunes, 2007; Mubeen *et al.*, 2019; Oladipupo *et al.*, 2022). These allelochemicals may act as oral or contact toxicants, repellents, feeding deterrents, microbicides, or even fumigants (Raina *et al.*, 2007). Several researchers (e.g., Laks *et al.*, 1988; Tascioglu *et al.*, 2012) have demonstrated the feasibility of applying extractives isolated from durable tree species to termite-susceptible woods as natural wood preservatives. However, this has not been proven to be commercially viable, likely due to the complexity of these natural compounds and the associated costs of extraction and purification or synthesis.

Compounds, in particular essential oils, extracted from a wide variety of plants have been evaluated for possible use in pest management (Oladipupo *et al.*, 2022). Studies with *Coptotermes* species have included large-scale plant screenings (e.g., Fokialakis

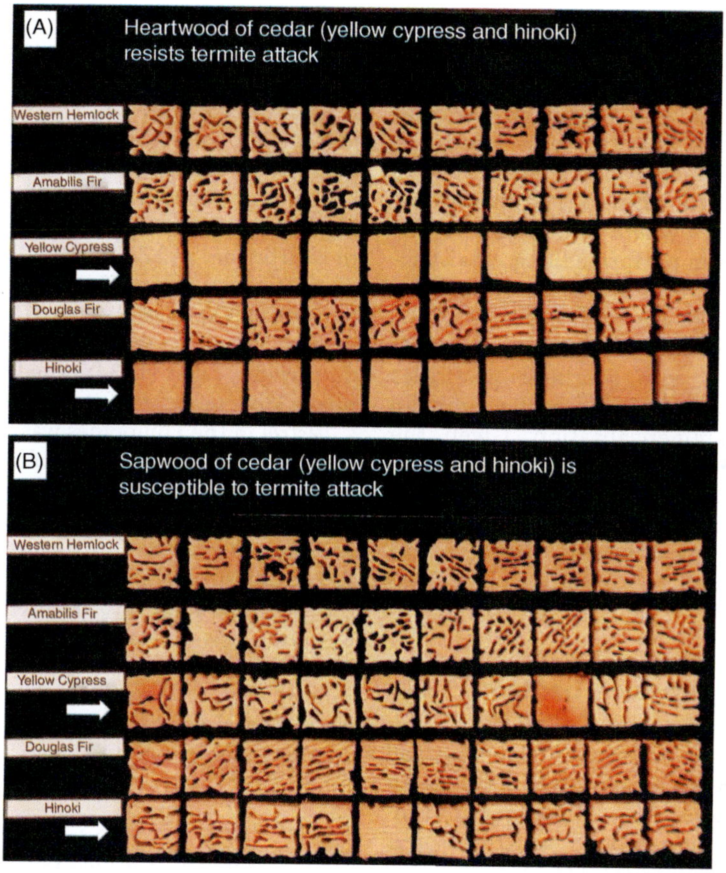

Fig. 14.1. Comparison of *Cototermes formosanus* Shiraki feeding on the heartwood (A) and sapwood (B) of five Pacific Rim timbers in a 28-day, no-choice laboratory test. *Chamaecyparis nootkatensis* (D. Don) Spach (yellow cypress, or Alaskan cedar) and *Chamaecyparis obtusa* (Siebold and Zucc.) Endl. (hinoki, or Japanese cypress) have durable heartwood, but the sapwood of both species is readily attacked.

et al., 2006), evaluations of compounds extracted from plants in a particular geographic region with reputed medicinal or insecticidal benefits (Elango *et al.*, 2012), focused studies of a specific plant genus or species (Zhu *et al.*, 2001b; Chieng *et al.*, 2008; Abdullah *et al.*, 2015), and evaluations of a particular class of extractives such as monoterpenoids (Cornelius *et al.*, 1997; Zhu *et al.*, 2001a). Toxic and behavioral chemicals produced by ants have also been evaluated against *C. formosanus*, both to understand ant–termite interactions and as a source of possible novel termiticides (Cornelius and Grace, 1994a,b; Cornelius *et al.*, 1995, 1997).

Despite many laboratory investigations of plant and insect allelochemicals over many years (e.g., Ahmad *et al.*, 2021; Mishra *et al.*, 2021), no commercial products targeting subterranean termites appear to have been developed to date. In the U.S., an orange-oil product is available for localized treatment of drywood termites (Kalotermidae) but has very limited efficacy in controlling soil-dwelling termites (Raina *et al.*, 2007) and is not sold for this purpose. Unfortunately, the current situation with allelochemicals is similar to that with termite biological control (Chouvenc *et al.*, 2011), in that a large number of laboratory studies have not led to successful field studies or to the sort of overall understanding of biological activity that is necessary for commercial development (Oladipupo *et al.*, 2022). Moreover, as with biological control studies, the laboratory bioassays conducted with these natural products frequently have little to no relationship to field conditions. In the few cases where more realistic bioassays have been performed (e.g., Maistrello *et al.*, 2001), processing/manufacturing costs and the relatively lower efficacy of naturally occurring bioactive compounds in comparison with synthetic pesticides (Osbrink *et al.*, 2005) may be limiting factors in commercialization. Given the lack of commercial progress, it seems that further research in this field should address these fundamental issues before continuing screening for additional allelochemical candidates.

14.3 Physical barriers

Physical barriers to termite tunneling and penetration are alternatives to soil insecticide applications for *Coptotermes* prevention. Such barriers around structures offer the advantages of greater longevity than soil treatments and of minimizing or even eliminating the need for insecticide use. Oi (2022) described the development and current status of these products in the U.S. and Australia.

A solid concrete foundation slab with no cracks or unprotected plumbing/conduit apertures could qualify as at least a partial termite barrier (Lenz and Runko, 1994). However, an elevated concrete building foundation, even when coupled with a treated or naturally durable wood post or sill plate upon it, clearly does not constitute an effective barrier to termite attack on the structure. In a study replicating Japanese building construction, *C. formosanus* was observed tunneling over concrete post supports, and over and through floor posts of both Japanese cypress (*C. obtusa*) and hiba arborvitae (*Thujopsis dolabrata* (Thunb. ex L. f.) Siebold & Zucc.), to reach the susceptible wood above (Grace, 2000).

Perhaps the oldest form of a physical barrier to subterranean termites is the metal "termite shield" (Horner *et al.*, 1934). A sheet of galvanized steel or copper flashing is placed over the top of the concrete foundation or pier and beneath the wooden bottom sill plate or support posts with the exposed edges of the flashing angled downward. Although Snyder and Zetek (1934) felt that such metal flashing was effective in Panama in preventing *Coptotermes* spp. from tunneling upward into the structure (although not downward), Light (1934) noted that a shield might "discourage" termite tunneling but that *C. gestroi* in the Philippines was quite capable of building over the edges of the shield. In fact, pest management professionals are well aware that these metal shields, although still in common use today, may function as an aid to termite inspection (if installed correctly) but not as a barrier to termite tunnel construction upward from the soil.

Several true physical barriers to subterranean termites have been developed, however. These can be categorized as particle barriers, steel mesh, and plastic membranes.

14.3.1 Particle barriers

Ebeling and Pence (1957) first reported that sand and crushed volcanic cinders screened to a specific size range prevented tunneling by *Reticulitermes hesperus* Banks and suggested that these materials could be used as barriers in building construction. While conducting tests of *C. formosanus* tunneling through various aggregates used in construction in Hawai'i, Tamashiro *et al.* (1987, 1991) noted that crushed basalt in a size range of 1.7–2.4 mm prevented tunneling. This discovery resulted in the first commercially available particle barrier, Basaltic Termite Barrier (BTB), marketed by Ameron HC&D (now HC&D) in 1987 and adopted into the Uniform Building Code of the City and County of Honolulu in 1989 as an alternative to pre-construction soil insecticide treatments (Yates *et al.*, 2000) (Figs 14.2, 14.3).

Su and Scheffrahn (1992) subsequently found that quartz and coral sand barriers in

Fig. 14.2. A simple Basaltic Termite Barrier installation for a home with a concrete slab foundation in Hawai'i.

Fig. 14.3. Basaltic Termite Barrier installation for a structure with a prestressed concrete slab foundation. In this more sophisticated method of installation, the particle barrier is held in place by metal forms.

the range of 2.00–2.36 mm and 2.36–2.80 mm were most effective in preventing penetration by field populations of *C. formosanus*, although a barrier of 1.7–2.36 mm allowed only slight penetration. Sornnuwat *et al.* (1995) also reported 1.7–2.4 mm gravel particles to be effective against *C. formosanus*, while the smaller *C. gestroi* was deterred by particles in the range of 1.2–2.4 mm. This size range was also found to be effective against *C. gestroi* in the Philippines (Acda and Ong, 2005; Acda, 2013). Glass shards have also been reported to prevent tunneling by *C. formosanus* (Pallaske and Igarashi, 1991), and by *Coptotermes acinaciformis* (Froggatt) and *Coptotermes lacteus* (Froggatt) (Shiday and French, 2011). The abrasive desiccant diatomaceous earth was found to be ineffective against *C. formosanus* (Grace and Yamamoto, 1993), although Ahmed *et al.* (2018) reported increased *Coptotermes heimi* (Wasmann) mortality when diatomaceous earth was mixed with organic soil amendments.

Crushed granite in this same size range of 1.7–2.4 mm was evaluated in Australia against *Coptotermes* spp. and approved for use as Granitgard™ in 1993 (French and Ahmed, 1993; Ewart *et al.*, 1997; French *et al.*, 2003). More recently, the TERM® Particle Barrier, a barrier of crushed and screened quartz aggregate, has been marketed in the United States in conjunction with a plastic membrane termite barrier as part of a system of waterproofing and termite barrier products (Polygard, 2022a). Basaltic Termite Barrier was sold directly to construction contractors and homeowners with suggested design details provided but no performance warranty, which impeded acceptance (Yates *et al.*, 2000); while Granitgard™ and TERM® Particle Barrier both require installation by approved pest management companies and offer performance warranties.

Key factors with a particle barrier are particle hardness, size, mass, and angularity. Particles must be too hard to be crushed by termite mandibles or by handling during installation, too large to be gripped and moved by termites, but small and heavy enough with sufficiently sharp edges to pack well and eliminate interstitial spaces that could permit termite movement (Tamashiro *et al.*, 1991; Ewart *et al.*, 1997; Keefer *et al.*, 2013).

Architectural design and care in installation are important with particle barriers. The layer of barrier product must be consistently thick, unbroken, and uncontaminated with soil or unscreened gravel in order to be effective (Yates *et al.*, 2000). Contractors in Hawai'i employed special metal forms to hold Basaltic Termite Barrier in place prior to pouring the foundation slab for military housing. It is also worth noting that termites may not be able to tunnel through the material, but they can walk over the surface of it (Ahmed *et al.*, 2004); and if it is not packed tightly against any adjacent concrete surfaces, they may be able to move upward through the resulting gap (Sornnuwat *et al.*, 1995). Careful installation and protection of the edges of the material and of any penetrations for plumbing or conduit are important with any termite barrier, and both Granitgard™ and TERM® offer additional physical or chemical products to supplement their particle barriers and facilitate installation.

14.3.2 Steel mesh

Termimesh (Fig. 14.4) is a high-grade stainless-steel mesh invented in Australia (Lenz and Runko, 1994). The mesh currently uses a proprietary TMA725 stainless-steel alloy that is said to have superior corrosion resistance to 316 grade stainless steel (Termimesh Australia, undated). According to the manufacturer, the product is currently available in Australia, Hawai'i, Singapore, Japan, the continental U.S., France, and Vietnam (Termimesh Australia, undated). At least one alternative steel mesh product, TermShield® (TermShield Australia, 2022), using a proprietary steel alloy said to outperform 316 grade, is available.

Although two different mesh aperture sizes are available in Australia to accommodate different termite species (Oi, 2022), the aperture size of 0.66 mm × 0.45 mm effectively excludes *C. acinaciformis* (Lenz and Runko, 1994) and *C. formosanus* (Grace *et al.*, 1996). Lenz *et al.* (1997) reported that *C. acinaciformis*

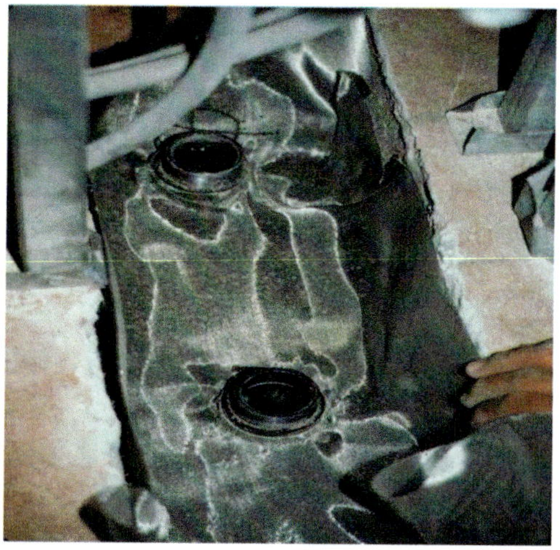

Fig. 14.4. Termimesh installed around pipes to prevent termite attack through a plumbing penetration in a concrete slab foundation.

could not penetrate cracks smaller than 1.5 mm in width, and Ewart *et al.* (1991, referenced by Grace *et al.*, 1996) found that mature workers of *C. formosanus* could not penetrate circular holes smaller than 1.4 mm in diameter. Sealing the edges of the material and tightly securing it around pipes or conduit penetrating the concrete foundation slab are critical steps in installation (Grace *et al.*, 1996) and add-on products are available to facilitate this (Oi, 2022).

In Hawai'i, one of the most significant uses of Termimesh has been as "socks" placed around the butt end of new utility poles to prevent termite attack. Although Termimesh requires that authorized contractors install the product in buildings, prefabricated "socks" are currently sold for this use in the U.S. (Termistop, 2019).

14.3.3 Membranes

Building codes often require installation of a waterproof membrane over the underlying substrate prior to pouring the concrete foundation. Manufacturers have leveraged this requirement to produce polyethylene membranes containing a repellent pyrethroid insecticide which can serve the dual purpose of moisture control and termite prevention, or in some situations, termite prevention alone. Two of the oldest such insecticide-containing membranes are Termifilm® (Adkalis, 2022), a French product used in many parts of the world, and the Australian product Kordon® (Bayer Environmental Science, 2004). Other products of this type are also available in Australia (Oi, 2022).

Su *et al.* (2004) demonstrated the efficacy of a polyethylene membrane impregnated with lambda-cyhalothrin (Wege *et al.*, 2003) against *C. formosanus*, although this particular product is no longer available (Oi, 2022). Both marketing and adoption of insecticide-impregnated membranes are challenging in the U.S., since insecticides are generally required by state and federal regulations to be handled by licensed pest management professionals, but building materials such as waterproofing membranes are typically purchased and installed by construction contractors.

A recent development with termite-resistant membranes has been the marketing of non-insecticidal plastic sheets reported by the manufacturers to act purely as physical barriers to termite penetration, such as TERM® Underslab Water/Termite Barrier (Polygard, 2022b) and Pango® Wrap Termite/Vapor Barrier (Stego Industries, 2021). Research has been conducted on these new materials, as indicated on the manufacturers' websites, but no published efficacy data or information on field performance are available at this point. As Oi (2022) has indicated, non-chemical pest control products in the U.S. fall into a regulatory gap that can make it difficult to accurately evaluate claims of structural protection

As with particle barriers and steel mesh, care in installation to keep the membrane barrier intact and seal all edges and areas around pipes and conduits is critically important. Add-on products such as adhesives, tapes and clamps are generally available from the manufacturers to facilitate this.

14.4 Biological control

Biological control can be achieved through the actions of predators, parasites, or pathogens.

The extensive literature on biological control of termites was reviewed by Culliney and Grace (2000) with special reference to *C. formosanus*, and by Chouvenc *et al.* (2011), who provided a critical analysis of the reasons why the field had largely not progressed beyond laboratory experiments. Reviews of more recent research on potential biocontrol agents are also available (Qasim *et al.*, 2015; Ahmad *et al.*, 2021; Mishra *et al.*, 2021).

Unfortunately, while the large amount of laboratory research conducted on this topic may well have scientific merit, from the point of view of termite management the field has not changed since the critical review of Chouvenc *et al.* (2011). Laboratory infectivity and toxicity studies are plentiful, but field studies are still virtually non-existent, and no practical methods of termite control have yet resulted from these research efforts. Bypassing termite social and individual immunity at the colony level is an elusive goal (Chouvenc and Su, 2010), and further screening for biological control candidates may remain fruitless for the foreseeable future. While some interesting work is being conducted in paratransgenesis (Sethi *et al.*, 2014; Tikhe *et al.*, 2016), it remains to be seen whether this approach can reach a commercial stage, given the potential costs and the regulatory climate around genetically engineered organisms.

14.4.1 Predators and parasites

Termites as a group have a wide variety of opportunistic and specialist predators, both vertebrates and invertebrates. Alates and termite species with foragers that travel on the soil surface are particular targets (Logan *et al.*, 1990), but a few predators attack termites in the nest (Scheppe, 1970; Johnson and Hagen, 1981). With *Coptotermes* species, ants are the most obvious predators. In Hawai'i, the ants *Ochetellus glaber* (Mayr.) and *Pheidole megacephala* (Fabricius) are often found occupying termite galleries in formerly infested wood when collecting *C. formosanus* in the field. However, *C. formosanus* is able to escape predation by walling off areas where these ants are encountered and retreating to underground refugia (Cornelius and Grace, 1995; Chouvenc *et al.*, 2015).

Despite a great deal of collection effort, no insect parasitoids have been found for *C. formosanus* (Grace, 2003). In fact, with rare exception (e.g., Disney, 1986), parasitism is virtually unknown in termites. Evidence also suggests that parasitoids, even if they were found, may be of limited value in controlling populations of concealed pests in general (Gross, 1991). Thus, to date there is no reason to dispute the conclusion of Culliney and Grace (2000) that there is little potential for the use of predators or parasitoids to manage *C. formosanus*.

14.4.2 Nematodes

Although nematodes have been marketed commercially in the past for subterranean termite control (Anonymous, 1985; Quarles, 1995), they are ill-suited to control of Rhinotermitidae, particularly *Coptotermes* species. Very high population densities of entomophagous nematodes are required for control even under laboratory conditions (Fujii, 1975; Wang *et al.*, 2002) and they have very limited mobility to traverse the diffuse gallery systems of subterranean termites (Baïmey *et al.*, 2017). Termites appear to avoid contact with nematodes (Epsky and Capinera, 1988), remove them from each other by grooming (Mankowski *et al.*, 2005), will "wall-off" infected individuals (Fujii, 1975), and nematodes are killed by the high temperatures that can be found in *Coptotermes* spp. nests (Khan *et al.*, 2016; Baïmey *et al.*, 2017).

14.4.3 Fungi

Fungal pathogens, particularly *Metarhizium anisopliae* (Metsch.) Sorokin and *Beauveria bassiana* (Balsamo) Vuillemin have been the focus of most research reports suggesting the feasibility of termite biological control (Chouvenc *et al.*, 2011). As with nematode infections, subterranean termites appear to successfully employ a variety of defense mechanisms to prevent an epizootic (Chouvenc and Su, 2010), particularly grooming (Hussain and Tian, 2013), cellular encapsulation (Chouvenc *et al.*, 2009), gut antifungal activity (Yanagawa and Shimizu, 2007), and immune response (Hussain and Wen, 2011). The very few field evaluations with *Coptotermes* species of pathogenic fungi (e.g., Lai, 1977) were considered to be either inconclusive or negative by Chouvenc *et al.* (2011).

With the loss of cyclodiene pesticides for subterranean termite control, enthusiasm for biological control was high in the 1990s. Despite the absence of successful field studies, a commercial formulation of *M. anisopliae* was marketed in the U.S. (Quarles, 1995; Grace, 1997; Rath, 2000), but is no longer available.

Although numerous laboratory tests continue to be published with optimistic suggestions for further development of the fungal strains evaluated, it is difficult to see these research efforts resulting in successful termite management options until there is a much better understanding of termite defense mechanisms and ways that pathogens might bypass such mechanisms (Chouvenc *et al.*, 2011). It is encouraging to see some researchers taking this approach (Syazwan *et al.*, 2021). The exception, as with plant allelochemicals, may be uses of fairly crude preparations of locally available materials in rural and less developed regions of the world, in the absence of more costly and likely more effective commercial products (Sahayaraj, 2018).

14.4.4 Viruses

Viruses are underreported and perhaps understudied in termites (Chouvenc *et al.*, 2011). A virus similar to acute bee paralysis virus (ABPV) was isolated from *C. lacteus* (Gibbs *et al.*, 1970), but Chouvenc *et al.* (2011) found only five published reports of laboratory tests with viruses against termites, none of which were Rhinotermitids. Since 2011, this number appears to have increased by one, with a reportedly successful laboratory evaluation of a nuclear polyhedrosis virus (NPV) against *C. formosanus* (Zhang and Mo, 2014, referenced in Ahmad *et al.*, 2021).

14.4.5 Bacteria

As with other entomopathogens, bacteria have been found to kill termites, including *Coptotermes* species (Osbrink *et al.*, 2001), in laboratory toxicity tests. Wang and Henderson (2013) reported that *C. formosanus* was susceptible to *Bacillus thuringiensis* (Berliner), but that bacterial growth was suppressed both by the presence of other bacteria in the environment and by the termites themselves through an unknown mechanism. In other laboratory investigations, the addition of lufenuron, a chitin synthesis inhibitor, has been reported to suppress *C. formosanus* resistance to *Pseudomonas aeruginosa* (Schroeter) Migula (Wang *et al.*, 2013); and immune suppressants increased *C. formosanus* susceptibility to *Serratia marcescens* Bizio (Connick *et al.*, 2001).

Grace and Ewart (1996) found that *C. formosanus* would readily feed upon recombinant *Pseudomonas fluorescens* Migula and suggested that bacteria could be used to deliver genetically engineered toxins. Interestingly, the only reported field evaluation of bacteria for control of *Coptotermes* spp. has been with transgenic *Enterobacter cloacae* (Jordan) Hormaeche and Edwards expressing a toxin from a bacterium associated with entomopathogenic nematodes (Zhang *et al.*, 2010). The authors reported success in eliminating *C. formosanus* field populations with bait stations, although it is difficult to exclude possible avoidance of the stations by foraging termites (Chouvenc *et al.*, 2011). This field test built upon a previous lab study (Zhao *et al.*, 2008) which demonstrated that recombinant *E. cloacae* transformed to express an insecticidal protein were readily transferred among termites and induced high mortality. This study, in turn, was an extension of the concept of paratransgenesis for termite control introduced by Husseneder and Grace (2005). This transgenic

"Trojan horse" concept has also been pursued with a yeast genetically engineered to express a protozoacidal lyctic peptide when ingested by subterranean termites (Husseneder and Collier, 2009; Sethi *et al.*, 2014; see Chapter 8 of this volume). In order to minimize the possibility of transfer of the transgenic microbe to nontarget species, Tikhe *et al.* (2016) demonstrated that an engineered termite-specific bacterium, *Trabulsiella odontotermitis* Chou *et al.*, was also readily fed upon by *C. formosanus* and transferred among nest mates. Economics and the regulatory climate with respect to genetically engineered organisms will likely determine whether this novel approach to microbial control will move from laboratory to field tests and commercial development.

14.5 Conclusions

Termite management continues to be a dynamic field of research. Several commercially successful physical barriers have been developed as alternatives to soil insecticide treatments, and research is continuing with plastic membranes, since such products are much easier to market and distribute outside of their local area of production than particle barriers, and potentially less expensive and simpler to install than steel mesh barriers. Naturally deterrent woods are used in some situations as alternatives to preservative-treated wood, but that use is limited due to cost, availability, and concerns over tropical deforestation. There continues to be a great deal of research conducted on biologically active phytochemicals and on the use of entomopathogens against termites, particularly in the tropical and subtropical regions where *Coptotermes* species are found, but commercial applications still appear to be limited. Although much of this research continues to be laboratory toxicity screening, a number of researchers have called for and are focusing on achieving the more comprehensive understanding of biological mechanisms, particularly with respect to termite field biology and ecology, that is needed to develop practical alternative methods of *Coptotermes* management.

References

Abdullah, F., Subramanian, P., Ibrahim, H., Malek, S.N.A., Lee, G.S. *et al.* (2015) Chemical composition, antifeedant, repellent, and toxicity activities of the rhizomes of galangal, *Alpinia galanga* against Asian subterranean termites, *Coptotermes gestroi* and *Coptotermes curvignathus* (Isoptera: Rhinotermitidae). *Journal of Insect Science* 15, 1–7.

Acda, M.N. (2013) Evaluation of lahar barrier to protect wood structures from Philippine subterranean termites. *Philippine Journal of Science* 142, 21–25.

Acda, M.N. and Ong, H.B. (2005) Use of volcanic debris as physical barrier against the Philippine milk termite (Isoptera: Rhinotermitidae). *Sociobiology* 46, 117–130.

Adkalis (2022) Termifilm. Available at: https://www.adkalis.com/en/export/products-offer/termifilm-uv/ (accessed 17 August 2022).

Ahmad, F., Fouad, H., Liang, S.-Y., Hu, Y. and Mo, J.-C. (2021) Termites and Chinese agricultural system: applications and advances in integrated termite management and chemical control. *Insect Science* 28, 2–20.

Ahmed, B.M., French, J.R. and Vinden, P. (2004) Review of remedial and preventative methods to protect timber in service from attack by subterranean termites in Australia. *Sociobiology* 44, 1–16.

Ahmed, S., Hassan, B. and Farooq, M.U. (2018) Effect of biofertilizers and diatomaceous earth on life and movement of subterranean termites under laboratory conditions. *International Journal of Tropical Insect Science* 38, 348–352.

Anonymous (1985) Update: nematodes vs. termites. *The IPM Practitioner* 7(8), 1–2.

Baïmey, H., Zadji, L., Afouda, L., Fanou, A., Kotchofa, R. *et al.* (2017) Searching for better methodologies for successful control of termites using entomopathogenic nematodes. In: Shah, M.M. and Mahamood, M. (eds) *Nematalogy – Concepts, Diagnosis and Control*. IntechOpen, London, pp. 53–83.

Bayer Environmental Science (2004) Kordon® termite barrier reference manual. Available at: https://www.pestcontrol.org.au/kordon-product-manual.pdf (accessed 16 August 2022).
Carter, F.L. and de Camargo, C.R.R. (1983) Testing antitermitic properties of Brazilian woods and their extracts. *Wood and Fiber Science* 15, 350–357.
Chieng, T.C., Assim, Z.B. and Fasihuddin, B.A. (2008) Toxicity and antitermite activities of the essential oils from *Piper sarmentosum*. *Malaysian Journal of Analytical Sciences* 12, 234–239.
Chouvenc, T. and Su, N.-Y. (2010) Apparent synergy among defense mechanisms in subterranean termites (Rhinotermitidae) against epizootic events: limits and potential for biological control. *Journal of Economic Entomology* 103, 1327–1337.
Chouvenc, T., Su, N.-Y. and Robert, A. (2009) Cellular encapsulation in the eastern subterranean termite, *Reticulitermes flavipes* (Isoptera), against infection by the entomopathogenic fungus *Metarhizium anisopliae*. *Journal of Invertebrate Pathology* 101, 234–241.
Chouvenc, T., Su, N.-Y. and Grace, J.K (2011) Fifty years of attempted biological control of termites – analysis of a failure. *Biological Control* 59, 69–82.
Chouvenc, T., Mullins, A.J. and Su, N.Y. (2015) Territorial status-quo between the big-headed ant (Hymenoptera: Formicidae) and the Formosan subterranean termite (Isoptera: Rhinotermitidae). *Florida Entomologist* 98, 157–161.
Connick, W.J. Jr, Osbrink, W.L.A., Wright, M.S., Williams, K.S., Daigle, D.J. *et al.* (2001) Increased mortality of *Coptotermes formosanus* (Isoptera: Rhinotermitidae) exposed to eicosanoid biosynthesis inhibitors and *Serratia marcescens* (Eubacteriales: Enterobacteriaceae). *Environmental Entomology* 30, 449–455.
Cornelius, M.L. and Grace, J.K. (1994a) Behavioral responses of the Formosan subterranean termite (Isoptera: Rhinotermitidae) to semiochemicals of seven ant species. *Environmental Entomology* 23, 1524–1528.
Cornelius, M.L. and Grace, J.K. (1994b) Semiochemicals extracted from a dolichoderine ant affect the feeding and tunneling behavior of the Formosan subterranean termite (Isoptera: Rhinotermitidae). *Journal of Economic Entomology* 87, 705–708.
Cornelius, M.L. and Grace, J.K. (1995) Laboratory evaluations of interactions of three ant species with the Formosan subterranean termite (Isoptera: Rhinotermitidae). *Sociobiology* 26, 291–298.
Cornelius, M.L., Grace, J.K., Ford, P.W. and Davidson, B.S. (1995) Toxicity and repellency of semiochemicals extracted from a dolichoderine ant (Hymenoptera: Formicidae) to the Formosan subterranean termite (Isoptera: Rhinotermitidae). *Environmental Entomology* 24, 1263–1269.
Cornelius, M.L., Grace, J.K. and Yates III, J.R. (1997) Toxicity of monoterpenoids and other natural products to the Formosan subterranean termite. *Journal of Economic Entomology* 90, 320–325.
Culliney, T.W. and Grace, J.K. (2000) Prospects for the biological control of subterranean termites (Isoptera: Rhinotermitidae), with special reference to *Coptotermes formosanus*. *Bulletin of Entomological Research* 90, 9–21.
Delate, K.M. and Grace, J.K. (1995) Susceptibility of neem to attack by the Formosan subterranean termite, *Coptotermes formosanus* Shiraki (Isoptera: Rhinotermitidae). *Journal of Applied Entomology* 119, 93–95.
Delate, K.M, Grace, J.K., Armstrong, J.W. and Tome, C.H.M. (1995) Carbon dioxide as a potential fumigant for termite control. *Pesticide Science* 44, 357–361.
Diba, F., Hadary, F., Panjaitan, S.D. and Yoshimura, T. (2013) Electromagnetic waves as non-destructive method to control subterranean termites *Coptotermes curvignathus* Holmgren and *Coptotermes formosanus* Shiraki. *Procedia Environmental Sciences* 17, 150–159.
Ding, W. and Hu, X.P. (2010) Antitermitic effect of the *Lantana camara* plant on subterranean termites (Isoptera: Rhinotermitidae). *Insect Science* 17, 427–433.
Disney, R.H.L. (1986) Two remarkable new species of scuttle-fly (Diptera: Phoridae) that parasitize termites (Isoptera) in Sulawesi. *Systematic Entomology* 11, 413–422.
Ebeling, W. and Pence, R.J. (1957) Relation of particle size to the penetration of subterranean termites through barriers of sand or cinders. *Journal of Economic Entomology* 50, 690–692.
Elango, G., Abdul Rahuman, A., Kamaraj, C., Bagavan, A., Abduz Zahir, A. *et al.* (2012) Efficacy of medicinal plant extracts against Formosan subterranean termite, *Coptotermes formosanus*. *Industrial Crops and Products* 36, 524–530.
Epsky, N.D. and Capinera, J.L. (1988) Efficacy of the entomogenous nematode *Steinernema feltiae* against a subterranean termite, *Reticulitermes tibialis* (Isoptera: Rhinotermitidae). *Journal of Economic Entomology* 81, 1313–1317.

Ewart, D.M., Tamashiro, M., Grace, J.K. and Ebesu, R.H. (1991) Minimum foraging aperture and particle barriers for excluding the Formosan subterranean termite. Unpublished poster presentation, *Annual Meeting of the Entomological Society of America*, Reno, Nevada (referenced by Grace *et al*., 1996).

Ewart, D.M., Rawlinson, E.R., Tolsma, A.D., Irvin, G.C. and Wiggins, P.R. (1997) *Development of a Granitgard® particulate termite barrier for use in tropical Australia*. Document no. IRG /WP 97-10190. International Research Group on Wood Preservation, Stockholm.

Fokialakis, N., Osbrink, W.L.A., Mamonov, L.K., Gemejieva, N.G., Mims, A.B. *et al*. (2006) Antifeedant and toxicity effects of thiophenes from four *Echinops* species against the Formosan subterranean termite, *Coptotermes formosanus*. *Pest Management Science* 62, 832–838.

French, J.R.J. and Ahmed, B. (1993) *Termite physical barriers: is retrofitting with Granitgard and option?* Document no. IRG/WP 93-40011. International Research Group on Wood Preservation, Stockholm.

French, J.R.J., Ahmed, B. and Trajstman, A. (2003) Laboratory and field evaluation of granite aggregate as a physical barrier against subterranean termites of the genus *Coptotermes* spp. (Isoptera: Rhinotermitidae). *Sociobiology* 42, 129–149.

Fujii, J.K. (1975) Effects of an entomogenous nematode, *Neoaplectana carpocapsae* Weiser, on the Formosan subterranean termite, *Coptotermes formosanus* Shiraki, with ecological and biological studies on *C. formosanus*. PhD dissertation. University of Hawaii at Mānoa, Honolulu, Hawaii.

Gibbs, A.J., Gay, E.J. and Wetherly, A.H. (1970) A possible paralysis virus of termites. *Virology* 40, 1063–1065.

Grace, J.K. (1996) Susceptibility of compressed bagasse fiber to termite attack. *Forest Products Journal* 46(9), 76–78.

Grace, J.K. (1997) Biological control strategies for termites. *Journal of Agricultural Entomology* 14, 281–289.

Grace, J.K. (2000) *Termite attack on susceptible lumber above naturally durable support posts*. Document no. IRG/WP 00-10370. International Research Group on Wood Preservation, Stockholm.

Grace, J. K. (2003) Approaches to biological control of termites. *Sociobiology* 41,115–121.

Grace, J.K. (2005a) *Termite response to agricultural fiber composites: bagasse*. Document no. IRG/WP 05-10549. International Research Group on Wood Protection, Stockholm.

Grace, J.K. (2005b) *Termite response to agricultural fiber composites: hemp*. Document no. IRG/WP 05-10548. International Research Group on Wood Protection, Stockholm.

Grace, J.K. and Ewart, D.M. (1996) Recombinant cells of *Pseudomonas fluorescens*: a highly palatable encapsulation for delivery of genetically engineered toxins to subterranean termites (Isoptera: Rhinotermitidae). *Letters in Applied Microbiology* 23, 183–186.

Grace, J.K. and Tome, C.H.M. (2005) Resistance of the Indonesia woods bangkirai (*Shorea laevis*) and merbau (*Intsia palembanica*) to Formosan subterranean termite attack. *Sociobiology* 45, 503–509.

Grace, J.K. and Yamamoto, R.T. (1993) Diatomaceous earth is not a barrier to Formosan subterranean termites (Isoptera: Termitidae). *Sociobiology* 23, 25–30.

Grace, J.K. and Yamamoto, R.T (1994) Natural resistance of Alaska-cedar, redwood, and teak to Formosan subterranean termites. *Forest Products Journal* 44(3), 41–45.

Grace, J.K., Goodell, B.S, Jones, W.E., Chandhoke, V. and Jellison, J. (1992) Evidence for inhibition of termite (Isoptera: Rhinotermitidae) feeding by extracellular metabolites of a wood decay fungus. *Proceedings of the Hawaiian Entomological Society* 31, 249–252.

Grace, J.K., Yates III, J.R., Tome, C.H.M. and Oshiro, R.J. (1996) Termite-resistant construction: use of a stainless steel mesh to exclude *Coptotermes formosanus* (Isoptera: Rhinotermitidae). *Sociobiology* 28, 365–372.

Grace, J.K., Wong, A.H.H. and Tome, C.H.M (1998) *Termite resistance of Malaysian and exotic woods with plantation potential: laboratory evaluation*. Document no. IRG/WP 98-10280. International Research Group on Wood Preservation, Stockholm.

Gross, P. (1991) Influence of target pest feeding niche on success rates in classical biological control. *Environmental Entomology* 20, 1217–1227.

Hapukotuwa, N.K. and Grace, J.K. (2011) Preferences of *Coptotermes formosanus* Shiraki and *Coptotermes gestroi* (Wasmann) (Blattodea: Rhinotermitidae) among three commercial wood species. *Insects* 2, 499–508.

Horner, A.C., Bowe, E.E., Putnam, W. and Chase, G.E. (1934) Buildings. In: Kofoid, C.A. (ed.) *Termites and Termite Control*, 2nd edn. University of California Press, Berkely, California, pp. 599–642.

Hussain, A. and Tian, M.-Y. (2013) Germination pattern and inoculum transfer of entomopathogenic fungi and their role in disease resistance among *Coptotermes formosanus* (Isoptera: Rhinotermitidae). *International Journal of Agriculture and Biology* 15, 319–324.

Hussain, A. and Wen, S.Y. (2011) Induction of immune response among Formosan subterranean termites, *Coptotermes formosanus* Shiraki (Rhinotermitidae: Isoptera). *African Journal of Microbiology Research* 6, 995–1000.
Husseneder, C. and Collier, R.E. (2009) Paratransgenesis in termites. In: Bourtzis, K. and Miller, T.A. (eds) *Insect Symbiosis, Volume 3*. CRC Press, Boca Raton, Florida, pp. 361–376.
Husseneder, C. and Grace, J.K. (2005) Genetically engineered termite gut bacteria (*Enterobacter cloacae*) deliver and spread foreign genes in termite colonies. *Applied Microbiology and Biotechnology* 68, 360–367.
Johnson, J.B. and Hagen, K.S. (1981) A neuropterous larva uses an allomone to attack termites. *Nature* 238, 506–507.
Keefer, T.C., Zollinger, D.G. and Gold, R.E. (2013) Evaluation of aggregate particles as a physical barrier to prevent subterranean termite incursion into structures. *Southwestern Entomologist* 38, 447–464.
Khan, M.A., Ahmad, W., Paul, B., Paul, S., Khan, Z. *et al.* (2016) Entomopathogenic nematodes for the management of termites. In: Hakeem, K.R., Akhtar, M. and Abdullah, S.N.A. (eds) *Plant, Soil and Microbes: Volume 1– Implications in Crop Science*. Springer International Publishing, Cham, Switzerland, pp. 317–352.
La Fage, J.P. (1984) A review of the bait-toxicant method of termite control: in search of the holy grail. *Proceedings of the Canadian Wood Preservation Association* 5, 77–81.
Lai, P.Y. (1977) Biology and ecology of the Formosan subterranean termite, *Coptotermes formosanus*, and its susceptibility to the entomogenous fungi, *Beauveria bassiana* and *Metarhizium anisopliae*. PhD dissertation. University of Hawaii at Mānoa, Honolulu, Hawaii.
Laks, P.E., Putnam, L.J. and Pruner, M.S. (1988) Natural products in wood preservation. *Proceedings of the Canadian Wood Preservation Association* 9, 105–123.
Lenz, M. and Runko, S. (1994) Protection of buildings, other structures and materials in ground contact from attack by subterranean termites (Isoptera) with a physical barrier – a fine mesh of high grade stainless steel. *Sociobiology* 24, 1–16.
Lenz, M., Schafer, B., Runko, S. and Glossop, L. (1997) The concrete slab as part of a termite barrier system: response of Australian subterranean termites to cracks of different width in concrete. *Sociobiology* 30, I03–118.
Light, S.F. (1934) The termite fauna of the Philippine Islands and its economic significance. In: Kofoid, C.A. (ed.) *Termites and Termite Control*, 2nd edn. University of California Press, Berkely, California, pp. 347–350.
Logan, J.W.M., Cowie, R.H. and Wood, T.G. (1990) Termite (Isoptera) control in agriculture and forestry by non-chemical methods: a review. *Bulletin of Entomological Research* 80, 309–330.
Maistrello, L., Henderson, G. and Laine, R.A. (2001) Efficacy of vetiver oil and nootkatone as soil barriers against Formosan subterranean termite (Isoptera: Rhinotermitidae). *Journal of Economic Entomology* 94, 1532–1537.
Mankowski, M.E., Kaya, H.K., Grace, J.K. and Sipes, B. (2005) Differential susceptibility of subterranean termite castes to entomopathogenic nematodes. *Biological Science and Technology* 15, 367–377.
Mishra, P., Verma, M., Jha, S., Tripathi, A., Pandey, A. *et al.* (2021) Biological approaches of termite management: a review. *Current Botany* 12, 121–131.
Morris, P.I., Grace, J.K. and Troughton, G.E. (1999) *Preliminary indications of the natural durability of spruce bark board*. Document no. IRG/WP 99-10312. International Research Group on Wood Preservation, Stockholm.
Mubeen, M., Iftikhar, Y., Ahmar, S., Aqleem, A., Hassan, F. *et al.* (2019) Management of termites using wood extractives and microbes – a mini review. *SSR Institute International Journal of Life Sciences* 5, 2410–2419.
Ngee, P.-S., Tashiro, A., Yoshimura, T., Jaal, Z. and Lee, C.-Y. (2004) Wood preference of selected Malaysian subterranean termites. *Sociobiology* 43, 535–550.
Nobre, T. and Nunes, L. (2007) Non-traditional approaches to subterranean termite control in buildings. *Wood Material Science and Engineering* 2, 147–156.
Oi, F. (2022) A review of the evolution of termite control: a continuum of alternatives to termiticides in the United States with emphasis on efficacy testing requirements for product registration. *Insects* 13(50), 1–28. doi: 10.3390/insects13010050
Oladipupo, S.O., Hu, X.P. and Appel, A.G. (2022) Essential oils in urban insect management – a review. *Journal of Economic Entomology* 115(5), 1375–1408. doi: 10.1093/jee/toac083
Osbrink, W.L.A., Williams, K.S., Connick Jr, W.J., Wright, M.S. and Lax, A.R. (2001) Virulence of bacteria associated with the Formosan subterranean termite (Isoptera: Rhinotermitidae) in New Orleans, LA. *Environmental Entomology* 30, 443–448.

Osbrink, W.L.A., Lax, A.R. and Cantrell, C.L. (2005) Comparison of natural and conventional insecticides against Formosan termites (Isoptera: Rhinotermitidae). In: Lee, C.-Y. and Robinson, W.H. (eds) *Proceedings of the Fifth International Conference on Urban Pests*. Perniagaan Ph'ng, Malaysia, pp. 213–223.

Pallaske M. and Igarashi, A. (1991) *Glass splinters as physical barriers: optimized material properties in use with and without insecticidal pre-treatment minimizes environmental contaminations*. Document no. IRG/WP 1476. International Research Group on Wood Preservation, Stockholm.

Polygard (2022a) Stone particle termite barrier data sheet. Available at: https://www.polyguardproducts.com/term/library/data-sheets-and-guide-specs/ (accessed 15 August 2022).

Polygard (2022b) Underslab water/termite barrier data sheet. Available at: https://www.polyguardproducts.com/term/library/data-sheets-and-guide-specs/ (accessed 15 August 2022).

Prefabworld Bali International (2014) Special Hawaii applications. Available at: https://www.prefabworldbali.com/special-hawaii-applications (accessed 15 August 2022).

Qasim, M, Lin, Y., Fang, D. and Wang, L. (2015) Termites and microbial control strategies. *South Asia Journal of Multidisciplinary Studies* 1(8), 1–27.

Quarles, W. (1995) New technologies for termite control. *The IPM Practitioner* 17(5/6), 1–9.

Ragon, K.W., Nicholas, D.D. and Schultz, T.P. (2008) Termite-resistant heartwood: the effect of the non-biocidal antioxidant properties of the extractives (Isoptera: Rhinotermitidae). *Sociobiology* 52, 47–54.

Raina, A., Bland, J., Doolittle, M., Lax, A., Boopathy, R. *et al.* (2007) Effect of orange oil extract on the Formosan subterranean termite (Isoptera: Rhinotermitidae). *Journal of Economic Entomology* 100, 880–885.

Rath, A.C. (2000) The use of entomopathogenic fungi for control of termites. *Biocontrol Science and Technology* 10, 563–581.

Sahayaraj, K. (2018) Bio-intensive integrated management of termites. In: Khan, M.A. and Ahmad, W. (eds) *Termites and Sustainable Management, Volume 2 – Economic Losses and Management.* Springer International Publishing, Cham, Switzerland, pp. 165–180.

Scheppe, W. (1970) Invertebrate predation on termites of the African savanna. *Insectes Sociaux* 17, 205–218.

Sethi, A., Delatte, J., Foil, L. and Husseneder, C. (2014) Protozoacidal trojan-horse: use of a ligand-lytic peptide for selective destruction of symbiotic protozoa within termite tuts. *PLoS ONE* 9(9), e106199.

Shiday, B.M.A. and French, J.R.J. (2011) Termite foraging behaviour and biological response against sintered glass screening as a potential graded physical barrier. *African Journal of Environmental Science and Technology* 5, 327–336.

Snyder, T.E. and Zetek, J. (1934) The termite fauna of the canal zone, Panama, and its economic significance. In: Kofoid, C.A. (ed.) *Termites and Termite Control*, 2nd edn. University of California Press, Berkely, California, pp. 342–346.

Sornnuwat, Y., Vongkaluang, C., Yoshimua, T, Tsunoda, K. and Takahashi, M. (1995) Tunneling of subterranean termites, *Coptotermes gestroi* Wasmann and *Coptotermes formosanus* Shiraki, into gravel barriers. *Japanese Journal of Environmental Entomology and Zoology* 7, 13–19.

Stego Industries (2021) Pango® Wrap termite/vapor barrier data sheet. Available at: https://www.stegoindustries.com/products/pango-wrap (accessed 17 August 2022).

Stirling, R., Morris, P.I. and Grace, J.K. (2015) Prediction of the decay and termite resistance of western red cedar heartwood. *Forest Products Journal* 65(3/4), 84–92.

Su, N.-Y. and Scheffrahn, R.H. (1992) Penetration of sized-particle barriers by field populations of subterranean termites (Isoptera: Rhinotermitidae). *Journal of Economic Entomology* 85, 2275–2278.

Su, N.-Y. and Tamashiro, M. (1986) Wood consumption rate and survival of the Formosan subterranean termite (Isoptera: Rhinotermitidae) when fed one of six woods used commercially in Hawaii. *Proceedings of the Hawaiian Entomological Society* 23, 109–113.

Su, N.-Y., Ban, P. and Scheffrahn, R.H. (2004) Polyethylene barrier impregnated with lambda-cyhalothrin for exclusion of subterranean termites (Isoptera: Rhinotermitidae) from structures. *Journal of Economic Entomology* 97, 570–574.

Syazwan, S.A., Lee, S.Y., Sajap, A.S., Lau, W.H., Omar, D. and Mohamed, R. (2021) Interaction between *Metarhizium anisopliae* and its host, the subterranean termite *Coptotermes curvignathus* during the infection process. *Biology* 10(263), 1–20.

Tamashiro, M., Yates, J.R. and Ebesu, R.H. (1987). The Formosan subterranean termite in Hawaii: problems and control. In: Tamashiro, M. and Su, N.-Y. (eds) *Biology and Control of the Formosan Subterranean Termite*. HITAHR Res. Ext. Series 083. University of Hawaii, Honolulu, pp. 15–22.

Tamashiro, M., Yates, J.R., Yamamoto, R.T. and Ebesu, R.H. (1991) Tunneling behavior of the Formosan subterranean termite and basalt barriers. *Sociobiology* 19, 163–170.

Tascioglu, C., Yalcin, M., de Troya, T. and Sivrikaya, H. (2012) Termiticidal properties of some wood and bark extracts used as wood preservatives. *BioResources* 7, 2960–2969.

Teak Bali (2019) Hawaii projects. Available at: https://www.teakbali.com/projects/hawaii-projects/ (accessed 15 August 2022).

Termimesh Australia (undated) The composition of Termimesh stainless steel termite mesh. Available at: https://www.termimesh.com.au/news/2019/08/06/product-composition (accessed 17 August 2022).

Termistop (2019) Utility protection. Available at: http://www.termistopusa.com/termistop-products?category=Utility+Protection (accessed 17 August 2022).

TermShield Australia (2022) About TermShield. Available at: https://termshield.com.au/about-termshield/ (accessed 17 August 2022).

Tikhe, C.V., Martin, T.M., Howells, A., Delatte, J. and Husseneder, C. (2016) Assessment of genetically engineered *Trabulsiella odontotermitis* as a 'Trojan Horse' for paratransgenesis in termites. *BMC Microbiology* 16(202), 1–11.

Wang, C. and Henderson, G. (2013) Evidence of Formosan subterranean termite group size and associated bacteria in the suppression of entomopathogenic bacteria, *Bacillus thuringiensis* subspecies *israelensis* and *thuringiensis*. *Annals of the Entomological Society of America* 106, 454–462.

Wang, C.L., Powell, J.E. and Nguyen, K. (2002) Laboratory evaluation of four entomopathogenic nematodes for control of subterranean termites (Isoptera: Rhinotermitidae). *Environmental Entomology* 31, 381–387.

Wang, C., Henderson, G. and Gautam, B.K. (2013) Lufenuron suppresses the resistance of Formosan subterranean termites (Isoptera: Rhinotermitidae) to entomopathogenic bacteria. *Journal of Economic Entomology* 106, 1812–1818.

Wege, P.J., McClellan, W.D., Bywater, A.F. and Hoppe, M.A. (2003) IMPASSE™ termite barrier: new preconstruction termite barrier technology. *Sociobiology* 41, 169–176.

Wong, A.H.H., Grace, J.K. and Kirton, L.G. (1998) *Termite resistance of Malaysian and exotic woods with plantation potential: field evaluation*. Document no. IRG/WP 98-10289. International Research Group on Wood Preservation, Stockholm.

Yanagawa, A. and Shimizu, S. (2007) Resistance of the termite, *Coptotermes formosanus* Shiraki to *Metarhizium anisopliae* due to grooming. *BioControl* 52, 75–85.

Yanagawa, A., Kajiwara, A., Nakajima, H., Quéméner, E. D.-L., Steyer, J.-P. *et al.* (2020) Physical assessments of termites (Termitidae) under 2.45 GHz. *Scientific Reports* 10(5197), 1–9.

Yates, J.R. III, Grace, J.K. and Reinhardt, J.N. (2000) Installation guidelines for the Basaltic Termite Barrier: a particle barrier to Formosan subterranean termites. *Sociobiology* 35, 1–16.

Youngquist, J.A., Krzysik, A.M., English, B.W., Spelter, H.N. and Chow, P. (1996) Agricultural fibers for use in building components. In: *The Use of Recycled Wood and Paper in Building Applications, Proceedings 7286*. Forest Products Society, Madison, Wisconsin, pp. 123–134.

Zhang, S. and Mo, J.C. (2014) Effect of *Autographa californica* nuclear polyhedrosis virus suspension concentrate against the workers of *Coptotermes formosanus* Shiraki. *Forest Pest and Disease* 3, 18–21.

Zhang, P.B., Yan, X., Qiu, X.H. and Han, R.C. (2010) Application of transgenic *Enterobacter cloacae* with the insecticidal tcdA1B1 genes for control of *Coptotermes formosanus* (Isoptera: Rhinotermitidae) in the field. *Sociobiology* 56, 27–37.

Zhao, R.H., Han, R.C., Qiu, X.H., Yan, X. and Liu, X. (2008) Cloning and heterologous expression of insecticidal-protein-encoding genes from *Photorhabdus luminescens* TT01 in *Enterobacter cloacae* for termite control. *Applied and Environmental Microbiology* 74, 7219–7226.

Zhu, B. C., Henderson, G., Chen, F., Fei, H. and Laine, R. A. (2001a) Evaluation of vetiver oil and seven insect-active essential oils against the Formosan subterranean termite. *Journal of Chemical Ecology*, 27, 1617–1625.

Zhu, B.C.R., Henderson, G., Chen, F., Maistrello, L. and Laine, R.A. (2001b) Nootkatone is a repellent for Formosan subterranean termite. *Journal of Chemical Ecology* 27, 523–531.

15 IPM and area-wide population management

Nan-Yao Su
Entomology and Nematology Department, Ft. Lauderdale Research and Education Center, University of Florida, nysu@ufl.edu

Abstract
Traditional IPM was developed for management of agricultural pests, and the economic injury level (EIL) is its unifying principle. For most of non-agricultural pest species, however, EIL is either zero or cannot be determined. Instead of EIL, a conceptual framework for termite IPM emphasizes the cost–benefit aspect of termite management programs. Baits are considered ideal for termite IPM because they reduce termite damage potential (= benefit) through colony elimination, and because they use less potent insecticides at a smaller quantity to minimize health and environmental risks (= cost). Area-wide projects with baits capable of eliminating colony(s) have been tried in several areas, and successful projects are those with clearly defined objectives, simple managerial structure with one person in charge, and employing control measures with data to support the colony elimination claims. Failed projects are usually plagued by political issues and complicated managerial structures.

15.1 Introduction

The publication of Stern *et al.* (1959) marked the beginning of integrated pest management (IPM) which has been widely practiced for the management of agricultural pests. The unifying principle of IPM is to maintain pest populations below the economic injury level (EIL) (Pedigo and Higley, 1992). EIL is defined as the "lowest population density that will cause economic damage", which is further defined as "the amount of injury which will justify the cost of artificial control measures" (Stern *et al.*, 1959, p. 86). Early attempts to formulate IPM programs by many in non-agricultural entomology, however, were mostly unsuccessful. One reason for their failures is the challenge to derive a meaningful EIL for pest species that do not damage agricultural commodities with predictable monetary values. These include medical, ornamental, landscape, household, and structural pests. This chapter reviews their challenges and re-examines the underlying philosophy of IPM as defined by Stern *et al.* (1959), which led to the proposed conceptual framework for termite IPM by Su and Scheffrahn (1998). Examples of termite IPM projects and related area-wide projects to manage subterranean termites are discussed.

 Biology and Management of the Formosan Subterranean Termite and Related Species (eds Nan-Yao Su and Chow-Yang Lee)
DOI: 10.1079/9781800621596.0015

15.2 Agricultural IPM

Stern *et al.* (1959) listed factors contributing to the rise of insect pest species in agricultural fields, namely: (1) the artificial monoculture; (2) invasion of exotic pests in the environment in the absence of their natural enemies; and (3) lower threshold in crop quality (e.g., appearance) acceptable to marketplaces. To alleviate the increasing pest pressure, more synthetic insecticides were used that led to another set of problems, including: (1) insecticide resistance; (2) outbreak of secondary pests; (3) migration of pest species into areas where the natural enemies were removed by pesticide applications; (4) insecticide contamination of food products; (5) insecticide exposure for farm workers and wildlife; and (6) legal ramifications.

Recognizing that natural enemies are essential elements of environmental resistance to keep pest populations in check, Stern *et al.* (1959) argued that chemical applications are "only needed at those times and places where natural control is inadequate" (p. 89). They proposed that biological and chemical control must be integrated to reduce pest populations below EIL. From the economic perspective, they emphasized that when synthetic insecticides are used, "the damage from the pest species must be sufficiently great to cover not only the cost of insecticidal treatment but also the possible deleterious effects ... on ecosystem" (p. 88). EIL is the threshold of pest density that can be tolerated and is calculated as: EIL = $C/VIDK$, where C = cost of management action per production unit, V = market value per production unit, I = injury units per pest, D = damage per injury unit, and K = proportional reduction in pest attack due to management action (Pedigo and Higley, 1992). For agricultural entomology, price unit of a commodity, be it cotton, soybean, corn, or livestock, can be used to calculate V, but it is almost impossible to provide a reliable V when the pests do not damage agricultural commodities, or when commodities are ornamental plants with unpredictable values.

15.3 Non-agricultural IPM and EIL

Ornamental and landscape pests damage agricultural commodities, but it is still challenging to adopt EIL for management of these pests, because the V value can be unpredictable. For a nurseryman, for example, a market value of a plant can be assigned for V, but for a homeowner who manages pests of ornamental plants of their property, the economic concern alone does not account for their V values (Pedigo *et al.*, 1986; Potter, 1986; Raupp *et al.*, 1988). Early on, Olkowski (1974) proposed the aesthetic injury level (AIL), defined as the lowest pest density to cause aesthetic injury, for decision making in management of ornamental pests. AIL was successfully adopted for pest management of ornamental plants such as turfgrass (Hellman *et al.*, 1982), and the American arborvitae (Raupp *et al.*, 1989), but it remained difficult to assign an objective and useful V value to derive AIL (Pedigo *et al.*, 1986), and the concept received limited support from the ornamental plant industry (Robinson, 1996).

For household pests such as cockroaches, attempts were made to determine AIL by conducting surveys of residents in public housing projects (Wood *et al.*, 1981; Zungoli and Robinson, 1984). Their results showed that AIL varied depending on individual attitude, and there is no single AIL that is acceptable for all housing or business situations, which questions the applicability of AIL for a practical IPM program for cockroaches. Pinto and Kraft (2000) proposed the action thresholds (ATs) for some pests in school IPM programs, including five ants per room, one (indoor) or three (outdoor) bees, two cockroaches per room, three houseflies per room, etc. The AT is equivalent to the economic threshold as described by Stern *et al.* (1959), but these ATs were mostly educated guesses rather than data-based values.

The above examples illustrated challenges facing non-agricultural entomologists in their attempts to derive EIL for IPM programs where the target insects do not damage commodities, or the values of commodities are unpredictable, and termites are no exception.

15.4 Termite IPM

15.4.1 Conceptual framework

There are two types of treatments for subterranean termite control: pre-construction treatment and post-construction (or remedial) treatment. A pre-construction treatment is mandated by building codes or required by financial institutes that issue the mortgage for new buildings. For pre-construction treatments, ~0.25–5 kg insecticide AI (active ingredient) is sprayed onto sub-slab soil of a house before the erection of building foundations. In recent years, baits are used in lieu of soil termiticide applications in some states. Because the pre-construction treatment is required regardless of the presence of termites, EIL = 0 and AT > EIL, resulting in the "inversion relationship" between EIL and AT as found in many urban and quarantine pests (Stejskal, 2002). Remedial treatments are done only when homeowners notice the signs of termite infestations, including alates (Fig. 15.1A), dispersal flight exit slits (Fig. 15.1B), damaged wood materials (Fig. 15.1C), and foraging tubes entering the building from soil (Fig. 15.1D) (see also Chapter 10 of this volume). When these signs of infestation are found, treatments are applied to remedy the situation in almost all cases, either by homeowners themselves or by professional applicators, hence EIL ~0. It is apparent that treatment decisions in subterranean termite control have little to do with EIL, AIL or AT, but this does not imply that IPM for termites is impossible to formulate.

Recognizing that EIL as proposed by Stern *et al.* (1959) is a decision-making tool, not the aim of an IPM program, Su and Scheffrahn (1998) argued that "it is probably unrealistic to apply the cost-benefit model developed for agricultural IPM directly to an urban pest management program"; however, "the underlying concept of IPM to use a cost-effective approach to solve pest problems ... is the same" (p. 9). With that premise, Su and Scheffrahn (1998)

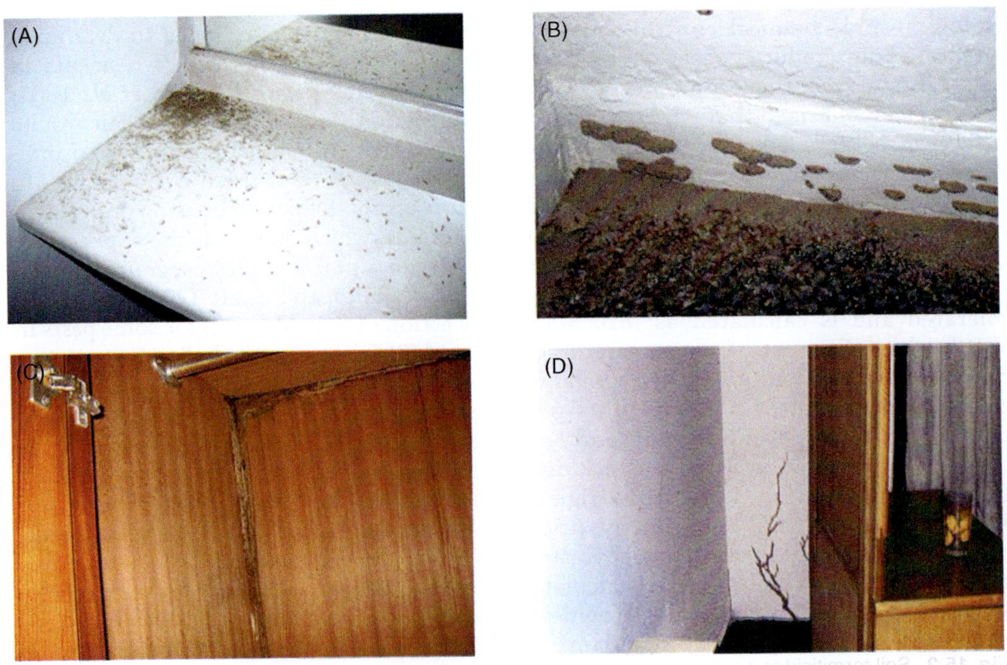

Fig. 15.1. External signs of termite infestations include alates (A), dispersal flight exit slits (B), damage wood materials (C), and foraging tubes (D).

concluded that the objective of a termite IPM program should be the protection of a home by using the most cost-effective method(s). In formulating the conceptual framework of termite IPM, they analyzed the cost–benefit structure of a termite control business. When a homeowner agrees to pay a fee (*F*) to a termite control firm to control a termite infestation, the potential damage by termites (*D*) has to exceed the fee, or $D > F$ (Su and Scheffrahn, 1998), and the goal of a control measure is to reduce or eliminate *D*. The fee includes operation costs (*T*) of a termite control firm plus profit (*P*), or $F=T+P$. The total operation costs $T = l+m+e+b+o$, where l = labor cost, m = material cost, e = equipment cost, b = liability, and o = overheads, hence $D > l+m+e+b+o+P$. Two variables in this simple equation, *D* and *b*, however, are almost impossible to determine.

Potential damage *D* does not include damage that has already occurred by the time the homeowner noticed the sign of infestation. Instead, *D* represents the future damage (and associated cost) the existing termite population may cause if no control measure is taken. The challenge here is that it is nearly impossible to project the future damage based on the observation of the initial signs of termite infestation (Fig. 15.1). However, other variables such as termite population pressure may be used as a proxy to represent *D*, i.e., more termites in the vicinity of a home will increase *D*. On the other hand, if a control measure can reduce or eliminate termite pressure, *D* can be reduced or brought to zero. Of the two major control methods for management of subterranean termites, soil barrier treatments are intended to exclude soilborne termites from the structures without affecting termite populations in the soil (Chapter 12 of this volume), hence *D* is generally unchanged. Baits, on the other hand, are intended to eliminate termite colony(s) beneath or near a house (Chapter 11 of this volume), which eliminates *D*.

Liabilities (*b*) in the termite control business may include financial loss due to termite reinfestations, and health and environmental risks associated with insecticide application. The issue with reinfestation can be minimized with the introduction of baits that eliminate termite colony(s) to drastically reduce termite damage potential (*D*). However, it is difficult to predict the long-term health and environmental risk potentials when applying a large quantity of potent, persistent liquid insecticides in soil near a home (Fig. 15.2). Between the 1950s and 1980s, for example, chlordane was heavily used for soil treatment against subterranean termites. A study was conducted in 1992–2001 to measure chlordane concentrations in whole fish from 648 stream sites nationwide, and the results showed a strong correlation between termiticidal use of chlordane in termite-infested areas and its concentration

Fig. 15.2. Soil termiticides are injected into holes drilled through concrete foundation (A) and sprayed onto trenches around a house (B). Long-term risks of applying a large quantity of potent and persistent insecticides in soil near a home is nearly impossible to predict.

in whole fish (Nowell *et al.*, 2009). It was evident that chlordane persisted in underwater sediments and wildlife 10–20 years after being applied, but it was not anticipated by the termite control industry who applied this soil termiticide. Although the precise figure is difficult to predict, health and environmental risks associated with insecticide application are correlated with the quality and quantity of pesticide being used, i.e., the application of more toxic and persistent insecticide at a larger quantity tends to increase the risk, whereas the application of less toxic and persistent insecticide at a smaller quantity reduces the risk.

The above simple cost–benefit analysis suggests that damage potential (*D*) and health and environmental risks (*b*) have to be reduced or eliminated, in order to achieve termite IPM.

15.4.2 Biological control

One major component of IPM concept as proposed by Stern *et al.* (1959) is to use natural enemies to suppress pest populations, and to apply chemical insecticides only when necessary. Many biological control options are available for non-agricultural pests, including those of ornamental and medical importance (Osborne and Landa, 1992; Lacey and Orr, 1994), but no such option exists for termite control. Many laboratory studies that tested the potential of biological control against termites reported promising results, especially with entomopathogenic fungi, *Metarhizium anisopliae* (Metchnikoff) Sorokin (now *Metayrhizium robertsii*) and *Beauveria bassiana* (Bals.-Criv.) Vuill., but there are no successful reports in field trials using these fungi (Chouvenc *et al.*, 2011a) or any other biocontrol agents (see Chapter 14 of this volume). Being soil insects, subterranean termites have evolved with defensive mechanisms to prevent fungi-induced epizootic (Chouvenc and Su, 2010), including grooming (Hussain and Tian, 2013), gut inhibition (Chouvenc *et al.*, 2009a, 2010), and cellular encapsulation (Yanagawa and Shimizu, 2007; Chouvenc *et al.*, 2009b, 2011b) (see Chapter 14 of this volume). Despite continuing laboratory studies that report potentials of biological control agents, none address the issue of bypassing termite defense mechanisms at cellular, individual or colony levels, and it appears that the biological control option will not be available in the near future (Culliney and Grace, 2000; Chouvenc *et al.*, 2011a, see Chapter 14 of this volume).

15.4.3 Cultural control

One option to reduce termite damage potential (*D*) is to remove termite-conducive conditions near a structure. Wood touching or buried in soil (Fig. 15.3A) provides an easy entry for subterranean termites to a house, and such wood–soil contact must be avoided. For the same reason, many building codes dictate that only treated wood be used if it comes into contact with ground. Similarly, wood piles such as firewood should be kept above ground level to avoid soil contact.

Subterranean termites favor wet conditions, and tunneling activities are known to increase in soil with higher moisture (Su and Puche, 2003). Elimination of water sources near building foundations can reduce termite activity. These include creating a soil grading to direct water away from the foundation or avoid accumulation of water near foundation soil from air-conditioning units (Fig. 15.3B). If a sprinkler is installed near a building foundation, a cover may be installed to prevent it from spraying water onto building walls (Fig. 15.3C). Overgrown vegetation adjacent to buildings (Fig. 15.3D) has to be trimmed as it tends to keep the soil moist and prevent adequate inspection of termite foraging tubes entering a building.

The above includes just some of many cultural control measures that can be taken to correct termite-conducive conditions, thereby reducing termite damage potential (*D*) (see Chapter 10 of this volume).

15.4.4 Soil treatment vs bait

Although both soil termiticide treatments and baits share the common goal of structural

Fig. 15.3. One option to reduce termite damage potential is to remove termite-conducive conditions such as avoiding wood–soil contact (A), avoiding water accumulation near foundation soil (B), diverting sprinkler water away from foundation wall (C), and trimming of overgrown vegetation adjacent to buildings.

protection, the mechanisms by which to achieve this goal differ. Soil termiticide treatments were used to exclude soilborne termites from a structure. According to the Termidor® label (80% fipronilor, BASF Corp.), for a pre-construction treatment, the sub-slab soil is treated with (at least) 0.06% water mixture at (the minimum) rate of 4.1 L/m^2, and soil adjacent to foundation walls is treated at the rate of 5.0 L/m (at 0.6 m depth). To treat an average home of 230 m^2 with a total foundation wall distance of 60 m (assuming a square foundation), at least ~1,243 L of liquid insecticide is required. Even at the minimum rate, the total amount of insecticide AI used for a pre-construction treatment of a single-family home amounts to 0.745 kg, or 32.4 kg/hectare (Fig. 15.4A), which is ~15 times higher than insecticide usage rate of 2.17 kg/hectare in agriculture (Pimentel and Levitan, 1986). At lower doses, fipronil may kill termites slowly and allow exposed termites to move away from treated soil before the onset of death, but only at the maximum distance of 2–5 m (Su, 2005; Saran and Rust, 2007). Soil treatments with fipronil may eliminate small *R. flavipes* colonies located near treated soil (Vargo and Parman, 2012), but it did not eliminate large colonies of *R. flavipes* (Ripa *et al.*, 2007) and had little effect on much larger colonies of *C. formosanus* (Chouvenc, 2018; Su, 2019).

Contrary to soil termiticides, baits are used to eliminate the source of subterranean termite infestation, i.e., colony(s) in the vicinity of a house (Fig. 15.4B). A large amount of literature has reported successful eliminations of subterranean termite colonies following the application of chitin-synthesis inhibitor (CSI) baits (see Chapter 11 of this volume). Elimination of subterranean

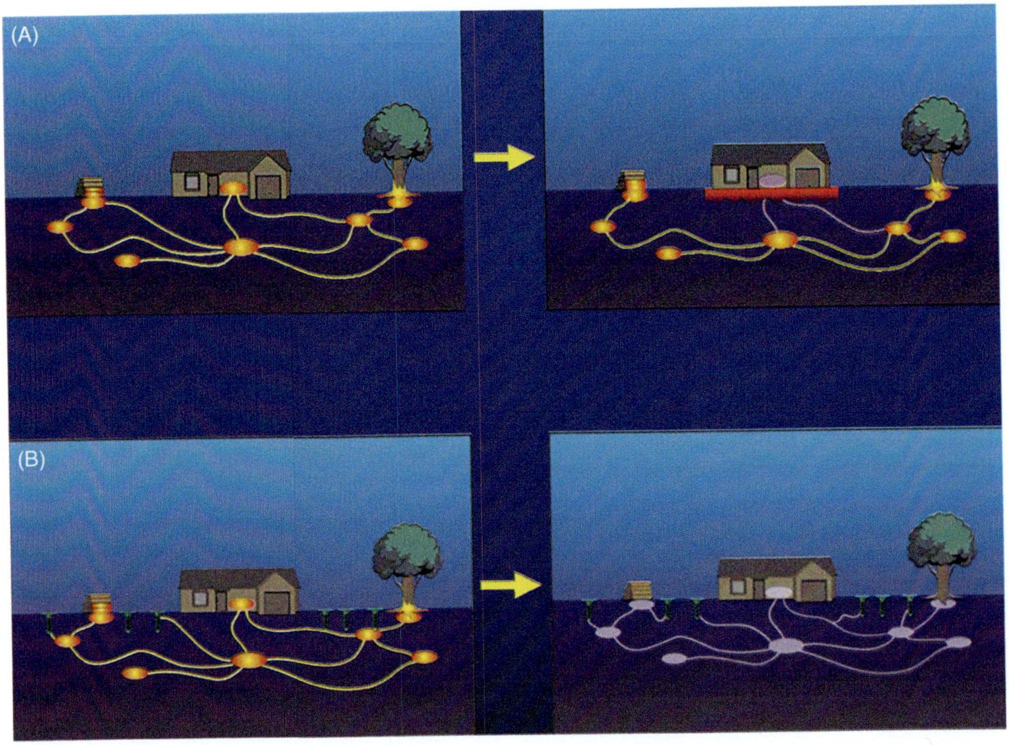

Fig. 15.4. Soil termiticides are applied at a rate of ~32.4 kg/hectare to exclude soilborne termites from a structure, but most of the termites in a colony survive (A). For a single-family home, bait AIs are used at a rate of 0.043 kg/hectare to eliminate termite colony(s) in soil (B), which is ~753 times less pesticide use than soil termiticide. Moreover, soil termiticide such as fipronil is more toxic (rat oral LD_{50}: 97 mg/kg) than a bait AI such as noviflumuron (rat oral LD_{50}: > 3,160 mg/kg). Because baits use less toxic insecticide (33-fold) at a lower rate (753-fold) than soil termiticides, the risks associated with bait applications is greatly reduced, resulting in a 24,849-fold less environmental impact by baits than soil termiticide treatments.

termite colony(s) will remove or reduce termite damage potential (*D*), making baits a better option for termite IPM than soil insecticide treatment. In field trials of bait efficacy studies, typically less than 1 g of CSI is needed to eliminate termite colony(s), hence the insecticide application rate for baits amounts to ~0.043 kg/hectare. Moreover, fipronil is more toxic (rat oral LD_{50}: 97 mg/kg) than a CSI such as noviflumuron (rat oral LD_{50}: > 3,160 mg/kg). Because baits use less toxic insecticide (33-fold) at a lower rate (753-fold) than soil insecticide treatments, the risk (*b*) associated with bait applications is greatly reduced, resulting in a 24,849-fold less environmental impact by baits than soil termiticide applications. A bait's ability to reduce termite damage potential (*D*) by using lower risk (*b*) measures makes it a better fit to be employed in termite IPM than soil insecticide treatments, and baits have been used in a large-scale IPM demonstration project in China as well as area-wide projects in the U.S. and elsewhere.

15.5 A large-scale IPM demonstration project in China

Of the six subterranean termite pest species in China, *C. formosanus* and three *Reticulitermes* spp., including *Reticulitermes flaviceps* (Oshima), *Reticulitermes chinensis* Snyder, and *Reticulitermes speratus* (Kolbe), are the most serious pests of houses and structures (Huang *et al.*, 1989). Termites are found in ~40% of the land area of China and it was estimated that 30–90% of buildings south of the Yangzhu River sustained damage from subterranean termites (Anonymous, 2005). Between the 1960s and 2000s, control of subterranean termites relied heavily on two chlorinated hydrocarbons: chlordane and mirex (dechlorane). During the 3-year period in 2000–2003, for example, 450–820 tons of chlordane and 9–31 tons of mirex were produced for subterranean termite control in China and neighboring countries (Anonymous, 2005; Su *et al.*, 2012).

In 2001, the Stockholm Convention on POPs (Persistent Organic Pollutants) was adopted under the auspices of the United Nations Environmental Program to eliminate these POPs because they "possess toxic properties, resist degradation, bioaccumulate and are transported ... across international boundaries and deposited far from their place of release, where they accumulate in terrestrial and aquatic ecosystems" (Anonymous, 2019). Along with persistent industrial chemicals and insecticides, chlordane and mirex were listed as the POPs targeted for elimination.

China ratified the Convention in 2004, and officially launched a project entitled, "Demonstration of Alternatives to Chlordane and Mirex in Termite Control" in 2006. Due to its expertise in POPs, World Bank was put in charge of implementing the project, and the Chinese Ministry of Environmental Protection was responsible for executing the project. Participating provinces were Anhui, Jiangsu, and Hunan, which provided the co-financing. Following the Convention Article 5e to "promote the use of best environmental practices", World Bank suggested the introduction of IPM instead of merely replacing chlordane and mirex with other liquid soil termiticides, and IPM became the underlying theme of the overall project objective. With this background, a group of Chinese termite experts adopted the conceptual framework of termite IPM as proposed by Su and Scheffrahn (1998) and drafted the *IPM Operation and Training Manual* (Anonymous, 2007).

In 2007, two inception workshops were held to introduce participants in the Stockholm POPs Convention to the project objectives, risks associated with POPs used for termite control, proposed project activities, IPM concepts, baiting technologies, and IPM projects elsewhere. The participants were government officials at national and provincial levels, and termite control personnel from over 80 termite control stations (TCSs) in Hunan and Anhui provinces. Three "train-the-trainers" workshops were held in the same year to familiarize the participants with the contents of the newly drafted *IPM Operation and Training Manual*, and the associated technologies, especially the installation of the monitoring–baiting system. The target audience was the 149 trainers, including 50 from 19 TCSs of Hunan, 39 from 18 TCSs of Jiangsu, and 60 from 20 TCSs of Anhui. Two vendors, Ensystex China (Shanghai, China) and Dow AgroSciences China (Shanghai, China), were invited to demonstrate their respective bait products, Extera™ Termite Control Baiting System and Sentricon® Colony Elimination System, respectively, in the field to provide participants with hands-on experiences in station assembly, drilling with power tools, and installation. After the 3-day workshop at each site, the trainers went back to their respective TSCs and organized additional training workshops to share their experiences with other termite control applicators who could not attend the workshops. Through this arrangement, by the end of 2010 over 5,000 termite control personnel (operation managers, termite control applicators, and clerical staff) in the three provinces were informed of the contents of the IPM training manual and trained with termite-baiting technologies.

Between 2008 and 2011, ~680,000 in-ground (IG) bait stations were installed at 55,354 buildings and/or sites in three participating provinces, which replaced 372 tons of chlordane. Approximately 150,000 above-ground (AG) stations were also used in some of these buildings with active infestations, which replaced ~163 kg of mirex. Moreover, closures of chlordane manufacturing plants removed 303 tons of chlordane produced by China (as of 2005). In total, the demonstration project reduced 675 tons of chlordane that otherwise would have been used in China. The performance of bait systems in China was comparative with other regions (Su *et al.*, 2012). Bait systems employed through the demonstration project covered only 15% of subterranean termite control in the three participating provinces, and as of 2011 the remaining 85% still relied on soil termiticides. There were challenges such as lack of understanding of the termite IPM concept among the participants and public (due to insufficient educational opportunities), unrealistic estimates for bait station procurement, and government authority's insistence on using domestic bait products that lacked efficacy data. Despite these shortcomings, the project was the first large-scale termite IPM project, and it is expected that the concept of using a lower-risk control option, be it a bait or anything else, to reduce termite damage potential will be adopted in China, the most populated nation in the world.

15.6 Area-wide (AW) management projects with subterranean termites

15.6.1 Current termite control business model and AW management program

The current business model for subterranean termite control is mostly based on the contract between a homeowner and a termite control firm to protect a single structure. AW pest management is a more sustainable approach for IPM program than a house-by-house approach (Klassen, 2007), but this option has rarely been considered. The following are examples of AW projects including those for *C. formosanus* and *Reticulitermes* species.

15.6.2 Golden Beach project, Florida, USA

The town of Golden Beach (~247 acres) is a small community of 350 houses situated at the northeastern corner of Dade County, Florida. Adjacent to its north is the town of Hallandale, where the first record of *C. formosanus* was collected in 1980 (Su and Scheffrahn, 1987). By 1999, when the town council adopted a resolution for full cooperation with the University of Florida to conduct an AW management project, numerous houses had been severely infested (Su *et al.*, 2004). IG monitoring stations and light traps were used to monitor *C. formosanus* activity between 2001 and 2004, but despite the town's resolution for a full collaboration, only 35% (116 out of 334 houses) cooperated. Of the 116 residents who responded to the survey, 112 indicated that they had their properties treated for termites, and 52 (46%) used baits whereas 35 (31%) used liquid termiticides. The remaining provided no information. Additional baiting systems with Recruit® II (0.5% hexaflumuron) were installed at the request of some residents but most of the properties received no treatments and did not provide access for monitoring. Because termite activity data of the entire town were not available, a Bait Impact Index (BII) that accounted for the amount of bait applied and distances between IG monitoring stations and bait stations were used to demonstrate a strong correlation between bait application and decline of termite activity (Su *et al.*, 2004). Although termite activity in 30 IG monitoring stations (note that these stations covered only a portion of the town) declined to near-zero by the summer of 2003, the mean number of *C. formosanus* alates collected in the spring of 2004 was not significantly different from 2001. Results of BII suggested that hexaflumuron baits eliminated some colonies but there remained other *C. formosanus* populations that were not monitored or treated and were unaffected. The lack of complete participation by the residents is the primary reason for the failure of this AW project, and it was terminated in 2004.

15.6.3 Cerro Navia project, Santiago, Chile

Since its first discovery in the metropolitan Santiago, Chile, in 1986, *R. flavipes* had spread to at least 13 cities in the 5th (Valparaiso) and 6th (O'Higgins) regions of that country by 2004 (Su *et al.*, 2006). During the same period, *R. flavipes* infestations were found in all the 31 municipalities of metropolitan Santiago, and one of the most severely infested was the low-income community of Cerro Navia (Smith *et al.*, 2006). Residents of six selected city blocks in Cerro Navia (~5.2 acres) were contacted by the municipality government in 1999 for initial site surveys, and 57 of 108 (53%) home sites agreed to participate in the project. The other 51 homes showed no interest, because they did not have termites, but they provided access for inspection to confirm the absence of termites.

Monitoring stations that comprised PVC tubes (5 cm diameter and 30 cm long) containing two pieces of radiata pine (*Pinus radiata* D. Don, 1 × 1 × 2.5 cm) were installed in all accessible sites, and wood consumption rates (g wood/day) were recorded to represent termite activity. After installing 130 Sentricon® stations in 2000, Recruit® II baits were placed in 53 active stations of the three northern blocks in 2001, and the other three blocks on the south were used as the untreated control. Following the consumption of hexaflumuron baits, termite activity declined steadily throughout 2001 from the baited three northern blocks, and by mid-2002, no termite activity was recorded from most of the monitoring stations and Sentricon® stations, whereas termites remained active in the three untreated southern blocks. In mid-2003, Recruit® II baits were applied in all active Sentricon® stations of all six blocks, and by October 2004, all termite activity had ceased. As with the Golden Beach project, there was a strong correlation between BII and decline in termite activity in monitoring stations, which demonstrated that more baits applied nearer to a monitoring station had more impact in reducing termite activity. The test area remained termite free when it was last surveyed in May 2005.

The project demonstrated that applications of hexaflumuron baits could eliminate all detectable termite activity in a large area, and factors for the success in the Cerro Navia project include: (1) that the project was carried out by a single pest control firm with a small crew of highly trained technicians; and (2) the good cooperation from the residents. Although the covered area (~5.2 acres) was relatively small, it was anticipated that the model could be applied for a large AW project.

15.6.4 Chatsworth project, California, USA

Termite infestation had been an ongoing problem for the 2,600 residents of a condominium complex of 139 buildings (~90 acres) located in the city of Chatsworth, a suburb of Los Angeles (Getty *et al.*, 2007). A termite control firm was contracted to install 7,327 Sentricon® station in October–December 2001. These stations were monitored monthly after the completion of installation. In January 2002, only a few showed termite activity, but by February, termites (*Reticulitermes hesperus* Banks, and one "undescribed" *Reticulitermes* species) were found in Sentricon® stations of 41% of building sites. In March 2002, a total of 160 stations harbored new termite activity. Six months after the initial installation, termites were detected in station(s) on 90% of building sites.

In 2002, a total of 878 (out of 7,327) Sentricon® stations harbored termite activity and were baited with 0.5% hexaflumuron (Recruit® II) (Getty *et al.*, 2007). Per label prescription, another 423 auxiliary stations were installed next to active stations, totaling 7,750 Sentricon® stations being installed at the site. Following the bait application through 2002, termite activity steadily declined and by September–December, only 10–20 stations with new termite activity were recorded. As temperatures climbed in the spring of 2003, termite activity rebounded (~40 stations with new activity). Then the newly registered 0.5% noviflumuron baits (Recruit® III) were applied in these stations. As with the previous year, termite activity steadily declined in 2003 and only a small number of newly active stations were found from August 2003 until December 2004, when the project was terminated.

Because of the large area, it took three months to install 7,327 Sentricon® stations, but the monthly monitoring of these stations was the "responsibility" of one technician, according to Getty *et al.* (2007). It is probably unrealistic for a single technician to perform the monthly monitoring of > 7,327 stations, during which she/he also had to apply baits in some stations and install auxiliary stations. Assuming a 20-working-day month, the technician had to inspect 366 stations daily, and for an 8 h working day she/he had only 1.3 min to spend per station, and this tight schedule did not even include the additional 423 auxiliary stations. This seemingly impossible schedule was probably relieved by the addition of more personnel on site. Project records suggested that there were multiple assistants under the supervision of the technician who is solely responsible for the monthly monitoring (M. Haverty, personal communication). Another possible reason for being able to monitor the large number of stations is the use of Sentricon® System with ESP™ (Electronic Sensing and Protection) that became available in 2001. An electronic sensor is embedded in the monitor of a Sentricon® station and is activated by a handheld interrogator to detect the sensor breakage, which may indicate the presence of termites in the station. With the use of the ESP™ system, stations with intact sensors were not opened, which would drastically reduce the number of stations to be manually inspected. By placing the responsibility of on-site monitoring, baiting, and installation of auxiliary stations on a single technician, the managerial structure of the operation is simplified, and the success of this AW project could be attributed to the simple operation and managerial structure.

15.6.5 French Quarter project, Louisiana, USA

"Operation Full Stop" was a federally funded program to manage *C. formosanus* populations in the city of New Orleans, Louisiana, and the major target was the 314-acre French Quarter. The objective was to "reduce the existing termite pressure in the French Quarter in order to minimize potential for future damages" (Guillot *et al.*, 2010, p. 313). The project began its first phase in 1998 to treat a 15-block (~0.1 × 0.1 km) area, or Area I, near the center of the Quarter, followed in 2002 by a 20-block area (Area II) surrounding Area I, a 19-block area (Area III) to the northeast in 2004, a stretch of land (1.7 km × 0.2 km) including railroad and Mississippi River levee (Area IV) to the southeast in 2003–2004, and a 22-block area to the northwest in 2006–2007 (Area V) (Fig. 15.5).

Property owners were informed of the project and asked to enter a contract with a termite control firm to treat their properties with either non-repellent termiticides and/or bait systems that were registered and approved. As a result, treatments (AI, number) being chosen in Area I were Sentricon® (hexaflumuron, 355), Termidor® (fipronil, 4), Premise® (imidacloprid, 42), and other products (9) including Exterra® (diflubenzuron), Firstline® (sulfluramid) and Subterfuge® (hydramethylnon). Sentricon®, Exterra®, Firstline®, and Subterfuge® were baiting systems, whereas Termidor® and Premise® were non-repellent soil termiticides. It should be noted that by 1998, studies had shown that metabolic inhibitors such as hydromethylnon or sulfluramid baits were ineffective in eliminating *C. formosanus* colonies (Su, 1982; Su *et al.*, 1995), yet because these bait products were EPA "registered" and "approved" by Louisianan Structural Pest Control Commission, they were allowed to be used in this AW project.

The number of property owners choosing Termidor® increased to 47 in Area II (2002–2003), 103 in Area III (2004–2005), and 133 in Area V (2006–2007), but those choosing Premise® declined to 28 in Area II, and 11 in Area III. In 2006 when the project was expanded to Area V, none chose Premise®. The number of private properties treated with Sentricon® declined from 355 in 1998 (Area I) to 219 in 2006 (Area V). In 2003, the active ingredient (AI) of Sentricon® baits was changed from hexaflumuron to noviflumuron (Recruit® III).

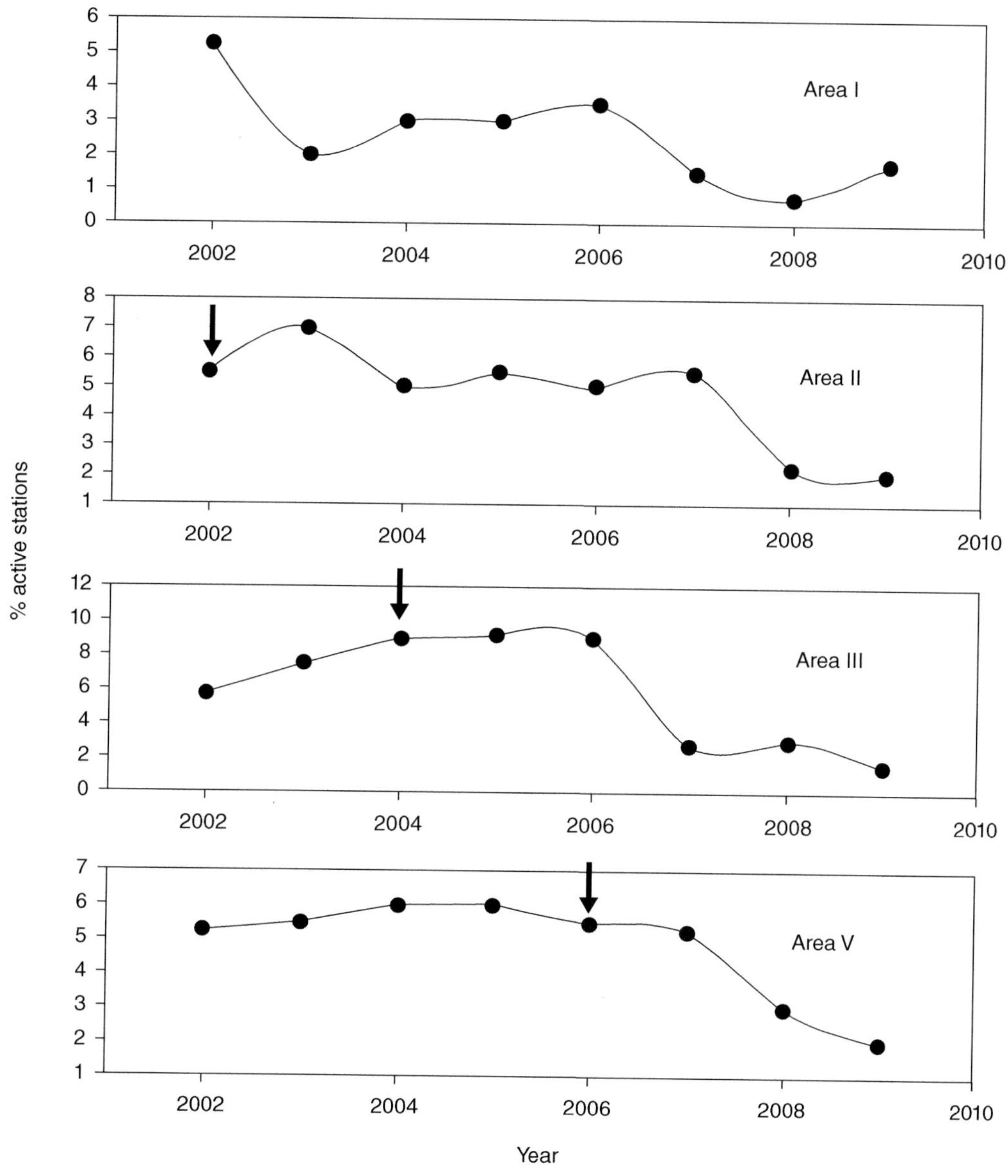

Fig. 15.5. Number of active in-ground (IG) stations for the French Quarter area-wide project that took place between 1998 and 2009. Area I with 410 properties received treatments in 1998, whereas treatments (arrows) began in 2002, 2004, and 2006 for Areas II (388 properties), III (403 properties), and V (360 properties), respectively. The 2006 decision to apply noviflumuron baits (Recruit® IV) in selected independent IG monitoring stations led to the decline of termite activity in 2007–2009 in all areas. Based on data from Guillot *et al.* (2010).

In-ground (IG) stations (modified Sentricon® stations with wooden slats inserted in holes bored through concrete pavement) were installed on the perimeter of each block to monitor *C. formosanus* activity in soil. Unmodified Sentricon® stations with wooden slats were also used on the levee for the same purpose, totaling 600 monitoring stations. Sticky traps were hung under streetlamps at intersections to measure the overall termite activity.

Four years after the 410 properties of Area I received the treatments, *C. formosanus* were found in 5.25% of the ~227 IG monitoring stations (based on Figs 3 and 7 of Guillot *et al.*, 2010), which amounts to 12 active stations (Fig. 15.5). The activity level declined to 2% in 2003 but remained 3–3.5% (6–8 active stations) through 2006. Area II (388 properties) was treated in 2002, but *C. formosanus* activity levels did not change (5–7%, or 3–4 stations) for another 5 years (Fig. 15.5). Due to the lack of apparent treatment impact on *C. formosanus* activity levels even 6 years after the commencement of the project, two actions were taken in 2004, including: (1) to carry out a thorough inspection of treated properties to uncover reinfestations for re-treatments; and (2) to employ Sentricon® system in public open spaces such as Jackson Square of Area I, and the railroad and levee of Area IV. In 2004, the project was expanded to Area III of 403 properties, but as with the Areas I and II, *C. formosanus* activity did not change substantially (~9%, or 7–8 active stations) through 2006 (Fig. 15.5). Again in 2006 another measure was taken to change the course of the project by applying noviflumuron baits (Recruit® IV of Sentricon® system) in "selected" independent IG monitoring stations. In the same year, the project was expanded to Area V of 360 properties. With the application of noviflumuron baits, *C. formosanus* colonies connecting to these IG monitoring stations were probably eliminated, leading to the drastic decline of termite activity after 2007–2009 in all areas (Fig. 15.5).

Alate data of Guillot *et al.* (2010) included those of Area VI (a 17-block read on the southwest side of French Quarter) that did not receive the project-sponsored treatments. There was a general decline in the number of alates caught in the sticky traps from 2003 to 2006 in Areas I, II, III, V, and VI, but the number rebounded slightly in 2007 and remained the same through 2009. The mean alate catch in 2009 was 213 ± 122 (SE), which was ~60% reduction from the peak year catch of 532 ± 353 in 2003. The final count of active IG monitoring stations in 2009 was 1.81 ± 0.21% (or 8.3 ± 0.96 stations), which was a 67% reduction from the 2002 activity level (5.44 ± 0.21%, or 24.9 ± 0.96 stations), which was remarkably similar to ~60% reduction of alate data. It should be noted, however, that the decline of active IG stations was, in part, due to the 2006 decision to apply noviflumuron baits in some monitoring stations.

In 2012, Operation Full Stop was terminated due to lack of government funding. A ~60% reduction of *C. formosanus* activity level after 11 years is not exactly a success, and there are many reasons for the less than satisfactory outcome. Guillot *et al.* (2010) listed the building construction practice in French Quarter as one factor contributing to lingering *C. formosanus* activity. They suggested that *C. formosanus* aerial colony(s) in shared walls of adjacent buildings escaped the treatments as different pest control firms were put in charge of separate properties, but none treated the walls. This explanation is contradictory, because the lack of decline in *C. formosanus* activity was recorded from IG monitoring stations, which cannot be attributed to "aerial colonies" that have no connection to the ground. The most likely reason behind the persistent *C. formosanus* IG activity 6 years after the initiation of the project in 1998 is the inability of some treatments to eliminate termite colony(s), including Termidor®, Premise®, Exterra®, Firsline®, and Subterfuge® that accounted for ~24% of treatments in Areas I, II, and III, 1998–2004 (based on Table 1 of Guillot *et al.*, 2010). *Coptotermes formosanus* colonies that received these treatments probably survived and continued to be detected from nearby IG monitoring stations (Fig. 15.5).

Because of the non-repellency, early on it was suggested that Termidor® (fipronil) and Premise® (imidacloprid) may act like liquid baits to impact termite colonies (Kard, 2001; Thorne and Breisch, 2001; Potter and Hillery, 2002; Ibrahim *et al.*, 2003; Wagner, 2003; Hu, 2005). Shelton and Grace (2003), however, concluded that a high concentration of fipronil (> 10 ppm) was needed for a successful horizontal transfer of lethal dose, but at such dose the donors may be killed too fast for a substantial transfer. Saran and Rust (2007) reached the same conclusion that the western subterranean termites, *R. hesperus*, exposed to fipronil were too severely impaired to act as donors and the horizontal transfer was not a major factor contributing to field efficacy of fipronil. A field study by Osbrink *et al.* (2005) showed that Premise® treatment had no effect on termites found in monitoring stations within 1–3 m of treatment. Results of a field study to compare efficacy of four treatments, fipronil, cypermethrin (a repellent soil termiticide), hexaflumuron bait and sulfluramid bait, against *R. flavipes* in Chile showed that no change in termite activity was observed in sites treated with sulfluramid bait, cypermethrin or fipronil soil termiticides 2 years after treatments, whereas termite activity in hexaflumuron-treated sites declined to zero at 1 year post-treatment (Ripa *et al.*, 2007). Laboratory studies with extended foraging arenas confirmed that fipronil treatments did not eliminate even small laboratory colonies of *C. formosanus* (Su, 2005; Chouvenc, 2018). As for Exterra® which uses a CSI (diflubenzuron) bait, Su and Scheffrahn (1993) reported that this CSI deterred *C. formosanus* feeding at a low concentration of 2 ppm while a concentration of > 1,000 ppm is required to cause significant mortality and concluded that diflubenzuron is probably ineffective as a bait toxicant (see Chapter 11 of this volume). A field study also concluded that diflubenzuron "did not have any observable effect on populations of either *C. formosanus* or *R. flavipes*" (Osbrink *et al.*, 2011, p. 1011). Collectively these laboratory and field studies indicated that the non-repellent termiticides such as fipronil and imidacloprid, or bait toxicants such as diflubenzuron, sulfluramid, and hydramethylnon, could not eliminate *C. formosanus* colonies, and should not have been used in an AW project if the objective is to reduce termite damage potential of a large area.

15.6.6 Armstrong Park project (Phase V), Louisiana, USA

Louis Armstrong Park is a 32-acre area adjacent to the northwest side of the French Quarter. The Armstrong Park project is part of Operation Full Stop to study the population dynamics of *C. formosanus* under the AW management scheme. The project was conducted in five phases from 1999 to 2015. During the first phase (1999–2001), > 13 *C. formosanus* colonies were identified by mark–recapture protocol (Messenger and Su, 2005) and genotyping (Husseneder *et al.*, 2003). In Phase II (2001–2002), three *C. formosanus* colonies were selected for elimination to observe inter-colony responses, and the results showed that vacant territories of baited colonies were immediately occupied by neighboring colonies (Messenger *et al.*, 2005). Hexaflumuron baits and later noviflumuron baits were used to eliminate all detectable *C. formosanus* colonies in the park in Phase III (2002–2003) (Mullins *et al.*, 2011), after which the progression of the recovery of *C. formosanus* colonies was recorded from 2004 to 2009 in Phase IV (Su *et al.*, 2016a) (see also Chapter 5 of this volume).

The objective of Phase V (2010–2015) was to demonstrate a sustainable AW management of *C. formosanus* populations in the park with a durable bait, Recruit® HD (Su *et al.*, 2016b). Before 2010, baiting systems comprised a monitoring–baiting procedure, in which IG stations were routinely monitored (monthly to quarterly) and AI-containing baits were applied only when termites were found in stations. The routine monitoring included traveling to sites, locating and opening IG stations, removing and replacing wood monitors when necessary, and installing baits when termites were detected. With the availability of electronic sensors (ESP™ system), locating and opening of the stations became less tedious, as shown in the Chatsworth project (Getty *et al.*, 2007) and Phase III of the Armstrong project (Mullins *et al.*, 2011), but other aspects of the routine monitoring procedure remained time-consuming and costly. In 2010 when the durable bait Recruit® HD was introduced to the marketplace, the monitoring segment of the earlier monitoring–baiting protocol was removed, and baits containing 0.5% noviflumuron were placed in IG stations during the initial installation (see Chapter 11 of this volume). Field aging studies showed no significant degradation of noviflumuron in Recruit® HD after 5 years (Eger *et al.*, 2014). The current label for Recruit® HD prescribes (at least) an annual on-site monitoring, and Recruit® HD baits are replaced only when more than one-third of the bait has been consumed. With the less frequent monitoring requirement, it was hypothesized that a durable bait such as Recruit® HD can provide economically sustainable management of *C. formosanus* in a large area (Su *et al.*, 2016b).

Phase V began in 2009 after termite populations fully recovered in the park and reached the carrying capacity (Su *et al.*, 2016a). Termite activity in the park had been monitored monthly by one technician (with one or two assistants when necessary) from 808 Sentricon® stations that contained wood monitors equipped with ESP® sensors. By April 2009, 14 *C. formosanus* colonies and one *R. flavipes* colony were identified with mark–recapture protocol and microsatellite genotyping (Su *et al.*, 2016a). For Phase V, an ESP® sensor was incorporated in Recruit® HD bait to further ease the monitoring of baited stations, and in accordance with label instruction, HD baits were not removed even after the cessation of termite activity. In September 2010, when termites were detected from 71 Sentricon® stations, Recruit® HD with ESP® sensors were applied in 13 selected stations of five *C. formosanus* colonies (Fig. 15.6A). Baits were applied in a relatively small number of active stations, due in part to the limited availability of Recruit® HD baits in 2010. Following the bait application, termite activity steadily declined, and by July 2011, activities of stations belonging to five baited colonies ceased. As more HD baits became available and more sites became accessible (due to the completion of renovations to repair damages resulting from hurricane Katrina of 2005), a series of active stations were successively baited from October 2011 to July 2012, and by December 2012, termites were found in none of the

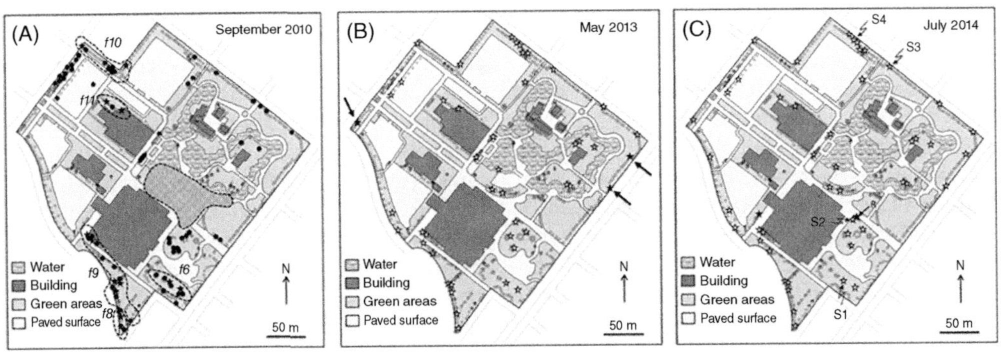

Fig. 15.6. Termite activity in Armstrong Park had been monitored by using 808 Sentricon® stations (small open circles) containing wood sensors equipped with ESP® sensors (A). In September 2010, termites were detected from 71 Sentricon® stations (solid circles), and Recruit® HD (high density, a durable bait with 0.5% noviflumuron) baits were applied in 13 selected stations (solid stars) of four *C. formosanus* colonies, *f6, f8, f9, f10* and *f11*. Dotted lines depict colony boundaries (A). After applications of Recruit® HD baits from October 2010 to July 2012, both *C. formosanus* and *R. flavipes* colonies in the park were eliminated by December 2012, and the absence of termite activity lasted 6 months until May 2013, when *C. formosanus* were found in three stations (arrows) at the park border (B), which suggested that following the eliminations of all termite colonies in the park, *C. formosanus* colonies outside the park began to invade the vacated territories in the park. Blank stars depict stations containing Recruit® HD baits without termite activity. The three colonies at the border were baited and eliminated by August 2013, and termite activity ceased until July 2014 when one station each was occupied by *R. flavipes* (solid star, arrow R) and *C. formosanus* (solid star) (C). Termites in both stations were small individuals and probably belong to incipient colonies founded by alates of 2013 or 2014. Seven interception and elimination (IE) events were recorded from stations S1–S4.

previously active 71 stations. The absence of termite activity lasted 6 months, which was the longest since the Armstrong Park project began in 1998. The results suggested that all termite colonies in the park, both *C. formosanus* and *R. flavipes*, were eliminated by the applications of Recruit® HD baits from October 2010 to July 2012. In May 2013, three stations became active with *C. formosanus*, and all of them were found at the park border (Fig. 15.6B). The lack of termite activity in any of the IG stations inside the park further confirmed that all termite colonies in the park were eliminated by December 2012, and *C. formosanus* colonies bordering the park began to invade the vacated territories in the park in May 2013. Following the application of Recruit® HD baits in the three (plus one in June 2013) newly active stations, termite activity ceased in August 2013 until August 2014, when two small incipient colonies (one each of *C. formosanus* and *R. flavipes*) were found in the park (Fig. 15.6C). These were apparently founded by alates that flew in during the swarming season of 2013 or 2012. The two stations were baited, and termite activity ceased in 2 months.

The Recruit® HD label prescribes that, following a colony elimination, baits are left in the stations (unless > 1/3 bait consumption and new baits are installed) to intercept and eliminate reinvading colony(s). Table 15.1 summarizes seven interception and elimination (IE) events observed from four IG stations (Fig. 15.6C, S1–S4) in Armstrong Park, September 2010–August 2012 (Su *et al.*, 2016b). A colony was considered eliminated when the activity ceased for > 2 months or stations were taken over by another species or colony(s) after bait consumption. Microsatellite genotyping was used to confirm colony origin (Vargo and Henderson, 2000; Owens *et al.*, 2012). Three IG stations, S1, S2, and S4, were occupied alternatively by another species following bait consumption, and in some instances invading colonies occupied the stations as soon as the baited colonies were eliminated, for example the *C. formosanus* colony that appeared in S1 in February 2012 only a month after a *R. flavipes* colony was eliminated. In S2, the *R. flavipes* colony invaded the station the same month (March 2012) that the previous occupant of a *C. formosanus* colony was eliminated (Table 15.1). Field and laboratory studies suggested that adjacent colonies were constantly interacting with each other through intersecting underground tunnels between them, and as soon as a colony is negatively affected by bait applications, the neighboring colony readily moves in to take over the vacant territory (Messenger *et al.*, 2005; Bernard *et al.*, 2017). The series of IE events were probably the result of the close inter-colony interactions.

After the cessation of one incipient colony each of *C. formosanus* and *R. flavipes* (Fig. 15.6 C) in September 2015, monthly monitoring resumed, but no termite activity was observed from the 808 IG stations for another 12 months and the project was terminated in September 2016.

Table 15.1. Date of installation, cessation of termite activity and subsequent interceptions of Recruit® HD baits in Sentricon® stations with multiple interception and elimination (IE) events[a]

Station ID	Initial HD installation (species)	1st inactive	1st interception (species)	2nd inactive	2nd interception (species)	3rd inactive
S1	Sep. 2010 (*Cf*)	Oct. 2010	Oct. 2011 (*Rf*)	Jan. 2012	Feb. 2012 (*Cf*)	Mar. 2012
S2	Nov. 2011 (*Rf*)	Dec. 2011	Jan. 2012 (*Cf*)	Mar. 2012	Mar. 2012 (*Rf*)	May 2012
S3[b]	Oct. 2011 (*Cf*)	Dec. 2011	Mar. 2012 (*Cf*)	Apr. 2012	Jul. 2012 (*Cf*)	Aug. 2012
S4	Nov. 2011 (*Cf*)	Dec. 2011	Mar. 2012 (*Rf*)	Jun. 2012	–	–

[a]*Cf*, *C. formosanus*; *Rf*, *R. flavipes*

[b]*C. formosanus* found in October 2011 and those found in July 2012 are of different colony origins according to microsatellite genotyping. Colony origin of *C. formosanus* found in March 2012 could not be identified due to insufficient number of sampled termites; however, it was considered a different colony from the one found in October 2011, and the one found in April 2012 because of > 2 months of cessation time post-eliminations.

15.6.7 College Station project, Texas, USA

This small-scale AW project was conducted in two field sites (~1 acre each) in College Station, Texas (Shults *et al.*, 2021). Each site had an open pasture adjacent to a wooded area with no structure. One hundred IG stations of Advance® Termite Bait System (BASF Corp.) were installed at 5 m equal distance for each site in 2016 to detect and monitor termite activity. Between 2016 and 2018, 12 colonies each of *Reticulitermes* species (*R. flavipes* and *Reticulitermes virginicus* (Banks)) were identified at both sites, according to microsatellite genotyping. After the colony survey in 2018, Trelona® baits (0.5% novaluron) were applied in one site, whereas the other site was used as the untreated control (blank baits). After baiting for 12 months, termite activity in all IG stations at the treated site ceased in June 2019, whereas those at untreated site remained active.

One shortcoming with the College Station project is the small size of the "area." An AW project is intended to manage subterranean termite populations to protect multiple structures in a large area, instead of the traditional structure-by-structure approach (Klassen, 2007). Because it is not uncommon to find a single-family home occupying a one-acre land in suburb or rural areas, this project resembled more of an attempt to manage multiple colonies of subterranean termites in a one-acre residential site than a typical AW project. Notwithstanding the small area, the project is worth mentioning because it is the first to use another CSI, novaluron, to eliminate multiple colonies of subterranean termites in the field.

15.6.8 Factors affecting the outcomes of AW projects

Successful AW projects listed above (Cerro Navia project, Chatsworth project, Armstrong Park project, and College Station project) share the following: (1) the project had a clear goal, i.e., to reduce or eliminate termite colonies in target areas; (2) a control measure with data to support the colony elimination claim was used to achieve the goal; and (3) a single person was put in charge of the project to carry out the daily works. No political agenda interfered with these successful AW projects. The failure with the Golden Beach project, on the other hand, can be attributed the lack of political leadership in persuading enough residents to cooperate with the project. With participation by only 35% of town residents, the project was doomed to fail from the start.

Political consideration to allow many treatment options, including those that have no field data of colony elimination (Termidor®, Premise®, Exterra®, Firsline® and Subterfuge®), to be employed was the main reason for the failure of the French Quarter project. Operation Full Stop was billed as an AW project, but in reality, it was a large-scale house-by-house treatment. The decision to allow each property owner to select their own treatment option was based more on a political consideration than a technical one. Because most property owners were not familiar with the pros and cons of available treatment options, they usually followed the recommendations of pest control firms that may use a variety of treatments, and this operation structure resulted in the use of treatments that allowed many *C. formosanus* colonies to survive. It may be understandable that a government agency such as USDA-ARS could not prohibit the use of registered and approved treatments. If the overall objective of the project was to reduce termite damage potential in a large area, however, a focused approach was needed to enforce the deployment of treatment measures capable of colony elimination. Another factor contributing to the less than satisfactory outcome of Operation Full Stop was the complicated managerial structures that involved many parties, i.e., researchers from government agencies and multiple universities, pest control firms, and property owners. Each of these parties has its own interest and agenda, and some conflict with others. As pointed out by Hendrichs *et al.* (2007), conflicting interest, complex managerial structure, and political interference often plague a large AW project.

References

Anonymous (2005) Pest Management Plan (E1285, v. 1). Prepared for the Global Environmental Facility (GEF) project, *China: Demonstration of Alternatives to Chlordane and Mirex in Termite Control*. Guangdong Entomological Institute. Global Environmental Facility; World Bank, Washington, DC. (http://www.gefonline.org/projectDetails.cfm?projID2359)

Anonymous (2007) *IPM Operation and Training Manual*. Chinese Termite Control Expert Committee, Stockholm Convention Implementation Office, Ministry of Environmental Protection, Beijing, China

Anonymous (2019) *Stockholm Convention on Persistent Organic Pollutants (POPs). Text and annexes* (revised). United Nations Environment Programme (UNEP), Geneva. Available at: http://chm.pops.int/Convention/ConventionText/tabid/2232/Default. aspx (accessed 8 September 2023)

Bernard, S., Osbrink, W. and Su, N.-Y. (2017) Response of the Formosan subterranean termite to neighboring con-specific populations after baiting with noviflumuron. *Journal of Economic Entomology* 110, 575–583. doi: 10.1093/jee/tow317

Chouvenc, T. (2018) Comparative impact of chitin synthesis inhibitor baits and non-repellent liquid termiticides on subterranean termite colonies over foraging distances: colony elimination versus localized termite exclusion. *Journal of Economic Entomology* 111, 2317–2328. doi: 10.1093/jee/toy210

Chouvenc, T., and Su, N.-Y. (2010) Apparent synergy among defense mechanisms in subterranean termites (Rhinotermitidae) against epizootic events – The limits and potential for biological control. *Journal of Economic Entomology* 103, 1327–1337.

Chouvenc, T., Su, N.-Y. and Robert, A. (2009a) Inhibition of *Metarhizium anisopliae* in the alimentary tract of the eastern subterranean termite *Reticulitermes flavipes*. *Journal of Invertebrate Pathology* 101, 130–136.

Chouvenc, T., Su, N.-Y. and Robert, A. (2009b) Cellular encapsulation in the eastern subterranean termite, *Reticulitermes flavipes* (Isoptera), against infection by the entomopathogenic fungus *Metarhizium anisopliae*. *Journal of Invertebrate Pathology* 101, 234–241.

Chouvenc, T., Su, N.-Y. and Robert, A. (2010) Inhibition of the fungal pathogen *Metarhizium anisopliae* in the alimentary tracts of five termite (Isoptera) species. *Florida Entomologist* 93, 467–469

Chouvenc, T., Su, N.-Y. and Grace, K. (2011a) Fifty years of attempted biological control of termites – Analysis of a failure. *Biological Control* 59, 69–82. doi: 10.1016/j.biocontrol.2011.06.015

Chouvenc, T., Su, N.-Y. and Robert, A. (2011b) Differences in cellular encapsulation of six termite (Isoptera) species against infection by the entomolpathogenic fungus *Metarhizium anisopliae*. *Florida Entomologist* 94, 389–397. doi: 10.1653/024.094.0302

Culliney, T.W. and Grace, J.K. (2000) Prospects for the biological control of subterranean termites (Isoptera: Rhinotermitidae), with special reference to *Coptotermes formosanus*. *Bulletin of Entomological Research* 90, 9–21.

Eger, J.E., Hamm, R.L., DeMark. J.J., Chin-Heady, E., Tolley, M.P. *et al.* (2014) Durability of a novel durable bait for control of subterranean termites (Isoptera: Rhinotermitidae): results of five-year field aging studies. *Journal of Economic Entomology* 107, 1201–1205.

Getty, G.M., Solek, C.W., Sbragia, R.J. and Haverty, M.I. (2007) Large-scale suppression of subterranean termite community using the Sentricon® Termite Colony Elimination System: A case study in Chatsworth, California, USA. *Sociobiology* 50, 1041–1050.

Guillot, F., Ring, D.R., Lax, A.R., Morgan, A., Brown, K. *et al.* (2010) Area-wide management of the Formosan subterranean termite, *Coptotermes formosanus* Shiraki (Isoptera: Rhinotermitidae), in the New Orleans French Quarter. *Sociobiology* 55, 311–338.

Hellman, J.L., Davidson, J.A. and Holmes, J. (1982) Urban integrated pest management in Maryland. In: Niemeyzk, H.D. and Joyner, B.G. (eds) *Advance in Turfgrass Entomology*. Hammer Graphics, Piqua, Ohio, pp. 31–38.

Hendrichs, J., Kenmore, P., Robinson, A.S. and Vreysen, M.J.B. (2007) Areawide integrated pest management (AW-IPM): Principles, practices and prospects. In: Vreysen, M.J.B., Robinson, A.S. and Hendrichs, J. (eds) *Area-wide Control of Insect Pests: From Research to Field Implementation*. Springer, Dordrecht, Netherlands, pp. 3–33.

Hu, X.P. (2005) Evaluation of efficacy and nonrepellency of indoxacarb and fipronil-treated soil at various concentrations and thickness against two subterranean termites (Isoptera: Rhinotermitidae). *Journal of Economic Entomology* 98, 509–517.

Huang, F.-S., Li, G.-X. and Zhu, S.-M. (1989) *The Taxonomy and Biology of Chinese Termites*. Tianze Press, Xian, China.

Hussain, A. and Tian, M.-Y. (2013) Germination pattern and inoculum transfer of entomopathogenic fungi and their role in disease resistance among *Coptotermes formosanus* (Isoptera: Rhinotermitidae). *International Journal of Agriculture and Biology* 15, 319–324.

Husseneder, C., Grace, J.K., Messenger, M.T., Vargo, E.L. and Su, N.-Y. (2003) Describing the spatial and social organization of Formosan subterranean termite colonies in Armstrong Park, New Orleans. *Sociobiology* 41, 61–65.

Ibrahim, S.A., Henderson, G. and Fei, H. (2003) Toxicity, repellency, and horizontal transmission of fipronil in the Formosan subterranean termite (Isoptera: Rhinotermitidae). *Journal of Economic Entomology* 96, 461–467.

Kard, B. 2001) Gulfport studies stay the course. *Pest Control* 69, 30–33.

Klassen, W. (2007) Introductory remarks. In: Vreysen, M.J.B., Robinson, A.S. and Hendrichs, J. (eds) *Area-wide Control of Insect Pests: From Research to Field Implementation*. Springer, Dordrecht, Netherlands, pp. vii–ix.

Lacey, L.A. and Orr, B.K. (1994) The role of biological control of mosquitoes in integrated vector control. *American Journal of Medical Hygiene* 50, 97–115.

Messenger, M.T. and Su, N.-Y. (2005) Colony characteristics of the Formosan subterranean termite in Louis Armstrong Park, New Orleans, Louisiana. *Journal of Economic Entomology* 40, 268–279.

Messenger, M.T., Su, N.-Y., Husseneder, C. and Grace, J.K. (2005) Elimination and reinvasion studies with *Coptotermes formosanus* (Isoptera: Rhinotermitidae) in Louisiana. *Journal of Economic Entomology* 98, 916–929.

Mullins, A.J., Su, N.-Y. and Owens, C. (2011) Reinvasion and colony expansion of *Coptotermes formosanus* (Isoptera: Rhinotermitidae) after areawide elimination. *Journal of Economic Entomology* 104, 1687–1697. doi: 10.1603/EC11036

Nowell, L.H., Crawford, C.G., Gilmore, R.J., Nakagaki, N., Stone, W.W. *et al.* (2009) Regression models for explaining and predicting concentrations of organochlorine pesticides in fish from streams in the United States. *Journal of Environmental Toxicology and Chemistry* 28, 1346–1358.

Olkowski, W. (1974) A model ecosystem management program. *Proceedings of Tall Timbers Conference on Ecological Animal Control by Habitat Management* 5, 103–117.

Osborne, L.S. and Landa, Z. (1992) Biological control of whiteflies with entomolopathogenic fungi. *Florida Entomologist* 75, 456–471.

Osbrink, W.L.A., Cornelius, M.L. and Lax, A. (2005) Effect of imidacloprid soil treatments on the occurrence of Formosan subterranean termite, *C. formosanus* Shiraki (Isoptera: Rhinotermitidae), in independent monitors. *Journal of Economic Entomology* 98, 2160 –2168.

Osbrink, W.L.A., Cornelius, M.L. and Lax, A. (2011) Areawide field study on effect of three chitin synthesis inhibitor baits on populations of *Coptotermes formosanus* and *Reticulitermes flavipes* (Isoptera: Rhinotermitidae). *Journal of Economic Entomology* 104, 1009–1017. doi: 10.1603/EC10217

Owens, C.B., Su, N.-Y., Husseneder, C., Riegel, C. and Brown, K.S. (2012) Molecular genetic evidence of Formosan subterranean termite (Isoptera: Rhinotermitidae) colony survivorship after prolonged inundation. *Journal of Economic Entomology* 105, 518–522. doi: 10.1603/EC11150

Pedigo, L.P. and Higley, L.G. (1992) The economic injury level concept and environmental quality. *American Entomologist* 38, 12–21.

Pedigo, L.P., Hutchins, S.H. and Higley, L.G. (1986) Economic injury levels in theory and practice. *Annual Review of Entomology* 31, 341–368.

Pimentel, D. and Levitan, L. (1986) Pesticides: Amounts applied and amounts reaching pests. *Bioscience* 36, 86–91.

Pinto, L.J. and Kraft, S.K. (2000) Action thresholds in school IPM programs. *Supplemental Materials for Integrated Pest Management – IPM Training Manual*. Maryland Department of Agriculture, Pesticide Regulation Section, Annapolis, Maryland.

Potter, D.A. (1986) Urban landscape pest management. In: Bennett, G.W. and Owens, J.M. (eds) *Advances in Urban Pest Management*. Van Nostrand Reinhold, New York, pp. 219–252.

Potter, M.H. and Hillery, A.E. (2002) Exterior-targeted liquid termiticides: an alternative approach to managing subterranean termites (Isoptera: Rhinotermitidae) in buildings. *Sociobiology* 39, 373–405.

Raupp, M.J., Davidson, J.A., Koehler, C.S., Sadof, C.S. and Reichelderfer, K. (1988) Decision-making considerations for pests causing aesthetic damage. *Bulletin of the Entomological Society of America* 34, 27–32. doi: 10.1093/besa/34.1.27

Raupp, M.J., Davidson, J.A., Koehler, C.S., Sadof, C.S. and Reichelderfer, K. (1989) Economic and aesthetic injury levels and thresholds for insect pests of ornamental plants. *Florida Entomologist* 72, 403–407.

Ripa, R., Luppicini, P., Su, N.-Y. and Rust, M.K. (2007) Field evaluation of potential area-wide control strategies against the invasive eastern subterranean termite (Isoptera: Rhinotermitidae) in Chile. *Journal of Economic Entomology* 100, 1391–1399.

Robinson, W.H. (1996) Integrated pest management in the urban entomology. *American Entomologist* 42, 76–78.

Saran, R.K. and Rust, M.K. (2007) The toxicity, uptake, and transfer efficiency of fipronil in western subterranean termites (Isoptera: Rhinotermitidae). *Journal of Economic Entomology* 100, 495–508.

Shelton, T.G. and Grace, J.K. (2003) Effects of exposure duration on transfer on nonrepellent termiticides among workers of *Coptotermes formosanus* Shiraki (Isoptera: Rhinotermitidae). *Journal of Economic Entomology* 96, 456–460.

Shults, P., Richardson, S., Eyer, P.-A., Chura, M., Barreda, H. *et al.* (2021) Area-wide elimination of subterranean termite colonies using a novaluron bait. *Insects* 12, 192. doi: 10.3390/insects12030192

Smith, J., Su, N.-Y. and Escoba, R. (2006) An area-wide population management project for the invasive eastern subterranean termite (Isoptera: Rhinotermitidae) in a low-income community in Santiago, Chile. *American Entomologist* 52, 253–260.

Stejskal, V. (2002). Inversion relationship between action threshold and economic/aesthetic injury level for the control of urban and quarantine pests. *Anzeiger für Schädlingskunde* 75, 158–160.

Stern, V.M., Smith, R.F., van den Bosch, R. and Hagen, K.S. (1959) The integrated control concept. *Hilgardia* 29, 81–101.

Su, N.-Y. (1982) An ethological approach to the remedial control of the Formosan subterranean termite, *Coptotermes formosanus Shiraki*. Ph.D. dissertation, University of Hawaii, Honolulu, Hawaii.

Su, N.-Y. (2005) Response of the Formosan subterranean termites (Isoptera: Rhinotermitidae) to baits or nonrepellent termiticides in extended foraging arenas. *Journal of Economic Entomology* 98, 2143–2152.

Su, N.-Y. (2019) Development of baits for management of subterranean termites. *Annual Review of Entomology* 64, 115–130. doi: 10.1146/annurev-ento-011118112429

Su, N.-Y. and Puche, H. (2003) Tunneling activity of subterranean termites (Isoptera: Rhinotermitidae) in sand with moisture gradients. *Journal of Economic Entomology* 96, 88–93.

Su, N.-Y. and Scheffrahn, R.H. (1987) Current status of the Formosan subterranean termite in Florida. In: Tamashiro, M. and Su, N.-Y. (eds) *Biology and Control of the Formosan Subterranean Termite*. College of Tropical Agriculture and Human Resources, University of Hawaii, Honolulu, Hawaii, pp. 27–31.

Su, N.-Y. and Scheffrahn, R.H. (1993) Laboratory evaluation of two chitin synthesis inhibitors, hexaflumuron and diflubenzuron, as bait toxicants against Formosan and eastern subterranean termites (Isoptera: Rhinotermitidae). *Journal of Economic Entomology* 86, 1453–1457.

Su, N.-Y. and Scheffrahn, R.H. (1998) A review of subterranean termite control practices and prospects for integrated pest management programs. *Integrated Pest Management Reviews* 3, 1–13.

Su, N.-Y., Scheffrahn, R.H. and Ban, P.M. (1995) Effects of sulfluramid-treated bait blocks on field colonies of the Formosan subterranean termite (Isoptera: Rhinotermitidae). *Journal of Economic Entomology* 88, 1343–1348.

Su, N.-Y., Ban, P. and Scheffrahn, R.H. (2004) Use of a bait impact index to assess the effects of bait application against populations of the Formosan subterranean termite (Isoptera: Rhinotermitidae) in a large area. *Journal of Economic Entomology* 97, 2029–2034.

Su, N.-Y., Ye, W., Ripa, R., Scheffrahn, R.H. and Giblin-Davis, R.M. (2006) Identification of Chilean *Reticulitermes* (Isoptera: Rhinotermitidae) inferred from three mitochondrial gene DNA sequences and soldier morphology. *Annals of Entomological Society of America* 99, 352–363.

Su, N.-Y., Lagnaoui, A., Wang, Q., Li, X., and Tan, S. (2012) A demonstration project of Stockholm POPs Convention to replace chlordane and mirex with IPM for termite control in China. *Journal of Integrated Pest Management* 4, 1–8. doi: 10.1603/IPM12020

Su, N.-Y., Guidry, E., Mullins, A.J. and Cotonne, C. (2016a) Reinvasion dynamics of subterranean termites (Isoptera: Rhinotermitidae) following the elimination of all detectable colonies in a large area. *Journal of Economic Entomology* 109, 809–814. doi: 10.1093/jee/tow018

Su, N.-Y., Guidry, E. and Cotonne, C. (2016b) Sustainable management of subterranean termite populations (Isoptera: Rhinotermitidae) in Armstrong Park, New Orleans, with durable baits. *Journal of Economic Entomology* 109, 1326–1332. doi: 10.1093/jee/tow051

Thorne, B.L. and Breisch, N.L. (2001) Effects of sublethal exposure to imidacloprid on subsequent behavior of subterranean termite *Reticulitermes virginicus* (Isoptera: Rhinotermitidae). *Journal of Economic Entomology* 94, 492–498.

Vargo, E.L. and Henderson, G. (2000) Identification of polymorphic microsatellite loci in the Formosan subterranean termite *Coptotermes formosanus* Shiraki. *Molecular Ecology* 9, 1919–1952.

Vargo, E. and Parman, V. (2012) Effect of fipronil on subterranean termite colonies (Isoptera: Rhinotermitidae) in the field. *Journal of Economic Entomology* 105, 523–532. doi: 10.1603/EC11155

Wagner, T. (2003) U.S. Forest Service termiticide tests. *Sociobiology* 41, 131–141.

Wood, F.E., Robinson, W.H., Kraft, S.K. and Zungoli, P.A. (1981) Survey of attitudes and knowledge of public housing residents toward cockroaches. *Bulletin of Entomological Society of America* 27, 9–13.

Yanagawa, A. and Shimizu, S. (2007) Resistance of the termite, *Coptotermes formosanus* Shiraki to *Metarhizium anisopliae* due to grooming. *BioControl* 52, 75–85.

Zungoli, P.A. and Robinson, W.H. (1984) Feasibility of establishing an Aesthetic Injury Level for German cockroach pest management programs. *Environmental Entomology* 13, 1453–1458.

16 Asian subterranean termite, *Coptotermes gestroi*

Kok-Boon Neoh[1] and Chow-Yang Lee[2]
[1]*Department of Entomology, National Chung Hsing University, Taiwan, neohkokboon@nchu.edu.tw;* [2]*Department of Entomology, University of California, Riverside, USA, chowyang.lee@ucr.edu*

Abstract

The Asian subterranean termite, *Coptotermes gestroi* (Wasmann), is a successful urban exploiter endemic to Southeast Asia. Over the past century, anthropogenic activity has spread this species far beyond its native range. Besides the Formosan subterranean termite, *Coptotermes formosanus* Shiraki, *C. gestroi* is the most economically important and invasive subterranean termite species. Despite its pest status, the lack of robust diagnostic characters in *Coptotermes* has resulted in many taxonomic challenges. Such ambiguity of taxonomical status has impeded an in-depth understanding of the species. This chapter presents the life history of *C. gestroi*, foraging and tunneling behaviors, and survival and swarming activity under different environments. This information is relevant to understand how the species thrives in the urban environment and achieves its pest status. The chapter also discusses the use of liquid termiticides and termite baiting and other methods against this species, the predicted future spread, and future research directions.

16.1 Introduction

Coptotermes gestroi (Wasmann) is an important urban exploiter. It is a highly invasive termite pest species in the urban environment (Fontes and Milano, 2002; Klangkaew *et al.*, 2002; Kirton and Brown, 2003; Kirton and Azmi, 2005; Lee *et al.*, 2007; Ab Majid and Ahmad, 2009; Albuquerque *et al.*, 2012; Arif *et al.*, 2019; Gazal *et al.*, 2019; Novita *et al.*, 2020; Huang *et al.*, 2022). Being a habitat generalist with nesting and feeding behavior plasticity, it is able to exploit a range of nest sites (from tree trunks to buildings) and food resources (from living trees to processed wood materials). *Coptotermes gestroi* also thrives at different levels of landscape disturbance (Neoh *et al.*, 2016, 2017). Coupled with fewer competitors and predators in cities relative to forests (Chiu *et al.*, 2016), this might explain why they are more commonly found in urban areas compared with forest or agricultural settings (Lee *et al.*, 2007; Takematsu and Vongkaluang, 2012; Zhang and Evans, 2017).

Advancement of transportation and international trade resulted in the spread of *C. gestroi* beyond its native range limits. *Coptotermes gestroi* has been dispersed by international movements of timber, timber products, solid wood-packing material, and termite-infested boats and vessels (Scheffrahn and Su, 2005; Scheffrahn and Crowe, 2011; Dhang, 2014; Nagaraju *et al.*, 2021). It has been found in 24 countries (Li *et al.*, 2013). *Coptotermes gestroi* has spread across

 Biology and Management of the Formosan Subterranean Termite and Related Species (eds Nan-Yao Su and Chow-Yang Lee)
DOI: 10.1079/9781800621596.0016

the Pacific Ocean to Polynesia and Micronesia and transported west across the Indian Ocean to Mauritius and Réunion Islands. It is found in the Caribbean islands and Americas. *Coptotermes gestroi* also expands northward in Taiwan, Florida, and southern China (Evans *et al.*, 2013; Huang *et al.*, 2022).

16.2 Origin, distribution, and synonym species

Coptotermes gestroi is native to Southeast Asia (Kirton and Brown, 2003; Li *et al.*, 2009). Wasmann (1896) first described *C. gestroi* using a soldier specimen. The species was first described from Myanmar (formerly Burma) and was believed to occur only in Thailand and northeast India (Roonwal and Chhotani, 1962). Holmgren (1911) described *Coptotermes havilandi* Holmgren from an alate. Kirton and Brown (2003) synonymized *C. gestroi* and *C. havilandi*, and since the latter was described after *C. gestroi*, it was assigned as the junior synonym.

The number of setae around the fontanelle has been used as a main feature to distinguish *C. gestroi* from *C. formosanus* in the U.S. and Taiwan (Scheffrahn *et al.*, 1990; Li *et al.*, 2009). Soldiers of *C. gestroi* possess a pair of setae around the fontanelle, while *C. formosanus* has two pairs of setae (Scheffrahn *et al.*, 1990). However, this feature is not a robust diagnostic character for morphological species identification, as varying numbers of setae may be found in incipient colonies (Chouvenc *et al.*, 2014). *Coptotermes* spp. in the native range, in general, have one pair of fontanelle setae, but some variations were also noted in *C. gestroi* (2–4 setae), *C. sepangensis* (2–3 setae), and *C. curvignathus* (1–4 setae) (Wikantyoso *et al.*, 2021). Instead of solely relying on the head capsule, Wikantyoso *et al.* (2021) proposed that additional characteristics, such as pronotum setae and postmentum, should be included.

The diversity of *Coptotermes* in the native range (Chouvenc *et al.*, 2016a) is far more complicated than for those in the New World (Scheffrahn *et al.*, 2015). Head capsule shape and size have been used as primary key species identifiers within the genus of *Coptotermes* (Kirton and Brown, 2003). However, Wikantyoso *et al.* (2021) reported that three extreme head-shape variations of *C. gestroi* were observed in nature, and this characteristic, to a certain extent, overlapped with those of *C. curvignathus*. This has also resulted in many taxonomic synonyms such as *Coptotermes pacificus* Light, *Coptotermes javanicus* Kemner in Indonesia, and *Coptotermes parvulus* Holmgren in India (Li *et al.*, 2013). Molecular techniques have enabled phylogenetic analyses on relatively close relationships within the *Coptotermes* genus. Over the past two decades, several Asian *Coptotermes* species were synonymized with *C. gestroi* (Chouvenc *et al.*, 2016a), including *Coptotermes vastator* Light (Yeap *et al.*, 2007), *Coptotermes cochlearus* Xia & He, *Coptotermes longignathus* Xia & He, *Coptotermes obliquus* Xia & He, *Coptotermes yaxianensis* Li, and *Coptotermes menglunensis* Tsai & Huang in China (Li *et al.*, 2011; Ke *et al.*, 2022).

This highly invasive species became established in the New World through human-aided transportation (Kirton and Brown, 2003). *Coptotermes gestroi* was first reported from the New World in the cities of southeast Brazil: Rio de Janeiro in 1923, and Santos, Sao Paulo in 1934 (Fontes and Milano, 2002). The spread of *C. gestroi* was later reported in Marquesas Island, French Polynesia in 1932, the Lesser Antilles, Caribbean islands in 1937, Mauritius and Réunion Island in 1936 and 1957, respectively, in Indian Ocean Islands (Evans *et al.*, 2013). DNA sequence comparison of mitochondrial genes suggested their origin of invasion in southeast Brazil to be Malaysia and Thailand (Martins *et al.*, 2010). In 2010, *C. gestroi* infestations were reported in three new localities in Campo Grande (state of Mato Grosso do Sul) in west-central Brazil, and Itajaí (state of Santa Catarina) and Porto Alegre (state of Rio Grande do Sul) in southeast Brazil via land transportation (Martins *et al.*, 2010). In 2000, *C. gestroi* was reported on the west coast of North America (Manzanillo, Colima, Mexico, 19° 02′ N, 104° 19′ W) (Ferraz and Méndez-Montiel, 2004).

In China, Xia and He (1986) first described *C. gestroi* (previously known as

C. obliquus) in the Hainan province of China. The species was found in Yunnan province (Huang *et al.*, 2000). It is believed that the distribution of *C. gestroi* also may have expanded to Guangxi province (Li *et al.*, 2011). In Taiwan, *C. gestroi* was first recorded in Kenting, Taiwan, in 2003 (Tsai and Chen, 2003), although it is now suspected that both *C. gestroi* and *C. formosanus* may have been in existence in Taiwan for more than 100 years (Li *et al.*, 2010a). The distribution of *C. gestroi* populations was known to be restricted to the Tropic of Cancer to 23° North (Su *et al.*, 2017). Until recently, the distribution of *C. gestroi* has been expanding to the northernmost region of Taiwan (25° N) (Su *et al.*, 2017; Huang *et al.*, 2022). Population structure analysis demonstrated that *C. gestroi* populations in Hainan and Taiwan were likely introduced from the Philippines (Li *et al.*, 2009; Yeap *et al.*, 2011; Li *et al.*, 2012) and shared a similar genetic structure with those in Hawai'i (Li *et al.*, 2009). However, another study using the cox1 gene revealed that populations from China were closely related to populations from the eastern and southern coasts of Peninsular Malaysia (Cheng *et al.*, 2014).

Coptotermes gestroi (previously known as *C. vastator* in early records in Hawai'i) was first intercepted in Hawai'i in 1918 in a shipment of banana stumps from Manila, Philippines (Roonwal and Chhotani, 1962). It was later found infesting a single house in the Kapahulu neighborhood of Honolulu, Oahu, Hawai'i, in 1963 (Weesner, 1965). *Coptotermes gestroi* was not reported in Hawai'i again until 1999 when it infested two residential structures in the former military housing community of Kalaeloa, a structure nearby Ewa Beach, seagrape trees (*Coccoloba uvifera* [L.] L.) at Ewa Beach Park, and a building at Hickam Airforce Base, all on the island of Oahu (Woodrow *et al.*, 2001). According to the last survey in 2012, *C. gestroi* alates were collected only in Kalaeloa (Tong *et al.*, 2017). However, pest control professionals subsequently found two infested homes in nearby Makakilo (J.K. Grace, Hawai'i, 2023, personal communication).

In Florida, *C. gestroi* was first reported in Miami in 1996, where it was collected from a storefront and a church (Su *et al.*, 1997). It was detected in Key West in 1999 and Lauderhill, Broward County, Florida, in 2001 (Scheffrahn and Su, 2005). The *C. gestroi* colony in Key West was established via shipborne and land transport from the West Indies (Scheffrahn and Su, 2005; Scheffrahn and Crowe, 2011). The samples from Key West likely originated from the indigenous populations in Malaysia (Jenkins *et al.*, 2007; Zhang and Evans, 2017). Termite records from 1990 to 2015 in South Florida showed that its distribution has since expanded (Chouvenc *et al.*, 2016b). In 2023, *C. gestroi* was found further north in Tampa (27.9506° N), Florida (PMP, 2023). For the most current distribution of *C. gestroi* in Florida, the information can be found at: https://flrec.ifas.ufl.edu/termites-in-florida/termite-distribution/

Coptotermes gestroi was discovered in Lautoka and Labasa in Fiji in late 2009. It was believed to have been introduced via infested wooden packaging materials (crates and pellets) from Asian countries 20–30 years before its discovery in 2009 (Chand *et al.*, 2018). The *C. gestroi* infestation was said to be treated using a fipronil termite bait application (Padayachi *et al.*, 2021). It remains unknown whether the colonies were successfully eliminated.

C. gestroi was first reported in Europe (Italy) in 2011 from a yacht that previously visited Panama, Bahamas, Barbados, and Jamaica from 2007 to 2009. However, no other infestations in the surrounding areas were reported and it is not known to be established in Italy (Ghesini *et al.*, 2011). Finally, *C. gestroi* was accidentally imported with plant material in a tropical garden in Germany (Plarre, 2018).

The distribution of *C. gestroi* is in the latitude range of 27.95° N to 26° S. However, it is not uncommon to find dead *C. gestroi* specimens collected or intercepted beyond the northernmost regions. For instance, Jenkins *et al.* (2007) collected desiccated samples in Cleveland, Ohio (41° N) that appeared to have been transported from

Singapore. Venkatesan *et al.* (2021) reported that imagos of *C. gestroi* were collected from wooden packaging imported from Harrisonburg, Virginia (38° N), to Southern India. However, it is questionable that the samples could have originated from Virginia. Unfortunately, no details were provided in the publication other than the shipment's origin.

16.3 Life history of *C. gestroi*

A new colony typically begins after a nuptial flight event. After landing on the ground and shedding their wings, the search for a partner begins. Females lead their partners to a nest site where mating occurs. The nest site selection is not random. Alates actively searched for a suitable soil substrate, and it has been shown that a sand–soil mixture was preferred over pure sand. The number of dealated termites also increased with the preferred soil substrate (Lima *et al.*, 2006). For a detailed illustration of the *Coptotermes* life cycle, refer to Fig. 4.2 in Chapter 4 of this volume.

The percentage of successful colony establishment in the laboratory was ~50% (Ferrax and Cancello, 2004). The quality of paired females and males also determines the success rate of colony establishment. Chouvenc (2019) demonstrated that successful incipient termite colonies had larger reproductive pairs than those in the failed colonies, indicating that high-quality alates with higher weight may have a slight advantage in increasing colony fitness at the incipient stage. The first batch of eggs (1–3 eggs per day) are laid by the primary queen within 4–14 days, depending on the types of substrate and environment conditions where the colony is found (Costa-Leonardo and Barsotti, 1998; Ferraz and Cancello, 2004).

The **egg** incubation period ranges from 30 to 70 days (Costa-Leonardo and Barsotti, 1998; Ferraz and Cancello, 2004). The first and second **larval instars** each take ~13 days, while the third instar takes ~35 days until the successive molt. The first and second instar larvae are immobile and fed by the king and queen. Female and male reproductives deplete most of their internal resources within 9 months (Chouvenc, 2019). The development of the termite, whether it will undergo the neuter or imaginal lines, is determined after the second molt (Ferraz and Cancello, 2004).

The **worker** caste has five instars (Barsotti and Costa-Leonardi, 2005). The third instar and onwards are functional true workers with sclerotizing head capsules, which actively participate in brood caring, food searching, and colony maintenance with trail pheromones (Ferraz and Cancello, 2004). In an incipient *C. gestroi* colony, the first batch of workers plays a pronounced role in ensuring colony productivity and survival. A minimum of ten workers would ensure the colony's growth (Chouvenc and Su, 2017a). Any mishap to the first-batch workers would have a detrimental effect on overall colony fitness, because raising a large brood is costly to a small worker force (Chouvenc and Su, 2017a; Chouvenc, 2022).

Soldiers of *C. gestroi* collected from a mature field colony had 14 (Chouvenc and Su, 2014) or 15 antennal segments (Ferraz and Cancello, 2004) and were twice the size of those in the laboratory with 12 antennal segments, which are derived from 2nd instar larvae in an incipient colony (Ferraz and Cancello, 2004). The presence of nanitic soldiers (defined as soldiers from incipient colonies) is due to accelerated development in incipient colonies, in which the nanitic soldiers have two fewer molting events than those produced in mature colonies (Chouvenc and Su, 2014). *Coptotermes gestroi* soldiers from mature colonies characteristically display a single pair of setae at the base on the fontanelle (Scheffrahn and Su, 2005), but nanitic soldiers from incipient colonies display a varying number and distribution of setae (Chouvenc *et al.*, 2014). Chouvenc *et al.* (2017a) examined the number of setae of soldiers and generated an asymmetry index. The index considers three aspects of the soldier setae asymmetry: the departure from the symmetry of paired setae; unpaired setae from expected pairs; and extra setae from the expected pairs. Their findings revealed that, unlike the soldiers in mature colonies, soldiers in incipient colonies showed asymmetry in the setae distribution area.

Nymphs develop into alates after undergoing six instars (Albino and Costa-Leonardo, 2011). The young nymphs have light brown and unpigmented eyes. The nymphs of *C. gestroi*, especially the fourth and fifth instars, actively participate in foraging activity throughout the year. Nymphs are produced primarily in mature colonies for dispersal (Albino and Costa-Leonardo, 2011). Costa-Leonardo *et al.* (2004) reported that orphaned young colonies (1–3 years old) produced nymphs more often than orphaned incipient colonies.

Some nymphs develop into brachypterous **neotenics** and do not disperse but remain within the parental queen-less colony as replacement reproductives after the primary(s) is dead (Costa-Leonardo *et al.*, 2004). These individuals may derive from the first nymphal instar (N1) (Costa-Leonardo *et al.*, 2004). However, none of these neotenics observed in either study was reproductively functional (Costa-Leonardo *et al.*, 2004; Chouvenc *et al.*, 2015a). Non-functional neotenics' ovaries contain ~10 ovarioles. The oocyte development of each neotenic varies; some individuals reach the oocyte development that resembles that of the nymphs, in which only previtellogenic oocytes were observed. On the other hand, some non-functional neotenics have vitellogenic oocytes of V stages, a stage before reaching a terminal stage (Costa-Leonardo *et al.*, 2022). This variation in oocyte maturation in non-functional neotenics could be due to: (1) non-functional neotenics derived from an early nymphal instar; or (2) the delayed maturation of the oocyte in a queen-right colony (Costa-Leonardo *et al.*, 2022). It is suggested that the presence of non-functional neotenics could be a vestigial trait and that they could regain their reproductive functionality in the presence of suitable resources (Chouvenc *et al.*, 2015a).

Caste development fates are determined based on environmental factors experienced during postembryonic development, such as nutritional and seasonal factors (Noirot, 1991; Fei and Henderson, 2002; Liu *et al.*, 2005; Scharf *et al.*, 2007). These stimuli are transferred to totipotent larvae through social interactions such as allogrooming or trophallactic exchange between workers (Tarver *et al.*, 2011). During the incipient phase of colony development, the worker force is extremely vital. As more workers are produced, more workforce would be available to provide for the primary reproductives, the brood, and nest maintenance. The soldier is equally important to maintain the caste proportion as well as to initiate their very first foraging activity. The first presoldier could appear in an incipient colony at 2 months on a filter paper (Ferraz and Cancello, 2004). However, soldier production was delayed by ~6 months when an incipient colony was initiated in the sawdust (Ferraz and Cancello, 2004). Substrate types affect the energy investment in nest maintenance and tunneling; hence in situations where more investment is required in nest construction and maintenance, soldier production will be affected. When facing worker force constraint and colony investment in soldiers is costly, this could reduce worker numbers and overall growth of the colony. This may explain why the success rate of incipient colonies is low (< 50%) (Chouvenc *et al.*, 2015b). In short, the optimal energy allocation for caste investment in the first offspring and nest maintenance by the incipient colony is vital to the colony's success during the first year and its subsequent growth rate. For a detailed discussion on developmental pathways and caste determination mechanisms in *Coptotermes*, please refer to section 4.3 in Chapter 4.

16.4 The reproductive caste

The king of *C. gestroi* has testes that consist of ~5–9 lobes per testis, and the lobes significantly increase in size as the king ages. The number of testicular lobes is difficult to determine in nymphs and alates, as they are not well-developed, and the lobes are compactly grouped. Similarly, the testicular development of neotenics is characterized by an intermediate stage between alates and kings. Unlike those in the king, the testicular lobes of neotenics are not well defined. In nymphs, seminal vesicles are small, and spermatozoa are absent until the third stage of the nymphal instar, indicating that spermatogenesis may begin as early as in 3rd instar nymphs. The spermatozoa are present in the seminal vesicles after the imaginal molt in alates or neotenics. The presence of spermatozoa in the seminal vesicles of male neotenics indicates their reproductive capability (Laranjo *et al.*, 2019).

The male alates possess a low volume of protein granules in the secretion present in the lumen of the seminal vesicles. The protein granules progressively increase in 6-month-old kings and 1-year-old kings. These protein secretions are composed of glycoproteins that may provide nutrition for the spermatozoa, reduce the receptivity or attractiveness of females, possess antimicrobial properties, and/or improve the fecundity of females (Laranjo *et al.*, 2019).

On the other hand, 4th instar nymphs, neotenics, alate females, 80-day-old and 2-year-old queens of *C. gestroi* showed variation in the number of ovarioles and oocyte development. In the 4th nymphal instars, 4–5 ovarioles were strongly compacted (Costa-Leonardo *et al.*, 2022). In non-functional neotenics, ovaries were composed of about ten ovarioles. Ovariole number in alates ranged from 4 to 5 per ovary (Costa-Leonardo *et al.*, 2022). Oocyte maturation in alates varied among individuals. Some females exhibited only stage V vitellogenic oocytes, similar to those observed in neotenics, whereas other alates showed stage VI terminal oocytes, resembling those from functional queens (Costa-Leonardo *et al.*, 2022). This implies that some alates may not fully mature in reproduction during the swarming event. However, ovarian maturation may be facilitated by a male and/or quality food availability (Brent and Traniello, 2001a; Brent and Traniello, 2001b).

A new (80-day-old) queen of *C. gestroi* may possess 5–6 ovarioles with fully developed vitellogenic oocytes, while a 2-year-old queen may possess as many as 35 ovarioles (Costa-Leonardo *et al.*, 2022). The secretion of proteins and polysaccharides in the spermathecal lumen is observed as early as the alate stages, even before copulation; however, it increases in mated queens (da Silva and Costa-Leonardo, 2022). This indicates spermatheca activation occurs before pairing, but its secretory activity intensifies to nourish and provide energy for the stored spermatozoa after mating (da Silva and Costa-Leonardo, 2022).

16.5 Relationship with endosymbiont

Coptotermes gestroi harbors five species of specialized wood-eating protists, namely *Pseudotrichonympha leei* del Campo & Keeling, *Holomastigotoides bigfooti* Jasso-Selles & Gile, *H. batututi* Jasso-Selles & Gile, *Cononympha skunkapei* Jasso-Selles & Gile, and *Con. monstrummogolloni* Jasso-Selles & Gile (Jasso-Selles *et al.*, 2020). The protist community structure in the hindgut of *C. gestroi* shares a high similarity with that in sympatric *C. formosanus* (Jasso-Selles *et al.*, 2020). The protist abundance of *C. gestroi* kings and queens peaks on either day 35 or 60 during the biparental care period and progressively decreases once the first batch of workers emerges. By day 400, the protozoan abundance of kings and queens is either similar to or less than that observed within unmated dealates (Velenovsky *et al.*, 2021). This indicates the dynamic nature of protist abundance in *C. gestroi* reproductives. In addition, the protozoan community in alates is often incomplete, but biparental colony foundation often provides sufficient mutualist inoculum to the incipient colony, as each protist species needs to be present in only one of the parents to be transmitted to the next generation (Velenovsky *et al.*, 2023).

Symbiotic protists are often perceived as the main endosymbionts for lignocellulosic biomass degradation. The contribution of lower termites' free-living gut microbial flora in the decomposition of lignocellulosic biomass is under-studied. Do *et al.* (2014) employed Illumina-based de novo sequencing to determine the free-living bacterial community in the gut of *C. gestroi* and to mine genes related to lignocellulose degradation from the gut bacteria. They found ~1,460 species with 12 most abundant bacterial orders, namely Spirochaetales,

Lactobacillales, Bacteroidales, Clostridiales, Enterobacteriales, Pseudomonades, Synergistales, Desulfovibrionales, Xanthomonadales, Burkholderiales, Bacillales, and Actinomycetales. The metagenome of the termite gut microbiota contained 125,431 protein-coding sequences, of which 36,477 had a predicted enzyme classification and 65,536 had a functional prediction. The predicted functions of gut bacteria may be related to carbohydrate metabolism, cellulose, and hemicellulose degradation (Do *et al.*, 2014). In particular, endo-β-1,4-glucanase and β-glucosidase were the most crucial glycoside hydrolase families related to lignocellulose degradation in worker and alate castes (Leonardo *et al.*, 2011). Some exo-glucanases, endo-glucanases, and classical xylanases associated with protists were also found in *C. gestroi* (Franco Cairo *et al.*, 2016). A set of pro-oxidant, antioxidant, and detoxification enzymes (PADs) is also found in termites. Among the PADs, aldo–keto reductases (AKRs) were highly expressed in the foregut of worker *C. gestroi* (Franco Cairo *et al.*, 2016; Tramontina *et al.*, 2017) and are likely involved in the digestion of a lignocellulosic diet (Tramontina *et al.*, 2017).

The parasitic sporozoan class Gregarinidea (or Gregarinea) can also be readily found in *C. gestroi*, with a 0.6–3.3% prevalence in foraging workers (Costa-Leonardo *et al.*, 2008). Gregarines occur as cyst(s) in the digestive system of *C. gestroi*. Infected workers generally have higher weights than those not infected, but no evidence showed that gregarines affect colony fitness (Costa-Leonardo *et al.*, 2008).

16.6 Food searching

Workers of *C. gestroi* first initiate food exploration. When food is located, workers return to the nest, laying a trail for recruiting nestmates to the food source at speed onefold higher than during food exploration (Arab *et al.*, 2012). More soldiers are recruited along with the workers to the food source during the first phase of the recruitment (Arab *et al.*, 2012). To date, (*Z*,*Z*,*E*)-3,6,8-dodecatrien-1-ol has been identified as the common component in the trail pheromone of workers of *C. gestroi* (Arab *et al.*, 2004). The different levels of workers' response toward trail pheromones between the exploration and recruitment phases might be titer dependence.

There have been a number of studies comparing tunneling patterns of different *Coptotermes* species. All these studies, however, were carried out in the laboratory and may not reflect the actual field situation. Unlike the laboratory experiments that were conducted under controlled conditions and standardized soil type, there is a multitude of factors such as soil density, soil moisture, soil type, underground obstacles, presence of rock barrier, etc., that potentially could influence the tunneling activity of the termites and how they find their food sources in the field. The age and the state of the colony would also influence the colony's food search behavior.

Yeoh and Lee (2007a) compared the tunneling activity of three native *Coptotermes* species of Southeast Asia, namely *C. gestroi*, *C. curvignathus* Holmgren, and *C. kalshoveni* Kemner. In large Petri dish bioassays, *C. gestroi* constructed moderately branched tunnels compared with *C. curvignathus*. Both *C. gestroi* and *C. curvignathus* made an average of six attack routes to access the test wood block. By contrast, Yeoh and Lee (2007a) reported that *C. kalshoveni* showed the least tunneling activity, in which most tunnels radiated from the point of origin and were straight with less branching. Only two attack routes were detected for *C. kalshoveni*. The tunneling pattern of *C. kalshoveni* is similar to that of *R. flavipes*, which built long and narrow primary tunnels (Puche and Su, 2001). However, the wood consumption rate of *C. gestroi* was one-third lower than that of *C. kalshoveni* (Yeoh and Lee, 2007a). The high tunnel branching and comparatively low *in situ* wood consumption rate suggest that rather than focusing on a foraging site, *C. gestroi* workers would use their foraging effort between foraging sites to locate more potential feeding sites to spread the risk of foraging (Lima and Costa-Leonardo, 2014). Unlike

C. gestroi, *C. formosanus* foragers display sequential foraging behavior, focusing their foraging efforts on the first food they encounter (Delaplane and La Fage, 1987).

Hapukotuwa and Grace (2012b) reported that *C. gestroi* constructed narrow, highly branched tunnels. Tunnels of *C. gestroi* showed more spatial dispersion than those of *C. formosanus*, and this species constructed a larger number of tunnels compared with *C. formosanus*. Primary tunnels randomly branched out from the point of origin. The tunnels were distributed in a way that maximized the angular distance between the tunnels and, thus, minimized the placement of tunnels in the explored area to avoid food search redundancy (Arab and Costa-Leonardo, 2005). This strategy may allow *C. gestroi* workers to maximize the *in situ* local food search. Extensive branching is probably an energy-saving strategy to increase food search and transportation efficiency (Lee *et al.*, 2007). On the contrary, Su and Lee (2023) investigated the tunnel geometry between *C. gestroi* and *C. formosanus* and found that *C. formosanus* constructed more abundant and complex tunnel networks and intercepted more woods than the latter species, which uses the distance searching strategy by constructing linear and long primary tunnels. It was suggested that this explains why none of the in-ground stations were intercepted by *C. gestroi* (Su *et al.*, 2023).

16.7 Factors affecting the tunneling behavior of *C. gestroi*

16.7.1 Temperature and moisture

Temperature influences the tunneling activity of *C. gestroi* (Cao and Su, 2014). The mean number of tunnels and mean total tunnel length increased with elevated temperature (15–20°C) (Arab and Costa-Leonardo, 2005). More tunneling activity of *C. gestroi* occurred at higher sand moisture levels (Arab and Costa-Leonardo, 2005; Kulis *et al.*, 2008). Feeding and survival of *C. gestroi* were significantly affected when termites were kept in a confined area with a 5% soil moisture level for 7 days (Kulis *et al.*, 2008). Hence, high soil moisture can be considered conducive to termite infestation (Su and Puche, 2003).

Forager *C. gestroi* workers are highly resistant to water stress. Janei *et al.* (2015) restricted water access by *C. gestroi* workers for periods of 3, 15, and 30 h. They observed that although forager workers lost body mass over time, no mortality occurred after the workers were rehydrated at 30 h. Wong and Lee (2010a) used twelve Petri dishes, which were interconnected to one another with Tygon tubes containing sand with different moisture levels of 5, 10, 15, 20, and 25% wt:wt. Although intense tunneling activity was found in dishes with a 20% moisture level, *C. gestroi* also tunneled extensively in the dry sand (0% moisture content) (Wong and Lee, 2010a). The average tunneling distance in the dish with 0% moisture content was statistically similar to that of the sand with moisture content up to 15% (Wong and Lee, 2010a). The occurrence of *C. gestroi* workers in dry sand indicated that *C. gestroi* could survive and forage well under adverse conditions, at least for a short period. *Coptotermes gestroi* workers possess many thin trichomes on the surface of the hypopharynx to facilitate water intake (da Silva and Costa-Leonardo, 2017). It is known that *C. formosanus* workers moisten dry soil using water from their salivary reservoirs (water sacs) and transport the moist soil to increase humidity (Gautam and Henderson, 2014). Santos *et al.* (2010) reported that soil moisture was negatively correlated with the damage rate of *C. gestroi* in the field. They showed that feeding at wood stakes by *C. gestroi* was negatively correlated with relative humidity, soil moisture, and rainfall.

16.7.2 Structural anomaly

Another factor that affects *C. gestroi* tunneling behavior is anomaly in the tunneling surface when obstacles are present. In a laboratory test in which twisted wires were introduced into bi-dimensional glass arenas as obstacles, *C. gestroi* foragers displayed more vigorous initial tunneling behavior than in arenas without obstacles (Lima and Costa-Leonardo, 2012). *Coptotermes gestroi* foragers constructed long primary tunnels along the twisted wire. The result is consistent with *C. formosanus*, which, when foragers encountered the edge of the arena, concentrated tunneling at the edge until the arena was circumnavigated (Campora and Grace, 2001). Constructing the tunnel along the anomaly would facilitate the tunnel excavation process. However, less effort was put into secondary, tertiary, or quaternary tunnel construction to increase their *in situ* foraging area (Lima and Costa-Leonardo, 2012). This thigmotaxic behavior is pervasive in buildings, where pipes, wires, and concrete may function as physical cues for all subterranean termite foragers, guiding foragers to locate distant food resources.

16.7.3 Preformed tunnel

In the presence of a preformed tunnel, Lima and Costa-Leonardo (2012) documented that the construction of the primary tunnel was less attempted, but more numerous and lengthy quaternary tunnels were observed compared with arenas without preformed tunnels. This may indicate that rather than constructing a new primary tunnel, *C. gestroi* workers would utilize the existing tunnels to access distant foraging sites while increasing the *in situ* foraging network to attain equivalent levels of foraging efficiency.

16.7.4 Termite carcasses

In *C. gestroi*, tunnel construction by workers was affected by the presence of soldier carcasses in tunnels. More primary tunnels were constructed to divert their foraging route from contact with carcasses. Such increased effort in energy investment in the initial tunnel is apparent when the workers encounter conspecific soldiers, compared with encountering heterospecific soldiers (Lima and Costa-Leonardo, 2012). Such observation may have important implications in termite management programs, particularly when fast-killing insecticides are used, which mainly target foraging workers and soldiers. The rapid onset of the mortalities of workers and soldiers in the tunnels will cause the termites to avoid contact with nestmate carcasses by sealing off the tunnel (Chouvenc, 2018). The condition may trigger more primary tunnel construction and lead to more locations of infestation in a given premises.

16.7.5 Food

Besides the impact of structural anomalies (Campora and Grace, 2001), soil moisture gradient, and temperature (Arab and Costa-Leonardo, 2005), the presence of wood on tunnel distribution and pattern of *C. gestroi* is negligible (Campora and Grace, 2001; Puche and Su, 2001; Arab and Costa-Leonardo, 2005). This contrasts with that observed with the family Termitidae (Araujo *et al.*, 2011).

Previous findings (Campora and Grace, 2001; Arab and Costa-Leonardo, 2005; Yeoh and Lee, 2007a; Hapukotuwa and Grace, 2012b) indicate heterogeneity in the tunneling behavior of *Coptotermes* species. This may be explained by the difference of mechanoreceptive sensilla in *C. gestroi* workers, in which the worker contains numerous short-peg sensilla (tIII) but a low number of long flexible-peg sensilla (tIb, tIc) as compared with *C. formosanus*. The lack of long flexible-peg sensilla may affect the sensitivity of *C. gestroi* to stimuli (Wikantyoso *et al.*, 2022). Other unexplored mechanisms underlying variations in the tunneling behavior of *C. gestroi*, such as the basal metabolic rate of termites and colony structure, could be involved.

16.8 Environmental parameters affecting foraging and survival

Coptotermes gestroi is active at 21.9–39.5°C (Patel *et al.*, 2019), and its wood consumption rate peaks at 35°C (Cao and Su, 2016). *Coptotermes gestroi* is sensitive to cold, cannot survive at 10°C, and has the narrowest temperature range compared with *C. formosanus*, *R. speratus*, and *R. flavipes* (Cao and Su, 2016). Therefore, the foraging activity of *C. gestroi* is dynamic and changes over time according to extrinsic factors such as climate effects. In regions close to subtropics, the temperature is the major climatic factor affecting the foraging activity of *C. gestroi* colonies. In southeastern Brazil (22° 23′ S), more wooden stakes were attacked in summer than in winter (Arab *et al.*, 2005). Similarly, in Hawai'i (21° 19′ N), *C. gestroi* was generally more active in the late spring and summer months (Hapukotuwa and Grace, 2012a). Precipitation and soil moisture are not major factors driving the foraging activity of *C. gestroi* in the region (Arab *et al.*, 2005; Hapukotuwa and Grace, 2012a). By contrast, in southeast Asia, where the temperature is constant, high precipitation during the rainy season triggers the foraging activity of *C. gestroi* (Sornnuwat *et al.*, 1996).

While climatic factors account for its foraging pattern, natural competition with other termites determines its foraging populations. The foraging population of *C. gestroi* is varied and dependent on locality. A mature colony may contain sizable foraging populations in a locality where *C. gestroi* is predominant in the urban area. For instance, the foraging population of *C. gestroi* may contain 0.11–4.18 million workers, with foraging territory of 172.5–5235 m^2 (Sornnuwat *et al.*, 1996; Arab *et al.*, 2005; Ab Majid and Ahmad, 2015). On the contrary, the foraging population size is relatively small in some localities where both *C. gestroi* and *C. formosanus* colonies coexist. Chiu *et al.* (2016) reported that *C. gestroi* populations at a forest site in Taiwan contained 0.12–0.20 million individuals. Competition for food with other local termite species and predator pressure from ants have been suggested to contribute to reduced foraging populations (Chiu *et al.*, 2016). These results are in line with those of Uchima and Grace (2003), who reported that the foraging population of *C. gestroi* colonies in Kalaeloa, Oahu, ranged from 0.19 to 0.68 million individuals with foraging territories encompassing 10.5–287.2 m^2. However, the authors dismissed that direct competition from *C. formosanus* limits the foraging population size and colony boundary. Habitat suitability for *C. gestroi* may be a factor in constraining its population size, colony boundary, and distribution (Uchima and Grace, 2003; Chiu *et al.*, 2016; Tong *et al.*, 2017).

In both its native range and introduced regions, *Coptotermes gestroi* is known to attack and kill live urban trees (Lee, 2014; Zorzenon and Campos, 2015; Chiu *et al.*, 2016; Chouvenc and Foley, 2018). In Malaysia, Lee (2014) reported that *C. gestroi* is highly destructive to urban trees that were planted for beautification of urban landscapes, with infestation rates as high as 87.5% of the trees surveyed. Similar observations were made in its introduced region. Chouvenc and Foley (2018) surveyed and reported *C. gestroi* caused feeding damage to a wide variety of trees in southeastern Florida, particularly the Slash pine, *Pinus elliottii* Engelm.

16.9 Agonistic behavior

Agonistic behavior is an aggressive response when encountering non-nestmates, unrelated conspecifics, and heterospecifics. The behavior usually results in acceptance or rejection by killing unrelated individuals. *Coptotermes gestroi*'s agonistic behaviors, such as head-drumming and jerking, are important (Hertel *et al.*, 2011) to maintain colony integrity and increase colony fitness. This will indirectly avoid various costs associated with food competition, intrusions by conspecifics, the transmission of parasites and disease, and reproduction by the intruders. The outcome of competition may affect the distribution and predominance of *C. gestroi*. *Coptotermes gestroi* is highly aggressive in its native range (Wong and Lee, 2010b) compared with those in the introduced regions (Uchima and Grace, 2009), although another study in the introduced region reported the opposite outcome (Li *et al.*, 2010b).

There is a caveat in these all-Petri-dish bioassays as they may not reflect the natural situation. These bioassays usually force confrontation among termites. However, there have been contrasting results when behavioral assays were performed using foraging arenas, in which termites may display a series of behavioral repertoires to avoid injury and death from direct combat. For instance, Li *et al.* (2010b) reported that tunnel interceptions are common for *C. gestroi* when paired in arenas with *C. formosanus*. Both species quickly buried the connected tunnel at multiple locations to minimize agonistic interactions. This may explain contradicting findings of Uchima and Grace (2009), who demonstrated that in the foraging competition experiment, neither *C. gestroi* nor *C. formosanus* monopolized a food source or affected each other's feeding rate. In contrast, when *C. gestroi* was introduced to *Heterotermes tenuis* (Hagen), Lima and Costa-Leonardo (2006) showed that *C. gestroi* was dominant in occupying the foraging territory, suggesting a high efficiency of this introduced species in exploring food sources, owing to its higher soldier-to-worker ratio.

In terms of intraspecific encounters, *C. gestroi* colonies from Hawai'i displayed a low level of aggression (< 16% mortality) in confrontation using the Petri-dish bioassays (Uchima and Grace, 2009). The result corroborates that of Chouvenc and Su (2017c) that *C. gestroi* laboratory colonies started by alates collected from Florida showed no sign of agonism when two colonies were paired. This lack of aggregation in the *C. gestroi* collected from Hawai'i and Florida could be due to reduced genetic diversity, resulting from a genetic bottleneck in these introduced regions.

The lack of intraspecific agonism among *C. gestroi* workers may create a precondition for two independent colonies to fuse. When 2-year-old separate laboratory colonies of *C. gestroi* were merged, the colonies were fused (Guaraldo and Costa-Leonardo, 2009). Gordon *et al.* (2020) reported that the younger colonies had a less-defined cuticular hydrocarbon identity, hence allowing a high acceptance threshold for non-nestmate conspecifics. However, not all foreign reproductives were fully accepted. About 80% of conspecific female reproductives of *C. gestroi* were accepted, while only one male reproductive managed to survive after the colony merged, indicating that the host colony identified the conspecific reproductives as foreigners. By contrast, the host colonies fully accepted the foreign reproductives when the host's biological parents were absent (Guaraldo and Costa-Leonardo, 2009).

Accepting a conspecific mature colony may not always be as straightforward as cue recognition between sterile castes. In termites, besides the similarity of template cues between the colonies, conspecific colony acceptance is determined by the cost of accepting a conspecific colony to the overall colony fitness. Although the interpretation of the link between the high affinity of colony fusion and genetic bottleneck experienced by populations in the introduced region provided by Guaraldo and Costa-Leonardo (2009) and Uchima and Grace (2009) may be correct, some important issues remain obscure. An intruding colony is rejected when the cost of accepting a foreign reproductive is greater than the benefit of accepting a conspecific reproductive. Two scenarios may cause conspecific colony rejection, the first being when the intruding colony has a higher nymph ratio than the host colony. This is because raising nymphs to the alate stage is energetically costly, and the host colony may not be ready to bear the additional cost (Matsuura and Nishida, 2001). Second, foreign reproductives are likely to be rejected when the host colony contains nymphs, because reproduction with replacement reproductives produced within the colony allows gene propagation (Neoh *et al.*, 2012a). This implies that the acceptance of a conspecific colony in the mature stage may have a temporal or seasonal association.

16.10 Swarming activity

Swarming is influenced by several ecological factors. Temperature, air humidity, rainfall, light intensity, wind velocity, atmospheric pressure, and electrical properties in the atmosphere play roles in initiating flight activity (Nutting, 1969). The alate flight season of different termite species is region-specific.

In its native range of Southeast Asia, *C. gestroi* flight activity occurs throughout the year, with peak activity between January and June (Neoh and Lee, 2009b). Most flights were confined at atmospheric pressures of 1,009–1,010 hPa, 27–28°C, and 83–84% relative humidity (RH). In the Indian subcontinent, *C. gestroi* alates swarmed at the temperature range of 21–44°C and 42–90% RH in June–August (Nasir and Akhtar, 2011; Harit *et al.*, 2014). In the southern hemisphere, swarming occurred at 17–3°C, 35–69% RH, and 0–15 mm precipitation. Flight activity occurred almost all year, peaking from August to October at the beginning of spring (Ferraz and Cancello, 2001). In all situations, the swarming activities occurred during the rainy season, but rain may not necessarily be required during the flight day (Ferraz and Cancello, 2001; Neoh and Lee, 2009b; Nasir and Akhtar, 2011; Harit *et al.*, 2014).

In contrast, disparate flight patterns were observed in regions where *C. gestroi* and *C. formosanus* distribution overlapped in Florida, Taiwan, and Hawai'i. Chouvenc *et al.* (2017b) documented that the swarming activity of *C. gestroi* in Florida peaked from mid-February to late April when the temperature at the time of the flights ranged from 21.1°C to 28.6°C, while the swarming activity of sympatric species *C. formosanus* began from early April to late June at 24.2–29.1°C. Other factors, such as wind, atmospheric pressure, and relative humidity, did not affect the flights (Chouvenc *et al.*, 2017b). In Taiwan, the dispersal flight season of *C. gestroi* started in mid-March and peaked in mid-April, while the flight of *C. formosanus* started a month later, in late April or early May, with a major swarm occurring in late May (Huang *et al.*, 2022). *Coptotermes gestroi* alates were observed in Hawai'i in April–July (Tong *et al.*, 2017). In general, *C. gestroi* alates swarm about a month before the peak of *C. formosanus* swarming (Tong *et al.*, 2017). The disparate flight periods and strategies exhibited by *C. gestroi* and *C. formosanus* strongly indicate that the reproductive isolation of these two sympatric termites is temporally mediated. This might be an adaptive strategy to ensure successful colony foundation. However, *C. gestroi* and *C. formosanus* are now known to mate and produce viable hybrids (Chouvenc *et al.*, 2015c; Su *et al.*, 2017; Li *et al.*, 2023) (also see Chapter 17 of this volume for more details). Since colony nutrient demand increases during the alate production season, disparate flight seasons could evince a niche partition that reduces colony competition, as shown in the fungus-growing termites *Macrotermes* spp. in Malaysia (Neoh and Lee, 2009a). This also indicates the comparatively high plasticity of this species and predisposes them as successful invasive species globally.

16.11 Prevention and control of *C. gestroi*

The preceding two subsections discussed the use of liquid termiticide and baiting for *C. gestroi* control. Detailed discussions on liquid termiticides and baiting can be found in Chapters 12 and 11, respectively. For other preventive and control methods, such as the use of wood preservatives, biological agents, botanicals, and nonchemical approaches, please refer to Chapters 13 and 14.

As detailed in Chapter 12, four types of termiticides are used against subterranean termites (see Section 12.4 and Table 12.2). Types I, II, and III are used as liquid termiticides, while Type IV toxicants are used in baits. Types I and II termiticides are those that are highly repellent (pyrethroids) and less repellent (organophosphates and cyclodienes), respectively. In contrast, Type III termiticides are non-repellent (diamides, neonicotinoids, oxadiazine, phenylpyrazole, and pyrroles). Types II and III are dose-dependent termiticides, while Type IV is dose-independent. Type IV consists of chitin synthesis inhibitors.

16.11.1 Liquid termiticides

Liquid termiticides have been used for termite control for more than 120 years and remain a popular preventive measure. This method involves creating a chemical barrier in the soil to protect building structures from termite attacks. The conventional method of using termiticides, such as pyrethroids and organophosphates, could only provide limited protection when poor treatment regimens, homeowner's post-treatment activities, soil erosion, and burrowing vertebrates resulted in gaps that allow termites free access into the building structure (Hu, 2011). On the other hand, in soil treated with non-repellent termiticides, the compounds are undetectable by termites, hence providing insecticide exposure to the termites when they tunnel through the treated zones. Manufacturers often advocate that, at recommended rates, this method allows termites to acquire a lethal dose of toxicant and transfer it to some healthy nestmates via grooming and trophallaxis, potentially leading to colony decline or suppression (Potter, 1999), although it has never been documented in *Coptotermes* spp.

Unlike *C. formosanus*, only limited studies have evaluated the performance of liquid termiticides on *C. gestroi* (Table 16.1). All published studies were carried out in the laboratory. Yeoh and Lee (2007b) evaluated bifenthrin (repellent termiticide) in the laboratory and found the tunneling activity of *C. gestroi* was limited to the untreated zone. In the same study, using glass-tube assay, the tunneling activity of *C. gestroi* was significantly reduced in soil treated with bifenthrin at 10 ppm. Since the compound is repellent and fast-acting, horizontal transfer was ruled out (Shelton *et al.*, 2005) unless at very low concentration, i.e., via cannibalism of moribund-poisoned *C. gestroi* workers or cadaver handling by healthy nestmates (da Silva *et al.*, 2019).

Non-repellent termiticides such as chlorfenapyr, chlorantraniliprole, and indoxacarb showed delayed action against *C. gestroi* (Yeoh and Lee, 2007b). In a glass-tube assay, *C. gestroi* tunneled through chlorfenapyr, chlorantraniliprole, and indoxacarb-treated sand, resulting in an impairment of tunneling activity. At the lower concentrations of indoxacarb and chlorantraniliprole, the transfer efficiency of the termiticides among nestmates is positively associated with either insecticide exposure duration or donor–recipient ratio. Long exposure time and/or a high donor–recipient ratio is required to ensure maximal uptake by recipient workers. *Coptotermes gestroi* donors treated with 100 ppm indoxacarb for at least 10 min significantly decreased the survival time of *C. gestroi* to 7 days (Neoh *et al.*, 2014b). Almost all *C. gestroi* workers survived post-exposure to 50 ppm chlorantraniliprole after 24 h, and the mean mortality of *C. gestroi* only reached 76% after 3 days. Chlorantraniliprole impairs normal muscle contraction and leads to feeding cessation of *C. gestroi* workers (Neoh *et al.*, 2012b). Hence, chlorantraniliprole-treated termites were more regularly attended by healthy termites (Neoh *et al.*, 2014a). A study showed that donors (chlorantraniliprole-treated termites) must be > 30% of the donor–recipient mixture to achieve 100% mortality of the recipient workers within 10 days (Neoh *et al.*, 2012b, 2014b). In contrast, a higher concentration of chlorfenapyr of more than 250 ppm is required for donor exposure to allow efficient transfer among healthy *C. gestroi* (Neoh *et al.*, 2014b).

Fipronil is highly toxic to *C. gestroi*, resulting in 100% mortality after 1 week, even at the lowest concentration of 1 ppm. A minimum of 1 ppm of fipronil-treated soil significantly reduced tunneling activity by up to 13% (Yeoh and Lee, 2007b). Tunneling activity was absent when the *C. gestroi* forager workers were exposed to concentrations up to 50 ppm (Yeoh and Lee, 2007b). Neoh *et al.* (2012b) documented that when *C. gestroi* workers were exposed to 60 ppm fipronil based on concentration on the label rate for field application for 6 h, the mortalities of *C. gestroi* donors recorded within 24 h were 100%. Despite high mortality of donor workers, substantial mortality (~73.5%) of recipient workers occurred (Neoh *et al.*, 2012b). The lethal transfer among *C. gestroi* workers was affected by the donor–recipient ratio but not donor exposure time (Neoh *et al.*, 2014b). *Coptotermes gestroi* workers treated with fipronil produced stomodeal and proctodeal fluids that were lethal to the healthy termites (Neoh *et al.*, 2014a).

The bioavailability of termiticides, adsorption of termiticides in soils, transport of termiticides in soils under precipitation, and degradation of termiticides varied with

Table 16.1. Laboratory and field evaluations of liquid termiticides against *C. gestroi*.

Class	Insecticide	Evaluation method	Findings	Reference
Diamide	chlorantraniliprole	Glass tube tunneling method at rates of 1, 10, 50, 100 ppm in sand, and Petri dish (treated/ untreated zones) method at 100 ppm	% cumulative tunneling distance ranged from 11.9 (100 ppm) to 100% (1, 10 ppm); tunneling activity in untreated: treated zones = 2.2: 1.6	Yeoh and Lee (2007b)
	chlorantraniliprole	Topical assay, horizontal transfer, bioavailability in 2 soil types (sandy clay, sand), feeding assay	LD_{50} = 41.9 ng/insect; exposure duration and donor: recipient ratio affect the mortality of recipient termites; mortality after exposure in sandy clay was significantly lower than in sand; exposed termites for 1 h to 50 ppm ceased to feed.	Neoh *et al.* (2012b)
	chlorantraniliprole	Effects on mutual interactions using rubidium as tracer	Mutual interaction between treated and untreated termites was not affected by their nutritional status; termites exposed to chlorantraniliprole were more likely to cease feeding and then undergo starvation	Neoh *et al.* (2014a)
	chlorantraniliprole	Horizontal transfer at different donor–recipient ratios, and exposure times (5, 10, 30 min) at rate of 100 ppm in sand	Mean survival time ranged from 350.3–430.7 h (5 min), 345.2–390 h (10 min), 217.8–307.3 h (30 min)	Neoh *et al.* (2014b)
	cyantraniliprole	Horizontal transfer at different donor: recipient ratios, and exposure times (5, 10, 30 min) at rate of 100 ppm in sand	Mean survival time ranged from 13–147.1 h (5 min), 12.8–90 h (10 min), 13.3–30.2 h (30 min)	Neoh *et al.* (2014b)
Neonicotinoid	imidacloprid	Glass tube tunneling method at rates of 1, 10, 50, 100 ppm in sand, and Petri dish (treated/ untreated zones) method at 25 ppm	% cumulative tunneling distance ranged from 0 (100 ppm) to 69.3% (1 ppm); tunneling activity in untreated: treated zones = 1.4: 1.0.	Yeoh and Lee (2007b)
	imidacloprid	Horizontal transfer at different donor: recipient ratios, and exposure times (5, 10, 30 min) at rate of 25 ppm in sand	Mean survival time ranged from 385.3–569.0 h (5 min), 221.9–487.9 h (10 min), 29.5–310.8 h (30 min)	Neoh *et al.* (2014b)
	thiamethoxam	Treated sand in Petri dish, horizontal transfer	50 ppm caused 100% mortality in 24 h, 2.5 ppm caused 98.4% mortality in 14 days; 25 ppm at donor: recipient ratio of at least 2:5 caused 100% mortality of recipient termites in 3 days	Acda (2014a)

Oxadiazine	indoxacarb	Glass tube tunneling method at rates of 1, 10, 50, 100 ppm in sand, and Petri dish (treated/ untreated zones) method at 100 ppm	% cumulative tunneling distance ranged from 2.0 (100 ppm) to 100% (1 ppm); tunneling activity in untreated: treated zones = 1.6: 1.4	Yeoh and Lee (2007b)
	indoxacarb	Topical assay	LD_{50} = 4.042 ng/insect	Neoh *et al.* (2012b)
	indoxacarb	Horizontal transfer at different donor: recipient ratios, and exposure times (5, 10, 30 min) at rate of 100 ppm in sand	Mean survival time ranged from 200.4–249.1 h (5 min), 137.5–181.3 h (10 min), 75.1–159.1 h (30 min)	Neoh *et al.* (2014b)
Phenylpyrazole	fipronil	Glass tube tunneling method at rates of 1, 10, 50, 100 ppm in sand, and Petri dish (treated/ untreated zones) method at 11 ppm	% cumulative tunneling distance ranged from 19.2 (100 ppm) to 87.7% (1 ppm); tunneling activity in untreated: treated zones = 4.0: 1.0	Yeoh and Lee (2007b)
	fipronil	Topical assay, bioavailability in 2 soil types	LD_{50} = 0.865 ng/insect; sandy clay treated with fipronil killed 73.5% recipient termites at 1 d	Neoh *et al.* (2012b)
	fipronil	Effects on mutual interactions using rubidium as tracer	Fipronil-treated termites were rarely attended by untreated workers, possibly due to the treated termites died too quickly; however, Rb transfer from fipronil-treated termites to starving untreated termites was threefold greater compared the exchange with fed untreated termites	Neoh *et al.* (2014a)
	fipronil	Horizontal transfer at different donor: recipient ratios, and exposure times (5, 10, 30 min) at rate of 11 ppm in sand	Mean survival time ranged from 142.5–215.5 h (5 min), 64.1–185.4 h (10 min), 38.5–103.2 h (30 min)	Neoh *et al.* (2014b)
	fipronil	Application (80 ppm) at treatment chamber at 12 m extended foraging arena	Termite activity remained stable throughout 84 days experimental period, except at locations near the treatment where large number of carcasses accumulated	Chouvenc (2018)
Pyrethroid	bifenthrin	Glass tube tunneling method at rates of 1, 10, 50, 100 ppm in sand and Petri dish (treated/ untreated zones) method at 30 ppm.	% cumulative tunneling distance ranged from 0.5 (100 ppm) to 100% (1 ppm); tunneling activity in untreated: treated zones = 3.2: 0	Yeoh and Lee (2007b)
	bifenthrin	Topical assay	LD_{50} = 0.128 ng/insect	Neoh *et al.* (2012b)

Continued

Table 16.1. Continued.

Class	Insecticide	Evaluation method	Findings	Reference
	bifenthrin	Horizontal transfer at different donor: recipient ratios, and exposure times (5, 10, 30 min) at rate of 100 ppm in sand	Mean survival time ranged from 215.3–513.5 h (5 min), 74.3–199 h (10 min) and 56.2–81.1 h (30 min)	Neoh *et al.* (2014b)
Pyrrole	chlorfenapyr	Glass tube tunneling method at rates of 1, 10, 50, 100 ppm in sand, and Petri dish (treated/ untreated zones) method at 30 ppm	% cumulative tunneling distance ranged from 29.8 (100 ppm) to 100% (1 ppm); tunneling activity in untreated: treated zones = 2.6: 1.8	Yeoh and Lee (2007b)
	chlorfenapyr	Topical assay	LD_{50} = 10.01 ng/insect	Neoh *et al.* (2012b)
	chlorfenapyr	Horizontal transfer at different donor: recipient ratios, and exposure times (5, 10, 30 min) at rate of 100 ppm in sand	Mean survival time ranged from 200.4–539.8 h (5 min), 438.2–535.6 h (10 min), 307.8–481.9 h (30 min)	Neoh *et al.* (2014b)

different soil types. A soil texture survey along structures infested with *C. gestroi* in Malaysia found that most *C. gestroi* infestations occurred at locations with sandy loam (Ab Majid and Ahmad, 2013). Besides, termiticide performance may vary depending on the type and dosage of chemical used, termite species, and soil types. Typically, manufacturers only recommend one label rate, irrespective of termite species, colonies, and localities. However, not all termite species and colonies behave uniformly or share a similar insecticide susceptibility. Neoh *et al.* (2012b) reported that a mere 0.9 ng fipronil was sufficient to cause 50% mortality in *C. gestroi* at 24 h, which was one- to fourfold lower than the amount required to achieve the same effect in *C. formosanus* (Ibrahim *et al.*, 2003). Assuming the rate of fipronil used was based on its performance against *C. formosanus*, using the same rate against *C. gestroi* will result in rapid onset of mortality of the exposed termites, leading to shorter tunneling distance and limited horizontal transfer to healthy nestmates. The accumulation of termite carcasses near and at the treated zone will result in a secondary repellent effect.

16.11.2 Termite baits

Baiting using paper- or alpha-cellulose containing type IV termiticide (chitin synthesis inhibitors) (CSIs) are increasingly popular as a post-construction remedial treatment approach against *C. gestroi*, especially in Southeast Asia. Unlike soil termiticide treatment, which requires a large quantity of active ingredients, baiting uses a minimal amount of active ingredients, mostly ~1 g, to eliminate a termite colony. This drastically reduces the impact on non-target soil organisms. However, to achieve colony elimination, the termites must feed on the baits and share the baits with nestmates.

The main goal of an effective baiting system is to deliver the toxicant to the colony fast and in sufficient quantity. Several characteristics contribute to the success of baiting against *C. gestroi*. First, unlike the fungus-growing termites (Neoh *et al.*, 2011; Lee *et al.*, 2014), *C. gestroi* readily consumes the bait and shares it with nestmates within a short period. Second, the active food-seeking behavior, high efficiency in food transportation (Cao and Su, 2014), and high food consumption rate (Cao and Su, 2016) in *C. gestroi* allow significant amounts of bait to be transferred to the colony and circulated among nestmates. Third, the tendency of cannibalism (da Silva *et al.*, 2019), in which worker termites cannibalize the newly intoxicated nestmates, allows secondary transmission of a toxicant to other healthy workers.

Several extrinsic factors may affect termite feeding before they receive an adequate lethal dose of the bait toxicant. These include disturbances, predators (e.g. ants), and mold in the bait (Wong and Lee, 2010c). Despite this, only a minimal amount (4.8–5.5 mg) of bait toxicant (e.g., noviflumuron) is required to eliminate colonies in laboratory experiments (Chouvenc and Su, 2017b; Gordon *et al.*, 2022).

The period required from baiting to colony elimination is another factor of interest. A summary of published field studies of baits on *C. gestroi* is shown in Table 16.2. It takes approximately 1–4 months to suppress or eliminate colonies of *C. gestroi* (Lee, 2002, 2007b; Sajap *et al.*, 2009; Chouvenc and Su, 2017b; Rojo and Acda, 2019; Chouvenc, 2021; Gordon *et al.*, 2022). Lee (2002) evaluated 0.5% hexaflumuron bait against field colonies of *Coptotermes travians* (= *C. havilandi sensu* Tho (1992), which is now synonymized as *C. gestroi*) and eliminated termite activity after 49–62 days. In another study, Lee (2007b) eliminated field colonies of *C. gestroi* using 0.5% and 1.0% bistrifluron in 6–8 weeks and 4–5 weeks, respectively. The total bistrifluron required to eliminate the colonies ranged from 1.1 g to 8.3 g. Umar and Ab Majid (2020) tested the performance of different baiting procedures on field *C. gestroi* colonies with 0.1% chlorfluazuron in Malaysia. The authors demonstrated that installing chlorfluazuron bait in only one of the active monitoring stations was sufficient to cause colony elimination in 4–8 weeks (Umar and Ab Majid, 2020), utilizing ~270 mg of the active ingredient.

Table 16.2. Field performance of termite baits against *C. gestroi*.

Location	% active ingredient	Amount of active ingredient used (g)	In-ground (IG) or above-ground (AG)	Period taken to colony elimination	Reference
Malaysia	0.5% hexaflumuron	0.92–1.46	IG	34–44 days	Lee (2002)[a]
Malaysia	0.5% hexaflumuron	0.15–1.09	AG	35–62 days	Sajap *et al.* (2002)[a]
Malaysia	0.1% noviflumuron	0.02–0.14	AG	46–78 days	Lee (2007a)
Malaysia	0.5% noviflumuron	0.04–0.70	AG	41–80 days	Lee (2007a)
Malaysia	0.5% noviflumuron	0.10–0.80	AG	35–46 days	Sajap *et al.* (2005)
Malaysia	0.5% bistrifluron	1.06–3.23	AG	6–8 weeks	Lee (2007b)
Malaysia	1.0% bistrifluron	1.73–8.34	AG	4–5 weeks	Lee (2007b)
Little Cayman	0.5% hexaflumuron	0.06–0.68	IG, AG	3–5 months	Su *et al.* (2000)
Taiwan	0.5% noviflumuron	< 0.16	IG	3–4 weeks	Chiu *et al.* (2016)
Philippines	0.1% chlorfluazuron	0.50–2.37	IG, AG	6–13 weeks	Garcia *et al.* (2007)
Philippines	0.1% chlorfluazuron	0.36–0.56	AG	8–12 weeks	Rojo and Acda (2019)
Malaysia	0.1% chlorfluazuron	0.22–0.30	IG	4–8.6 weeks	Umar and Majid (2020)
Florida	0.5% noviflumuron	n/a	AG	3–13 weeks	Su *et al.* (2023)

[a]Tested on *C. gestroi* (*Coptotermes travians* sensu Tho)

In Southeast Asia, baiting of *C. gestroi* infestations in structures is mainly carried out using above-ground (AG) baits (> 90%) and not in-ground (IG) baits. This is for several reasons. First, many termite species besides *C. gestroi* could be found foraging around structures. They include *Globitermes sulphureus* (Haviland), *Macrotermes gilvus* (Hagen), *Macrotermes carbonarius* (Hagen), *M. crassus*, *Ancistrotermes pakistanicus* (Ahmad) (Lee, 2007a; Lee *et al.*, 2007). With the exception of *M. gilvus*, the other species are rarely found infesting structures. However, they can be found infesting IG monitoring stations, decreasing the baiting success of *C. gestroi* using IG bait.

Unlike the interception rate of *C. formosanus* in the IG monitoring station (~17%) (Su, 1994), the interception rate of *C. gestroi* in the IG monitoring station in Southeast Asia is low (2–4%). Even then, it may take several months before the termites intercept the IG monitoring stations. Hence, if the infestation is visible, pest management professionals will resort to using AG baits to save time, labor, and cost. Also, unlike in the U.S., where single-family homes are typical, most people who live in regions where *C. gestroi* is the primary termite pest reside in high-rise apartments or linked (terrace) houses with no perimeter space. This rules out the possibility of installing IG monitoring stations for baiting.

Su *et al.* (2023) reported that none of the 83 IG monitoring stations installed in soil was intercepted by *C. gestroi* colonies in Florida, despite the infestations being close to the stations. They were, however, able to eliminate the infestations using AG baits that were installed on infested trees. This implied that the present practice of baiting *C. formosanus* using IG stations may not be feasible in the management of *C. gestroi*. Regular inspection of structures and trees, and the use of AG baits when the infestation is detected, is likely the most viable option in managing *C. gestroi*.

20-Hydroxyecdysone (20E) is an insect steroid hormone that coordinates with juvenile hormone (JH) to regulate the ecdysis (molting) and metamorphosis of arthropods. Excessive ingestion by insects can induce hyperecdysonism, which leads to premature mortality. Carnohan and Su (2018) exposed *C. gestroi* workers to different concentrations of 20E in combination with noviflumuron in no-choice assays. They reported 78% mortality on days 13–14 after the insects were exposed to 2000 ppm of 20E. The delayed intoxication of *C. gestroi* workers allowed them to return to the central nest and spread the toxicant to other nestmates. Like CSIs, 20E affects termites' molting process, but the lethal time is much shorter than the CSIs. However, it was found in choice assays that *C. gestroi* workers ceased feeding on treated paper discs containing 20E concentrations of 500–2000 ppm (Carnohan *et al.*, 2021). Hence, incorporating additives to mask the presence of 20E appears crucial for future bait development.

The potential of using insecticides such as boric acid and fipronil was also evaluated. Casarin *et al.* (2009) impregnated corrugated paper with different concentrations of boric acid and fipronil and introduced it to *C. gestroi* workers. The study showed that boric acid concentrations between 2000 and 3000 ppm caused approximately 90–100% mortality of *C. gestroi* workers at 35 days, while fipronil concentrations between 0.01 and 0.0001 ppm resulted in 100% termite mortality after 2 weeks of exposure (Casarin *et al.*, 2009). However, this laboratory test was only limited to direct contact evaluation, and no experiments of horizontal transfer between nestmates were conducted. Judging from the Type III characteristics of these two toxicants, it is unlikely that they will be suitable toxicants to be used in termite bait.

16.11.3 Other control methods

Other control methods that had been tested on *C. gestroi* were entomopathogenic fungi such as *Isaria fumosorosea* Wize and *Metarhizium anisopliae* (Maketon *et al.*, 2007; Jessica *et al.*, 2019) and botanical extracts (Himmi *et al.*, 2013; Acda, 2014b; Kadir, 2017; Bakaruddin and Ab Majid, 2019). The performance of these methods to control

C. gestroi in the field remains unproven. There are also prevention methods against *C. gestroi*, such as the use of durable or resistant wood for construction (Ngee *et al.*, 2004; Kadir and Hale, 2012; Roszaini *et al.*, 2013), processed wood-particle board for flooring and furniture (Hashim *et al.*, 2009; Syamani *et al.*, 2011; Acda and Cabangon, 2013; Tarasin and Rattanapun, 2019), and durable polymer plastics as plastic cable sheathings (Lenz *et al.*, 2012, 2013).

16.12 Future spread of *C. gestroi*

Local and global expansions of *C. gestroi* are mainly dependent on anthropogenic activity. Based on species distribution models, the habitat suitable for *C. gestroi* colony establishment is larger than the distribution recorded to date (Li *et al.*, 2013). Those suitable areas with no *C. gestroi* recorded so far include most countries in South Asia, northern Australia, surrounding areas of the Gulf of Mexico, Central America, northeastern South America (northeast of Andes Mountains), central and southeastern Africa, and Madagascar (Li *et al.*, 2013). By 2050–2070, *C. gestroi* is expected to expand by < 5% in its distribution range (Buczkowski and Bertelsmeier, 2017). In particular, rising temperatures in the event of global climate change are predicted to cause upward shifts of *C. gestroi*'s distribution from tropical regions to subtropical regions, as evident from the northward spread of *C. gestroi* in Florida (Chouvenc *et al.*, 2016b; PMP, 2023) and in Taiwan (Huang *et al.*, 2022). Li *et al.* (2013) also predicted that continuous urbanization and increasing human population would increase its introduction frequency and cause further extension, especially in fast-developing tropical and subtropical countries. This is in accordance with the species distribution prediction that urban areas in southeast Florida seem to have an increased probability of *C. gestroi* presence (Tonini *et al.*, 2014). In contrast, despite a high introduction risk to temperate cities in Europe, the U.S., northeastern China, and Japan, successful colonization and naturalization of *C. gestroi* in these regions are unlikely, due to low temperatures (Li *et al.*, 2013).

16.13 Conclusion and future research questions

Coptotermes gestroi is a highly invasive species that has become increasingly economically important beyond its native range. Similar to *C. formosanus*, its dispersal has been associated with anthropogenic activity. The current control strategies of *C. gestroi* rely on soil treatment using liquid termiticides and baiting. The use of AG baits is highly effective in managing *C. gestroi*, more so than IG baits, due to the low interception rate of the latter in the ground monitoring stations and also the frequent presence of other non-pest subterranean termite species (especially in Southeast Asia) that may disrupt the IG baiting process.

Most liquid termiticides and baits were originally developed and tested in the U.S. and Europe on *Reticulitermes* spp. and *C. formosanus*, which were the major termite species. Many of these products were subsequently marketed in other parts of the world, assuming they would also work against *C. gestroi* and other termite species, which may not necessarily be true. Su *et al.* (2023) revealed the low interception rate of *C. gestroi* in IG termite monitoring stations (0%) compared with that of *C. formosanus* (~17%) (Su, 1994), possibly due to different tunneling behavior between the two species. Despite being sympatric species in some regions, the susceptibility of *C. gestroi* and *C. formosanus* to termiticides differs, sometimes by several-fold (Ibrahim *et al.*, 2003; Neoh *et al.*, 2012b). Considering how widespread *C. gestroi* is now, it is imperative for termiticide manufacturers to consider the importance of this species during the development of future termiticides and delivery systems.

Most published studies on *C. gestroi* were conducted on populations that were introduced to Brazil, the U.S., and Taiwan. These introduced populations may have undergone genetic bottlenecks, unlike those in their native range (Southeast Asia). Agonistic behavior assays have shown a lack of aggression between *C. gestroi* colonies in Hawai'i and in Florida. To what extent these introduced populations differ from their native counterparts, especially in biology and ecology, would be an exciting research direction. Some questions arise. Could the recent hybridization between *C. formosanus* and *C. gestroi* found in introduced regions happen because these populations have gone through a genetic bottleneck? Would a similar phenomenon be observed if native populations of *C. formosanus* and *C. gestroi* were paired? How does genetic diversity in native populations affect the production of hybrids?

While invasive termite species such as *C. formosanus* and *C. gestroi* are often found in urban areas in their introduced regions, will it also affect the natural environment, such as forests, in these introduced regions? Chiu *et al.* (2016) reported the presence of *C. gestroi* in the forest margins in Taiwan, while *C. formosanus* was found invading forest patches in South Carolina and Louisiana, U.S. (Evans *et al.*, 2019; Evans, 2021). This raises another question. In its native range, *C. gestroi* feeds on a wide variety of timbers that differ from the ones used in constructions in the introduced regions. The tree species in its introduced areas also differ from those in its native range. Are there any changes in the abundance and diversity of gut microbiota in those populations in introduced regions that enable them to thrive beyond their native range?

Lastly, it remains unknown how climate change impacts (rising temperature, rising atmospheric CO_2, and irregular rainfall patterns) could affect the destructive intensity of *C. gestroi*. The rising atmospheric CO_2 level is known to reduce the nitrogen and overall mineral concentrations in plants (Loladze, 2014). This will lead to reduced nitrogen in trees grown for construction timbers. With less nitrogen in wood, the wood density will be reduced (Pretzsch *et al.*, 2018; Beets *et al.*, 2001). This may increase the feeding of termites and other wood-destroying insects to compensate for the reduced nitrogen content, leading to more significant damage to the timber. As the climate warms, termite feeding rate (Cao and Su, 2016), activity, and abundance will also increase (Zanne *et al.*, 2022). Could these changes potentially lead to greater damage by *C. gestroi* in the future? This certainly warrants further investigation.

References

Ab Majid, A.H. and Ahmad, A.H. (2009) Termites infestation selected from premises in Penang, Seberang Prai & Sungai Petani, Malaysia. *Malaysian Applied Biology Journal* 38, 37–48.

Ab Majid, A.H. and Ahmad, A.H. (2013) Termite infestation associated with type of soil in Pulau Pinang, Malaysia (Isoptera: Rhinotermitidae). *Tropical Life Sciences Research* 24, 91–95.

Ab Majid, A.H. and Ahmad, A.H. (2015) Define colony number of subterranean termites *Coptotermes gestroi* (Isoptera: Rhinotermitidae) in selected infested structures. *Sains Malaysiana* 44, 211–216.

Acda, M.N. (2014a) Toxicity and transmission of thiamethoxam in the asian subterranean termite *Coptotermes gestroi* (Isoptera: Rhinotermitidae). *Journal of Insect Science* 14, 222. doi: 10.1093/jisesa/ieu084.

Acda, M.N. (2014b) Repellent effects of annona crude seed extract on the Asian subterranean termite *Coptotermes gestroi* Wasmann (Isoptera:Rhinotermitidae). *Sociobiology* 61, 332–337.

Acda, M.N. and Cabangon, R.J. (2013) Termite resistance and physico-mechanical properties of particleboard using waste tobacco stalk and wood particles. *International Biodeterioration & Biodegradation* 85, 354–358.

Albino, E. and Costa-Leonardo, A.M. (2011) Nymphs in foraging populations of *Coptotermes gestroi* (Isoptera: Rhinotermitidae). *Animal Biology* 61, 427–439.

Albuquerque, A.C., Matias, G., Couto, A., Oliveira, M.A.P., and Vasconcellos, A. (2012) Urban termites of Recife, Northeast Brazil (Isoptera). *Sociobiology* 59, 183–188.

Arab, A. and Costa-Leonardo, A.M. (2005) Effect of biotic and abiotic factors on the tunneling behavior of *Coptotermes gestroi* and *Heterotermes tenuis* (Isoptera : Rhinotermitidae). *Behavioural Processes* 70, 32–40.

Arab, A., Costa-Leonardo, A.M., Batista-Pereira, L.G., dos Santos, M.G., Correa, A.G. *et al.* (2004) Trail-pheromone specificity of two sympatric termites (Rhinotermitidae) from southeastern brazil. *Sociobiology* 43, 377–387.

Arab, A., Costa-Leonardo, A.M., Casarin, F.E., Guaraldo, A.D., and Chaves, R.C. (2005) Foraging activity and demographic patterns of two termite species (Isoptera : Rhinotermitidae) living in urban landscapes in southeastern Brazil. *European Journal of Entomology* 102, 691–697.

Arab, A., Carollo Blanco, Y,. and Costa-Leonardo, A.M. (2012) Dynamics of foraging and recruitment behavior in the Asian subterranean termite *Coptotermes gestroi* (Rhinotermitidae). *Psyche: A Journal of Entomology* 2012, 1–7. Article ID 806782. doi: 10.1155/2012/806782

Araujo, A.P.A., de Araujo, F.S., and Desouza, O. (2011) Resource suitability affecting foraging area extension in termites (Insecta, Isoptera). *Sociobiology* 57, 271–283.

Arif, A., Putri, G., Muin, M. (2019) Hazard mapping of subterranean termite attacks in Makassar City, South Sulawesi, Indonesia. *Insects* 11, 31. doi: 10.3390/insects11010031

Bakaruddin, N.H. and Ab Majid, A.H. (2019) Efficacy of several plants extracts on the tunneling activity and survival of subterranean termites (*Coptotermes gestroi* and *Globitermes sulphureus*). *Tropical Life Sciences Research* 30, 33–56.

Barsotti, R.C. and Costa-Leonardi, A.M. (2005) The caste system of *Coptotermes gestroi* (Isoptera : Rhinotermitidae). *Sociobiology* 46, 87–103.

Beets, P.N., Gilschrist, K., and Jeffreys, M.P. (2001) Wood density of radiata pine: effect of nitrogen supply. *Forest Ecology and Management* 145, 173–180.

Brent, C.S. and Traniello, J.F.A. (2001a) Influence of sex-specific stimuli on ovarian maturation in primary and secondary reproductives of the dampwood termite *Zootermopsis angusticollis*. *Physiological Entomology* 26, 239–247.

Brent, C.S. and Traniello, J.F.A. (2001b) Social influence of larvae on ovarian maturation in primary and secondary reproductives of the dampwood termite *Zootermopsis angusticollis*. *Physiological Entomology* 26, 78–85.

Buczkowski, G. and Bertelsmeier, C. (2017) Invasive termites in a changing climate: A global perspective. *Ecology and Evolution* 7, 974–985. doi: 10.1002/ece3.2674.

Campora, C.E. and Grace, J.K. (2001) Tunnel orientation and search pattern sequence of the Formosan subterranean termite (Isoptera: Rhinotermitidae). *Journal of Economic Entomology* 94, 1193–1199.

Cao, R.X. and Su, N.Y. (2014) Tunneling and food transportation activity of four subterranean termite species (Isoptera: Rhinotermitidae) at various temperatures. *Annals of the Entomological Society of America* 107, 696–701.

Cao, R.X. and Su, N.Y. (2016) Temperature preferences of four subterranean termite species (Isoptera: Rhinotermitidae) and temperature-dependent survivorship and wood-consumption rate. *Annals of the Entomological Society of America* 109, 64–71.

Carnohan, L. and Su, N.Y. (2018) Mortality time trends of three subterranean termite species when exposed to 20-Hydroxyecdysone, alone and in combination with noviflumuron. *Journal of Economic Entomology* 111, 2312–2316.

Carnohan, L., Lee, S.B., and Su, N.Y. (2021) Concentration-dependent feeding deterrence to 20-Hydroxyecdysone for three subterranean termite species (Blattodea: Rhinotermitidae). *Insects* 12, 218; doi: 10.3390/insects12030218

Casarin, F.E., Costa-Leonardo, A.M. and Bueno, O.C. (2009) Laboratory Assessment of two active ingredients for control of *Coptotermes gestroi* (Isoptera: Rhinotermitidae). *Sociobiology* 54, 787–797.

Chand, R., Jokhan, A., Charan, H., Raj, K. and Singh, P. (2018) Threats posed by Asian subterranean termites in the Fiji Islands and their potential controls: A review. *New Zealand Plant Protection* 71, 129–139.

Cheng, S., Thinagaran, D., Mohanna, S.Z.M. and Noh, N.A.M. (2014) Haplotype-habitat associations of *Coptotermes gestroi* (Termitoidae: Rhinotermitidae) from mitochondrial DNA genes. *Environmental Entomology* 43, 1105–1116.

Chiu, C.I., Yeh, H.T., Tsai, M.J. and Li, H.F. (2016) Naturalization and control of *Coptotermes gestroi* (Blattodea: Rhinotermitidae) in a Taiwanese forest. *Journal of Economic Entomology* 109, 1317–1325.

Chouvenc, T. (2018) Comparative impact of chitin synthesis inhibitor baits and non-repellent liquid termiticides on subterranean termite colonies over foraging distances: Colony elimination versus localized termite exclusion. *Journal of Economic Entomology* 111, 2317–2328.

Chouvenc, T. (2019) The relative importance of queen and king initial weights in termite colony foundation success. *Insectes Sociaux* 66, 177–184.

Chouvenc, T. (2021) Subterranean termite (*Coptotermes gestroi* (Blattodea: Rhinotermitidae)) colony elimination through exposure to a novaluron csi bait formulation in laboratory. *Journal of Economic Entomology* 114, 1249–1255.

Chouvenc, T. (2022) Eusociality and the transition from biparental to alloparental care in termites. *Functional Ecology* 36, 3049–3059.

Chouvenc, T. and Foley, J.R. IV (2018) *Coptotermes gestroi* (Wasmann) (Blattodea [Isoptera]: Rhinotermitidae), a threat to the southeastern Florida urban tree canopy. *Florida Entomologist* 101, 79–90.

Chouvenc, T. and Su, N.Y. (2014) Colony age-dependent pathway in caste development of *Coptotermes formosanus* Shiraki. *Insectes Sociaux* 61, 171–182.

Chouvenc, T. and Su, N.Y. (2017a) Irreversible transfer of brood care duties and insights into the burden of caregiving in incipient subterranean termite colonies. *Ecological Entomology* 42, 777–784.

Chouvenc, T. and Su, N.Y. (2017b) Subterranean termites feeding on CSI baits for a short duration still results in colony elimination. *Journal of Economic Entomology* 110, 2534–2538.

Chouvenc, T. and Su, N.Y. (2017c) Testing the role of cuticular hydrocarbons on intercolonial agonism in two subterranean termite species (*Coptotermes*) and their hybrids. *Insectes Sociaux* 64, 347–355.

Chouvenc, T., Basille, M., Li, H.F. and Su, N.Y. (2014) Developmental instability in incipient colonies of social insects. *PLoS ONE* 9, 11, e113949. doi: 10.1371/journal.pone.0113949

Chouvenc, T., Mullins, A.J. and Su, N.Y. (2015a) Rare production of nymphs in an Asian subterranean termite (Isoptera: Rhinotermitidae) incipient colony. *Florida Entomologist* 98, 972–973. doi: 10.1653/024.098.0327

Chouvenc, T., Basille, M. and Su, N.Y. (2015b) The production of soldiers and the maintenance of caste proportions delay the growth of termite incipient colonies. *Insectes Sociaux* 62, 23–29.

Chouvenc, T., Helmick, E.E. and Su, N.Y. (2015c) Hybridization of two major termite invaders as a consequence of human activity. *PLoS ONE* 10, e0120745.

Chouvenc, T., Li, H.F., Austin, J., Bordereau, C., Bourguignon, T. *et al.* (2016a) Revisiting *Coptotermes* (Isoptera: Rhinotermitidae): a global taxonomic roadmap for species validity and distribution of an economically important subterranean termite genus. *Systematic Entomology* 41, 299–306.

Chouvenc, T., Scheffrahn, R.H. and Su, N.Y. (2016b) Establishment and spread of two invasive subterranean termite species (*Coptotermes formosanus* and *C. gestroi*; Isoptera: Rhinotermitidae) in metropolitan southeastern Florida (1990–2015). *Florida Entomologist* 99, 187–191.

Chouvenc, T., Basille, M. and Su, N.Y. (2017a) Role of accelerated developmental pathway and limited nurturing capacity on soldier developmental instability in subterranean termite incipient colonies. *Insectes Sociaux* 64, 477–483.

Chouvenc, T., Scheffrahn, R.H., Mullins, A.J. and Su, N.Y. (2017b) Flight phenology of two *Coptotermes* species (Isoptera: Rhinotermitidae) in Southeastern Florida. *Journal of Economic Entomology* 110, 1693–1704.

Costa-Leonardo, A.M. and Barsotti, R.C. (1998) Swarming and incipient colonies of *Coptotermes havilandi* (Isoptera, Rhinotermitidae). *Sociobiology* 31, 131–142.

Costa-Leonardo, A.M., Arab, A. and Casarin, F.E. (2004) Neotenic formation in laboratory colonies of the termite *Coptotermes gestroi* after orphaning. *Journal of Insect Science* 4. 10. doi: 10.1093/jis/4.1.10.

Costa-Leonardo, A.M., Casarin, F.E. and Constantini, J.P. (2008) Record of a gregarine (Apicomplexa : Neogregarinida) in the abdomen of the termite *Coptotermes gestroi* (Isoptera, Rhinotermitidae). *Journal of Invertebrate Pathology* 97, 114–118.

Costa-Leonardo, A.M., Janei, V. and da Silva, I.B. (2022) Comparative reproductive biology of pre-, imaginal, and neotenic castes of the Asian termite *Coptotermes gestroi* (Blattaria, Isoptera, Rhinotermitidae). *Bulletin of Entomological Research* 112, 827–836.

da Silva, I.B. and Costa-Leonardo, A.M. (2017) Hypopharynx in termites: Morphological and functional aspects. *Micron* 101, 186–196.

da Silva, I.B. and Costa-Leonardo, A.M. (2022) Mating mediates morphophysiological changes in the spermathecae of *Coptotermes gestroi* queens. *Entomologia Experimentalis et Applicata* 171, 361–373.

da Silva, L.H.B., Haifig, I., and Costa-Leonardo, A.M. (2019) Facing death: How does the subterranean termite *Coptotermes gestroi* (Isoptera: Rhinotermitidae) deal with corpses? *Zoology* 137, 125712.

Delaplane, K. and La Fage, J. (1987) Variance in feeding on equivalent wood blocks by the Formosan subterranean termite in laboratory choice tests. *Sociobiology* 13, 227–233.

Dhang, P. (2014) Managing subterranean termites in ships and yachts across Philippine ports. *International Pest Control* 56, 262–263.

Do, T.H., Nguyen, T.T., Nguyen, T.N., Le, Q.G., Nguyen, C. *et al.* (2014) Mining biomass-degrading genes through Illumina-based de novo sequencing and metagenomic analysis of free-living bacteria in the gut of the lower termite *Coptotermes gestroi* harvested in Vietnam. *Journal of Bioscience and Bioengineering* 118, 665–671.

Evans, T.A. (2021) Predicting ecological impacts of invasive termites. *Current Opinion in Insect Science* 46, 88–94.

Evans, T.A., Forschler, B.T. and Grace, J.K. (2013) Biology of invasive termites: A worldwide review. *Annual Review of Entomology* 58, 455–474.

Evans, T.A., Forschler, B.T. and Trettin, C.C. (2019) Not just urban: The Formosan subterranean termite, *Coptotermes formosanus*, is invading forests in the Southeast USA. *Biological Invasions* 21, 1283–1294.

Fei, H. and Henderson, G. (2002) Formosan subterranean termite (Isoptera: Rhinotermitidae) wood consumption and worker survival as affected by temperature and soldier proportion. *Environmental Entomology* 31, 509–514.

Ferraz, M.V. and Cancello, E. (2001) Swarming behavior of the economically most important termite, *Coptotermes havilandi* (Isoptera: Rhinotermitidae), in southeastern Brazil. *Sociobiology* 38, 683–694.

Ferraz, M.V. and Cancello, E.M. (2004) Strategies on the developmental biology of incipient colonies of *Coptotermes gestroi* (Isoptera: Rhinotermitidae) in different substrates. *Sociobiology* 44, 109–122.

Ferraz, M.V. and Méndez-Montiel, J.T. (2004) First record of a subterranean termite, *Coptotermes havilandi* Holmgren (Isoptera: Rhinotermitidae), on the west coast of North America (Mexico). *Acta Zoológica Mexicana* 20, 39–43.

Fontes, L.R. and Milano, S. (2002) Termites as an urban problem in South America. *Sociobiology* 40, 103–152.

Franco Cairo, J.P., Carazzolle, M.F., Leonardo, F.C., Mofatto, L.S., Brenelli, L.B. *et al.* (2016) Expanding the knowledge on lignocellulolytic and redox enzymes of worker and soldier castes from the lower termite *Coptotermes gestroi*. *Frontiers in Microbiology* 7, 1518.

Garcia, C.M., Giron, M.Y. and Broadbent, S.G. (2007) Termite baiting system: a new dimension of termite control in the Philippines. *Proceedings of the 38th International Research Group on Wood Preservation*, Wyoming, USA, p. 12. https://www.irg-wp.com/irgdocs/details.php?618a4a51-8d2b-45b3-bcb6-901120a48366

Gautam, B.K. and Henderson, G. (2014) Water transport by *Coptotermes formosanus* (Isoptera: Rhinotermitidae). *Environmental Entomology* 43, 1399–1405.

Gazal, V., Bailez, O. and Viana-Bailez, A.M. (2019) Termite (Isoptera) survey in urban area in Northern of Rio de Janeiro State, Brazil. *Revista Colombiana De Entomologia* 45, e7813. doi: 10.25100/socolen.v45i0.7813.

Ghesini, S., Puglia, G. and Marini, M. (2011) First report of *Coptotermes gestroi* in Italy and Europe. *Bulletin of Insectology* 64, 53–54.

Gordon, J.M., Šobotník, J. and Chouvenc, T. (2020) Colony-age-dependent variation in cuticular hydrocarbon profiles in subterranean termite colonies. *Ecology and Evolution* 10, 10095–10104.

Gordon, J.M., Velenovsky, J.F. and Chouvenc, T. (2022) Subterranean termite colony elimination can be achieved even when only a small proportion of foragers feed upon a CSI bait. *Journal of Pest Science* 95, 1207–1216.

Guaraldo, A.C. and Costa-Leonardo, A.M. (2009) Preliminary fusion testing between whole young colonies of *Coptotermes gestroi* (Isoptera: Rhinotermitidae). *Sociobiology* 53, 767–774.

Hapukotuwa, N.K. and Grace, J.K. (2012a) Comparison of *Coptotermes formosanus* and *Coptotermes gestroi* (Blattodea: Rhinotermitidae) field sites and seasonal foraging activity in Hawaii. *Sociobiology* 59, 203–212.

Hapukotuwa, N.K. and Grace, J.K. (2012b) *Coptotermes formosanus* and *Coptotermes gestroi* (Blattodea: Rhinotermitidae) exhibit quantitatively different tunneling patterns. *Psyche: A Journal of Entomology* 2012, Article ID 675356. doi: 10.1155/2012/675356

Harit, A.K., Gajalakshmi, S. and Abbasi, S.A. (2014) Swarming of the termite *Coptotermes gestroi* in north-eastern Puducherry. *Zoology and Ecology* 24, 62–69. doi: 10.1080/21658005.2014.886407

Hashim, R., Boon, J.G., Sulaiman, O., Kawamura, F. and Lee, C.Y. (2009) Evaluation of the decay resistance properties of *Cerbera odollam* extracts and their influence on properties of particleboard. *International Biodeterioration & Biodegradation* 63, 1013–1017.

Hertel, H., Hanspach, A. and Plarre, R. (2011) Differences in alarm responses in drywood and subterranean termites (Isoptera: Kalotermitidae and Rhinotermitidae) to physical stimuli. *Journal of Insect Behavior* 24, 106–115.

Himmi, S.K., Tarmadi, D., Ismayati, M. and Yusuf, S. (2013) Bioefficacy performance of neem-based formulation on wood protection and soil barrier against subterranean termite, *Coptotermes gestroi* Wasmann (Isoptera: Rhinotermitidae). *Procedia Environmental Sciences* 17, 135–141. doi: 10.1016/j.proenv.2013.02.021
Holmgren, N. (1911) Ceylon – Termiten. In: Escherich, K. (ed.) *Termitenleben auf Ceylon. Neu Studien zur Sociologie der tiere Zugleich ein Kapitel Kolonialer Forstentomologie*. Gustav Fischer, Jena, Germany, pp. 184–212.
Hu, X.P. (2011) Liquid termiticides: their role in subterranean termite management. In: Dhang, P. (ed.) *Urban Pest Management: An Environmental Perspective*. CABI, Wallingford, UK, pp. 114–132.
Huang, F., Zhu, S., Ping, Z., He, X., Li, G. and Gao, D. (2000) *Fauna Sinica, Vol. 17, Insecta Isoptera*. Science Press, Beijing, China.
Huang, S.Y., Chiu, C.I., Tsai, Y.Y., Li, W.J., Wu, C.C. and Li, H.F. (2022) Nationwide termite pest survey conducted in Taiwan as a citizen science project. *Journal of Economic Entomology* 115, 1650–1658.
Ibrahim, S.A., Henderson, G. and Fei, H. (2003) Toxicity, repellency, and horizontal transmission of fipronil in the Formosan subterranean termite (Isoptera: Rhinotermitidae). *Journal of Economic Entomology* 96, 461–467.
Janei, V., Lima, J.T. and Costa-Leonardo, A.M. (2015) Rehydration after water stress in forager workers of *Coptotermes gestroi* (Wasmann) (Blattaria: Rhinotermitidae). *Neotropical Entomology* 44, 301–307.
Jasso-Selles, D.E., De Martini, F., Velenovsky, J.F., Mee, E.D., Montoya, S.J. *et al.* (2020) The complete protist symbiont communities of *Coptotermes formosanus* and *Coptotermes gestroi*: Morphological and molecular characterization of five new species. *Journal of Eukaryotic Microbiology* 67, 626–641. doi: 10.1111/jeu.12815
Jenkins, T.M., Jones, S.C., Lee, C.Y., Forschler, B.T., Chen, Z.B. et al. (2007) Phylogeography illuminates maternal origins of exotic *Coptotermes gestroi* (Isoptera : Rhinotermitidae). *Molecular Phylogenetics and Evolution* 42, 612–621.
Jessica, J.J., Peng, T.L., Sajap, A.S., Lee, S.H. and Syazwan, S.A. (2019) Evaluation of the virulence of entomopathogenic fungus, *Isaria fumosorosea* isolates against subterranean termites *Coptotermes* spp. (Isoptera: Rhinotermitidae). *Journal of Forestry Research* 30, 213–218.
Kadir, R. (2017) Toxic effects of three selected Malaysian timbers plant extracts against subterranean termites. *Maderas-Ciencia Y Tecnologia* 19, 417–432.
Kadir, R. and Hale, M.D. (2012) Comparative termite resistance of 12 Malaysian timber species in laboratory tests. *Holzforschung* 66, 127–130.
Ke, Y.L., Zhang, S.J. and Li, Z.Q. (2022) Phylogenetic relationships of Chinese *Coptotermes* (Blattodea: Isoptera: Rhinotermitidae) termites and a new synonym inferred from morphological data. *Journal of Entomological Science* 57, 64–81.
Kirton, L.G. and Azmi, M. (2005) Patterns in the relative incidence of subterranean termite species infesting buildings in peninsular Malaysia. *Sociobiology* 46, 1–15.
Kirton, L.G. and Brown, V.K. (2003) The taxonomic status of pest species of *Coptotermes* in Southeast Asia: Resolving the paradox in the pest status of the termites, *Coptotermes gestroi*, *C. havilandi* and *C. travians* (Isoptera : Rhinotermitidae). *Sociobiology* 42, 43–63.
Klangkaew, C., Inoue, T., Abe, T., Takematsu, Y., Kudo, T. et al. (2002) The diversity and abundance of termites (Isoptera) in the urban area of Bangkok, Thailand. *Sociobiology* 39, 485–493.
Kulis, J., Sajap, A.S., and Loong, C. (2008) Effect of moisture and relative humidity on survival and feeding activity of the Asian subterranean termite, *Coptotermes gestroi* (Isoptera : Rhinotermitidae). *Sociobiology* 52, 579–587.
Laranjo, L.T., Haifig, I. and Costa-Leonardo, A.M. (2019) A Glycoproteinaceous secretion in the seminal vesicles of the termite *Coptotermes gestroi* (Isoptera: Rhinotermitidae). *Insects* 10, 428. doi: 10.3390/insects10120428
Lee, C.C., Neoh, K.B. and Lee, C.Y. (2014) Colony size affects the efficacy of bait containing chitin synthesis inhibitor against the fungus-growing termite. *Journal of Economic Entomology* 107, 2154–2162.
Lee, C.Y. (2002) Control of foraging colonies of subterranean termites, *Coptotermes travians* (Isoptera: Rhinotermitidae) in Malaysia using hexaflumuron baits. *Sociobiology* 39, 411–416.
Lee, C.Y. (2007a) Perspective in Urban Insect Pest Management in Malaysia. Vector Control Research Unit, Universiti Sains Malaysia, Penang, Malaysia.
Lee, C.Y. (2007b) Evaluation of bistrifluron as an above-ground bait against the Asian subterranean termite, *Coptotermes gestroi* (Wasmann) in Malaysia. In: *Proceedings of the 4th Conference of the Pacific-Rim Termite Research Group*, pp. 8–11.
Lee, C.Y. (2014) Urban forest insect pests and their management in Malaysia. *Formosan Entomologist* 33, 207–214.

Lee, C.Y., Vongkaluang, C. and Lenz, M. (2007) Challenges to subterranean termite management of multi-genera faunas in Southeast Asia and Australia. *Sociobiology* 50, 213–221.

Lenz, M., Creffield, J.W., Evans, T.A., Kard, B., Vongkaluang, C. *et al.* (2012) Resistance of polyamide and polyethylene cable sheathings to termites in Australia, Thailand, USA, Malaysia and Japan: A comparison of four field assessment methods. *International Biodeterioration & Biodegradation* 66, 53–62.

Lenz, M., Kard, B., Creffield, J.W., Evans, T.A., Brown, K.S. *et al.* (2013) Ability of field populations of *Coptotermes* spp., *Reticulitermes flavipes*, and *Mastotermes darwiniensis* (Isoptera: Rhinotermitidae; Mastotermitidae) to damage plastic cable sheathings. *Journal of Economic Entomology* 106, 1395–1403.

Leonardo, F.C., da Cunha, A.F., da Silva, M.J., Carazzolle, M.F., Costa-Leonardo, A.M. *et al.* (2011) Analysis of the workers head transcriptome of the Asian subterranean termite, *Coptotermes gestroi*. *Bulletin of Entomological Research* 101, 383–391.

Li, H.F., Ye, W.M., Su, N.Y. and Kanzaki, N. (2009) Phylogeography of *Coptotermes gestroi* and *Coptotermes formosanus* (Isoptera: Rhinotermitidae) in Taiwan. *Annals of the Entomological Society of America* 102, 684–693.

Li, H.F., Su, N.Y. and Wu., W.J. (2010a) Solving the hundred-year controversy of *Coptotermes* taxonomy in Taiwan. *American Entomologist* 56, 222–227.

Li, H.F., Yang, R.L. and Su, N.Y. (2010b) Interspecific competition and territory defense mechanisms of *Coptotermes formosanus* and *Coptotermes gestroi* (Isoptera: Rhinotermitidae). *Environmental Entomology* 39, 1601–1607.

Li, H.F., Fujisaki, I. and Su, N.Y. (2013) Predicting habitat suitability of *Coptotermes gestroi* (Isoptera: Rhinotermitidae) with species distribution models. *Journal of Economic Entomology* 106, 311–321.

Li, H.F., Chen, G.Y., Huang, S.Y., Lin, M.D., Chouvenc, T. and Ching, Y.H. (2023) Hybrids of two destructive subterranean termites established in the field, revealing a potential for gene flow between species. *Research Square* (pre-print). doi: 10.21203/rs.3.rs-2801694/v1

Li, Z.-Q., Liu, B.R., Li, Q.J., Xiao, W.L. and Zhong, J.H. (2011) Two new synonyms of *Coptotermes gestroi* (Wasmann) (Isoptera: Rhinotermitidae) in China. *Sociobiology* 58, 449–455.

Li, Z., Liu, B., Zeng, W. and Zhong, J. (2012) Genetic variation in *Coptotermes gestroi* (Isoptera: Rhinotermitidae) populations from Hainan, Taiwan, and the Philippines. *Journal of Agricultural and Urban Entomology* 28, 57–65.

Lima, J.T. and Costa-Leonardo, A.M. (2006) Interspecific interactions between *Coptotermes gestroi* and *Heterotermes tenuis* (Isoptera : Rhinotermitidae) under laboratory conditions. *Sociobiology* 48, 759–770.

Lima, J.T. and Costa-Leonardo, A.M. (2012) Tunnelling behaviour of the Asian subterranean termite in heterogeneous soils: Presence of cues in the foraging area. *Animal Behaviour* 83, 1269–1278.

Lima, J.T. and Costa-Leonardo, A.M. (2014) Foraging in subterranean termites (Isoptera: Rhinotermitidae): How do *Heterotermes tenuis* and *Coptotermes gestroi* behave when they locate equivalent food resources? *Bulletin of Entomological Research* 104, 525–533.

Liu, Y., Henderson, G., Mao, L. and Laine, R.A. (2005) Seasonal variation of juvenile hormone titers of the Formosan subterranean termite, *Coptotermes formosanus* (Rhinotermitidae). *Environmental Entomology* 34, 557–562.

Loladze, I. (2014) Hidden shift of the ionome of plants exposed to elevated CO_2 depletes minerals at the base of human nutrition. *eLife* 3, e02245.

Maketon, M., Sawangwan, P. and Sawatwarakul, W. (2007) Laboratory study on the efficacy of *Metarhizium anisopliae* (Deuteromycota : Hyphomycetes) in controlling *Coptotermes gestroi* (Isoptera : Rhinotermitidae). *Entomologia Generalis* 30, 203–218.

Martins, C., Fontes, L.R., Bueno, O.C. and Martins, V.G. (2010) *Coptotermes gestroi* (Isoptera: Rhinotermitidae) in Brazil: possible origins inferred by mitochondrial cytochrome oxidase II gene sequences. *Genome* 53, 65--657.

Matsuura, K. and Nishida, T. (2001) Colony fusion in a termite: What makes the society “open”? *Insectes Sociaux* 48, 378–383.

Nagaraju, D.K., Kalleshwaraswamy, C.M., Iyyanar, D., Singh, M., Jain, R.K. *et al.* (2021) First interception of two wood feeding potential invasive *Coptotermes* termite species in India. *International Journal of Tropical Insect Science* 41, 1043-1052.

Nasir, S. and Akhtar, M. (2011) Swarming of termites (Isoptera) in the Mianwali district, Pakistan. *Sociobiology* 58, 151–164.

Neoh, K.B. and Lee, C.Y. (2009a) Flight activity of two sympatric termite species, *Macrotermes gilvus* and *Macrotermes carbonarius* (Termitidae: Macrotermitinae). *Environmental Entomology* 38, 1697–1706.

Neoh, K.B. and Lee, C.Y. (2009b) Flight activity and flight phenology of the Asian subterranean termite, *Coptotermes gestroi* (Blattodea: Rhinotermitidae). *Sociobiology* 54, 521–530.

Neoh, K.B., Jalaludin, N.A. and Lee, C.Y. (2011) Elimination of field colonies of a mound-building termite *Globitermes sulphureus* (Isoptera: Termitidae) by bistrifluron bait. *Journal of Economic Entomology* 104, 607–613.

Neoh, K.B., Indiran, Y., Lenz, M. and Lee, C.Y. (2012a) Does lack of intraspecific aggression or absence of nymphs determine acceptance of foreign reproductives in *Macrotermes*? *Insectes Sociaux* 59, 223–230.

Neoh, K.B., Hu, J., Yeoh, B.H. and Lee, C.Y. (2012b) Toxicity and horizontal transfer of chlorantraniliprole against the Asian subterranean termite *Coptotermes gestroi* (Wasmann): effects of donor: recipient ratio, exposure duration and soil type. *Pest Management Science* 68, 749–756.

Neoh, K.B., Lee, C.C. and Lee, C.Y. (2014a) Effects of termiticide exposure on mutual interactions between the treated and untreated workers of the Asian subterranean termite *Coptotermes gestroi*. *Pest Management Science* 70, 240–244.

Neoh, K.B., Yeoh, B.H. and Lee, C.Y. (2014b) Mortality patterns in *Coptotermes gestroi* (Blattodea: Rhinotermitidae) following horizontal transfer of nonrepellent and repellent insecticides: Effects of donor: recipient ratio and exposure time. *Journal of Economic Entomology* 107, 1563–1572.

Neoh, K.B., Bong, L.J., Muhammad, A., Itoh, M., Kozan, O. *et al.* (2016) The impact of tropical peat fire on termite assemblage in Sumatra, Indonesia: Reduced complexity of community structure and survival strategies. *Environmental Entomology* 45, 1170–1177.

Neoh, K.B., Bong, L.J., Muhammad, A,. Itoh, M., Kozan, O. *et al.* (2017) The effect of remnant forest on insect successional response in tropical fire-impacted peatland: A bi-taxa comparison. *PLoS ONE* 12, e0174388. doi: 10.1371/journal.pone.0174388

Ngee, P.S., Tashiro, A., Yoshimura, T., Jaal, Z. and Lee, C.Y. (2004) Wood preference of selected Malaysian subterranean termites (Isoptera : Rhinotermitidae, Termitidae). *Sociobiology* 43, 535–550.

Noirot, C. (1991) Caste differentiation in Isoptera: Basic features, role of pheromones. *Ethology Ecology & Evolution* 3, 2–7.

Novita, N., Amiruddin, H., Ibrahim, H., Jamil, T.M., Syaukani, S. *et al.* (2020) Investigation of termite attack on cultural heritage buildings: A case study in Aceh Province, *Indonesia. Insects* 11, 385. doi: 10.3390/insects11060385

Nutting, W.L. (1969) Flight and colony foundation. In: Krishna, K. and Weesner, F.M. (eds) *Biology of Termites*. Academic Press New York, pp. 233-282.

Padayachi, B., Prasad, N. and Rounds, P. (2021) Status of Asian subterranean termites in the sugarcane belt of Fiji. In: *Proceedings of the Australian Society of Sugar Cane Technologists, 20–23 April 2021*, Bundaberg, Australia, pp. 540–542.

Patel, J.S., Chouvenc, T. and Su, N.Y. (2019) Temperature preference of two invasive subterranean termite species and their hybrids (Blattodea: Rhinotermitidae: Coptotermes). *Journal of Economic Entomology* 112, 2888–2893.

Plarre, R. (2018) Termiten in Deutschland – wo gibt es denn so was? *Sitzungsberichte der Gesellschaft Naturforschender Freunde zu Berlin* 53, 83–90. https://opus4.kobv.de/opus4-bam/frontdoor/index/index/year/2018/docId/44055

PMP (2023) Asian subterranean termite, *Coptotermes gestroi*. *Pest Management Professional* 91(5), 54. https://www.mypmp.net/2023/05/11/asian-subterranean-termite-coptotermes-gestroi/

Potter, M.F. (1999) The changing face of termite control. II. *Pest Control Technology* 27, 33–34, 36, 38–39, 42, 90.

Pretzsch, H., Biber, P., Schütze, G., Kemmerer, J. and Uhl, E. (2018). Wood density reduced while wood volume growth accelerated in Central European forests since 1870. *Forest Ecology and Management* 429, 589–616.

Puche, H. and Su, N.Y. (2001) Tunnel formation by *Reticulitermes flavipes* and *Coptotermes formosanus* (Isoptera: Rhinotermitidae) in response to wood in sand. *Journal of Economic Entomology* 94, 1398–1404.

Rojo, M.J.A. and Acda, M.N. (2019) Elimination of the Asian subterranean termite *Coptotermes gestroi* (Blattodea: Rhinotermitidae) using above-ground chlorfluazuron baits. *Philippine Agricultural Scientist* 102, 90–93.

Roonwal, M.L. and Chhotani, O.B. (1962) Termite fauna of Assam region, Eastern India. *Proceedings of the National Institute of Sciences of India, Part B, Biological Sciences* 28, 281–406.

Roszaini, K., Rafeadah, R. and Dahlan, J.M. (2013) Durability of Malaysian timbers against the Asian subterranean termite *Coptotermes gestroi* Wasmann. *Journal of the Institute of Wood Science* 19, 16–21.

Sajap, A.S., Jaafar, M.A. and Ouimette, D. (2002) Above-ground baiting for controlling subterranean termite, *Coptotermes travians* (Isoptera: Rhinotermitidae) in Selangor, Peninsular Malaysia. *Sociobiology* 39, 345–350.

Sajap, A.S., Lee, L.C., Ouimette, D. and Jaafar, M.A. (2005) Field evaluation of noviflumuron for controlling Asian subterranean termite, *Coptotermes gestroi* (Isoptera: Rhinotermitidae). In: Lee, C.Y. and W.H. Robinson (eds) *Proceedings of the 5th International Conference on Urban Pests*. P&Y Design Network, Penang, Malaysia, pp. 239–241.

Sajap, A.S., Lee, L.C. and Shah, Z.M. (2009) Elimination of subterranean termite colonies with hexaflumuron in an improved bait matrix, preferred textured cellulose (PTC). *Sociobiology* 53, 891–901.

Santos, M.N., Teixeira, M.L.F., Pereira, M.B. and Menezes, E.B. (2010) Environmental factors influencing the foraging and feeding behavior of two termite species (Isoptera: Rhinotermitidae) in natural habitats. *Sociobiology* 55, 763–777.

Scharf, M.E., Buckspan, C.E., Grzymala, T.L. and Zhou, X. (2007) Regulation of polyphenic caste differentiation in the termite *Reticulitermes flavipes* by interaction of intrinsic and extrinsic factors. *Journal of Experimental Biology* 210, 4390–4398.

Scheffrahn, R.H. and Crowe, W. (2011) Ship-borne termite (isoptera) border interceptions in Australia and onboard infestations in Florida, 1986–2009. *Florida Entomologist* 94, 57–63.

Scheffrahn, R.H. and Su, N.Y. (2005) Distribution of the termite genus *Coptotermes* (Isoptera: Rhinotermitidae) in Florida. *Florida Entomologist* 88, 201–203.

Scheffrahn, R.H., Su, N.Y. and Diehl, B. (1990) Native, introduced, and structure-infesting termites of the Turks and Caicos islands, B.W.I. (Isoptera: Kalotermitidae, Rhinotermitidae, Termitidae). *Florida Entomologist* 73, 622–627.

Scheffrahn, R.H., Carrijo, T.F., Krecek, J., Su, N.Y., Szalanski, A.L. *et al.* (2015) A single endemic and three exotic species of the termite genus *Coptotermes* (Isoptera, Rhinotermitidae) in the New World. *Arthropod Systematics & Phylogeny* 73, 333–348.

Shelton, T.G., Bell, C.D. and Wagner, T.L. (2005) Lack of transfer of permethrin among nestmates of *Reticulitermes flavipes* in laboratory trials (Isoptera: Rhinotermitidae). *Sociobiology* 45, 69–75

Sornnuwat, Y., Tsunoda, K., Yoshimura, T., Takahashi, M. and Vongkaluang, C. (1996) Foraging populations of *Coptotermes gestroi* (Isoptera: Rhinotermitidae) in an urban area. *Journal of Economic Entomology* 89, 1485–1490.

Su, N.Y. (1994) Field evaluation of a hexaflumuron bait for population suppression of subterranean termites (Isoptera: Rhinotermitidae). *Journal of Economic Entomology* 87, 389–397.

Su, N.Y. and Lee, S.B. (2023) A comparison of tunnel geometry between the Formosan subterranean termite and Asian subterranean termite (Blattodea: Rhinotermitidae). *Pest Management Science* (in press).

Su, N.Y and Puche, H. (2003) Tunneling activity of subterranean termites (Isoptera: Rhinotermitidae) in sand with moisture gradients. *Journal of Economic Entomology* 96, 88–93.

Su, N.Y., Scheffrahn, R.H., and Weissling, T. (1997) A new introduction of a subterranean termite, *Coptotermes havilandi* Holmgren (Isoptera: Rhinotermitidae) in Miami, Florida. *Florida Entomologist* 80, 408–411.

Su, N.Y., Ban, P.M. and Scheffrahn, R.H. (2000) Control of *Coptotermes havilandi* (Isoptera: Rhinotermitidae) with hexaflumuron baits and a sensor incorporated into a monitoring and baiting program. *Journal of Economic Entomology* 93, 415–421.

Su, N.Y., Chouvenc, T. and Li, H.F. (2017) Potential hybridization between two invasive termite species, *Coptotermes formosanus* and *C. gestroi* (Isoptera: Rhinotermitidae), and its biological and economic implications. *Insects* 8, 14. doi: 10.3390/insects8010014.

Su, N.Y., Mullins, A., and Chouvenc, T. (2023) Elimination of structural and tree infestations of the Asian subterranean termite, Coptotermes gestroi (Wasmann) (Blattodea: Rhinotermitidae) with noviflumuron baits in above-ground stations. *Journal of Economic Entomology* 116, 909–915.

Syamani, F.A., Subiyanto, B. and Massijaya, M.Y. (2011) Termite resistant properties of sisal fiberboards. *Insects* 2, 462–468. doi: 10.3390/insects2040462

Takematsu, Y. and Vongkaluang, C. (2012) A taxonomic review of the Rhinotermitidae (Isoptera) of Thailand. *Journal of Natural History* 46, 1079–1109.

Tarasin, M. and Rattanapun, W. (2019) Termite resistance of *Melaleuca cajuputi* wood treated with citric acid. *Agriculture and Natural Resources* 53, 662–666.

Tarver, M.R., Schmelz, E.A. and Scharf, M.E. (2011) Soldier caste influences on candidate primer pheromone levels and juvenile hormone-dependent caste differentiation in workers of the termite *Reticulitermes flavipes*. *Journal of Insect Physiology* 57, 771–777.

Tho, Y.P. (1992) *Termites of Peninsular Malaysia.* Malayan Forest Records No. 36. Forest Research Institute of Malaysia, Kuala Lumpur.

Tong, R.L., Grace, J.K., Mason, M., Krushelnycky, P.D., Spafford, H. and Aihara-Sasaki, M. (2017) Termite species distribution and flight periods on Oahu, *Hawaii. Insects* 8, 58. doi: 10.3390/insects8020058

Tonini, F., Divino, F., Lasinio, G.J., Hochmair, H.H. and Scheffrahn, R.H. (2014) Predicting the geographical distribution of two invasive termite species from occurrence data. *Environmental Entomology* 43, 1135–1144.

Tramontina, R., Cairo, J., Liberato, M.V., Mandelli, F., Sousa, A., *et al.* (2017) The *Coptotermes gestroi* aldo-keto reductase: a multipurpose enzyme for biorefinery applications. *Biotechnology for Biofuels* 10, 4. doi: 10.1186/s13068-016-0688-6

Tsai, C.C. and Chen, C.S. (2003) First record of *Coptotermes gestroi* (Isoptera: Rhinotermitidae) from Taiwan. *Formosan Entomologist* 23, 157–161.

Uchima, S.Y. and Grace, J.K. (2003) Characteristics of *Coptotermes vastator* (Isoptera: Rhinotermitidae) colonies on Oahu, *Hawaii. Sociobiology* 41, 281–288.

Uchima, S.Y. and Grace, J.K. (2009) Interspecific agonism and foraging competition between *Coptotermes formosanus* and *Coptotermes gestroi* (Blattodea: Rhinotermitidae). *Sociobiology* 54, 765–776.

Umar, W. and Ab Majid, A.H. (2020) Efficacy of minimum application of chlorfluazuron baiting to control urban subterranean termite populations of *Coptotermes gestroi* (Wasmann) (Blattodea: Rhinotermitidae). *Insects* 11, 569. doi: 10.3390/insects11090569.

Velenovsky, J.F., Gile, G.H., Su, N.Y. and Chouvenc, T. (2021) Dynamic protozoan abundance of *Coptotermes* kings and queens during the transition from biparental to alloparental care. *Insectes Sociaux* 68, 33–40.

Velonovsky, J.F., Martini, F.D., Hileman, J.T., Gordon, J.M., Su, N.Y., Gile, G.H. and Chouvenc, T. (2023) Vertical transmission of cellulolytic protists in termites is imperfect, but sufficient, due to biparental transmission. *Symbiosis* (in press).

Venkatesan, T., Kalleshwaraswamy, C.M., Gupta, A. and Ashika, T.R. (2021) Intrusion pathway of invasive Asian subterranean termite, *Coptotermes gestroi* (Wasmann) from the Neotropics into the Indian mainland. *Current Science* 120, 1778–1781.

Wasmann, E. (1896) Viaggio di Leonardo Fea in Birmania e regioni vicine. lxxii. Neue Termitophilen und Termiten aus Indian. *Annali del Museo civico di storia naturale di Genova* 36, 613–630.

Weesner, F. (1965) *The Termites of the United States*. The National Pest Control Association, Elizabeth, New Jersey.

Wikantyoso, B., Imai, T., Himmi, S.K., Yusuf, S., Hata, T. and Yoshimura, T. (2022) Ultrastructure and distribution of sensory receptors on the nonolfactory organs of the soldier caste in subterranean termite (*Coptotermes* spp.). *Arthropod Structure & Development* 70, 101201. doi: 10.1016/j.asd.2022.101201

Wikantyoso, B., Tseng, S.P., Himmi, S.K., Yusuf, S. and Yoshimura, T. (2021) Morphometric analysis of *Coptotermes* spp. soldier caste (Blattodea: Rhinotermitidae) in Indonesia and evidence of *Coptotermes gestroi* extreme head-capsule shapes. *Insects* 12, 477. doi: 10.3390/insects12050477.

Wong, N. and Lee, C.Y. (2010a) Influence of different substrate moistures on wood consumption and movement patterns of *Microcerotermes crassus* and *Coptotermes gestroi* (Blattodea: Termitidae, Rhinotermitidae). *Journal of Economic Entomology* 103, 437–442.

Wong, N. and Lee, C.Y. (2010b) Intra- and interspecific agonistic behavior of the subterranean termite *Microcerotermes crassus* (Isoptera: Termitidae). *Journal of Economic Entomology* 103, 1754–1760.

Wong, N. and Lee, C.Y. (2010c) Effects of disturbance and the presence of termite and other invertebrate carcasses at feeding sites on the behavior of the subterranean termite *Microcerotermes crassus* (Blattodea: Termitidae). *Sociobiology* 55, 353–367.

Woodrow, R., Grace, J.K. and Higa, S. (2001) Occurrence of *Coptotermes vastator* (Isoptera: Flhinotermitidae) on the Island of Oahu, Hawaii. *Sociobiology* 38, 667–673

Xia, K. and He, X. (1986) Study on the genus *Coptotermes* from China (Isoptera: Rhinotermitidae). *Contributions from Shanghai Institute of Entomology* 6, 157–182.

Yeap, B.K., Othman, A.S., Lee, V.S. and Lee, C.Y. (2007) Genetic relationship between *Coptotermes gestroi* and *Coptotermes vastator* (Isoptera : Rhinotermitidae). *Journal of Economic Entomology* 100, 467–474.

Yeap, B.K., Othman, A.S. and Lee, C.Y. (2011) Genetic analysis of population structure of *Coptotermes gestroi* (Isoptera: Rhinotermitidae) in native and introduced populations. *Environmental Entomology* 40, 470–476.

Yeoh, B.H. and Lee, C.Y. (2007a) Tunneling activity, wood consumption and survivorship of *Coptotermes gestroi*, *Coptotermes curvignathus* and *Coptotermes kalshoveni* (Isoptera : Rhinotermitidae) in the laboratory. *Sociobiology* 50, 1087–1096.

Yeoh, B.H. and Lee, C.Y. (2007b) Tunneling responses of the Asian subterranean termite, *Coptotermes gestroi* in termiticide-treated sand (Isoptera : Rhinotermitidae). *Sociobiology* 50, 457–468.

Zanne, A.E., Flores-Moreno, H., Powell, J.R., Cornwell, W.K., Dalling, J.W. *et al.* (2022) Termite sensitivity to temperature affects global wood decay rates. *Science* 377, 1440–1444.

Zhang, M.P. and Evans, T.A. (2017) Determining urban exploiter status of a termite using genetic analysis. *Urban Ecosystems* 20, 535–545.

Zorzenon, F. and Campos, A. (2015) Subterranean termites in urban forestry: Tree preference and management. *Neotropical Entomology* 44, 180–185.

17 Hybridization between *Coptotermes formosanus* and *Coptotermes gestroi*

Thomas Chouvenc[1] **and Hou-Feng Li**[2]
[1]*Entomology and Nematology Department, Ft. Lauderdale Research and Education Center, University of Florida, tomchouv@ufl.edu;* [2]*Department of Entomology, National Chung Hsing University, Taiwan, houfeng@nchu.edu.tw*

Abstract
Hybridization is a biological process that may allow for the exchange of genetic information between lineages, with potential ecological and evolutionary consequences. In areas where the Formosan subterranean termite, *Coptotermes formosanus* Shiraki, and the Asian subterranean termite, *Coptotermes gestroi* (Wasmann), coexist, it was demonstrated that alates produced from mature colonies of both species would fly simultaneously at the same location, allowing opportunities for mating interactions. The absence of pre-zygotic barriers between the two species results in courtship confusion, which leads to heterospecific colony foundation. In Florida and in Taiwan, hybrid colonies were obtained in the laboratory, and it was confirmed that heterospecific mating can result in viable hybrid *Coptotermes* colonies. Since 2019, hybrid alates have been found in the field during their dispersal flight seasons, both in Florida and in Taiwan, and one established hybrid colony was found in the field in Taiwan. All castes produced by hybrid colonies were found to display some trait anomalies, which can either be intermediate, cumulative, or overlapping with the two parental species, or can depart away from both parental species. As F1 and advanced hybrids (> F1 generations) of alates were found in the field, it implies that there is potential gene flow between the two species in geographic areas where they overlap. However, laboratory colonies displayed poor F2 fertility, suggesting a potentially limited gene flow between the two populations. The current knowledge about hybrid *Coptotermes* is still in its infancy, and an ongoing investigation will further reveal the possible short-term and long-term implications of this biological entity.

"The interaction between the two species could prove to be interesting." (Henry A. Bess, 1970)

17.1 Introduction

Global anthropogenic activities have resulted in the alteration of habitats and in the change of geographic ranges for many species with invasive capabilities (Vitousek *et al.*, 1997; Frishkoff *et al.*, 2016; Lenoir *et al.*, 2017; Sardain *et al.*, 2019; Wallingford *et al.*, 2020). Biological invasions often have negative ecological and economic impacts, as some invasive organisms can destabilize ecosystems, critically reduce agricultural yields, or impact human health and infrastructures (Lowe *et al.*, 2000; Perrings *et al.*, 2002; Diagne *et al.*, 2021). Biological invasions may also result in situations where congeneric species that have been geographically isolated over evolutionary time may now be found in sympatry, potentially providing opportunities for hybridization and gene flow among species from heterospecific mating events (Crispo *et al.*, 2011; Chunco,

 Biology and Management of the Formosan Subterranean Termite and Related Species (eds Nan-Yao Su and Chow-Yang Lee)
DOI: 10.1079/9781800621596.0017

2014; San Jose *et al.*, 2023). Such events have increasingly been reported in recent times, especially among insect invaders (Fournier and Aron, 2021; Arce-Valdés and Sánchez-Guillén, 2022).

Heterospecific mating events may occur when pre-zygotic barriers between two species did not form during allopatric speciation and when courtship and mating phenologies still overlap sufficiently (Crispo *et al.*, 2011; Chouvenc *et al.*, 2015). However, while post-zygotic isolation mechanisms may result in maladaptive phenotypes, potentially owing to genetic incompatibilities and outbreeding depression (Frankham *et al.*, 2011; Lowe *et al.*, 2015; Todesco *et al.*, 2016; Bourret *et al.*, 2022), persisting genetic compatibilities with intermediate or transgressive phenotypes may result in viable, vigorous, and fertile hybrid individuals (Birchler *et al.*, 2003; Dittrich-Reed and Fitzpatrick, 2013). As a result, hybridization events can lead to various scenarios, from displacement of native species (Huxel, 1999; Buggs, 2007; Muhlfeld *et al.*, 2017), to absence of hybrid vigor (Burke and Arnold, 2001; Mallet, 2007), hybrid zones with degrees of gene flow among populations (Mallet, 2005), long-term evolutionary consequences such as hybrid speciation, and sources of genetic innovations (Abbott *et al.*, 2013, 2016; Moran *et al.*, 2021).

Hybridization events between an invasive and a native species have commonly been reported (Mooney and Cleland, 2001; Chunco, 2014), and it was estimated that ~10% of animal species may have gone through a history of hybridization (Mallet, 2005). However, reports of hybridization between two invasive species in non-native ranges are relatively rare (Shoemaker *et al.*, 1996; Gaskin and Kazmer, 2009; Diedericks *et al.*, 2018). In the past decade, the report of a potential hybridization between two *Coptotermes* invaders in Florida (Chouvenc *et al.*, 2015) has raised concerns about the potential gene flow and economic consequences of a putative hybrid termite population (Su *et al.*, 2017). *Coptotermes formosanus* Shiraki and *Coptotermes gestroi* (Wasmann) are two of the most successful termite invaders in the world (Evans *et al.*, 2013; Chouvenc *et al.*, 2016a; Evans, 2021; Blumenfeld *et al.*, 2021) with major economic and ecological impacts in their respective native and invasive ranges, making these two particular subterranean termite species responsible for a significant portion of the ~US$40 billion global annual termite cost to society (Rust and Su, 2012). Evidence of heterospecific mating behaviors between *C. gestroi* and *C. formosanus* was first observed in Florida in 2013 (Chouvenc *et al.*, 2015). It was then also confirmed in Taiwan in 2019 (Chen, 2022), where both species co-occur, although *C. formosanus* is native to the area, while *C. gestroi* has at least a 100-year-old history of invasion on the island (Li *et al.*, 2010a). Therefore, the prediction by Bess (1970) that *"The interaction between the two species could prove to be interesting,"* while made in the context of comparing the two *Coptotermes* species' destructive power and competitiveness as two invaders in Hawai'i (Uchima and Grace, 2009; Li *et al.*, 2010b), has now taken on a whole new meaning that Bess (1970) may not have foreseen at the time.

This chapter aims to provide the current knowledge and biological context about the interaction between *C. gestroi* and *C. formosanus* and their hybrid. First, the geographic distribution overlap and dispersal flight phenology of the two species are described. Second, we provide an overview of the lack of interspecific mating barriers which has led to an absence of pre-zygotic isolation. Third, observations from hybrid laboratory colonies have allowed for an investigation of the similarity and differences between hybrid colonies and their respective parental species. Fourth, the implication of the recent detection of hybrid *Coptotermes* in the field is discussed. Finally, hybrid *Coptotermes* colonies provide opportunities to test novel termite research questions, and an overview of ongoing and future research possibilities is covered.

17.2 Overlapping geographic distribution and dispersal flight phenology

Coptotermes formosanus and *C. gestroi* are the only two species within the genus that have spread far beyond the constraints of their respective native ranges thanks to human activity (see Chapters 2, 3, and 16 of this volume) (Chouvenc *et al.*, 2016a). While *C. formosanus* is primarily restricted to areas of warm temperate and subtropical climates, *C. gestroi* has spread to many tropical localities around the world (Li *et al.*, 2013). Therefore, their respective geographic distributions are almost completely mutually exclusive as a result of their different environmental requirements (Cao and Su, 2016), except for narrow overlap in a few locations. Such geographic overlaps can be found on Oahu Island (Hawai'i), southwest Taiwan, and in southeast Florida (Su *et al.*, 2017) (Fig. 17.1A), and are primarily restricted to highly urbanized areas (Chouvenc *et al.*, 2016b; Huang *et al.*, 2022). As a result, some communities within these areas are under the risk of infestation by both species. In addition, these areas have been expanding over the past few decades, as it was shown that, in southeastern Florida, both species are expanding their invasive range, increasing their area of overlap over time (Chouvenc *et al.*, 2016b; Su *et al.*, 2017). Similarly in Taiwan, *C. gestroi* has recently expanded its invasive range northward, gaining ground on localities historically infested with *C. formosanus* (Huang *et al.*, 2022).

In all known overlapping ranges, distinct dispersal flight periods were observed for alates of each species (Scheffrahn and Su, 2005). In Florida, the *C. gestroi* flight activity peaks early March, while the *C. formosanus* flight activity peaks around late April to early May (Chouvenc *et al.*, 2017a). In Taiwan, *C. gestroi* flight activity peaks early–mid-April, while the *C. formosanus* flight activity peaks around mid–late May (Huang *et al.*, 2022). In both locations, the peaks of flight activity differ between the two species, but the dispersal flight seasons can last for up to ~2 months for each species (Fig. 17.1B,C), with multiple flight events throughout the season (see Chapter 4 of this volume). In Broward County, Florida, flight season overlap has been observed every year since 2010 in southeast Florida during late March to early April (Chouvenc *et al.*, 2017a). In 2013, the first direct observation of alates of both species collected in a single light trap confirmed that *C. formosanus* and *C. gestroi* alates flew at the same place and at the same time (Chouvenc *et al.*, 2015) despite a slight temporal delay in phenology, i.e., early twilight for *C. gestroi*, late twilight for *C. formosanus* (see Chapter 4 of this volume). Such dispersal season overlap was also confirmed in Taiwan (Chen, 2022; Huang *et al.*, 2022). In 2019–2021, during

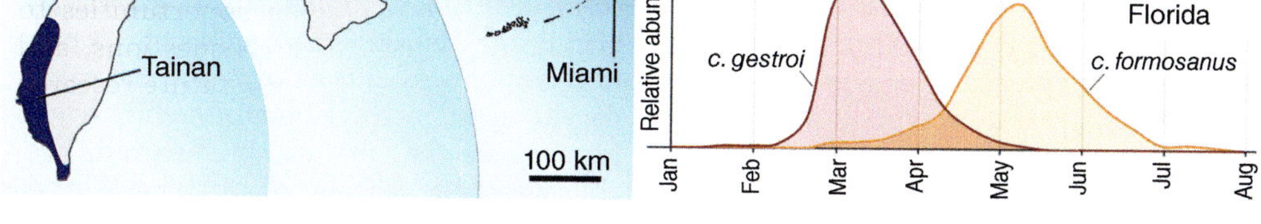

Fig. 17.1. (A) Blue areas represent known distribution overlap for *C. formosanus* and *C. gestroi*, from left to right, in Taiwan, Hawai'i, and Florida (map generated from the University of Florida Termite Collection, and from the National Chung Hsing University Termite Collection, used with permission). (B) Dispersal flight season for *C. formosanus* and *C. gestroi* in Taiwan (data from the National Chung Hsing University Termite Collection, used with permission). (C) Dispersal flight season for *C. formosanus* and *C. gestroi* in Florida (data from the University of Florida Termite Collection, used with permission)

Coptotermes dispersal flight season in Taichung City, Taiwan, multiple simultaneous dispersal flight events of *C. formosanus* and *C. gestroi* were recorded (Chen, 2022). It is now confirmed that in areas with overlapped distributions, multiple potential simultaneous flight events have occurred each year since about 2010 in Florida, and most likely for many decades in Taiwan. While the dispersal flight seasons of the two *Coptotermes* species also partially overlap in Hawai'i, simultaneous flight events have not been confirmed there (Tong *et al.*, 2017).

17.3 Tandem behavior confusion

During simultaneous dispersal flight events, it was observed that males and females of both species readily engage in tandem run behaviors in field conditions, regardless of the species of the tandem partner (Chouvenc *et al.*, 2015; Chen, 2022). As *Coptotermes* alates drop their wings in search of a potential mate, dealates engage in a tandem run when a male and a female encounter, as the female assumes the role of the leader followed by the male (Mizumoto *et al.*, 2020). Like most termite species, *Coptotermes* females produce a sex pheromone from their tergal glands which allows the male to maintain contact with the female during the tandem run. In both *C. formosanus* and *C. gestroi*, females produce dodecatrienol (see Chapter 7 of this volume), implying that the two species use the same semiochemical as a sex pheromone to maintain tandem runs (Chouvenc *et al.*, 2020). However, *C. formosanus* female tergal glands are bigger and produce ~10 times the quantity of sex pheromone when compared with *C. gestroi* female tergal glands (Chouvenc *et al.*, 2020). As a result, *C. gestroi* males primarily engage in tandem runs with *C. formosanus* females in a choice assay, while *C. formosanus* males primarily engage in tandem runs with females of their own species (Chouvenc *et al.*, 2015, 2020; Mizumoto *et al.*, 2021).

Such courtship behavior confusion results in opportunities for a female of one species to initiate a claustral chamber and establish an incipient colony with a male of the other species under field conditions (Chouvenc *et al.*, 2017a, 2020). While pairings of *C. formosanus* female × *C. gestroi* male are likely to occur readily during simultaneous dispersal flights, it remains possible for pairings of *C. gestroi* female × *C. formosanus* male to form, despite its relative rarity (Chouvenc *et al.*, 2020; Mizumoto *et al.*, 2021). It is therefore possible for colonies of both mating combinations to potentially establish in the field.

17.4 F1 hybrid *Coptotermes* in the laboratory

In the initial absence of the discovery of established hybrid field colonies, incipient colonies were established in the laboratory with heterospecific mating pairs, to evaluate hybrid colony viability and biological characteristics. Colonies established by a *C. formosanus* female × *C. gestroi* male pair are referred as "Hybrid F," while colonies established by a *C. gestroi* female × *C. formosanus* male pair are referred as "Hybrid G," where the letter indicates the species of the female. Colonies of both hybrid mating combinations were successfully established in laboratory conditions in Florida and Taiwan (Chouvenc *et al.*, 2015; Chen, 2022; Patel *et al.*, 2023a). The initial study only obtained a low number of colony replicates and first suggested a potential high hybrid vigor, as it was observed that populations of hybrid colonies could grow up to about twice the number of individuals within the first year, when compared with both parental species (Chouvenc *et al.*, 2015). However, it was later confirmed that *C. formosanus*, *C. gestroi*, Hybrid F, and Hybrid G had equivalent population growth within the first 4 years of colony development in the laboratory, with a wide intraspecies (i.e., mating combination) variability over the years in colony size and in soldier ratio (Patel *et al.*, 2023a).

Wood consumption rates were similar among colonies from all mating combinations (Patel *et al.*, 2019a, 2020) and temperature-dependent activity of both hybrid types was more like that of *C. gestroi* than *C. formosanus* (Patel *et al.*, 2019b). In addition, temperature-dependent survivorship of both hybrid types covered the temperature tolerance of both parental species (15–35°C), suggesting that hybrid colonies have the potential to establish and survive in climate conditions of either parental species (Patel *et al.*, 2019a). From a semiochemical perspective, the cuticular hydrocarbon profiles of both *Coptotermes* hybrid types were intermediate to the two parental species profiles (Chouvenc and Su, 2017). Remarkably, despite similar cuticular hydrocarbon profiles, Hybrid F and Hybrid G colonies display high degrees of agonism (Chouvenc and Su, 2017).

Thus far, studies have shown that many traits in *Coptotermes* hybrids are either intermediate, cumulative, or overlapping with the two parental species. However, some traits deviate from either of the parental species, including soldier morphology, molting rates, and construction behavior. Soldiers produced by hybrid *Coptotermes* colonies can be morphologically differentiated from their two parental species with a few distinct qualitative trait differences, despite having a wide overlap in quantitative morphometric traits (Patel, 2020; Chen, 2022; Patel *et al.*, 2023b). The most reliable characteristic for identification is the fontanelle area, including the shape of the fontanelle and number of setae around the fontanelle (Fig. 17.2). The fontanelle shape is trianguliform (arched dorsally) in *C. formosanus*, subcircular (sometimes with soft a hexagonal edge) in *C. gestroi*, and horizontally ellipsoid in Hybrid F and Hybrid G. There are two pairs of setae beside the fontanelle in *C. formosanus*, while there is a single pair of setae in *C. gestroi*. In Hybrid F and Hybrid G, there are usually two pairs of setae as in *C. formosanus*, although one or two setae may sometimes be missing, reflecting degrees of developmental instability in hybrid soldiers (Chouvenc *et al.*, 2017b). In addition, the cuticle suture besides the soldier postmentum (postmental sulcum) displays few heavily sclerotized striations in *C. gestroi*, no striation at all (smooth) in *C. formosanus*, and several faint striations in Hybrid F and Hybrid G. Finally, hybrid soldiers tend to have longer mandibles than parental species (Fig. 17.3), especially in Hybrid F (Patel *et al.*, 2023b). The ability of hybrid soldiers to bite and to produce their frontal gland secretions appears to be partially impaired, raising questions regarding their relative defensive effectiveness (Patel *et al.*, 2023b).

With an apparent lack of heterosis in hybrid *Coptotermes* colonies, it was also observed that some individual workers within hybrid colonies (both Hybrid F and Hybrid G) displayed sluggish locomotion, while workers from colonies of the parental

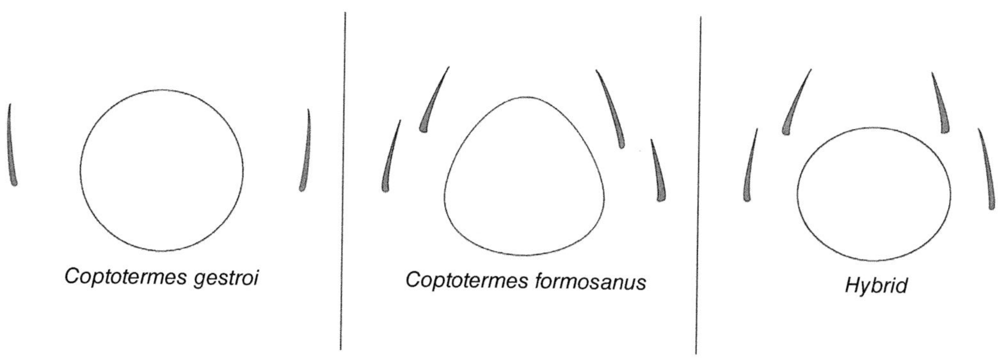

Fig. 17.2. Typical shape of the fontanelle in soldiers of *Coptotermes gestroi* (subcircular), *Coptotermes formosanus* (trianguliform), and *Coptotermes* hybrids (horizontally ellipsoid), with relative position and count of setae beside the fontanelle. However, shapes and setae counts vary across colonies, and from the Florida and Taiwan source material (T. Chouvenc.)

Fig. 17.3. Soldiers from a Hybrid F colony (laboratory-reared colony, 4-year-old, *C. formosanus* female × *C. gestroi* male mating combination). Note the elongated mandibles and the relatively parallel position of the mandibles at rest. (Photo: Thomas Chouvenc.)

species did not (Lee *et al.*, 2020). In addition, it was determined that the average molting rate of workers in hybrid colonies was lower than in parental species (Lee *et al.*, 2020). Developmental issues may therefore emerge in some hybrid workers, as they may be stuck in intermolt indefinitely, resulting in their reduced locomotive ability over time, and potentially in reduced foraging efficiency (Lee *et al.*, 2020, 2022). As some atypical (= transgressive) phenotypes were observed in workers and soldiers of hybrid colonies (i.e., phenotypes not intermediate or not within the range of variation of the two parental species), it suggests that there may be some alteration of gene expression resulting from the interaction of the genetic backgrounds of the two parental lineages, producing developmental anomalies. Finally, the construction behavior of the hybrids deviates from both parental species. While *C. gestroi* and *C. formosanus* build typical sponge-like structure carton nests (Chouvenc *et al.*, 2013), which appear similar between the two species, carton nests in Hybrid F and Hybrid G colonies lack the typical sponge-like structure (Patel *et al.*, 2023b). Instead, the nest construction by hybrid colonies appears to be the superimposition of mud tubes on top of each other, resulting in a layer cake-style carton building up over time (Patel *et al.*, 2023b).

To date, no hybrid *Coptotermes* laboratory-reared colonies have produced nymphs or alates, although they occasionally produced nymphoid secondary reproductives, even in the presence of both primary reproductives (Patel *et al.*, 2023a). In the absence of known parthenogenesis in *Coptotermes* (see Chapter 4 of this volume), such viable and potentially functional secondary reproductives (nymphoids) are most likely the product of sexual reproduction between the heterospecific queen and king. The presence of hybrid secondary reproductives implies that the nymphal pathway (see Chapter 4 of this volume) is possible in hybrid individuals. It is therefore theoretically possible for hybrid colonies to mature and engage in nymph differentiation toward alate production ($L_2 \rightarrow N_1 \rightarrow \ldots \rightarrow A$).

17.5 Hybrid detection in the field

The ability of hybrid colonies to establish in the field, mature, and produce viable alates has remained elusive since the first report of the potential hybridization between the two species (Chouvenc *et al.*, 2015; Patel *et al.*, 2023a). Su *et al.* (2017) suggested that intensive survey efforts at specific locations that have a long history of invasion of *C. gestoi* in areas with pre-existing *C. formosanus* populations in Taiwan may reveal field-established hybrid colonies. In 2019, *Coptotermes* alates with atypical morphologies were collected during dispersal flight events from a preliminary survey in Taichung, Taiwan (Chen, 2022). The atypical morphology includes differential antennal spot shape and color when compared with parental species, and also an intermediate abdominal color between *C. gestroi* and *C. formosanus* (Li *et al.*, 2023). Molecular analysis from subsamples of atypical morphology alates confirmed them to be hybrid *Coptotermes* alates (Chen, 2022). In 2019, 3,295 *C. formosanus alates*, 13,956 *C. gestroi* alates, and 335 hybrid alates were collected in Taiwan. In 2020, 4,416 *C. formosanus* alates, 1,380 *C. gestroi* alates, and 748 hybrid alates were collected. In 2021, 2,118 *C. formosanus* alates, 1,220

C. gestroi alates, and 135 hybrid alates were collected. The peak of the dispersal flight season of hybrid alates appears to be intermediate to the two parental species (Li *et al.*, 2023). Both Hybrid F and Hybrid G alates were collected, implying that both heterospecific mating combinations readily occurred in the field in Taiwan, and both produced viable F1 alates (Chen, 2022; Li *et al.*, 2023). In addition, while collected *Coptotermes* alates with atypical morphologies were mainly F1 generations, a handful of advanced hybrids (> F1 generations) were also collected (Chen, 2022; Li *et al.*, 2023). Such observation revealed that F1 hybrid alates can be viable and fertile, raising the possibility for potential gene flow between the two species in Taiwan. However, laboratory colonies established in Taiwan with hybrid alates revealed, with few exceptions, poor fertility regardless of the origin of the mating partner (Li *et al.*, 2023). Such preliminary field and laboratory observations may potentially cast doubt on the relative success of F1 hybrid alates as successful colony founders in the field. Therefore, while potential gene flow between the two *Coptotermes* species is possible, it may be relatively limited and needs to be investigated further.

In Florida, two morphologically atypical alates (both females) were collected in late March 2020. Unfortunately, both samples were lost during the Covid-19 pandemic shutdown. In 2021, 87 female alates (no males) displaying atypical morphologies and matching the description from the atypical alates from Taiwan were collected and were genetically confirmed to be hybrid F1 (Chouvenc *et al.*, in preparation). During the 2022 dispersal flight season, 102 hybrid (65 females, 37 males) alates were collected (compared with > 26,500 *C. gestroi* alates and > 7,500 *C. formosanus* alates during the same period). Preliminary observation using hybrid alates to establish laboratory colonies with mating partners of all possible mating combinations supported the potentially low fertility of F1 hybrid alates, as previously observed in Taiwan, although some viable F2 colonies were observed (Li *et al.*, 2023; Chouvenc *et al.*, in preparation). Anecdotally, both in Taiwan and in Florida, hybrid dispersal flights were biased toward a high female ratio, which raises questions on males' ability to properly develop, survive, and/or to fly during dispersal events.

The first confirmation of mating confusion in the field between *C. gestroi* and *C. formosanus* in Florida was made in 2013 (Chouvenc *et al.*, 2015), and the first detection of hybrid alates in 2020. Because it may take 5–8 years for *Coptotermes* to reach colony maturity (Chouvenc and Su, 2014) (Chapter 4 of this volume), these data suggest that F1 hybrid colonies successfully established in the field in Florida around the same time when courtship behavior confusion was detected. The light trap procedure was identical every year since 2013 for all dispersal flight events, i.e., same location, same duration, and same setup (Chouvenc *et al.*, 2017a). Therefore, the progressive increase of hybrid alates collected from 2020 to 2022 in standardized conditions may imply that one or more mature hybrid colonies were in the process of increasing their F1 alate productivity; this hypothesis will be further investigated in the near future. It appears that the transition from a pre-hybrid time to a post-hybrid era in Florida was observed in real time.

Finally, during the May 2020 dispersal season, the first case of a field-established hybrid *Coptotermes* colony was confirmed in Taichung (Chen, 2022). This colony was found dispersing from an infested tree (NCHU Termite Collection TW11376, TW11393) with visible mud tubes on the trunk. Guarding the mud tubes where alates were emerging, soldiers were found and were able to secrete defensive secretions from their frontal gland when disturbed by the observer. However, some of these soldiers had wing buds, a trait not found in soldiers of typical *Coptotermes* species (Fig. 17.4). Such morphology most likely resulted from developmental issues, where a co-expression of two different developmental trajectories (nymph and soldier) (see Chapter 4 of this volume) were observed in an individual, further supporting the idea that hybrid *Coptotermes* individuals may go through various degrees of developmental instability resulting in morphological aberrations. Soldier–nymph intercaste individuals are relatively rare across other termite species under field

Fig. 17.4. Abnormal soldier morphology (soldier–nymph intercaste) found in a field-established *Coptotermes* hybrid colony in Taichung, Taiwan. Some of the soldiers had wing buds. (Picture: Chen, G.-Y., with permission; modified from Chen, 2022, with permission.)

conditions (Lefeuve and Thorne, 1984; Roisin and Pasteels, 1986), but can be artificially produced in the laboratory with juvenoids (Howard, 1979; Lelis and Everaerts, 1993; Miura *et al.*, 2003; Scharf *et al.*, 2003; Miura, 2005; Cesar *et al.*, 2019). In *Coptotermes*, the only mention of a soldier–nymph individual can be found from a laboratory experiment with fenoxycarb (Jones, 1984). In the case of the hybrid *Coptotermes* field colony discovered in Taiwan, soldier–nymph intercastes were readily produced by a mature field colony. However, such winged soldier morphology was not observed from hybrid *Coptotermes* laboratory colonies, raising the possibility that such abnormal morphology may only emerge in mature colonies, where the nymphal pathway is active. As the Taiwan and Florida termite research teams continue to collect additional information, a more comprehensive understanding of the biology of hybrids between the two *Coptotermes* species will be determined in the years to come.

17.6 Ongoing and future research

The field collection of *Coptotermes* hybrid alates in Taiwan and in Florida during dispersal flights, and the discovery of a mature hybrid *Coptotermes* colony in Taiwan, confirmed that *C. gestroi* and *C. formosanus* have the ability to form viable heterospecific colony foundations and reach colony maturity to produce F1 hybrid alates. Undoubtedly, the study of hybrid *Coptotermes* is still in its infancy, as the research into their biology and evolutionary implications is ongoing. The initial studies on laboratory-reared *Coptotermes* hybrid colonies confirmed that both mating combinations (Hybrid F and Hybrid G) result in viable colonies, and hybrid individuals display many traits that are similar, intermediate, or a combination of both their parental species. However, some hybrid traits are transgressive and deviate from either one or both of the parental species. There is an opportunity to study how gene interactions can result in atypical phenotypes, such as alate and soldier morphology, worker molting rate, caste regulatory processes, or nest construction behaviors. The genetic divergence between *C. formosanus* and *C. gestroi* is ~18 million years (Bourguignon *et al.*, 2016), and while there are no pre-zygotic barriers between the two species, it seems that such evolutionary time in allopatry has resulted in partial post-zygotic limitations that may lead to developmental alterations and partial F1 fertility.

Hybrid termites can provide unique biological perspectives and can be a useful tool to study potential evolutionary innovation (Yashiro *et al.*, 2021) in an evo-devo context (Toth and Rehan, 2017). In *Coptotermes* hybrids, novel traits may emerge or offer opportunities to study aspects of termite biology that would not be possible otherwise. One ongoing project is to use protist lineages from *C. gestroi* and *C. formosanus* (Jasso-Selles *et al.*, 2020) to test the biparental symbiotic acquisition hypothesis in hybrid colonies, and to determine how different lineages of protists may – or may not – persist over generations of colonies (Velenovsky, 2022).

Finally, while the concern of a negative economic impact from the establishment of a hybrid *Coptotermes* population remains valid (Su *et al.*, 2017; Patel *et al.*, 2023a), one may consider that the already established *C. gestroi* and *C. formosanus* populations in several areas represent a far greater threat to manmade structures and trees. The presence of hybrid *Coptotermes* in the field should not distract from the already existing threat from both parental species (Chouvenc *et al.*, 2016b; Huang *et al.*, 2022). While gene flow between populations of the two species may have long-term evolutionary consequences, it may take decades before such impacts can be measured. However, as *C. gestroi* populations have been moving northward in the past few decades in Taiwan, partially attributable to the ongoing climate change, it may also be speculated that some of the northernmost *C. gestroi* population may carry some *C. formosanus* genes from past hybridization events allowing them to survive and thrive in areas historically too cold for them (Cao and Su, 2016). Further investigations in Taiwan and in Florida will continue to monitor hybrid *Coptotermes* activity in the laboratory and in the field in the hope that some of these outstanding questions could be answered.

References

Abbott, R., Albach, D., Ansell, S., Arntzen, J.W., Baird, S.J. *et al.* (2013) Hybridization and speciation. *Journal of Evolutionary Biology* 26, 229–246.

Abbott, R.J., Barton, N.H. and Good, J.M. (2016) Genomics of hybridization and its evolutionary consequences. *Molecular Ecology* 25, 2325–2332.

Arce-Valdés, L.R. and Sánchez-Guillén, R.A.S. (2022) The evolutionary outcomes of climate change-induced hybridization in insect populations. *Current Opinion in Insect Science* 54, 100966.

Bess, H.A. (1970) Termites of Hawaii and the oceanic islands. In: Krishna K. and Weesner F.M. (eds) *Biology of Termites*. Vol. 1., Academic Press, New York, pp. 449–476.

Birchler, J.A., Auger, D.L. and Riddle, N.C. (2003) In search of the molecular basis of heterosis. *The Plant Cell* 15, 2236–2239.

Blumenfeld, A.J., Eyer, P.A., Husseneder, C., Mo, J., Johnson, L.N. *et al.* (2021) Bridgehead effect and multiple introductions shape the global invasion history of a termite. *Communications Biology* 4, e.196.

Bourguignon, T., Lo, N., Šobotník, J., Sillam-Dussès, D., Roisin, Y. *et al.* (2016) Oceanic dispersal, vicariance and human introduction shaped the modern distribution of the termites *Reticulitermes*, *Heterotermes* and *Coptotermes*. *Proceedings of the Royal Society B: Biological Sciences* 283, 20160179.

Bourret, S.L., Kovach, R.P., Cline, T.J., Strait, J.T. and Muhlfeld, C.C. (2022) High dispersal rates in hybrids drive expansion of maladaptive hybridization. *Proceedings of the Royal Society B: Biological Sciences* 289, 20221813.

Buggs, R.J.A. (2007) Empirical study of hybrid zone movement. *Heredity* 99, 301–312.

Burke, J.M. and Arnold, M.L. (2001) Genetics and the fitness of hybrids. *Annual Review of Genetics* 35, 31–52.

Cao, R. and Su, N.-Y. (2016) Temperature preferences of four subterranean termite species (Isoptera: Rhinotermitidae) and temperature-dependent survivorship and wood-consumption rate. *Annals of the Entomological Society of America* 109, 64–71.

Cesar, C.S., Giacometti, D., Costa-Leonardo, A.M. and Casarin, F.E. (2019). Drywood pest termite *Cryptotermes brevis* (Blattaria: Isoptera: Kalotermitidae): A detailed morphological study of pseudergates. *Neotropical Entomology* 48, 822–833.

Chen, G.Y. (2022) Hybridization between Formosan subterranean termite and Asian subterranean termite in Taiwan. MSc dissertation, National Chung Hsing University, Taichung, Taiwan.
Chouvenc, T. and Su, N.-Y. (2014) Colony age-dependent pathway in caste development of *Coptotermes formosanus* Shiraki. *Insectes Sociaux* 61, 171–182.
Chouvenc, T. and Su, N.-Y. (2017) Testing the role of cuticular hydrocarbons on intercolonial agonism in two subterranean termite species (*Coptotermes*) and their hybrids. *Insectes Sociaux* 64, 347–355.
Chouvenc, T., Efstathion, C.A., Elliott, M.L. and Su, N.-Y. (2013) Extended disease resistance emerging from the faecal nest of a subterranean termite. *Proceedings of the Royal Society B: Biological Sciences* 280, 20131885.
Chouvenc, T., Helmick, E.E. and Su, N.Y. (2015) Hybridization of two major termite invaders as a consequence of human activity. *PLoS ONE* 10, e0120745.
Chouvenc, T., Li H.-F., Austin, J., Bordereau, C., Bourguignon, T. *et al.* (2016a) Revisiting *Coptotermes* (Isoptera: Rhinotermitidae): a global taxonomic road map for species validity and distribution of an economically important subterranean termite genus. *Systematic Entomology* 4, 299–306.
Chouvenc, T., Scheffrahn, R.H. and Su, N.-Y. (2016b) Establishment and spread of two invasive subterranean termite species (*Coptotermes formosanus* and *C. gestroi*; Isoptera: Rhinotermitidae) in metropolitan southeastern Florida (1990–2015). *Florida Entomologist* 99, 187–191.
Chouvenc, T., Scheffrahn, R.H., Mullins, A.J. and Su, N.-Y. (2017a) Flight phenology of two *Coptotermes* species (Isoptera: Rhinotermitidae) in southeastern Florida. *Journal of Economic Entomology* 110, 1693–1704.
Chouvenc, T., Basille, M. and Su, N.-Y. (2017b) Role of accelerated developmental pathway and limited nurturing capacity on soldier developmental instability in subterranean termite incipient colonies. *Insectes Sociaux* 64, 477–483.
Chouvenc, T., Sillam-Dussès, D. and Robert, A. (2020) Courtship behavior confusion in two subterranean termite species that evolved in allopatry (Blattodea, Rhinotermitidae, *Coptotermes*). *Journal of Chemical Ecology* 46, 461–474.
Chouvenc, T., Helmick, E.E., Velenovsky IV, J.F., Lee, S.-B., Su N.-Y, Li, H.-F. [in prep.] Evidence of field establishment of hybrid *Coptotermes* in Florida.
Chunco, A.J. (2014) Hybridization in a warmer world. *Ecology and Evolution* 4, 2019–2031.
Crispo, E., Moore, J.S., Lee-Yaw, J.A., Gray, S.M. and Haller, B.C. (2011) Broken barriers: Human-induced changes to gene flow and introgression in animals: An examination of the ways in which humans increase genetic exchange among populations and species and the consequences for biodiversity. *BioEssays* 33, 508–518.
Diagne, C., Leroy, B., Vaissière, A.C., Gozlan, R.E., Roiz, D., Jarić, I., Salles, J.M., Bradshaw, C.J. and Courchamp, F. (2021) High and rising economic costs of biological invasions worldwide. *Nature* 592, 571–576.
Diedericks, G., Henriques, R., von der Heyden, S., Weyl, O.L. and Hui, C. (2018) Sleeping with the enemy: introgressive hybridization in two invasive centrarchids. *Journal of Fish Biology* 93, 405–410.
Dittrich-Reed, D.R. and Fitzpatrick, B.M. (2013) Transgressive hybrids as hopeful monsters. *Evolutionary Biology* 40, 310–315.
Evans, T.A. (2021) Predicting ecological impacts of invasive termites. *Current Opinion in Insect Science* 46, 88–94.
Evans, T.A., Forschler, B.T. and Grace, J.K. (2013) Biology of invasive termites: a worldwide review. *Annual Review of Entomology* 58, 455–474.
Fournier, D. and Aron, S. (2021) Hybridization and invasiveness in social insects – the good, the bad and the hybrid. *Current Opinion in Insect Science* 46, 1–9.
Frankham, R., Ballou, J.D., Eldridge, M.D., Lacy, R.C., Ralls, K., Dudash, M.R. and Fenster, C.B. (2011) Predicting the probability of outbreeding depression. *Conservation Biology* 25, 465–475.
Frishkoff, L.O., Karp, D.S., Flanders, J.R., Zook, J. *et al.* (2016) Climate change and habitat conversion favour the same species. *Ecology Letters* 19, 1081–1090.
Gaskin, J.F. and Kazmer, D.J. (2009) Introgression between invasive saltcedars (*Tamarix chinensis* and *T. ramosissima*) in the USA. *Biological Invasions* 11, 1121–1130.
Howard, R.W. (1979) Termites and juvenile hormone analogues: a review of methodology. *Sociobiology* 4, 269–278.
Huang, S.-Y., Chiu, C.-I, Tsai, Y.-Y., Li, W.-J., Wu, C.-C. *et al.* (2022) Nationwide termite pest survey conducted in Taiwan as a citizen science project. *Journal of Economic Entomology* 115, 1650–1658.
Huxel, G.R. (1999) Rapid displacement of native species by invasive species: effects of hybridization. *Biological Conservation* 89, 143–152.

Jasso-Selles, D.E., De Martini, F., Velenovsky IV, J.F., Mee, E.D., Montoya, S.J. *et al.* (2020) The complete protist symbiont communities of *Coptotermes formosanus* and *Coptotermes gestroi*: morphological and molecular characterization of five new species. *Journal of Eukaryotic Microbiology* 67, 626–641.

Jones, S.C. (1984) Evaluation of two insect growth regulators for the bait-block method of subterranean termite (Isoptera: Rhinotermitidae) control. *Journal of Economic Entomology* 77, 1086–1091.

Lee, S.B., Chouvenc, T., Patel, J. and Su, N.Y. (2020) Altered mobility and accumulation of inefficient workers in juvenile hybrid termite colonies. *Frontiers in Ecology and Evolution* 8, 589762.

Lee, S.-B., Chouvenc, T., Mizumoto, N., Mullins, A. and Su, N.-Y. (2022) Age-based spatial distribution of workers is resilient to worker loss in a subterranean termite. *Scientific Reports* 12, e7837.

Lefeuve, P. and Thorne, B.L. (1984) Nymph–soldier intercastes in *Nasutitermes lujae* and *N. columbicus* (Isoptera; Termitidae). *Canadian Journal of Zoology* 62, 959–964.

Lelis, A.T. and Everaerts, C. (1993) Effects of juvenile hormone analogues upon soldier differentiation in the termite *Reticulitermes santonensis* (Rhinotermitidae, Heterotermitinae). *Journal of Morphology* 217, 239–261.

Lenoir, J., Hattab, T. and Pierre, G. (2017) Climatic microrefugia under anthropogenic climate change: implications for species redistribution. *Ecography* 40, 253–266.

Li, H.-F., Su, N.-Y. and Wu, W.-J. (2010a) Solving the hundred-year controversy of *Coptotermes* taxonomy in Taiwan. *American Entomologist* 56, 222–227.

Li, H.-F., Yang, R.-L. and Su, N.-Y. (2010b) Interspecific competition and territory defense mechanisms of *Coptotermes formosanus* and *Coptotermes gestroi* (Isoptera: Rhinotermitidae). *Environmental Entomology* 39, 1601–1607.

Li, H.-F., Fujisaki, I., & Su, N.-Y. (2013) Predicting habitat suitability of *Coptotermes gestroi* (Isoptera: Rhinotermitidae) with species distribution models. *Journal of Economic Entomology* 106, 311–321.

Li, H.-F., Chen, G.-Y., Huang, S.-Y. Lin, M.-D. Chouvenc, T. and Ching, Y-H. (2023) Hybrids of two destructive subterranean termites established in the field, revealing a potential for gene flow between species. Available at: https://www.researchsquare.com/article/rs-2801694/v1 (accessed 29 October 2023).

Lowe, S., Browne, M., Boudjelas, S. and De Poorter, M. (2000) *100 of the world's worst invasive alien species: a selection from the global invasive species database*. Invasive Species Specialist Group, Auckland, 12 pp.

Lowe, W.H., Muhlfeld, C.C. and Allendorf, F.W. (2015) Spatial sorting promotes the spread of maladaptive hybridization. *Trends in Ecology & Evolution* 30, 456–462.

Mallet, J. (2005) Hybridization as an invasion of the genome. *Trends in Ecology & Evolution* 20, 229–237.

Mallet, J. (2007) Hybrid speciation. *Nature* 44, 279–283.

Miura, T. (2005) Developmental regulation of caste-specific characters in social-insect polyphenism. *Evolution & Development* 7, 122–129.

Miura, T., Koshikawa, S. and Matsumoto, T. (2003) Winged presoldiers induced by a juvenile hormone analog in *Zootermopsis nevadensis*: implications for plasticity and evolution of caste differentiation in termites. *Journal of Morphology* 257, 22–32.

Mizumoto, N., Rizo, A., Pratt, S.C. and Chouvenc, T. (2020) Termite males enhance mating encounters by changing speed according to density. *Journal of Animal Ecology* 89, 2542–2552.

Mizumoto, N., Lee, S.B., Valentini, G., Chouvenc, T. and Pratt, S.C. (2021) Coordination of movement via complementary interactions of leaders and followers in termite mating pairs. *Proceedings of the Royal Society B: Biological Sciences* 288, 20210998.

Mooney, H.A. and Cleland, E.E. (2001) The evolutionary impact of invasive species. *Proceedings of the National Academy of Sciences* 98, 5446–5451.

Moran, B.M., Payne, C., Langdon, Q., Powell, D.L., Brandvain, Y. *et al.* (2021) The genomic consequences of hybridization. *Elife* 10, e69016.

Muhlfeld, C.C., Kovach, R.P., Al-Chokhachy, R., Amish, S.J., Kershner, J.L. *et al.* (2017) Legacy introductions and climatic variation explain spatiotemporal patterns of invasive hybridization in a native trout. *Global Change Biology* 23, 4663–4674.

Patel, J.S. (2020) Biology of the hybrid termites of *Coptotermes formosanus* and *Coptotermes gestroi* (Blattodea: Rhinotermitidae). Doctoral dissertation, University of Florida, 128 pp.

Patel, J.S., Tong, R.L., Chouvenc, T. and Su, N.-Y. (2019a) Comparison of temperature-dependent survivorship and wood-consumption rate among two invasive subterranean termite species (Blattodea: Rhinotermitidae: *Coptotermes*) and their hybrids. *Journal of Economic Entomology* 112, 300–304.

Patel, J.S., Chouvenc, T. and Su, N.-Y. (2019b) Temperature preference of two invasive subterranean termite species and their hybrids (Blattodea: Rhinotermitidae: *Coptotermes*). *Journal of Economic Entomology* 112, 2888–2893.

Patel, J.S., Lee, S.B., Chouvenc, T. and Su, N.-Y. (2020) Inferring termite colony size using wood consumption in subterranean termites (Blattodea: Rhinotermitidae) in laboratory-rearing conditions. *Journal of Economic Entomology* 113, 905–910.

Patel, J.S., Lee, S.B., Chouvenc, T., and Su, N.-Y. (2023a) Equivalent colony growth of hybrids of two invasive Coptotermes species can threaten urban areas. *Journal of Economic Entomology.* [In press.] doi: 10.1093/jee/toad020

Patel, J.S., Chouvenc, T., Wu, D.C.-C., Li, H.-F. *et al.* (2023b) Phenotypic differences among mating combinations of two *Coptotermes* species and their hybrids (Blattodea: Rhinotermitidae) *Journal of Insect Science*, In press.

Perrings, C., Williamson, M., Barbier, E.B., Delfino, D., Dalmazzone, S. *et al.* (2002) Biological invasion risks and the public good: an economic perspective. *Conservation Ecology* 6, 1–7.

Roisin, Y. and Pasteels, J.M. (1986) Differentiation of worker-derived intercastes and precocious imagoes after queen removal in the Neo-Guinean termite *Nasutitermes princeps* (Desneux). *Journal of Morphology* 189, 281–293.

Rust, M.K. and Su, N.-Y. (2012) Managing social insects of urban importance. *Annual Review of Entomology* 57, 355–375.

San Jose, M., Doorenweerd, C. and Rubinoff, D. (2023) Genomics reveals widespread hybridization across insects with ramifications for species boundaries and invasive species. *Current Opinion in Insect Science* 58, 101052.

Sardain, A., Sardain, E. and Leung, B. (2019) Global forecasts of shipping traffic and biological invasions to 2050. *Nature Sustainability* 2, 274–282.

Scharf, M.E., Ratliff, C.R., Hoteling, J.T., Pittendrigh, B.R. and Bennett, G.W. (2003) Caste differentiation responses of two sympatric *Reticulitermes* termite species to juvenile hormone homologs and synthetic juvenoids in two laboratory assays. *Insectes Sociaux* 50, 346–354.

Scheffrahn, R.H. and Su, N.-Y. (2005) Distribution of the termite genus *Coptotermes* (Isoptera: Rhinotermitidae) in Florida. *Florida Entomologist* 88, 201–203.

Shoemaker, D.D., Ross, K.G. and Arnold, M.L. (1996) Genetic structure and evolution of a fire ant hybrid zone. *Evolution* 50, 1958–1976.

Su, N.-Y., Chouvenc, T. and Li, H.-F. (2017) Potential hybridization between two invasive termite species, *Coptotermes formosanus* and *C. gestroi* (Isoptera: Rhinotermitidae), and its biological and economic implications. *Insects* 8, 14.

Todesco, M., Pascual, M.A., Owens, G.L., Ostevik, K.L., Moyers, B.T. *et al.* (2016) Hybridization and extinction. *Evolutionary Applications* 9, 892–908.

Tong, R.L., Grace, J.K., Mason, M., Krushelnycky, P.D., Spafford, H. *et al.* (2017) Termite species distribution and flight periods on Oahu, Hawaii. *Insects* 8, 58.

Toth, A.L. and Rehan, S.M. (2017). Molecular evolution of insect sociality: an eco-evo-devo perspective. *Annual Review of Entomology* 62, 419–442.

Uchima, S.Y. and Grace, J.K. (2009) Interspecific agonism and foraging competition between *Coptotermes formosanus* and *Coptotermes gestroi* (Blattodea: Rhinotermitidae). *Sociobiology* 54, 765–776.

Velenovsky, J.F. IV (2022) The protist communities of colonies established by conspecific and heterospecific *Coptotermes* alate pairs confirm some and reveal other aspects of aspects of symbiosis in lower termites. PhD Dissertation, University of Florida, 172 pp.

Vitousek, P.M., D'Antonio, C.M., Loope, L.L., Rejmanek, M. and Westbrooks, R. (1997) Introduced species: a significant component of human-caused global change. *New Zealand Journal of Ecology* 21, 1–16.

Wallingford, P.D., Morelli, T.L., Allen, J.M., Beaury, E.M., Blumenthal, D.M. *et al.* (2020) Adjusting the lens of invasion biology to focus on the impacts of climate-driven range shifts. *Nature Climate Change* 10, 398–405.

Yashiro, T., Tea, Y.K., Van Der Wal, C., Nozaki, T., Mizumoto, N. *et al.* (2021) Enhanced heterozygosity from male meiotic chromosome chains is superseded by hybrid female asexuality in termites. *Proceedings of the National Academy of Sciences* 118, e2009533118.

18 Using *Coptotermes* for laboratory experiments: field collection, laboratory rearing, and bioassay visualization

Thomas Chouvenc
Entomology and Nematology Department, Ft. Lauderdale Research and Education Center, University of Florida, tomchouv@ufl.edu

Abstract

Coptotermes laboratory research has historically relied on the ability to manipulate groups of termites for experimental protocols. However, there is an accumulation of scientific literature on subterranean termite research with questionable interpretative power, often stemming from poor initial termite vigor or inappropriate experimental approaches. Moving forward, it is imperative that empirical assays adopt strict rules on acquiring, maintaining, manipulating, and observing live groups of termites, to ensure that studies produce biologically meaningful, repeatable, comparable, and relevant results. This chapter provides a scientific philosophy for how *Coptotermes* may be approached as a biological model for research purposes, and compiles decades of experience from various laboratories around the world. It covers: (1) limitations that are inherent to subterranean termite research, with an emphasis on mistakes and pitfalls to avoid when first approaching *Coptotermes*; (2) protocols to establish termite field colonies as a perennial source of biological material; (3) rearing protocols in laboratory conditions using *Coptotermes* alates as an alternative reliable source for termites; and (4) protocols to observe subterranean termites with an emphasis on the use of planar arenas. This general overview provides universal guidelines toward a standardization of empirical *Coptotermes* research, while opening new avenues and opportunities for current and future laboratories.

"The social behavior of termites is the most important biological factor to be considered in their breeding and testing." G. Becker (1969)

18.1 Introduction

Over the past six decades of *Coptotermes* research, studies have used laboratory experiments to investigate termite behavioral and physiological processes, to evaluate material resistance to termite damage, and to determine the impact of pest control approaches on termite survival. For example, extensive screening efforts to evaluate pesticide efficacy on *Coptotermes* species alone have resulted in the accumulation of more than 2,750 publications (1960–2022), although only a fraction of them resulted in successful commercial pest control products. One of the reasons for the accumulation of literature about termite control with little-to-no practical application is often the inappropriateness of the protocols used to test hypotheses about termite biology. Becker (1969)

 Biology and Management of the Formosan Subterranean Termite and Related Species (eds Nan-Yao Su and Chow-Yang Lee)
DOI: 10.1079/9781800621596.0018

highlighted the importance of uniform testing conditions for comparable and repeatable results when studying termites, but also warned that the biological relevancy of an experiment may be limited when the social behavior of termites is not taken into account. Lenz (1986) further emphasized the importance of optimal experimental conditions toward minimizing the stress imposed on termites in order to obtain any meaningful results, and Lenz (2009) later reiterated these concerns with palpable levels of frustration.

These repeated warnings were often overlooked, as many researchers continued to test chemical compounds, microbial agents, plant extracts or other miscellaneous biocides, using small groups of low-vigor termites in small containers, which did not provide environmental conditions that would have allowed termites to properly display the broad range of their behavioral repertoire. Such testing conditions fundamentally depart from what makes *Coptotermes* such a challenging pest: large colonies with long foraging distances in soils (King and Spink, 1969; Su and Scheffrahn, 1988) and a complex reproductive demography (Shimizu, 1962; Higa, 1981; Vargo and Husseneder, 2009; Chouvenc and Su, 2014; Chouvenc *et al.*, 2022) (Chapter 4 of this volume). Consequently, *Coptotermes* control research is crowded with studies that are, in essence, candidly preliminary. They have often reported "highly promising" results for future potential applications, while most of them should have instead concluded with a variation of: "we just killed a few bugs in a jar" (Chouvenc *et al.*, 2011a). Similarly, many research groups continued to attempt investigations of various behavioral and physiological aspects of termite biology, while neglecting the inherent sociobiological components of termite groups (as pointed out by several authors, including: Thorne and Haverty, 1991; Arquette *et al.*, 2006; Lenz, 2009; Miramontes and DeSouza, 2008; Chouvenc, 2018a).

Nonetheless, preliminary screenings investigating any aspect of *Coptotermes* biology still often require a "quick and dirty" Petri dish-style experiment. Such preliminary observations can have informative value toward setting up a protocol that considers relevant aspects of the termite's biology, so that results can be interpreted in their appropriate biological context (Becker, 1969). Unfortunately, in most instances, such follow-up experiments were never conducted (and studies were eventually published "as is") owing to several limiting factors inherent to termite research: (1) limited access to resources; (2) limited access to biological material; (3) limited visibility; and (4) limited knowledge.

18.1.1 Limited access to resources

One of the primary limiting factors for setting up *Coptotermes* experiments that would consider their large populations, extensive foraging distances, and a complex colony demography is the resources necessary to scale protocols up to their biologically relevant context, while still being replicable. These resources can be narrowed down to financial ones, which, by extension, also limits time (working hours) and laboratory space. Financial resources, which are often limited in academic set-ups, often dictate the number of people who could contribute their time and skills to set up large-scale or technical experiments that involve the manipulation of large numbers of termites within expensive and extensive experimental contraptions (Su, 2005a; Chouvenc, 2018b; Gordon *et al.*, 2022). Alternatively, the decision to restrict termite group sizes within small containers for experimentation is often made out of convenience (Lenz, 2009), and such shortcuts may be more prevalent when laboratory resources are limited. Although it is important to acknowledge that termite research laboratories may have limited access to resources to establish such large-scale protocols, some approaches discussed herein may allow for an optimization of the resources required to proceed toward these protocols.

18.1.2 Limited access to biological material

The number of termites and types of caste required to perform a study may vary. However, there is a consensus that termites can display unique behaviors and relatively high survivorship only when held in relatively large groups and in minimal stress conditions (Becker, 1961; Becker, 1970; Schultze-Dewitz, 1962; Lenz, 1986; Rosengaus *et al.*, 1998; Miramontes and DeSouza, 2008; Lenz, 2009; Chouvenc *et al.*, 2011b; Gao *et al.*, 2012; Du *et al.*, 2016; Lin *et al.*, 2023), which may provide a social context that is as close as possible to field colonies (Su *et al.*, 1982; Su, 2005a; Du *et al.*, 2017a; Chouvenc, 2018b). Therefore, termite research laboratories must ensure a steady supply of healthy termites to enable the set-up of relatively large experiments when attempting to emulate field conditions (Chouvenc, 2018a, 2022a). There are two primary ways to obtain the biological material necessary for a perennial termite research program: (1) periodically collecting foraging populations from established field colonies; and (2) rearing laboratory colonies from alates. Both methods have their advantages and disadvantages, which are discussed below.

18.1.3 Limited visibility

One of the major issues encountered while studying subterranean termites is the difficulty of observing them continuously, while providing them with environmental conditions as close as possible to field conditions. Their cryptic nesting habits and extensive subterranean foraging abilities make it challenging to observe colony-wide variables, both for field colonies and laboratory colonies. While Petri-dish experiments are convenient and allow for some variable levels of visibility, they fail to provide an environment conducive for *Coptotermes* to display their extensive behavioral repertoire and for them to thrive throughout the duration of the experiment (Chouvenc *et al.*, 2011b; Du *et al.*, 2016). Over the past 20 years, the use of planar arenas, which provide soil-like foraging conditions while allowing for continuous observation, have fundamentally transformed how multiple aspects of *Coptotermes* behavior and biology are understood (Campora and Grace, 2009; Bernard *et al.*, 2017; Du *et al.*, 2017a; Su *et al.*, 2017; Chouvenc, 2018a; Kakkar *et al.*, 2018). In addition, extensive foraging ability is an inherent trait of *Coptotermes* (Su and Scheffrahn, 1988; Su *et al.*, 2017), and appropriately moistened soil conditions during experimentation can provide enhanced survival over long periods of observation (Chouvenc *et al.*, 2011b). This chapter also discusses how the use of planar arenas can be normalized in laboratories, and how versatile they can be for a wide range of experimental protocols.

18.1.4 Limited knowledge

This issue is probably the most pervasive across termite biology and control research and can be expressed in many ways. (1) Too often, results from various bioassays are presented with unacceptably low control survival over short periods of observation, preventing any reliable interpretation when considering that termites were set in lethally stressful situations (Becker, 1969). Unfortunately, this implies that little thought was given to providing favorable conditions for termite survival, which inherently stems from not knowing or caring how to keep termites alive and healthy throughout the duration of the experiment (Lenz, 1986, 2009). This can be remedied by optimizing experimental protocols, by referring to the existing literature that has solved such issues, and by avoiding mistakes made in the past, as detailed by Lenz (2009). (2) *Coptotermes* control literature has accumulated a vast backlog of poorly relevant publications which claim that a "new" approach for killing termites is "very promising," while only killing small groups of termites in a Petri dish, and most often never expanding the study further. Unfortunately, the ongoing flow of these publications has resulted in a positive publication bias that has led to overoptimism, which inherently perpetuated itself (Chouvenc *et al.*, 2011a). This could be avoided if a rigorous dive into the existing *Coptotermes* literature is made, so as to be familiar with termite biology in general and then the idiosyncrasies of the biology of *Coptotermes*, such as their inherent molting process and their long foraging distances. Laboratories should be incentivized to emulate established experimental set-ups that already addressed biological relevancy limitations and could realistically be scaled up to field assay. (3) Similar to other aspects of termite biology, many studies fail to take into account the critical parameters of termites' social organization, which often results in oversimplified interpretations. The social context of termite colonies defines their ability to respond to environmental changes, and hence experiment protocols, and observations from a small subgroup of a termite colony rarely hold when scaling up to colony-wide observations (Becker, 1969; Miramontes and DeSouza, 2008; Gordon *et al.*, 2022).

18.1.5 Toward optimized laboratory protocols

The non-mutually exclusive limitations described above have hindered the progress of empirical termite biology research for decades. Such limitations can be solved by: (1) taking into consideration the complexity of the termite social context and nesting habits; (2) investing resources efficiently within the laboratory, so that a series of questions can be answered while limiting their inherent costs; (3) establishing a cost-effective and reliable source of biological material; and (4) improving termite survivorship and visibility in experimental assays. This book seeks to provide a comprehensive background on what makes *Coptotermes* both a unique termite pest and also a unique termite model system to study. This final chapter specifically covers aspects of laboratory experimental approaches with *Coptotermes* and provides a general incentive for current and future laboratories to take time in establishing appropriate protocols toward answering questions within their relevant biological context.

18.2 Collection of large numbers of termites from field populations

Collecting a whole *Coptotermes* colony with a complete brood, foragers, and reproductives from the field is virtually impossible. Mature *Coptotermes* colonies can reach several million individuals, with foraging distances that can spread underground for up to 100 m, and with a cryptic nesting habit that makes the localization of their central nest impractical (Greaves, 1962; King and Spink, 1969; Su and Scheffrahn, 1988; Vargo *et al.*, 2003; Ruan *et al.*, 2015; Chouvenc *et al.*, 2022). Historically, the monitoring of subterranean termite activity in the soil has used wooden stakes (or equivalent) driven in the ground in a grid pattern, to obtain a preliminary estimation of the foraging territory of an established *Coptotermes* colony, which often involved mark–recapture protocols (Lai, 1977; Su *et al.*, 1984; Su and Scheffrahn, 1988; Messenger and Su, 2005; Crossland and Su, 2006) to confirm interconnection among feeding sites by a single termite colony.

Ehrhorn (1934) reported one of the first uses of wooden stakes to detect (and attempt to treat) a *Coptotermes* population, and Sands (1972) later mentioned the use of such wooden monitoring stakes (referred to as "bait sampling method") in various studies on tropical subterranean termite species. La Fage *et al.* (1973) popularized this monitoring approach with the iconic use of large grids of toilet paper rolls, laid on the surface of the Arizona desert, to monitor *Heterotermes* and *Gnathamitermes* populations. However, the toilet paper roll approach to monitor and collect termites is often not possible with *Coptotermes*, especially in urban areas. In addition, while wooden stake protocols may allow for a long-term monitoring of termite activity, they do not permit the collection of large numbers of foragers. Indeed, extracting termites from an infested piece of sound wood is challenging, owing to subterranean termites initially feeding on the spring wood, resulting in termites being sandwiched within layers of wood, which leads to most of them being crushed during the destructive extraction process. In addition, wooden stakes are often too small to obtain large numbers of termites at any given time. This issue was eventually solved by the implementation of a "bucket trap" method, initially established by Tamashiro *et al.* (1973). Bucket traps allow for periodically collecting several thousand termites at a time from field-established colonies. The aim of establishing such trap locations is to ensure a reliable source of biological material, while minimizing labor cost, processing efforts, and stress imposed on termites.

18.2.1 Finding suitable collection sites

To obtain a reliable source of termites for years, or even decades (Chouvenc *et al.*, 2022), several conditions should be met before implementing bucket traps. First, established mature *Coptotermes* colonies must be detected, either by inspecting trees for potential *Coptotermes* infestation, or through systematic surveys with wooden stakes. Colonies detected on private properties are often not good candidates, owing to the incentive from homeowner to eliminate colonies for long-term structural protection. However, some privately owned properties with no risk for structural damage may still provide opportunities, as long as the private or corporate owner agrees to use the site for termite research purposes. Alternatively, public parks or public areas under various administrative jurisdictions can be used as long-term collecting sites, as long as permission is provided by the administrator, or an agreement is reached. Second, colonies must be located in areas that are readily accessible by laboratory technicians, as commuting distances between the laboratory and established colony field sites can put a strain on financial and time resources. Therefore, the periodical monitoring frequency of multiple field sites at various locations should be optimized, by establishing a routine bucket trap inspection protocol that minimizes commuting time among sites. Finally, the field sites must be subjected to the least possible disturbance (primarily anthropogenic, but also from wildlife and environmental extremes). Termite foraging populations can permanently abscond from established bucket traps because of excess disturbances. Once established, bucket traps should be out of sight and/or out of reach of the general public or any other third party, to prevent unwelcomed tampering.

18.2.2 Establishing bucket traps at a field location

Once identified through the initial wooden stakes survey or turning over existing downed woody debris in a field site with an active *Coptotermes* foraging population, in-ground station bucket traps can be established (Tamashiro *et al.*, 1973) (see Chapter 5 of this volume). This method for collecting *Coptotermes* in the field from in-ground stations has since been emulated, adapted, and modified by several laboratories around the world (Su and Scheffrahn, 1986; Ewart *et al.*, 1992; Grace *et al.*, 1995; Sornnuwat *et al.*, 1996; Osbrink *et al.*, 1999; Su, 2001; Acda, 2004; Arab *et al.*, 2005; Chiu *et al.*, 2016; Chouvenc *et al.*, 2022). The concept is to provide a semi-permanent rigid structure in the ground that protects the termite feeding site, but that can easily be accessed by an operator with minimal disturbance. To do so, a hole is dug around the initial infested wooden stakes, or where termites are observed entering downed woody debris on the ground (Su and Scheffrahn, 1986). A plastic bucket, the bottom of which has been cut off (to allow for termite access), is then placed in the hole and with the lid is flushed with the surface of the ground (Fig. 18.1A). At all times, a food source must remain present in the bucket to maintain termite activity. The food source can be a preassembled bundle of wood (for monthly collection protocols, as described below), rolls of moistened corrugated cardboard (for collection within a few days after installation), or

Fig. 18.1. Method to collect large numbers of *Coptotermes* from field-established colonies, adapted from Tamashiro *et al*. (1973). (A) Bucket trap (photograph courtesy of Aaron Mullins). (B) Wood bundle preparation (photograph: Thomas Chouvenc). (C) Wood bundle collected from an in-ground bucket trap after 3 weeks in the field (photograph: Thomas Chouvenc). (D) Above-ground field trap attached to a tree (image courtesy of Hou-Feng Li). (E) Termite processing protocol in the laboratory (photograph: Thomas Chouvenc). (F) Termites after being processed, clean from wood and nest debris (photograph: Thomas Chouvenc).

any equivalent contraption that provides a feeding matrix with high surface access (Evans and Gleeson, 2006). Moistened cardboard rolls allow for rapid termite extraction in case of emergency needs and rapid collection of large numbers of termites for an experiment (La Fage *et al*., 1983). However, preassembled bundles of wood or durable feeding matrix are preferred for long-term termite collection protocols. Finally, using two food sources in a bucket and alternating their periodical replacement while leaving one of them undisturbed in the bucket may minimize chances for the foraging population to abscond during the collection process.

Although the size of the bucket and the dimensions of the wood bundle used in this trapping method are at the discretion of each laboratory, they must remain practical for a technician to optimize productivity, i.e., the number of termites collected over time within the constraints of available resources. Initially from Su and Scheffrahn (1986), and fully redescribed by Chouvenc *et al.* (2022), the standard protocol of the University of Florida subterranean termite laboratory (referred hereon as the "UF Termite Lab") for more than 35 years has used semi-flexible plastic buckets (18 cm diameter × 20 cm height) (Fig. 18.1A). Bundles of wood can be assembled from five or more (15 × 10 × 0.3 cm) pieces of wood (*Picea* sp.) and the wood must be pre-emptively cured to allow undesirable wood volatiles to fade away. Wood pieces must each be separated by fine plastic straws or equivalent, bundled by tied-up plastic-coated metal wires that compress the straws, leaving ~2 mm gaps between each wood layer, creating an equivalent "termite space" (Chouvenc *et al.*, 2011b) which allows for maximal surface access to the termites. Prior to assembly, the wood can first be oven-dried to obtain initial dry weight of the bundle, in case it involves a wood consumption study. When a roll of corrugated cardboard or wood bundle (Fig. 18.1B) is ready to be placed inside an active bucket trap, it must first be soaked in water, shortly before installation in the bucket trap.

18.2.3 Collecting termites from field bucket traps

First, the frequency of termite collection should be established, which may vary depending on colony size, colony health, termite species, and seasonality (Waller and La Fage, 1987; Grace *et al.*, 1995; Chouvenc *et al.*, 2022). Leaving a wood bundle in the bucket trap for too short of a time can minimize the potential number of termites actively feeding on it at the time of collection, while leaving it for too long may result in full wood consumption and site abandonment by the foraging termite population. In the case of the use of cardboard rolls, it may only remain for a few days before being fully consumed, potentially losing the opportunity to acquire several thousand termites. All preparations, which include soaking the feeding matrix with water, must be made at proper distances from the active bucket trap, to minimize ground vibration. For the same reason, approach to the trap must be gentle. Lid removal and the extraction of the already established feeding matrix must be done swiftly, to reduce the time of disturbance to just a few seconds. Carefulness during the collection process is critical, as workers rapidly abscond from the trap while soldiers aggregate near the disturbance (Gautam and Henderson, 2012a; Wang *et al.*, 2016; Sillam-Dussès *et al.*, 2023), resulting in the collection of a relatively small group of termites with an aberrant soldier ratio. Retrieved wood bundles/cardboard rolls can then be placed in individual containers and brought back to the laboratory for processing (Fig. 18.1C). Once the wood/cardboard is replaced, the lid is placed back on the bucket trap and it must be concealed using surrounding leaf litter or soil around the traps, to prevent tampering from any third party, and where only the technical staff monitoring the traps would know their exact locations (with additional GPS records for the books). It is important that multiple bucket traps are installed within the foraging territory of a single *C. formosanus* colony to increase the sampling yield, where mark–recapture methods can be used to confirm initial interconnectivity of the bucket traps (Lai, 1977; Su and Scheffrahn, 1988; Husseneder *et al.*, 2005; Messenger and Su, 2005; Crosland and Su, 2006). Subsequent wooden stakes can be installed in the vicinity to further expand the numbers of available bucket traps at active feeding sites, to take into account the potential for shifting foraging territory of a colony over time and casual loss of activity (Chouvenc *et al.*, 2022). Finally, in addition to in-ground bucket traps, above-ground traps can be installed for *Coptotermes* activity in trees or structural infestations, as used in above-ground bait formulations (Su *et al.*, 1997; Yates and Grace, 2000). An above-ground contraption containing moist corrugated cardboard or a wood bundle (Chiu *et al.*, 2016) can be attached to the site of an active infestation, carton material, or mud tube (Fig. 18.1D), and may also serve as a source of termites for laboratory use.

18.2.4 Processing termites collected from the field

This phase is a critical point in the procedure, where termites are the most likely to be stressed and injured. Carelessness during this phase inevitably results in mediocre termite survival, making such processed batches of termites unusable. There are many ways that termites collected from the field can be processed in the laboratory, but among the potential protocols that would result in separating termites from the cellulose matrix and debris, Tamashiro *et al.* (1973) established, to date, the most efficient and most gentle method to obtain large numbers of healthy termites, fully separated from debris. The concept tapped into the inherent subterranean termite behavior, where individual termites move away from dry locations toward a moist refuge, which equates to a termite "self-sorting" method. Tamashiro *et al.* (1973) used wooden moist tongue depressors as a ladder for termites to climb on moist blocks of wood stacked on each other with thumb-tags beneath each wood block to create space for termites to crawl in. To ensure that termites moved to wood blocks, the tip of the tongue depressor has to touch the side wall of the wood block. Once the contraption is set, termites can then progressively move from the debris, toward the wood block refuge (Fig. 18.1E). Repeatedly used wood blocks appear to aggregate more termites quickly but avoid using a set of wood blocks for different subterranean termite species.

Once the termite transfer is completed, the blocks can then be gently tapped in a clean container or tray (Fig. 18.1F), from which termites can be further sorted for the need of the experiment (e.g., caste, instar). As an alternative to the Tamashiro *et al.* (1973) method, a simplified version of this concept is to apply wet paper towels on top of the layer of debris and termites in the tray, as individuals can anchor themselves on the towel within a few minutes. The towels can then be gently shaken into a new tray, where termites are at this point fully separated from the debris. The process can be repeated multiple times, until the vast majority of termites are sorted out. The few remaining termites that are still mixed with debris can either be sorted out with a mouth aspirator, a soft entomological forceps, or can be discarded. Finally, instead of being used directly for experiments, groups of termites can be stored in containers with moist layers of wood or other cellulosic material, from a few days to a couple of weeks prior to experimentation, depending on vigor and storing conditions.

18.2.5 Aspects to consider when using processed termites collected from the field

After being collected from field traps, termite vitality often decreases over time, especially if not maintained in optimal conditions. Therefore, termites kept in groups for over a month should not be used for experiments, as control mortality at this point would not be acceptable. As a generally accepted standard, survival of at least 80% of termites from control groups should be observed by the end of the experiment. It is also important to keep in mind that the temperature at which the termites are maintained may affect their feeding activity and their molting rate (Cao and Su, 2016; Kakkar *et al.*, 2016). Temperatures used for maintenance and experiments vary in the literature, often from 21°C to 28°C, and humidity inside rearing vials must remain above 90% RH. In addition, the collection of large numbers of termites using the approach described above can provide the opportunity for a laboratory to acquire additional metrics about the quality of the obtained biological material. For example, the data collection for the average weight of individuals and the caste composition of a group of foragers from field traps can directly inform on the demographic status and shifts in reproductive activity over time, directly informing also on the overall health of the field colony of origin (Chouvenc *et al.*, 2022).

One of the advantages of using field-collected termites from the method outlined above is that once field sites are established, it can provide a reliable source of "fresh" biological material from field *C. formosanus* mature colonies. With a bucket trap protocol implemented at multiple sites, it can also provide termites from distinct colonies, to ensure the replicability of experiments. A planned routine from a technical member of staff to monitor, maintain, and collect all active traps on a determined rotation schedule therefore ensures a reliable supply of healthy and vigorous termites for laboratory use. However, this method has limitations. As mentioned previously, laboratory resources can be thin, and establishing an extensive network of field bucket traps requires the skill of a trained applicator and an infrastructure to maintain it. This means that such field programs come at a cost, which includes salaries, vehicles, gas, temperature-controlled incubators, and miscellaneous consumables (wood, containers, etc.). In addition, one of most common shortfalls of using groups of termites collected from field foraging sites is that they primarily represent workers and soldiers from a relatively old demography within the colony (Su *et al.*, 2017; Chouvenc *et al.*, 2022; Lee *et al.*, 2022). Thus field-collected termites do not represent a "colony" so much as a portion of the "foraging population of a colony." This implies that experiments using field-collected termites cannot consider the effect of the brood (eggs, larvae), primary reproductives (queen and king), and relatively young cohorts of foragers and caretakers (Du *et al.*, 2017a) in their experimental design, therefore ignoring important aspects of the complexity of the social interactions found in mature *Coptotermes* colonies. One such aspect was highlighted by Kakkar *et al.* (2018) and Chouvenc (2018b) concerning the re-interpretation of the results from Su (2005a) on the importance of the brood and central nest in *Coptotermes* on pest management strategies.

18.3 Rearing termite colonies from foundation to maturity in the laboratory

This approach fundamentally departs from the field collection approach in methodology, in resources needed, and in its purpose. Laboratory colony rearing can be a rather resource-consuming approach that requires a dedicated and perennial laboratory workforce. Ultimately, this approach attempts to emulate the complete life cycle of termites within the constraints of laboratory conditions (Osbrink *et al.*, 2016). If successful, it allows for the mass rearing of termites in standardized conditions, and for minimizing environmental fluctuations, providing a supply of biological material at a scale that no other approaches would permit (Chouvenc, 2022; Gordon *et al.*, 2022). Laboratory rearing of *Coptotermes* colonies must take into account various challenges as colonies are established in artificial conditions and are then reared over many years through a demanding maintenance regime. Alates must be collected from the field during each seasonal *Coptotermes* dispersal flight, so that incipient colonies can be established in the laboratory annually to maintain rearing continuity. Subsequently, colonies must be maintained to be allowed to grow toward maturity over many years. The following sections provide a detailed protocol of the rearing routine followed at the UF Termite Lab since 2013, but variations from this protocol can be implemented depending on the needs and limitations of other laboratories working with *Coptotermes*. The protocol includes three steps: (1) collecting termite alates from field dispersal flights; (2) establishing incipient colonies in the laboratory; and (3) rearing termite colonies toward maturity through a diligent maintenance regimen.

18.3.1 Collection of termite alates from field dispersal flights

From a locality with known field-established populations of *Coptotermes*, alates can be obtained to initiate the establishment of laboratory colonies. The most convenient way to obtain a large number of alates from various colonies of origin is to use a light trap during dispersal flight events (Chouvenc *et al.*, 2017). Of course, only species that disperse around dusk or at night can be trapped using this method, which applies to *C. gestroi* and *C. formosanus*, as they fly shortly after sunset. Extracting alates directly from field nest material is impractical and the alates' degree of maturity and vigor toward laboratory colony establishment may be questionable and should be avoided if possible (personal observation). Therefore, a light trap designed to collect large numbers of live flying termite alates remains a critical element to start laboratory colonies. To my knowledge, there are no commercially available light traps specifically designed for this purpose, or that would fulfill the function. Modification of monitoring light traps for other insects may be an affordable solution. However, it is relatively inexpensive to build one, and the size of the contraption can be adapted at the convenience of the user. The complete material required for a functional light trap for termite alate collection can be assembled with as little as a container, two panels, one panel stand, and a light source. The concept is to create a trap with a funnel design that allows termites to fall into it but would prevent them from flying out (Fig. 18.2A), which is adapted from Peppuy (1999).

The panels must be installed on a stand at an offset angle to create a linear funnel entrance, leaving a space of no more than 1.5 cm (Fig. 18.2B). Panels must be smooth so that termite alates cannot have any grip and will readily slide down the funnel from gravity. Panels must also be opaque so that, when attracted by the light, it is not through the panels, which would prevent access to the funnel area. Black or white acrylic laminate (~3.2 mm = 1/8 inch) panels works well, and are easy to assemble and maintain. A light source must be placed above for alates to direct their flights toward the trap. Urban lights may compete with the trap, so it is important to provide a strong alternative. While *C. formosanus* alates are apparently attracted to 400–420 nm wavelength (Yamano, 1987), a bright floodlight (white spectrum) with high output (> 3,000 lumen, LED only to prevent accumulation of heat) can be used as a light source and can also serve to anchor the two panels in case of sudden strong wind events (Fig. 18.2C). An additional blacklight (UV spectrum, LED or fluorescent) source is recommended, as various *Coptotermes* species may be attracted to different light frequencies. For example, when lights were both combined, it reliably attracted *C. gestroi* and *C. formosanus* alates (Chouvenc *et al.*, 2017). The stand must be built in a way that allows for the two panels to be at the appropriate angle, but must also cover the container it sits on, to serve as a lid and prevent termites from escaping (Fig. 18.2D). In addition, the container receiving all alates must have smooth walls to prevent termites from crawling out and should be large enough for termite alates to fly further within the contained space, which may facilitate the alate self-dealation process (Raina *et al.*, 2003a). Note that if the trap is located in an area with strong ant pressure from the ground, a plastic tarp may first be installed on the ground surface prior to installing the light trap container, so as to prevent ant raids within the trap itself.

Once termite alates have flown into the trap (Fig. 18.2E) and alate flight density in the surrounding environment is thinning out, light sources must be turned off to prevent a large number of undesirable insects from entering the trap. Beetles, moths, ant imagoes and other miscellaneous flying insects also tend to be attracted to the light trap and there is often a short window of time (5–10 min) between the moment *Coptotermes* alates stop flying in the environment and the moment other insects increase their flight activity toward the trap. Therefore, turning off the lights at the right time can avoid an unnecessary (often messy and time consuming) sorting process between termite alates and other insects. Once termite alates are captured in the light trap, they can be transferred by disassembling the trap (away from any light sources, if possible, to prevent an unwanted alates exodus) and gently dumping the alate load from the light trap container into a smaller portable and closable container. The transportable container should first be accommodated with stacked layers of moist corrugated cardboard. This step is critical, because not only does it incentivize termites to dealate (loss of wings) and hide within the holes of the carboard, but also they can be stored in such conditions until the next morning with a high survival rate. Finally, in the case of a large dispersal flight event, the alates should

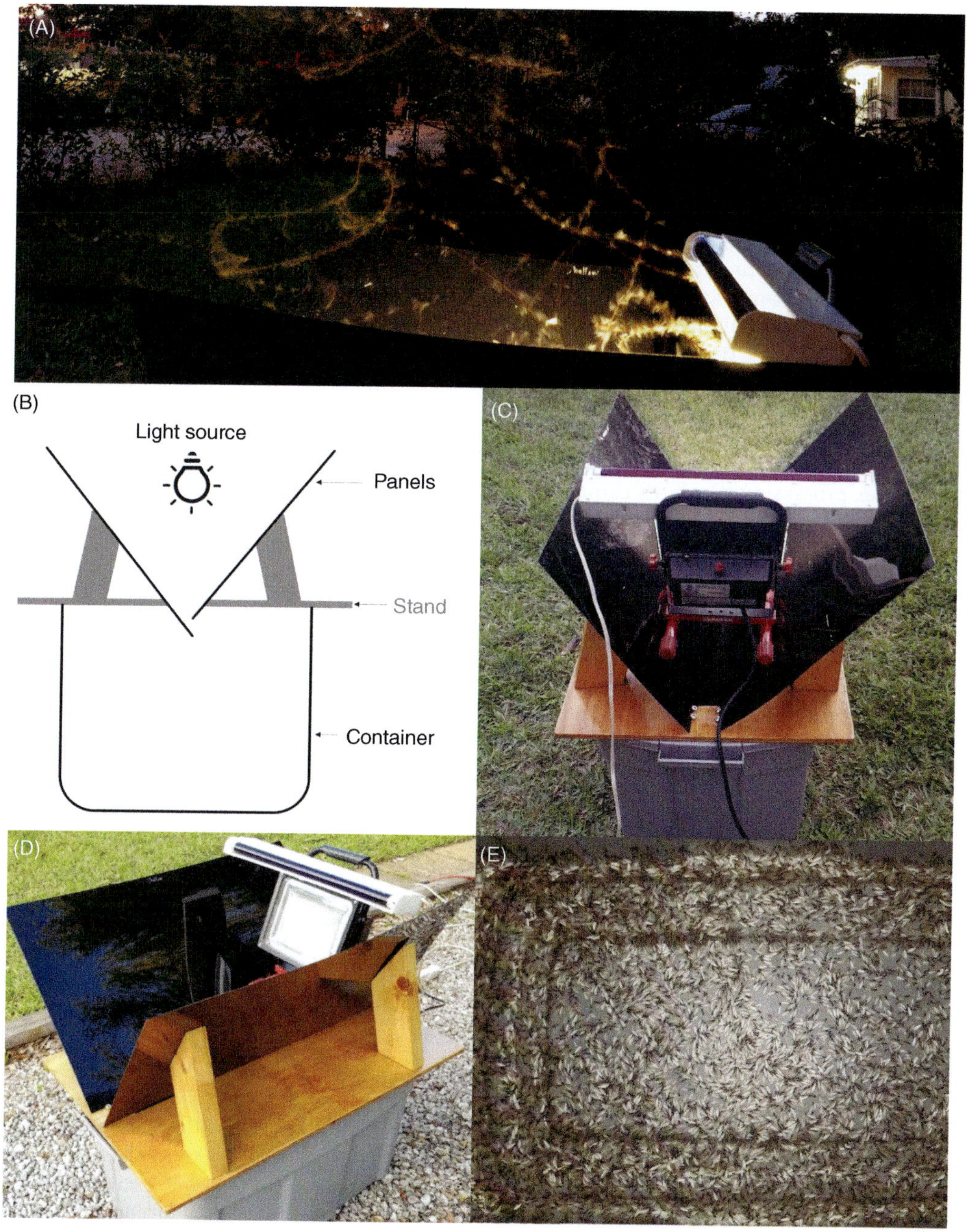

Fig. 18.2. Example of a light trap for collecting *Coptotermes* alates during dispersal flight events. (A) Termite alates flying toward the light trap. (B) The two panels must be offset to create a linear funnel that prevents termites from flying back out. (C) Light sources can be placed on top, anchoring the panels against potential wind gust events. (D) The panel stand serves as a lid for the container. (E) Accumulation of alates in the trap after a large dispersal flight event (here, ~7,000 *Coptotermes gestroi* alates). (All photographs: Thomas Chouvenc.)

be split in multiple containers, because termites tend not to dealate if their density remains too high when in the container. It is important to provide the best conditions for most alates to drop their wings, otherwise it makes the sorting process impractical the following day.

While incipient colony establishment could be initiated immediately after the dispersal flight if all preparations were made ahead of time, it is time-consuming and often requires laboratory equipment (see next section). Therefore, since the location of dispersal flights may be remote from the laboratory, *Coptotermes* alates can be kept overnight in the moist corrugated cardboard and be brought to the laboratory the next morning for further processing. Dealates can then be separated out from the cardboard and must be sorted out by sex. Females' last abdominal segments are fused ventrally, forming a sternal plate that is not present in males (Park *et al.*, 2004; Chouvenc *et al.*, 2017). In addition, females produce a sex pheromone from a pair of tergal glands, located dorsally on the 9th and 10th abdominal segments which elicit males to form a tandem behavior. The female leads and the male follows (Chouvenc *et al.*, 2020; Mizumoto *et al.*, 2021a), facilitating sex identification from a group of active dealates. Once sorted, the number of individuals of each sex can be used to determine the sex ratio of the dispersal flight, and the average weight of each sex (which both vary greatly throughout the dispersal flight season) (Chouvenc *et al.*, 2017; Chouvenc, 2019). Using subsamples of individuals and the total mass of termite alates collected, it is possible to estimate the number of alates collected from large dispersal events without having to sort all dealates individually. Then, if not used immediately, groups of sorted dealates may be kept in Petri dishes with a moisture source (moist filter paper, sand, or cardboard) for a few hours (< 6 h) before being used for colony foundation. However, it is important that the whole process, from field collection of alates to laboratory colony foundation, is completed within ~20 h of alates collection, and the sooner the better. Colony foundation success rate plumets drastically when dealates are used after 24 h from collection and their use for incipient colonies should be avoided after one day of collection.

18.3.2 Establishing incipient colonies in the laboratory

Colony foundation in the laboratory with *Coptotermes* dealates has been attempted and replicated over time in several laboratories, with various approaches (King and Spink, 1974; Higa, 1981; Huang and Chen, 1984; Costa-Leonardo and Barsotti, 2001; Fei and Henderson, 2003; Morales-Ramos and Rojas, 2003; Raina *et al.*, 2003b; Ferraz and Cancello, 2004; Chouvenc and Su, 2014). *Coptotermes formosanus* and *C. gestroi* are among the most successful invasive termite species (Evans *et al.*, 2013), where incipient colonies can readily establish in boats and later produce alates as a propagule pressure for establishing in novel geographic locations (Hochmair and Scheffrahn, 2010; Su, 2013a; Chouvenc *et al.*, 2016) (Chapter 2 of this volume). Such observation revealed that alates of either species can readily establish new colonies in challenging conditions (Mullins and Su, 2018; Chouvenc, 2019), which directly translates to them being relatively easy to rear in laboratory conditions (Osbrink *et al.*, 2016). In addition, as alates can be collected in relatively large numbers using the protocol described above, incipient colonies can be mass produced to mitigate the inherent foundation failure rate. By standardizing the production of incipient colony rearing units each year, a laboratory should be able to sustain optimal rearing capabilities over time.

Incipient *Coptotermes* colonies have minimal requirements to be self-sufficient within the first year of their life, with little-to-no maintenance required. If the founding pair of dealates have access to wood, to a reliable moisture source, to soil rich in organic matter, and to appropriate space, incipient colonies can successfully survive the first few critical months of foundation (Chouvenc, 2022b) and thrive toward further colony growth. The following protocol for rearing units was optimized at the UF Termite Lab over the past decade, as initially adapted from Higa (1981). A rearing unit is composed of a transparent plastic cylindrical vial, 8 cm height × 2.5 cm diameter, internal volume of 37.5 cm^3 (10 drams), (bigger containers should be avoided for incipient colonies) with 6 g of moistened organic soil material at the bottom. The organic soil should be first sifted to a coarse powder and mixed with a small amount of water (below saturation) to ease the introduction into the vials, and then should be slightly compacted in the vial with a dowel to leave room for the other elements. Four blocks of wood, *Picea* sp. (5 cm × 0.5 cm × 0.5 cm) are positioned vertically, and one additional wood block (10 cm × 0.5 cm × 0.5 cm) is placed inside the vial, along the vertical side, sticking out on the top. A 3% agar solution (autoclaved) is then poured into the vial, leaving a 2 cm space at the top of the vial (Fig. 18.3A), as a 500 ml solution flask should allow for the preparation of ~40 vials. It may take several hours for the agar to cool down and solidify. The dealate sorting process described above can take place during this down time, so as to streamline the colony foundation process in the shortest possible time.

When the agar is solidified, the long wood block is pulled from the vial to leave a hole in the agar, providing direct access to the soil at the bottom, to the wood and to the agar. A male and a female dealate *Coptotermes* can then be placed in the vial, where they can initiate a claustral chamber in the preformed access hole (Fig. 18.3B). A perforated plastic cap is then placed on the top to allow for aeration (with a thin safety pin), but to limit desiccation and prevent escape over time. Rearing units are then stored at 26–28°C in larger container for ~10 months (Fig. 18.3C). During this time, a small amount of water may be added to vials that show signs of dryness (shriveled agar), but otherwise they do not require much attention. If the container is kept closed, the relative humidity of the room or incubator is irrelevant, because the internal space of the vials can maintain > 99% RH air moisture over time if the air exchange from the vials to the environment is limited. At ~10 months of growth, colonies can reach > 150 individuals and often start escaping from their rearing vial by chewing through the lid. It is often necessary to transfer the colonies to bigger containers prior to 10 months for two primary reasons. First, it prevents escapes, which often result in colonies fusing with each other (which should be avoided) (Fig. 18.3D), leading most individuals from involved colonies to dry out, because the air exchange with the outside environment is now too high. Second, the transfer to larger containers (see next section) can be time-consuming and it is preferable to complete the transfer of all viable colonies before the start of the upcoming dispersal flight season, so that critical and time-sensitive tasks do not occur all at once.

Although this *Coptotermes* foundation protocol is now standard at the UF Termite Lab, it can be altered according to the need of experiments or rearing conditions. For example, rectangular vials instead of cylindrical ones can be used to improve visibility of termite behaviors within the first few months of incipient colonies, allowing for behavior recording (Chouvenc and Lee, 2021). In addition, the use of soils rich in organic matter is critical for sustaining initial colony growth (Mullins and Su, 2018). *Coptotermes* colonies require supplementary dietary nitrogen from soils to optimize the production of new individuals, as the use of plain mineral sand instead of organic soils results in stalling any further colony growth (Mullins *et al.*, 2021). Finally, it is preferable not to autoclave the soil prior to being used for rearing units (personal observation). If autoclaved, the soil in rearing units can rapidly be overwhelmed by fast-growing saprophytic fungi such as *Aspergillus*, *Trichoderma*, *Paecilomyces*, and *Penicillium*, most likely acquired from contamination during the rearing unit assembly or from alates themselves, and which may negatively impact incipient *Coptotermes* colonies (Chouvenc *et al.*, 2013a). Instead, non-autoclaved soils readily come with a complex biotic assemblage of microorganisms that buffers the system from being taken over by a single strain of fast-growing fungus or bacterium (Chouvenc *et al.*, 2018). Laboratories may have access to various commercially available organic soils (organic potting soil, topsoil, peat moss mixtures, etc.), which are often heterogeneous even within brands. It may be necessary to test them individually or in combination to assess what would work best for initial colony survival and colony

Fig. 18.3. Laboratory colony rearing protocol. (A) Rearing vials, with the agar drying. (B) Incipient colony where the king and queen established the claustral chamber in the preformed hole. (C) Storage solution for incipient colonies. (D) Escape of colonies around ~10 months of growth. (E) One-year-old colony transferred to 1.5 l container. (F) Colonies transferred to larger containers toward maturity, with (on top shelves) 13 l containers with 3–5-year-old colonies of *Coptotermes* and (on the bottom) three "cubes" containing > 5-year-old *Coptotermes* colonies. (All photographs: Thomas Chouvenc.)

growth in the rearing units, so that adjustment can be made in the following years if necessary. Finally, initial rearing substrate can be species or genus specific, as an incipient colony rearing protocol optimized for *Coptotermes* may not necessarily be the best for *Reticulitermes* which may require different rearing conditions (Mitaka *et al.*, 2023).

18.3.3 Colony growth and laboratory maintenance

As incipient colonies reach critical carrying capacity in their initial rearing unit (~10 months), it is necessary to transfer them individually to larger rearing containers. At the UF Termite Lab, rearing conditions were standardized using ~1.5 L containers (Pioneer Plastics, Dixon, Kentucky, USA, 17.5 × 12.5 × 7 cm^3, model 049C, hard plastic with lid). A 2 cm layer of moistened organic soil is first deposited at the bottom of the container (no need to be sifted) and two moistened *Picea* sp. wood blocks (15 × 4 × 1.5 cm^3) are stacked on one side of the container. On the other side, the initial cylindrical rearing vial containing an incipient colony is placed horizontally on the surface of the soil (lid is first removed). While it is possible to extract all individuals of a whole incipient colony from the initial rearing vial for a complete census and metrics acquisition protocol (for internal records or for an experiment), it must first be relocated in a similar rearing vial (with soil, wood, and agar) prior to the transfer, so that the colony can reacclimatize itself in a contained space for several days. If processed termites are instead directly transferred to the open space of the new 1.5 L container, the mortality rate becomes unacceptable for rearing purpose, and should be avoided. It is important that incipient colonies are left to explore their new rearing environment on their own from their initial vial, and to let the termites colonize the new soil and wood by themselves over time, so as to minimize the stress during the transfer. Finally, an additional step may be considered to improve colony survival and growth between the 37.5 mL vial and 1.5 L container. The 37.5 mL is often too small for the royal chamber to persist over time after the transfer to the 1.5 L containers, leading to the royal couple establishing between two layers of wood in the main container, increasing the chance of killing them by accident during maintenance. To minimize such issue, the ~10-month-old colonies can first be processed and transferred to a medium-sized vial (9.5 cm height × 4.5 cm diameter, or 40 drams), with soil, wood, and a thin layer of agar (~2 cm thick), where the queen and king often remain safely for the years to come. Colonies can be left for several additional months in such an intermediate vial, prior to being transferred to a 1.5 L container as described above (Fig. 18.3E).

After transfer to 1.5 L containers, colonies' health and growth primarily rely on simple routine maintenance and inspection protocol of all colonies for the years to come (~2-week rotation cycle). (1) Colonies must have access to wood at all times. If a colony is about to run out of food, wood blocks can be added as needed to the rearing container. (2) Colonies must be continuously maintained in moist conditions. Water can be regularly added to the container with a spray bottle each time the colony is inspected during the rotation cycle. (3) While the containers are not air-tight, opening the lid of the container every other week may help prevent suffocation from the potential accumulation of metabolic gas. (4) Termite colonies must be kept captive within their individual container. Daily visual inspection can detect mud tubes coming out of lids that can be propped open by termites over time. Such escape can be facilitated if during the maintenance routine, some soil or frass particles are inadvertently left jammed within the space between the container and the lid, which allows termites to wedge additional particles to crank the lid open. If unmonitored, termites can either expand the mud tube and reach the substrate, leading to an unpleasant surprise on that fateful Monday morning, the termites fall out of the container, slowly depopulating the colony and leading escaped termites to unavoidable desiccation, or to sustain the laboratory's resident spider population.

Space eventually becomes a limiting factor again, and the lack of space and food resources can lead to physiological stress to the colony, making them unsuitable for experimentation (Osbrink *et al.*, 2016; Lin *et al.*, 2023). Within 2.5 years post-foundation, colonies usually reach 5,000 to 10,000 termites. At this point, colonies stop growing if kept in such conditions, and escaping events may become frequent. Therefore, over the years, colonies must be transferred to progressively larger containers to accommodate further colony growth. Again, the protocol at the UF Termite Lab was streamlined using 13 L containers (30.5 × 45.7 × 15.2 cm^3) (Carlisle, Oklahoma City, Oklahoma, model 1061107 StorPlus Clear Food

Storage Box and Lid). This container model possesses a rim that allows for a space between the external border of the container and the lid. Such a design provides two important features for the rearing protocol biweekly routine. First, clamps can be installed on the four sides, to secure the lid and limit escape events (Fig. 18.3F). Second, if the colony initiates an escape, it first forms a mud tube ring in the space between the lid and the container rim, providing a delay of ~1 week for the maintenance staff to detect it and prevent a full colony escape. Between 2013 and 2023 such protocol has allowed the UF Termite Lab to grow thousands of *Coptotermes* colonies, while keeping them safely contained within their dedicated space.

Colonies can then continue to grow in such conditions and reach a maximum capacity of ~60,000 termites within 4 years post-foundation (Chouvenc and Su, 2017a). If colonies are maintained in these containers, they stop further growth and can be maintained as they are for several years, prior to use in experiments, as standardized-size colonies (Gordon *et al.*, 2022). At this point, colonies can be accumulated in a rearing room, where a dedicated space must be allocated. Colonies maintained in limited containers like these are not in the condition to fully mature, which could be analogous to a colony development stasis (Osbrink *et al.*, 2016) as space is now the primary limiting factor. However, if provided additional space (~1 m^3) and food, colonies can reach > 1 million termites within 5 years post-foundation and seasonally produce alates (personal observation) (Fig. 18.3F), therefore completing the full life cycle within laboratory conditions. While seasonality of alate production may not be achieved within constant laboratory temperatures, conditions can be temporarily altered by changing the room temperature to "winter" (e.g., from 28°C to 18°C), or by temporarily moving colonies to a cooler location. In addition, colonies that are allowed to reach such a size have a tangible metabolic output, as the inside of the container often displays temperatures that are several degrees higher than the rearing chamber, and, more problematic, are subject to suffocation if the rearing container is closed. Therefore, contrary to the 1.5 L and 13 L containers described above, such large colonies must remain "open lid" at all times, and mud tube constructions by the active colony must be cleared periodically through the laboratory maintenance routine. Alternatively, a "water moat" designed on the rearing container can prevent escapes and provide a permanent source of water for the colony.

18.3.4 Aspects to consider when rearing *Coptotermes* laboratory colonies

When compared with termites collected from foraging field sites, such a colony laboratory rearing protocol offers a series of advantages. First, it provides a high degree of replicability by using whole termite colonies, offering a quasi-unlimited supply of biological material. Second, it has often been stressed how a portion of a foraging population does not reflect the behavior and demographic complexity of whole termite colonies and that observations obtained in conditions that are far detached from the biological relevancy of a *Coptotermes* nest must be interpreted with caution (Becker, 1969; Thorne and Haverty, 1991; Lenz, 2009; Chouvenc, 2018a). Laboratory-reared colonies bypass such limitations by providing a complete biological entity for experimentation (queen, king, brood, comprehensive demographic profiles of workers and soldiers) which allows for testing hypotheses that were previously not testable (see Kakkar *et al.*, 2018; Gordon *et al.*, 2022). In addition, using a reproductive excluder (Lee *et al.*, 2019), colonies of various ages can be used to study the importance of distance from the "central nest" (Lee *et al.*, 2022). Whole colonies can also be connected to a foraging arena system (see next section for more details), simply by drilling a hole on one of the sides of the rearing container. Finally, large numbers of termites can be harvested from mature laboratory colonies (Fig. 18.3F) using moist corrugated cardboard rolls (~80,000 termites within 1 h), allowing for the rapid set-up of an impromptu experiment, and also providing replication if multiple colonies are used, with little to no impact on the health and functioning of the sampled mature colonies.

However, rearing millions of termites in laboratory conditions from initial colony foundation to maturity comes with a series of limitations that research laboratories must take into consideration as colonies successfully grow. First, space rapidly becomes a primary limiting factor. Colony-rearing containers must often be stacked vertically to optimize their footprint in the rearing facility, and there can only be a limited number of colonies that may be allowed to be transferred in large containers toward maturity. Second, inspection and maintenance protocols are time-consuming. It may take several full-time employees to ensure that laboratory-reared colonies are kept healthy, growing, and contained, by carefully following the rearing protocol described above. Salaries for such a dedicated taskforce must be accounted for in the laboratory research budget. Third, this rearing protocol is resource-demanding. Wood must be acquired, cured, processed, and cut to usable pieces, which involves the use of power tools, with potential liabilities. Organic soil must be repeatedly resupplied to growing colonies and it must be ensured that the soil does not come with traces of pesticides. Containers of various sizes for different colony ages must be systematically resupplied to accommodate the accumulation of colonies over the years. All of these resources are inherent “consumables” and must be budgeted accordingly.

Fourth, the colony foundation process remains the most sensitive part of the rearing protocol, and unacceptable levels of colony establishment failure can easily occur if attention to details is neglected during the preparation of the initial vials. A goal of 80% colony survival should be aimed at, and lower survival rates would indicate that issues occurred during the laboratory preparation of incipient colony vials and should be remedied in the next colony-making cycle. The inherent ~20% colony establishment failure can be due to various factors that are out of the control of the laboratory, such as inbreeding (the mating pair may be coming from the same colony of origin), incomplete symbiotic gut profile (Michaud *et al.*, 2020; Velenovsky, 2022), and relatively poor quality of founders (Chouvenc, 2019, 2022b).

Fifth, if successful in rearing *Coptotermes* with the accumulation of thousands of colonies, a laboratory may run into colony-tracking issues. The laboratory must therefore implement a colony-labeling procedure from the beginning with a unique identity label for each colony, which can then be associated with a log of maintenance events and observations over the years of rearing. Sixth, laboratory-reared colonies often display workers and soldiers that are noticeably smaller than field-collected colonies. This can be due to a direct access to a relatively young cohort of foragers (Chouvenc *et al.*, 2022), but could also be the result of the colony missing access to a resource that has yet to be determined.

Finally, rearing millions of laboratory-kept termites is a “labor of love” and dedication. One must not forget that these colonies are inherent biological material that is destined toward scientific research and will most likely be killed in the process. Such colonies must not become frivolous exotic pets with no other goal than being a vanity project to the research laboratory.

18.4 Use of planar arenas in laboratory experimentation

Subterranean termites such as *Coptotermes* have evolved extensive foraging abilities in soil conditions, which allows them to connect feeding sites using long underground galleries and above-ground shelter tubes (King and Spink, 1969; Su and Scheffrahn, 1988). Ironically, most of the literature concerned with experiments on the biology of subterranean termites often ignored such important aspects, resulting in limited biological relevancy and interpretive power (Becker, 1969; Lenz, 1986). Often, termites were tested in Petri-dish style or “jar” experiments (Lenz, 2009) which offered suboptimal survival conditions with limited observation possibilities, and a limited biological context.

One of the main challenges in studying subterranean termites that still remains is the difficulty to keeping them alive for the duration of an experiment (maintaining high vigor), while being simultaneously observable. Both aspects have proven to be extensively problematic, often limiting research abilities (Becker, 1969; Lenz, 2009).

Alternatively, the use of "planar arenas" unequivocally provides one of the best compromises that allows for optimal observation of behaviors while maintaining relatively high survival in soil foraging conditions (Chouvenc *et al.*, 2011b). With such foraging planar arenas, subterranean termites can display their inherent tunneling behaviors (Bardunias and Su, 2005; Li and Su, 2009), which then enables the deposition of fecal material within the gallery system to further their ability to establish a homeostatic environment found in field nest conditions (Chouvenc *et al.*, 2013b). As stated by Becker (1969):

> "Soil-dwelling species need space for moving and gallery building as part of their natural behavior and in long-term tests, Rhinotermitidae should be provided with soil to satisfy their building instinct."

Having better control of their environment, subterranean termites can display their complete behavioral repertoire while still being observable at all times (Du *et al.*, 2016), which inherently facilitates high survival rates. Despite their versatility, ease of use, and experimental potentials, the use of planar arenas for subterranean termite research has been relatively limited (Chouvenc *et al.*, 2011b). This section will first go over the historical use of planar arena for subterranean termite research, then describe the basic concept which can be adapted by any laboratory for their experimental use, and finally further advocate for their universal use, especially with the advent of novel technologies making it accessible and customizable to all termite researchers.

18.4.1 Historical use of planar arenas for subterranean termite research

The first planar arena designs by Adamson (1941), Lüscher (1949), Becker (1965), and McMahan (1966) allowed them to observe incipient colonies or small groups of termites, reared between two glass plates, while providing them with wood and soil sandwiched between the two plates (= "glass plate termitary", Becker, 1969). Despite being an elegant solution for temporary observations, the concept did not receive traction for decades, presumably because of the limitation and challenges of working with glass. The commercial advent of transparent thermoplastic sheets (Plexiglas® and others) may have allowed for a change in attitudes, as Robson *et al.* (1995) were among the first to offer a preliminary study of subterranean termite foraging behavior in the laboratory using vertical plastic planar arenas. In the following decade, only a dozen studies improved on the concept by implementing changes (horizontal arenas) and by increasing the complexity of the designs (Reinhard *et al.*, 1997; Hedlund and Henderson, 1999; Pitts-Singer and Forschler, 2000; Campora and Grace, 2001, 2004; Puche and Su, 2001; Su and Puche, 2003; Tucker *et al.*, 2004; Su *et al.*, 2004; Bardunias and Su, 2005; Su, 2005b). Ultimately, Su (2005a) bypassed the size limitation of the arena by connecting a network of planar arenas using tubing to simulate long foraging distance, while being able to observe most of the termites at all times (= "extended foraging arenas").

In the decade following Su (2005a), the UF Termite Lab and collaborators extensively used various iterations of planar arenas and established a robust body of knowledge concerning *Coptotermes* foraging behavior and its implication on the success of *Coptotermes* as challenging structural pest insects (Lee *et al.*, 2007, 2008a,b,c,d,e, 2009; Li and Su, 2008, 2009; Chouvenc *et al.*, 2008, 2011b, 2011c, 2012, 2013a,b, 2015; Bardunias and Su, 2009a,b, 2010a,b; Lee and Su, 2009, 2010, 2011; Su and Lee, 2009; Yang *et al.*, 2009; Bardunias *et al.*, 2010; Jeon *et al.*, 2010; Ku *et al.*, 2010, 2012a,b, 2013; Li *et al.*, 2010; Chouvenc and Su, 2012; Su, 2013b; Cao and Su, 2014). During that time, a call for a universal use of planar arenas was made (Chouvenc *et al.*, 2011b), so as to progressively shift experimental designs away from over-simplified approaches. However, relatively few other laboratories used planar arenas for subterranean termite research during this period (Nobre *et al.*, 2007; Whitman and Forschler, 2007; Campora and Grace, 2009; Gautam and Henderson, 2012b; Hapukotuwa and Grace, 2012a,b; Sim *et al.*, 2012, 2015; Cornelius and Gallatin, 2015; Cornelius and Osbrink, 2015; Li *et al.*, 2015).

After 2015, the UF Termite Lab continued to expand *Coptotermes* behavioral research by connecting whole laboratory-reared colonies to planar arenas or by introducing whole colonies in planar arenas (Du *et al.*, 2016, 2017a,b; Kakkar *et al.*, 2016, 2017, 2018; Bernard *et al.*, 2017; Chouvenc and Su, 2017b; Su *et al.*, 2017; Chouvenc, 2018b, 2020; Kakkar and Su, 2018; Lee *et al.*, 2019, 2020a,b, 2021; Gazdick 2023). Finally, other laboratories studying subterranean termite foraging behaviors also used planar arena protocols (Lee *et al.*, 2016; Brossette *et al.*, 2017, 2019; Sim and Lee, 2017; Yanagihara *et al.*, 2018; Mizumoto *et al.*, 2020, 2021b; Konishi and Matsuura, 2021; Janowiecki and Vargo, 2022; Pailler *et al.*, 2022; Zhang *et al.*, 2022), leading to an ongoing expansion of their use by independent research teams around the world.

18.4.2 General concept for protocols using planar arenas

Planar arenas consist of two plates separated by a uniform distance, creating an area where termites can readily move and display their behaviors while being observable at all times (Fig. 18.4A). The two panels must be made of transparent thermoplastic and the panel thickness must be sufficient to prevent warping. For relatively small arenas (< 10 × 10 cm), panels may be ~2 mm thick (Fig. 18.4B). However, for larger protocols, panels ~6 mm thick should be used. To create a space between the two panels, plastic laminates (~1.5 mm thick) can be used for the external borders and for additional spacers within the arena to ensure uniformity of the distance between the two panels across the arena. Internal spacers must be used in conjunction with external screws by drilling holes in the panels and the spacers. The number of spacers must be adjusted according to the size of the arena (see for example: Puche and Su, 2001; Su, 2005a; Chouvenc *et al.*, 2008, 2011b). Laminate borders can either be anchored with glue on one of the panels, or can be assemble using screws, by pre-drilling the laminate and the panels at predetermined locations (Fig. 18.4C).

Wood pieces of ~1.5 mm thickness can be installed in the arena prior to introducing the "foraging" matrix. While organic soil could be used to fill the arena, its use is inconvenient for two primary reasons: (1) the darkness of the soil creates a contrast that makes photography and video recording difficult for observation, defeating the purpose of the arena to observe termites; and (2) organic soils tend to be hydrophobic, rendering the moistening of the arena almost impossible. Instead, white sand particles should be used (pre-emptively sifted to ~0.3–0.4 mm for particle homogeneity) to fill the arena. For small arenas with glued laminate borders, a rolling pin can be used to establish a moistened soil layer (1 ml water per 5 g of sand), and the arena can be assembled using binder clips. In such small set-ups, use of hot glue to seal the surrounding of the arena can prevent rapid desiccation and improve the longevity of the experiment (Chouvenc *et al.*, 2011b). For larger and thicker arenas, the arena should be first assembled, leaving a single access point opened on one side to introduce dry sand using a linear funnel. Once the arena is filled with sand, the last laminate border can be assembled, with all screws tightened properly. Through pin holes drilled at strategic locations on the top panel of the arena, a water drip can be installed to slowly introduce the moisture into the arena, which can be stopped when the sand reaches water saturation (Fig. 18.4D). Once the arena is fully assembled, termites can then be introduced into the system using an introduction chamber (Chouvenc, 2020), or through an interface piece that connects tubing from a contained termite group to the planar arena or to connect arenas in a network (Su, 2005a; Chouvenc, 2018b) (Fig. 18.4E).

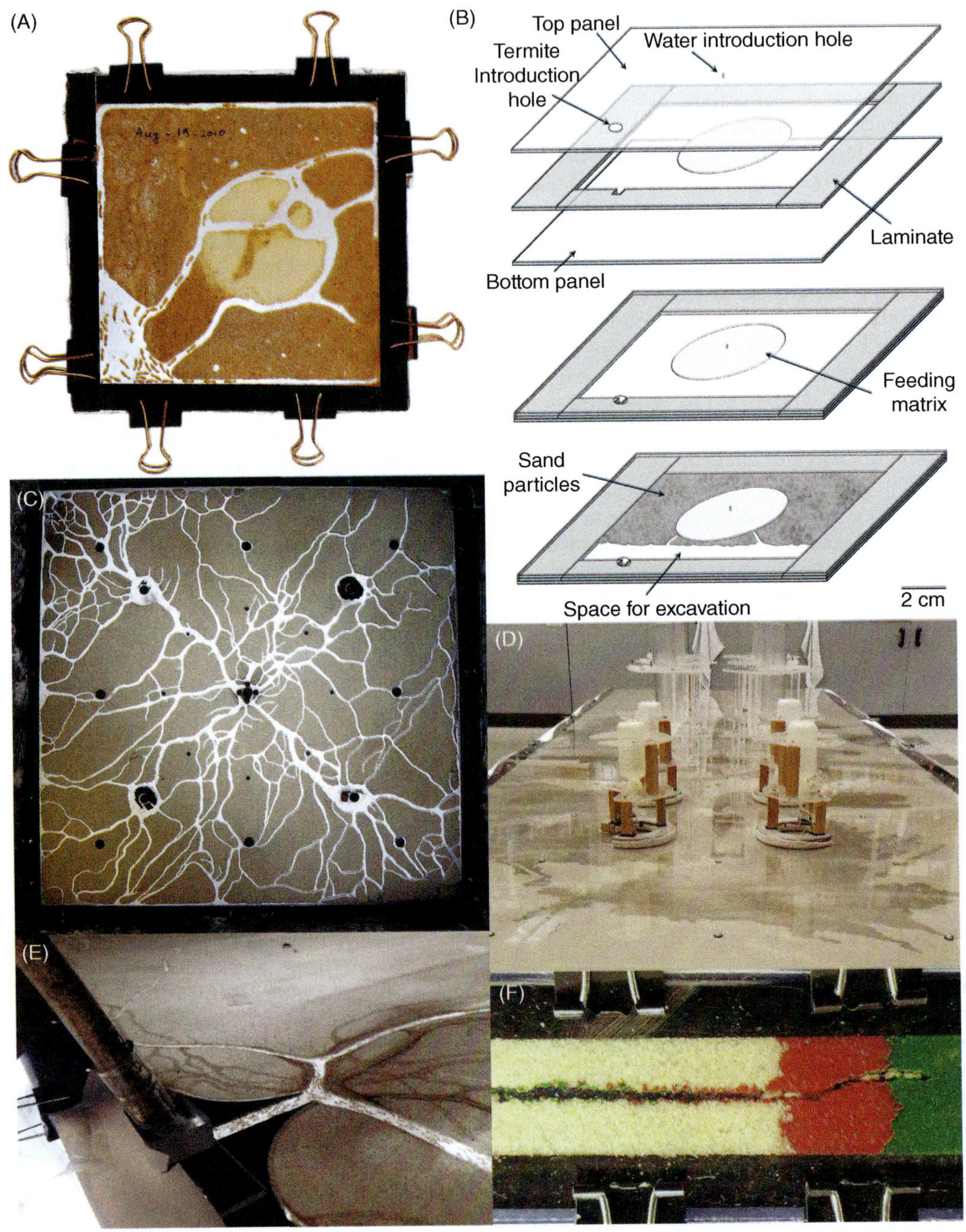

Fig. 18.4. Planar arenas for termite observation and experimentation. (A) Small arena (10 × 10 cm) (photograph: Thomas Chouvenc). (B) Panels assembly example, where the wet sand can be incorporated with a rolling pin. (C) Large arena (1 × 1 m) with spacers and borders assembled with screws throughout the contraption (photograph courtesy of Nan-Yao Su). (D) Introduction of moisture in dry-sand-assembled arenas with water drippers through pre-drilled holes (photograph: Thomas Chouvenc). (E) Connection piece from a round tubing to a planar arena allows for connection between sets (photograph: Thomas Chouvenc). (F) Use of colored sand to keep track of excavation behavior (photograph courtesy of Sang-Bin Lee).

18.4.3 Aspects to consider when using planar arenas

The size and design of the arena should be adapted to the experimental purpose and to the number of termites used. For example, planar arenas as small as 10 × 10 cm may hold between 50 and 100 individuals, while arenas as large as 50 cm × 50 cm may hold up to 2,000 individuals (Chouvenc *et al.*, 2011b; Chouvenc, 2020). Access to resources can be temporarily manipulated using movable sliding parts (Su, 2005a) or excluders (Lee *et al.*, 2019). The use of mini-chambers can be exploited for rapid behavioral observations and repeatability (Chouvenc and Su, 2017b; Lee *et al.*, 2020a). When filling the arena with sand (dry or wet method mentioned above), foraging sand expansion resulting from excavation should be taken into account (Li and Su, 2008; Bardunias and Su, 2009a) by (1) initially leaving an empty space if the arena is a closed system (Chouvenc *et al.*, 2011b), or (2) allowing the excavated sand to be deposited in chambers external to the arena (Yang *et al.*, 2009).

If the termites used in the experiment are relatively small, the thickness of the space between the two panels of the arena may become problematic, as the height of termite tunnels may be smaller than the distance between the two panels. As a result, a layer of sand may be maintained, potentially obscuring direct observation. To bypass this issue, thinner laminate may be used for the border pieces and spacers (either using a different origin/brand of laminate, or by sanding all laminate pieces homogeneously, which is not always convenient). Alternatively, an LED light can be placed under the arena, to incentivize the termites to tunnel in the upper layer of the sand, providing direct visual to the observer. Conversely, observation from the bottom panels can be done using mirrors for ease of access. Different-colored sand particles can be used (Fig. 18.4F) to monitor foraging activity over time (Lee *et al.*, 2007; Li and Su, 2009; Lee *et al.*, 2020b). Also, planar arenas can provide an opportunity to track individual behaviors if termites are marked using paint or dye markers (Mizumoto *et al.*, 2020, 2021b)

As highlighted throughout this section, the use of planar arenas offers an experimental versatility that provides opportunities to study unique aspects of subterranean termite biology and behavior. While glass panels may have long been a limitation for their use, the access to commercial transparent thermoplastic products should have readily resulted in their extensive use. However, between 1995 and 2015, their use remained limited, with relatively simple designs. In hindsight, the extensive use of planar arenas by the UF Termite Lab primarily relied on the technical skill of a single laboratory employee (Ronald Pepin) who was able to produce precision work with plastic materials. Such skill may not have readily been available to other laboratories during that time, and the cost of large panels of transparent thermoplastic may have further limited their use. However, with the recent advent of desktop laser-cutter machines at an affordable price, the mass production of intricate arena designs has become widely available and accessible to all. It is therefore expected that, in the near future, subterranean termite research will continue to rely increasingly on planar arenas.

One common detracting argument against the use of planar arena is that it "only" provides a "2-dimensional" environment for the foraging termites, not fully providing "natural" foraging conditions from the field. Such an argument often ignores the prime advantage of such designs: total visibility, which rapidly eclipses any negative aspects of their use. Such an argument is also fallacious, because foraging termites are navigating within a gallery system from a central nest to a foraging site, which can be the equivalent of moving within a single dimension. In addition, "3D" tunneling geometry is maintained when using planar arenas (Bardunias and Su, 2005). The concept of "up and down" is strongly anthropocentric because of our own movement limitation from gravity. Foraging termites are mostly detached from such constraints, as a wall and ceiling (for us) are simply an accessible surface for termites to walk on. In addition, while the use of planar arenas possesses its own limitation, it provides compelling advantages that no other approaches can match. As listed in this chapter, the extensive body of work (2005–2020) generated using planar arenas has fundamentally changed how *Coptotermes* biology and behavioral complexity are understood and how control methods impact whole colonies (Su, 2005a; Bernard *et al.*, 2017; Du *et al.*, 2017b; Su *et al.*, 2017; Kakkar *et al.*, 2018; Chouvenc, 2018b) (see Chapter 5 of this volume). The final argument is that such arenas are too "artificial" to represent natural field conditions, but detractors have always failed to suggest a superior alternative approach while providing visibility for behavioral observations and high termite survivorship. The field of subterranean termite behavioral research has long moved on from such detractors, as no better alternative was proposed to obtain a compromise between high visibility and biologically relevant results.

18.5 Conclusion

Coptotermes research will remain an important field of study because this termite genus will continue to be a major economic pest to structures and will impact more and more people as *C. formosanus* and *C. gestroi* (and potentially others) will continue to spread as successful invaders (Evans *et al.*, 2013; Chouvenc *et al.*, 2016). In addition, the phylogenetic position of the *Heterotermes–Reticulitermes–Coptotermes* clade (i.e., "Heterotermidae"; see Chapter 4 of this volume), which is the sister clade to Termitidae, makes it a relevant biological model to study evolutionary processes in termites (Chouvenc *et al.*, 2021). Finally, the relative ease of rearing *Coptotermes* furthers the argument that it can be used as a primary biological model. This chapter provides an overview of a series of advice and warnings for what can be done or what should be avoided by current and future subterranean termite research groups. Such groups can benefit from the experience of various laboratories (described herein), so as to avoid common mistakes, and to strive toward a sustainable research program. By strategically investing in skills and equipment, optimizing biological material acquisition protocols, streamlining experimental design, and observing complex behaviors, a wide range of opportunities become possible for a research team. While most of the protocols described herein have been developed over time primarily in Hawai'i, Louisiana, and Florida, each suggested approach can be altered and customized for the need of one's research program. However, this information provides the key elements to start or to strengthen a subterranean termite research program and it should be carefully considered when investigating new research avenues. The techniques described in this chapter provide a general overview of the research approaches that may be adopted or modified to produce sound and biologically relevant results when working with *Coptotermes*.

References

Acda, M.N. (2004) A method to trap and rapidly segregate subterranean termites (Isoptera: Rhinotermitidae) for laboratory bioassays. *Sociobiology* 44, 255–260.

Adamson, A.M. (1941) Laboratory technique for the study of living termites. *Ecology* 22, 411–414.

Arab, A., Costa-Leonardo, A.M., Casarin, F.E., de Camargo Guaraldo, A. and Chaves, R.C. (2005) Foraging activity and demographic patterns of two termite species (Isoptera: Rhinotermitidae) living in urban landscapes in southeastern Brazil. *European Journal of Entomology* 102, 691–697.

Arquette, T.J., Champagne, D.E., Brown, M.R. and Forschler, B.T. (2006) Evaluation of novel and traditional measures for vigor of laboratory-cultured termites, *Reticulitermes flavipes* (Kollar). *Journal of Insect Physiology* 52, 51–66.

Bardunias, P. and Su, N.-Y. (2005) Comparison of tunnel geometry of subterranean termites (Isoptera: Rhinotermitidae) in "two-dimensional" and "three-dimensional" arenas. *Sociobiology* 45, 679–686.

Bardunias, P. and Su, N.-Y. (2009a) Opposing headings of excavating and depositing termites facilitate branch formation in the Formosan subterranean termite. *Animal Behaviour* 78, 755–759.

Bardunias, P. and Su, N.-Y. (2009b) Dead reckoning in tunnel propagation of the Formosan subterranean termite (Isoptera: Rhinotermitidae). *Annals of the Entomological Society of America* 102, 158–165.

Bardunias, P., and Su, N.-Y. (2010a) Queue size determines the width of tunnels in the Formosan subterranean termite (Isoptera: Rhinotermitidae). *Journal of Insect Behavior* 23, 189–204.

Bardunias, P. and Su, N.-Y. (2010b) Change in tunnel heading in the Formosan subterranean termite (Isoptera: Rhinotermitidae) in response to movement through opened space. *Annals of the Entomological Society of America* 103, 449–454.

Bardunias, P., Su, N.-Y. and Yang, R.-L. (2010) Behavioral variation among tunnelers in the Formosan subterranean termite. *Journal of Asia-Pacific Entomology* 13, 45–49.

Becker, G. (1961) Beiträge zur prüfung und beurteilung der natürlichen dauerhaftigkeit con holz gegen termitin. *Holz Roh-u. Werkstoff* 19, 278–290.

Becker, G. (1965) Über die haltung von termiten im laboratorium. *Zeitschrift für angewandte Zoologie* 52, 385–398.

Becker, G. (1969) Rearing of termites and testing methods used in the laboratory. In: Krishna, K. and Weesner, F.M. (eds) *Biology of Termites*, Vol. 1. Academic Press, New York, pp. 351–385.

Becker, G. (1970) Grundlagen für eine Norm zur Laboratoriumsprüfung mit erdbewohnenden Termiten. *Materials Testing* 12, 293–299.

Bernard, S., Osbrink, W. and Su, N.-Y. (2017) Response of the Formosan subterranean termite to neighboring con-specific populations after baiting with noviflumuron. *Journal of Economic Entomology* 110, 575–583.

Brossette, L., Bagneres, A.G., Millot, A., Blanchard, S., Dupont, S. and Lucas, C. (2017) Termite's royal cradle: does colony foundation success differ between two subterranean species? *Insectes Sociaux* 64, 515–523.

Brossette, L., Meunier, J., Dupont, S., Bagnères, A.G. and Lucas, C. (2019) Unbalanced biparental care during colony foundation in two subterranean termites. *Ecology and Evolution* 9, 192–200.

Campora, C.E. and Grace, J.K. (2001) Tunnel orientation and search pattern sequence of the Formosan subterranean termite (Isoptera: Rhinotermitidae). *Journal of Economic Entomology* 94, 1193–1199.

Campora, C.E. and Grace, J.K. (2004) Effect of average worker size on tunneling behavior of Formosan subterranean termite colonies. *Journal of Insect Behavior* 17, 777–791.

Campora, C.E. and Grace, J.K. (2009) Comparison of tunneling in the laboratory and field by the Formosan subterranean termite, *Coptotermes formosanus* (Isoptera: Rhinotermitidae). *Sociobiology* 53, 389–401.

Cao, R. and Su, N.-Y. (2014) Tunneling and food transportation activity of four subterranean termite species (Isoptera: Rhinotermitidae) at various temperatures. *Annals of the Entomological Society of America* 107, 696–701.

Cao, R. and Su, N.-Y. (2016) Temperature preferences of four subterranean termite species (Isoptera: Rhinotermitidae) and temperature-dependent survivorship and wood-consumption rate. *Annals of the Entomological Society of America* 109, 64–71.

Chiu, C.-I., Yeh, H.-T., Tsai, M.-J. and Li, H.-F. (2016) Naturalization and control of *Coptotermes gestroi* (Blattodea: Rhinotermitidae) in a Taiwanese forest. *Journal of Economic Entomology* 109, 1317–1325.

Chouvenc, T. (2018a) So, you want to study termite control? Step 1: raise a few million termites. *Entomology Today*, October 31, 2018. Available at: https://entomologytoday.org/2018/10/31/study-termite-control-raise-millions/ (accessed 8 September 2023)

Chouvenc, T. (2018b) Comparative impact of chitin synthesis inhibitor baits and non-repellent liquid termiticides on subterranean termite colonies over foraging distances: colony elimination versus localized termite exclusion. *Journal of Economic Entomology* 111, 2317–2328.

Chouvenc, T. (2019) The relative importance of queen and king initial weights in termite colony foundation success. *Insectes Sociaux* 66, 177–184.

Chouvenc, T. (2020) Limited survival strategy in starving subterranean termite colonies. *Insectes Sociaux* 67, 71–82.

Chouvenc, T. (2022a) Not enough bugs, too many bugs – scaling up. *American Entomologist* 68, 32–33.

Chouvenc, T. (2022b) Eusociality and the transition from biparental to alloparental care in termites. *Functional Ecology* 36, 3049–3059.

Chouvenc, T. and Lee, S.-B. (2021) Queen egg laying and egg hatching abilities are hindered in subterranean termite colonies when exposed to a chitin synthesis inhibitor Bait Formulation. *Journal of Economic Entomology* 114, 2466–2472.

Chouvenc, T. and Su, N.-Y. (2012) When subterranean termites challenge the rules of fungal epizootics. *PloS ONE* 7, e34484.

Chouvenc, T. and Su, N.-Y. (2014) Colony age-dependent pathway in caste development of *Coptotermes formosanus* Shiraki. *Insectes Sociaux* 61, 171–182.

Chouvenc, T. and Su, N.-Y. (2017a) Subterranean termites feeding on CSI baits for a short duration still results in colony elimination. *Journal of Economic Entomology* 110, 2534–2538.

Chouvenc, T. and Su, N.-Y. (2017b) Testing the role of cuticular hydrocarbons on intercolonial agonism in two subterranean termite species (*Coptotermes*) and their hybrids. *Insectes Sociaux* 64, 347–355.

Chouvenc, T., Su, N.-Y. and Elliott, M.L. (2008) Interaction between the subterranean termite *Reticulitermes flavipes* (Isoptera: Rhinotermitidae) and the entomopathogenic fungus *Metarhizium anisopliae* in foraging arenas. *Journal of Economic Entomology* 101, 885–893.

Chouvenc, T., Su, N.-Y. and Grace, J.K. (2011a) Fifty years of attempted biological control of termites–Analysis of a failure. *Biological Control* 59, 69-82.

Chouvenc, T., Bardunias, P., Li, H.-F., Elliott, M.L. and Su, N.-Y. (2011b) Planar arenas for use in laboratory bioassay studies of subterranean termites (Rhinotermitidae). *Florida Entomologist* 94, 817–826.

Chouvenc, T., Elliott, M.L. and Su, N.-Y. (2011c) Rich microbial community associated with the nest material of *Reticulitermes flavipes* (Isoptera: Rhinotermitidae). *Florida Entomologist* 94, 115–116.

Chouvenc, T., Efstathion, C.A., Elliott, M.L. and Su, N.-Y. (2012) Resource competition between two fungal parasites in subterranean termites. *Naturwissenschaften* 99, 949–958.

Chouvenc, T., Bardunias, P., Efstathion, C.A., Chakrabarti, S., Elliott, M.L., Giblin-Davis, R. and Su, N.-Y. (2013a) Resource opportunities from the nest of dying subterranean termite (Isoptera: Rhinotermitidae) colonies: a laboratory case of ecological succession. *Annals of the Entomological Society of America* 106,771–778.

Chouvenc, T., Efstathion, C.A., Elliott, M.L. and Su, N.-Y. (2013b) Extended disease resistance emerging from the faecal nest of a subterranean termite. *Proceedings of the Royal Society B: Biological Sciences* 280, 20131885.

Chouvenc, T., Mullins, A.J. and Su, N.-Y. (2015) Territorial status-quo between the big-headed ant (Hymenoptera: Formicidae) and the Formosan subterranean termite (Isoptera: Rhinotermitidae). *Florida Entomologist* 98, 157–161.

Chouvenc, T., Scheffrahn, R.H. and Su, N.-Y. (2016) Establishment and spread of two invasive subterranean termite species (*Coptotermes formosanus* and *C. gestroi*; Isoptera: Rhinotermitidae) in metropolitan southeastern Florida (1990–2015). *Florida Entomologist* 99, 187–191.

Chouvenc, T., Scheffrahn, R.H., Mullins, A.J. and Su, N.-Y. (2017) Flight phenology of two *Coptotermes species* (Isoptera: Rhinotermitidae) in southeastern Florida. *Journal of Economic Entomology* 110, 1693–1704.

Chouvenc, T., Elliott, M.L., Šobotník, J., Efstathion, C.A. and Su, N.-Y. (2018) The termite fecal nest: a framework for the opportunistic acquisition of beneficial soil *Streptomyces* (Actinomycetales: Streptomycetaceae). *Environmental Entomology* 47, 1431–1439.

Chouvenc, T., Sillam-Dussès, D. and Robert, A. (2020) Courtship behavior confusion in two subterranean termite species that evolved in allopatry (Blattodea, Rhinotermitidae, *Coptotermes*). *Journal of Chemical Ecology* 46, 461–474.

Chouvenc, T., Šobotník, J., Engel, M.S. and Bourguignon, T. (2021) Termite evolution: mutualistic associations, key innovations, and the rise of Termitidae. *Cellular and Molecular Life Sciences* 78, 2749–2769.

Chouvenc, T., Ban, P.M. and Su, N.-Y. (2022) Life and death of termite colonies, a decades-long age demography perspective. *Frontiers in Ecology and Evolution* 10, 911042.

Cornelius, M.L. and Gallatin, E.M. (2015) Task allocation in the tunneling behavior of workers of the Formosan subterranean termite, *Coptotermes formosanus* Shiraki. *Journal of Asia-Pacific Entomology* 18, 637–642.

Cornelius, M.L. and Osbrink, W.L. (2015) Effect of chemical cues on the foraging and tunneling behavior of Formosan subterranean termites (Isoptera: Rhinotermitidae). *Midsouth Entomologist* 8, 1–9.

Costa-Leonardo, A.M. and Barsotti, R.C. (2001) Growth patterns of incipient colonies of *Coptotermes havilandi* (Isoptera, Rhinotermitidae) initiated in the laboratory from swarming alates. *Sociobiology* 37, 551–561.

Crosland, M.W.J. and Su, N.-Y. (2006) Mark–recapture without estimating population sizes: a tool to evaluate termite baits. *Bulletin of Entomological Research* 96, 99–103.

Du, H., Chouvenc, T., Osbrink, W.L.A. and Su, N.-Y. (2016) Social interactions in the central nest of *Coptotermes formosanus* juvenile colonies. *Insectes Sociaux* 63, 279–290.

Du, H., Chouvenc, T., Osbrink, W.L.A. and Su, N.-Y. (2017a) Heterogeneous distribution of castes/instars and behaviors in the nest of *Coptotermes formosanus* Shiraki. *Insectes Sociaux* 64, 103–112.

Du, H., Chouvenc, T. and Su, N.-Y. (2017b) Development of age polyethism with colony maturity in *Coptotermes formosanus* (Isoptera: Rhinotermitidae). *Environmental Entomology* 46, 311–318.

Ehrhorn, E.M. (1934) The termites of Hawaii, their economic significance and control, and the distribution of termites by commerce. In: Kofoid *et al.* (eds) *Termites and Termite Control*. University of California Press, Berkeley, California, pp. 321–333.

Evans, T.A. and Gleeson, P.V. (2006) The effect of bait design on bait consumption in termites (Isoptera: Rhinotermitidae). *Bulletin of Entomological Research* 96, 85–90.

Evans, T.A., Forschler, B.T. and Grace, J.K. (2013) Biology of invasive termites: a worldwide review. *Annual Review of Entomology* 58, 455–474.

Ewart, D.M., Grace, J.K., Yamamoto, R.T. and Tamashiro, M. (1992) Hollow stakes for detecting subterranean termites (Isoptera: Rhinotermitidae). *Sociobiology* 20, 17–22.

Fei, H.X. and Henderson, G. (2003) Comparative study of incipient colony development in the Formosan subterranean termite, *Coptotermes formosanus* Shiraki (Isoptera, Rhinotermitidae). *Insectes Sociaux* 50, 226–233.

Ferraz, M.V. and Cancello, E.M. (2004) Strategies on the developmental biology of incipient colonies of *Coptotermes gestroi* (Isoptera: Rhinotermitidae) in different substrates. *Sociobiology* 44, 109–122.

Gao, Q., Bidochka, M.J. and Thompson, G.J. (2012) Effect of group size and caste ratio on individual survivorship and social immunity in a subterranean termite. *Acta Ethologica* 15, 55–63.

Gautam, B.K. and Henderson, G. (2012a) Escape behavior of the Formosan subterranean termite (Isoptera: Rhinotermitidae) in response to disturbance. *Journal of Insect Behavior* 25, 70–79.

Gautam, B.K. and Henderson, G. (2012b) Laboratory study of the influence of substrate type and temperature on the exploratory tunneling by Formosan subterranean termite. *Insects* 3, 629–639.

Gazdick, K. (2023) Foraging behavior of Coptotermes formosanus and its ability to intercept in-ground bait. University of Florida, MS thesis, 59 pp.

Gordon, J.M., Velenovsky, J.F. and Chouvenc, T. (2022) Subterranean termite colony elimination can be achieved even when only a small proportion of foragers feed upon a CSI bait. *Journal of Pest Science* 95, 1207–1216.

Grace, J.K., Yamamoto, R.T. and Tamashiro, M. (1995). Relationship of individual worker mass and population decline in a Formosan subterranean termite colony (Isoptera: Rhinotermitidae). *Environmental Entomology* 24, 1258–1262.

Greaves, T. (1962) Studies of foraging galleries and the invasion of living trees by *Coptotermes acinaciformis* and *C. brunneus* (Isoptera). *Australian Journal of Zoology* 10, 630–651.

Hapukotuwa, N.K. and Grace, J.K. (2012a) Do tunnel patterns of *Coptotermes formosanus* and *Coptotermes gestroi* (Blattodea: Rhinotermitidae) reflect different foraging strategies? *Sociobiology* 59, 189–202.

Hapukotuwa, N.K. and Grace, J.K. (2012b) *Coptotermes formosanus* and *Coptotermes gestroi* (Blattodea: Rhinotermitidae) exhibit quantitatively different tunneling patterns. *Psyche* 2012, 675356.

Hedlund, J.C. and Henderson, G. (1999) Effect of available food size on search tunnel formation by the Formosan subterranean termite (Isoptera: Rhinotermitidae). *Journal of Economic Entomology* 92, 610–616.

Higa, S.Y. (1981) Flight, colony foundation, and development of the gonads of the primary reproductives of the Formosan subterranean termite, *Coptotermes formosanus* Shiraki. PhD Dissertation, University of Hawaii, Honolulu, 173 pp.

Hochmair, H.H. and Scheffrahn, R.H. (2010) Spatial association of marine dockage with land-borne infestations of invasive termites (Isoptera: Rhinotermitidae: *Coptotermes*) in urban South Florida. *Journal of Economic Entomology* 103, 1338–1346.

Huang, L.W. and Chen, L.L. (1984) Biology and colony development of *Coptotermes formosanus* Shiraki. *Acta Entomologica Sinica* 27, 64–69.

Husseneder, C., Messenger, M.T., Su, N.Y., Grace, J.K. and Vargo, E.L. (2005) Colony social organization and population genetic structure of an introduced population of Formosan subterranean termite from New Orleans, Louisiana. *Journal of Economic Entomology* 98, 1421–1434.

Janowiecki, M.A. and Vargo, E.L. (2022) Effect of soldiers on collective tunneling behavior in three species of *Reticulitermes* (Blattodea: Rhinotermitidae). *Insectes Sociaux* 69, 237–245.

Jeon, W., Kang, S.-Y., Su, N.-Y. and Lee, S.-H. (2010) A constraint condition for foraging strategy in subterranean termites. *Journal of Insect Science* 10.146, 1–8.

Kakkar, G. and Su, N.-Y. (2018) Molting drives mortality in foraging populations of Formosan subterranean termites (Isoptera: Rhinotermitidae) baited with a chitin synthesis inhibitor, noviflumuron. *Pest Management Science* 74, 219–224.

Kakkar, G., Chouvenc, T., Osbrink, W. and Su, N.-Y. (2016) Temporal assessment of molting in workers of Formosan subterranean termites (Isoptera: Rhinotermitidae). *Journal of Economic Entomology* 109, 2175–2181.

Kakkar, G., Osbrink, W., Mullins, A. and Su, N.Y. (2017) Molting site fidelity in workers of Formosan subterranean termites (Isoptera: Rhinotermitidae). *Journal of Economic Entomology* 110, 2512–2517.

Kakkar, G., Osbrink, W. and Su, N.-Y. (2018) Molting site fidelity accounts for colony elimination of the Formosan subterranean termites (Isoptera: Rhinotermitidae) by chitin synthesis inhibitor baits. *Scientific Reports* 8, e1259.

King, E.G. Jr and Spink, W.T. (1969) Foraging galleries of the Formosan subterranean termite, *Coptotermes formosanus*, in Louisiana. *Annals of the Entomological Society of America* 62, 536–542.

King, E.G. Jr and Spink, W.T. (1974) Laboratory studies on the biology of the Formosan subterranean termite with primary emphasis on young colony development. *Annals of the Entomological Society of America* 67, 953–958.

Konishi, T. and Matsuura, K. (2021) Royal presence promotes worker and soldier aggression against non-nestmates in termites. *Insectes Sociaux* 68, 15–21.

Ku, S.-J., Su, N.-Y. and Lee, S.-H. (2010) Directional selection by the subterranean termite *Coptotermes formosanus* Shiraki (Isoptera: Rhinotermitidae) at tunnel intersections. *Entomological Science* 13, 363–366.

Ku, S.-J., Su, N.-Y. and Lee, S.-H. (2012a) Measurement of time taken by the Formosan termite, *Coptotermes formosanus*, to pass tunnel intersections. *Journal of Insect Science* 12.23, 1–7.

Ku, S.-J., Jeon, W., Su, N.-Y. and Lee, S.-H. (2012b) Analysis of the responses of termites to tunnel irregularity. *Insectes Sociaux* 59, 549–555.

Ku, S.-J., Su, N.-Y. and Lee, S.-H. (2013) Movement efficiency and behavior of termites (Isoptera) in tunnels with varying pore sizes. *Florida Entomologist* 96, 810–817.

La Fage, J.P., Nutting, W.L. and Haverty, M.I. (1973) Desert subterranean termites: a method for studying foraging behavior. *Environmental Entomology* 2, 954–956.

La Fage, J.P., Su, N.-Y., Jones, M.J. and Esenther, G.R. (1983) A rapid method for collecting large numbers of subterranean termites from wood. *Sociobiology* 7, 305–309.

Lai, P.Y. (1977) Biology and ecology of the Formosan subterranean termite, *Coptotermes formosanus* and its susceptibility to the entomogenus fungi, *Beauveria bassiana* and *Metarhizium anisopliae*. PhD Thesis, University of Hawaii.

Lee, S.-B., Chouvenc, T. and Su, N.-Y. (2019) A reproductives excluder for subterranean termites in laboratory experiments. *Journal of Economic Entomology* 112, 2882–2887.

Lee, S.-B., Chouvenc, T., Patel, J. and Su, N.-Y., (2020a) Altered mobility and accumulation of inefficient workers in juvenile hybrid termite colonies. *Frontiers in Ecology and Evolution* 8, 589762.

Lee, S.-B., Su, N.-Y., Song, H.-S. and Lee, S.-H. (2020b) Minimizing moving distance in deposition behavior of the subterranean termite. *Ecology and Evolution* 10, 2145–2152.

Lee, S.-B., Chouvenc, T. and Su, N.-Y. (2021) Differential time allocation of foraging workers in the subterranean termite. *Frontiers in Zoology* 18, e61.

Lee, S.-B., Chouvenc, T., Mizumoto, N., Mullins, A. and Su, N.-Y. (2022) Age-based spatial distribution of workers is resilient to worker loss in a subterranean termite. *Scientific Reports* 12, e7837.

Lee, S.-H., and Su N.-Y. (2009) The influence of branching tunnels on subterranean termites' foraging efficiency: Considerations for simulations. *Ecological Informatics* 4, 152–155.

Lee, S.-H. and Su, N.-Y. (2010) A novel approach to characterize branching network: Application to termite tunnel patterns. *Journal of Asia-Pacific Entomology* 13, 117–120.

Lee, S.H. and Su, N.-Y. (2011) Territory size distribution of Formosan subterranean termites in urban landscape: Comparison between experimental and simulated results. *Journal of Asia-Pacific Entomology* 14, 1–6.

Lee, S.-H., Bardunias, P. and Su, N.-Y. (2007) Optimal length distribution of termite tunnel branches for efficient food search and resource transportation. *Biosystems* 90, 802–807.

Lee, S.-H., Bardunias, P. and Su, N.-Y. (2008a) Rounding a corner of a bent termite tunnel and tunnel traffic efficiency. *Behavioural Processes* 77, 135–138.

Lee, S.-H., Bardunias, P., Su, N.-Y. and Yang, R.-L. (2008b) Behavioral response of termites to tunnel surface irregularity. *Behavioural Processes* 78, 397–400.

Lee, S.-H., Yang, R.-L. and Su, N.-Y. (2008c) Tunneling response of termites to a pre-formed tunnel. *Behavioural Processes* 79, 192–194.

Lee, S.-H., Bardunias, P. and Su, N.-Y. (2008d) Two strategies for optimizing the food encounter rate of termite tunnels simulated by a lattice model. *Ecological Modelling* 213, 381–388.

Lee, S.-H., Yang, R.-L. and Su, N.-Y. (2008e) Surface irregularity induced-tunneling behavior of the Formosan subterranean termite. *Behavioural Processes* 78, 473–476.

Lee, S.-H., Su, N.-Y. and Lee, M. (2009) Why is the number of primary tunnels of the formosan subterranean termite, *Coptotermes formosanus* Shiraki (Isoptera: Rhinotermidae), restricted during foraging? *Journal of Asia-Pacific Entomology* 12, 151–154.

Lee, S.-H., Sim, S. and Song, H.-S. (2016) Path selection by the termite *Reticulitermes speratus kyushuensis* (Isoptera: Rhinotermitidae) at a bifurcation node of Y-shaped tunnel. *Journal of Asia-Pacific Entomology* 19, 497–501.

Lenz, M. (1986) Principles behind the laboratory assessment of materials with subterranean termites. Recent perspectives and shifts in emphasis. *Material und Organismen* 21, 123–137.

Lenz, M. (2009) Laboratory bioassays with subterranean termites (Isoptera). The importance of termite biology. *Sociobiology* 53, 573–595.

Li, H.-F. and Su, N.-Y. (2008) Sand displacement during tunnel excavation by the Formosan subterranean termite (Isoptera: Rhinotermitidae). *Annals of the Entomological Society of America* 101, 456–462.

Li, H.-F. and Su, N.-Y. (2009) Buccal manipulation of sand particles during tunnel excavation of the Formosan subterranean termite (Isoptera: Rhinotermitidae). *Annals of the Entomological Society of America* 102, 333–338.

Li, H.-F., Yang, R.-L. and Su, N.-Y. (2010) Interspecific competition and territory defense mechanisms of *Coptotermes formosanus* and *Coptotermes gestroi* (Isoptera: Rhinotermitidae). *Environmental Entomology* 39, 1601–1607.

Li, H., Yang, M., Chen, Y., Zhu, N., Lee, C.-Y., Wei, J.-Q. and Mo, J. (2015) Investigation of age polyethism in food processing of the fungus-growing termite *Odontotermes formosanus* (Blattodea: Termitidae) using a laboratory artificial rearing system. *Journal of Economic Entomology* 108, 266–273.

Lin, S., Pen, I. and Korb, J. (2023) Effect of food restriction on survival and reproduction of a termite. *Journal of Evolutionary Biology* 36, 542–549.

Lüscher, M. (1949) Continuous observation of termites in laboratory cultures. *Acta Tropica* 6, 161–165.

McMahan, E.A. (1966) Studies of termite wood-feeding preferences. *Studies* 19, 239–250.

Messenger, M.T. and Su, N.-Y. (2005) Colony characteristics and seasonal activity of the Formosan subterranean termite (Isoptera: Rhinotermitidae) in Louis Armstrong park, New Orleans, Louisiana. *Journal of Entomological Science* 40, 268–279.

Michaud, C., Hervé, V., Dupont, S., Dubreuil, G., Bézier, A.M., Meunier, J., Brune, A. and Dedeine, F. (2020) Efficient but occasionally imperfect vertical transmission of gut mutualistic protists in a wood-feeding termite. *Molecular Ecology* 29, 308–324.

Miramontes, O. and DeSouza, O. (2008) Individual basis for collective behaviour in the termite, *Cornitermes cumulans*. *Journal of Insect Science* 8, p.22.

Mitaka, Y., Akino, T. and Matsuura, K. (2023) Development of a standard medium for culturing the termite *Reticulitermes speratus*. *Insectes Sociaux* (in press). doi: 10.1007/s00040-023-00907-6

Mizumoto, N., Bardunias, P.M. and Pratt, S.C. (2020) Complex relationship between tunneling patterns and individual behaviors in termites. *The American Naturalist* 196, 555–565.

Mizumoto, N., Lee, S.B., Valentini, G., Chouvenc, T. and Pratt, S.C. (2021a) Coordination of movement via complementary interactions of leaders and followers in termite mating pairs. *Proceedings of the Royal Society B* 288, 20210998.

Mizumoto, N., Gile, G.H. and Pratt, S.C. (2021b) Behavioral rules for soil excavation by colony founders and workers in termites. *Annals of the Entomological Society of America* 114, 654–661.

Morales-Ramos, J.A. and Rojas, M.G. (2003) Nutritional ecology of the Formosan subterranean termite (Isoptera: Rhinotermitidae): growth and survival of incipient colonies feeding on preferred wood species. *Journal of Economic Entomology* 96, 106–116.

Mullins, A. and Su, N.-Y. (2018) Parental nitrogen transfer and apparent absence of N2 fixation during colony foundation in *Coptotermes formosanus* Shiraki. *Insects* 9.37, 1–8.

Mullins, A., Chouvenc, T. and Su, N.-Y. (2021) Soil organic matter is essential for colony growth in subterranean termites. *Scientific Reports* 11, e21252.

Nobre, T., Nunes, L. and Bignell, D.E. (2007) Tunnel geometry of the subterranean termite *Reticulitermes grassei* (Isoptera: Rhinotermitidae) in response to sand bulk density and the presence of food. *Insect Science* 14, 511–518.

Osbrink, W.L.A., Woodson, W.D. and Lax, A.R. (1999) Population of Formosan subterranean termite, *Coptotermes formosanus* (Isoptera: Rhinotermitidae), established in living urban trees in New Orleans, Louisiana, USA. In: *Proceedings, 3rd International Conference on Urban Pests*. Graficke zavody hronov. Prague, Czech Republic, pp. 19–22.

Osbrink, W.L.A., Cornelius, M.L. and Showler, A.T. (2016) Bionomics and formation of “Bonsai” colonies with long-term rearing of *Coptotermes formosanus* (Isoptera: Rhinotermitidae). *Journal of Economic Entomology* 109, 770–778.

Pailler, L., Matte, A., Groseiller, A., Eyer, P.A., Ruhland, F. and Lucas, C. (2022) High exploration behavior of termite propagules can enhance invasiveness. *Frontiers in Ecology and Evolution* 10, e840105

Park, Y.-I., Bland, J.M. and Raina, A.K. (2004) Factors affecting post-flight behavior in primary reproductives of the Formosan subterranean termite, *Coptotermes formosanus* (Isoptera: Rhinotermitidae). *Journal of Insect Physiology* 50, 539–546.

Peppuy, A. (1999) Termites du nord Vietnam, communication chimique et isolement spécifique. PhD dissertation, Université Pierre et Marie Curie, Paris VI. 146 pp.

Pitts-Singer, T.L. and Forschler, B.T. (2000) Influence of guidelines and passageways on tunneling behavior of *Reticulitermes flavipes* (Kollar) and *R. virginicus* (Banks) (Isoptera: Rhinotermitidae). *Journal of Insect Behavior* 13, 273–290.

Puche, H. and Su, N.-Y. (2001) Tunnel formation by *Reticulitermes flavipes* and *Coptotermes formosanus* (Isoptera: Rhinotermitidae) in response to wood in sand. *Journal of Economic Entomology* 94, 1398–1404.

Raina, A.K., Bland, J.M., Dickens, J.C., Park, Y.-I. and Hollister, B. (2003a) Premating behavior of dealates of the Formosan subterranean termite and evidence for the presence of a contact sex pheromone. *Journal of Insect Behavior* 16, 233–245.

Raina, A., Park, Y.-I. and Florane, C. (2003b) Behavior and reproductive biology of the primary reproductives of the Formosan subterranean termite (Isoptera: Rhinotermitidae). *Sociobiology* 41, 37–48.

Reinhard, J., Hertel, H. and Kaib, M. (1997) Systematic search for food in the subterranean termite *Reticulitermes santonensis* De Feytaud (Isoptera, Rhinotermitidae). *Insectes Sociaux* 44, 147–158.

Robson, S.K., Lesniak, M.G., Kothandapani, R.V., Traniello, J.F.A., Thorne, B.L. and Fourcassié, V. (1995) Nonrandom search geometry in subterranean termites. *Naturwissenschaften* 82, 526–528.

Rosengaus, R.B., Maxmen, A.B., Coates, L.E. and Traniello, J.F. (1998) Disease resistance: a benefit of sociality in the dampwood termite *Zootermopsis angusticollis* (Isoptera: Termopsidae). *Behavioral Ecology and Sociobiology* 44, 125–134.

Ruan, G., Song, X., Hu, Y., Han, N. and Zhang, D. (2015) Foraging activities of *Coptotermes formosanus* in subtropical areas in China. *Journal of Economic Entomology* 108, 701–706.

Sands, W.A. (1972) Problems in attempting to sample tropical subterranean termite populations. *Ekologia Polska* 20, 23–31.

Schultze-Dewitz, G. (1962) Abhängigkeit des versuchsergebnisses bei der termitenprüfung von der grösse der prüfgemeinschaft. *Symp. Genet. Biol. Ital.* 11, 24–31

Shimizu, K. (1962) Analytical studies on the vitality of colonies of the Formosan subterranean termite, *Coptotermes formosanus* Shiraki. I. Analysis of the strength of vitality. *Bulletin of the Faculty of Agriculture, Miyazaki University* 8, 106–110.

Sillam-Dussès, D., Jandák, V., Stiblik, P., Delattre, O., Chouvenc, T. *et al.* (2023). Alarm communication predates eusociality in termites. *Communications Biology* 6, 83.

Sim, S. and Lee, S.-H. (2017) Direction selection of termites at a skewed T-shaped tunnel junction. *Journal of Asia-Pacific Entomology* 20, 61–64.

Sim, S., Ku, S.-J. and Lee, S.-H. (2012) Directional selection by termites at a branching node created by a ballpoint pen. *Journal of Asia-Pacific Entomology* 15, 447–450.

Sim, S., Kang, S.-H. and Lee, S.-H. (2015) Using hidden Markov models to characterize termite traveling behavior in tunnels with different curvatures. *Behavioural Processes* 111, 101–108.

Sornnuwat, Y., Tsunoda, K., Yoshimura, T., Takahashi, M. and Vonckaluang, C. (1996) Foraging populations of *Coptotermes gestroi* (Isoptera: Rhinotermitidae) in an urban area. *Journal of Economic Entomology* 89, 1485–1490.

Su, N.-Y. (2001) Studies on the foraging of subterranean termites (Isoptera). *Sociobiology* 37, 253–260.

Su, N.-Y. (2005a) Response of the Formosan subterranean termites (Isoptera: Rhinotermitidae) to baits or nonrepellent termiticides in extended foraging arenas. *Journal of Economic Entomology* 98, 2143–2152.

Su, N.-Y. (2005b) Directional change in tunneling of subterranean termites (Isoptera: Rhinotermitidae) in response to decayed wood attractants. *Journal of Economic Entomology* 98, 471–475.

Su, N.-Y. (2013a) How to become a successful invader. *Florida Entomologist* 96, 765–769.

Su, N.-Y. (2013b) Estimating population size of large laboratory colonies of the Formosan subterranean termite using the capture probability equilibrium. *Journal of Economic Entomology* 106, 2442–2447.

Su, N.-Y. and Lee, S.-H. (2009) Tunnel volume regulation and group size of subterranean termites (Isoptera: Rhinotermitidae). *Annals of the Entomological Society of America* 102, 1158–1164.

Su, N.-Y. and Puche, H. (2003) Tunneling activity of subterranean termites (Isoptera: Rhinotermitidae) in sand with moisture gradients. *Journal of Economic Entomology* 96, 88–93.

Su, N.-Y. and Scheffrahn, R.H. (1986) A method to access, trap, and monitor field populations of the Formosan subterranean termite (Isoptera: Rhinotermitidae). *Sociobiology* 12, 299–304.

Su, N.-Y. and Scheffrahn, R.H. (1988) Foraging population and territory of the Formosan subterranean termite (Isoptera: Rhinotermitidae) in an urban environment. *Sociobiology* 14, 353–360.

Su, N.-Y., Tamashiro, M., Yates, J.R. and Haverty, M. (1982) Effect of behavior on the evaluation of insecticides for prevention of or remedial control of the Formosan subterranean termite. *Journal of Economic Entomology* 75, 188–193.

Su, N.-Y., Tamashiro, M., Yates, J.R. and Haverty, M.I. (1984) Foraging behavior of the Formosan subterranean termite (Isoptera: Rhinotermitidae). *Environmental Entomology* 13, 1466–1470.

Su, N.-Y., Ban, P.M. and Scheffrahn, R.H. (1997) Remedial baiting with hexaflumuron in above-ground stations to control structure-infesting populations of the Formosan subterranean termite (Isoptera: Rhinotermitidae). *Journal of Economic Entomology* 90, 809–817.

Su, N.-Y., Stith, B.M., Puche, H. and Bardunias, P. (2004) Characterization of tunneling geometry of subterranean termites (Isoptera: Rhinotermitidae) by computer. *Sociobiology* 44, 471–483.

Su, N.-Y., Osbrink, W., Kakkar, G., Mullins, A. and Chouvenc, T. (2017) Foraging distance and population size of juvenile colonies of the Formosan subterranean termite (Isoptera: Rhinotermitidae) in laboratory extended arenas. *Journal of Economic Entomology* 110, 1728–1735.

Tamashiro, M., Fujii, J.K. and Lai, P.Y. (1973) A simple method to observe, trap, and prepare large numbers of subterranean termites for laboratory and field experiments. *Environmental Entomology* 2, 721–722.

Thorne, B.L. and Haverty, M.I. (1991) An assessment of the potential uses of agonistic behaviors in termite control. In: Haverty, Michael I.; Wilcox, W. Wayne (technical coordinators) *Proceedings of the symposium on current research on wood-destroying organisms and future prospects for protecting wood in use; September 13, 1989; Bend, Oregon. Gen. Tech. Rep. PSW-GTR-128*. Pacific Southwest Research Station, Forest Service, US Department of Agriculture; Berkeley, California, pp. 24–27 (Vol. 128).

Tucker, C.L., Koehler, P.G. and Oi, F.M. (2004) Influence of soil compaction on tunnel network construction by the eastern subterranean termite (Isoptera: Rhinotermitidae). *Journal of Economic Entomology* 97, 89–94.

Vargo, E.L. and Husseneder, C. (2009) Biology of subterranean termites: insights from molecular studies of *Reticulitermes* and *Coptotermes*. *Annual Review of Entomology* 54, 379–403.

Vargo, E.L., Husseneder, C. and Grace, J.K. (2003) Colony and population genetic structure of the Formosan subterranean termite, *Coptotermes formosanus*, in Japan. *Molecular Ecology* 12, 2599–2608.

Velenovsky, J.F. IV (2022) The protist communities of colonies established by conspecific and heterospecific *Coptotermes* alate pairs confirm some and reveal other aspects of symbiosis in lower termites. Doctoral dissertation, University of Florida, 172 pp.

Waller, D.A. and La Fage, J.P. (1987) Food quality and foraging response by the subterranean termite *Coptotermes formosanus* Shiraki (Isoptera: Rhinotermitidae). *Bulletin of Entomological Research* 77, 417–424.

Wang, C., Henderson, G., Gautam, B.K., Chen, J. and Bhatta, D. (2016) Panic escape polyethism in worker and soldier *Coptotermes formosanus* (Isoptera: Rhinotermitidae). *Insect Science* 23, 305–312.

Whitman, J.G. and Forschler, B.T. (2007) Observational notes on short-lived and infrequent behaviors displayed by *Reticulitermes flavipes* (Isoptera: Rhinotermitidae). *Annals of the Entomological Society of America* 100, 763–771.

Yamano, K. (1987) *Physical control of the Formosan subterranean termite, Coptotermes formosanus Shiraki*. Research extension series-College of Tropical Agriculture and Human Resources, University of Hawaii, *Cooperative Extension Service (USA)* 83, 43–47.

Yanagihara, S., Suehiro, W., Mitaka, Y. and Matsuura, K. (2018) Age-based soldier polyethism: old termite soldiers take more risks than young soldiers. *Biology Letters* 14, 20180025.

Yang, R.-L., Su, N.-Y. and Bardunias, P. (2009) Individual task load in tunnel excavation by the Formosan subterranean termite (Isoptera: Rhinotermitidae). *Annals of the Entomological Society of America* 102, 906–910.

Yates, J.R. III and Grace, J.K. (2000) Effective use of above-ground hexaflumuron bait stations for Formosan subterranean termite control (Isoptera: Rhinotermitidae). *Sociobiology* 35, 333–356.

Zhang, L., Zhang, J., Wang, C., He, Y., Wen, X., Xu, Z. and Wang, C. (2022) Toxicological, behavioral, and horizontal transfer effects of cycloxaprid against Formosan subterranean Termites (Blattodea: Rhinotermitidae). *Journal of Economic Entomology* 115, 1240–1250.

19 More Questions to Answer

Nan-Yao Su

Science is a dynamic field characterized by a never-ending quest for knowledge, where each answer invariably gives rise to a fresh set of questions. Throughout this book, numerous chapters have explored intriguing topics and presented potential avenues for future research. However, it is worth noting that not all chapter authors had the opportunity to address the unresolved questions in their respective fields. This concluding chapter presents a concise compilation of questions contributed by those authors who were unable to include their own "need to be answered" enquiries within their chapters. These questions serve as promising starting points for further exploration and discovery in their respective domains.

19.1 Sulfuryl fluoride fumigation dosage for *Coptotermes* spp.

Rudolf H. Scheffrahn

Sulfuryl fluoride (SF) fumigation for *Coptotermes formosanus* Shiraki and related species is a rapid method of aerial colony eradication. The most common application is for boat fumigations either in water or in drydock. The labels for both SF products, Vikane™ and Zythor ™, specify four times the dosage required for drywood termite eradication. It is unknown how the FST 4× dosage came to be. La Fage *et al.* (1983) intended to simulate field conditions by conducting SF diffusion trials through nest carton. Their bioassay used water-saturated carton discs (dry carton did not retard SF diffusion) 8.9 cm thick, glued inside PVC pipes. Fifty FST workers were placed in the void on one side of the carton before capping the pipe ends. The other void was injected with SF at 24 h accumulated dosages of 0, 79, 105, or 131 mg-h/L for a total pipe volume of 3.8 l. After 24 h at 30°C, mean (±SD) worker mortality was 3.3±2.1, 62.7±37.1, 95.3±11.4, and 100±0 per cent, respectively. For comparison, the laboratory dosage for *Incisitermes minor* (Hagen) is 52 mg-h/L at 25°C. The La Fage *et al.* (1983) study brings to light that the 4× SF label rate for FST is excessive and has no basis. Furthermore, the study did not simulate field conditions but was an unrealistic worst-case scenario for the following reasons: (1) in the field, the termites are not enveloped by solid carton as they require diffusion of ambient, oxygen-laden air; (2) nest carton contains open galleries leading to foraging sites, and SF diffuses along these; (3) nest carton, although high in

DOI: 10.1079/9781800621596.0019

moisture content, is not water saturated; and (4) latent mortality was not considered after the fumigant exposure period. Therefore, the correct SF dosage for FST must yet be determined from field fumigations of aerial infestations.

Reference

La Fage, J.P., Jones, M., Lawrence, T. and O'Melia, F.C. (1983) Vikane® gas fumigant successful against Formosan termite in laboratory tests. *Down to Earth* (The Dow Chemical Company) 39, 27–32.

19.2 Questions regarding colony foundation

Thomas Chouvenc

1. What is the true mortality rate of alates during dispersal flight? Chapter 4 speculated that only < 0.01% of alates may successfully establish a colony, but actual numbers remain challenging to determine. Such mortality rate is important because it is a relevant factor on the ability of invasive *Coptotermes* to rapidly colonize a new location, once established.

2. In incipient colonies, what triggers the shift to alloparental care? Once the incipient colony shifts from biparental to alloparental care, do the queen and king stop providing caregiving to the new cohort of offspring because: (1) they no longer have the resources to take care of them (physiological limit); (2) they no longer have the ability to display brood care behaviors (behavioral shift); or (3) they simply do not need to provide brood care (social context)? Such a question can be difficult to answer but can further reveal how deeply rooted such caregiving shift is within the emergence of eusociality in termites. For this, a comparative approach across termite taxa may be necessary.

3. During colony foundation, if alates provide different lineage of mutualistic protozoa, how can such symbiosis be stable over evolutionary times if the different lineages must compete for survival in the gut of those host offspring, each time a colony is established?

19.3 Questions regarding colony development

Thomas Chouvenc

1. Do L_1 and L_2 larvae engage in a fasting period prior to molting, and how short is the complete molting process when compared with workers (~14 days)?

2. How does L_2 initiate the $L_2 \rightarrow N_1$ differentiations? What are the socio-environmental factors opening up the pathway, and what are the molecular regulatory processes involved in the development toward the imaginal line? *Coptotermes* has a rigid bifurcated development, and in the absence of ergatoid reproductives, the opening up of the $L_2 \rightarrow N_1$ path can only be tested in mature colonies, which are inherently difficult to work with in control laboratory set-ups. It is also unknown if in *Coptotermes* such N_1 differentiation is triggered at the L_2 instar, or if epigenetic regulators are involved in L_1 or even pre-oviposition (transgenerational epigenetic regulation).

3. Do nymphs (N_{3+}) that have moved toward foraging sites to feed on their own return to the central part of the nest when it is time for them to molt, as observed in workers?

4. How long is the molting cycle of an individual and what influences the timing of molting? The molting cycle of workers can be as short as 30 days and as long as 6 months, according to various observations. Such discrepancy can emerge from the age of workers (the older the worker, the longer the intermolt period). However, it is unclear what are the factors (internal and socio-environmental) that are triggering the timing of the molting process in *Coptotermes*.

5. What nymphal instars can become secondary reproductives? The nymphal development toward secondary reproductive (N_n→SR) has not been investigated in depth in *Coptotermes*. It is unclear which nymphal instars are the most prone to become secondary reproductives, and what triggers such differentiation.

19.4 Colony demographic considerations

Thomas Chouvenc

1. What are the short- and long-term impacts of inbreeding in *Coptotermes* colonies that go through successions and accumulation of secondary reproductives? The longevity of workers produced in colonies under such reproductive conditions may be shorter than colonies under single-family conditions but remains to be investigated. Similarly, alates produced by extended colonies appear to be smaller in body size than in single-family colonies. The potential trade-off between quality and quantity of alates for colonies under different reproductive strategies remains to be confirmed.
2. Do soldiers have a function beyond defense in *Coptotermes*? Soldiers have protozoa in their gut and therefore can contribute to the digestion process through trophallaxis of food particles and could be part of the social stomach. In addition, they may serve as water reservoirs, and help move moisture within the gallery system, from the central nest to new foraging sites. Such secondary functions, among others, remain to be investigated.
3. Do workers display an intermolt polyethism? Beyond age polyethism, it is possible that workers perform different tasks throughout their intermolt period, as mandible sharpness potentially decreases over time, and task efficiency may change over time.
4. Is *Coptotermes* a central nester? *Coptotermes* possess a "central nest" where the primary reproductives and their brood are found, and where workers return for molting. However, colonies with an accumulation of many secondary reproductives apparently can establish reproduction at satellite locations, potentially changing the demographic flow of individuals within the colony over time, and changing, from a behavioral perspective, the colony as a "satellite nester." If true, then where do workers return to when it is time to molt, if the nest is "decentralized" at this point?
5. What triggers the definitive loss of reproduction toward colony senescence, with an absence of replacement of reproductive? Why are colonies not "forever"?

19.5 How do termites create a direct tunnel leading to food after its discovery?

Nan-Yao Su

Subterranean termites exhibit remarkable tunneling behavior as they diligently search for food in or on soil, guided by a set of rules. Various factors, such as moisture gradients and semiochemicals permeated in soil, can influence the geometry of these tunnels. However, the propagation of tunnels is generally determined by a Global Away Vector aided by path integration to navigate away from the tunnel's origin (Bardunias and Su, 2009).

Available data suggest that termites navigating through the intricate gallery system possess the ability to geo-reference their current position relative to the tunnel's origin, be it the nest or a food source (Michael *et al.*, 2023). Yet, it remains unclear whether this information is communicated to nestmates and, if so, by whom. When encountering a new food source, do the first termites to find it return to the nest and communicate its position to their peers? Alternatively, do they simply lay down a trail-following pheromone for others to navigate the existing tunnels?

Even if the latter scenario holds true, where a trail-following pheromone guides other termites to the discovered food, it fails to explain how shortcuts or direct tunnels to the food source are typically established (Michael *et al.*, 2023). To propagate a new tunnel directly from the tunnel origin to a recently discovered food source, the tunnelers must possess some knowledge of its general position. Therefore, it is crucial to investigate whether the termites that discover new food initiate the tunneling process directly towards the source, with other nestmates following their lead. Alternatively, it is worth exploring whether information about the general position of the food source is communicated to nestmates, akin to the well-known waggle dance observed in honeybees nearly 80 years ago.

Understanding the mechanisms underlying the communication and decision-making processes within termite colonies regarding tunnel construction and navigation can shed light on the remarkable co-operative behavior exhibited by these social insects. Exploring whether termites possess sophisticated communication methods similar to the waggle dance could unravel fascinating insights into their collective intelligence and decision-making abilities.

References

Bardunias, P. and Su, N.-Y. (2009) Dead reckoning in the tunnel propagation of the Formosan subterranean termite (Isoptera: Rhinotermitidae). *Annals of the Entomological Society of America* 102, 158–165.

Michael, M., Chouvenc, T., Su, N.-Y. and Lee, S.-B. (2023) Finding shortcuts through collective tunnel excavations in a subterranean termite. *Behavioral Ecology* 34, 354–362. doi: 10.1093/beheco/arad007

19.6 Relationship between termite tunnel geometry and distribution of food in or on soil

Nan-Yao Su

The laboratory study by Robson *et al.* (1995) suggested that *Reticulitermes flavipes* (Kollar) create tunnels to intercept “wood on the soil surface” that is distributed randomly and clumped together. The simulation study by Lee *et al.* (2006) found that the tunnel geometry of *C. formosanus* is optimized for efficiently encountering clumped food sources rather than randomly or uniformly distributed ones. However, there is currently no available data on the natural distribution of termite food sources such as logs, branches, or decayed wood debris, and whether this distribution varies across environments. To validate the findings of laboratory and simulation studies, further investigation into the distribution patterns of these termite food sources in their natural habitats is required.

References

Lee, S.-H., Bardunias, P. and Su, N.-Y. (2006) Food encounter rates of simulated termite tunnels with variable food size/distribution pattern and tunnel branch length. *Journal Theoretical Biology* 243, 493–500.

Robson, S.K., Lesniak, M.G., Kothandapani, R.V. and Traniello J.F.A. (1995) Nonrandom search geometry in subterranean termites. *Naturwissenschaften* 82, 526–528.

19.7 How do soldiers digest semi-digested wood (cellulose) given by workers?

Gaku Tokuda, Shuji Itakura, and Nathan Lo

Termite workers forage and eat wood. The partially digested wood and gut metabolites are then consumed by other caste members, including soldiers. However, the digestion and energy metabolism of soldiers is still poorly understood. Such knowledge is important for understanding how the whole termite colony is maintained from an energy balance point of view.

19.8 Function of gene homologs of endogenous cellulase

Gaku Tokuda, Shuji Itakura, and Nathan Lo

Recent studies have shown that each termite genome encodes four or more homologs of endogenous cellulase genes. At least one of these is phylogenetically distantly related to digestive cellulases and is uniformly expressed at low levels regardless of the body part, sex, or caste (Shigenobu *et al.*, 2022). Since such a gene homolog is also found in non-cellulolytic insects such as hymenopterans (Shigenobu *et al.*, 2022), elucidation of its function would be relevant to the general mechanism in insects.

Reference

Shigenobu, S., Hayashi, Y., Watanabe, D., Tokuda, G., Hojo, M.Y. *et al.* (2022) Genomic and transcriptomic analyses of the subterranean termite *Reticulitermes speratus*: gene duplication facilitates social evolution. *Proceedings of the National Academy of Sciences of the USA* 119, e2110361119.

19.9 Sensory cues in social behavior

Qian Sun

Subterranean termites perform complex and sophisticated social behaviors with a division of labor among castes. While alates develop compound eyes and respond to visual cues, most individuals heavily rely on chemical and tactile cues to perceive the social environment. However, few chemical cues are characterized (see Table 7.1), and little is known about the nature of tactile cues in *C. formosanus*. The social cues that modulate major colony activities, including reproduction, caste differentiation, foraging, and nestmate recognition, remain to be fully identified. Future research should also determine the biological activities of identified compounds isolated from the exocrine glands of termites and their nest and investigate the potential integration of different sensory cues in the modulation of social behavior. Determining the sensory mechanisms will fill a major knowledge gap in the chemical and behavioral ecology of *Coptotermes* and may facilitate the development of termite monitoring and control tactics through behavioral manipulation.

19.10 Molecular mechanism for communication among termite gut microbes

Claudia Husseneder

What is the molecular mechanism, the universal language (so to speak), allowing the four levels of the quadripartite relationship (termite, protists, prokaryotes, and bacteriophages) to communicate with each other to negotiate the common living conditions, trade-offs, collaboration, and competition in the termite gut?

19.11 Impact of climate change on termiticide performance

Chow-Yang Lee and Kok-Boon Neoh

High moisture and temperature are known to accelerate the deterioration of termiticides. With climate change, termiticide performance will likely be impacted. As climate warms, termite activity and abundance will rise (Zanne *et al.*, 2022). Long-term drought may cause the production of thicker cuticular hydrocarbon (CHCs) in termites as a response to prevent desiccation. This has been observed in *C. formosanus* (Richardson, 2022) and may reduce insecticide penetration. Besides, enhanced CHC synthesis has also been associated with the overexpression of cytochrome P450 genes implicated in insecticide detoxification (Chen *et al.*, 2019). It is interesting to investigate how these climate factors potentially affect the future of termiticide usage.

References

Chen, N., Pei, X.J., Li, S., Fan, Y.L. and Liu, T.X. (2019) Involvement of integument-rich CYP4G19 in hydrocarbon biosynthesis and cuticular penetration resistance in *Blattella germanica*. *Pest Management Science* 76, 215–226.

Richardson, S.J. (2022) Soil moisture mediated behavioral and physiological alterations of *Coptotermes formosanus* and *Reticulitermes flavipes* (Blattodea: Rhinotermitidae). MS thesis. Louisiana State University, Baton Rouge, Louisiana, 60 pp.

Zanne, A.E., Flores-Moreno, H., Powell, J.R., Cornwell, W.K., Dalling, J.W. *et al.* (2022) Termite sensitivity to temperature affects global wood decay rates. *Science* 377, 1440–1444.

19.12 Questions regarding hybrid termites

Thomas Chouvenc

1. Will hybrid *Coptotermes* establish and become a source of gene flow between *C. formosanus* and *C. gestroi*, in Taiwan and in Florida? Even if F1 alates are produced, it remains unclear if an admixture of genetic material within local population will result in phenological changes. For example, could a *C. gestroi* population that historically received genetic material from *C. formosanus* spread further geographically in colder regions?

2. Similarly, if F1 alates are viable and partially fertile, what is the true potential for introgressive hybridization, and would such introgression be asymmetrical between the two parental species?

3. Hybrid colonies display workers and soldiers with altered morphology, physiology, and behaviors. Such phenotypic alterations could be used as a model to investigate regulatory processes involved in molting, caste differentiation, and nest construction. In addition, hybrid colonies provide an opportunity to test the potential conflict between lineages of symbionts from distinct origin.

Index

Note: The page numbers in italics and bold represents figures and tables respectively.

Printed and bound by CPI Group (UK) Ltd, Croydon, CR0 4YY

13/09/2026

14961109-0004